Advances in Non-volatile Memory and Storage Technology

Woodhead Publishing Series in Electronic and Optical Materials

Advances in Non-volatile Memory and Storage Technology

Second Edition

Edited by

Blanka Magyari-Köpe
Yoshio Nishi

ELSEVIER

WP
WOODHEAD
PUBLISHING
An imprint of Elsevier

Woodhead Publishing is an imprint of Elsevier
The Officers' Mess Business Centre, Royston Road, Duxford, CB22 4QH, United Kingdom
50 Hampshire Street, 5th Floor, Cambridge, MA 02139, United States
The Boulevard, Langford Lane, Kidlington, OX5 1GB, United Kingdom

Library of Congress Cataloging-in-Publication Data
A catalog record for this book is available from the Library of Congress

British Library Cataloguing-in-Publication Data
A catalogue record for this book is available from the British Library

ISBN: 978-0-08-102584-0

For information on all Woodhead publications visit our
website at https://www.elsevier.com/books-and-journals

Publisher: Matthew Deans
Acquisition Editor: Kayla Dos Santos
Editorial Project Manager: Emma Hayes
Production Project Manager: Joy Christel Neumarin
    Honest Thangiah
Cover Designer: Miles Hitchen

Typeset by SPi Global, India

Working together
to grow libraries in
developing countries

www.elsevier.com • www.bookaid.org

Contents

Contributors

Stefano Ambrogio IBM Research-Almaden, San Jose, CA, United States

Y. Ando Tohoku University, Sendai, Japan

G. Bersuker The Aerospace Corporation, Los Angeles, CA, United States

Chong Bi Department of Materials Science and Engineering; Department of Electrical Engineering, Stanford University, Stanford, CA, United States

Philippe Blaise CEA LETI Minatec Campus, Grenoble, France

B. De Salvo CEA-LETI Minatec Campus, Grenoble, France

Jonas Deuermeier CENIMAT/i3N, Department of Materials Science, Faculty of Sciences and Technology, Universidade NOVA de Lisboa and CEMOP/UNINOVA, Campus de Caparica, Caparica, Portugal

Regina Dittmann Peter Grünberg Institute (PGI-7), Forschungszentrum Jülich GmbH and JARA-FIT, Jülich, Germany

T. Endoh Tohoku University, Sendai, Japan

S. Fukami Tohoku University, Sendai, Japan

D.C. Gilmer Nantero, Inc., Woburn, MA, United States

Ludovic Goux Imec, Kapeldreef, Leuven, Belgium

T. Hanyu Tohoku University, Sendai, Japan

Michel Harrand CEA LETI Minatec Campus, Grenoble, France

Susanne Hoffmann-Eifert Peter Grünberg Institute (PGI-7), Forschungszentrum Jülich GmbH and JARA-FIT, Jülich, Germany

Hyunsang Hwang Department of Materials Science and Engineering, Pohang University of Science and Technology, Pohang, South Korea

Cheol Seong Hwang Department of Materials Science and Engineering, and Inter-University Semiconductor Research Center, Seoul National University, Seoul, South Korea

Daniele Ielmini Dipartimento di Elettronica, Informazione e Bioingegneria, Politecnico of Milan and IU.NET, Milan, Italy

S. Ikeda Tohoku University, Sendai, Japan

Asal Kiazadeh CENIMAT/i3N, Department of Materials Science, Faculty of Sciences and Technology, Universidade NOVA de Lisboa and CEMOP/UNINOVA, Campus de Caparica, Caparica, Portugal

H. Koike Tohoku University, Sendai, Japan

Yunmo Koo Department of Materials Science and Engineering, Pohang University of Science and Technology, Pohang, South Korea

Luca Larcher Applied Materials, Reggio Emilia, Italy

Seokjae Lim Department of Materials Science and Engineering, Pohang University of Science and Technology, Pohang, South Korea

Massimo Longo CNR—Institute for Microelectronics and Microsystems—Unit of Agrate Brianza, Italy

Y. Ma Tohoku University, Sendai, Japan

Stephan Menzel Peter-Grünberg-Institut (PGI-7), Forschungszentrum Jülich GmbH and JARA-FIT, Jülich, Germany

Rivu Midya Department of Electrical and Computer Engineering, University of Massachusetts, Amherst, MA, United States

Thomas Mikolajick NaMLab gGmbH; Institute of Semiconductors and Microsystems, TU Dresden, Dresden, Germany

Gabriel Molas CEA LETI Minatec Campus, Grenoble, France

Cécile Nail CEA LETI Minatec Campus, Grenoble, France

H. Ohno Tohoku University, Sendai, Japan

Andrea Padovani Applied Materials, Reggio Emilia, Italy

Jaehyuk Park Department of Materials Science and Engineering, Pohang University of Science and Technology, Pohang, South Korea

Paolo Pavan Dipartimento di Ingegneria "Enzo Ferrari", Università di Modena e Reggio Emilia, Modena, Italy

L. Perniola CEA-LETI Minatec Campus, Grenoble, France

Francesco Maria Puglisi Dipartimento di Ingegneria "Enzo Ferrari", Università di Modena e Reggio Emilia, Modena, Italy

Mingyi Rao Department of Electrical and Computer Engineering, University of Massachusetts, Amherst, MA, United States

Noriyuki Sato Department of Materials Science and Engineering; Department of Electrical Engineering, Stanford University, Stanford, CA, United States

H. Sato Tohoku University, Sendai, Japan

R. Shirota National Chiao Tung University, Hsinchu, Taiwan

Jeonghwan Song Department of Materials Science and Engineering, Pohang University of Science and Technology, Pohang, South Korea

D. Suzuki Tohoku University, Sendai, Japan

Navnidhi Kumar Upadhyay Department of Electrical and Computer Engineering, University of Massachusetts, Amherst, MA, United States

D. Veksler The Aerospace Corporation, Los Angeles, CA, United States

E. Vianello CEA-LETI Minatec Campus, Grenoble, France

Shan X. Wang Department of Materials Science and Engineering; Department of Electrical Engineering, Stanford University, Stanford, CA, United States

Zhongrui Wang Department of Electrical and Computer Engineering, University of Massachusetts, Amherst, MA, United States

Rainer Waser Peter-Grünberg-Institut (PGI-7), Forschungszentrum Jülich GmbH and JARA-FIT, Jülich; Institut für Werkstoffe der Elektrotechnik (IWE 2), RWTH Aachen, Aachen, Germany

J. Joshua Yang Department of Electrical and Computer Engineering, University of Massachusetts, Amherst, MA, United States

Part One

Progress in nonvolatile memory research and application

OxRAM technology development and performances

1

Ludovic Goux
Imec, Kapeldreef, Leuven, Belgium

1.1 Introduction

1.1.1 Nonvolatile memory applications

1.1.1.1 Storage-class memory (SCM)

In today's computational systems, memories are categorized into volatile memory and nonvolatile memory (NVM) technologies.

Due to the ever increasing demand for more memory capacity, planar NAND Flash has been scaled down to below 20-nm feature size. Concomitantly, three-dimensional (3D) vertical NAND Flash has been developed as a Bit-Cost-Scalable (BiCS) solution and allows today entering the Terabyte era. As a result, the NVM market is by far dominated by NAND Flash technology, and the forecast is that the future of NAND will be NAND [1].

In the volatile memory category, the main technologies are the static RAM (SRAM) and dynamic RAM (DRAM), which are higher-speed and higher-performance technologies, however, exhibit poor scalability. Their role in a central process unit (CPU) is to store data that require immediate access while NAND Flash or hard-disk drive (HDD) store information that is not required immediately but for available future usage [2].

The problem arises when transferring data from DRAM to NAND: the overall performance of the system is limited by the huge latency gap between these two technologies. This gap has been virtually fitted with architectural solutions to increase the data access speed but at the expense of complex system design and increased chip area. In recent years, researchers have started exploring the possibility of novel memory concepts to improve the existing memory hierarchy. The concept of storage class memory (SCM) has been proposed, aiming to fill the access time gap between the "memory-memory" and the "storage-memory."

As a "bridge" technology between DRAM and Flash, the main requirements for SCM are intermediate between DRAM and Flash and should be cost effective.

In short, SCM should be enabled by a nonvolatile, cheap, and scalable technology having clearly better reliability (write endurance and retention) than Flash. At the lead in this future booming market, Intel-Micron announced the 3D X-point Memory in 2015 [3] and launched products in 2017. Although not officially confirmed by Intel-Micron, it is generally agreed in the memory community that 3D X-point is based on phase-change memory technology.

Advances in Non-volatile Memory and Storage Technology. https://doi.org/10.1016/B978-0-08-102584-0.00001-2

As will be detailed in the following sections, the filamentary resistive RAM (RRAM) technology has also proven to be a promising candidate in this growing market.

1.1.1.2   Internet of Things (IoT)—Embedded memory

Another booming market is the concept of Internet of Things (IoT), which is defined as "intelligent connectivity of smart devices by which objects can sense one another and communicate" [4].

The advent of billions of connected devices is creating new opportunities and huge markets for the emerging memories. In fact, the market for IoT devices has been projected to more than 3.5 trillion in 2020.

In a typical electronic system, logic and NVM components are fabricated separately due to the incompatibility in integration flow. To accommodate the exponential growth of IoT devices, chip costs have to be reduced.

This drives the need for developing new embedded memory technology where a chip can contain both logic and memory components to lower the cost and save space on the printed circuit boards. This is referred to as system-on-a-chip (SoC). The existing SoC chip uses NOR Flash as embedded memory. However, to integrate embedded NOR Flash in 28 nm node and below, up to 15 extra photomasks need to be added in the overall fabrication process, which makes embedded NOR Flash an extremely expensive technology, all the more for more advanced logic nodes. Therefore, an alternative CMOS-compatible and low-cost embedded memory technology would be highly desired to feed the IoT market as well as other embedded markets.

In this respect too, the RRAM technology is a strong contender, as will be shown next.

1.1.2   Resistive RAM technology

To fulfill the requirements of these new applications, various new memory device concepts have been proposed and studied. The prominent concepts are spin-transfer torque magnetic RAM (STT-MRAM), phase-change RAM (PCRAM), and RRAM or ReRAM. These technologies, due to limited maturity, are categorized as emerging memories.

The RRAM category is the name of a group of memory technologies characterized by an electrically reversible resistive switching functionality between a low resistance state (LRS) and a high resistance state (HRS). One of the most attractive advantages of RRAM is the low-cost integration allowed by the combination of CMOS-friendly materials within a simple two-terminal device structure, which typically consists of a dielectric layer sandwiched between two metal electrodes.

Between the numerous mechanisms potentially at the origin of resistive-switching effects, nano-ionic transport and redox-reaction mechanisms taking place at the nanometer scale [5] have been clearly identified as accounting for the switching functionality of various systems.

In most of the RRAM devices reported [6–8], the resistive switching property originates from the growth and shrinkage of a nanoscaled conductive filament (CF) in the dielectric layer. These devices are generally referred to as filamentary RRAM devices, whose characteristics mainly depend on the CF. In particular, due to the nanoscale character of the switching, filamentary RRAM has been so far perceived as highly scalable [5,9], and has thus been developed intensively.

This is, for example, the case of electrochemical metallization memory (ECM) and valence change memory (VCM) cell concepts.

The ECM concept has led to the development of the conductive bridge RAM technology (CBRAM), whereby the filament consists of metallic species injected from an active electrode, typically Ag or Cu, into the dielectric layer [10].

On the other hand, the VCM concept is at play in the Oxide switching RAM technology (OxRAM), whereby the dielectric layer is an insulating oxide material through which a CF of oxygen-vacancy defects (V_o) is electrically created [11]. This technology generally uses a transition metal oxide (TMO) material, typically HfO_2, Ta_2O_5, or TiO_2 [11,12].

On the other hand, devices with nonfilamentary resistive switching mechanism have also been reported [13,14]. These devices exhibit area-dependent current flow, which is not observed in filamentary RRAM after the filament has been created. Here the resistive switching is achieved by modulation of the effective tunneling barrier thickness at the oxide-metal interface, as induced electrically by the uniform motion of V_o defects. An advantage of this concept is the scalability of the operating current with the device area.

Between these different concepts, this chapter is dedicated to the filamentary OxRAM technology, and we specifically focus on the works achieved at imec in this field over the last decade.

In Section 1.2, we first review how the initial developments moved from the unipolar to the bipolar switching concept, and we describe typical structures, fabrication flows, testing procedures, and electrical characteristics of bipolar devices that mainstream nowadays.

In Section 1.3, we focus on material developments allowing substantial improvements in memory performances, and finally, in Section 1.4, we address key reliability challenges to address in the future.

1.2 History and basics of filamentary OxRAM

Resistive switching and negative differential resistance phenomena were first reported in the 1960s in rather thick oxide materials, for example, Al_2O_3 [15] and NiO [16] layers. This early period of research has been comprehensively reviewed in Refs. [17,18].

Since the 2000s, there has been a renewed interest in resistive switching in TMO systems, driven by the potential industrial application as RRAM. In this period, the developments focused on NiO and TiO_2 systems exhibiting unipolar switching operation [19,20], which means that the same voltage polarity may be used both for set switching to LRS and reset switching to HRS. This unipolar operation received large consideration due to possible integration with a two-terminal selector element, such as a diode, which in turn held the promise of dense memory integration potential. Goux and Spiga [19] give a recent review of unipolar-switching developments.

1.2.1 From unipolar to bipolar concept

The unipolar switching originates from a distinct class of switching mechanism, which is called thermochemical memory (TCM) mechanism. Indeed, contrary to bipolar VCM switching mechanisms where the motion of oxygen ionic species is dominated by drift, the TCM switching is dominated by thermal-controlled diffusion and redox reaction of active oxygen species.

This major microscopic difference between VCM and TCM mechanisms reflects in different memory characteristics.

1.2.1.1 Unipolar operation mechanism

Unipolar switching is usually observed in simple metal–insulator–metal (MIM) structures, where the "I" and "M" elements are typically constituted of an oxide layer and noble metallic layers respectively.

Fresh cells are most often in a very high resistive state requiring the application of a large "forming" voltage (V_f), which is generally regarded as a sort of electrical-breakdown [21,22] allowing to turn the cell for the first time in LRS. After this so-called "forming" conditioning step, the cell may be reversibly reset-switched to HRS and set-switched to LRS, as illustrated in Fig. 1.1A–C.

The reset switching operation corresponds to the rupture of the CF, resulting in a drop of conductivity. Let us assume first that the CF consists of a V_O-chain. Upon reset operation, the current density through the CF has been shown to reach high values, generating the necessary thermal energy to activate the migration of oxygen from O-rich regions outside the CF toward O-deficient regions inside the CF. According to this picture, the reset switching corresponds to the local reoxidation of the CF. It requires moving back the forming-generated O-species to V_O sites, and as the switching is unipolar in nature, this process is expected to take place primarily laterally by diffusion mechanisms. In agreement with this scenario, we have evidenced by means

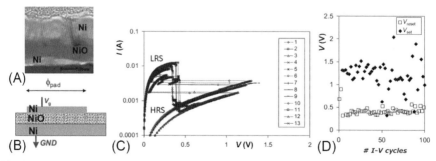

Fig. 1.1 (A) Cross-sectional TEM images and (B) schematic structures of Ni\NiO\Ni cells; (C) Consecutive unipolar *I-V* sweeps obtained on the same cell; (D) Extracted V_{set} and V_{reset} as a function of the *I-V* cycles [21].

of TEM and EELS characterization a significant increase of the O/Ni ratio after reset operation in TiN\NiO\Ni cells [23]. Assuming now that the CF consists of a metallic Ni-chain the thermal dissolution of Ni upon reset switching might be a more realistic mechanism for CF rupture. In this scenario too the diffusion mechanism is expected to occur sideways from inside the Ni-rich CF toward outside Ni-poor regions. In both situations, the physics of the reset operation is overall expected to depend on the spatial distribution of defects, on local fields as well as on temperature profiles.

1.2.1.2   Variability and endurance of unipolar switching

Some NiO cells have been reported to exhibit coexistence of unipolar and bipolar resistive switching properties [21]. While NiO is intrinsically a thermochemical material, the disordered/defective microstructure of the layers proved favorable to ionic drift and electrochemical redox mechanisms, allowing thus selecting reversibly any switching mode between unipolar and bipolar [21]. Note that the forming step itself is sufficient to induce disorder and facilitate ionic drift in high-quality dielectrics. For example, coexisting unipolar and bipolar operation modes were obtained after forming operation in TiN\HfO$_2$\Pt cells where a high-quality amorphous HfO$_2$ layer was prepared by ALD [11,24].

The switching variability was investigated in various reports [20,21]. In general, the switching-voltage distribution is wider for unipolar than for bipolar switching. Indeed, as the reset mechanism is a self-accelerated process for unipolar mechanisms, it may be argued that the programming of the HRS state is difficult to control, leading to different cycle-to-cycle (C2C) R_{HRS} and subsequent V_{set} parameters, as observed in Fig. 1.1D [21]. Other reports also suggested that the variability degrades in a configuration of multiple-CF network [25], wherein the switching "hot" spot is likely to move within the network from cycle to cycle.

Hence, not surprisingly, write-endurance lifetime is clearly shorter for unipolar than for bipolar switching. Between the invoked origins are the larger temperatures required for reset compared to bipolar switching, possibly inducing electro-migration effects [26], the gradual loss of species involved in the radial diffusion, or simply the drift-induced loss of species inherent to the use of a single programming polarity [11].

Overall, unipolar OxRAM suffered from high operating voltage, large switching variability, as well as limited endurance. Hence, it gradually lost interest concomitantly with the substantial improvement of bipolar OxRAM technologies by the years 2010.

Nowadays, modern OxRAM devices are implemented in bipolar mode where the growth or shrinkage of the conductive filament is achieved by relying on a voltage controlled ion migration assisted by temperature and electric field enabling faster set and reset operations. The rest of the chapter will be dedicated to the development of this concept.

1.2.2   Bipolar switching concept

1.2.2.1   Device structure

In addition to memory performance limitations, unipolar switching OxRAM showed the major drawback of involving difficult-to-integrate materials, for example, noble metals required in the model system Pt/NiO/Pt to avoid parasitic oxidation during the thermal-induced reset operation.

In contrast, the drift-dominated mechanisms involved in bipolar VCM allow using CMOS-friendly electrode materials like TiN or TaN.

Regarding the oxide layer, excellent performances have been demonstrated using mature atomic-layer deposition (ALD) oxide layers like HfO_2 [7,27] or Ta_2O_5 [28,29], which are considered today as the mainstream oxide families of OxRAM.

However, the key success of bipolar OxRAM lies elsewhere, that is in the electrical asymmetry of the structure, as required for a bipolar functionality. We evidenced by Internal Photoelectron Emission (IPE) that some low work-function metallic layers like Hf, Ti, Ta inserted at one interface of the MIM stack are very appropriate to induce such required asymmetry. Fig. 1.2 shows the substantial decrease of electron barrier of ~1 eV from TiN to Hf electrode [30]. This effect is due to the oxygen scavenging by Hf leading to the development of an oxygen deficient hafnia interlayer between Hf and HfO_2. A resulting substochiometric oxide interlayer is formed, often referred to as oxygen-exchange-layer (OEL) [31]. The role of this layer will be to "exchange" oxygen species with the filament in the oxide layer during the bipolar switching operation. On the other hand, the TiN opposite electrode is considered inert to oxygen in the first approximation. Hence, the O-scavenging layer of Hf induces an oxygen vacancy profile along the oxide thickness and will be the main knob allowing to tune the bipolar switching functionality and characteristics of the device. Both the *nature* and the *thickness* of this layer will be key to the device performances, as will be shown in the following sections.

As best O-scavenging materials, in general, the use of "mother" metals in $Ta_2O_5\backslash$ Ta [29] or $HfO_2\backslash Hf$ [27] is preferred and shows robust stack stability after integration thermal budget. And in practice, up to today the $TiN\backslash HfO_2\backslash Hf\backslash TiN$ and $TiN\backslash Ta_2O_5\backslash$

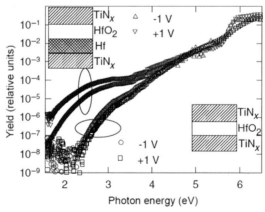

Fig. 1.2 Internal Photo-Emission (IPE) characterization of $TiN_x/HfO_2/TiN_x$ and $TiN_x/Hf/HfO_2/TiN_x$ samples, showing the logarithmic plots of the electron quantum yield as a function of photon energy. The signal observed under positive +1 V and negative 1 V bias corresponds to electron IPE from the bottom and top metal electrode, respectively [30].

Ta\TiN are considered as mainstream OxRAM devices and will be the baselines discussed in the following sections.

1.2.2.2   Integration and scaling

In this section, we describe a typical 1-Transistor/1-Resistor (1T1R) integration flow used to fabricate TiN\HfO$_2$\Hf\TiN and TiN\Ta$_2$O$_5$\Ta\TiN in a crossbar configuration, and where the "T" element is the selector device addressing the "R" OxRAM device (Fig. 1.3A).

After the front-end-of-line (FEOL) processing of this select transistor (nMOSFET), the OxRAM device is fabricated as follows: A 100-nm-thick TiN is deposited by physical-vapor-deposition (PVD) as a bottom electrode (BE) where the electrical connection to the transistor is realized by tungsten (W) plugs. After BE patterning SiO$_2$ is deposited and polished by chemical–mechanical polishing (CMP) down to the TiN surface. This step ensures flat BE surface and is thus critical for device performance. Subsequently, the active HfO$_2$ layer, or Ta$_2$O$_5$ respectively, is deposited by ALD, typically using HfCl4, or TaCl5 resp., as precursor, and H$_2$O as the oxidant, followed by the PVD of the Hf, respectively. Ta, O-scavenging metal layer and a TiN top electrode (TE). Then, the oxide\TE stack is patterned via lithography and etch steps, where a typical device size of $40 \times 40 \, nm^2$ is formed in imec process flow. Finally, passivation modules using Si$_3$N$_4$ and SiO$_2$ layers deposited by CVD and contact formation steps are carried out.

Fig. 1.3B depicts the cross-section TEM image of a pristine TiN\HfO$_2$\Hf device. By means of patterning trimming processes, sub-10nm devices are realized. As observed on Fig. 1.3B the Hf layer is laterally oxidized during process flow, which reduces further the active device size [27].

1.2.2.3   Typical device testing and characteristics

The role of the select transistor is not only to isolate and address the cell but also to limit more efficiently the current overshoot during forming and set, as compared to the

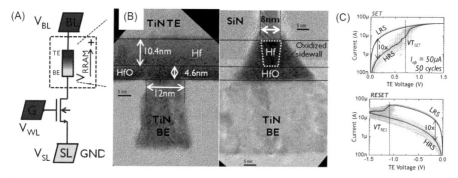

Fig. 1.3 (A) Schematic representation of the so-called 1T1R configuration; (B) cross section along BE and TE of a minimum-sized ($8 \times 12 \, nm^2$) reference 5 nm-HfO$_2$\10 nm-Hf RRAM stack; the HfO$_2$ layer is amorphous, Hf is crystalline, oxides sidewalls at TE are further reducing effective cell size; (C) typical set and reset I-V traces obtained on the cells [27].

current compliance (CC) function of a commercial semiconductor parameter analyzer (SPA) [22]. In a 1T1R scheme, the transient current flowing through the OxRAM cell is controlled by fixing an appropriate gate voltage (V_g), which is a key aspect for reliable characterization of the device because the CF properties depend on this transient current [19].

The set operation is carried out by applying a positive voltage to the selected cell Bit-Line (BL) while the Source-Line (SL) is grounded. Conversely, for reset operation, a negative voltage is applied to the BL. In a well-designed 1T1R structure with controlled stray capacitance, the maximum current reached during reset is linked to the saturation current imposed by the transistor during the set operation. In any case, in practice, larger V_g may be applied during the reset operation because this step does not need to limit the current. For state resistance readout, the transistor is fully open and $V_{read} = 0.1\,V$ is applied to the BL.

Using this methodology Fig. 1.3C shows typical set/reset I-V traces obtained after forming on TiN\HfO$_2$\Hf cells and using a V_g amplitude allowing a maximum operating current $I_{op} = 50\,\mu A$ during set, as imposed by the saturation level of the transistor.

Forming

As mentioned, before memory operation the cells need to undergo a "forming" step, whereby the conductive filament is formed through the oxide layer.

In Ref. [27] we studied the scaling behavior of the forming voltage (V_f) down to 10-nm TiN\HfO$_2$\Hf cell size. We observe that while amorphous HfO$_2$ maintains a well-behaved scaling vs area, polycrystalline HfO$_2$ shows abrupt dispersion and median V_f increase below ~40 nm size, which is attributed to the abrupt decrease of grain-boundary density [27]. Therefore, in order to keep V_f control with scaling to sub-10 nm size, amorphous oxide layers are preferred, allowing uniform weak electrical path and controllably low V_f.

We also observed that the forming operation is also highly controlled by the engineered O-scavenging layer. Fig. 1.4 shows for TiN\Ta$_2$O$_5$\TE cells that V_f may be drastically decreased by changing Ta to Ti TE O-scavenging layer [29]. This is due to the larger O-affinity of Ti, resulting in a lower formation energy (E_f) of V_o in Ta$_2$O$_5$, and which in turn induces a lower V_f. As can be observed in Fig. 1.4B, the breakdown voltage (V_{bd}) data extracted by applying opposite voltage polarity are higher than V_f, which is due to a larger energy required for the drift of oxygen anionic species toward TiN as compared to the O-scavenging layer.

Interestingly, a thicker Ta layer or the use of a substochiometric TaOx TE layer also allows to tune E_f and thus V_f. Actually, the nature and thickness of the scavenging layer tune the oxygen chemical potential at the interface, as will be detailed in the following sections.

Note also that the scavenging is so strong for Ta$_2$O$_5$\Ti cells that the top part of the Ta$_2$O$_5$ layer is fully depleted after integration (Fig. 1.4C) and the switching operation is poor. This effect confirms that the mother-metal Ta is preferred to preserve device integrity. Using Ta (10 nm) TE, excellent V_f uniformity is observed for a Ta$_2$O$_5$ thickness range down to 3 nm, which allows to controllably limit $V_f < 1.5\,V$ for device sizes down to 20 nm [29].

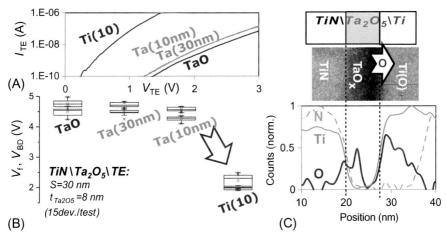

Fig. 1.4 Median preforming *I-V* traces (A), and V_f and V_{bd} data of 30-nm-size Ta_2O_5 devices having different TE materials, showing strong scavenging difference between Ta- and Ti-based TE; (C) XTEM and EELS profiles on as-integrated devices, showing a huge Ti-induced O-soaking from the Ta_2O_5 layer [29].

Filament shape and scaling

The forming step is expected to result in a V_o-rich filamentary region having asymmetric $[V_o]$ profile increasing from the TiN to the Hf interface [32,33]. This forming scenario is more likely to occur than the virtual-cathode growth model (growth of V_o defects from the cathode side) because the V_o formation energy (E_f) will decrease closer to the scavenging layer, resulting in larger $[V_o]$, and the mobility of the O-anionic species will increase with the increase of $[V_o][33]$. Hence the filament will gradually form from the scavenging layer toward the "inert" TiN electrode. Based on this scenario, it is expected that the forming step results in a conical filament having its narrow constriction closer to TiN.

Likewise, the reset switching is expected to take place "naturally" closer to the TiN anode due to: (i) the lower V_o density to recover; (ii) the O-blocking role of the TiN interface during reset-induced O-anionic drift; and (iii) the larger O-mobility within O-deficient HfO_2 materials.

A few years after proposing this scenario, we performed a 3D Conductive Atomic-Force-Microscopy (C-AFM) tomography experiment on a TiN\HfO₂\Hf cell [34]. This technique allows, after electrical forming of the device, to record the current flowing through the scanning conductive tip while scrapping gradually the device materials. In this "scalpel" mode, the tip is brought in contact with the sample and mechanically scrapes a layer of the material to expose the layer underneath. The C-AFM imaging mode is used in-between successive scalpel steps to generate 2D images of the conduction map. After complete device scrapping, a 3D tomogram is reconstructed by interpolation of 2D images. Fig. 1.5 shows the result for TiN\HfO₂\Hf cells programmed using a forming current of $50\,\mu A$. The experiment confirmed the conical shape of the filament, with its widest part near the Hf TE (7 nm) and the narrowest region near the TiN BE (3 nm), as shown in Fig. 1.5C.

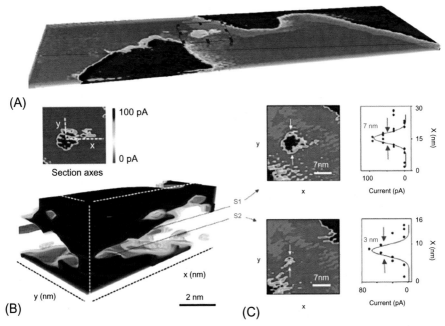

Fig. 1.5 3D observation of a conductive filament by C-AFM tomography technique in TiN\
HfO₂\Hf cell. (A) Tomographic reconstruction of the crossbar memory cell visualized by
volume rendering after top electrode removal (scan size 800×200 nm); (B) 2D zoom into
the region containing the CF and observation by volume rendering and isosurface at fixed
threshold *(blue shape)* for the CF under investigation in 5-nm-thick HfO₂ (scale bar 2 nm).
The conductive filament is shown in double cross section (axes according to inset). The
low-current contribution in the tomogram is suppressed to enhance the contrast of the highly
conductive features. (C) 2D observation of the CF section-planes (left panels) and C-AFM
spectra (right panels) to determine effective CF size [34].

This experiment also evidenced that OxRAM technology may be scaled below
10-nm cell size.

Set/reset switching

After forming, the OEL will serve as a V_o reservoir supplying V_o species during set
switching to LRS and absorbing them back during reset to HRS. The actual set/reset
operations will translate geometrically by the modulation of the constriction width
close to the TiN BE, which will, in turn, modulate the electrical resistance level.
Electrically, this translates, similarly to the unipolar switching, into an HRS resistance
level intermediate between LRS and pristine resistance levels.

Note that *only* the constriction will impact the device resistance, so it should be
understood that the OEL element extends from over the rest of the filament. This OEL
element created during forming contains all V_o species required for set/reset operation,
meaning no additional V_o defects are created.

As set/reset voltages (V_{set}/V_{reset}) drop locally on the defective constriction, their amplitude is lower than V_f. In practice, to better characterize the intrinsic switching properties, it is convenient to use a structure having an integrated load resistor as a current limiter (in a so-called 2R structure) [35]. In this scheme, we typically apply triangular pulses to one electrode of the 2R device structure. The applied voltage is acquired on one channel of a high-bandwidth oscilloscope while the current is read out by connecting the opposite electrode in series to the 50-Ω impedance-matched input resistance of the oscilloscope and converting into current the voltage generated on this shunt resistor.

Fig. 1.6 shows typical set/reset I-V curves after subtracting the voltage dropped on the load resistor. For the set switching, we observe that after overcoming a "trigger" set voltage, the voltage snap-backs to a transient V_{trans} voltage, and the transient current rises at this constant V_{trans}, which corresponds to the gradual growth of the constriction. Interestingly, the transient reset is onset after applying an opposite voltage having the same amplitude of V_{trans}, which is followed by a voltage snap-forward due to voltage redistribution during constriction narrowing. Contrary to set, reset operation will thus be self-limited.

V_{trans} was observed to depend neither on the operating current nor on geometrical factors [35]. Hence, V_{trans} constitutes the intrinsic transient switching voltage parameter of the device, which we relate to the V_o mobility measure during switching. Consistently, we also observed that V_{trans} varies in the range 0.4 –0.7 V (Fig. 1.7A and B) depending on the scavenging material although the constriction sits at the opposite side [29]. This effect is due to the "remote" character of the scavenging [30], which means that the scavenging layer modulates the oxygen chemical potential up to the opposite electrode (Fig. 1.7C). Therefore, a strong scavenging layer, such as Hf, will lower the chemical potential and thus increase V_o mobility. In agreement, HfO$_2$\Hf cells exhibit lower V_{reset} parameter than Ta$_2$O$_5$\Ta cells [29].

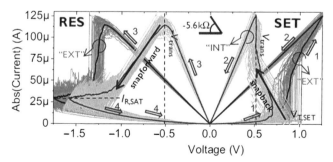

Fig. 1.6 DC cycling on 2R structure ($RA = 5.68$ kΩ) showing the "extrinsic" I-V trace *(blue)* and "intrinsic" *(red)*, both lines are mean over 50 cycles. The abrupt current drop in "extrinsic" RESET (not observed in 1R and 1T1R structures) can be linked to voltage accelerated RESET consequence of the large overvoltage ($|V$-$V_{trans}|$) already present at RESET onset [35].

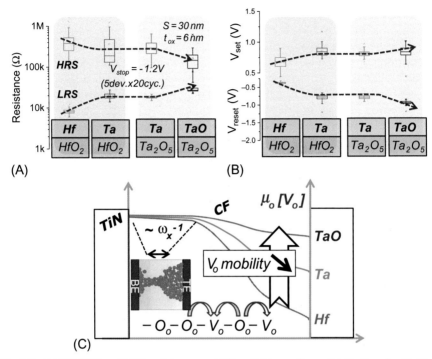

Fig. 1.7 (A) DC HRS and LRS resistance, and (B) switching-voltage distributions for Ta_2O_5 and HfO_2 devices, showing the tuning of switching parameters by the TE material, irrespective of the dielectric material; (C) Schematic representation of the oxygen chemical-potential profile along the CF, tunable by the electrodes and thus allowing modulation of the V_o mobility for fixed CF geometry (ωx) [29].

Quantum point-contact conduction and hour-glass model

We developed both a Quantum Point Contact (QPC) conduction model [36] and a so-called Hour-Glass (HG) model [37] allowing to describe the complete set of experimental electrical properties of our devices.

The QPC model is based on the Landauer-Büttiker formulation which treats the transport as a transmission problem for electrons at the Fermi level. In this approach, the conduction is controlled by the CF constriction length and width, which are inversely proportional to parameters ωx and ωy, respectively, which are determined themselves by a parabolic potential barrier along the filament and a transversal parabolic potential well, respectively, defining a saddle surface in the x, y, E-space (Fig. 1.8A). Since the transport is assumed to be ballistic inside the QPC, the applied potential drops only at the two CF interfaces of the constriction.

Using this QPC model, we have proposed an HG electrical switching model aiming to describe the atomistic movement of the ions while retaining analytical tractability. The proposed model proved successful to model the electrical switching characteristics and statistical fluctuations of our HfO_2 and Ta_2O_5-based devices.

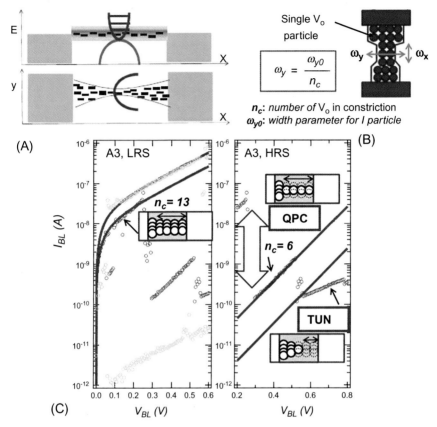

Fig. 1.8 (A) Energy diagram (top) and spatial schematic drawing (bottom) of a filament, assuming many interacting V_o defects forming a conduction band; (B) schematic representation of the CF constriction made of nc defect particles and whose length and width are determined by ωx and ωy geometry parameters; (C) QPC modeling exercise of I-V traces obtained in the course of a reset operation performed on a TiN\ Al$_2$O$_3$\Hf (A3) cell operated at $I_{op} = 1$ μA; dots are experimental I-V$_s$ while solid red lines are nc-dependent QPC traces [38].

This switching model consists of five major components with specific assumptions:

1. *Conduction* model: The current is controlled by QPC at the constriction
2. *Structural* model: The model calculates the number of available V_{os} ("absence" of oxygen ions) constituting the CF after forming. Based on this, the CF is abstracted into three reservoir systems: (i) a top reservoir (TR) extending from the OEL to the constriction interface, (ii) a bottom reservoir (BR) near the constriction and BE interface, and (iii) a current limiting constriction (C). The V_o particles may move from TR to C and to BR or the other way around, like sand moves in an hourglass so that at any time the total number of particles ntotal = nBR + nTR + nc. The number of particles in the constriction nc is the integer state variable which is related to the QPC geometrical parameters of the constriction ωx and ωy (Fig. 1.8B).
3. *Kinetic* model: This model describes the dynamic change of nc where four processes need to be considered: particle emission from C to TR or BR (reducing nc) and emission from TR or BR to C (increasing nc)

4. *Thermal* model: The temperature calculation is based on classical Joule heating applied in the maximum field zones of the CF, and is expressed as $T = \text{Tambient} + a.V.I.R_{th}/nc$, where R_{th} is the effective thermal resistance of the CF. Note that the narrower the constriction, the higher the temperature increases.

5. *Stochastic* component: Finally, to address the stochastic character of the switching, the time constants of V_o emissions are considered as mean values of the exponential time constant distribution. A Monte-Carlo algorithm is used for the set/reset transient modeling with the inclusion of statistical variation.

In our developments and investigations, we used the QPC model for estimating the constriction geometries, and we used the HG model to estimate the local temperature, the statistical variability of the switching parameters, as well as the reliability failures.

Memory performances

During switching, the motion of oxygen species is driven by a thermally assisted drift mechanism. Thanks both to the high temperatures reached locally by the high current density flowing through the narrow constriction, as well as the very short distance required for V_o emission in and out of the constriction, set/reset operations are very fast. Typically the switching transient is in the order of the nanosecond, and the required switching time for stable filament set and reset lies in the range of a few nanoseconds [7, 29]. In Ref. [29] we showed functional switching of TiN\Ta$_2$O$_5$\Ta cells using voltage pulses in the nanosecond range for pulse amplitudes $V_p < 2$ V (Fig. 1.9A). Using these conditions >1 billion write cycles may be performed without failure (Fig. 1.9B).

Similar switching and endurance characteristics may be observed on HfO$_2$\Hf cells [7, 27,39]. In Ref. [27], we showed, however, that endurance lifetime is decreased to ~1 million cycles for devices scaled below 10 nm.

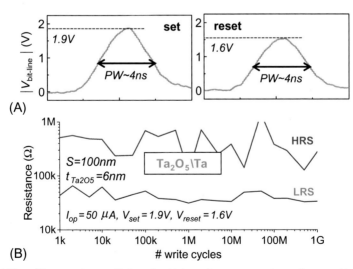

Fig. 1.9 (A) oscilloscope traces of ultra-short triangular programming pulses used to assess write endurance properties of TiN\Ta$_2$O$_5$\Ta devices; (B) 1e9 unverified write cycles reliably obtained using $PW < 5$ ns and $V_p < 2$ V [29].

With respect to switching energy, it may be fairly approximated by multiplying the transient current by $V_{trans} \sim 0.5$ V (Fig. 1.6). Hence, for an operating current of 100 µA, typical set energy and power lie in the range of ~1 pJ and 50 µW, respectively. As the maximum reset current is similar to the maximum set current [35], the reset energy is similar to the set energy.

Overall these results showing sub-10 nm scalability, low switching energy and time, as well as good endurance potential, have triggered intense development efforts to further improve device performances. The next section will detail a few aspects of key material developments.

1.3 Bipolar OxRAM material stack development

1.3.1 Optimization of the scavenging layer for retention

As mentioned in the previous section, the O-affinity of scavenging material controls the oxygen chemical potential (μ_o) at the interface with the oxide. In turn, it controls the forming and switching voltages [40,41]. It also affects the mobility of V_o species involved in the switching. Indeed, using the same operating conditions, stronger scavenging layer results in larger Memory Window (MW = RHRS/RLRS) (Fig. 1.7B), or alternatively lower switching voltage is observed in such case for a same resistive window.

To a lower extent than the *nature* of the material, the *thickness* of the scavenging layer also allows tuning the chemical potential profile through the stack. The beneficial effect of this thickness engineering is specifically observed on the retention characteristics. In Ref. [40] we made an extensive retention study of TiN\Ta$_2$O$_5$\Ta cells having either 10-nm-(Ta10 cells) or 30-nm-thick Ta scavenging layer (Ta30 cells).

We observed that for 10-nm-thick Ta the LRS resistance state (RLRS) distribution has slightly shifted to larger resistances after 10 days at 250°C while the RHRS distribution has hardly moved. On the other hand, and very interestingly, for 30-nm-thick Ta the situation is reversed, meaning that it is the RHRS distribution that has shifted to lower values after baking while RLRS did not move (Fig. 1.10A and B). Both for Ta10 and Ta30 cells, the baking-time dependency of the loss followed a slow power law, such that the state distributions did not move significantly more between 10 and 60 days at 250°C (Fig. 1.10C). The power law is due to a distribution of time constants related to local fluctuations of energy barriers for defect motion.

We attributed these retention results to the modulation of μ_o at the CF\Ta interface by the Ta thickness. In other words, the thicker Ta layer in Ta30 cells increases slightly the heat of formation of Ta$_2$O$_5$ at the interface, favoring thus slightly more the CF reduction as compared to Ta10 cells, and resulting in stabilized LRS. This view is also in agreement with the slightly reduced V_f measured for Ta30.

On the other hand, oxygen drifts back to the CF during reset, which lowers the V_o concentration (V_o) closer to the TiN interface and results thus in a locally richer oxygen environment leading to larger μ_o. This effect thus offsets in HRS the balanced chemical profile met in LRS, leading to less stable HRS. A symmetric description holds for Ta10 cells.

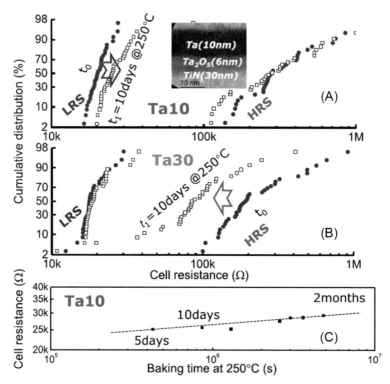

Fig. 1.10 Impact of the Ta thickness on the LRS and HRS retention behaviors of TiN\Ta$_2$O$_5$\Ta cells, showing excellent tunability between (A) more-stable HRS (obtained for thinner Ta), and (B) more-stable LRS (obtained for thicker Ta); inset of (A) shows an XTEM image of the as-integrated Ta10 cell; (C) baking-time dependency of the median resistance of LRS distribution for Ta10 cells [40].

Importantly, these results emphasized the dominant role of the "vertical" diffusion and local redox mechanisms between the CF and the scavenging material over bulk diffusion of V_o species.

Note that a severe baking temperature of 250°C is used in Ref. [40]. When, however, a milder temperature of <150°C is used, retention stability is achieved both for LRS and HRS states on the median level, while only retention tail losses are observed. In Ref. [42] we studied the origin of tail losses for TiN\HfO$_2$\Hf cells and confirmed that the tail bits originate from low activation-energy (E_a~0.5 eV) diffusing species, which are understood as metastable Oxygen (O)-ions moving along the CF between the constriction and the scavenging layer. We will come back to this study in Section 1.4.

1.3.2 Optimization of the bottom electrode for endurance

In Ref. [43] we investigated the endurance characteristics of TiN\Ta$_2$O$_5$\Ta cells as a function of set/reset pulse amplitudes and pulse width. We showed that the use of a standard 10-ns-long reset pulse of 1.5 V resulted in endurance lifetime of 1 million cycles. From this point, we showed that the increase of both the reset voltage V_{reset} and pulse width PW_{reset} resulted in a drastic decrease of the lifetime, while the set pulse had a minor impact on endurance behavior [43].

On the other hand, HG simulations of the CF temperature during switching revealed that the CF temperature is much larger during reset than during set. The main reasons are as follows: (i) during set transient the voltage snap-backs to a lower standby voltage (V_{trans}~0.5 V) while most of the voltage redistributes to the select transistor; (ii) the CF temperature is calculated from the power dissipation in the constriction, which consists of less nc defects in HRS than in LRS, and is thus larger in HRS, resulting in larger temperature rise during reset transient.

Furthermore, HG simulations showed large temperature fluctuations during reset, which is related to the nc fluctuations due to the competition between reset-induced V_{o}-emission and thermal-induced V_{o}-capture in the narrow HRS constriction, which also accounts for the current noise typically observed during reset and is a signature of this dynamic balance [44].

Hence, endurance lifetime appeared to be controlled by the large temperature peaks occurring during reset, and we showed it could be increased to 100 million cycles by using ns-short reset pulses of moderate amplitude [28, 43].

The chemical origin of the thermal-induced degradation lies in the chemical interaction between the oxide and the TiN BE. Indeed, in the previous section, we described the conical shape of the CF, having its "hot" constriction close to the TiN BE. And although TiN may be considered as an "inert" electrode for bipolar functionality demonstration, its nonideal character is revealed by endurance stress. We calculated by ab initio the free enthalpy of oxidation ΔG of BE materials by Oxygen (O) from the CF. Large ΔG ~6 eV is obtained for ideal TiN crystal, however, it drops to -2 eV for pure Ti (Fig. 1.11A). In PVD-deposited TiN layers, any point defects, such as nitrogen vacancies (V_{N}) or Ti dangling bonds will have the same effect of locally decreasing ΔG and will result in TiN oxidation.

Hence, gradual TiN oxidation by these microscopic mechanisms is likely induced by the large reset field and temperature during endurance, leading finally to failure to set (failure B in Fig. 1.11) due to difficult O-detrapping from TiN, which is the failure mode observed for "strong" reset conditions.

On the other hand, endurance stress using "softer" conditions with lower reset field and temperature may be insufficient for TiN oxidation, however, sufficient for O in-diffusion as interstitials within TiN or simply through defects like grain boundaries, which would lead finally to failure to reset (failure A in Fig. 1.11).

From ab initio, the use of a Ru BE would result not only in restored large ΔG but also in reduced O in-diffusion (Fig. 1.11B), which should improve the immunity to failures B and A, respectively. Note that the crystal structures used in the calculation for Ru are hexagonal closed packed and cubic for TiN. In agreement, experimental endurance tests on Ru\Ta$_2$O$_5$\Ta devices confirmed the ab initio predictions, as shown in Fig. 1.11F. In Ref. [28] we showed further that even for large temperatures and fields associated with a reset pulse of -2.3 V, no failure was observed before 1 billion endurance cycles.

Using the same reasoning, the use of oxide materials having a lower enthalpy of metal oxidation is predicted to result in improved robustness to O-injection into the

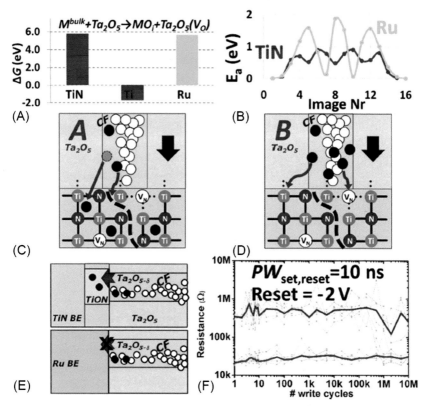

Fig. 1.11 Ab initio calculation of the energy barrier required for O (A) to react (ΔG) with BE materials and (B) to diffuse (E_a) in TiN and Ru; (C) and (D) schematics describing possible degradation mechanisms accounting for failure modes A and B, where O is shown in black. (E) Schematics depicting suppressed O-induced BE degradation during switching using Ru. (F) Endurance characteristics obtained on Ru\Ta$_2$O$_5$\Ta devices tested in standard condition except for $V_{reset} = -2$ [43].

BE material. As expected, TiN\HfO$_2$\Hf cells showed increased endurance lifetime in various cycling conditions, as compared to TiN\Ta$_2$O$_5$\Ta [28].

Note overall that the guideline to use Ru as BE material not only helps improve the endurance lifetime but also benefits the MW because the use of Ru allows using safely larger reset voltage pulses.

1.3.3 Optimization of the oxide for lower operating current

For storage-class memory application, the OxRAM device is required to be integrated with a two-terminal selector device in a very dense array configuration. In order to minimize the voltage drop on the array access lines, as well as to comply with the maximum current supplied by the selector, the OxRAM device is required to operate at low current. To this purpose, the optimization of the oxide element is a key aspect.

1.3.3.1   Optimization of Al₂O₃\HfO₂ system

In Refs. [32,33,38,45]. we showed that an enhanced stack asymmetry helped improve the device performances at lower operating current. This is achieved through oxygen chemical potential profile engineering from BE to TE. In this respect, the incorporation of Al element close to the BE proved most beneficial, as will be detailed in this part.

We observed by IPE characterization that the remote scavenging induced by the Hf layer down to the BE interface is suppressed when we change the oxide from HfO_2 to an Al_2O_3 layer processed in the same conditions [45]. This result indicates that V_o defects are less easily created in the Al_2O_3 layer interfacing the BE, which is supported also by ab initio calculations (Fig. 1.12) [46] and translates in a larger oxygen chemical potential.

Therefore we designed a $TiN\backslash Al_2O_3\backslash HfO_2\backslash Hf$ stack where the thickness of each layer was engineered to enhance the μ_o profile through the stack [33]. This stack allows to electrically form conical filaments having a *narrower* constriction close to BE, which thus shows larger state resistances [33]. The advantage of programming controllably in particular, a larger HRS resistance is that the cell may be operated at lower current I_{op}, resulting in a large LRS resistance, while still showing potentially a large MW, as shown in Fig. 1.12B for $I_{op} = 10\,\mu A$.

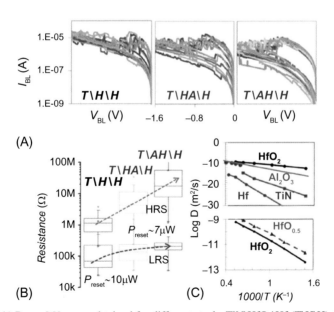

Fig. 1.12 (A) Reset I-V traces obtained for different stacks TiN\HfO₂\Hf (T\H\H), TiN\HfO₂\ Al₂O₃\Hf (T\HA\H), and TiN\ Al₂O₃\HfO₂\Hf (T\AH\H), using forming and set current of $10\,\mu A$; (B) distribution of LRS and HRS resistances extracted at $V_{BL} = -20\,mV$ from 100 I-V traces (5 cells × 20 cycles); (C) ab initio calculations of O-diffusion coefficient in different materials [33].

In Refs. [38, 45], we engineered further the TiN\ Al_2O_3\$HfO_2$\Hf stack for optimizing memory performances obtained using $I_{op} < 10\,\mu A$. We evidenced that the sub-$10\,\mu A$ operation is largely affected by the defect chemistry of the dielectric materials. In particular, we found that data retention is affected when the constriction is located within the Al_2O_3 layer, due to the strong O–Al affinity, and we, therefore, engineered the stack for controlling the constriction location at the Al_2O_3\$HfO_2$ interface in order to optimize the trade-off between retention and MW. The improved control on CF shape and constriction also proved beneficial on the switching variability.

1.3.3.2   Al incorporation in HfO₂ system

Operating OxRAM cells at lower currents generate narrower CF constrictions, and thus less nc defects (Fig. 1.8C), which results in weaker LRS data retention [47].

One effective solution to improve this key parameter of LRS retention is to inhibit the V_o diffusion from the constriction toward the oxide bulk matrix. This is not addressed satisfactorily when the constriction sits at the Al_2O_3\HfO_2 interface because V_o species may readily diffuse from the constriction toward the HfO_2 layer. To better address this issue we also developed an ALD process where Al_2O_3 cycles are alternated with HfO_2 cycles in order to incorporate Al uniformly to the HfO_2 layer during deposition and generate a Hf-Al-O ternary compound [27,48], wherein the diffusion of oxygen is slowed down, resulting in further improved LRS retention [48].

1.3.3.3   H incorporation in Ta₂O₅ system

There are other possible engineering knobs to improve the LRS retention at low current. For example, in Ref. [49] we improved the LRS stability of TiN\Ta_2O_5\Ta cells through NH3 treatment of the Ta_2O_5 layer. From electrical and ab initio analyses we suggested that this improvement originated from the formation of O–H complexes, which both stabilize V_o defects and slow down the diffusion of oxygen species, resulting in retarded retention loss.

1.3.3.4   Gd₂O₃ system

Inspired by the supposed beneficial impact of H-species incorporation in oxide layers [49], we considered hygroscopic oxide materials for OxRAM application, and we developed TiN\Gd_2O_3\Hf cells. As expected we evidenced a high level of H species in the Gd_2O_3 layer by ToF-SIMS [50].

As compared to HfO_2 and Ta_2O_5 baselines, Gd_2O_3-based cells clearly showed larger MW at low current as well as long endurance lifetime. Fig. 1.13A shows typical endurance properties achieved on our 40 nm-size cells, showing no failure up to 1e12 write cycles at $50\,\mu A$ [50]. Such long endurance lifetime is to our knowledge a record achievement in the case of TiN BE, which shows that OxRAM properties do not systematically exhibit a MW/endurance trade-off (Fig. 1.13B).

As for the large MW, it is actually a result of the programming to larger HRS resistance level, which we attributed to the improved constriction oxidation process allowed by the involvement of OH– species during reset.

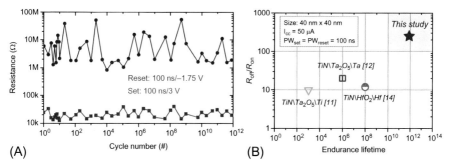

Fig. 1.13 (A) 1e12 Write endurance lifetime obtained on a 40-nm-size TiN\Gd$_2$O$_3$\Hf cell operated at $I_{cc} = 50\,\mu A$; (B) Endurance benchmarking between different OxRAM stacks fabricated by imec using the same test vehicle; all data are collected from 40 nm × 40 nm cells with the same programming PW and I_{cc}; between all different materials, the TiN\Gd$_2$O$_3$\Hf cell *(blue star)* offers best switching performance both in terms of endurance and MW [50].

This improved reset efficiency also allowed to generate promising properties at low $I_{op} = 10\,\mu A$, which we reported in Ref. [51].

1.4 Bipolar OxRAM key reliability concerns

1.4.1 *Switching variability*

In Ref. [38,45], we evidenced that the behavior of the cells deviates considerably when I_{op} is decreased below $15\,\mu A$, which we associated to an incomplete voltage snapback during forming/set operations and a change of switching regime from the progressive to the soft-breakdown regime (SBD). Due to this change of regime, we showed in particular that sub-μA operation is largely affected by the defect chemistry of the dielectric material in the SBD regime due to the increased impact of single-V_o defects over subband characteristics.

We also show that the switching variability originates from both (i) the large impact of single V_o-defect variation on the current level controlled by only a few particles (nc < 15) in the QPC constriction, and (ii) the variations of geometry parameters (both ωy and ωx) of the constriction itself (Fig. 1.8C).

In order to address the respective contributions of these variability causes, we carried out an extensive switching variability characterization and modeling of TiN\HfO$_2$\Hf devices in Ref. [52].

Firstly, by comparing C2C and device-to-device (D2D) resistance distribution parameters (median μ and dispersion σ), we showed that once extrinsic processing factors are under control the switching variability is mainly intrinsically generated by the single switching event, which is stochastic. In other words, it does not depend on the cell size and is marginally affected by geometrical factors or by the microstructure of the HfO$_2$ layer, but it is dominated by the CF properties themselves which are altered cycle after cycle.

Endurance tests also confirmed the uncorrelation between states produced by set and reset from cycle to cycle, confirming the stochastic character of the switching.

Secondly, we evidenced two distinct populations in the resistance distribution (Fig. 1.14A):

1. a population (A) obeying a lognormal distribution, and which originates from the width (ωy) variation of the CF constriction
2. a bent component (B) of the LRS distribution, which can be observed above $30\,k\Omega$ and with a slope matching the HRS distribution slope

Using Monte-Carlo simulations, we reproduced these variability results (Fig. 1.14B and C) and we were able to associate the population (A) to the fluctuation of nc, the

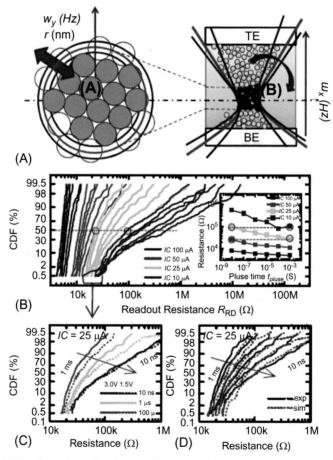

Fig. 1.14 (A) Physical origin of resistance dispersion: fluctuations of nc defining the constriction radius dominate the variability in region (A), while fluctuations in the constriction geometry (ωx, ωy) become the dominant component in (B); (B) D2D resistance distribution for different compliances and pulse width PW (10 ns to 1 ms); (C) LRS distribution at $I_{cc} = 25\,\mu A$ for different V_{set}; (D) Measured and simulated distribution for $I_{cc} = 25\,\mu A$ [52].

number of (equally sized) V_o defects defining *the radius* of the CF narrowest cross section, while the population (B) was related to the change of *constriction shape* (ωx and ωy fluctuations) occurring for fixed nc.

These results indicate that OxRAM devices operated at lower currents will intrinsically generate large state resistance distribution, mainly accounted for by the fluctuations of the constriction geometry. This is a serious intrinsic limitation of the technology.

1.4.2   Program instability

The well-known "Write-verify" method appears as an obvious solution to circumvent the large switching variability of OxRAM, and indeed it allows to straighten the distributions when the cell resistance is read out shortly after programming [51].

However, in Ref. [53] we uncovered the dynamics of the state fluctuations and evidenced a program instability phenomenon. This instability clearly questions the relevance of verifying methods.

In this work, we used cumulative or noncumulative incremental step pulse programming (ISPP) verify methods to open a memory window between 20 and 200 kΩ using 20-nm-size TiN\HfO$_2$\Hf cells. The reset adaptive algorithm was carried out by stepping the Bit-Line (BL) pulse voltage amplitude, while for the set algorithm it is the gate voltage amplitude that was stepped up in order to control an increased I_{op} from 30 µA to a maximum of 70 µA. Resistance readout takes place immediately after each programming pulse with a sampling time of 100 µs, and both algorithms stop as soon as the target resistance is reached.

Fig. 1.15 shows the effectiveness of such algorithms in straightening the LRS and HRS distributions. However, a different scenario arises if the same resistance is read out again after a delay of 1 s: the distribution appears to revert back to its single pulse equivalent.

A more complicated and counter-intuitive picture emerges if a subpopulation of 10% in the upper part, the center part or the lower part of the ISPP verified distribution is isolated and compared after 1 s: it is observed that all subpopulations have spread out to equally wide distributions while retaining at the same time their respective median resistance level (Fig. 1.15C and D). These regenerated distributions also share a similar shape to the one produced by a single set/reset pulse programming. Interestingly, even the bits initially programmed in the lowest 10% LRS resistance subgroup may, after 1 s, be found in the opposite side of the distribution.

This phenomenon is referred to as resistance *programming instability* [53].

We used a specific procedure to monitor the resistance evolution using a set of 9 log-spaced readout pulses between 100 µs and 1 s, and we quantified the correlation loss against the time-from-program ΔTP. We observed that the correlation ρ shows the same rate of decay when plotted against $\log(\Delta TP)$ implying a process that uniformly degrades in each time-decade elapsed from programming. We verified that this effect neither depends on the read voltage up to 0.5 V (to discard possible disturb origin) nor on program pulse width and trailing edge time (to discard possible thermal effects). Only the use of stronger programming current is able to reduce the rate of correlation loss [53].

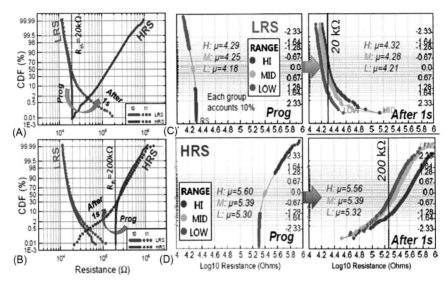

Fig. 1.15 Program instability: (A) distribution of programmed resistances produced by Set- and (B) Reset-adaptive algorithms; while initially successful, after 1 s the verify level is lost and the distributions revert back to their single pulse value; (C) evolution of isolated subpopulation (10% each) for set and (D) Reset programming algorithm; after 1 s, all groups regenerate similar distributions; note that, while initial placement is lost, each group maintains its median [53].

From the analysis of the resistance distribution changes between all consecutive sampling intervals, we evidenced a bimodal behavior, where only one of the two populations (~5% of states) significantly contributes to the cell instability through large resistance jumps. When such a large jump occurs, states with low or high resistance tends to be pulled back to the center of the distribution. Furthermore, the magnitude of the jump is also related to the distance from the median, behaving like a (damped) spring-mass system.

From a physical point of view, this can be interpreted using the HG switching model. Each programming event creates a different nc number of current controlling conductive defects in the constriction. This number remains constant in time (fixing the median resistance level), while the shape of the constriction containing this fixed number of defects changes over time (see schematic representation in Fig. 1.14A). The constriction shape, which is described by ωx and ωy and is related to the microscopic arrangement of the defects, tends to *rearrange itself around a rest position*. The exact physics behind remains unclear but may be attributed to the mechanical stress relaxation or the structural relaxation immediately after programming.

To summarize, the origin of the program instability lies in the CF shape fluctuation over time for a fixed nc. On the other hand, the switching variability previously described may be understood as the *convolution of the nc dispersion from C2C and the intrinsic constriction shape fluctuations*.

From a device viewpoint, the program instability issue substantially impairs the possibility of adaptively controlling the resistance level, and poses a serious obstacle to the adoption of filamentary OxRAM for low-current applications due to strong switching variability.

1.4.3   Retention tails

As mentioned earlier, lower program currents result in lower nc in the CF constriction, which results in weaker LRS retention properties [47]. Although we demonstrated in Ref. [48] that Al incorporation may suppress the bulk diffusion of V_o particles out of the constriction, the low activation-energy (E_a ~0.5 eV) diffusing O-ions species moving along the CF between the constriction and the scavenging layer will always generate a retention tail loss to some extent [42].

In Ref. [42] we investigated TiN\HfO$_2$\Hf cells in which programming knobs may allow to mitigate this issue. We found out that not only the I_{op} amplitude but also the set *PW* greatly affect the LRS retention.

We firstly showed that the LRS state resistance programmed using $PW_{set} = 200$ ns at 100 µA is lower than the resistance programmed using $PW_{set} = 100$ ms at 50 µA. In spite of the shorter set pulse used for the former conditions, this result is in agreement with the higher nc obtained for higher I_{op}. However, we observed that the LRS retention loss was clearly larger in the former programming condition than in the latter, clearly indicating that this LRS far-end tail formation is not dominantly determined by nc in the constriction. Rather, it is highly improved using long set pulses, which is effective to drift O-ion species from the constriction toward far in the OEL element. To the contrary, for short set pulses, these species remain in the neighborhood of the constriction and readily recombine with V_o defects in constriction, generating LRS tail loss [42]. These findings were also supported by HG simulations [42, 54].

This work showed that retention tails may be mitigated, however, at the expense of long write times that might not be affordable from product viewpoint.

1.4.4   Impact of programming history

In Ref. [55] we carried out further extensive retention study at 150°C, considering also the impact of conditioning programming history of the cell prior to the set program pulse. In particular, we confirmed the substantial impact of the reset pulse width prior to set programming, as well as the duration between these pulses (Fig. 1.16A).

Based on the Program Instability phenomenon (Fig. 1.15), the resistance of the programmed bits may jump to any resistance value described by the entire single-pulse resistance state distribution. Therefore, the correlation to the programmed resistance gets lost very fast, and thus longer-duration retention studies need to consider the single-pulse resistance distribution as a whole. In our methodology, the retention loss considers the evolution of a given percentile of the distribution with time, instead of particular bits. As shown in Fig. 1.16B, we determined the time-to-fail as the baking time at which $R_{10\%} > 2 \times R_{10\%}$, initial, where $R_{10\%}$ stands for the resistance of the 10th percentile, as chosen in order to focus the study on the tail bits.

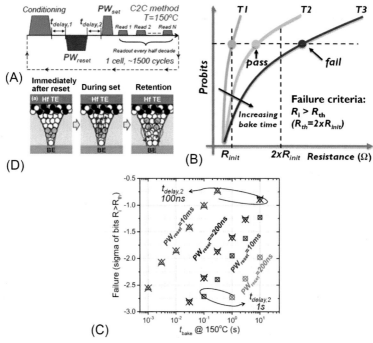

Fig. 1.16 (A) Detailed programming scheme using multiple readout pulses after set programming up to 10 s and repeated over 1500 cycles; the impacts of PW_{reset} and delay durations ($t_{delay,1}$ and $t_{delay,2}$) on LRS stability are studied; (B) Metrics used to quantify the LRS retention failure bits: at each sampling time t_i, the resistance R_i of each quantile is compared to its corresponding R_{th} (where $R_{th} = 2 \times R_{init}$), and the failure is defined when $R_i > R_{th}$; (C) Summary of failure populations for different PW_{reset} and $t_{delay,2}$ on TiN\HfO₂\Hf cells; (D) Cartoon explaining that insufficient delay $t_{delay,2}$ after reset leaves O-ions in the neighborhood of the constriction, leading to LRS failure bits [55].

Fig. 1.16C shows that shorter PW_{reset} prior to set allows decreasing the failing LRS population. The origin of this effect is as follows: a short reset allows decreasing nc effectively without drifting a large quantity of O-ion species in the neighborhood of the constriction, which would account for LRS failure bits (Fig. 1.16D). Subsequently, during set programming, this resulting low amount of O-ion species will be evacuated more efficiently far from the constriction.

Similarly, we evidenced a beneficial impact of the delay $t_{delay,2}$ after the reset pulse (Fig. 1.16C), which can be understood as follows: after the reset pulse, the O-ion species drifted close to the constriction gradually diffuse back toward the scavenging layer. Therefore, a longer $t_{delay,2}$ will result in a more pronounced depletion of these species close to the constriction, which will thus improve the retention properties (Fig. 1.16C).

Using the optimum programming conditions based on these learnings, which was applied to the optimized imec stack, we have demonstrated excellent retention properties of LRS and HRS states after 30 days at $T = 150°C$ [55].

Overall, these works allowed to obtain improved LRS retention properties using (i) long set write, (ii) short reset pulse prior to set, and (iii) long delay between these pulses.

Again, from product design viewpoint this sensitivity to program history issue may be accommodated by appropriate algorithms, however, at the expense of write latency.

1.5 Conclusion and outlook

To conclude, OxRAM technology has undoubtedly made tremendous progress over the past decade. The investigations carried out at imec have allowed proving sub-10 nm scalability, low-power, and reliable operation in the current range of $100 \mu A$, using a write pulse width ~100 ns, in which conditions >1e12 write endurance and >30 days retention stability at 150°C are achieved.

In this chapter we have also shown some guidelines for further improvements of the device stacks, showing the key impacts of the bottom electrode, oxide and scavenging layer materials, and thickness. We have also pointed out serious concerns with respect to the intrinsic variability, program instability and retention tail loss issues, which will be strong obstacles to overcome for low-current ($<50 \mu A$) applications.

Under the current situation, it is not surprising that high-current embedded applications are the first segments where OxRAM made it up to the product. OxRAM has indeed been commercialized for the first time in 2015 as embedded in microcontrollers by Panasonic [56].

Further developments in the embedded area with 1T1R structure will need to focus on the lowering of the write voltage to keep compatibility with the core transistor voltage along with the node scaling to 28 nm or even 20 nm.

The main challenges of OxRAM in the future are obviously for high-density storage-class memory application, where the variability and retention issues need to be tackled and solved.

SCM applications indeed require reasonably low operating current levels, for which unacceptable variability and retention characteristics are systematically observed. In addition, these memory characteristics will need to be investigated as operated in combination with a two-terminal selector device, as required for the high-density purpose. Not only the intrinsic variability of such two-terminal selector will add to the OxRAM memory element, but the control of the ON current will not be as efficient as for a select transistor, which may further degrade the dispersion and window closure of the cell.

Therefore, a large emphasis will be required in the design and optimization of an appropriate selector system. Regarding the memory element, a solution may also arise from the identification of a system allowing a strong increase of the single-pulse memory window in the whole distribution. In all these areas, device physics and material science will be key expertise centers and will define the future of this technology.

References

[1] C. Lee, Conventional memory technologies: DRAM, in: IEDM Short Course-Memory Technologies for Future Systems, 2015. IEEE.

[2] M.R. Zargham, Computer Architecture: Single and Parallel Systems, Prentice-Hall, Inc., Upper Saddle River, NJ, 1996.

[3] 3D XPoint™ Technology Revolutionizes Storage Memory, www.youtube.com. (video, infomercial), Intel.

[4] M. Albert, 7 Things to Know about the Internet of Things and Industry 4.0, 2015.

[5] R. Waser, R. Dittmann, G. Staikov, K. Szot, Redox-based resistive switching memories—nanoionic mechanisms, prospects, and challenges, Adv. Mater. 21 (2009) 2632–2663.

[6] H.S.P. Wong, H.Y. Lee, S. Yu, Y.S. Chen, Y. Wu, P.S. Chen, B. Lee, F.T. Chen, M.J. Tsai, Metal-Oxide RRAM, Proc. IEEE 100 (June 2012) 1951–1970.

[7] B. Govoreanu, et al., 10 x 10nm2 Hf/HfOx crossbar resistive RAM with excellent performance, reliability and low-energy operation, in: 2011 IEEE International Electron Devices Meeting (IEDM), Dec 2011, 2011, pp. 31.6.1–31.6.4.

[8] H.Y. Lee, et al., Low power and high speed bipolar switching with a thin reactive Ti buffer layer in robust HfO2 based RRAM, in: 2008 IEEE International Electron Devices Meeting, Dec 2008, 2008, pp. 1–4.

[9] K.M. Kim, D.S. Jeong, C.S. Hwang, Nanofilamentary resistive switching in binary oxide system; a review on the present status and outlook, Nanotechnology 22 (2011) 254002. 1–17.

[10] L. Goux, I. Valov, Electrochemical processes and device improvement in conductive bridge RAM cells, Phys. Status Solidi A 213 (2) (2016) 274–288.

[11] L. Goux, P. Czarnecki, Y.Y. Chen, L. Pantisano, X.P. Wang, R. Degraeve, B. Govoreanu, M. Jurczak, D.J. Wouters, L. Altimime, Evidences of oxygen-mediated resistive-switching mechanism in TiN\HfO2\Pt cells, Appl. Phys. Lett. 97 (2010) 243509.

[12] D. Ielmini, Resistive switching memories based on metal oxides: mechanisms, reliability and scaling, Semicond. Sci. Technol. 31 (6) (2016) 063002.

[13] B. Govoreanu, et al., Vacancy modulated conductive oxide resistive RAM (VMCO-RRAM): an areascalable switching current, self-compliant, highly nonlinear and wide on/offwindow resistive switching cell, in: 2013 IEEE International Electron Devices Meeting, Dec 2013, pp. 10.2.1–10.2.4.

[14] Luo, et al., IEDM, 2017.

[15] T.W. Hickmott, Low-frequency negative resistance in thin anodic oxide films, J. Appl. Phys. 33 (1962) 2669–2682.

[16] J.F. Gibbons, W.E. Beadle, Switching properties of thin NiO films, Solid State Electron. 7 (1964) 785.

[17] G. Dearnaley, D.V. Morgan, A.M. Stoneham, A model for filament growth and switching in amorphous oxide films, J. Non-Cryst. Solids 4 (1970) 593–612.

[18] D.P. Oxley, Electroforming, switching and memory effects in oxide thin films, Electrocompon. Sci. Technol. 3 (1977) 217–224.

[19] L. Goux, S. Spiga, Unipolar resistive switching mechanisms, in: Resistive Switching—From Fundamentals of Nanoionic Redox Processes to Memristive Device Applications, Wiley-VCH, 2015.

[20] Baek, et al., IEDM, 2005.

[21] L. Goux, J.G. Lisoni, M. Jurczak, D.J. Wouters, L. Courtade, C. Muller, Coexistence of the bipolar and unipolar resistive-switching modes in NiO cells made by thermal oxidation of Ni layers, J. Appl. Phys. 107 (024512) (2010) 1–7.

[22] L. Goux, J.G. Lisoni, X.P. Wang, M. Jurczak, Wouters, Optimized Ni oxidation in 80-nm contact holes for integration of forming-free and low-power Ni/NiO/Ni memory cells, IEEE Trans. Electron Devices 56 (10) (2009) 2363–2368.

[23] L. Goux, R. Degraeve, J. Meersschaut, B. Govoreanu, D.J. Wouters, S. Kubicek, M. Jurczak, Role of the anode material in the unipolar switching of TiN\NiO\Ni cells, J. Appl. Phys. 113 (2013) 054505.

[24] L. Goux, Y.-Y. Chen, L. Pantisano, X.-P. Wang, G. Groeseneken, M. Jurczak, D.J. Wouters, On the gradual unipolar and bipolar resistive switching of TiN\HfO2\Pt memory systems, Electrochem. Solid-State Lett. 13 (6) (2010) G54–G56.

[25] Y.S. Kim, J.S. Kim, J.S. Choi, I.R. Hwang, S.H. Hong, S.O. Kang, B.H. Park, Resistive switching behaviors of NiO films with controlled number of conducting filaments, Appl. Phys. Lett. 98 (2011) 192104.

[26] D. Ielmini, F. Nardi, C. Cagli, Universal reset characteristics of unipolar and bipolar metal-oxide RRAM, IEEE Trans. Electron Devices 58 (10) (2011) 3246–3253.

[27] A. Fantini, L. Goux, A. Redolfi, R. Degraeve, G. Kar, Y.Y. Chen, M. Jurczak, Lateral and vertical scaling impact on statistical performances and reliability of 10nm TiN/Hf(Al)O/ Hf/TiN RRAM devices, in: Symposium on VLSI Technology Digest of Technical Papers, 2014, pp. 242.

[28]. C.Y. Chen, L. Goux, A. Fantini, A. Redolfi, S. Clima, R. Degraeve, Y.Y. Chen, G. Groeseneken, M. Jurczak, "Understanding the impact of programming pulses and electrode materials on the endurance properties of scaled Ta2O5 RRAM cells", IEDM14, 355

[29] L. Goux, A. Fantini, A. Redolfi, C.Y. Chen, F.F. Shi, R. Degraeve, Y.Y. Chen, T. Witters, G. Groeseneken, M. Jurczak, Role of the ta scavenger electrode in the excellent switching control and reliability of a scalable low-current operated TiN\Ta2O5\ta RRAM device, in: Symposium on VLSI Technology Digest of Technical Papers, 2014, pp. 130.

[30] V.V. Afanas'ev, A. Stesmans, L. Pantisano, S. Cimino, C. Adelmann, L. Goux, Y.Y. Chen, J.A. Kittl, D. Wouters, M. Jurczak, TiNx/HfO2 interface dipole induced by oxygen scavenging, Appl. Phys. Lett. 98 (2011) 132901.

[31] S. Clima, et al., First-principles thermodynamics and defect kinetics guidelines for engineering a tailored RRAM device, J. Appl. Phys. 119 (22) (2016) 225107.

[32] L. Goux, A. Fantini, G. Kar, Y.-Y. Chen, N. Jossart, R. Degraeve, S. Clima, B. Govoreanu, G. Lorenzo, G. Pourtois, D.J. Wouters, J.A. Kittl, L. Altimime, M. Jurczak, Ultralow sub-500nA operating current high-performance TiN\Al2O3\HfO2\Hf\TiN bipolar RRAM achieved through understanding-based stack-engineering, in: Symposium on VLSI Technology Digest of Technical Papers, 2012, pp. 159.

[34] U. Celano, L. Goux, R. Degraeve, et al., Imaging the three-dimensional conductive channel in filamentary-based oxide resistive switching memory, Nano 15 (12) (2015) 7970–7975.

[35] A. Fantini, D.J. Wouters, R. Degraeve, L. Goux, L. Pantisano, G. Kar, Y.-Y. Chen, B. Govoreanu, J.A. Kittl, L. Altimime, M. Jurczak, Intrinsic switching behavior in HfO2 RRAM by fast electrical measurements on novel 2R test structures, in: IMW, 2012, pp. 45.

[36] R. Degraeve, P. Roussel, L. Goux, D. Wouters, J. Kittl, L. Altimime, M. Jurczak, G. Groeseneken, Generic learning of TDDB applied to RRAM for improved understanding of conduction and switching mechanism through multiple filaments, in: IEDM, 2010, pp. 632–635. 978-1-4244-7419-6/10/$26.00.

[37] R. Degraeve, A. Fantini, S. Clima, B. Govoreanu, L. Goux, Y.Y. Chen, D.J. Wouters, P. Roussel, G.S. Kar, G. Pourtois, S. Cosemans, J.A. Kittl, G. Groeseneken, M. Jurczak, L. Altimime, Dynamic 'hour glass' model for SET and RESET in HfO2 RRAM, in: Symposium on VLSI Technology Digest of Technical Papers, 2012, pp. 75.

[38]   L. Goux, N. Raghavan, A. Fantini, R. Nigon, S. Strangio, R. Degraeve, G. Kar, Y.Y. Chen,
       F. De Stefano, V.V. Afanas'ev, M. Jurczak, On the bipolar resistive-switching characteris-
       tics of Al2O3- and HfO2-based memory cells operated in the soft-breakdown regime, J.
       Appl. Phys. 116 (2014) 134502.
[39]   Y.Y. Chen, B. Govoreanu, L. Goux, R. Degraeve, A. Fantini, G.S. Kar, D.J. Wouters,
       G. Groeseneken, J.A. Kittl, M. Jurczak, L. Altimime, Balancing SET/RESET pulse for >
       1010 endurance in HfO2/Hf 1T1R bipolar RRAM, IEEE Trans. Electron Devices 59 (12)
       (2012) 3243.
[40]   L. Goux, A. Fantini, Y.Y. Chen, A. Redolfi, R. Degraeve, M. Jurczak, Evidences of
       electrode-controlled retention properties in Ta2O5-based resistive-switching memory
       cells, ECS Solid State Lett. 3 (11) (2014) Q79–Q81.
[41]   Y.Y. Chen, L. Goux, S. Clima, B. Govoreanu, R. Degraeve, G.S. Kar, A. Fantini,
       G. Groeseneken, D.J. Wouters, M. Jurczak, Endurance/retention trade-off on HfO2/metal
       cap 1T1R bipolar RRAM, IEEE Trans. Electron Devices 60 (3) (2013) 1114.
[42]. C.Y. Chen, A. Fantini, L. Goux, R. Degraeve, S. Clima, A. Redolfi, G. Groeseneken,
       M. Jurczak, "Programming-conditions solutions towards suppression of retention tails of
       scaled oxide-based RRAM", IEDM15, 261
[43]   C.Y. Chen, L. Goux, A. Fantini, S. Clima, R. Degraeve, A. Redolfi, Y.Y. Chen,
       G. Groeseneken, M. Jurczak, Endurance degradation mechanisms in TiN\Ta2O5\ta resis-
       tive random-access memory cells, Appl. Phys. Lett. 106 (2015) 053501.
[44]   R. Degraeve, A. Fantini, N. Raghavan, Y.Y. Chen, L. Goux, S. Clima, S. Cosemans,
       B. Govoreanu, D.J. Wouters, P. Roussel, G.S. Kar, G. Groeseneken, M. Jurczak, Modeling
       RRAM set/reset statistics resulting in guidelines for optimized operation, in: Symposium
       on VLSI Technology Digest of Technical Papers, 2013, pp. T98.
[33]   L. Goux, A. Fantini, B. Govoreanu, G. Kar, S. Clima, Y.-Y. Chen, R. Degraeve,
       D.J. Wouters, G. Pourtois, M. Jurczak, Asymmetry and switching phenomenology in
       TiN\(Al2O3)\HfO2\Hf systems, ECS Solid State Lett. 1 (4) (2012) P63–P65.
[45]   L. Goux, A. Fantini, R. Degraeve, N. Raghavan, R. Nigon, S. Strangio, G. Kar, D.J. Wouters,
       Y.Y. Chen, M. Komura, F. De Stefano, V.V. Afanas'ev, M. Jurczak, Understanding of the
       intrinsic characteristics and memory trade-offs of sub-µA filamentary RRAM operation,
       in: Symposium on VLSI Technology Digest of Technical Papers, 2013, pp. T162.
[46]   S. Clima, K. Sankaran, Y.Y. Chen, A. Fantini, U. Celano, A. Belmonte, L. Zhang, L. Goux,
       B. Govoreanu, R. Degraeve, D.J. Wouters, M. Jurczak, W. Vandervorst, S. De Gendt,
       G. Pourtois, RRAMs based on anionic and cationic switching: a short overview, Phys.
       Status Solidi RRL (2014) 1–11.
[47]   Y.Y. Chen, M. Komura, R. Degraeve, B. Govoreanu, L. Goux, A. Fantini, N. Raghavan,
       S. Clima, L. Zhang, A. Belmonte, A. Redolfi, G.S. Kar, G. Groeseneken, D.J. Wouters,
       M. Jurczak, Improvement of data retention in HfO2/Hf 1T1R RRAM cell under low op-
       erating current, in: IEDM, 2013, pp. 252–255.
[48]   A. Fantini, L. Goux, S. Clima, R. Degraeve, A. Redolfi, C. Adelmann, G. Polimeni,
       Y.Y. Chen, M. Komura, A. Belmonte, D.J. Wouters, M. Jurczak, Engineering of Hf1-
       xAlxOy amorphous dielectrics for high-performance RRAM applications, in: IMW,
       2014.
[49]   L. Goux, J.Y. Kim, B. Magyari-Kope, Y. Nishi, A. Redolfi, M. Jurczak, H-treatment im-
       pact on conductive-filament formation and stability in Ta2O5-based resistive-switching
       memory cells, J. Appl. Phys. 117 (2015) 124501.
[50]   C.Y. Chen, L. Goux, A. Fantini, A. Redolfi, G. Groeseneken, M. Jurczak, Doped Gd-O
       based RRAM for embedded application, in: IMW, 2016.

[51] C.Y. Chen, L. Goux, A. Fantini, A. Redolfi, G. Groeseneken, M. Jurczak, Low-current operation of novel Gd2O3-based RRAM cells with large memory window, Phys. Status Solidi A (2016) 1–5.

[52] A. Fantini, L. Goux, R. Degraeve, D.J. Wouters, N. Raghavan, G. Kar, A. Belmonte, Y.Y. Chen, B. Govoreanu, M. Jurczak, Intrinsic switching variability in HfO2 RRAM, in: 5th International Memory Workshop (IMW), 2013, pp. 30–33.

[53]. A. Fantini, G. Gorine, R. Degraeve, L. Goux, C.Y. Chen, A. Redolfi, S. Clima, A. Cabrini, G. Torelli, M. Jurczak, "Intrinsic program instability in HfO2 RRAM and consequences on program algorithms", IEDM15, 169

[54] R. Degraeve, C.Y. Chen, U. Celano, A. Fantini, L. Goux, D. Linten, G.S. Kar, Quantitative retention model for filamentary oxide-based resistive RAM, Microelectron. Eng. 178 (2017) 38–41.

[55] C.Y. Chen, A. Fantini, R. Degraeve, A. Redolfi, G. Groeseneken, L. Goux, G.S. Kar, Statistical investigation of the impact of program history and oxide-metal interface on OxRRAM retention, in: IEDM, 2016, pp. 99–102.

[56] Website of Panasonic (Microcontrollers), https://na.industrial.panasonic.com/products/ semiconductors/microcontrollers/8-bitlow-power-microcomputers-mn101l-series.

Metal-oxide resistive random access memory (RRAM) technology: Material and operation details and ramifications

2

G. Bersuker, D.C. Gilmer†, D. Veksler*
*The Aerospace Corporation, Los Angeles, CA, United States, †Nantero, Inc., Woburn, MA, United States

2.1 Introduction

Continued progress in the various types of memory systems, such as high-density memory cell arrays, dynamic random access memories (DRAM), NAND, storage class memory (SCM), advanced embedded-type applications, etc., depend on continued advances toward low operation current and voltage range while increasing density and speed. In addition, the rising requirements for reducing power consumption for mobile application and convenience of use have increased the efforts and focus toward technology development of nonvolatile memories [1]. Desirable memory stack characteristics include, among others, area-independency (scalability) and high density, fast programming, low voltage and power operations, good retention and endurance, and three-dimensional (3D) integration capability.

One of the most interesting nonvolatile memory (NVM) applications, and intensely developing, is neuromorphic computing (NC): conventional microelectronic systems based on von Neumann hardware rely on energy-intensive off-chip memory storage that is a severe limitation for data-driven computing. Neuromorphic inspired architectures, which combine both computation and memory through an array of neuron-like elements with synapse-like connections, are aimed to address the shortcomings of modern computing and achieve the requirements of next generation of artificial intelligence (AI) systems. These architectures can be implemented using conventional complementary metal-oxide-semiconductor (CMOS) devices and their combination with NVM [2–4]. To achieve both faster operations and power reduction compared to the CPU, digital memory devices should be replaced with high-density analog NVMs, with resistances representing synaptic weights. For analog resistance changes, an NVM with continuously variable resistance, such as resistive random access memory (RRAM), is advantageous. Desirable characteristics of analog devices for NC include fast low-power programming of multiple analog levels and gradual and symmetric resistance changes in both set and reset operations. These requirements impose strict limitations on the applicable NVM technologies. Therefore, properties of RRAM cell stack materials discussed below also

Advances in Non-volatile Memory and Storage Technology. https://doi.org/10.1016/B978-0-08-102584-0.00002-4

consider the advantages and limitations they provide to meet the application-specific performance and fabrication requirements for NC.

The resistance switching random access memory (RRAM) technology presents an attractive option for the above-listed NVM systems, due to its demonstrated potential for low-complexity/high-density/high-speed/low-cost/low-energy nonvolatile operation and prospective ability to satisfy the requirements of advanced scaled memory system types [5, 6].

RRAM employs a memory storage mechanism through forming atomic defects

Within a large family of the metal-oxide-based resistance switching memory schemes, a common characteristic is that their operating mechanisms involve either rearranging of the atomic structure of the dielectric material (rendering it conductive) or movement of atoms in the dielectric (resulting in the formation of a conductive path), as opposed to the current incumbent memory technologies based on the electron storage. The variety of resistive switching memory types currently under consideration are discussed in other chapters of this book. Below we identify RRAM stack properties with respect to the abovementioned requirements applied to metal-oxide materials.

The RRAM concept in NVM technology is based on controllably varying conduction through an insulating layer. In its basic configuration, a RRAM cell is composed of an insulating film sandwiched between conducting electrodes forming a metal-insulator-metal (MIM) structure. Effective resistance of the insulating layer can be switched between high and low levels (set/reset operations) by changing an applied voltage polarity/magnitude (as well as other factors, such as temperature, light, etc.). This resistance change modifies the current passing through the device that defines a memory state [5, 6].

Current through an insulating dielectric layer is generally determined by the transport of injected electrons via as-fabricated and generated electrically active defect sites combining contributions of multiple parallel conductive paths formed by either isolated defects (metal ions/vacancies) supporting electron hopping conductivity, or links/clusters of accumulated defects of semiconducting or metallic properties [7, 8]. Resistance change memory switching then involves generating/activating and passivating/deactivating a certain number of defects (preferably the same number in each switching cycle) through movement/rearrangement of atoms/ions/bonds.

Area independency/scalability: Filamental RRAM

At any given defect density, the number of current paths through the dielectric, in the virgin or fresh state, is proportional to the device area, and consequently the total current is area dependent. In addition, the current magnitude tends to fluctuate from device to device due to randomness of the initial distribution of vacancies/ions. However, cell area dependency is eliminated when the current is dominated by a single conductive path (conductive filament (CF)). The CF provides an ultimate scaling advantage since it is only limited to the active filament size, which potentially may be as small as a few nm. The specific mechanisms in filament-type switching depend on the materials (dielectric and metal electrodes) employed in the fabrication of the memory cell and may involve more than one type of a conduction mode. Filament-based metal-oxide RRAM implemented with a variety of transition-metal-oxides (i.e., HfO_2, ZrO_2, Ta_2O_5, and TiO_2) has received considerable attention due to demonstrated nanosecond, low-power (<pJ) switching with high (up to 10^{12} cycles) endurance and retention of more than 10 years, using fab friendly and simple binary oxides and metal electrodes [5–20].

Controllable resistance states: Ultrafast (pulsed) operations

A stable preferential conduction path is known to form through oxide films subjected to electrical stress: under the applied voltage, a current abruptly increases at some point in time indicating the occurrence of a dielectric breakdown (BD) resulting in the formation of a CF.

A serious issue with the BD-driven forming process is that its outcome, that is, the resistance of the formed filament, is essentially uncontrollable: the postforming current tends to exceed a targeted value (a so-called current overshoot) that reflects an excessively low-resistance state (LRS) resistance. This uncontrolled LRS varies from device to device, and this variation is unacceptable for multicell memory array operations. The forming operation, therefore, requires employing a current limiter which prevents the BD current from exceeding a targeted value. A current compliance limit that regulates the LRS resistance was shown to be effectively controlled by a transistor in-series with the memory cell [21, 22]. The configuration combining one RRAM cell and one transistor is referred to as 1T1R. However, the transistor incorporation increases the footprint of memory arrays, and complicates 3D integration.

Thus, in metal-oxide devices, a critical issue for practical RRAM implementation in ultimate scaled configurations is determining overshoot-free forming and switching conditions that do not rely on implementing the current compliance (to control maximum current value), while remaining consistent with high-frequency, low-power operations. In order to remove dependency on compliance current control, one needs to consider the BD mechanism responsible for an uncontrollable (runaway) process.

Forming a CF in metal oxides involves the creation of a metal-rich region that requires (a) breaking metal-oxygen bonds followed by (b) out-diffusion of released oxygen ions. Both processes are temperature and electric field driven. Temperature is increased in the vicinity of existing electrically active vacancies: electron transport was shown to proceed effectively via electron localization/delocalization at the vacancy sites that is accompanied by energy dissipation [23]. Continuing oxygen vacancy generation in each other's vicinity leads to a higher electron flow (higher current—see Fig. 2.1), which further increases local temperature and, consequently, the vacancy generation rate. At this point, BD enters a runaway phase of a positive feedback process of self-accelerated vacancy generation [24]. In this phase, an extremely fast generation rate (<ns) does not allow the timely removal of the applied voltage upon reaching the targeted current (resistance) value to stop continued filament growth. Thus, limiting BD runaway is imperative for achieving compliance-free forming.

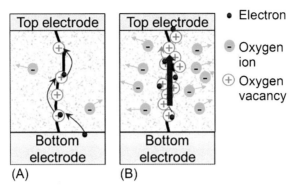

Fig. 2.1 Schematics of the filament formation process. (A) Electrons hopping via positively charged oxygen vacancies along the grain boundary assist in releasing oxygen ions and generating new vacancies. (B) Current increase induces a run-away process of the oxygen vacancy/ion pair generation.

To avoid an onset of the runaway vacancy generation, observed as a fast uncontrollable current increase along the extending filament, the forming process should be active only for a sufficiently short time, as then the overall energy dissipation can remain small and only a limited number of conductive defects can be generated under the selected briefly applied voltage. With this "time" control, a CF of the targeted resistance may be formed gradually, via a sequence of repeatedly applied extra-short voltages of sufficiently low amplitude. The effectiveness of such a stress regime was demonstrated by employing the fast pulse technique [25, 26].

Material features enabling well-controlled forming: Preexisting preferential conductive path

Forming a filament crossing the entire metal-oxide dielectric thickness under such restrictive time-voltage conditions becomes more feasible when it proceeds via a preexisting continuous electron transport path. Indeed, in this case, the temperature increase caused by energy dissipation during the electron transport (in particular, its hopping between the vacancies, Fig. 2.1) induces the generation of multiple new electrically active defects (vacancies) throughout the path length. Consequently, the filament can grow through the entire thickness of the film even under extremely short voltage pulses. On the other hand, forming in stoichiometric dielectric materials requires much higher applied voltages (electric fields) causing higher propensity toward BD runaway and usually resulting in subsequently poor resistance switching due to significant out-diffusion of the released oxygen ions.

Thus, initial, as-fabricated, conductive paths propagating through the dielectric film define possible precursor sites of the upcoming CF. To form a filament switchable at low currents (preferable in the case of large cross-bar networks), the initial path should be sufficiently resistive. In this respect, certain polycrystalline films, in particular, HfO_2, are promising since their leakage current paths are determined primarily by their multigrain structure, specifically grain boundaries (GBs) where the oxygen diffusion along and extraction from GBs can proceed efficiently. In thin hafnia films, few grains boundaries partially oriented across the film thickness constitute effective filament precursor paths.

Some of the advantages of HfO_2 for metal-oxide filament-based RRAM, besides the proven "fab-friendly" track record, are the extensive research performed developing gate stacks for advanced logic using metal-gate high-k with HfO_2 toward manufacturability and productization, assisting the detailed understanding of HfO_2 conductivity, and BD mechanisms [27–30]. In addition, HfO_2 has one of the stronger oxygen affinities for transition metals and thus, thermodynamically speaking, is relatively stable and compatible with most of the commonly used fab-friendly electrodes such as TiN, TaN, W, etc. [31–36].

The above considerations point to a strong link between materials structural features and memory cell operations: material properties impose certain requirements on electrical conditions to achieve targeted cell characteristics. Below, we discuss structural changes induced by the electron transport processes that drive CF forming and resistance switching operations in hafnia-based RRAM.

This chapter focuses on the HfO_2-based RRAM system, discussing optimization of the device properties via material engineering and developed atomistic models for possible mechanisms involved in the observed repeatable resistance switching. Some of the processes contributing to resistive switching may be active in other metal-oxide RRAM systems; however, due to the material-specific nature of RRAM characteristics (in particular, relative oxygen affinities, valance states, and atom diffusivities), any conclusions or comparisons to other material systems must be made with care. Here below we first discuss RRAM memory operational characteristics under conventional (dc, pulse) operation conditions (together with details on material structure, modeling, and transport simulation results) including information regarding read current instability (random telegraph noise (RTN)) and then conclude with circuitry operation-relevant ultrashort pulse data demonstrating dramatic effect on cell performance.

2.2   Operational characteristics of HfO_2-based RRAM

The resistance change behavior for bipolar operation of filament-based RRAM switching, such as observed for HfO_2, is shown in Fig. 2.2A, where an ohmic low resistance "ON" state (LRS) and a nonohmic high resistance "OFF" state (HRS), can be related to the formed CF and its rupture, respectively (Fig. 2.2B). To describe the material changes in the dielectric associated with this type of resistive switching, we must start with the CF formation, where a microscopic description of the CF features that enable memory operations have been proposed [7].

The CF forming process in HfO_2 can be discussed in terms of dielectric BD, an abrupt formation of a localized region between the electrodes, within which the dielectric composition becomes more oxygen-deficient, rendering this region conductive (the CF formation). The associated abrupt conductance change for this CF formation is observed during the DC voltage sweep or ramp of the RRAM dielectric "fresh state" (Fig. 2.3). This figure also indicates the typical DC switching operation for filament-based RRAM, as noted on the graph with:

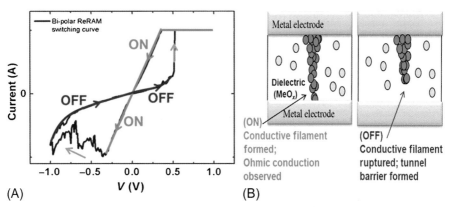

(A)   (B)

Fig. 2.2 (A) Characteristic DC properties of the on (LRS) and off (HRS) states; and (B) related physical picture for filament-based RRAM. *Circles* represent oxygen vacancies, and their propagation through the dielectric may form a conductive path.

1. *Forming*: a dielectric BD event leading to a CF; the voltage of the abrupt BD (forming) is termed V_F.
2. *Reset operation*: where the device changes from the LRS to the HRS state due to rupture of the CF.
3. *Set operation*: where the device changes from the HRS to the LRS state due to reformation of the CF path.

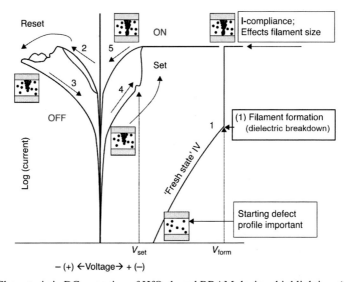

Fig. 2.3 Characteristic DC operation of HfO_2-based RRAM devices highlighting: (1) Forming (a dielectric breakdown event leading to formation of a conductive filament (CF)); (2) reset operation, where the device changes from the LRS to the HRS state due to rupture of the CF; (3) the DC return sweep after reset is also shown; (4) set operation, where the device changes from the HRS to the LRS state due to reformation of the CF path; and (5) the DC return sweep after set is also shown.

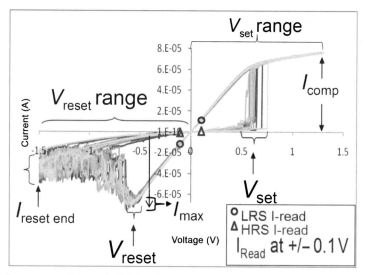

Fig. 2.4 A summary of the typical DC switching operation parameter terms for the bipolar operated RRAM devices; V_{set}, V_{reset}, V_{reset}-range, I_{comp}, I_{max}, LRS I-read, HRS I-read. V_{reset} is defined by the onset of current reduction trend, V_{reset} range is the maximum voltage magnitude applied during the reset operation. Note that in DC operations, "V_{reset} range" is usually referred to as the reset voltage.

A summary of the typical DC switching operation parameters for the bipolar operated RRAM devices is shown in Fig. 2.4. The well-behaved devices can repeatedly switch between the LRS and HRS states for billions of cycles, with the corresponding switching operations being reset (LRS → HRS), and set (HRS → LRS), respectively [12, 15, 20]. Note also that "anode" is defined to be the positive voltage electrode and "cathode" defined to be the negative voltage electrode for any applied operational bias. Detailed discussion on the switching mechanisms and modeling of the various switching operations are shown below.

2.2.1 Sources of conductive filament variability

Lack of precise control of the filament formation, perceived to be a random process, introduces variability, both device-to-device (even when fabrication process-related variability can be neglected) and cycle-to-cycle, into the switching characteristics of this class of devices. A reliable figure of merit (FoM) of the properties of the formed filament (e.g., its size and resistivity) is found to be the maximum current during the first voltage ramped reset operation after forming (I_{max}). Devices in which the I_{max} values are above the compliance current limit (I_{comp}) used during forming, exhibit lower LRS resistance values (leading to greater power consumption during reset), and a higher variation in the HRS resistance during subsequent device cycling [22]. In extreme cases, the devices with higher I_{max} vs I_{comp} can fail to reset. Since a greater I_{max} likely reflects a larger effective filament cross section (with lower resistivity) indicative of higher power

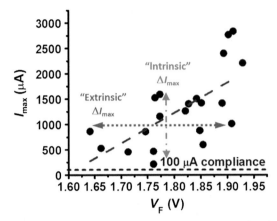

Fig. 2.5 Dependence of the maximum first reset current (I_{max}) on the forming voltage, V_F, resulting from the voltage ramp forming process. Two contributions to I_{max} variation can be noted: (1) due to random distribution of the V_F values ("extrinsic" variation); and (2) at any given V_F value ("intrinsic" variation).

consumption during memory cell operation, a forming process resulting in lower I_{max} values is highly desirable. As illustrated in Fig. 2.5, the device-to-device variation of I_{max} values can be roughly decomposed into two major contributions: "extrinsic" variation is due to the random distribution of the forming voltage values (V_F) associated with structural variability of the samples, and "intrinsic" variation is due to additional factors determining the conduction characteristics of the filament created at a given V_F.

A high V_F distribution correlates to the variability of the subsequent I_{max}, primarily due to related parasitic capacitance magnifying the released energy at V_F (see the following section). A constant voltage forming (CVF, which includes also pulse forming) eliminates (by definition) the "extrinsic" V_F distributional variability contribution to I_{max}, and when performed at low voltages, can also reduce the "intrinsic" contribution. However, only when the parasitic capacitance is substantially small (~<100 fF), can the variability from any V_F distribution be mitigated.

2.2.2 Variability caused by lack of filament growth control

2.2.2.1 Conductivity overshoot

In the conventional RRAM scheme, the filament growth is considered to proceed up to the point when the current reaches a predefined compliance value controlled by a current limiter, for example, a transistor in-series to the cell (see Section 2.2.3). It is generally assumed that the resistance of a formed conductive path is determined by the maximum current the path sustains at a given voltage. Therefore, no further filament growth requiring an increase of the current is expected. However, a continuing constant current flow during forming heats up a large dielectric volume surrounding the current path; the temperature remains high under a reached compliance (maximum) current flow. High surrounding temperature along with the

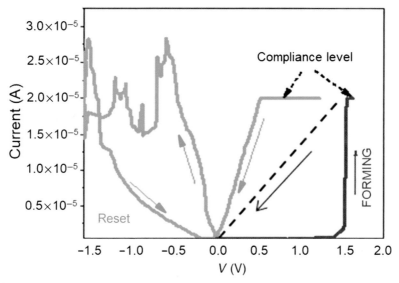

Fig. 2.6 Conductivity overshoot observed in monolithic 1T1R device after constant voltage forming at 1.6 V, 0.02 mA current compliance, in subsequent voltage back-sweep and reset operations. Ohmic current in the −0.4 to 0.4 V range is higher than the current *(broken line)* expected based on the compliance limit, indicating lower filament resistance.

electric field (though reduced) induces continuing bond breakage and associated filament growth even though the current remains constant (no current overshoot). As a result, when a current flow continues, the actual filament resistance is getting lower than that extracted from the measured forming current-voltage (*I-V*) data (the voltage across the cell drops). This phenomenon is called a conductivity overshoot [37] to emphasize a time-dependent process lowering the filament resistance that cannot be controlled by limiting the current magnitude increase. Such overshoot results in higher operation currents (Fig. 2.6) that can contribute to cell variability, similarly to the effect of parasitics discussed below.

To form a stable filament of a targeted resistivity, it is important to maintain a voltage-temperature balance between the rates of oxygen ion generation and their diffusion away from the forming filament and temperature of the surrounding oxide. In particular, forming at a higher applied voltage may generate vacancies much faster than the released oxygen ions can diffuse away. After the forming process ends, a number of the close-by ions may then move back to the filament, driven by ions repulsion at elevated temperature, reoxidizing it and increasing its resistance [38, 39]. On the other hand, at lower forming voltages, longer forming times generate high temperature throughout the surrounding dielectric region enabling vacancy generation under a constant current and low-voltage drop (conductivity overshoot), which reduces the filament resistance below a targeted value. Similar considerations are applicable to the set-reset switching operations. Overall, longer forming times and higher voltages result in larger filaments.

Since a conductivity overshoot depends on both the generated temperature and overall duration of a forming process, such overshoot can be mitigated by using a short pulse technique, which closer matches actual circuitry operations: it allows temperature to dissipate during the intervals between pulses, preventing high-temperature buildup in the dielectric around the conductive path and, thus, continuing bond breakage and oxygen ion out-diffusion (see Section 2.6).

As a positive side effect of the pulse operations, short periods of high temperature during forming limit out-diffusion of the released oxygen ions, some of which can then be driven back to the filament during a reset operation leading to a partial reoxidation of a certain filament section and, subsequently, its less conductive state [40]. In a subsequent set operation, a current flow raises temperature in this more resistive section of the filament breaking the oxygen bonds there and restoring postforming filament resistivity. Such spatially localized switching process involves a smaller number of moving atoms making it more energy/time efficient (it will be discussed elsewhere).

2.2.2.2   Current overshoot

Due to extremely fast dielectric BD process (<ns, Fig. 2.24), an external parametric analyzer cannot timely terminate an applied voltage to prevent the current growing above the target value I_{comp}. Similar effects are caused by excessive parasitic capacitance (C_p) in RRAM device connections that may result in the maximum current through the RRAM device during forming to surpass the current compliance limit, I_{comp}, as a result of the AC path arising from the C_p, $I_{max} = I_{comp} + I_{Cp}$ [22, 41]. The transient (overshoot) current, I_{Cp}, induced by this parasitic capacitance is proportional to dV/dt, the rate of the voltage decrease across the cell during the dielectric BD instance (the final, ultrafast current-runaway forming phase), as given by $I_{Cp} = C_p \cdot dV/dt$.

The peak value of I_{Cp} is observed to match well to the I_{max} (Fig. 2.7), indicating that the preset compliance limit determined by the external current compliance limiter in series with the RRAM cell is not always effective in controlling the current overshoot, thus introducing an uncontrolled random component in the forming process.

The value of C_p can be reduced through proper device integration (see below), while smaller V_F reduces overshoot (Section 2.3). Suppressing overshoot not only leads to the formation of filaments of smaller cross sections but also improves LRS and HRS variability (Fig. 2.8) during the subsequent device cycling [22]. However, in using the CVF method with a low V_F, the forming process takes too long to be practical for industrial applications. To take advantage of the low-voltage CVF method and its ability to reduce device variability, it was proposed [19, 22] to perform forming at elevated temperatures that drastically reduces time-to-forming. This "hot" forming method has demonstrated lower LRS and HRS variability and greater memory-windows in subsequent room-temperature switching operations (Figs. 2.9A and B). The switching mechanism with respect to "hot" forming is discussed in further detail in Section 2.3.

2.2.3   1T1R test vehicle

An imbedded (integrated) transistor in series with the memory cell (integrated 1T1R, Fig. 2.10) can better control the current passing through the cell during the filament

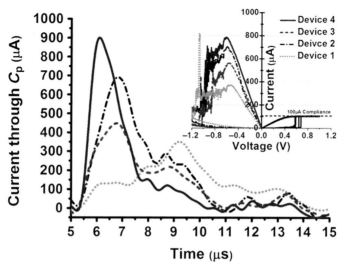

Fig. 2.7 Examples of the overshoot currents and the corresponding I_{max} values (the "intrinsic" variation illustrated in Fig. 2.3). The transient currents $I = C_p \cdot dV/dt$ (charging of the parasitic capacitance C_p) is estimated using the dV/dt values as measured by the oscilloscope (see the measurement setup schematic in the inset), for four devices during the forming event under the identical CVF condition of 1.7V with the 100μA compliance. Inset: the first reset/set of the corresponding devices.

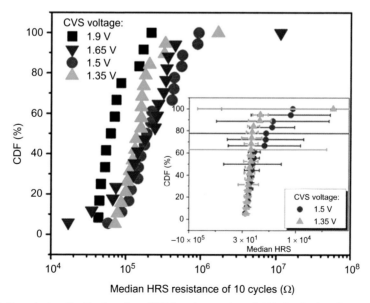

Fig. 2.8 Cumulative distribution plot of HRS resistance (at −0.1V) for devices formed under different CVF conditions (constant voltage stress (CVS)). Each point corresponds to the median HRS resistance obtained from 10 set/reset cycles on a given device. The inset depicts the HRS median and standard deviation for the devices formed at 1.5 and 1.35V CVF.

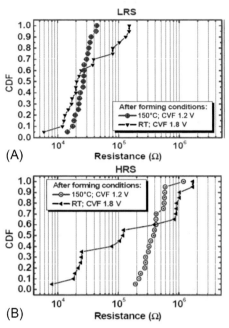

Fig. 2.9 Distributions of (A) LRS and (B) HRS resistance values for devices formed under constant voltage forming (CVF) conditions of 150°C/1.2 V and room temperature (RT)/1.8 V. Each symbol represents a resistance value averaged over 15 consecutive set/reset cycles at room temperature.

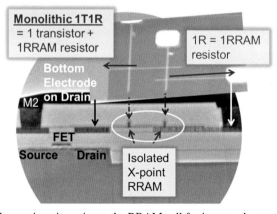

Fig. 2.10 Example transistor in-series to the RRAM cell for improved current compliance.

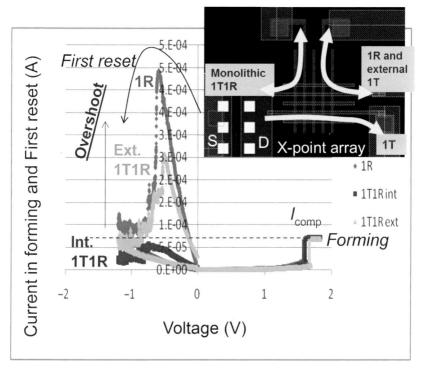

Fig. 2.11 Impact of parasitics on reset current. DC *I-V* curves during forming and first reset on 100 nm × 100 nm crossbar TiN/Ti-OEL/HfO$_{2-x}$/TiN RRAM stack for the test structure shown in inset when current compliance can be limited by monolithic (integrated) transistor or externally connected transistor configuration or externally connected parametric analyzer. The integrated transistor configuration may result in no current overshoot above the forming compliance value during reset [12, 13].

forming step (and set operations). When employing appropriately designed integrated 1T1R structures, the parasitic capacitance can be expected to be less than approximately 50–100 fF, which greatly improves current compliance control and minimizes or eliminates overshoot issues [12, 13, 21]. In addition, when using an integrated transistor to accurately control the filament formation and set operations (dielectric BDs), the intrinsic properties of the RRAM devices can more clearly be revealed. The parasitic capacitance is limited in general by using a smaller area drain, smaller area bottom and top electrodes of the RRAM element, smaller area probe pads, and thicker isolation layers. Through use of a specially designed test structure (Fig. 2.11), which allows for testing of the same RRAM element with various configurations of integrated 1T1R, an external parametric analyzer or an external transistor to limit operation current, clearly demonstrates that only the integrated 1T1R substantially minimizes the parasitic capacitance (as manifested by elimination of the I_{max} current overshoot above I_{comp}) [13]. However, it does not prevent the conductivity overshoot (see Fig. 2.12).

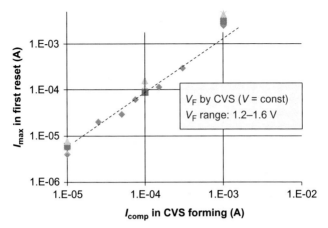

Fig. 2.12 The correlation between the maximum reset current and current compliance values used in the constant voltage stress (CVS) forming in the integrated transistor case. A linear correlation demonstrates effectiveness of the integrated 1T1R configuration in controlling current overshoot in a lower I_{comp} range. However, overshoot is observed at higher I_{comp} value of 10^{-3} A (conductivity overshoot, see Section 2.2.2.1). The experimental setup for dC and real-time ac characterization are discussed in Refs. 12, 13.

2.3   Modeling forming and switching processes

2.3.1   Properties of HfO$_2$ dielectric assisting filament formation

The properties of the initial CF created by the forming process essentially determine the device switching characteristics and therefore, we start by considering the forming process. Forming an intrinsic (no ion injection from the adjacent layers) conductive path in metal oxides requires generating a metal-rich region propagating through the film thickness. The forming process involves applying a sufficient bias across a dielectric resulting in the formation of a CF, which generally exhibits ohmic or near-ohmic type conduction (as follows from observed linear I-V dependency and increasing resistance with temperature). These metallic characteristics indicate that the filament is represented by the Hf-rich/oxygen-deficient region in the dielectric. Thus, the filament formation is associated with breaking metal-oxygen bonds followed by out-diffusion of released oxygen ions that requires temperature assistance overcoming certain energy barriers.

We start by identifying the morphological properties of the HfO$_2$ film assisting formation of the CF. For this task, conductive atomic force microscopy (C-AFM) was found to be an effective technique [40, 42, 43], allowing a nanometer-resolved characterization of the electrical and topographical properties of the gate oxides [44]. When the C-AFM tip-sample system is biased, a current flows through the structure so that along with the surface topographical features the electrical properties of devices can distinguish individual nanocrystals and allow profiling of much of the current both along and across the GB (depression in the topographical profile). Results demonstrate that the current through the crystalline dielectric preferably flows along the GBs

(Fig. 2.13), consistent with scanning tunneling microscopy (STM) measurements [45]. Dielectric BD induced by the continuously applied voltage also preferentially occurs at the GB sites.

Ab initio calculations of the GB structures in monoclinic HfO_2 were employed to identify the GB properties responsible for the current flow. Despite the boundary stability, segregation of vacancies to the boundary is thermodynamically favorable [46]. The formation energy and properties of vacancies segregated to the boundary are modified as their concentration increases. Neutral O-vacancies (Vo) can achieve spacing at the GB down to about 3.5 A, while positively charged vacancies (V+) can be placed with spacing as close as about 5 A.

At the low vacancy concentration, individual defect states are introduced into the HfO_2 bandgap. Higher vacancy density at GB leads to the creation of a conductive subband, confined to the GB region, due to overlapping of the localized Hf d-states in the HfO_2 bandgap. This subband effectively constitutes a percolation path for the current flow, in agreement with the C-AFM and STM results, which show a perfect match between the maps of the leakage current and GBs.

2.3.2 Grain boundaries as conductive filament precursor

Electrical transport along the GBs can be successfully described by the multiphonon trap-assisted tunneling (TAT) [47] via the vacancy sites. The leakage current through the TiN/HfO$_2$/TiN capacitor was simulated [48, 49], considering both direct tunneling (DT) and TAT components of the gate current. The TAT current, which is shown to be

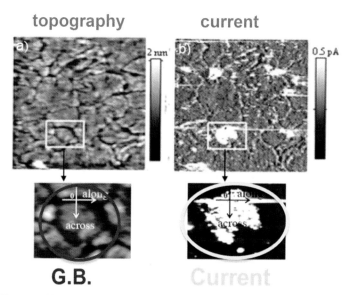

Fig. 2.13 Topographical (A) and current (B) images of the 5-nm HfO$_2$ film obtained with C-AFM. Multiple scans resulted in the breakdown spots formed at the sites of the grain boundaries and expended into the surrounding grains.

the dominant component in the substoichiometric hafnia discussed here, is calculated by taking into account contributions from the single- and multitrap conductive paths by the defects randomly placed along the GBs. The current driven through a conductive path, I, is determined by the slowest trap in the path as

$$I = e / \left(\tau_c + \tau_e \right) \tag{2.1}$$

where e is the electron charge and τ_c and τ_e are the time constants associated with the electron capture and emission, respectively, by the slowest trap. The calculations of $\tau_{c(e)}$ take into account the electron tunneling to/from the oxide trap and lattice relaxation associated with the electron trapping/detrapping [50–52], which causes displacements of the oxide atoms due to the Coulomb attraction between their nuclei and the electron. While the tunneling portion of the charge transport process is temperature independent, the lattice relation governed by the electron-phonon interaction generally strongly depends on temperature. This relaxation introduces a barrier to the electron trapping process which, within the high-temperature approximation, can be approximately described as $\exp(-E_b/kT)$, where the effective barrier is expressed by

$$E_b = \left(E_F - E_t \right) - E_r^2 / 4E_r \tag{2.2}$$

Here E_F is the electrode Fermi level and E_t and E_r are trap total energy and its relaxation energy, respectively. These trap characteristics, which identify the trap atomic structure and its charge state, can be extracted by matching voltage/temperature dependencies of the calculated and experimental currents.

The TAT description successfully reproduces current-voltage (I-V) temperature dependency in the crystalline hafnia (Fig. 2.14) [53]. The extracted values for the

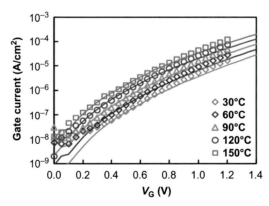

Fig. 2.14 Examples of measured (*symbols*) and simulated (*lines*) leakage current in TiN/5 nm-HfO$_2$/TiN capacitor at specified temperatures [53]. The oxygen vacancy defect parameters used in the simulations are as follows: the thermal ionization and relaxation energies are $E_T = 1.7$–2.7 eV and $E_{REL} = 1.19$ eV, respectively; and the vacancy density in GBs is $N_T = 10^{20}$ cm^{-3}; and the GB area is about 5% of the total device area.

energy of the contributing traps and their relaxation energy match those calculated for the V+ at the GBs [46], indicating that the TAT current is associated with the $(V^+ + e \rightarrow V^0 \rightarrow V^+ + e)$ process. Since the doubly positively charged O-vacancies [V(2+)], which can be present in hafnia, exhibit lowest diffusion activation energy (0.6–0.7 eV) [28, 46], their eventual segregation at the GBs leads to the formation of a conductive subband within the dielectric energy gap. By capturing the injected electrons, these vacancies are converted to a V^+ state, which is shown to support the TAT current. As discussed below, a current flow along the GBs facilitates dielectric BD at these locations, which results in the preferential formation of the CFs at the GBs in metal oxides.

To evaluate the effect of GBs on the memory cell characteristics, the forming/switching operations were performed in nanoscale by using the tip of a C-AFM probe as a top electrode of the MIM memory device (the biased substrate constitutes the bottom electrode) [42]. As seen in Fig. 2.15, higher conductivity of the GBs in the polycrystalline film is accompanied by lower forming (BD) voltages.

These low forming voltage sites at GBs exhibit repeatable resistive switching, while no switching is observed when the forming is induced on the grain sites (as well as in the case of an amorphous stoichiometric HfO_2), which exhibit significantly higher forming voltage values. These results indicate that resistive switching is benefited from higher oxygen deficiency of the dielectric that makes it less resistive and reduces the forming voltage. Indeed, devices fabricated using the stack of a HfO_2 dielectric and overlaying thin metal film that scavenges or "getters" oxygen from HfO_2 leading to the dielectric oxygen deficiency (HfO_{2-x}) as shown by the synchrotron X-ray photoelectron spectroscopy (XPS) data (Fig. 2.16), demonstrate much better switching characteristics than those of lower oxygen deficiency (the gettering metal can be an early transitional metal (i.e., Ti, Zr, Hf, etc.)). These gettering metal layers are often termed oxygen exchange layers (OEL) in RRAM stacks, even though actual oxygen exchange may not be possible because oxygen ions extracted by the metal films (which assist defect creation in the dielectric), is generally irreversible [54]. It has been demonstrated that once crossed the metal electrode interface, oxygen has no drive to return back to the oxide. Only oxygen ions remaining in the dielectric, because they either did not have enough time to reach the OEL metal or got trapped by the positively charged vacancies (see below), can contribute to the filament reoxidation.

2.3.3 Oxygen gettering: Conditioning the conductive filament precursor

Limited spatial resolution of the available physical characterization techniques does not allow addressing the relationship between the oxygen gettering process and film morphology, although physical observations combined with electrical data have implicated HfO_2 GBs as susceptible toward reduction, even when not favored from bulk thermodynamic considerations [55]. However, certain aspects of the gettering process can be deduced directly from electrical measurement data, in particular from a trend between the leakage in fresh devices and their forming voltages, which is observed in devices of different areas and various combinations of oxide

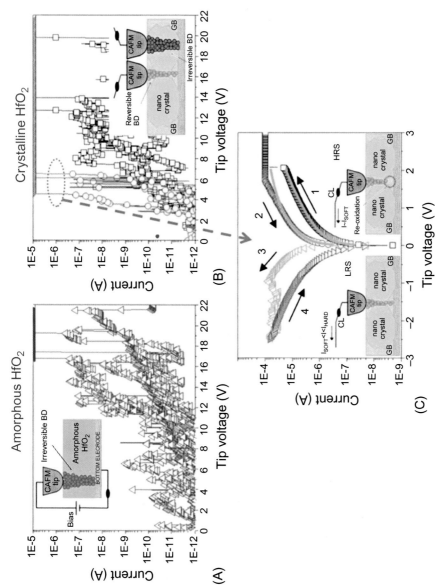

Fig. 2.15 Forming process at different random locations[42]: (A) Sample with nonannealed amorphous 3 nm HfO_2; (B) annealed polycrystalline 3 nm HfO_2, where two different I-V patterns associated with low-voltage and high-voltage forming can be distinguished; and (C) an example of switching observed at the low-voltage forming site in (B). The schematics indicate the probing location and measurement conditions.

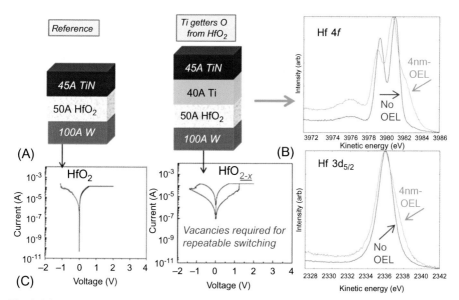

Fig. 2.16 Schematic of the stacks measured by the synchrotron XPS (at the National Synchrotron Light Source (NSLS) at Brookhaven National Lab): (A) The Hf 4f; and 3d 5/2 spectra indicating changes in hafnia composition with the insertion of Ti-OEL (B); and (C) switching characteristics of the corresponding stacks indicating improved RRAM switching with "the OEL-reduced" hafnia.

and metal-OEL thicknesses (Fig. 2.17) [56]. To explain such trends, we may consider that the oxygen scavenging proceeds more effectively along the GBs where the bonding of oxygen is known to be weaker [46] and the oxygen ion diffusion rate is higher [57]. The oxygen extraction from GBs is expected to occur essentially nonuniformly, with oxygen being more efficiently scavenged from the GB portion located in closer proximity to the metal layer. As a result, a GB portion next to the metal film is less resistive (due to higher concentration of the oxygen vacancies there), while a more stoichiometric section of the GB constitutes an HfO_2 dielectric barrier, albeit with higher density of oxygen vacancies than that of the regular, "un-gettered" oxide (Fig. 2.18).

Such stochastic process of oxygen gettering is expected to proceed differently in each GB due to variability of the GB characteristics (orientation, length, structure, etc.), resulting in a distribution of thicknesses of the dielectric barriers in the various GBs of each given device and, subsequently, a distribution of the magnitudes of leakage currents through these various GBs. The GB where the oxygen extraction was the most efficient (for whatever reasons) represents the preferential conductive path through the HfO_{2-x} film. The leakiest path is expected to have the thinnest barrier since the electron transport through the barrier controls the overall current. This thinnest barrier should be the first one (out of all the barriers in other less conductive paths) to be broken by the applied voltage during forming. Thus, a single structural feature—the thinnest GB barrier in a given device—is responsible for both the lowest

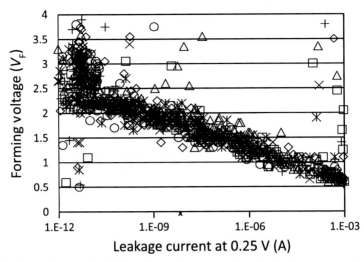

Fig. 2.17 Correlation trend of the forming voltages vs preforming leakage currents for various device sizes ($50 \times 50\,nm$–$30 \times 30\,\mu m$), HfO_{2-x} and OEL thicknesses of approximately (3.5–5 nm) and (2–4 nm), respectively, tested under identical conditions.

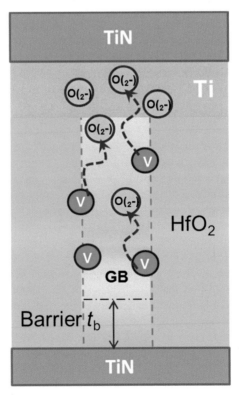

Fig. 2.18 Schematic illustrating oxygen ions O(2−) diffusing out from the grain boundary (GB) to the overlaying OEL during postdeposition anneal during postdeposition anneal leaving behind oxygen vacancies, v. The portion of the GB with high vacancy density exhibits low resistance, while the GB portion further away from the OEL remains more stoichiometric, thus constituting an effective dielectric barrier.

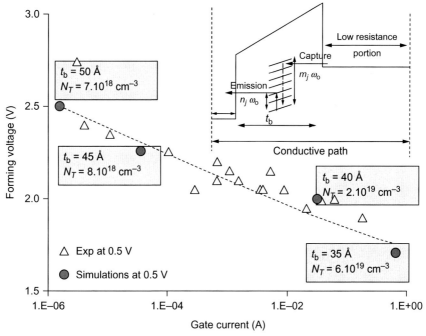

Fig. 2.19 Modeling the leakage current vs forming voltage in the stack TiN/5 nmHfO$_2$/X nmTi/ TiN with varying thickness, X, of the Ti layer at $V_g = 0.5$ V using the multiphonon trap-assisted tunneling (TAT) description for the electron transport via the traps in the remaining dielectric barrier, t_b, as shown by the schematic energy band diagram in the inset. The TAT model accounts for emission and absorption (denoted by the numbers m and n, respectively) of the phonons ($\hbar\omega_0$) that accompany the electron capture and emission, respectively, during its transfer via the given trap. Trap density, N_T, is specified in the graph for each barrier thickness. Traps are randomly distributed through the barrier volume. Experimental data for the corresponding devices are taken from Fig. 2.17.

BD voltage magnitude (i.e., the forming voltage) and highest leakage current (which is proportional to the total current through the device [56]) that explains their correlation.

This hypothesis was tested by modeling the leakage current and forming voltage (using the multiphonon TAT description), which was calibrated by matching the *I-V* dependencies in a variety of HfO$_2$ devices in a wide range of temperatures, from 6 to 400 K [58], with different ratios of the metal/dielectric thicknesses. The observed *I-V$_F$* trend for different ratios is successfully reproduced by varying the barrier thickness and densities of vacancies, which support the current through these barriers (Fig. 2.19). The simulations show that devices with higher ratios of the metal to oxide layer thicknesses have thinner barriers with slightly higher vacancy densities, as expected. Thus, one of the beneficial effects of oxygen gettering is a creation of highly conductive current paths, which exhibit low forming voltages—consistent with the above-discussed C-AFM results. However, the major effect is that the dielectric oxygen deficiency enables resistive switching phenomenon, as discussed below.

2.3.4 Mechanism of filament forming

We can now discuss the CF forming process based on the above-outlined structural and morphological dielectric features controlling the leakage current in the MIM stack. The forming process in HfO_2 can be considered in terms of dielectric BD, an abrupt formation of a conductive path between the electrodes. Dielectric BD represents a "weak link" event, the occurrence of which is determined by the moment when the first conductive path is formed; therefore, its characteristics are expected to follow the Weibull statistics [59], as is confirmed below.

2.3.4.1 Bond breakage

As a first step of the formation of an intrinsic conductive path in the dielectric material (without considering infusion of metal atoms from the electrodes), the oxygen-metal bonds have to be broken, which can be described by the Arrhenius dependency:

$$G(x,y,z) = G_0 \exp\left(-\frac{E_A - \beta E_{ox}(x,y,z,)}{kT(x,y,z)}\right) \qquad (2.3)$$

Here G is the probability of the bond breakage in a given location (x,y,z) at a given moment, and G_0 is the characteristic vibration frequency of the Hf–O bond. The barrier (E_A), corresponding to the breakage of the Hf–O bond, is lowered by $\beta E_{ox}(x,y,z)$, due to the polarization of the chemical bonds (as accounted for by β) by the external field $E_{ox}(x,y,z)$ [27, 30]. In many cases, the polarization factor is negligibly small [60] leaving the barrier too high for the bond-breaking process to be feasible. This barrier is significantly reduced, down to 1.12 eV, when this process takes place near a preexisting oxygen vacancy, especially if it trapped extra (injected) electrons [61]. However, electric field by itself cannot lower the barrier sufficiently to induce bond breakage under practically employed thermal conditions.

The critical contribution to oxygen release process comes from the elevation of local temperature at the oxygen site caused by the electron transport (trapping/detrapping) via the nearby vacancy. Within the TAT description, as discussed above, the current through the dielectric is associated with the injected electrons hopping between the vacancies. The process of the electron localization/delocalization at a vacancy site during the hopping is accompanied by the displacements of the surrounding lattice atoms (the lattice relaxation) [52] and associated release of phonons, which heat up nearby surrounding region [7]. Elevated local temperature intensifies the Hf–O bond vibrations (excites O atom toward the barrier top), increasing the probability of O going over the barrier into the interstitial position (bond breakage). Depending on the relative O-vacancy positions, in order to initiate an O release a number of electron trapping/detrapping events at the neighboring vacancy may be required that results in time dependency of the bond-breaking process.

Thus, each electron trapping/detrapping event is accompanied by the energy dissipation (P), in the form of phonons, around the trap location: $P = I/em\hbar\omega_0$, where I/e is the carrier flux through the trap (Eq. 2.1) and m is the net number of phonons of the frequency ω_0 emitted during the charge transfer. This energy dissipation leads to a temperature increase in the surrounding region, a value of which at each (x,y,z) location is determined by the distance from the charge transporting defect, frequency of the charge trapping/detrapping process, thermal conductivity of materials constituting the device, as well as overall device dimensions and geometry (which determine the boundary conditions) and, when the steady-state description is applicable, can be approximated by Joule heating expression:

$$P(x,y,z) = k_T \cdot \nabla^2 T(x,y,z) \tag{2.4}$$

where k_T is the oxide thermal conductivity. This approximation works well under the condition of continuous electron flux along the conductive path (in particular, GB) but it is not adequate for describing the temperature increase around an isolated electron trap experiencing sporadically electron trapping/detrapping events (it will be discussed elsewhere).

The presence of oxygen vacancies changes locally the dielectric constant, thus affecting the electric field distribution. Therefore, at any moment t, the field and temperature at a given location depend on the spatial distribution of the defects in the surrounding dielectric region, $n(x',y',z',t)$, generated prior to this moment in time, t:

$$E_{ox}(x,y,z,t) = f_{E_{ox}}\left\{n(x',y',z',t)\right\}, \; T(x,y,z,t) = f_T\left\{n(x',y',z',t)\right\} \tag{2.5}$$

where $f_{Eox}\{n(x',y',z',t)\}$, $f_T\{n(x',y',z',t)\}$ describe the functional dependencies of E_{ox} and T on the defects distribution, which cannot be generally expressed in a compact form (the E_{ox} and T maps can be obtained by numerically solving Poisson, charged carrier transport, and heat flow equations, accounting for a field direction, for a given vacancy distribution $n(x',y',z',t)$ — see simulations below in Section 2.3.6). Substituting Eq. (2.5) into Eq. (2.3), we see that the local bond breakage rate depends on a global spatial defect distribution that makes G essentially time dependent:

$$G(x,y,z,t) = G_0 \exp\left(-\frac{E_A - \beta f_{E_{ox}}\left\{n(x',y',z',t)\right\}}{k f_T\left\{n(x',y',z',t)\right\}}\right) \tag{2.6}$$

while the defect generation at each location (x,y,z) depends on the overall evolution of the generation rate $G(x,y,z,t)$ overtime at this location:

$$n(x,y,z,t) = \int_0^t G(x,y,z,t')\,dt' \tag{2.7}$$

As follows from Eq. (2.6) (consistent with the mechanism discussed following Eq. 2.3), the probability to generate new defects (vacancy) in the vicinity of the existing one is much higher than that in a lattice site spatially isolated from existing vacancies. This generation probability is growing with the density of defects in the region surrounding the (x, y, z) site (since both E_{ox} and T tend to increase with $n(x,y,z,t)$). At a certain moment, t_o, when a critical number of defects are generated close to each other (forming an oxygen vacancy cluster), the local field induces further barrier lowering while $T(x,y,z,t_o)$ drastically increases (see Eq. 2.4) due to a higher charge transport rate through this conductive defect cluster [53]. Under these conditions, the defect generation probability Eq. (2.6) increases sharply, triggering a positive feedback process of self-accelerated bond breakage described by the set of coupled Eqs. (2.6), (2.7).

The breakage of Hf–O bonds leads to creation of additional oxygen vacancies supporting the electron hopping, which further increases the electron transport through this dielectric region. Higher local current density increases temperature in the surrounding region, promoting generation of new defects nearby. At the same time, formation of lower resistance regions along the conductive path, due to higher local density of the generated vacancies, causes a greater share of the applied voltage to drop across the dielectric regions of higher resistance [53] which, in turn, enhances the vacancy generation rate there. This temperature-field-driven process of vacancy generation in the vicinity of a critical size defect cluster describes a runaway dielectric BD.

Upon breakage of the Hf–O bond, some of the oxygen ions diffuse out of the high-current region following the electric field direction, while others may recombine with the nearby vacancies and become subject to the possible subsequent bond breakage events. The oxygen release and its out-diffusion culminate in the formation of the oxygen deficient (hence metal (Hf)-rich) region through the dielectric, called the CF. Formation of a CF, which is accompanied by a change of the charge transport mechanism, constitutes a forming event usually observed in electrical measurements as an abrupt current increase by up to several orders of magnitude when governed by a runaway BD (an alternative option of the forming process avoiding BD runaway is discussed below in Section 2.6).

The described model suggests that the CF resulting from the above-described temperature-field-driven forming process might be of a cone-like shape. Indeed, more energy is dissipated during each electron-trapping event when it occurs at the traps located further away from the cathode due to acceleration of the injected electron by the applied electric field. In addition, the charge transfer rate (the characteristic times in Eq. 2.1) of the positively charged vacancies located further away from the cathode is also higher, because of their lower trapping barrier (Eq. 2.2), due to relatively high relaxation energy of these defects. Higher power dissipation, in turn, translates to higher rates of oxygen dissociation and out-diffusion in farther proximity from the cathode, which facilitates the formation of a wider oxygen deficient area. The transmission electron microscopy (TEM)-electron energy loss spectroscopy (EELS) data are consistent with the proposed cone-like shape of the filament (Fig. 2.20) [62], the filament cross section being observed to be larger with higher forming current compliance, in accord with electrical data [7].

2.3.5 Formation of vacancy cluster

The bond breakage discussed above represents the initial step in the CF formation process. The final CF characteristics depend on the diffusion of the released oxygen ions, and their spatial distribution in the dielectric at the end of the forming process, thus directly affecting the subsequently performed reset process. Obtaining this CF information requires detailed description of the kinetics of oxygen release and its diffusion during forming.

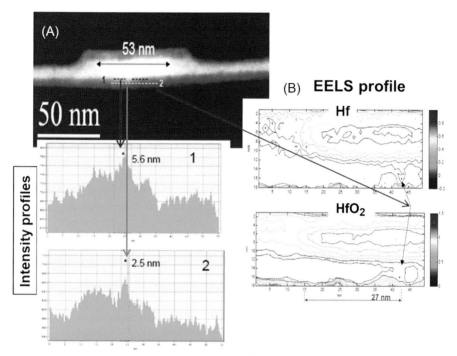

Fig. 2.20 TEM-EELS measurements on $50 \times 50 \, \text{nm}^2$ cross-bar device: (A) dark-field image showing a conical filament with an estimated top diameter of 5.6 nm and bottom diameter of 2.5 nm and (B) EELS spectra indicating a reduction of the HfO_2 content (bottom) and increase in the Hf content (top) in the filament region. HfO_2 and Hf coefficients determined from the fitting of the 3D spectral image acquired in the region outlined in the dark-field image.

When the Hf–O bonds "break," the O(2−) ion shifts to the interstitial lattice position leaving behind a double positively charged vacancy, V(2+). Due to a strong Coulomb coupling in the V(2+)-O(2−) Frenkel pair, released oxygen very quickly recombines with the vacancy. However, the dissociation process can proceed when a new O vacancy is created in the vicinity of a preexisting vacancy (thus forming nearest-neighbor Frenkel defects) [61]. In this case, a newly created Frenkel pair, O(2−)+V(2+), is stabilized by sharing electrons trapped by the next-neighbor vacancy, which can accept up to five injected electrons and be in (2−) state. A neutral charge state of a newly formed vacancy, V(2+)+2e=V(0), prevents its quick recombination with the interstitial O(2−). In this case, the recombination barrier of 0.83 eV is larger than O(2−) diffusion barrier (0.5–0.7 eV) indicating that oxygen ions most likely diffuses away. Such process accelerates (the binding energy per vacancy increases) when a new vacancy is formed next to a larger vacancy aggregate that leads to faster growth of the forming filament. The forming, therefore, proceeds more effectively when it initiates around a preexisting conductive path formed by vacancies, in particular, along the GB.

The above factors show that filament formation occurs more effectively via the process of vacancies generation in each other proximity rather than by vacancies diffusion and aggregation. While vacancies in a positively charge state can migrate [63], it is less likely that they maintain such state under electrons injection conditions. The additional factor complicating the diffusion-controlled forming is that positively charged vacancies are neither attracted nor driven by any external forces to each other (or to a vacancy cluster) to form/grow a filament.

Electron hopping between vacancies along GB increases the temperature along the path that promotes generation of additional vacancies within the GB and, hence, higher current [38]. Released negatively charged oxygen ions are moved out of GB to nearest interstitial sites in the surrounding oxide; lower oxygen ions energy at these sites (by about 0.4–0.5 eV) forms an additional barrier to their return to the GB (and subsequent recombination with vacancies there). Released oxygen ions migrate (over the barrier of approximately 0.5–0.7 eV) across the grains. In the runaway stage of filament formation, oxygen vacancies are generated in close proximity from Hf-rich clusters, which express metallic transport characteristics. With further current increase along the filament, the local temperature may reach extremely high values (about 800–1000°C). Singularities of the electric field at the rather abrupt Hf-rich clusters' interface with the adjoined dielectric material elevate the probability of the interfacial Hf–O bond breaking and releasing oxygen ions into the GB vicinity [64].

The process kinetic changes when a cluster of overlapping vacancies is eventually formed. The injected electrons captured by such a cluster are delocalized through the entire cluster volume and fully compensate positive charges associated with the oxygen vacancies. When a new vacancy is created at the interface between the metal-rich cluster and surrounding oxide, the vacancy charge is instantaneously neutralized by the electrons shared within a cluster. This significantly reduces the activation energy for releasing the O(2−) ion (down to about 2 eV) as compared to the formation of an isolated vacancy in the bulk of HfO_2 [53]. Thus, the formation of an initial vacancy cluster reduces the energy penalty for oxygen ion out-diffusion and assists the cluster growth process described by Eqs. (2.6), (2.7).

Material modeling results [64] indicate that oxygen deficient hafnia prefers to phase separates into a slightly oxidized hcp-Hf metal phase ($HfO_{0.2}$) and a stoichiometric oxide phase (HfO_2). In other words, highly oxygen deficient hafnia (a filament region, above the critical size of about 4 nm, with high density of vacancies) condenses/converges into the energetically preferred hcp-Hf metal structure during a forming operation accompanied by Joule heating and rising high temperature. Therefore, the CF formation and switching processes should be considered in terms of oxygen ions release, migration, and reoxidation of the Hf metal region. Describing it in terms of oxygen ions diffusion along the filament region is not adequate for metal materials: oxygen cannot move through a metal filament fast enough to explain reset/set kinetics; in particular, reset would only proceed via a layer-by-layer gradual filament oxidation starting from its interface with the overlaying oxygen-containing layer (OEL).

When a cluster of a certain critical size is created [64] in the cause of a dielectric BD (runaway) process its growth is expected to proceed at a very high rate (Fig. 2.24) forming a continuous CF through the entire dielectric thickness, it is usually manifested electrically by ohmic-type conductivity. As follows from the above discussion, a filament

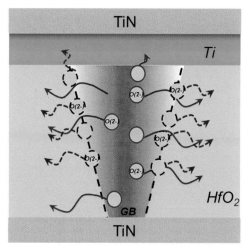

Fig. 2.21 Schematic of the proposed filament formation process. The filament growth is initiated at GB *(vertical blue dashed lines)* where oxygen bonding is weaker than in the bulk oxide and proceeds via the process of the O(2−) ions release from the filament outer surface.

cross-section extends preferentially via the oxygen ion release from the filament outer surface toward the surrounding oxide (Fig. 2.21): newly generated vacancies share electrons diffusing along the CF that prevents creation of V(+) sites, which would recombine with the released nearby O(−) ions.

According to the ab initio calculations [28, 29], the released interstitial O(2−) ions in HfO₂ can diffuse (by substituting the regular oxygen atoms) with the activation energy of $E_d = 0.5–0.7$ V. The O(2−) diffusion can then be described as proceeded via an O(2−) ion hopping between unoccupied interstitial positions with the rate:

$$R\left(x,y,z,t\right) = v * \exp\left[\left(-qE_{diff}\left(x,y,z,t\right)\right)/\left(k_B * T_{eff}\left(x,y,z,t\right)\right)\right] \qquad (2.8)$$

where $E_{diff}(x,y,z,t) = E_d - Q * \lambda/2 * E_{ox}(x,y,z,t)$, λ is the hopping distance (~O-O distance), Q is the ion charge, and v is the characteristic vibration frequency (Fig. 2.22). In addition to the applied field, the local field $E_{ox}(x,y,z,t)$ also includes the ion-ion and ion-vacancy Coulomb repulsion and attraction components, respectively, which depend on the spatial distributions of the O(2−) ions and vacancies (some of which are confined within the GBs) at any given moment, t.

2.3.6 Forming process simulations

The forming process can be approximated by two phases:

1. an initial stage dominated by the TAT conduction via preexisting vacancies, which is characterized by limited vacancy generation [65]; and
2. the final, subnanosecond, thermal-runaway phase, when vacancy generation proceeds effectively via a self-accelerated (positive-feedback) process [53].

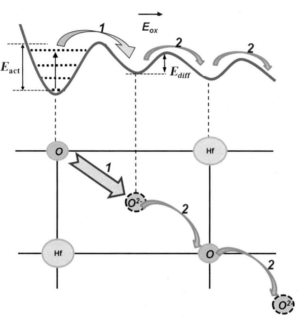

Fig. 2.22 Schematics of the energy and corresponding spatial diagrams describing the O^{2-} displacement into interstitial position; (1) due to a breakage of the O-atom bonds with neighboring Hf atoms and (2) O^{2-} ion diffusion along the applied field.

The discussion now focuses on the second, fast phase. The bond breakage and oxygen diffusion processes cannot be described in a closed analytical form because of their probabilistic nature and essentially nonlocal characteristics (Eqs. 2.6–2.8) and therefore, require the use of Monte Carlo simulations based on the rate equations [24, 53, 65]. The simulations start with constructing a unit memory cell containing the multilayer stack of metal electrodes, OEL, and dielectric materials. The metal-oxide dielectric layer (in this case, HfO_2) is initially randomly seeded with the carrier transporting defects (oxygen vacancies) distributed across the dielectric thickness and confined to the regions representing the GBs (their radius is calculated to be ~0.5 nm) [46]. The initial nonuniform vacancy distribution should reflect on the conditions expected in the result of the OEL layer gettering of O ions from the initial HfO_2 material: vacancies are preferentially located in the GBs [46, 56], closer to the metal OEL layer.

In the final runaway phase, which is characterized by the high density of vacancies within a conductive path where the TAT description of the charge transport is no longer valid, we consider the electron drift through a defect subband. Such electron transport can be described by the effective resistance approach [23, 24], when every elementary volume (bin) of the discretized dielectric space is assigned with an effective resistivity value, which depends on the oxygen vacancy concentration in this bin. At each iteration corresponding to a certain time interval, the program calculates the voltage across the RRAM (V_{RRAM}) taking into account the effective load resistor (R_{Load}), and then current and corresponding 3D local electric field and temperature maps at each

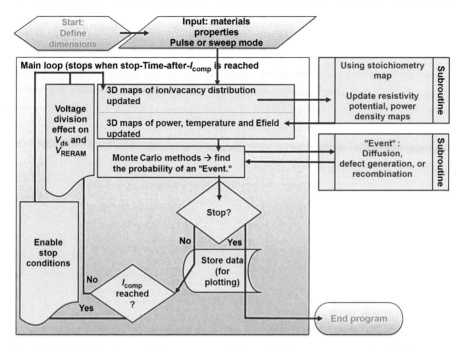

Fig. 2.23 The block scheme of the program, which models the processes controlling the HfO$_2$-RRAM operations. Main loop involves numerically solving Fourier heat transfer and charge continuity equations in each unit volume in the dielectric.

elementary bin (Fig. 2.23). Next, the probabilities of the vacancy generation (oxygen displacement to an interstitial position) and oxygen hopping (for the already shifted oxygen) are calculated for each oxygen ion throughout the entire volume. After each vacancy generation or ion-vacancy recombination event, the change in stoichiometry is translated into the change in the local resistivity of the corresponding bin, based on which a new field and temperature distribution is recalculated. The 3D maps of vacancy positions, ion positions, temperature, resistivity, electric field, and I-V curve data are generated at each iteration cycle.

Upon formation of a CF (the characteristics of which, specifically the cross section [7], depend on the current compliance limit), the resistance of the MIM device reduces usually by several orders of magnitude and, correspondingly, most of the applied voltage now drops over the external resistance.

The measured time for this voltage to change from low to high values defines an upper time length limit of the runaway phase of the forming process, which is experimentally estimated to be less than 3 ns (Fig. 2.24) (limited by the measurement setup resolution) [24]. The runaway time period, during which most of the oxygen ions are released, determines the duration of the directional ions diffusion under the influence of an applied field E_{ox}.

By reducing the effective diffusion barrier, the applied field during forming directs oxygen ions toward the anode electrode. Oxygen ions, which reach the metal layer

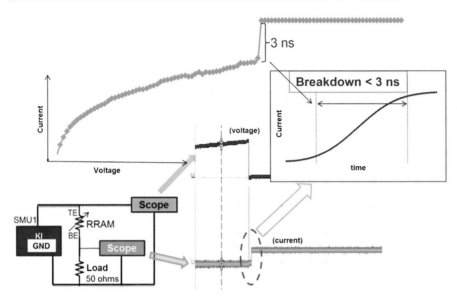

Fig. 2.24 The *I-V* characteristic of the RRAM device during the voltage ramp forming, measurement setup, and time dependencies of the voltages across the RRAM and load resistor during the voltage ramp. The inset shows the time dependency of the voltage transient.

(OEL) at the oxide/anode interface, oxidize it [66] and, therefore, cannot readily be extracted back to the CF and contribute to the subsequent reset process when an opposite bias (of a relatively small amplitude on the order of 1 V) is applied. However, the Coulomb repulsion between the oxygen ions slows down their transport along the direction of the external field, and a significant number of ions do not reach the metal layer during a very short transient (runaway) forming phase. These ions remain in the filament vicinity (in the interstitial positions), as illustrated in the example of a voltage ramp forming simulation (Fig. 2.25) and are available for a subsequent reset operation as discussed below. In a radial direction, the released ions move along the density/temperature gradient, their diffusion slowing down as the temperature decreases with larger distance from the filament.

According to the simulation results, higher forming voltages V_F increase the defect generation rate (Eq. 2.3), leading to shorter forming transient times (and potentially greater overshoot). By comparing radial and vertical distributions of the generated vacancies and diffused ions immediately after completion of the runaway phase, we find that under a higher forming voltage condition the oxygen ion and vacancy distributions significantly overlap (Fig. 2.26A′) indicating limited O(2−) diffusion. This is strikingly different from the low forming voltage case (Fig. 2.26B′), which exhibits a more spread-out radial distribution of the O(2−) ions due to a much longer duration of the transient (runaway) phase.

Simulations show that during the runaway phase, variations in the filament geometry caused by random vacancy generation at multiple sites along the conductive path induce local spikes in the electric field. These spikes reduce the activation energy for vacancy generation (Eq. 2.3), causing its rate to be higher than those of ion diffusion

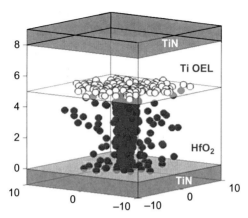

Fig. 2.25 Example of the simulated voltage ramp forming (Fig. 2.24). *Blue, red, and white dots* represent oxygen vacancies, interstitial oxygen ions, and oxygen ions, which oxidized the OEL, respectively. Device dimensions are in nanometer, the total ramp time t is in second and the final current I is in Amperes.

and recombination. Therefore, when the forming runaway phase is shorter, as in the higher voltage case, oxygen ions are generated at a high rate and do not have sufficient time to diffuse away from the filament. The use of higher ambient temperature accelerates the onset of the runaway phase in the trap generation process, thus enabling forming at lower voltage conditions to occur within a practically reasonable timescale.

The diffusion and recombination processes continue after the transient phase, when the forming current compliance limit is reached and the voltage across the device sharply decreases; it drops instead across an external resistance, while the current continues to flow at the compliance level. During this "relaxation" period (of no vacancy generation, but still higher temperature around the filament), more recombination events occur in the higher voltage forming case, since more ions are located in close proximity to the filament at the end of the vacancy generation period (Fig. 2.26A'). A lower forming voltage results in greater radial distribution or spread-out of oxygen ions during the relaxation phase. As a consequence, with higher forming voltages (at similar I_{comp}), the resulting filaments tend to exhibit higher resistance and are less stable (Fig. 2.26(d)).

2.3.6.1 Switching operations

Transition to the HRS by applying a reset voltage, where polarity is opposite to that of the forming voltage in bipolar operations, is caused by a rupture of the CF established during forming. To understand the major factors controlling HRS properties, we simulated the HRS currents and I-V temperature dependency for different reset (slow dc) conditions using the above-described multiphonon TAT method. The HRS experimental data are reliably reproduced when we account for the nonohmic character of the HRS current by assuming the existence of an approximate 1 nm dielectric barrier across the CF; the electron transport through this barrier, which includes both DT and TAT via the traps in the barrier, determines the HRS current [8]. As was reported [67], higher reset voltages result in higher HRS resistance, which is modeled as increased thickness

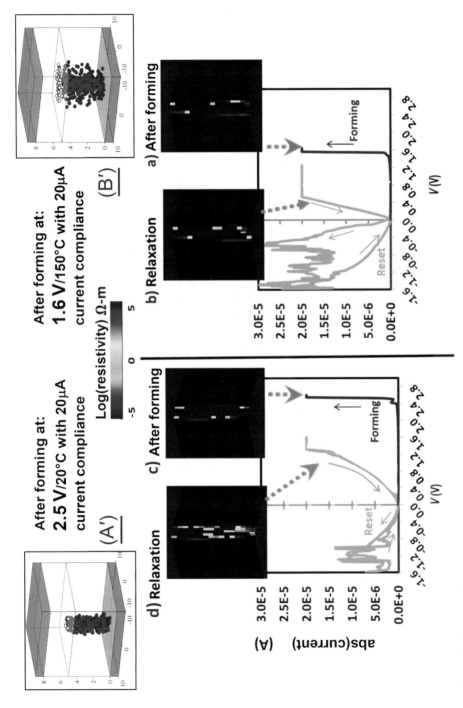

Fig. 2.26 See figure legend on next page.

of the dielectric barrier, up to 2.5 nm in the case of a slow DC reset [68], its band offset and k-value matching those of HfO_2. The temperature dependency of the HRS current is controlled by the TAT process via defects (oxygen vacancies) in the barrier layer (Fig. 2.27), with density 2×10^{21} cm^{-3} that is independent from the barrier thickness. The ionization energy of the contributing traps exhibits a wide distribution, 1.7–2.7 eV, matching the values obtained for the electrically active traps, which control the current in a fresh HfO_2 [53]. Therefore, it is reasonable to suggest that the CF rupture is associated with the reoxidation of a portion of the filament [8]. This can be achieved as the oxygen ions, which are out-diffused from the CF region during the forming, move back to the CF when the voltage polarity changes during reset operation.

Indeed, under the reset DC (long) stress bias, oxygen ions are expected to accumulate near the bottom electrode, which should present a good oxygen diffusion barrier to avoid the irreversible loss of oxygen during repeated switching cycles. In this respect, the TiN electrode appears to meet this requirement for HfO_2 RRAM dielectrics: according to the ab initio calculation [64], its crystal structure (ignoring its process-dependent polycrystallinity) does not have room to accommodate oxygen atoms, assuming the absence of N vacancies in TiN, and its free energy of oxide formation is much less favored compared to Hf. With sufficiently high density of O(2−) ions next to the bottom electrode, they tend to diffuse toward the filament, enabled by local temperature and driven by a density gradient and mutual Coulomb repulsion (Fig. 2.28). It is considered that the affinity of Hf and O-ions lowers the energy barrier for the O-ions hopping into the vacant site in the filament region. The filament works as an oxygen sink: oxygen ions readily bond to the available Hf atoms, thus creating room for the next oxygen ion to move toward the filament. However, ion diffusion in the direction away from the filament is suppressed due to a steep temperature reduction gradient: the lower temperature away from the filament reduces the ion diffusion rate (Eq. 2.8), which in turn slows down further out-diffusion of ions located closer to the filament due to ions' mutual repulsion.

Therefore, the spatial distribution of oxygen ions at the end of the forming process presents the initial boundary condition for modeling the reset process. This distribution, achieved under the forming conditions when the temperature at any given site in the oxide is much higher than that achieved during reset, determines the "supply" region, from which the O-ions might be driven back to the CF during the reset operation.

Fig. 2.26, Cont'd Effects of the forming voltage amplitude and ambient temperature on the resulting filament geometry. The experimental DC I-V curves during the forming (under the conditions as specified on the graphs) and subsequent voltage back-sweep and reset operations. (a), (b) and (c), (d) are the simulated 2D filament resistivity maps immediately after the forming and relaxation (at $V=0$ during ≥ 1 μs), respectively, and (A′) and (B′) are the corresponding simulated 3D spatial distribution of vacancies *(blue dots)* and oxygen ions in oxide *(red dots)* and in overlaying Ti metal film *(white dots)*. After the forming, the filament resistivity maps (a, c), are very similar (see also blue dots in (A′) (B′)). However, after a high-voltage forming, the filament exhibits a nonohmic conductance (see back-sweep), consistent with the simulated higher filament resistivity caused by a high rate of recombination between the vacancies and oxygen ions closely distributed around the filament, see A′ and d.

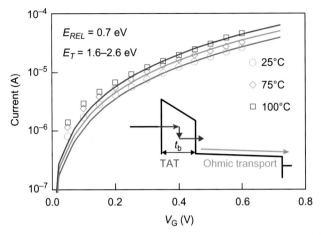

Fig. 2.27 Measured *(symbols)* and simulated *(lines)* temperature dependency of the HRS (postreset) DC current. Inset: schematic of the band diagram of the CF in HRS used for the current simulation. TAT via the traps (the trap density is $2.0 \times 10^{21}\,\mathrm{cm}^{-3}$) in the barrier layer (of 0.9 nm) is responsible for the temperature dependency of the HRS current shown.

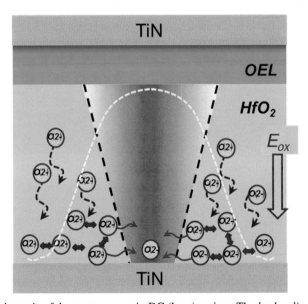

Fig. 2.28 A schematic of the reset process in DC (long) regime. The *broken lines and broken bell-shape curve* outline the filament and temperature radial profile, respectively. The O-ions diffuse toward the bottom electrode following the electric field, and laterally due to density gradients and ions Coulomb repulsion *(red arrows)*, and may reoxidize a portion of the filament next to the anode electrode.

Due to a very low O(2–) hopping barrier (0.7 eV), oxygen ion diffusion proceeds very effectively at room temperature, which would lead to an eventual uniform distribution of oxygen ions throughout the entire dielectric volume. This would result in oxygen-depletion of the dielectric region around the filament, thus disabling the reset process (Fig. 2.29A–D). However, we have to consider that the dielectric film generally contains a significant number of imperfections, specifically GBs in the case of polycrystalline hafnia, which are known to accumulate oxygen vacancies in a positively charged state, as discussed above [7]. These pinned positive charges, the density

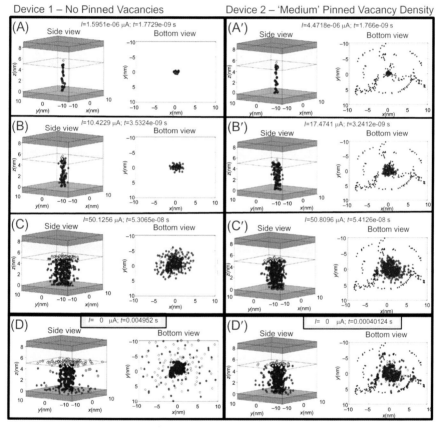

Fig. 2.29 An example of simulated forming process in 20×20 nm TiN/Ti/5 nm-HfO$_2$/TiN cell, $V_F = 1.7$ V, $I_{comp} = 50 \mu$A, considering no pinned vacancies (Device 1 (A–D)) or pinned vacancies (Device 2 (A′–D′)). The *red and white dots* represent oxygen ions in the interstitial positions in the dielectric, and those which diffused into the Ti metal layer and oxidized it, respectively. *Cyan dots* are the oxygen ions which diffused outside of the cell volume. *Blue and black dots* are oxygen vacancies generated during the forming process, and as-processed, pinned at the grain boundaries, respectively. A–A′ are the initial moment of stress, C–C′ are the final moment of forming, and D–D′ are the ion/vacancy distributions after the postforming relaxation for 0.0004 s. (In the 3D graphs, in order to provide a better view of the filament region, the vacancies at the grain boundaries are not shown.)

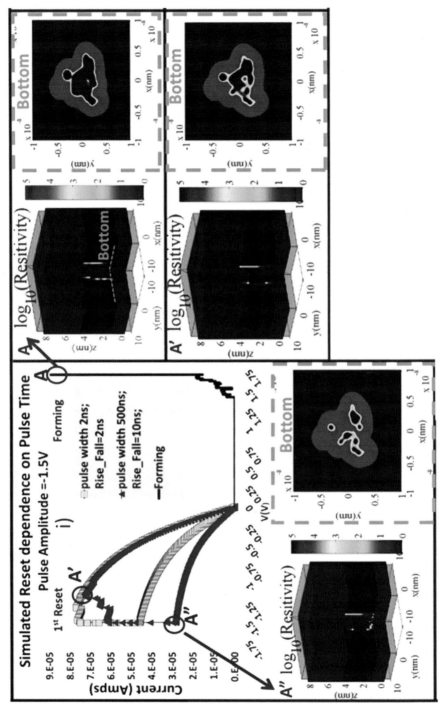

Fig. 2.30 See figure legend on next page

of which is dramatically increased as a result of oxygen gettering [56] (Fig. 2.16), attract negatively charged oxygen ions, preventing them from leaving the CF region surrounded by randomly distributed GBs [69]. The energy barrier associated with the vacancy segregation at GBs [46] prevents the recombination of the coupled (via Coulomb interaction) oxygen ions and pinned vacancies. These coupled oxygen ions remain within the limited region defined by the grains surrounding the CF (which itself encompasses the GB) (Fig. 2.29A′–D′), thus enabling the high endurance of the HfO_2-based devices. The simulations confirmed that the pinned vacancies, with a density determined by the degree of the hafnia oxygen deficiency, can effectively contain oxygen ions enabling repeatable set/reset operations [24, 39] (Fig. 2.30).

As seen in Fig. 2.30, the simulations demonstrate that a longer reset pulse (as well as higher reset voltage) [39] leads to reoxidation of a larger section of the CF, resulting in a less conductive HRS, consistent with the literature reports on the voltage vs time trade-off [70].

A switching from high to low resistive state (i.e., set operation toward LRS defined by the current compliance level) leads to the initial CF being restored. For this, the thin dielectric barrier formed during reset needs to be broken. Due to relatively high conductivity of the undisrupted portion of the "conductive" filament, most of the applied voltage during a set operation drops across the dielectric barrier. If this voltage (in particular, a pulse amplitude) is sufficiently high so that the electric field in this thin dielectric layer approaches the intrinsic dielectric strength of HfO_2 (estimated to be $\sim 4\,MV\,cm^{-1}$) [30], the Hf–O bond breakage proceeds rapidly, resulting in a release of the oxygen ions and restoration of the CF toward its postforming shape. The description of the set process (which occurs generally at lower voltages than forming), essentially follows that of the forming process, as discussed in Section 2.3.

2.4 Materials development: Engineering vacancy profiles for RRAM

2.4.1 Vacancy asymmetry and the filament-based RRAM model

As described in Section 2.3, oxygen vacancies in substoichiometric films may agglomerate (in polycrystalline film), by accumulating along the GBs [8], thus forming conductive paths through the dielectric. The presence of conductive paths assists with lowering the filament forming voltage, allowing for more controlled CF formation.

Fig. 2.30, cont'd The effect of reset pulse time on the barrier formation. A: simulated *I-V* characteristic during forming with the pulse of the amplitude $V_F = 1.7\,V$ for the device in Fig. 2.29A′–D′. The ions/vacancies distributions resulting from this forming are used as the initial conditions for the subsequently simulations of reset under two different conditions with different pulse widths and pulse rise/fall times, as specified in the legend. The 3D and 2D (corresponding to the device cross section near the bottom "anode" electrode) maps show resistivity after the forming (A) and during the reset at the end of the pulse rising time (A′), and at the end of the reset pulse width (A″).

The nonuniform cross section of the CF provides certain advantages for switching operations, with bipolar reset described as oxidation of the narrower end of the CF tip to form a thin dielectric barrier near the reset-anode, which for bipolar operation is also near the forming-cathode, leading to the HRS. The subsequent set operation (from HRS to LRS) is then the BD of this thin dielectric barrier, thus reforming the CF. In order for a device to demonstrate well-behaved RRAM switching, the forming and switching conditions should be selected according to the filament composition profile to achieve favorable postforming distribution of the oxygen ions, which subsequently contribute to reoxidation of the filament during reset.

The DC reset process in an HfO_2-based RRAM device (the atomic-level simulations of which are presented above) is illustrated in Fig. 2.31 [8]. Changes in the LRS during increasing reset bias are first observed as a deviation of the *I-V* curve linearity expected for the ohmic LRS due to temperature increase of the CF caused by joule heating (Fig. 2.31 inset shows simulated temperature increase with increasing bias). Eventually, upon continued bias increase and sufficient heat, oxidation of the CF tip begins, as observed by a "noisy" decrease in current with continued voltage increase. This fluctuating resistance reveals the competing processes of CF oxidation and generation of new oxygen-vacancy pairs in the electrically stressed dielectric barrier, which is growing during the reset. This model of a reset process suggests that the vacancy concentration profile asymmetry in the RRAM dielectric stack is critically important for repeatable switching [71, 72] as discussed below.

Robust switching in the transition metal oxides (TMO) RRAM devices are known to require some amount of oxygen deficiency of the dielectric film [41, 72]. It has also been demonstrated that performance of RRAM devices is improved by utilizing asymmetrical vacancy profiles of substoichiometric TMO films. These asymmetrical

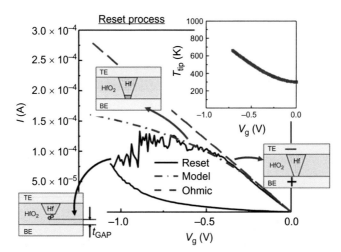

Fig. 2.31 Reset *I-V* sweep with schematics of the corresponding physical processes. The current deviates from ohmic, initially due to a resistance increase caused by increasing filament temperature. Reset occurs when temperature/voltage is sufficient for oxidation and oxygen is available.

vacancy profiles are typically formed using reactive-physical vapor deposition (PVD) [17, 41, 73], or through depositing a dielectric or metal and applying a subsequent posttreatment reduction or oxidation anneal [74–76], or with depositing reactive gettering-type metals (often termed OEL in RRAM stacks) onto the TMO [15, 16, 20, 77, 78]. Many of the various TMO-RRAM reports in the literature also consistently show bipolar operations with the reset bias applied having the reset-anode against the more oxygen rich or more stoichiometric part of the fabricated dielectric stacks [15, 16, 41, 72, 73, 75–77]. The reason the bipolar operation for TMO filament-type RRAM consistently has this polarity preference characteristic is related to the TMO asymmetric defect/vacancy profile, the CF asymmetry, and switching mechanism described in Section 2.3. Having less defects/vacancies near the "reset-anode" allows for a more effective reoxidation of the narrow CF tip in this region. When an opposite bipolar operational polarity is used to switch the RRAM device, too many defects/vacancies near the "reset-anode" can prevent effective oxidation of the CF tip [72].

2.4.2 Switching in asymmetric vacancy engineered RRAM

To further investigate the preferred bipolar operational biasing and vacancy asymmetry relationship, various substoichiometric HfO_{2-x} films with asymmetrical oxygen vacancy profiles were formed [72]. A thin OEL of titanium was deposited on top of HfO_2 and subsequently capped with TiN, creating an asymmetrical defect/vacancy profile with more defects/vacancies near the OEL side of the stack due to preferential gettering of oxygen from the HfO_2 grains and GBs nearer the OEL metal. Fig. 2.32A and B presents the switching behavior for the two possible bipolar biasing schemes (Fig. 2.32A) having reset-anode opposite the OEL (or more defects/vacancies) side, and Fig. 2.32B biased with reset-anode at the OEL (or more defects/vacancies) side). Operating in the bipolar mode with a bias polarity having the reset-anode against the

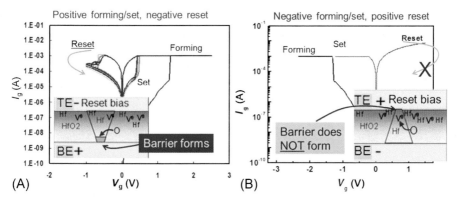

Fig. 2.32 Switching characteristics for the two bipolar operational biasing schemes applied to devices with asymmetric oxygen composition profile. When operational biasing has the reset anode near the more stoichiometric side, a dielectric barrier forms, successfully achieving reset (A). However, having the oxygen poor (higher vacancy concentration) near the reset anode prevents a dielectric barrier from forming and no reset is observed (B).

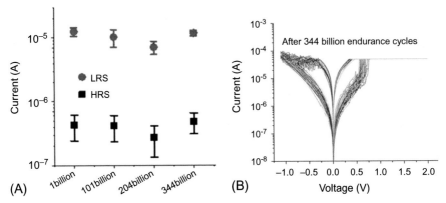

Fig. 2.33 Endurance data for the RRAM stack used in Figs. 2.19 and 2.20 (TiN[HfO$_{2-x}$/Ti-OEL]TiN). Averaged LRS and HRS reads from 50 DC cycles performed after 1, 101, 204, and 344 billion pulse endurance cycles using the set current compliance of $I_{comp}=60\,\mu A$; pulse set voltage = 1.5 V, pulse reset voltage = −1.3 V; pulse rise/fall time = 15 ns, and pulse width = 5 ns (A). The 50 DC cycles performed after 344 billion pulse cycles is shown in (B).

more stoichiometric part of the film (away from the OEL) results in well-behaved RRAM switching with well over 100 billion ac cycles demonstrated on dimensions down to 50 nm × 50 nm (Fig. 2.33). Operating the device opposite this preferred polarity often results in the failure to reset toward higher resistance states (Fig. 2.32B), due to difficulty in reoxidation of the CF because of the high defect/vacancy density near the reset-anode in this "nonpreferred" biasing case.

This oxygen vacancy asymmetry and preferred bipolar biasing is demonstrated further with a RRAM device using a highly asymmetrical HfO$_{2-x}$ film formed by oxidizing the top of an exceedingly substoichiometric HfO$_{2-x}$ film. The backside secondary ion mass spectrometry (SIMS) profile of the oxygen concentration for this type of stack is shown in Fig. 2.34, indicating the film has high defect/vacancy concentration near the bottom TiN electrode, and stoichiometric HfO$_2$ near the top TiN electrode. Note the defect profile for this stack (defects/vacancies near bottom electrode) is opposite to that of the engineered HfO$_{2-x}$/OEL described above (with defects/vacancies near the OEL at the top electrode). The ramifications of this opposite defect profile in the (top) oxidized HfO$_{2-x}$ with respect to the HfO$_{2-x}$/OEL (top "reduced") stack are that now the preferred bipolar operational biasing (i.e., with reset-anode against more stoichiometric side of the film) occurs with negative forming/set and positive reset voltages applied to top electrode.

In addition, when performing the preferred operational biasing for this highly asymmetric oxidized HfO$_{2-x}$, not only is over 10^8 ac endurance cycling and 100% DC switching yield observed (Fig. 2.35), but also operating with opposite the preferred bipolar biasing with reset-anode against the defective/vacancy-rich side of the film, resulted in 0% yield. Thus, this highly asymmetrical substoichiometric film unambiguously illustrates a relation of the operational biasing to oxygen vacancy profiles, in agreement with the described reset mechanism of CF tip oxidation. Having reset-anode against the defective/vacancy-rich part of the stack allows for too many

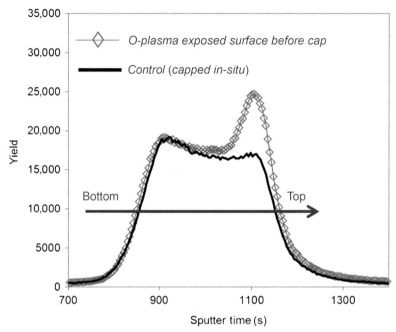

Fig. 2.34 Backside SIMS showing the oxygen profiles for highly substoichiometric HfO$_x$ with and without exposing the top surface to oxygen plasma before capping in situ with Pt (Pt used only for SIMS analysis).

available conduction paths and inefficient oxidation of the CF for HRS tunnel barrier formation, that is, no repeatable switching. Reset-anode against the more stoichiometric part and more than a hundred million of repeated cycling may be possible.

Comparing DC reset sweeps for the two cases of reset-anode against "good" (more-stoichiometric) or "bad" (more-oxygen vacancies) parts of the dielectric, indicate that even for the nonpreferred bias polarity of reset-anode against the bad vacancy-rich part of the dielectric, a similar deviation in current upon nearly reaching the forming compliance can be seen (Fig. 2.36). However, for the nonpreferred bias of the asymmetrical stack, the current does not continue a downward trend to HRS, but begins to increase and the device "breaks" (due to no compliance in reset, and increasing bias and current leading to a large filament). Thus, the reset case with too many defects near the reset-anode suppresses the ability for CF oxidation and denies a competition between filament oxidation and BD leading to catastrophic dielectric BD and reset failure. Engineering the oxygen vacancy profile to support the described reset and set mechanisms is required for RRAM performance optimization.

Similar inclination as the HfO$_x$-based system for a preferred bipolar operation, biasing and related engineered oxygen vacancy asymmetry is demonstrated for TaO$_x$-based RRAM. Fig. 2.37A shows DC behavior for a TaO$_x$-based stack consisting of a deposited thin Ta-OEL followed by a deposited thick highly substoichiometric TaO$_x$, and then a thin stoichiometric Ta$_2$O$_5$ layer. The preferred bipolar operation of this stack is as expected with reset-anode against the good, or more stoichiometric, Ta$_2$O$_5$ side

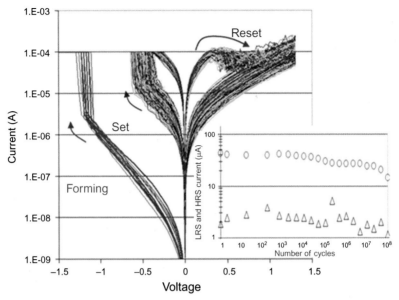

Fig. 2.35 The DC *I*-*V* characteristics of the forming and switching operations performed with the preferred operational biasing for highly asymmetrical oxidized HfO_{2-x} devices (with external compliance control). The data are collected on 35 die × 10 cycles (100% yield) each. Inset shows HRS and LRS read current during over 10^8 ac endurance cycling.

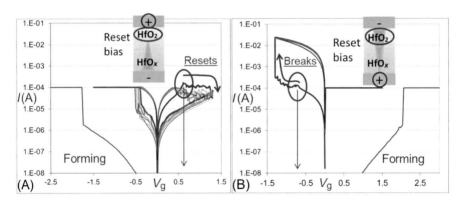

Fig. 2.36 Typical DC forming and cycling with asymmetric HfO_x for (A) preferred bipolar biasing (reset anode against "good" more stoichiometric HfO_2) vs (B) nonpreferred biasing (reset anode against "bad" vacancy rich HfO_x).

of the stack. Furthermore, a stack having near 1-nm stoichiometric HfO_2 deposited and capped with thick substoichiometric TaO_x and then a thin Ta-OEL (Fig. 2.37B) demonstrates similar switching as the RRAM stack with a single thicker HfO_2 with the top deposited OEL (Figs. 2.31–2.33), where the preferred reset-anode is opposite the vacancy-rich OEL/top electrode (and against the more stoichiometric side of the stack). The similar reset for these different RRAM material stacks also supports the

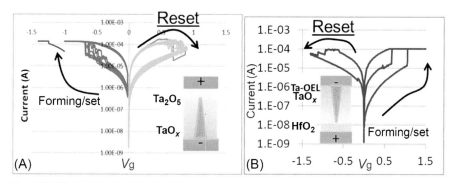

Fig. 2.37 The DC forming/cycling for (A) asymmetric (bottom\Ta\TaO$_x$\Ta$_2$O$_5$\top) with "preferred" negative forming/set and positive reset applied to top electrode, and (B) asymmetric (bottom\HfO$_2$\TaO$_x$\Ta-OEL\top) with positive forming/set and negative reset applied to top electrode.

reset mechanism of oxidizing the CF tip toward a thin (~1 nm) dielectric barrier [8], and comparing directly the devices for Fig. 2.37B and Figs. 2.31–2.33 which do have "more" HfO$_2$ at the reset-anode, they form the similar approximate 1 nm HfO$_2$ tunnel barrier with similar observed *I-V* characteristics.

2.4.3   Real-time monitoring of RRAM switching highlights operating mechanism

The relevant mechanisms for the bipolar RRAM operation can be highlighted by capturing changes in the device conductivity in real time during applied set or reset type ac bias to the RRAM stack [12, 13]. For instance, in general, a dielectric barrier limiting the current in the HRS should break similarly irrespective of polarity of the applied bias; however, there is a clear difference in the HRS when pulse voltages of the same pulse width and height but opposite polarities are applied [72]. For preferred set-type bias (set-anode bias near the defect/vacancy-rich side) applied to the device in the HRS, a clear abrupt change to a lower resistance state occurs (Fig. 2.38A): the dielectric barrier abruptly breaks (when the electric field across the barrier approaches the critical hafnia BD field value (>4 MV cm^{-1} subject to dielectric stoichiometry)), with the released negatively charged oxygen ions being pushed away from the barrier and toward the anode by the electric field. Due to lack of "available" oxygen pushed toward the break in the dielectric barrier, and a higher oxygen vacancy concentration (from asymmetrical oxygen vacancy engineering of the fabricated TMO stack) near the anode for this set-type bias, the reformation of the CF occurs readily. For preferred reset-type bias (reset-anode bias opposite the defect/vacancy-rich side) applied to the device in the HRS, many continuous pulses are required before any resistance change occurs (Fig. 2.38B). In addition, when the resistance does change, the read current is noisy. According to the above-discussed model, higher barrier stability against the "preferred" reset-bias stress results from competition between the Hf−O bond breakage in the dielectric barrier, and the near immediate reoxidation of the Hf atoms caused

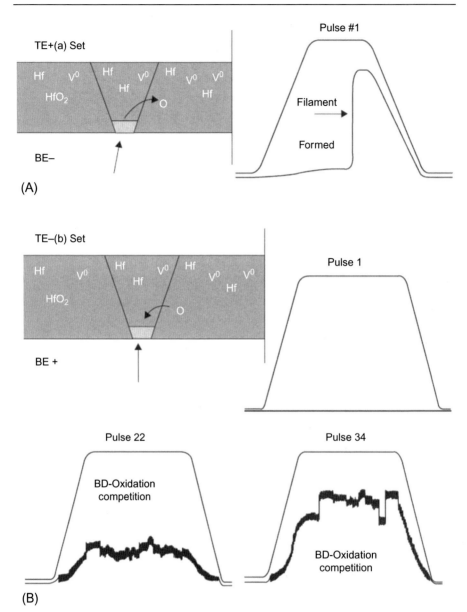

Fig. 2.38 (A) Set pulse (100 μs and 0.8 V) applied to HRS state of the RRAM stack (similar RRAM stack as used for Figs. 2.31–2.33). Immediate and singular resistance change to LRS is observed indicating reformation of the CF for the set bias. (B) Reset pulse (100 μs and − 0.8 V) applied to HRS state of the RRAM stack where many pulses are required for resistance change, and then the fluctuating resistance change is reflecting the competition between HRS breakdown and reoxidation (TE/BE, top/bottom electrode).

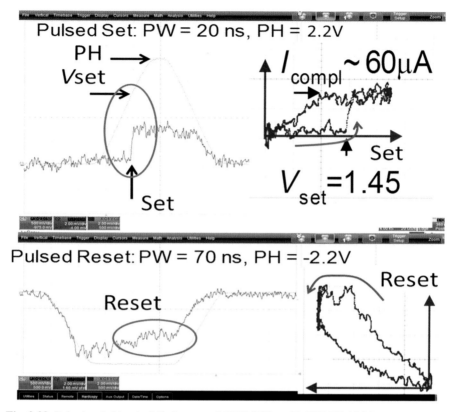

Fig. 2.39 Pulsed switching in fully integrated 1T1R HfO_{2-x}/Ti-OEL RRAM for $100\,nm \times 100\,nm$ cross-bar devices. *Smooth lines*: pulse voltage applied to the top electrode; *wavy lines*: voltage at the bottom electrode as read through the oscilloscope. Insets: pulsed *I-V* traces corresponding to the respective pulsed set and reset operations (PW/PH, pulse width/ height).

by a continued presence of the oxygen ions in their vicinity: the applied field pushes the ions against the electrode, thus preventing their fast out-diffusion from the filament region and providing ample available oxygen for filament tip oxidation.

Fig. 2.39 shows oscilloscope captured switching data for a pulsed width (PW) of 20 ns for set, and 70 ns for reset, for fully integrated 1T1R HfO_{2-x}-OEL RRAM in $100\,nm \times 100\,nm$ cross-bar devices (similar RRAM stack as used in Figs. 2.31–2.33) [72]. Switching for both set and reset between HRS and LRS in a similar ns time frame is clearly observed. In addition, for the preferred operational biasing, even at this ns time scale, the immediate and singular resistance change is observed for set operation, and more noisy increase/decrease fluctuations in resistance are observed for reset operation, due to the reset bias additionally having an effective competition between oxidation of the CF tip to form a dielectric barrier, and BD of this developing dielectric barrier.

Optimizing RRAM switching requires an understanding of how asymmetrical oxygen vacancy profiles and filament geometry assist the nature and preference of the bipolar biasing operational schemes. The goal is creating a vacancy profile that is advantageous for dielectric barrier formation (from CF tip oxidation) during reset, and dielectric barrier BD (filament reformation) in set operations, at conditions benefiting high-speed/low-power switching.

2.4.4   Performance and tunability

The reported results show that vacancy profile engineered HfO_{2-x}-based RRAM can demonstrate robust low-power operations, which appear to satisfy performance/endurance requirements for many applications of advanced memory systems. Key RRAM performance parameters, including LRS and HRS resistance, set/reset voltage and switching time, have been measured as a function of compliance current during set (I_{comp}), applied voltage range and pulse speed (dV/dt) using real-time ac methodology, as well as DC characterization [12, 13]. For this study, fully integrated $50\,nm \times 50\,nm$ cross-bar HfO_{2-x}-based 1T/1R RRAM devices with low parasitics were used. Unless noted differently, measurements were made on the same device, and trends were confirmed on multiple devices/wafers. An example of the DC I-V characteristic trends during HfO_{2-x}-RRAM set/reset operations with increasing set-I_{comp} levels is shown in Fig. 2.40 [13].

It has been demonstrated that capacitive parasitics can render the LRSs after forming to have a strong dependence on the forming voltage V_F [41]. The properties of HfO_{2-x}-based RRAM, when examined in devices using integrated 1T1R with low

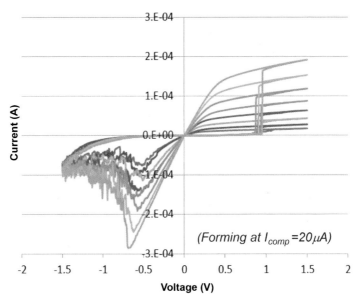

Fig. 2.40 DC set/reset cycles obtained on the same device at different I_{comp} in set during each cycle.

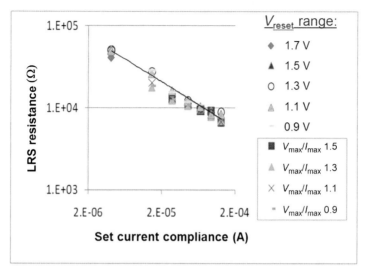

Fig. 2.41 DC RLRS for the integrated 1T1R (measured at 0.1 V) and $V_{reset\,max}/I_{reset\,max}$, as a function of I_{comp} during set and various V_{reset}-range.

parasitics, demonstrate that the LRS is primarily controlled by the I_{comp} during forming operation (Fig. 2.12) for filament resistance vs forming-I_{comp}. Similarly, in set operations, with the low parasitic integrated 1T1R, the operating LRS is primarily determined by I_{comp} during set (see Fig. 2.41 for set-LRS vs set-I_{comp} for a wide range of I_{comp}) [12, 13]. In addition, due to relatively low V_{set} values, any potential parasitic capacitance effects are generally small. The LRS current reveals a logarithmic dependency on I_{comp} in DC set operation, while the HRS current exhibits only a weak linear increase with these I_{comp} limits (Fig. 2.42) [12, 13]. These LRS and HRS resistance

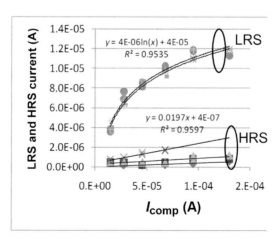

Fig. 2.42 Dependence of DC LRS and HRS read currents on I_{comp} in set obtained on the same device.

trends with I_{comp} result in a bell-shaped Ion/Ioff feature (Fig. 2.43) [12]. The smaller On/Off ratio with reduction of reset-range is due to thinner oxidized portion of the filament in reset, which leads to a lower HRS resistance. For increasing V_{reset}-range, a thicker and/or reset more robust dielectric barrier is formed having higher HRS resistance [68], which results subsequently in higher V_{set} required to break the "thicker" formed dielectric barrier (Fig. 2.44) [12].

Thus, to paraphrase these DC trend observations; the magnitude of I_{comp} during set is the major parameter controlling the LRS (via the filament cross section), whereas HRS (determined by the dielectric barrier at a given filament cross section) is mainly controlled by the V_{reset}-range. A more resistive HRS resulting from higher V_{reset}-range leads to a subsequently higher V_{set}.

The ac operating kinetics for these devices have been examined using the real-time operation/monitoring methodology, which can measure the current through the RRAM during pulsed forming and set/reset switching to achieve accurate determination of switching voltage, and set/reset switching time [12, 13]. The HfO_{2-x}-based RRAM devices can switch rapidly, down to the ps regime [25, 42].

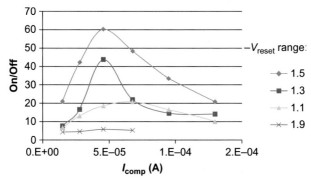

Fig. 2.43 DC Ion/Ioff ratio (at $I_{read}(V=0.1 \text{ V})$) as a function of I_{comp} during set for various V_{reset}-range.

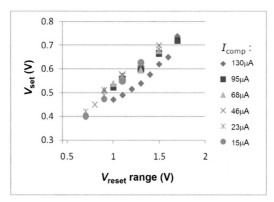

Fig. 2.44 Vt set as a function of V_{reset}-range. The measurements performed on same device at different values of I_{comp} in set.

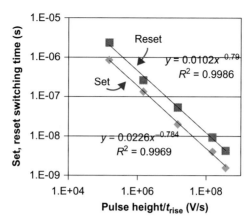

Fig. 2.45 Set and reset switching transition time (to the same resistance values) as a function of voltage pulse speed (pulse height/pulse rise time).

The data in Fig. 2.45 show that the faster the pulse (as defined by the pulse ramp rate: pulse voltage magnitude/pulse rise time), the faster the switching (between the LRS and HRS). However, the trade-off for decreased pulse rise time (faster pulsing) is an increase in the minimum set/reset voltages required to similarly change the device resistance state (Fig. 2.46) [12].

Although retention characteristics have not been discussed, reports of well-behaved retention (predicted beyond 10-year operation) above 200°C for HfO_{2-x}-based RRAM have been reported [16]. The promising retention behavior is based on the structure of the CF, and related structural changes, which for the case of HfO_{2-x}-based RRAM involves the movement of atoms as opposed to the "incumbent" charge-trap flash technologies that utilize the storage of electrons. Compared to the latter, the energy barrier of the atomic movement is higher, which impedes spontaneous structural rearrangements, thus preserving the material's resistive state.

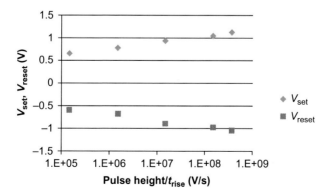

Fig. 2.46 Set and reset voltage as a function of pulse speed maintaining constant resistance values in set/reset switching.

In summary, operating performance, tunability, and intrinsic dependencies have been established between switching voltage and HRS/LRS resistance, which in turn depends on the current/voltage ranges. It is further established that HfO_{2-x}-based RRAM devices integrated with a low parasitics selector (as in 1T1R) can achieve low-power, high-speed, and high-endurance operations that appear to satisfy requirements for various memory applications.

2.5 Read current instability (random telegraph noise)

The read current in RRAM can be observed to exhibit stochastic digital-like changes of various amplitudes (Fig. 2.47) [14, 80–84]. This instability, called RTN, effectively reduces a memory window limiting further scaling of operational currents. In this section, we focus on analyzing the RTN-induced read instability amplitude in a statistically significant number of cycles of individual HfO_x-based RRAM cells (TiN/HfO_x/Ti-OEL/TiN RRAM), which is also representative of the switching in idealized arrays of identically formed cells.

2.5.1 Statistics of read instability

The maximum variation of the read current, called peak-to-peak (P-p) amplitude (Fig. 2.48) [79], is associated with the event when all possible random changes occur simultaneously producing the maximum current fluctuation, which constitutes the worst-case scenario from the standpoint of an erroneous read of the memory state. Therefore, the P-p amplitude can be used as a FoM for a quantitative evaluation of the read instability in both HRS and LRS. The sometime observed irreversible step-like

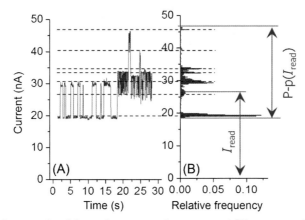

Fig. 2.47 (A) An example of the read current vs time trace; and (B) corresponding histogram of the read current values. Peaks in the histogram correspond to the RTN levels observed during the actual (limited) measurement time. I_{read} represents the mean DC current of the corresponding resistive state; P-p(I_{read}) is the figure of merit (FoM) of read instability [79].

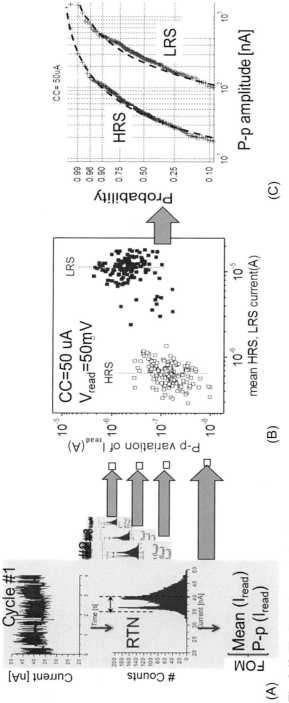

Fig. 2.48 The RTN data analysis: (A) I_{read} (averaged over read time) and P-p (peak-to-peak amplitude) are measured in HRS and LRS in each set/reset cycle; (B) distribution of I_{read} and P-p(I_{read}) values measured in each memory state over 200 set/reset cycles; and (C) cumulative distribution of the P-p amplitudes in (B) in the log scale. Both LRS and HRS distributions can be successfully fitted with the CDF of a normal distribution (thick dashed curves). In this example, $I_{comp} = 50\,\mu A$ and $V_{read} = 50\,mV$. Each symbol represents mean I_{read} (x-axis) and P-p amplitude (y-axis) of the RTN measured after each set/reset cycle [85].

changes in the read current, which can be caused by generation/annihilation of an atomic defect contributing to the charge transport [79], are not considered here.

By establishing a functional dependency of the distribution of the collected RTN data over statistically significant number of cycles, it is possible to assess the maximum P-p amplitude, which may get potentially realized in the given device. The distribution of the P-p amplitudes (on a log scale) is found to be fitted very well by the Gaussian probability density function (p.d.f.), yielding the most probable maximum noise amplitude and its dispersion value (Fig. 2.48). These two parameters define the entire P-p amplitude distribution in both LRS and HRS, allowing estimation of the probability of an RTN signal of any given amplitude to occur in this device. For successful reads of the resistive states, the memory window should be larger than the maximum noise amplitude. Therefore, the latter determines the minimum memory window necessary to keep the signal/noise ratio under a given limit.

This approach was employed to analyze the effect of RTN with scaling of the read current through scaling of the filament size [85]. Both I_{HRS} and I_{LRS} (at the same read voltage) are usually smaller in devices formed under lower current compliance (I_{comp}) limits, which lead to smaller filaments [23], while the P-p amplitude is also found to reduce with the I_{comp}. The fact that the noise reduces proportionally to the memory window (Fig. 2.49) is favorable for achieving lower operating currents by forming devices at lower I_{comp}.

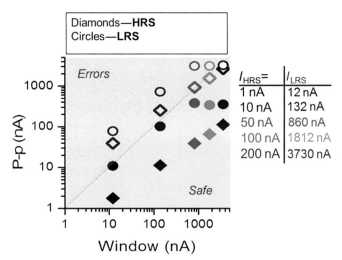

Fig. 2.49 The trends of the noise-induced variations of the read current (P-p) vs memory window trends for devices formed at different I_{comp} limits (ranging from 10 to 90 μA), which leads to different filament cross sections. *Diamonds*—HRS and *circles*—LRS. Values are given for 1σ *(closed symbols)* and 3σ *(open symbols)* of the noise amplitude distribution [79].

2.5.2 RTN origin

The RTN characteristics can be understood within the framework of the HRS/LRS description discussed in Section 2.3. In HRS, a current through the disrupted portion of the filament is shown to be supported by the trap-assisted tunneling process (Fig. 2.27) [7, 23]. Therefore, RTN of the read current in HRS is caused by the fluctuation of the number of traps contributing to TAT. As was discussed in Section 2.3, the TAT-active traps are the V+ oxygen vacancies in HfO_{2-x}, which are created by the activation of the V(2+) vacancies when they capture injected electrons [86]. The process responsible for RTN is thus stochastic activation/deactivation of electron trapping/detrapping in some of the electron transporting V+ defects (Fig. 2.50). Each reset/set cycle results in the CF being ruptured and restored, each time having a slightly different number and spatial distribution of V(2+) defects that translate to different characteristic times of the $V(2+) \rightarrow V+/V+ \rightarrow V(2+)$ processes and is subsequently reflected in variations read and RTN signals.

The RTN model has been verified by comparing its prediction to the measured P-p dependency on the forming I_{comp} conditions. First, by using a single set of the earlier reported trap characteristics [7], the I_{HRS} currents were calculated for various random spatial distributions of vacancies in the dielectric barrier in the devices formed under different I_{comp} conditions used to obtain the data in Fig. 2.49. The fitting of the calculated I_{HRS}, averaged over different randomly generated trap distributions, to the measured I_{HRS} data (averaged over 200 switching cycles) yielded a number of traps (N) required to reproduce the HRS read current for each I_{comp} limit. CFs of larger cross sections (i.e., formed at larger forming compliance currents) are found to contain a larger number of traps contributing to TAT transport, as expected (Fig. 2.50).

The extracted number of the traps contributing to I_{HRS} for each I_{comp} was used to check the feasibility of the RTN being controlled by the fluctuation of this number due to the spontaneous activation/deactivation $(V(2+) \rightarrow V^+/V^+ \rightarrow (V2+))$ of some of these traps. The activation/deactivation rates for N randomly distributed traps were calculated using the above-discussed multiphonon description. The Markov chain theory was then employed to calculate the probabilities, $P(s)$, of finding a system of N

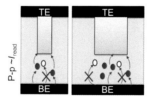

Fig. 2.50 Schematic of HRS RTN mechanism in a filament with (A) small and (B) large cross section. The HRS current is controlled by the TAT process via the traps in the dielectric barrier. Each trap can be either in an activated conductive (when it transfers an electron) or deactivated state. Trap activation is associated with its capturing of an electron (a relatively slow process). The TAT current is determined by the rates of capture/emission of a second electron by each active trap. RTN is associated with activation/deactivation of traps in TAT path.

vacancies in any of the 2^N states, $s = \{1, 2, \ldots, 2^N\}$, when each vacancy is either active (V+) or inactive (V(2+)) based on the abovementioned activation rates. Using these probabilities, calculated separately for devices formed at each given I_{comp}, the dispersion of the theoretically simulated I_{HRS} current fluctuations, comp with respect to its average value $\langle I_{HRS} \rangle$, can be evaluated:

$$\sigma = \sqrt{\sum_s P(s)\left(I_{HRS}(S) - I_{HRS}\right)^2} \tag{2.9}$$

Here σ is averaged over the various random trap distributions employed to simulate I_{HRS} at each I_{comp} limit. This current fluctuation, σ, should correlate to the mean value of the P-p distribution measured in devices formed under a given I_{comp} condition. Indeed, each cycle potentially realizes one of the possible distributions of N vacancies, and for each of these distributions, the most likely current fluctuations (=P-p value in a given cycle) are determined by the vacancies (at configuration s) with the higher activation/deactivation probabilities ($P(s)$). In the example in Fig. 2.51, the 2σ range, which represents about 98% of the possible I_{HRS} current fluctuations in any given set/reset cycle, resulted in an excellent agreement with the experimental (P-p) values obtained from 200 consecutive set/reset cycles. This agreement confirms that the HRS RTN can be associated with the traps in the oxide barrier in the filament.

In contrast to the HRS case, the traps responsible for RTN in LRS may reside outside of the filament (Fig. 2.52) and, therefore, do not directly support the electron current through the RRAM cell. To describe the dependency of the noise amplitude on I_{read}, the I_{read} fluctuations were modeled by assuming that they are associated with a change in a partial Coulomb-blockade of the electron flux through the CF caused by a change of a charge state of the nearby defect after it captured an electron. The CF section affected by the Coulomb blockade is located within a few Debye lengths

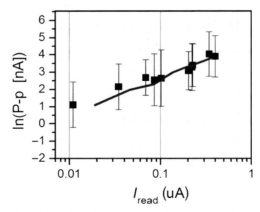

Fig. 2.51 HRS: Experimental P-p(I_{HRS}) vs I_{HRS} dependence, each data point represents a value averaged over 200 set/reset cycles. Data obtained from 10 devices formed at different I_{comp} (10–90 mA). Each I_{HRS} value is obtained in a device formed at the specific compliance current. Error bars show dispersion of P-p(I_{HRS}) values over 200 cycles for each device. *Solid line* represents the theoretical calculations of 2σ [85].

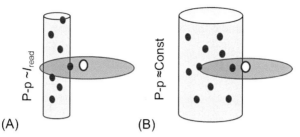

Fig. 2.52 Schematic representation of the LRS RTN mechanism in a filament of (A) small and (B) large cross section (small circles represent electrons and larger white circle represents a trap). Electron transport in the filament is described by a constant conductance value ($s = 25,000\,S\,m^{-1}$ was used in the simulations). RTN is caused by a Coulomb blockade of a portion of the filament by the charge associated with the electron trapping/detrapping at the trap in the vicinity of the filament. Electron density in the filament, $n = 10^{26}\,m^{-3}$, was assumed to theoretically reproduce experimental P-p(I_{read}) data (Fig. 2.53) [85].

from the charged trap. Theoretical values of RTN amplitudes for the CF of different sizes (which correspond to certain I_{read} values shown at the x-axis in Fig. 2.53) were calculated as P-p=$I_{read-c} - I_{read}$, where I_{read} and I_{read-c} are the simulated CF currents before and after the electron trapping at the adjacent defect. I_{read} is obtained by numerically solving Poisson and current continuity equations within the approximation of a uniform charge flow in the CF. Results in Fig. 2.53 show that the LRS P-p amplitude begins to scale with I_{read} only at its smaller values, when the trap effective blocking distance is becoming comparable with the entire filament cross section. Note that a reduction of P-p(I_{read}) may also be associated with a change in the filament composition when it was formed at low I_{comp} (as manifested by a nonohmic conductivity, indicating a transition in the charge transport to the TAT regime).

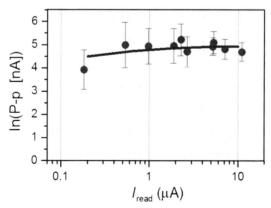

Fig. 2.53 LRS: The experimental P-p(I_{LRS}) vs I_{read} dependence *(symbols)* each value is averaged over 200 set/reset cycles. Error bars show the P-p(I_{LRS}) dispersion over 200 cycles. *Solid line*: theoretical fitting assuming, for simplicity, a uniform filament composition [85].

The HRS and LRS trends of the P-p amplitude vs the read current magnitude reflect on the difference between the transport mechanisms in these states. The RTN can be attributed to the activation and deactivation of the electron traps in or near the filament and the effect of these processes on the electron transport through the RRAM cell.

2.6 Multilevel ultrafast compliance-free RRAM switching

As discussed above, mobile NC systems present specific requirements that HfO_{2-x}-RRAM cells would have to satisfy to serve as microelectronic "synapses." The HfO_{2-x}-RRAM technology provides exceptional scalability and 3D integration capability, as well as high retention and endurance. However, sufficiently high-stability/low-variability memory states for the reliable gradual memory state modulation required in analog applications to imitate adaptive synaptic changes was not experimentally demonstrated using conventional (DC, pulse) test techniques.

Forming and switching of the CF in the HfO_{2-x}-based devices involves an atomic-level rearrangement it is an essentially stochastic process, which tends to lead to high site to site and cycling variability. The CF formation in HfO_{2-x}-RRAM is temperature and electric field driven (see Section 2.3.3) where local temperature increases in the vicinity of existing electrically active vacancies that stimulates faster generation of new oxygen vacancies/interstitial oxygen ions in the vicinity leading to a higher electron flow, which further increases local temperature, etc. This positive feedback loop is responsible for a runaway nature of the BD process when the vacancies generation rate becomes extremely fast (see Fig. 2.24), which does not allow to controllably stop continuing filament growth at the desired resistance value. Therefore, instead of attempting to control the filament resistance by limiting the current magnitude (which is still not effective even in 1T1R configuration due to conductivity overshoot, see Section 2.2.2.1), one may focus on limiting the delivered energy that drives the filament formation processes. This energy depends on the *duration* of the current flow, which is determined by the *time* the voltage drops across the device.

The effectiveness of such approach was demonstrated in the above-discussed HfO_{2-x}-RRAM cells. A controlled growth of a CF was achieved by applying ultrashort voltage pulses of sufficiently low amplitude that closely resembles circuitry operation conditions, without employing a current-limiting element (transistor (1T)) [25].

A measurement setup of the high-speed signal system is shown in Fig. 2.54. To accurately assess the energy released during short pulses, a robust procedure for de-embedding the parasitic displacement current from the current through the RRAM cell measured during forming has to be implemented (Fig. 2.55). Large displacement currents associated with the parasitic capacitances in a device and extremely short rise and fall times of the pulse are overshadowing the active current drawn by the RRAM cell during the pulse. The de-embedding procedure (its outcome is shown in Fig. 2.55B) is based on subtracting of the preforming current $I_{pre}(t)$, which has purely parasitic nature, from the postforming current, $I_{measured}(t)$, shown in Fig. 2.55A [25].

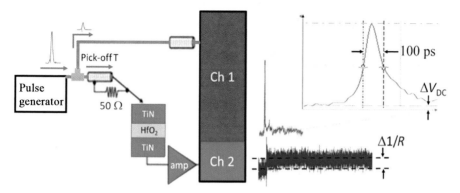

Fig. 2.54 High-speed pulse RRAM switching setup. In all, $50\,\Omega$ termination improves signal integrity by preventing pulse reflection from the cell and reducing RC constant. Channel 1 of the oscilloscope ($20\,\text{GHz}$ bandwidth) is used to measure voltage pulse (shown on the right-hand side) applied to the RRAM device. Small DC bias ($<100\,\text{mV}$) is superimposed on the voltage pulse to continuously monitor RRAM resistance. By varying the amplitude of switching pulse (from 0.5 to 5 V) different target resistance levels of RRAM can be reached. Channel 2 measures RRAM conductance ($1/R$) before and after the pulse ($\Delta 1/R$, plotted below the pulse signal) [87].

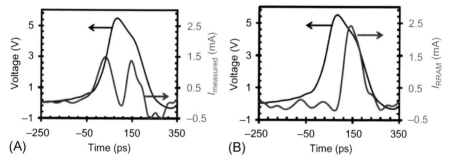

Fig. 2.55 Measured voltage and current pulse before (A) and after (B) de-embedding the parasitic current components. This step is necessary to accurately extract forming energy [25].

Maximum reset current (determined by the resistance (dimension) of the CF) depends on the pulse duration (Fig. 2.56A), and the forming energy delivered by the pulse to the RRAM cell, $E_{Forming}$ (Fig. 2.56B):

$$E_{Forming} = \int_{t_0}^{t_0 + t_{pulse}} V(t) I_{RRAM}(t)\, dt,$$

where $I_{RRAM}(t)$ is the current flowing through the filament during voltage pulse, t_0 is the beginning of the voltage pulse, and t_{pulse} is a pulse duration. The above expression can be generalized in case of multiple forming pulses

$$E_{Forming} = \sum_{i=1}^{N} \int_{t_i}^{t_i + t_{pulse}} V_i(t) I_i(t)\, dt,$$

where t_i is the beginning of the i voltage pulse and N is a number of pulses required to form the CF.

As follows, distribution of the postforming resistance values at a given pulse width (see Fig. 2.56A) is driven by the distribution of the forming energy, $E_{Forming}$, caused by device-to-device variability of the initial (preforming) precursor conductive paths rather than variability of the forming process itself.

Switching stability evaluation was performed on $200\,nm \times 200\,nm$ cross-point RRAM, with TiN\HfO$_{2-x}$(5.8 nm)\Ti\TiN stacks and *no* current-limiting elements, using programming pulses with fixed width (100 ps) and amplitude (Fig. 2.54) [25]. Switching endurance was evaluated using the developed Compliance-free UltraShort Smart Pulse Programming (CUSPP) [26]. In this CUSPP setup, the resistance of the RRAM states is monitored continuously by a low DC voltage, V_{read} ($-8\,mV$). This approach is adopted to provide a continuous recording of the resistance evolution during, and after programming. A fast comparator with response time of less than 5 ns was used to automatically (1) compare the instantaneous state of the RRAM vs the target resistance and (2) orchestrates ...-HRS-LRS-HRS-LRS-... cycling by switching the pulse polarity after reaching the target resistance. This approach allows for cycling RRAM resistance states at the rate of a few millions cycles per second, and supports the tuning of the pulse amplitude to allow the investigation of switching/forming via single pulse or multiple pulses as shown in Fig. 2.57A and B. While forming with ultrashort pulses requires higher pulse amplitudes, switching pulse amplitudes can be flexible, ranging within 0.5–2 V for both set and reset. Smaller amplitudes of set and reset pulses enable more gradual changes of RRAM resistance with a larger number of switching pulses. Smaller programming pulse amplitudes are also expected to have weaker effect the RRAM resistance state, thus making a cycle-to-cycle distribution of the resistances of a particular memory state tighter.

Fig. 2.58A shows the distributions of the HRS and LRS values for a representative RRAM cell measured over a few thousands switching cycles. It is worth noting that

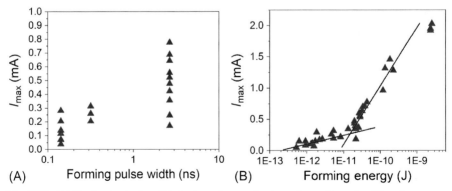

Fig. 2.56 (A) Device-to-device distribution of current values, I_{max}, after pulsed forming with varying pulse durations [25]. Dispersion in conductance of the filaments formed by the identical pulses is associated with the device-to-device variability of the initial precursor conductive paths. (B) Dependency of I_{max} on the energy released by a forming pulse. Above a certain critical energy value (near the intersection of the *solid lines*) the filament conductance increases at a much higher rate (as indicated by the *line slopes*).

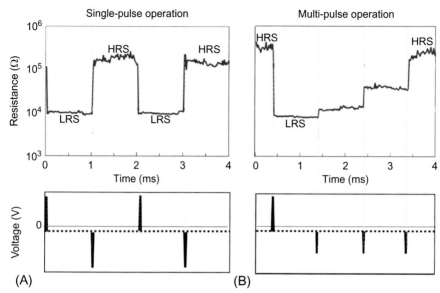

Fig. 2.57 By varying voltage pulse amplitude resistance switching can be performed in (A) digital or (B) analog regimes using larger single pulse or multiple smaller pulses, respectively. In the low panels schematic, the pulse width is 100 ps and pulse amplitudes are 1 V (digital switching) and 0.8 V (gradual (analog) switching) [37, 88].

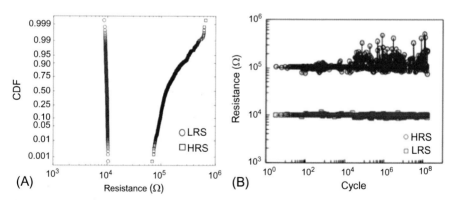

Fig. 2.58 Ultrafast pulse compliance-free switching of HfO_{2-x}-RRAM [88]. (A) Cycle-to-cycle distribution of HRS and LRS, measured over 10,000 consecutive cycles. Pulse duration is 100 ps, pulse amplitude is around 1 V for both set and reset. Resistance is measured immediately after programming. (B) Endurance: LRS and HRS resistances vs a number of switching cycles (in log scale). Maximum variation of HRS resistance is observed approximately every 10,000 cycles. Note that both variations amplitudes and their frequency remain constant within the entire switching cycles measurement, it implies that no permanent device degradation occurs and switching instability reflects random structural fluctuations (their origin will be discussed elsewhere).

the RRAM cell switches in every single cycle. Since the resistance is verified after each program pulse and, if needed, pulses are repeated until the target resistance is reached, there are no failed switching events using this method (called "program-verify"). Verification of the resistance state after programing allows for adjusting this state leading to a tighter resistance distribution, both cycle to cycle and device to device. The desired resistance value can be achieved through a consecutive application of up to several ultrashort pulses. After each pulse, the resistance value is verified and compared with the target resistance. This approach leads to resistance distributions being truncated below and above the target values for HRS and LRS, respectively. Setting the target HRS resistances to a different value changes the number of required switching pulses resulting in a shift of the mean of the resistance distribution toward the new target value. Although tight memory distribution may be desired for practical applications [37], these results set an important milestone on the path to implement hafnia-based RRAM in neuromorphic applications. Better resolution can be obtained using larger memory windows, or tightening the resistance distributions by further tuning the switching conditions, and/or material properties.

Stable RRAM operations were demonstrated by employing the CUSPP methodology (Fig. 2.58B), where HRS/LRS switching was performed for up to 10^8 cycles upon when the switching was terminated, with no degradation of the memory window [88].

The ultrafast pulse compliance-free switching approach demonstrates the HfO_{2-x}-RRAM capability of symmetric and gradual conductance change (Fig. 2.59), which is critical to neuromorphic systems [87]. As seen in Fig. 2.59, switching between multiple states is indeed almost linear, the read noise is low.

Although generally RRAM is experimentally subjected to various noise sources during read and write operations, their effect can be mostly eliminated by increasing the integration time, reducing operating/read voltages, etc. More important, the reported broadening of the resistance distributions in HfO_{2-x}-RRAM with time after programming [89], is not a concern in the ultrafast pulse compliance-free switching method relevant to circuit operations. In this approach, the resistance relaxation, which was considered a showstopper for metal-oxide RRAM, is not observed [90]. The advantage of the ultrafast pulse switching is associated with fast dissipation of heat emitted during extremely short RRAM programing [26], contrarily to the case with conventional (longer) switching pulse operations [89], where heat was accumulated in a relatively large volume of the device allowing for further postprogramming atomic structural rearrangements and related modifications of the RRAM conductance.

The duration of the heat dissipation was evaluated experimentally [87] by studying the effect of two consecutive programming pulses on the RRAM resistance (Fig. 2.60). Reducing the time interval between pulses below 2 ns resulted in significant amplification of their effect on resistance. It's consistent with theoretical expectations that, by briefly increasing local temperature, the first pulse magnifies the structural changes inducing by the subsequent pulse. This correlated impact of programming pulses on the HfO_{2-x}-RRAM conductance emulates modulations of the synaptic weight controlled by the time between the incoming signals in the spiking neuromorphic systems.

Thus, the compliance-free ultrafast pulse technique allows for the switching

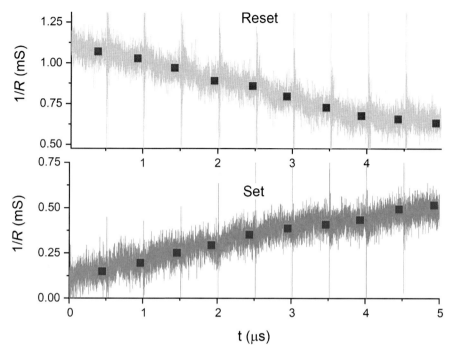

Fig. 2.59 Semilinear increase and decrease of the RRAM conductance (1/R) under the continuous set and reset pulses. Program pulse width is 100 ps. Conductance spikes seen at the moments of programing pulses are the measurement artifact. *Square marks* on the conductance traces correspond to averaged values of the RRAM conductance after each pulse [87].

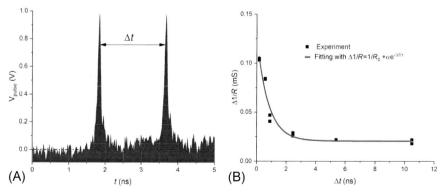

Fig. 2.60 Effect of the time interval between consecutive programming (reset) pulses on the filament conductance. (A) A pair of 100-ps voltage pulses applied to RRAM. The delay between pulses, Δt, is varied from 200 ps to 11 ns. (B) Conductance change vs pulses delay Δt. The measurements are repeated twice for each time delay to account for variations in switching between HRS and LRS states [87].

of HfO_{2-x} filamentary RRAM states in a way satisfying most of the requirements imposed by neuromorphic applications. This approach eliminates the need for a compliance-limiting device, excluding contributions from conductivity overshoot and parasitic capacitance. Precise control over the total released energy is achieved by applying ultrashort voltage pulses (in ≈ 100 ps to a few ns range) limiting the process duration. This also limits the active volume of the devices that contribute to switching with the size of the "hotspot" during the pulse not increasing more than a few nanometers around the filament. Smaller size of the active volume also reduces variability, through ultrafast dissipation of heat, freezing the ions in their places and not allowing for further changes of the filament size.

2.7 Conclusion

Detailed operational and intrinsic switching characteristics for hafnia-based RRAM have been presented, including performance and tunability along with materials/vacancy engineering ramifications. The entire set of reported experimental data and material structure modeling and transport simulation results are consistent with the physical picture of the hafnia-based RRAM operations, as caused by oxidation/reduction processes in the conducting filament formed in the dielectric. The outcome of the filament formation process—an oxygen-deficient region (which constitutes a CF) surrounded by the interstitial oxygen ions expelled from the filament region—establishes initial structural conditions for the subsequent reset and set operations. Kinetics of these redox processes can be considered by accounting for oxygen ions diffusion driven by the local electric fields and temperatures in the dielectric surrounding the filament. A microscopic description of these processes, the framework of which is presented here, directly links the device electrical and material characteristics, thus enabling improving device performance through optimization of the material compositional profile and operation conditions.

By nature of this process, its mechanism is material specific, meaning that it explicitly includes dependencies on dielectric morphology (crystallinity and GBs density), its stoichiometric profile (oxygen vacancy distribution), parameters of the electron-phonon coupling of the defects supporting electron transport in this material (e.g., ionization and relaxation energies of positively charged oxygen vacancies at GBs), processes of the oxygen-vacancy pair generation (metal-oxygen bond breakage), and diffusion of the released oxygen ions (hopping via substitutional oxygen sites). Although there are no "fitting parameters," it is important to realize that the values of all parameters used in the simulations are obtained outside the model, either from physical characterization data or material structure modeling (there is always some uncertainty in even these parameter values).

At any given moment during the device operation, the processes of ion hopping and ion-vacancy pair generation/recombination at each location in the device are determined by the ion/vacancy distribution in the entire active device region, since it affects the local electric fields and temperatures. Due to essentially nonlocal characteristics of these processes and their probabilistic nature, an analytical compact description for the physical changes associated with the device switching is not feasible. Developing a simplified compact model, which would retain all essential properties of the comprehensive Monte Carlo description, presents a challenging but highly valuable goal. Alternative approaches allowing a compact description are phenomenological models, which are usually focused on specific features of the switching phenomenon (e.g., dependency of reset time on the applied voltage), and utilize values of the employed parameters that are extracted by fitting the modeling results to the experimental data. These models can be helpful in explaining certain observations in a given material stack; however, they are generally less sensitive to the specifics of the material characteristics and therefore, may not be as useful in driving fabrication process improvement as is the atomistic model described in this chapter.

Conductance in metal-oxide films is determined by material structural characteristics, which can be locally reversibly changed under applied operation conditions. Consequently, cell switching properties are extremely sensitive to operation conditions; neglecting this factor led to a number of confusing conclusions on the capability of this technology. Polycrystalline oxides, in particular, hafnia-based RRAM cells, are promising since these materials have preexisting preferential conductive paths (specifically along the GBs), one of which gets converted to a switchable CF. Limitations observed in this class of materials are shown to be largely related to measurement artifacts, as was demonstrated by employing the compliance-free operations using an ultrafast pulse technique.

Using the developed models and resulting understanding of the primarily driving forces for RRAM switching and their ramifications toward material properties, materials engineering of the HfO_{2-x}-based RRAM was employed to optimize device performance. Based on the modeling results, the operating conditions leading to lower switching and device-to-device variability were identified. It is demonstrated that a substoichiometric dielectric film with an asymmetrical vacancy profile is advantageous for effective dielectric barrier formation during reset and dielectric barrier BD in set operations, which are the processes controlling the switching in the bipolar operated HfO_{2-x}-based memory cells. The highly scalable and robust (>300 billion cycle endurance) devices, which can operate at the ~sub-ns time scale at low power (~fJ), fabricated using materials already common to the integrated circuits industry, exhibit characteristics meeting a variety of NVM applications.

Regarding neuromorphic applications, the hafnia-based filamental devices can resolve the fundamental time-energy conundrum by delivering both ultrafast operations with low-energy consumption. The compliance-free test approach, which closely reproduces circuitry operation conditions, results in a more stable hafnia-RRAM switching compared to DC and conventional microsecond pulse operations. Besides improved cycle-to-cycle distribution, compliance-free switching results provide an additional benefit of removing the need to incorporate compliance transistors, thus, reducing the array footprint.

Through continued refinement of developed models and materials engineering, it is expected that these RRAM devices will be successfully implemented for high-volume manufacturing in many future commercial products.

Acknowledgment

This work was partially supported by the Aerospace Corporation's Internal Research and Development Program.

References

[1]   C.-Y. Lu, Future prospects of NAND flash memory technology: evolution from float-
      ing gate to charge trapping to 3D stacking, J. Nanosci. Nanotechnol. 12 (10) (2012)
      7604–7618.
[2]   M.J. Marinella, et al., Multiscale co-design analysis of energy, latency, area, and accuracy
      of a ReRAM analog neural training accelerator, IEEE J. Emerging Sel. Top. Circuits Syst.
      8 (1) (2018) 86–101.
[3]   S. Ambrogio, et al., Equivalent-accuracy power-efficient neuromorphic hardware accel-
      eration of neural network training using analog memory, Nature 558 (2018) 60.
[4]   H. Tsai, et al., Recent progress in analog memory-based accelerators for deep learning, J.
      Phys. D Appl. Phys. 51 (2018) 283001.
[5]   R. Waser, M. Aono, Nanoionics-based resistive switching memories, Nat. Mater. 6 (2007)
      833–840.
[6]   A. Chen, Ionic memory technology, in: V.V. Kharton (Ed.), Solid State Electrochemistry
      II: Electrodes, Interfaces and Ceramic Membranes, Wiley, 2011, pp. 1–26. [S.l.],
      (Chapter 1).
[7]   G. Bersuker, et al., Metal oxide resistive memory switching mechanism based on conduc-
      tive filament properties, J. Appl. Phys. 110 (12) (2011) 124518.
[8]   G. Bersuker, et al., Metal oxide RRAM switching mechanism based on conductive fila-
      ment microscopic properties, in: IEDM, Tech. Dig, 2010, pp. 19.6.1–19.6.4.
[9]   A. Sawa, Resistive switching in transition metal oxides, Mater. Today 11 (2008) 28–36.
[10] R. Waser, et al., Redox-based resistive switching memories—nanoionic mechanisms,
      prospects, and challenges, Adv. Mater. 21 (2009) 2632.
[11] I.G. Baek, et al., Highly scalable non-volatile resistive memory using simple binary oxide
      driven by asymmetric unipolar voltage pulses, in: IEDM, Tech. Dig, 2004, pp. 587–590.
[12] S. Koveshnikov, et al., Real-time study of switching kinetics in integrated 1T/HfO$_x$ 1R
      RRAM: Intrinsic tunability of set/reset voltage and trade-off with switching time, in:
      IEDM, Tech. Dig, 2012.
[13] S. Koveshnikov, et al., Development of NVM HfOx based 1T1R RRAM devices: needs
      and challenges for potential NAND replacement, in: (IGSTC) International Gate Stack
      Technology Symposium, 2012.
[14] M.-J. Lee, et al., A fast, high-endurance and scalable non-volatile memory device made
      from asymmetric Ta$_2$O$_5$/TaO$_2$ bilayer structures, Nat. Mater. 10 (2011) 625–630.
[15] H.Y. Lee, et al., Low-power and nanosecond switching in robust hafnium oxide resistive
      memory with a thin Ti cap, IEEE Electron Device Lett. 31 (1) (2010) 44–46.
[16] Y.S. Chen, et al., Challenges and opportunities for HfOx based resistive random access
      memory, in: IEDM, Tech. Dig., December, 2011, pp. 31.3.1–31.3.4.
[17] Y.B. Kim, et al., Bi-layered RRAM with unlimited endurance and extremely uniform
      switching, in: VLSI Technology Symposium, 2011, pp. 52–53.
[18] J.J. Yang, et al., Metal/TiO$_2$ interfaces for memristive switches, Appl. Phys. A 102 (2011)
      785–789.

[19] B. Butcher, et al., High endurance performance of 1T1R HfO based RRAM at low (20 μA) operative current and elevated (150 °C) temperature, in: IIRW IEEE Transactions in Device and Material Reliability (TDMR), October, 2011, pp. 146–150.

[20] B. Govoreanu, et al., 10 nm × 10 nm Hf/HfO$_x$ crossbar resistive RAM with excellent performance, reliability and low-energy operation, in: IEDM, Tech. Dig, 2011.

[21] K. Kinoshita, et al., Reduction in the reset current in a resistive random access memory consisting of NiOx brought about by reducing a parasitic capacitance, Appl. Phys. Lett. 93 (2008) 033506.

[22] A. Kalantarian, et al., Controlling uniformity of RRAM characteristics via the forming process, in: IRPS IEEE, Anaheim, CA, 2012.

[23] L. Vandelli, et al., Comprehensive physical modeling of forming and switching operations in HfO$_2$ RRAM devices, in: IEDM, Tech. Dig., December, 2011, pp. 17.5.1–17.5.4.

[24] B. Butcher, et al., Modeling the effects of different forming conditions on RRAM conductive filament stability, in: IEEE, International Memory Workshop (IMW), Monterey, 2013.

[25] P.R. Shrestha, et al., Energy control paradigm for compliance-free reliable operation of RRAM, in: Proc. 2014 IEEE International Reliability Physics Symposium, 2014, pp. MY.10.1–MY.10.4.

[26] D.M. Nminibapiel, et al., Characteristics of resistive memory read fluctuations in endurance cycling, IEEE Electron Device Lett. 38 (3) (2017) 326–329.

[27] G. Bersuker, Y. Jeon, H.R. Huff, Degradation of thin oxides during electrical stress, Microelectron. Reliab. 41 (12) (2001) 1923–1931.

[28] N. Capron, P. Broqvist, A. Pasquarello, Migration of oxygen vacancy in HfO$_2$ and across the HfO$_2$/SiO$_2$ interface: a first-principles investigation, Appl. Phys. Lett. 91 (19) (2007) 192905.

[29] S. Foster, A.L. Shluger, R.M. Nieminen, Mechanism of interstitial oxygen diffusion in Hafnia, Phys. Rev. Lett. 89 (2) (2002) 225901.

[30] J. McPherson, et al., Thermochemical description of dielectric breakdown in high dielectric constant materials, Appl. Phys. Lett. 82 (13) (2003) 2121.

[31] J.K. Hubbard, D.G. Schlom, Thermodynamic stability of binary oxides in contact with silicon, J. Mater. Res. 11 (11) (1996) 2757–2776.

[32] S. Stemmer, Stability of gate dielectrics and metal gate electrodes, in: International SEMATECH Gate Stack Engineering Working Group Symposium, Austin, TX, 2003.

[33] S. Stemmer, Thermodynamic considerations in the stability of binary oxides for alternative gate dielectrics in complementary metal-oxide-semiconductors, J. Vac. Sci. Technol. B Microelectron. Nanometer Struct. 22 (2) (2004) 791–800.

[34] V. Cosnier, et al., Understanding of the thermal stability of the hafnium oxide/TiN stack, Microelectron. Eng. 84 (9–10) (2007) 1886–1889.

[35] S. Guha, V. Narayanan, High-κ/metal gate science and technology, Mater. Res. 39 (2009) 181–202.

[36] J.K. Schaeffer, et al., Physical and electrical properties of metal gate electrodes on HfO$_2$ gate dielectrics, J. Vac. Sci. Technol. B Microelectron. Nanometer Struct. 21 (1) (2003) 11–17.

[37] G. Bersuker, et al., Toward reliable RRAM performance: macro- and micro-analysis of operation processes, J. Comput. Electron. 16 (4) (2017) 1085–1094.

[38] A. Padovani, et al., Microscopic modeling of HfOx RRAM operations: from forming to switching, IEEE Trans. Electron Devices 62 (2015) 1998–2006.

[39] B. Butcher, et al., Connecting the physical and electrical properties of Hafnia-based RRAM, in: IEDM, 2013.

[40] M. Porti, M. Nafria, X. Aymerich, Nanometer-scale analysis of current limited stresses impact on SiO$_2$ gate oxide reliability using C-AFM, IEEE Trans. Nanotechnol. 3 (1) (2004) 55–60.

[41] D.C. Gilmer, et al., Effects of RRAM stack configuration on forming voltage and current overshoot, in: IEEE-International Memory Workshop (IMW), 2011.

[42] M. Lanza, et al., Grain boundaries as preferential sites for resistive switching in the HfO$_2$ resistive random access memory structures, Appl. Phys. Lett. 100 (2012) 123508.

[43] L. Zhang, Y. Mitani, Structural and electrical evolution of gate dielectric breakdown observed by conductive atomic force microscopy, Appl. Phys. Lett. 88 (3) (2006) 032906.

[44] M. Porti, et al., Electrical characterization of stressed and broken down SiO$_2$ films at a nanometer scale using a conductive atomic force microscope, J. Appl. Phys. 91 (4) (2002) 2071–2079.

[45] K.S. Yew, et al., Nanoscale characterization of HfO2/SiO2 gate stack degradation by scanning tunneling microscopy, in: Proceedings of the International Conference on Solid State Devices and Materials (SSDM), Japan, 2009.

[46] K.P. McKenna, A.L. Shluger, Electronic properties of defects in polycrystalline dielectric materials, Microelectron. Eng. 86 (2009) 1751–1755.

[47] L. Larcher, Statistical simulation of leakage currents in MOS and flash memory devices with a new multiphonon trap-assisted tunneling model, IEEE Trans. Electron Devices 50 (2003) 1246.

[48] A. Padovani, Statistical modeling of leakage currents through SiO$_2$/high-κ dielectrics stacks for non-volatile memory applications, in: IEEE Int. Reliability Physics Symposium (IRPS), 2008, pp. 616–620.

[49] L. Vandelli, et al., Modeling temperature dependency (6–400K) of the leakage current through the SiO$_2$/high-κ stacks, in: ESSDERC, Sivilia, Spain, 2010, pp. 388–391.

[50] K. Huang, A. Rhys, Theory of light absorption and non-radiative transitions in F-centres, Proc. R. Soc. Lond. A 204 (1950) 406–423.

[51] C.H. Henry, D.V. Lang, Nonradiative capture and recombination by multiphonon emission in GaAs and GaP, Phy. Rev. B 15 (2) (1977) 989–1016.

[52] W.B. Fowler, et al., Hysteresis and Franck-Condon relaxation in insulator-semiconductor tunneling, Phy. Rev. B 41 (1990) 8313.

[53] L. Vandeli, et al., Microscopic modeling of electrical stress-induced breakdown in poly-crystalline hafnium oxide dielectrics, IEEE Trans. Electron Devices 60 (5) (2013) 1754–1762.

[54] A. O'Hara, A. Demkov, G. Bersuker, Assessing hafnium on hafnia as an oxygen getter, J. Appl. Phys. 115 (2014) 183703.

[55] D.C. Gilmer, et al., Compatibility of polycrystalline gate deposition with HfO$_2$ and Al$_2$O$_3$/HfO$_2$ gate dielectrics, Appl. Phys. Lett. 81 (2002) 1288.

[56] K.G. Young-Fisher, et al., Leakage current-forming voltage relation and oxygen gettering in HfO$_x$ RRAM devices, IEEE Electron Device Lett. 34 (2013) 750–752.

[57] L.V. Goncharova, et al., Diffusion and interface growth in hafnium oxide and silicate ultrathin films on Si(001), Phy. Rev. B 83 (2011) 1153329.

[58] L. Vandeli, et al., A physical model of the temperature dependence of the current through SiO$_2$/HfO$_2$ stacks, IEEE Trans. Electron Devices 58 (9) (2011) 2878–2887.

[59] S. Lombardo, et al., Dielectric breakdown mechanisms in gate oxides, J. Appl. Phys. 98 (12) (2005) 121301.

[60] M. Schie, et al., Field-enhanced route to generating anti-Frenkel pairs in HfO2, Phys. Rev. Mater. 2 (2018) 035002.

[61] S.R. Bradley, A.L. Shluger, G. Bersuker, Electron-injection-assisted generation of oxygen vacancies in monoclinic HfO$_2$, Phys. Rev. Appl. 4 (2015) 064008.

[62] S. Privitera, et al., Microscopy study of the conductive filament in HfO_2 resistive switching memory devices, Microelectron. Eng. 109 (2013) 75–78.

[63] S.R. Bradley, et al., Modelling of oxygen vacancy aggregates in monoclinic HfO_2: can they contribute to conductive filament formation? J. Phys. Condens. Matter 27 (2015) 41540.

[64] K.P. McKenna, et al., Optimal stoichiometry for nucleation and growth of conductive filaments in HfO_2, Model. Simul. Mater. Sci. Eng. 22 (2014) 025001.

[65] L. Larcher, et al., Microscopic understanding and modeling of HfO2 RRAM device physics, in: IEDM, Tech. Dig, 2012.

[66] M. Sowinski, et al., Hard X-ray photoelectron spectroscopy study of the electroforming in Ti/HfO_2-based resistive switching structures, Appl. Phys. Lett. 100 (2012) 233509.

[67] H.-L. Chang, et al., Physical mechanism of HfO2 random access memory, in: VLSI Symp. Tech. Dig. Based Bipolar Resistive, 2011.

[68] F.M. Puglisi, et al., An empirical model for RRAM resistance, IEEE Electron Device Lett. 99 (2013) 387–389.

[69] G. Bersuker, et al., Connecting RRAM performance to the properties of the hafnia-based dielectrics, in: ESSDERC, Bucharest, 2013.

[70] B. Gao, Pulse voltage dependent resistive switching behaviors of HfO2-based RRAM. Solid-state and integrated circuit technology (ICSICT), in: 10th IEEE International Conference, November, 2010, pp. 1145–1147.

[71] D.C. Gilmer, et al., Engineering metal-oxide based RRAM for high volume manufacturing, in: Non-Volatile Memory Technology Symposium (NVMTS), Singapore, 2012.

[72] D.C. Gilmer, et al., Asymmetry, vacancy engineering and mechanism for bipolar RRAM, in: IEEE, International Memory Workshop (IMW), Milano, 2012.

[73] J.E. Stevens, et al., Reactive sputtering of substoichiometric Ta_2O_5 for resistive memory applications, J. Vac. Sci. Technol. A 32 (2) (2014) 021501-1-6.

[74] C. Vallee, et al., Plasma treatment of HfO_2-based metal-insulator-metal resistive memories, J. Vac. Sci. Technol. A 29 (2011) 041512.

[75] J.-H. Hur, et al., Universal model for bipolar resistance random access memory (BReRAM) switching, in: IEDM, Tech. Dig, 2011.

[76] W.-C. Chien, et al., A multi-level 40 nm WO_x resistive memory with excellent reliability, in: IEDM Tech. Dig, 2011.

[77] T.-C. Chang, et al., Single atom redox reaction inducing resistance switching with Ti/HfO_2/TiN RRAM device, in: IEDM, Tech. Dig, 2011.

[78] P.-S. Chen, et al., Improved resistive switching of HfO_2/TiN stack with a reactive metal layer and annealing. Solid State Devices Mater. (2009) 444–445, https://doi.org/10.1143/JJAP.49.04DD18.

[79] D. Veksler, et al., Methodology for the statistical evaluation of the effect of random telegraph noise (RTN) on RRAM characteristics, in: IEDM, Tech. Dig. Dec, 2012, pp. 9.6.1–9.6.4.

[80] R. Soni, et al., Probing Cu doped Ge0.3Se0.7 based resistance switching memory devices with random telegraph noise, J. Appl. Phys. 107 (2010) 024517.

[81] M. Terai, et al., Resistance controllability of Ta2O/TiO stack ReRAM for low-voltage and multilevel operation, IEEE Electron Device Lett. 31 (3) (2010) 20–26.

[82] Y.H. Tseng, et al., Electron trapping effect on the switching behavior of contact RRAM devices through random telegraph noise analysis, in: IEDM, Tech. Dig, 2010, pp. 28.5.1–28.5.4.

[83] J.-K. Lee, et al., Extraction of trap location and energy from random telegraph noise in amorphous TiO_x resistance random access memories, Appl. Phys. Lett. 98 (2011) 143502.

[84] D. Ielmini, F. Nardi, C. Cagli, Resistance-dependent amplitude of random telegraph-signal noise in resistive switching memories, Appl. Phys. Lett. 96 (2010) 053503.

[85] D. Veksler, et al., Random telegraph noise (RTN) in scaled RRAM devices. in: IEEE Int. Reliability Physics Synposium (IRPS), 2013, https://doi.org/10.1109/IRPS.2013.6532101.

[86] G. Bersuker, et al., Grain boundary-driven leakage path formation in HfO_2 dielectrics, Solid State Electron. 65–66 (2011) 146–150.

[87] D. Veksler, et al., Synaptic weight modulation by controlling metal oxide RRAM switching, in: ECS/AiMES, 2018.

[88] D. Nminibapiel, The Efficacy of Programming Energy Controlled Switching in Resistive Random Access Memory (RRAM), (Ph.D. thesis) Old Dominion University, 2017. https://digitalcommons.odu.edu/ece_etds/21.

[89] A. Fantini, et al., Intrinsic program instability in HfO2 RRAM and consequences on program algorithms, in: IEDM Tech. Dig, 2015, pp. 7.5.1–7.5.4.

[90] D.M. Nminibapiel, et al., Impact of RRAM read fluctuations on the program-verify approach, IEEE Electron Device Lett. 38 (6) (2017) 736–739.

Advanced modeling and characterization techniques for innovative memory devices: The RRAM case

3

Francesco Maria Puglisi[a], Andrea Padovani[b], Paolo Pavan[a], Luca Larcher[b]
[a]Dipartimento di Ingegneria "Enzo Ferrari", Università di Modena e Reggio Emilia, Modena, Italy, [b]Applied Materials, Reggio Emilia, Italy

3.1 Introduction

The Semiconductor Industry Association (SIA), representing US leadership in semiconductor manufacturing, design, and research, revealed on February 5th, 2018 the industry's highest-ever annual sales and an increase of 21.6% compared to the 2016 total. "As semiconductors have become more heavily embedded in an ever-increasing number of products (from cars to coffee makers) and nascent technologies like artificial intelligence, virtual reality, and the Internet of Things have emerged, global demand for semiconductors has increased, leading to landmark sales in 2017 and a bright outlook for the long term," said John Neuffer, SIA president and CEO. "The global market experienced across-the-board growth in 2017, with double-digit sales increases in every regional market and nearly all major product categories. We expect the market to grow more modestly in 2018" [1].

Several semiconductor product segments stood out in 2017. Memory was the largest semiconductor category by sales with $124.0 billion in 2017, and the fastest growing, with sales increasing to 61.5%. Within the memory category, sales of DRAM products increased by 76.8% and sales of NAND flash products increased by 47.5%. Logic ($102.2 billion) and micro-ICs ($63.9 billion)—a category that includes microprocessors—rounded out the top three product categories in terms of total sales. Other fast-growing product categories in 2017 included rectifiers (18.3%), diodes (16.4%), and sensors and actuators (16.2%). Even without sales of memory products, sales of all other products combined increased by nearly 10% in 2017 [1].

One can easily derive from these data that the semiconductor memory sector represents a large part of the overall semiconductor market; memory revenue changes will have a large effect on the industry. From the economic point of view, favorable DRAM and NAND supply-demand dynamics have resulted in robust earnings and stock prices for memory manufacturers, but something might be changing. Indeed, additional memory capacity coming online in 2018, especially from China, is expected to soften memory prices. In turn, the impact of softened memory prices could affect

Advances in Non-volatile Memory and Storage Technology. https://doi.org/10.1016/B978-0-08-102584-0.00004-8

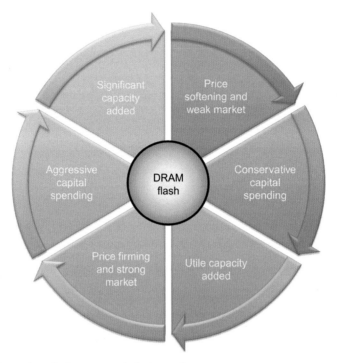

Fig. 3.1 The semiconductor memory industry cycle.

stock prices of memory manufacturers. Is this the "classic" cycle of development in semiconductors (Fig. 3.1), or has something changed?

In the last few years, we have observed that the number of memory suppliers has collapsed from 32 DRAM companies decades ago to just three today: Samsung, SK Hynix, and Micron. On the NAND side, there are four major IP holders: a joint venture between Micron and Intel; another alliance between Toshiba and Western Digital; Samsung; and SK Hynix.

Analysts say that despite these structural shifts, the memory segment's recent profitability may not continue indefinitely [1]. While DRAM players should continue to benefit from segment consolidation, they may have to contend with an oversupply, as well as competition from next-generation memory technologies. The NAND segment is seeing some shifts, including increasing moves to vertical and horizontal integration, that potentially offer opportunities to improve overall profitability. However, NAND faces new competition from entrants at the system level. The transition to 3-D NAND and next-generation nonvolatile memory (NVM) technologies may pose significant challenges for the NAND segment.

Besides these economic challenges, there are also some architecture and device challenges that are driving the changes in the semiconductor memory world. These challenges are making the research activity even more interesting and demanding. Semiconductor waves have been traditionally driven by disruptive applications. It is

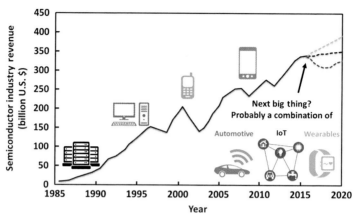

Fig. 3.2 Semiconductor industry revenues waves.

thought that the next wave will be likely driven by a mix of Internet of Things (IoT), wearable, and automotive applications, see Fig. 3.2. In all these fields, a common characteristic could be the "small dimensions": new architectures can possibly enter this market by providing optimized HW solutions, suitable for low power, low data transfer, low capacity, and low cost. This is also the reason why next-generation nonvolatile memories, in particular resistive random access memory (RRAM), are investigated in these new emerging scenarios for the many new opportunities they offer with respect to the large density of 3D NAND, which might not be the key parameter anymore.

New possibilities can be envisaged in *nonvolatile processing, memory-driven computing, neuromorphic computing, neural networks, matrix-array calculations, non-Boolean calculations*, and others to come. In all these scenarios, the unique features of RRAM devices are exploited to achieve a fast response, a dense array, a cheap (cheaper than 3D-NAND, at least) solution, and an intrinsically random physics, leading to an RRAM-centric vision [2].

In this framework, the multiscale modeling approach adopted by Larcher and coworkers and implemented in Ginestra [3–9] and presented here, offers a unique opportunity to understand the underlying physics and to exploit the fundamental material properties to study the feasibility of the solution, or to optimize the proposed circuit. In the following, we will sketch some of these scenarios.

3.1.1   Nonvolatile processing/non-Boolean calculations

Many applications in the field of IoT not only require low power consumption coupled with nonvolatile data storage, but also the capability to maintain data and system states in the absence of input power is essential for an energy harvesting powered computing device. Conventional processors [10] adopt an off-chip memory (such as Flash) for data backup. CMOS and the emerging NVM can now be integrated onto one die, thanks to the advances in IC design and processing. These emerging

NVM technologies, including ferroelectric random access memory (FRAM) [11], magnetic random access memory (MRAM) [12], phase-change memory (PCM) [13], and RRAM [14], offer a full range of benefits such as high density, low read/write energy, long endurance, and 3-D integration compatibility. Nevertheless, this is not a fundamental solution: bits held in flip-flops still have to be transferred into and read out from the centralized NVM in a sequential manner, resulting in significant energy and timing overheads.

An NVM element could be attached to the standard flip-flop—the component that holds data in a processor—to form a nonvolatile flip-flop (NVFF) and to realize in-place data backup and restore. The main idea behind nonvolatile processing (NVP) is to replace the time-consuming and energy-inefficient byte-by-byte global data migration with a localized full-parallel bit-to-bit transfer.

Aside from the emerging NVM technologies, the advent of emerging beyond-CMOS logic transistors, such as tunneling field-effect transistors (TFETs) and negative capacitance field-effect transistors (NCFETs), has also brought great opportunities toward a new paradigm of future low-power nonvolatile computing. These emerging devices could either show enhanced Boolean logic operation with higher energy efficiency or actually be harnessed to redesign existing computing methods by introducing features beyond CMOS Boolean logic.

3.1.2   Memory-driven computing

Today's computers perform to do jobs that scientists could not have envisioned 20 years ago, and we might need to change the actual architecture and move to a new paradigm that shifts processing from slow silicon to hyperfast memory. This approach is called Memory-Driven Computing (HPE e.g., reports a real prototype [15]).

Memory-Driven Computing is based on a new architecture giving every processor in a system access to a large amount of memory—a big change from today's systems where relatively small amounts of memory are tethered to each processor.

The perfect memory for Memory-Driven Computing combines the best features of today's memory and storage technologies, what is called today "storage class memory," and RRAM are the candidates for this position.

We also need a new element to make this new approach feasible and a communication vehicle that facilitates data transfer between the different elements of the computer system. Today, each computer component is connected using a different type of interconnect. Memory-driven computing proposes that every component is connected using the same high-performance interconnect protocol. Using photonics, in particular using microscopic lasers, we can funnel hundreds of times more data down an optic fiber ribbon using much less energy. The fibers are also tiny, making physical installation easier.

This advanced hardware needs also a specific and compatible software. Moreover, with a new computer design come new opportunities to embed security throughout software and hardware.

3.1.3 Neuromorphic computing/neural networks/array-matrix computing

A lot of interest has risen in very-large-scale integration hardware to mimic neurobiological architectures and, therefore, to obtain advanced computing systems with high efficiency [16]. This approach is called neuromorphic computing. The advantages of RRAM in high density, fast operation, low power consumption, and analog to biological synapse make it extremely attractive in neuromorphic system implementation.

It is worthwhile to mention that RRAM cross-point can be naturally used to implement the matrix–vector computation (or sum-of-product) [17], especially in neural networks in which sum-of-product computations are a fundamental mathematical component that can be implemented efficiently [18]. For example, basic computations in neural network implementation based on RRAM cross-point array have been proposed [17–19]. The synaptic weights in the neural network are represented by the conductance of RRAM cells. In this way, the cross-point array conducts both network configuration (storage) and computation. The output of the array will be sensed out by periphery circuitry and transferred to the recognition result.

A neuromorphic core with 64 k cells with phase change memory as a synaptic device was reported by IBM in 2015 [20]. This is the new neuromorphic chip released by IBM after the True-North system reported in 2011; very recently, scientists from HP Labs reported a dot-product engine based on the 1T1R cross-point array [21].

3.1.4 RRAM-centric design

Recently a lot of interest has risen on 3D vertical RRAM as a specific subset of nonvolatile processing and memory-driven computing for their intrinsic capability of being integrated into a real CMOS process. Authors in [2] describe the design principles and experimental demonstrations of in-memory logic operations with 1T-4R 3D vertical RRAM. Their design methodology is based on (i) the cost-effectiveness, high-density data storage, and manipulation capabilities of 3D RRAM device/circuit structure over conventional 2D RRAM structures; (ii) three different computation needs for in-memory operation schemes (half-VDD scheme, VDD/GND scheme, and 3D-LUT scheme); (iii) nonvolatile and cascadable operations, free from destructive read; and (iv) LUTs can be reconfigured in 3D memories during computation to help mitigate endurance limitations.

It is 3D RRAM technology that offers a high-capacity, high-bandwidth on-chip storage solution, as well as the capability toward fine-grained monolithic 3D integration with logic [22–24]. Crossbar Inc. has developed stand-alone 3D RRAM products targeting data storage solutions by integrating thousands of RRAM cells in a cross-point architecture [25], but the same architecture is proposed also for in-memory computing. As the milestone research directed by Intel Corp. and Micron Technology, the 3D XPoint technology with initial 128-Gb storage per die using the resistive switching technology was released in 2015 [26].

3.1.5 Needs for advanced modeling and characterization

This chapter is organized to stress the importance of modeling, characterization, and their interplay to accelerate RRAM technology development and to promote its full industrial exploitation. The next section will introduce the multiscale modeling platform that is used to connect material properties to electrical device performance, including variability and reliability. The potential of the platform in material screening, allowing predicting device performance from atomic structures of materials and defects, will be described in detail. In addition, the platform is used to confirm the understanding of the material properties through comparison with electrical curves. Nevertheless, executing either of the tasks requires a number of crucial material-related parameters as input. Some of these parameters can be extracted exploiting electrical "characterization for modeling" techniques described in the third section of this chapter. Finally, the chapter concludes by emphasizing how TCAD-assisted material-device co-design and optimization are essential to assess the performance of RRAM devices for application-specific targets.

3.2 A multiscale material-to-device modeling platform

Modeling the electrical response of electron devices by accounting for microscopic material properties is fundamental for the development of novel technologies for logic, memory, and neuromorphic computing. This is crucial especially for RRAM devices, whose operations are based on the interplay between the charge and ionic transports and on the material modifications induced by the applied voltage. In particular, connecting the material properties to the electrical device performances through physical modeling is crucial to accelerate the technology development in the RRAM domain, as it allows:

(i)   predicting the device performances from material and defect properties, including reliability;
(ii) understanding the ultimate device scaling limits in spite of the variability;
(iii) evaluating the process effects on devices and materials through the interpretation of electrical characterization data;
(iv) material/device co-design that enables material screening and device geometry optimization from electrical device specifications.

The implications of the material properties on the electrical behavior of RRAM devices are especially important for devices integrated into the Back-End of Line (BEOL), implementing cross-point memories, hybrid memory-in-logic solutions, and artificial synapses for computing architectures going beyond von Neumann [2, 10]. The RRAM performances, that is, fast switching times, low voltage/power operations, endurance and retention [3–9] are strongly affected by defect properties and material morphology (e.g., grains). Indeed, defects such as vacancies, interstitial ions, and dopants affect strongly the current flowing across transition metal oxides (TMOs), which is typically due to trap-assisted tunneling (TAT), that is the dominant conduction mechanism in wide-bandgap materials such as TMOs [7, 8, 27]. In addition, the generation, diffusion, and recombination of atomic species such as interstitials and vacancies severely affect the kinetics of forming and switching as well as the degradation dynamics of RRAM devices.

Handling properly this complex set of physical mechanisms requires a multiscale modeling approach to account for atomic material properties in the description of both electron transport and the bias-induced material modifications. Interestingly, this material-connected modeling will allow explaining some dilemmas associated with RRAM device operation and reliability that are: (i) the location of oxygen ions and vacancies in different states (i.e., the oxygen reservoir); (ii) the morphology and microscopic structure of the conductive filament (CF); and (iii) the excellent endurance shown by HfO_2- and TaO_x-based RRAM despite the relatively low number of ions or vacancies involved in the partial disruption and reconstruction of the CF.

Since the number of ions and vacancies involved in RRAM operations is typically low, kinetic Monte-Carlo (kMC) techniques are adopted in order to properly handle the device statistics and to describe the diffusion and generation processes of atomic species.

3.2.1   Physical mechanisms

The multiscale modeling platform has to describe comprehensively the operations of RRAM device, that is, from forming to switching including endurance and retention. The current has to be simulated in any device conditions, that is, from the pristine state up to set/reset, in both static and transient conditions. In parallel, the structural device and material modification, that is, the creation and modification of the CF taking place during forming, set, and reset operations must be considered. This requires to consistently account for a variety of charge transport mechanisms (i.e., TAT, direct and Fowler-Nordheim tunneling, drift, and diffusion), the associated power dissipation and local temperature increase, the atomic-level material modifications due to the distortion and breakage of atomic bonds, chemical reactions (e.g., redox), and diffusion of atomic species (e.g., vacancies, interstitials). For this reason, this modeling platform is comprised of two main portions, addressing the charge transport and the generation-diffusion of atomic species (e.g., vacancies, interstitials), as shown in Fig. 3.3. The material-dependent parameters needed for a physical description of charge and ion transport are derived from the interpretation of electrical characterization measurements and of ab initio calculations [28, 29].

The electric field and potential profiles across the device volume are calculated by solving Poisson's equation while accounting for the vacancy and interstitial charge, including charge trapping. This is crucial for a correct description of the kinetics of the forming and switching processes, which needs to account for the mutual Coulomb repulsion (attraction) between species of the same (opposite) charge polarity.

3.2.1.1   Charge transport mechanisms

Many charge transport mechanisms are taken into account, comprising TAT, Thermoionic Emission (TE), Poole-Frenkel (PF), and Drift/Diffusion (DD) through conduction and valence bands, as well as through bands originated by metal-rich regions formed within the insulating layers, that is, defect clusters [3–9].

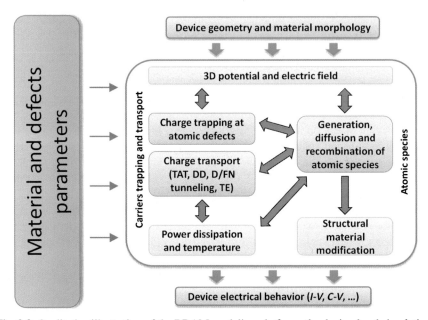

Fig. 3.3 Qualitative illustration of the RRAM modeling platform: the device-level simulation branch (enclosed in the *blue box*) includes modules addressing the charge (electron/hole) transport and generation and diffusion of atomic species. Besides defining the device geometry and morphology *(top box)*, the device simulator requires some material and defects parameters (thermal ionization and relaxation energies, diffusion barrier, and formation energies) calculated using ab initio methods (*green box* on the left).

In most of the RRAM devices reported in the literature, TAT is the dominant charge transport mechanism; in most of the binary and ternary oxides, charge transport is indeed assisted by defects, that is, in many cases (e.g., Transitional Metal Oxides, TMOs) oxygen vacancies, that dictate the electrical behavior of the RRAM device [3–9]. The TAT conduction is described by means of a nonradiative multiphonon TAT model that inherently considers the electron-phonon coupling [5–8] by accounting for the atomic lattice rearrangement, that is, displacement of the atoms surrounding the defect required to accommodate the trapped charge. This structural relaxation process heavily depends on the atomic structure of the defect, which is described by the relaxation energy, E_{REL}, [7, 8, 27]. The latter and the defect thermal ionization energy, E_T, which are the two key parameters affecting the TAT charge transport, are calculated by means of ab initio methods, that is, Density Functional Theory and Molecular Dynamics simulations, as illustrated in Fig. 3.3. E_{REL} and E_T values are typically calculated by considering an isolated defect despite their dependence on the defect density. Indeed, E_{REL} reduces at high defect densities in agreement with the Marcus theory of electron transfer [4, 30] because of the increasing electron delocalization among adjacent defects, which eventually lowers the effective E_{REL}. Moreover, at very high defect densities (e.g., in LRS), extended defects and/or defect clusters are formed (eventually culminating in the formation of a CF) that leads to the creation of a conductive defect subband where the charge transport occurs mainly through drift and diffusion. Indeed, in this regime, the TAT through individual defects is no longer a valid description of charge transport, and the Landauer approach needs to be adopted to correctly account for the delocalized electron flow that underlies the ohmic-like behavior observed in LRS.

The defect-assisted charge transport naturally results in localized power dissipation at the defect sites, which is self-consistently computed across the entire device volume. This is accomplished by including the charge carriers' energy released at both the defects (at every charge trapping event) and the lattice (due to inelastic scattering mechanisms, i.e., optical and acoustic phonons). The power dissipation leads to a localized temperature increase calculated by solving the Fourier's heat flow equation.

3.2.1.2   Generation and diffusion of atomic species

The generation and DD of defects (i.e., interstitial ions and vacancies) play a critical role, dictating RRAM devices response in all regions of operations, which are based on the intermixing and interactions between the electronic and ionic/atomic transport. Therefore, the generation and the diffusion of "mobile" atomic species must be consistently calculated to properly take into account the structural material modifications occurring during RRAM operations, that is, forming, set, and reset.

The generation of mobile atomic species mainly occurs at specific favorable locations within the device. Such locations are identified by the regions at which peaks of both the electric field and the temperature are found, as they heavily facilitate defects formation. Moreover, atomic species generation can also be enhanced at the interfaces due to the lower defect formation energies (compared to the bulk). Two main mechanisms are responsible for the generation of active atomic species: (i) the breakage of atomic bonds, enhanced by the local electric field and temperature profiles and locally favored by the possible presence of precursors [31] and other defects; and (ii) the oxidation/reduction chemical reactions that occur at the interfaces, as well favored by the local electric field and temperature.

The defects generation rate for all of the above-mentioned mechanisms is described by compact effective-energy formulas efficiently accounting for the microscopic mechanisms and material properties. A quite popular bond breakage model used especially for reliability projection is the thermochemical model proposed by McPherson in [32]. According to this model, the generation rate depends on a small set of microscopic parameters, such as the vibration frequency of the atomic bond, the molecule bond polarization factor, and the activation energy required to break the bond. These parameters are functions of the material properties and are calculated using ab initio methods [28, 29], as shown in Fig. 3.3. Extensions to this model have been proposed in order to effectively account for the active role played by precursors [31], atomic defects, and local charges that may favor the generation process. This leads to a very complex scenario, as these effects may be present simultaneously. All the above material parameters are derived from ab initio calculations, as illustrated in Fig. 3.3.

Once the atomic species are generated in the dielectric, the vacancies and the interstitial ions can move within the stack (also transitioning from layer to layer) driven by the internal electric field (in turn affected by the charge state of local atomic species and defects), and by the local temperature (itself affected by localized power dissipation at defect sites). The diffusion of such defect species is strongly accelerated by the local temperature (that has to be self-consistently calculated), and plays a critical role during the set/reset operations, as they result from the field-and temperature-driven motion and generation/recombination of defect species. Moreover, diffusion processes are crucial to describe reliability issues (i.e., retention and endurance). As a result of the filamentary nature of the switching mechanism, the number of defects involved in RRAM device operation is limited to some hundreds. Hence, kinetic Monte-Carlo approaches have been successfully employed to investigate the reasons for the relevant intrinsic variability and for the stochastic nature of the physical processes [3–9].

Due to the strong coupling among different processes occurring at defect sites (e.g., charge trapping dynamics influences, and is in turn influenced by the local potential and temperature profiles), every time an individual defect is generated, recombined, or moved from its position, the internal device conditions (e.g., trapped charge distribution, electric field, current, and temperature) are updated. This allows considering the entire dynamics of the physical processes responsible for RRAM operations and gaining insights into the role of the local temperature.

3.2.2 Modeling and simulation results

Combining all the above-mentioned physical mechanisms in a multiscale modeling platform allows directly connecting the electrical performance of RRAM devices to the fundamental material properties. Particularly, the modeling platform can be used to investigate the nature of the atomic species involved in the various mechanisms governing the behavior of RRAM devices, and can be exploited to simulate the whole switching cycle (including forming, set, and reset operations). Therefore, it can be used to predict the performance of RRAM device and proves crucial to boost the development of the RRAM technology focusing on the specific application (e.g., memory [14, 22, 23, 25], logic [2, 10, 15], neuromorphic computing [18–21], and security and authentication [33, 34]). Moreover, it helps in identifying the culprits for the reliability issues that affect RRAM devices [8, 14, 31, 32, 35–40].

3.2.2.1 Defect spectroscopy

Ultimately, to assess the device performance and reliability, it is compulsory to identify the nature and the properties (e.g., distribution in energy and space) of the atomic species involved in the physical processes governing RRAM operations. This requires a comprehensive characterization of the device in pristine conditions, which must include the analysis of the TMO properties (bandgap, dielectric constant, etc.) as well as a spectroscopy of the preexisting defects. The modeling platform presented here allows identifying the nature and properties of the defects controlling TAT charge transport and degradation in TMOs by connecting these properties with the results of electrical characterization. This capability is demonstrated by applying our multiscale modeling approach to the case of HfO_2-based RRAM devices.

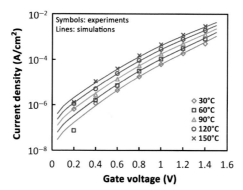

Fig. 3.4 Current *(symbols)* measured at different temperatures on a MIM structure with a 5-nm-thick HfO$_2$ layer and TiN top and bottom electrodes, along with corresponding simulations *(lines)*.

Fig. 3.4 shows the comparison between the current measured at different temperatures on a TiN/Ti/HfO$_2$(5 nm)/TiN device and the simulations results. Correctly reproducing both the voltage and the temperature dependencies of the measured IV curves allow identifying the spatial and energy distribution of the defects in the dielectric that are contributing to charge transport, Fig. 3.5. Simulations confirm that the defects involved in charge transport are more or less evenly distributed in space, with the defects located at approximately the middle of the dielectric layer being responsible for the vast majority of the TAT current. In addition, they show an approximately uniform energy distribution centered at $E_T = 2.4\,eV$ from the bottom of HfO$_2$ conduction band and exhibiting a spread of nearly 1 eV. Notably, by reproducing the current temperature dependence it is possible to extract the defects' relaxation energy ($E_{REL} = 1.02\,eV$), which is an important indicator of the atomic nature of the defect [3–9, 27]. The set

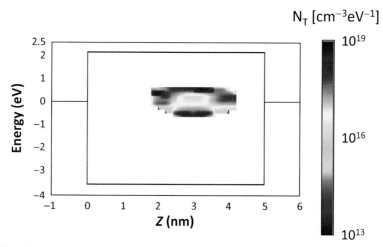

Fig. 3.5 Defect map extracted from the simulation of the IV characteristics in Fig. 3.4.

of energy parameters extracted by matching simulations to experimental data is in excellent agreement with the values found in the literature for positively charged oxygen vacancy defects in HfO_2 ($E_{REL} = 1\,eV$ and $E_T = 2.32\,eV$) [28, 29].

3.2.2.2   Forming simulations

The modeling platform described above is a powerful tool to investigate the kinetics of the physical processes governing RRAM operations. It can be used to simulate all the operations in static and transient conditions, that is, from forming to switching, and to investigate the dependence on material properties and external conditions (temperatures, voltage pulses) on the device electrical characteristics.

Fig. 3.6 shows the evolution of the currents simulated on a TiN/5 nm-HfO_x/TiO_x/ Ti RRAM device during ramped voltage forming (with a 10 μA current compliance), reset, and set operations. As can be seen, the simulated current exhibits all the distinctive features typically observed in experiments [4–6, 9, 14, 36, 39–41]: the abrupt current jump at the forming voltage, the gradual current reduction in reset, and the occurrence of set at a much lower voltage as compared to forming. To gain more insights into the mechanisms controlling RRAM operations, we monitored the evolution of temperature, potential, and generated oxygen vacancies/ions (V/I) during simulations. Figs. 3.7, 3.9, and 3.10 show these quantities at different phases of the simulated forming, reset, and set operations, respectively.

Fig. 3.7 shows the evolution of V/I (top), temperature (center), and potential (bottom) at different stages of the simulated forming operation (labeled as A, B, C, and D in Fig. 3.6). Prior to forming, the current is dominated by the TAT through preexisting defects, that is, mainly oxygen vacancies that accumulate preferentially at grain boundaries in the polycrystalline materials [4, 5, 41]. Since their density is relatively

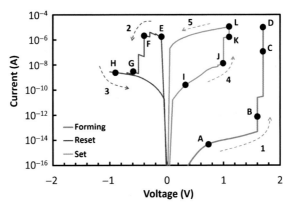

Fig. 3.6 Current-voltage characteristics simulated during (1) forming, (2)–(3) reset, and (4)–(5) set operations with a current compliance of 10 μA. The RRAM device stack is composed of a 6-nm-thick HfO_x layer, with the bottom electrode made of TiN and the top one made of Ti. In order to account for the O extracted from the hafnia, we considered a substoichiometric TiO_x layer formed at the interface between Ti and HfO_2. Labels (A–L) indicate the key points at different stages of forming, set, and reset operations.

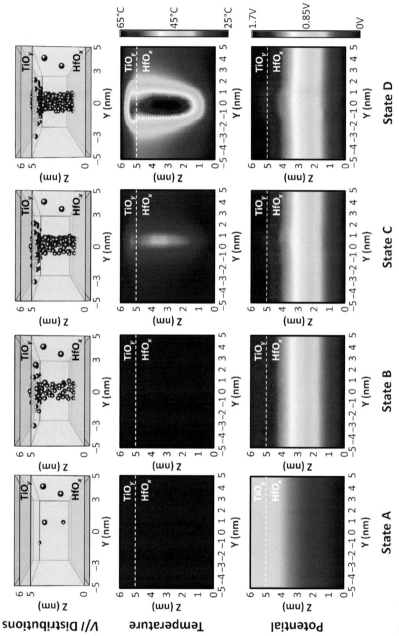

Fig. 3.7 From top to bottom: distribution of oxygen ions (*blue spheres*) and vacancies (*red spheres*), temperature profile and potential profile at different stages of the simulated forming operation (labeled A, B, C, and D in Fig. 3.6).

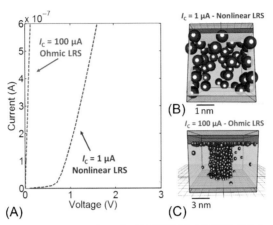

Fig. 3.8 (A) Simulated *I-V* characteristics of a TiN/Ti/HfO$_2$/TiN RRAM device in LRS formed with a current compliance of 1 μA (*red line*—nonlinear behavior) and 100 μA (*magenta line*—quasiohmic behavior). (B) Sketch of the CF morphology for the case in which the current compliance is 1 μA. (C) Sketch of the CF morphology for the case in which the current compliance is 100 μA.

low, the current flow does not lead to an appreciable increase of the power dissipation and of the local temperature (state A in Fig. 3.6). Moreover, the electric field across the oxide is also small (as $V_G < V_{FORM}$), leading to a low and uniform probability to generate new defects. As a result, few defects are generated randomly throughout the RRAM volume without any correlation and the current slowly increases with the applied voltage. When the voltage reaches V_{FORM} (around 1.7 V in Fig. 3.6), the bond stretching induced by the high electric field enhances the Hf–O bond breakage, leading to a significant creation of new V/I species elements (state B in Figs. 3.6 and 3.7). The generated vacancies support the TAT transport, thus increasing the current, the associated power dissipation and temperature [4–6], see state B in Fig. 3.6. This triggers a thermally driven positive feedback that quickly leads to the formation of the CF (states C and D in Figs. 3.6 and 3.7) [4–6]. The positive feedback and the defect generation process is only limited by the current compliance (typically imposed by either a series resistor or a transistor). The charge transport characteristics observed in LRS depend on the CF morphology, in turn controlled by the dynamics of the defect generation process, hence by the current compliance.

As illustrated in Fig. 3.8, it is possible to correctly reproduce the current in LRS on RRAM devices formed at low (e.g., 1 μA) and high (e.g., 100 μA) current compliance levels, corresponding to different CF morphology. Using low current compliance leads to the formation of a "weak" CF formed by a relatively low number of defects, where the charge transport occurs by a combination of TAT and DD, as reflected by the nonlinear LRS *I-V* characteristic. Conversely, a sufficiently high current compliance level leads to a CF formed by a large number of tightly packed defects that exhibits a quasiohmic (i.e., linear) LRS *I-V* curve. The oxygen ions generated during this field- and temperature-driven process diffuse away under the action of the applied electric field,

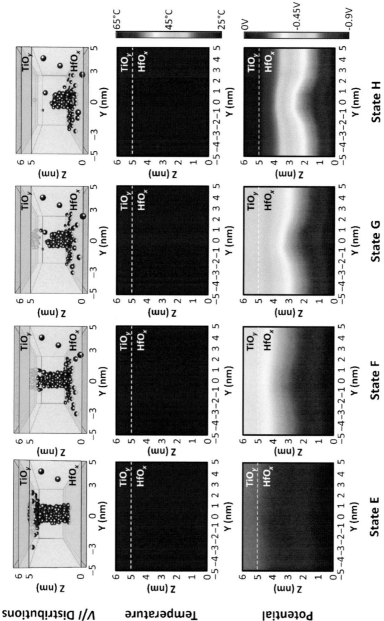

Fig. 3.9 From top to bottom: distribution of oxygen ions (*blue spheres*) and vacancies (*red spheres*), temperature profile and potential profile at different stages (labels E, F, G, and H) of the reset operation simulated in Fig. 3.6.

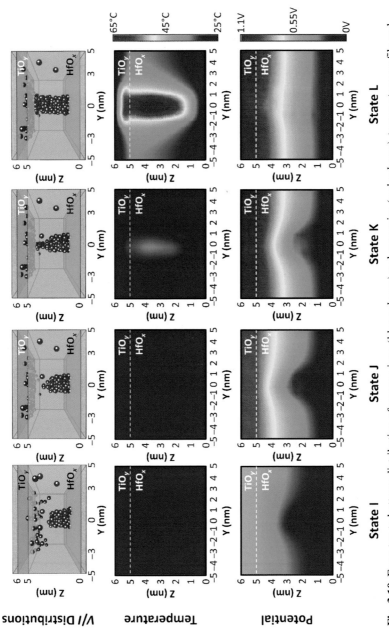

Fig. 3.10 From top to bottom: distribution of oxygen ions (*blue spheres*) and vacancies (*red spheres*), temperature profile and potential profile at different stages (labels I, J, K, and L) of the set operation simulated in Fig. 3.6.

temperature, and ions mutual Coulomb repulsion. The field-driven vertical motion dominates during forming, driving the ions toward the top electrode and into the overlying TiO_y layer that is typically formed during the fabrication process [23]. Oxygen ions accumulate into the TiO_y layer, creating the so-called "oxygen reservoir" that will provide the O ion supply for the subsequent switching operations (state D in Figs. 3.6 and 3.7).

3.2.2.3   Set and reset simulations

Starting from results of the simulations of the forming process (i.e., the CF), the platform is then used to simulate set and reset operations (i.e., cycling). This is a critically important aspect since RRAM switching characteristics strongly depend on the properties of the CF. A distinctive feature of the proposed framework with respect to the other approaches proposed in the literature [42, 43] is that no assumptions are made on the CF characteristics.

Fig. 3.9 shows the evolution of oxygen vacancies/ions (top), temperature (center), and potential (bottom) at different stages of the simulated reset operation (labeled as E, F, G, and H in Fig. 3.6). The O ions gathered during forming in the top TiO_y layer drift toward the bottom HfO_x/TiN interface during the negative voltage ramp (reset). During their motion toward the bottom electrode, O ions can reoxidize a portion of the CF, see states (F)–(H) in Fig. 3.9. This typically leads to the formation of a dielectric barrier (state H in Fig. 3.9), in this case located close to the TiO_y layer. This latter barrier is mainly responsible for the current reduction associated with the reset transition, Fig. 3.6. The remnant part of the CF preserves its metallic nature, that is, the potential will mostly fall across the dielectric barrier, see state H in Fig. 3.9. Finally, it is worth noting that the voltage and current levels reached during reset are not high enough to cause a significant increase in the local temperature, Fig. 3.9, suggesting that the operation is field-driven rather than temperature-driven. Lastly, the application of a positive voltage to the device in the HRS state (i.e., the set operation) results in the sudden current upsurge that switches the device back to the LRS as shown in Fig. 3.6. Fig. 3.10 shows the evolution of oxygen vacancies/ions (top), temperature (center), and potential (bottom) at different stages of the simulated set operation (labeled as I, J, K, and L in Fig. 3.6). The starting condition of the set simulation is represented by the device state obtained at the end of the reset operation.

The dynamics of the set operation is very similar to the one described for the initial forming. On the one hand, the positive voltage ramp applied to drive the device in the LRS promotes the diffusion of oxygen ions toward the TiO_y layer where they accumulate and recreate the "reservoir" needed for the following reset operation, see Fig. 3.10. On the other hand, because of the metallic nature of the remnant part of the CF, the applied voltage drops almost completely across the thin dielectric barrier inducing a high electric field (exceeding the critical breakdown field) also at relatively low voltages, as shown by the 2D potential profiles in Fig. 3.10 (especially states I and J). The high electric field strongly supports the Hf−O bond breakage process in the dielectric barrier region, leading to a significant generation of V/I at smaller voltages (1.1 V in Fig. 3.6) as compared to the forming operation, see states K and L in Fig. 3.10. Similarly to what discussed for forming, the generated vacancies increase the TAT current and the related power dissipation as well as the local temperature (states K and L in Fig. 3.10). This quickly triggers the thermally driven positive feedback that results in the restoration of the full CF, associated with the abrupt current jump observed at the set voltage, Fig. 3.6.

3.2.2.4 Reliability simulations: Cycling and endurance

Besides accounting for the mechanisms responsible for resistive switching and RRAM operations, the modeling platform can also be used to gain insights into RRAM device fundamental reliability issues, the most important of which is endurance (i.e., device failure during cycling). Simulations reveal that the kinetics of the set and reset processes (as well as the location of the dielectric barrier) are very sensitive to the anisotropic properties of diffusion kinetics of the oxidizing species. An efficient reset operation can be obtained only when the motion of the O ions occurs predominantly along the vertical direction (as expected due to the more relaxed material structure in the surroundings of the CF) though some radial ion movement is still required to maximize V/I recombination probability. In addition, ions diffusion must not be too fast to provide enough time for the ions to recombine with the CF vacancies. Under fast oxygen ions diffusion, the reset efficiency is reduced and the oxide barrier is formed preferentially at the bottom interface, Fig. 3.11. Finally, when the radial motion dominates, the reset operation is inefficient, hindering the opening of a dielectric barrier, Fig. 3.11.

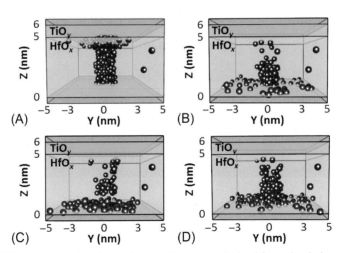

Fig. 3.11 Different oxygen ions/vacancies distributions as obtained from simulations (A) after forming and (B)–(D) after reset operation performed considering different oxygen ions diffusion properties (i.e., the energy barrier for diffusion). Results show a clear dependence of the reset state on oxygen ion diffusion; (B) the oxide barrier is formed in the proximity of the TE when O ions diffusion is relatively slow; (C) the oxide barrier is shown to form at the bottom electrode interface in the presence of a fast diffusion occurring also along the lateral directions; and (D) the inefficient reset operation is observed when the lateral diffusion dominates, indicating that the O ion diffusion is anisotropic. Note that, due to the exponential dependence of the process on the diffusion energy barrier, above differences are obtained even in the presence of relatively small variations (0.1–0.3 eV) of this important material-dependent parameter.

Similarly, during the set operation, the generated O ions must be able to diffuse toward the top TiO_y "reservoir" without experiencing a significant motion in the radial direction. A large lateral diffusion would inhibit the successful migration of O ions in the "reservoir," decreasing the efficiency of the following reset operation. In such conditions, the cycling endurance of the device will be negatively affected, causing a significant reduction of the memory window and, possibly, the device failure after a few switching cycles, Figs. 3.12 and 3.13.

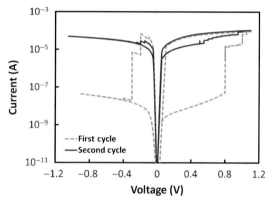

Fig. 3.12 Current-voltage characteristics simulated during two consecutive set-reset cycles under a current compliance of $100\,\mu A$. The RRAM device stack is composed of a 6-nm-thick HfO_x layer, with the bottom electrode made of TiN and the top one made of Ti. In order to account for the O extracted from the hafnia, we considered a substoichiometric 1-nm-thick TiO_x layer formed at the interface between Ti and HfO_2.

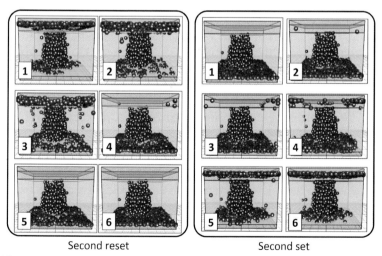

Second reset   Second set

Fig. 3.13 Evolution of oxygen ions *(blue spheres)* and vacancies *(red spheres)* at different phases (from 1 to 6) of the reset (left) and set (right) operations during the second cycle as simulated in Fig. 3.12.

3.3   Material and device characterization for modeling

The characterization of materials is of fundamental relevance for modeling purposes. Most of the RRAM devices in the literature are based on the presence of defects and/ or impurities in a dielectric layer and their ability to (i) greatly affect the electrical properties of the device, and (ii) determine permanent (though reversible) structural changes responsible for the resistive switching phenomenon. It is hence imperative to quantify meaningful defect parameters that rule over (i) their electrical response and (ii) their motion and generation/recombination. Since these properties affect the behavior of the device, it is possible to devise experimental procedures to extract their values or, at least, to identify a range of plausible values for these parameters. In the following, we review some of the advanced "characterization for modeling" techniques that can be used to extract parameter values for simulation and TCAD-assisted engineering.

3.3.1   Characterization for charge transport

Exploiting the capabilities of a simulation platform to drive the engineering of RRAM devices requires as a first step to be able to understand the mechanisms behind charge transport. Its accurate modeling must consider all the most relevant physical properties of the materials, as well as the presence of imperfections and nonidealities (e.g., structural defects and exogenous atomic species). Reproducing the electrical characteristics of the simulated device structure requires simultaneously considering many different charge transport phenomena (i.e., carriers drift and diffusion, direct tunneling, Fowler-Nordheim tunneling, TAT, and thermionic emission). Specifically, the current conduction through the insulating stacks employed in RRAM devices is typically dominated by TAT [7]. The main parameters affecting the current assisted by defects are their relaxation energy and their thermal ionization energy. In the following, we show how it is possible to extract the values of such parameters starting from simple measurements based on (i) the temperature dependence of the TAT leakage current (in pristine devices or in devices in HRS); and (ii) the voltage dependence of the statistical parameters of random telegraph noise (RTN) fluctuations observed in HRS.

3.3.1.1   Temperature dependence of the static I-V characteristics

The leakage current simulations performed through the multiphonon TAT model allows reproducing the experimental current by taking into account the contributions of a relatively large number of defects randomly located at different positions and energy levels inside the dielectric layer. Nevertheless, they require as input the thermal ionization and relaxation energies of the defects. Since these parameters determine the dynamics of the energy exchange between the defects and the lattice, they also set the temperature dependence of the TAT leakage current measured in the device [4, 7, 8]. However, the complexity of the TAT formalism (in terms of formulae to calculate the capture and the emission times) hinders the extraction of these parameters from experimental data. Nevertheless, it is possible to formulate an extraction procedure by introducing appropriate simplifying assumptions: (i) we neglect the trap-to-trap transitions among defects (this holds for relatively thin layers, i.e., <7 nm); (ii) defects are all assumed to have the same energy level. Under these assumptions, the formulae for the calculation of capture and emission times are strongly simplified,

$$\tau_c^{-1} = c_0 N_c e^{\left(-\frac{x_t}{\lambda_c}\right)} e^{\left(-\frac{E_c}{k_B T}\right)} \quad \tau_e^{-1} = c_0 N_c e^{\left(-\frac{t_{ox}-x_t}{\lambda_e}\right)} e^{\left(-\frac{E_e}{k_B T}\right)}$$

$$\lambda_c = \left(2\sqrt{\phi_B 2m^* q / \hbar^2}\right)^{-1} \quad \lambda_e = \left(2\sqrt{E_T 2m^* q / \hbar^2}\right)^{-1}$$

$$E_c = \frac{\left(E_{REL} - \Delta E\right)^2}{4 E_{REL}} \quad \Delta E = E_T - \phi_B + qV\frac{x_t}{t_{ox}} \quad E_e = \frac{E_{REL}}{4}$$

with exponential functions replacing the complex multiphonon transition and tunneling probabilities [44]. In the equations defining the capture and emission times, the first exponential term represents the tunneling probability through the potential barrier between the cathode and the trap, and the trap and the anode. The second exponential accounts for the effect of lattice rearrangement associated with the phonon exchange at the defect site when an electron is captured into (emitted from) a defect. In this simplified formalism, x_t is the trap distance from the cathode; E_T and E_{REL} are the thermal ionization and the relaxation energies of the defect; t_{ox} is the dielectric barrier thickness; m^* is the effective electron mass; q is the elementary charge; $\hbar$ is the reduced Plank's constant; k_B is the Boltzmann constant; N_c is the density of states at the bottom of the conduction band; λ_c and λ_e are the characteristic electron tunneling lengths calculated through the Wentzel-Kramers-Brillouin (WKB) approximation (neglecting the potential drop across the barrier); φ_B is the energy barrier at the interface between the cathode and the dielectric; and V is the applied voltage. E_c and E_e are the associated thermal activation barriers [7], with ΔE being the difference between the electron energy and the defect ground state. In this scenario, while E_E is independent of the applied voltage (it depends only on the relaxation energy of the defect), E_C shows a quadratic dependence on V. Both capture and emission processes contribute to the carrier TAT time, hence both contribute to the leakage current and affect its activation energy.

$$I_{TAT} = \frac{q}{\tau_e + \tau_c} = \frac{q}{P_{T,out}(V)\exp\left(-\frac{E_E}{k_B T}\right) + P_{T,in}(V)\exp\left(-\frac{E_C(V)}{k_B T}\right)}$$

The TAT current is maximum at defects with similar capture and emission times [7], that is, $\tau_c = \tau_e$. Indeed, defects lying too close to the electrodes will have strongly unbalanced capture and emission times. This is due to the fact that in and out tunneling probabilities $P_{T,in}(V)$ and $P_{T,out}(V)$ (which are calculated using the WKB method) depend exponentially on both the applied field and the tunneling length [7]. Hence, one of the two time constant is exponentially larger than the other one, which leads to a significantly higher total TAT time and a lower TAT current contribution. Thus, minimizing the total TAT time (i.e., maximizing the TAT current) requires having defects located approximately in the middle of the dielectric layer and suitably aligned with the Fermi level at the electrodes. So, the TAT current expression can be simplified as follows:

$$I_{TAT} \cong \frac{q}{2P_{T,in}(V)\exp\left(-\dfrac{E_C}{k_B T}\right)} \propto \exp\left(\frac{E_C(V)}{k_B T}\right)$$

The TAT current thermal activation energy equals the capture process one, that is, $E_A = E_c$, which is consistent with the reduction of the activation energy with the applied voltage, as shown in the lower left panel of Fig. 3.14. E_A is extracted experimentally from the Arrhenius plot of I-V curves measured at different temperatures, such as those in the upper right panel of Fig. 3.14. The relations between E_c, E_e, and the other quantities at play can be further simplified allowing the derivation of a simple formula connecting E_A to E_T and E_{REL}.

$$E_T(E_{REL}, V, \alpha) = \Phi_B - qV_{XT} + E_{REL} \pm 2\sqrt{E_A(V)E_{REL}} \quad V_{XT} = V\frac{x_T}{t_{OX}} = V\alpha$$

This formula establishes a simple relation between E_T and E_{REL}, which depends on the activation energy $E_A(V)$, on the applied voltage, and on the exact defect position

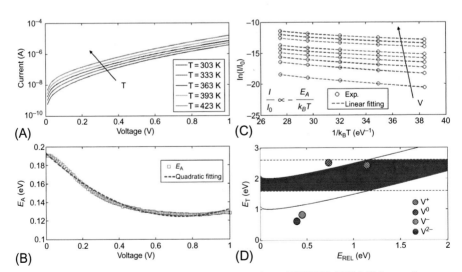

Fig. 3.14 (A) Experimental I-V vs T curves measured on a TiN/HfO$_x$/TiN MIM capacitor with a 5-nm-thick substoichiometric HfO$_x$ layer. (B) Arrhenius plot of data in (A) at different voltages and linear fittings. Notice that, to the extent of the extraction procedure, the value of I_0 (here set to one) is immaterial. (C) Extracted E_A vs V and quadratic fitting. (D) Extracted E_{REL} vs E_T map. The *red region* is the "safe zone" in which the colored striped symbols represents the known E_{REL}, E_T combinations for the V_o^+, V_o^0, V_o^-, and V_o^{2-} defect configurations in HfO$_2$ and HfO$_x$ [28] ($\varphi_B = 2.1\,\text{eV}$ for the TiN/HfO$_x$ interface).

through V_{XT}. Since the exact defect position is not known a priori, we should consider α values from 0 to 1, corresponding to defect positions ranging from the cathode to the anode interfaces. Thus, E_T (E_{REL}, V, α) plotted for different values of V, E_{REL}, and α identifies a region of the (E_T, E_{REL}) plane with defect characteristics that are consistent with the measured E_A. This region, delimited by the solid lines in the lower right panel of Fig. 3.14, can be further refined by considering that the defects mostly contributing to the TAT current are aligned with the Fermi level at the cathode, that is, E_T ranges from $\varphi_B - 0.5\,eV$ to $\varphi_B + 0.5\,eV$. This result, which holds for moderate voltages (0–1 V) independently on the bandgap, can be used to further limit the region of interest (see the dashed lines in Fig. 3.14), resulting in the red area shown in the lower right panel of Fig. 3.14. However, since E_{REL} and E_T are not known a priori, determining the defect properties (E_T, E_{REL}) requires coupling the region derived using this method with the results of ab initio calculations of the atomic defect structures. Fig. 3.14 shows both the region obtained with the proposed methodology and the (E_T, E_{REL}) values calculated with DFT methods for the oxygen vacancy defect (the most common defect in HfO_2) in different charge states. The best match is found for positively charged oxygen vacancy defects, in agreement with previous reports [7, 28, 31, 36, 41, 44–50].

This simple procedure takes advantage of easy measurements and processing to reveal important information about the defects' nature that is typically hard to retrieve. Indeed, E_{REL} and E_T values for defects are typically calculated by means of DFT analysis, with E_{REL} being often neglected in the literature, together with the leakage current temperature dependence. Therefore, if E_T is known, this procedure allows estimating a reasonable range for the value of E_{REL}. In the opposite case, the procedure can be used to estimate an initial guess for (E_{REL}, E_T) to be used in simulations. It must be noted that this procedure intrinsically assumes that the temperature dependence of the leakage current is dominated by the temperature dependence of the capture/emission dynamics. In this respect, careful attention must be paid when characterizing materials that display strong temperature dependence of the bandgap and/or of the electron affinity [51–53]. These parameters strongly affect the leakage current; therefore, experimental data must be corrected to properly account for these dependencies before the analysis.

3.3.1.2   *Random telegraph noise characterization*

RRAM devices are known to suffer from severe current fluctuations that typically appear in the form of RTN [35–37, 45–50, 54, 55]. In their simplest form, such fluctuations appear as random abrupt changes of the current between two discrete values (two-level RTN). Current fluctuations may also appear in more intricate forms, exhibiting many discrete levels (i.e., multilevel RTN). Although its origin is still discussed [45, 56], there is agreement about RTN being associated with the random activation and deactivation of defects assisting charge transport, for example positively charged oxygen vacancies in HfO_2 [45]. Two-level RTN is associated with the activation and deactivation of an individual defect while multilevel RTN is thought to be associated with the concurrent activation and deactivation of many defects [45]. In agreement with recent contributions [45], RTN could result from charge trapping and de-trapping at defects not directly contributing to charge transport (due to their slower capture/emission dynamics) and located in the proximity of those supporting TAT. The trapped charge at an additional "slow" defect alters the potential profile in its surroundings, hence changing the potential at the location of the TAT-supporting defect. This potential alteration could temporarily impede the electron transport through the TAT-supporting defect, which would be restored only after the charge emission from the additional "slow" defect. The effect of charge trapping/de-trapping together with the Coulomb interaction [45, 57–59], results in the observed RTN current fluctuations. As the RTN phenomenon in RRAM devices is identified as a defect-related alteration of the charge transport mechanism, its characterization can be helpful in (i) providing an independent confirmation of the charge transport mechanism; and (ii) extracting useful parameters related to the defects involved in the phenomenon.

The experimental procedure described hereafter has been performed on HfO_2 RRAM devices but can be performed on devices made up of any material system. Initially, the device is driven in HRS, which is characterized by the existence of a dielectric barrier within the CF in which charge transport is dominated by TAT. The thickness of the dielectric barrier can be estimated by reproducing the I-V curve in HRS by simulations. By applying a constant read voltage, RTN current fluctuations over time can be recorded on a single device, and this measurement can be repeated for different applied voltages. For each acquired time series, the statistical properties of the RTN fluctuations (the fluctuation amplitude, ΔI, the average capture, τ_c, and emission, τ_e, times) can be retrieved (Fig. 3.15). The extraction of such parameters from the RTN time series can be performed with many methods carefully described in the literature [60–62]. Among them, the Factorial Hidden Markov Model (FHMM) can be used to decompose the multilevel RTN into a superposition of two-level RTN fluctuations (components), Fig. 3.15. This allows separately retrieving the statistical properties of each component. Briefly, the FHMM is a machine-learning algorithm that considers the (potentially multilevel) RTN signal as a superposition of two-states Markov chains [38, 60, 61, 63] The latter are well suited to represent RTN signals, which are Markov (memoryless) processes with unobserved (hidden) states, that is, the discrete current levels. The details of this technique can be found in [60, 61].

The ΔI can be regarded as the effect of the electrical activation/deactivation of the defect assisting the TAT current, which holds under the assumption of a strong Coulomb interaction between the TAT-supporting defect and the additional "slow" defect. In this framework, the Coulomb effect induced by the charge trapped at the additional "slow" defect on the TAT-supporting one causes the complete suppression of its contribution to the overall TAT current, so that the ΔI can be identified with the current driven by the TAT-supporting defect when the "slow" defect is empty (i.e., no Coulomb interaction). It is hence possible to interpret the experimental ΔI vs V curve as the I-V curve related to an individual TAT-supporting defect, Fig. 3.16. If the relaxation energy of such defects is known, simulations can be performed to reproduce the experimental ΔI vs V curve, which allows extracting the position and energy of the TAT-supporting defect, x_t and E_T respectively. The successful matching between experiments and simulations provides an independent confirmation of the charge transport mechanism and allows identifying

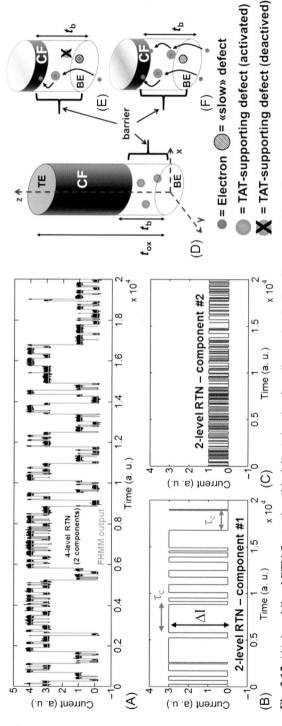

Fig. 3.15 (A) A multilevel RTN fluctuation (*black line*) showing four discrete levels (i.e., the superposition of two components) together with the FHMM fitting (*green line*). (B and C) The two components as retrieved by the FHMM. In (B) the statistical properties of a two-level RTN fluctuation (ΔI, τ_c, and τ_e) are evidenced. (D) Schematic picture of the device in HRS and (E and F) of the RTN mechanism in HRS, showing the defects' (E) de-activation, and (F) activation.

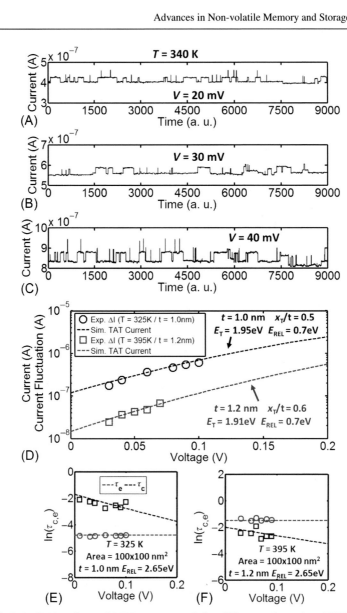

Fig. 3.16 (A-B-C) Experimental RTN traces recorded at different voltages on a TiN/HfO$_x$/TiN MIM capacitor with a 5-nm-thick substoichiometric HfO$_x$ layer. The statistical features (e.g., capture and emission time constants) of a single two-level RTN fluctuation are extracted through the FHMM analysis as a function of the applied voltage. (D) Experimental RTN ΔI *(symbols)* and simulated TAT current contributions *(solid line)* driven by an individual oxygen vacancy defect are shown as a function of the applied voltage on two devices with different barrier thickness. The simulation of the ΔI-V curves allows extracting the position of the individual oxygen vacancy defect within the barrier x_t, and its thermal ionization energy, E_T. t is the barrier thickness, extracted by simulating the I-V curve of the device in HRS. (E, F) Simulated *(dashed lines)* and experimental *(symbols)* RTN capture and emission times extracted at different voltages for the same devices considered in (D). The E_{REL} for the slow defect is 2.65 eV.

the vertical position of the TAT-supporting defect in the dielectric barrier, x_t, as shown in Fig. 3.16. Consistently with the expectations, the TAT-supporting defect is found to be in proximity of the middle of the barrier and its thermal ionization energy value is compatible with the typical values characterizing the positively charged oxygen vacancy. The voltage dependence of τ_c and τ_e extracted from RTN can then be reproduced by simulations, considering the charge trapping (modeled in the TAT framework) at the "slow" defect. For simplicity, the "slow" additional defect is supposed to be at the same vertical position as the TAT-supporting defect. Matching simulations and experiments require adjusting the relaxation energy considered for the "slow" defect, Fig. 3.16. In the case of HfO_2 RRAM devices, the extracted relaxation energy value is $E_{REL} = 2.65\,eV$, which agrees with that predicted for a neutral interstitial oxygen defect [29], being at the same time distant from that of any other TAT-supporting defect (e.g., oxygen vacancy in HfO_2) atomic structure [28]. Interestingly, in HfO_2, oxygen interstitials and oxygen vacancies are simultaneously generated by a bond breakage event [32], and their dynamics dictates the resistive switching mechanism [3–9]. Therefore, RTN characterization is shown to be a powerful technique to retrieve the microscopic parameters of defects involved in the switching mechanism. Moreover, the extraction of such parameters for different material system could help in identifying which material (if any) is less affected by RTN fluctuation. The latter can indeed be a nuisance in neuromorphic circuits, in which RTN fluctuations can cause temporary and unpredictable random variations of the synaptic weight associated with the device [3, 59]. Moreover, recently, it has been suggested [59, 64] to exploit the intrinsic randomness of RTN as an entropy source to realize RTN-based Physical Unclonable Functions and Random Number Generators. From this perspective, it is mandatory to characterize RTN and to extract the microscopic properties of related defects.

3.3.2   Characterization for structural material changes

The wide landscape of materials that are shown to be well suited for RRAM devices leads to many possibilities in terms of tailoring the electrical performance of such devices for a specific target. In this scenario, TCAD-driven device engineering is an attractive option due to its predictive capabilities. Nevertheless, dealing with the complexity of the physical mechanisms behind the CF formation and manipulation requires knowing specific properties of the materials and of the defect species involved in the resistive switching phenomenon, such as the activation energy and field acceleration factor for Frenkel pair (vacancy/ion) generation. Further, we illustrate the details of a powerful measurement technique based on time-dependent dielectric breakdown (TDDB) that can be useful to retrieve the main physical properties of stress-induced defect generation, crucial to describe forming and resistive switching.

3.3.2.1   Time-dependent dielectric breakdown

Experiments involving TDDB can be extremely useful to extract the properties that are necessary to describe defects generation. In this context, although many models exist in the literature to describe the breakdown phenomenon in dielectrics, it is desirable to rely on an approach that connects the atomistic material-related mechanisms to the macroscopic electrical behavior as retrieved from electrical characterization. Particularly, it is necessary to reproduce the observed voltage and temperature dependence of TDDB and its Weibull statistics. Among the models proposed in the literature, the thermochemical model for bond breaking proposed by McPherson [32] is one of the most popular. This model describes the stress-induced defect (i.e., Frenkel pair) generation rate, G, due to bond breaking as

$$G = G_0 e^{-\left(\frac{E_{A,G} - bF}{k_B T}\right)} \qquad b = p_0 \left(\frac{2+k}{3}\right)$$

where k_B is the Boltzmann's constant; T is the absolute temperature; F is the electric field; G_0 the bond vibration frequency [32]; $E_{A,G}$ is the zero-field activation energy for the generation process; b is the bond polarization factor related to the molecular dipole moment, p_0, and to the material relative permittivity, k. $E_{A,G}$ is extracted by assuming that the time-to-breakdown is proportional to the inverse of the generation rate, that is, $t_{BD} \approx G^{-1}$. This allows writing

$$E_{A,EFF} = E_{A,G} - bF = -k_B T \ln\left(\frac{1}{t_{BD} G_0}\right)$$

which allows the extraction of both $E_{A,G}$ and b from time-to-breakdown measurements at different stress electric fields, Fig. 3.17. This procedure requires using a significant number of samples to be investigated due to the stochastic character of this process and to the device-to-device variability of defects location in space and energy. Indeed, for each stress condition (F, T) many measurements on many devices must be performed in order to represent the time-to-breakdown Weibull statistics. Then, the time-to-breakdown statistics must be analyzed to extract the corresponding $t_{63\%}$ that defines t_{BD} for each (F, T) condition. Finally, the effective activation energy derived

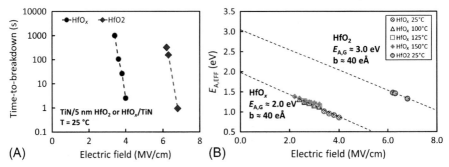

Fig. 3.17 (A) TDDB characteristics for TiN/5 nm HfO$_2$/TiN and TiN/5 nm HfO$_x$/TiN stacks as a function of the applied stress electric field, F. (B) Corresponding $E_{A,EFF}$ vs F plot allows extracting $E_{A,G}$ and b by linear fitting *(dashed lines)*. A reduction of the $E_{A,G}$ value is observed in HfO$_x$ samples as compared to HfO$_2$ samples.

from the experiments can be plotted as a function of F, see the right panel in Fig. 3.17, and both $E_{A,G}$ and b are easily extracted using a linear fitting [8]. Interestingly, high temperatures can be used to speed up the breakdown process at low applied field, as the extracted parameters do not show temperature dependence, Fig. 3.17.

The extracted $E_{A,G}$ parameter can be regarded as a value that effectively accounts for additional phenomena not directly included in the thermochemical model but effectively described by the thermochemical formalism. In particular, the charge injection and trapping at defect precursors and existing defects have been shown to accelerate the defect creation by significantly reducing $E_{A,G}$ in SiO_2 and HfO_2 [31]. Furthermore, also substoichiometry was shown to reduce $E_{A,G}$ as compared to one of the stoichiometric samples, see the right panel of Fig. 3.17. Notwithstanding the significant number of samples that have to be used, the information extracted through this technique is not only useful to describe the forming/set processes and their implications, but also to gain insight into the mechanisms that rule over HRS retention.

3.4 Conclusions

The focus of this chapter is on the relevance of advanced "modeling for characterization" techniques and "characterization for modeling" methods to achieve a fast and reliable material/device co-design for application-specific RRAM devices. The effectiveness of the proposed multiscale modeling platform relies on the interplay between these two procedures, making the material/device co-optimization fast and reliable. Indeed, the simulation platform requires a set of parameters that is not always available in the literature (either measured or calculated by ab initio methods) but it can be extracted by the proposed novel "characterization for modeling" techniques. The accuracy of the physics-based description of the charge transport mechanisms and of the structural changes occurring in RRAMs is confirmed by the excellent agreement that the multiscale model achieves with experimental results obtained on different material systems and device structures. This makes the platform a dependable tool for analysis and design, potentially shortening the time-to-market of RRAM-based solutions for several different applications.

References

[1] www.semiconductors.org.
[2] H. Li, T.F. Wu, S. Mitra, H.-S.P. Wong, Resistive RAM-centric computing: design and modeling methodology, IEEE Trans. Circuits Syst. 64 (9) (2017) 2263–2273.
[3] L. Larcher, A. Padovani, V. Di Lecce, Multiscale modeling of neuromorphic computing: from materials to device operations. in: IEEE International Electron Devices Meeting (IEDM), San Francisco, CA, 2017, 2017, pp. 11.7.1–11.7.4, https://doi.org/10.1109/IEDM.2017.8268374.
[4] A. Padovani, L. Larcher, O. Pirrotta, L. Vandelli, G. Bersuker, Microscopic modeling of HfOx RRAM operations: from forming to switching, IEEE Trans. Electron Devices 62 (6) (2015) 1998–2006.

[5]   L. Vandelli, et al., Comprehensive physical modeling of forming and switching opera-
      tions in HfO2 RRAM devices, in: Proc. IEEE Int. Electron Devices Meeting, December,
      2011, pp. 421–424.

[6]   L. Larcher, A. Padovani, O. Pirrotta, L. Vandelli, G. Bersuker, Microscopic understand-
      ing and modeling of HfO2 RRAM device physics, in: Proc. IEEE Int. Electron Devices
      Meeting, December, 2012, pp. 474–477.

[7]   L. Vandelli, A. Padovani, L. Larcher, R.G. Southwick III, W.B. Knowlton, G. Bersuker, A
      physical model of the temperature dependence of the current through SiO2/HfO2 stacks,
      IEEE Trans. Electron Devices 58 (9) (2011) 2878–2887.

[8]   A. Padovani, L. Larcher, G. Bersuker, P. Pavan, Charge transport and degradation in HfO2
      and HfOx dielectrics, IEEE Electron Device Lett. 34 (5) (2013) 680–682.

[9]   L. Vandelli, A. Padovani, L. Larcher, G. Bersuker, Microscopic modeling of electrical
      stress-induced breakdown in poly-crystalline hafnium oxide dielectrics, IEEE Trans.
      Electron Devices 60 (5) (2013) 1754–1762.

[10] F. Su, K. Ma, X. Li, T. Wu, Y. Liu, V. Narayanan, Nonvolatile processors: why is it trend-
      ing? in: Proc. of Design, Automation and Test in Europe (DATE), Lausanne (CH), 27–31
      March, 2017, pp. 966–971.

[11] J.F. Scott, Ferroelectric Memories, vol. 3, Springer Science & Business Media, 2013.

[12] J.-G. Zhu, Magnetoresistive random access memory: the path to competitiveness and
      scalability, Proc. IEEE 96 (11) (2008) 1786–1798.

[13] H.-S.P. Wong, et al., Phase change memory, Proc. IEEE 98 (12) (2010) 2201–2227.

[14] H.-S.P. Wong, et al., Metal–oxide RRAM, Proc. IEEE 100 (6) (2012) 1951–1970.

[15] https://news.hpe.com/memory-driven-computing-explained/.

[16] H. Li, Y. Chen, C. Liu, J.P. Strachan, N. Davila, Looking ahead for resistive memory
      technology, IEEE Consum. Electron. Mag. (2017) 94–107.

[17] M. Hu, H. Li, Q. Wu, G.S. Rose, Hardware realization of BSB recall function using mem-
      ristor crossbar arrays, in: Proc. Design Automation Conf. (DAC), 2012, pp. 498–503.

[18] C.C. Liu, Q. Yang, C. Zhang, H. Jiang, Q. Wu, H. Li, A memristor based neuromorphic engine
      with a current sensing scheme in artificial neural network application, in: Proc. 2017 22nd Asia
      and South Pacific Design Automation Conf. (ASP-DAC), Chiba, Japan, 16–19 January, 2017.

[19] C. Liu, B. Yan, C. Yang, L. Song, Z. Li, B. Liu, Y. Chen, H. Li, Q. Wu, H. Jiang, A spiking
      neuromorphic design with resistive cross- bar, in: Proc. Design Automation Conf. (DAC),
      2015, pp. 1–6.

[20] S. Kim, M. Ishii, S. Lewis, T. Perri, M. BrightSky, W. Kim, R. Jordan, G.W. Burr,
      N. Sosa, A. Ray, J.P. Han, C. Miller, K. Hosokawa, C. Lam, NVM neuromorphic core
      with 64k-cell (256-by-256) phase change memory synaptic array with on-chip neuron cir-
      cuits for continuous in-situ learning, in: Proc. 2015 IEEE Int. Electron Devices Meeting
      (IEDM), 2015, pp. 17.1.1–17.1.4.

[21] M. Hu, J.P. Strachan, Z. Li, E.M. Grafals, N. Davila, C. Graves, S. Lam, N. Ge, J. Yang,
      R.S. Williams, Dot-product engine for neuromorphic computing: programming 1T1M
      crossbar to accelerate matrix-vector multiplication, in: Proc. Design Automation Conf.
      (DAC), 2016, pp. 1–6.

[22] I.G. Baek, et al., Realization of vertical resistive memory (VRRAM) using cost effective
      3D process, in: IEDM Tech. Dig., December, 2011, pp. 737–740.

[23] H.-Y. Chen, S. Yu, B. Gao, P. Huang, J. Kang, H.-S.P. Wong, HfOx based vertical resistive
      random access memory for cost-effective 3D cross-point architecture without cell selec-
      tor, in: IEDM Tech. Dig. December, 2012, pp. 497–500.

[24] M.M. Shulaker, et al., Monolithic 3D integration of logic and memory: carbon nanotube
      FETs, resistive RAM, and silicon FETs, in: IEDM Tech. Dig., December, 2014, pp. 1–4.

[25] Stand-Alone 3D RRAM Products Targeting Data Storage Solutions, Crossbar Inc. Tech. Rep. [Online]. Available: http://www.crossbar-inc.com/products/3-D-rram/.

[26] J. Hruska, Intel, Micron Reveal Xpoint, a New Memory Architecture that Could Outclass DDR4 and NAND, Extreme-Tech. [Online]. Available: http://www.extremetech.com/extreme/211087-intel-micron-reveal-xpoint-a-new-memory-architecture-that-claims-to-outclass-both-ddr4-and-nand, 2015 July 28.

[27] M. Zhang, Z. Huo, Z. Yu, J. Liu, M. Liu, Unification of three multi-phonon trap-assisted tunneling mechanisms, J. Appl. Phys. 110 (2011) 114108.

[28] D. Muñoz Ramo, J.L. Gavartin, A.L. Shluger, G. Bersuker, Spectroscopic properties of oxygen vacancies in monoclinic HfO2 calculated with periodic and embedded cluster density functional theory, Phys. Rev. B 75 (2007) 205336.

[29] A.S. Foster, A.L. Shluger, R.M. Nieminen, Mechanism of interstitial oxygen diffusion in hafnia, Phys. Rev. Lett. 89 (22) (2012) 225901.

[30] R.A. Marcus, Electron transfer reactions in chemistry. Theory and experiment, Rev. Mod. Phys. 65 (1993) 599–610.

[31] A. Padovani, D.Z. Gao, A.L. Shluger, L. Larcher, A microscopic mechanisms of dielectric breakdown in SiO2 films: an insight from multi-scale modeling, J. Appl. Phys. 121 (2017) 155101.

[32] J. McPherson, J.-Y. Kim, A. Shanware, H. Mogul, Thermochemical description of dielectric breakdown in high dielectric constant materials, Appl. Phys. Lett. 82 (13) (2003) 2121–2123.

[33] R. Liu, H. Wu, Y. Pang, H. Qian, S. Yu, Experimental characterization of physical unclonable function based on 1 kb resistive random access memory arrays. IEEE Electron Device Lett. 36 (12) (2015) 1380–1383, https://doi.org/10.1109/LED.2015.2496257.

[34] A. Chen, Comprehensive assessment of RRAM-based PUF for hardware security applications. in: Proc. of IEEE International Electron Devices Meeting (IEDM), Washington D.C., USA, 2015, pp. 10.7.1–10.7.4, https://doi.org/10.1109/IEDM.2015.7409672.

[35] F.M. Puglisi, P. Pavan, A. Padovani, L. Larcher, G. Bersuker, RTS noise characterization of HfO$_x$ RRAM in high resistive state. Solid State Electron. 84 (2013) 160–166, https://doi.org/10.1016/j.sse.2013.02.023.

[36] F.M. Puglisi, L. Larcher, A. Padovani, P. Pavan, Bipolar resistive RAM Based on HfO$_2$: physics, compact modeling, and variability control. IEEE J. Emerging Sel. Top. Circuits Syst. 6 (2) (2016) 171–184, https://doi.org/10.1109/JETCAS.2016.2547703.

[37] S. Ambrogio, S. Balatti, A. Cubeta, A. Calderoni, N. Ramaswamy, D. Ielmini, Understanding switching variability and random telegraph noise in resistive RAM. in: IEEE Proc. of IEEE Proc. of International Electron Device Meeting, 9-11 December, 2013, pp. 31.5.1–4, https://doi.org/10.1109/IEDM.2013.6724732.

[38] F.M. Puglisi, Measuring and analyzing random telegraph noise in nanoscale devices: the case of resistive random access memories. in: 2017 17th Non-Volatile Memory Technology Symposium (NVMTS), Aachen, 2017, pp. 1–5, https://doi.org/10.1109/NVMTS.2017.8171308.

[39] A. Fantini, et al., Intrinsic switching variability in HfO$_2$ RRAM. in: 2013 5th IEEE International Memory Workshop, Monterey, CA, 2013, pp. 30–33, https://doi.org/10.1109/IMW.2013.6582090.

[40] Y.Y. Chen, et al., Improvement of data retention in HfO$_2$/Hf 1T1R RRAM cell under low operating current. in: 2013 IEEE International Electron Devices Meeting, Washington, DC, 2013, pp. 10.1.1–10.1.4, https://doi.org/10.1109/IEDM.2013.6724598.

[41] G. Bersuker, D.C. Gilmer, D. Veksler, P.D. Kirsch, L. Vandelli, A. Padovani, L. Larcher, K. McKenna, A. Shluger, V. Iglesias, M. Porti, M. Nafria, Metal oxide resistive memory switching mechanism based on conductive filament microscopic properties. J. Appl. Phys. 110 (2011) 124518, https://doi.org/10.1063/1.3671565.

[42] D. Ielmini, Modeling the universal set/reset characteristics of bipolar RRAM by field-
      and temperature-driven filament growth, IEEE Trans. Electron Devices 58 (12) (2011)
      4309–4317.
[43] S. Long, X. Lian, C. Cagli, L. Perniola, E. Miranda, M. Liu, J. Suñé, A model for the set
      statistics of RRAM inspired in the percolation model of oxide breakdown, IEEE Electron
      Device Lett. 34 (8) (2013) 999–1001.
[44]. L. Larcher, A. Padovani, P. Pavan, "Leakage current in HfO2 stacks: from physical to
      compact modeling", Nanotechnology 2012: Electronics, Devices, Fabrication, MEMS,
      Fluidics and Computational, (vol. 2) pp. 809–814. (ISBN: 978-1-4665-6275-2).
[45] F.M. Puglisi, L. Larcher, A. Padovani, P. Pavan, A. Complete Statistical, Investigation of
      RTN in HfO2-based RRAM in high resistive state. IEEE Trans. Electron Devices 62 (8)
      (2015) 2606–2613, https://doi.org/10.1109/TED.2015.2439812.
[46] F.M. Puglisi, P. Pavan, A. Padovani, L. Larcher, A study on HfO2 RRAM in HRS based
      on I–V and RTN analysis, Solid State Electron. 102 (2014) 69–75. ISSN 0038-1101
      https://doi.org/10.1016/j.sse.2014.06.001.
[47] F.M. Puglisi, P. Pavan, Guidelines for a reliable analysis of random telegraph noise in
      electronic devices. IEEE Trans. Instrum. Meas. 65 (6) (2016) 1435–1442, https://doi.
      org/10.1109/TIM.2016.2518880.
[48] R. Thamankar, N. Raghavan, J. Molina, F.M. Puglisi, S.J. O'Shea, K. Shubhakar,
      L. Larcher, P. Pavan, A. Padovani, K.L. Pey, Localized characterization of charge trans-
      port and random telegraph noise at the nanoscale in HfO2 films combining scanning
      tunneling microscopy and multi-scale simulations. J. Appl. Phys. 119 (2016) 084304,
      https://doi.org/10.1063/1.4941697.
[49] R. Thamankar, F.M. Puglisi, A. Ranjan, N. Raghavan, K. Shubhakar, J. Molina, L. Larcher,
      A. Padovani, P. Pavan, S.J. O'Shea, K.L. Pey, Localized characterization of charge trans-
      port and random telegraph noise at the nanoscale in HfO2 films combining scanning
      tunneling microscopy and multi-scale simulations. J. Appl. Phys. 122 (2) (2017) 024301,
      https://doi.org/10.1063/1.4991002.
[50] N. Raghavan, R. Degraeve, A. Fantini, L. Goux, S. Strangio, B. Govoreanu, D.J. Wouters,
      G. Groeseneken, M. Jurczak, Microscopic origin of random telegraph noise fluctuations
      in aggressively scaled RRAM and its impact on read disturb variability. in: Proc. of
      IEEE International Reliability Physics Symposium (IRPS), Anaheim, CA, April, 2013,
      pp. 5E.3.1–5E.3.7, https://doi.org/10.1109/IRPS.2013.6532042.
[51] P. Kumbhakar, A.K. Kole, C.S. Tiwary, S. Biswas, S. Vinod, J. Taha-Tijerina, U. Chatterjee,
      P.M. Ajayan, Nonlinear optical properties and temperature-dependent UV–vis absorption
      and photoluminescence emission in 2D hexagonal boron nitride Nanosheets. Adv. Opt.
      Mater. 3 (2015) 828–835, https://doi.org/10.1002/adom.201400445.
[52] F.M. Puglisi, et al., 2D h-BN based RRAM devices. in: 2016 IEEE International Electron
      Devices Meeting (IEDM), San Francisco, CA, 2016, pp. 34.8.1–34.8.4, https://doi.
      org/10.1109/IEDM.2016.7838544.
[53] C.B. Pan, Y.F. Ji, N. Xiao, F. Hui, K. Tang, Y. Guo, X. Xie, F.M. Puglisi, L. Larcher,
      E. Miranda, L.L. Jiang, Y.Y. Shi, I. Valov, P.C. McIntyre, R. Waser, M. Lanza, Coexistence
      of grain-boundaries-assisted Bipolar and threshold Resistive switching in multilayer hex-
      agonal boron nitride, Adv. Funct. Mater. 27 (2017) 1604811.
[54] F.M. Puglisi, L. Larcher, G. Bersuker, A. Padovani, P. Pavan, An empirical model for
      RRAM resistance in low-and high-resistance states. IEEE Electron Device Lett. 34 (3)
      (2013) 387, https://doi.org/10.1109/LED.2013.2238883.
[55] D. Veksler, G. Bersuker, L. Vandelli, A. Padovani, L. Larcher, A. Muraviev, B. Chakrabarti,
      E. Vogel, D.C. Gilmer, P.D. Kirsch, Random telegraph noise (RTN) in scaled RRAM de-

vices. in: Proceedings of the IEEE International Reliability Physics Symposium (IRPS), MY, April, 2013, pp. 10.1–4, https://doi.org/10.1109/IRPS.2013.6532101.

[56] T. Grasser, Stochastic charge trapping in oxides: from random telegraph noise to bias temperature instabilities. Microelectron. Reliab. 52 (2009) 39–70, https://doi.org/10.1016/j.microrel.2011.09.002.

[57] F.M. Puglisi, P. Pavan, L. Vandelli, A. Padovani, M. Bertocchi, L. Larcher, A microscopic physical description of RTN current fluctuations in HfOx. in: Proc. of IEEE International Reliability Physics Symposium (IRPS), Monterey, CA, 19-23 April, 2015, pp. 5B.5.1–5B.5.6, https://doi.org/10.1109/IRPS.2015.7112746.

[58] F.M. Puglisi, P. Pavan, L. Larcher, Random telegraph noise in HfOx resistive random access memory: from physics to compact modeling. in: 2016 IEEE International Reliability Physics Symposium (IRPS), Pasadena, CA, 2016, pp. MY-8-1–MY-8-5, https://doi.org/10.1109/IRPS.2016.7574624.

[59] F.M. Puglisi, N. Zagni, L. Larcher, P. Pavan, A new verilog-a compact model of random telegraph noise in oxide-based RRAM for advanced circuit design. in: 2017 47th European Solid-State Device Research Conference (ESSDERC), Leuven, 2017, pp. 204–207, https://doi.org/10.1109/ESSDERC.2017.8066627.

[60] F.M. Puglisi, P. Pavan, RTN analysis with FHMM as a tool for multi-trap characterization in HfOx RRAM. in: Proc. of IEEE International Conference on Electron Devices and Solid-State Circuits (EDSSC), 3–5 June 2013, 2013, pp. 1–2, https://doi.org/10.1109/EDSSC.2013.6628059.

[61] F.M. Puglisi, P. Pavan, Factorial hidden Markov model analysis of random telegraph noise in Resistive random access memories, ECTI Trans. Electr. Eng. Electron. Commun. 12 (1) (2014) 24–29.

[62] H. Awano, H. Tsutsui, H. Ochi, T. Sato, Multi-trap RTN parameter extraction based on Bayesian inference. in: International Symposium on Quality Electronic Design (ISQED), Santa Clara, CA, 2013, pp. 597–602, https://doi.org/10.1109/ISQED.2013.6523672.

[63] F.M. Puglisi, A. Padovani, L. Larcher, P. Pavan, Random telegraph noise: measurement, data analysis, and interpretation. in: 2017 IEEE 24th International Symposium on the Physical and Failure Analysis of Integrated Circuits (IPFA), Chengdu, 2017, pp. 1–9, https://doi.org/10.1109/IPFA.2017.8060057.

[64] T. Figliolia, P. Julian, G. Tognetti, A.G. Andreou, A true Random number generator using RTN noise and a sigma delta converter. in: 2016 IEEE International Symposium on Circuits and Systems (ISCAS), Montreal, QC, 2016, pp. 17–20, https://doi.org/10.1109/ISCAS.2016.7527159.

Mechanism of memristive switching in OxRAM

4

*Stephan Menzel**, Rainer Waser**, †*
*Peter-Grünberg-Institut (PGI-7), Forschungszentrum Jülich GmbH and JARA-FIT, Jülich, Germany, †Institut für Werkstoffe der Elektrotechnik (IWE 2), RWTH Aachen, Aachen, Germany

4.1 Introduction

In memristive switching devices, its resistance state can be switched between at least two different levels by applying appropriate voltage stimuli. In a binary device, the resistance changes between one low resistive state (LRS) and one high resistive state (HRS). The transition from the HRS to the LRS is called SET operation, and the reverse transition is called RESET. In principle, intermediate resistance states between the LRS and the HRS can be programmed, too. This enables storing more than one bit in a single device.

As the resistance change is nonvolatile, memristive switching devices can be potentially applied in future random access memories (RAM). Oxide-based RAMs (OxRAM), quite often also called resistive RAMs (RRAM), comprise a class of memristive switching devices that is based on the valence change mechanism (VCM). A VCM device comprises an oxide thin film sandwiched between two electrodes. Typically, one of the electrodes consists of a high work function inert metal and the other one is made of a low work function metal. The VCM mechanism relies on the motion of ionic defects within the oxide material and a concurrent valence change in the cation sublattice [1]. Oxygen defects are typically named as the relevant ionic defects, thus the name OxRAM, but in principle cation interstitials could induce the same effect [2]. As charged ionic defects are responsible for the resistance switching effect, the device operation is inherently bipolar, that is, SET and RESET operation occur at opposing voltage polarities. Before a VCM cell can be switched repetitively an electroforming step is typically required, during which the ionic defects are introduced into the oxide layer.

VCM cells can be divided into two groups by considering the spatial extent of the resistance switching. If the switching occurs over the complete area of the device, it is called homogenous or interfacial switching [3]. The more common switching phenomenon is restricted to a very confined region, and is called filamentary switching.

Besides the spatial extent, one can distinguish two different bipolar switching modes of VCM cells: Eightwise (8w) switching and counter-eightwise (c8w) switching [4–6]. To distinguish both modes, the reference electrode needs to be defined in the following manner. The reference electrode is the one at which (or close to which) the resistance switching takes place. The voltage is applied to this electrode whereas

Advances in Non-volatile Memory and Storage Technology. https://doi.org/10.1016/B978-0-08-102584-0.00005-X

the counter electrode is grounded. In an 8w switching system, the SET operation occurs with a positive polarity and the RESET with a negative polarity. In contrast, the system shows c8w switching if the device is set with a negative polarity and resets with a positive one. Please note that in the literature quite often the SET operation is related to a positive applied voltage, but, in fact, in most of the cases the device shows a c8w switching polarity. In these cases, the reference electrode is defined in the opposite way. A typical *I-V* characteristic showing the electroforming and the subsequent RESET and SET switching for filamentary c8w switching Pt/SrTiO$_3$/TiN cell is shown in Fig. 4.1A [7]. In this case, the switching takes place close to the high work function Pt electrode, which thus serves as reference electrode.

Besides 8w and c8w, also complementary switching (CS) has been observed in a single-layer oxide device when electronically similar electrodes are used [9–11]. In this case, the switching occurs at two different switching locations and c8w bipolar switching can be achieved at both electrodes by limiting the maximum current. When using the full dynamic voltage range one of the switching locations is in the HRS and the other one in the LRS, which results in the CS *I-V* curve illustrated in Fig. 4.1B. This behavior was initially shown by connecting to bipolar resistive switching devices in an antiserial manner, the complementary resistive switch [12].

The chapter focuses on the c8w filamentary switching mode, as it is the most common one. Nevertheless, many explanations can be transferred to the homogenous switching case. In this chapter, it is first discussed how ionic defects are introduced into the system. This so-called electroforming mechanism is required to facilitate the

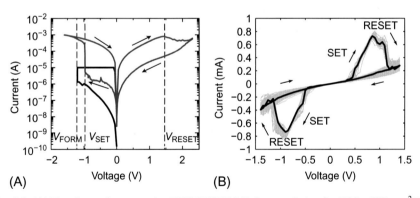

Fig. 4.1 (A) Forming and consecutive RESET/SET *I-V* characteristic of a (100×100) nm^2 Pt/SrTiO$_3$/TiN crossbar device. The forming voltage, the SET voltage and the RESET voltage are marked using dashed lines [7]. (B) Complementary resistive switching *I-V* characteristic of a Pt/Ta$_2$O$_5$/Ta/Pt micro crossbar ($3\,\mu m \times 3\,\mu m$) device [8].

Panel (A): Reprinted with permission from K. Fleck, C. La Torre, N. Aslam, S. Hoffmann-Eifert, U. Böttger, S. Menzel, Uniting gradual and abrupt SET processes in resistive switching oxides, Phys. Rev. Appl. 6 (2016) 064015, https://doi.org/10.1103/PhysRevApplied.6.064015. Copyright 2016 by the American Physical Society. Panel (B): Reprinted with permission from A. Schoenhals, S. Menzel, V. Rana, R. Waser, 3-bit Read Scheme for Single Layer Ta2O5 ReRAM. Non-Volatile Memory Technology Symposium (NVMTS), https://doi.org/10.1109/NVMTS.2014.7060845. © 2014 IEEE.

repetitive switching between different resistance states. To this end, two different electroforming models are critically reviewed: the generation of anti-Frenkel pairs in the
oxide bulk and the oxygen evolution reaction at the metal/oxide interface. Further,
it will be discussed how the introduced ionic defects modulate the electronic transport through the device stack. Section 4.4 outlines the current understanding of c8w
switching in VCM cells. Finally, open questions and the directions of further research
are addressed.

4.2 Electroforming mechanism

To facilitate the repetitive switching between different resistance states an initial electroforming step is typically required. This step results in a high concentration of the
ionic defects responsible for the switching process. When considering oxygen vacancies as the dominant type of defects, two different ways of generating these defects in
the oxide film have been proposed in the literature. The first one postulates the generation of anti-Frenkel pairs consisting of one oxygen interstitial O_i'' and one oxygen
vacancy $V_O^{\cdot\cdot}$ under high electric fields [13–19]. The corresponding reaction can be
written in Kröger-Vink notation as

$$O_O \rightarrow V_O^{\cdot\cdot} + O_i''. \tag{4.1}$$

This means an oxygen ion on a regular oxygen ion lattice site is released to an
interstitial site O_i'' under high electric field and leaves behind an oxygen vacancy $V_O^{\cdot\cdot}$,
which is doubly positively charged with respect to the regular lattice. This process is
illustrated in Fig. 4.2A. Alternatively, an oxygen exchange reaction at the anode is
proposed. In this case, an oxygen ion is removed from the oxide close to the interface
and oxygen (O_2) evolves. In the oxide, a doubly positively charged oxygen vacancy
and two electrons form. In Kröger-Vink notation this process can be written as

$$O_O \rightarrow \frac{1}{2}O_2 + V_O^{\cdot\cdot} + 2e^{\cdot}. \tag{4.2}$$

Fig. 4.2B illustrates the oxygen evolution reaction. Please note that oxygen can
be also extracted by oxidizing the anode material, which is the more likely reaction
when a non-noble metal serves as anode. In the following sections, these two forming
models are discussed in more detail.

4.2.1 Anti Frenkel-pair generation model

The anti-Frenkel pair (aFp) generation model was proposed by several groups and
implemented within kinetic Monte Carlo (KMC) frameworks to simulate the electroforming and switching processes in HfO_2 [13, 14, 16–18, 20], Ta_2O_5 [19] or SiO_2-
based VCM cells [21]. In the aFp electroforming model, $V_O^{\cdot\cdot}$-O_i'' anti-Frenkel pairs
are created within the bulk accompanied with a fast outward diffusion of the oxygen
interstitials. Typically, the oxygen vacancies are assumed to be immobile and they

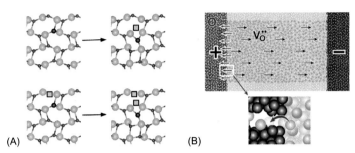

Fig. 4.2 (A) Formation of an $V_O^{\cdot\cdot}$-$O_i^{''}$ anti-Frenkel pair in the bulk of HfO$_2$ *(upper images)* and in the presence of a preexisting oxygen vacancy *(lower images)*. The large green balls represent the Hf ions, the small red balls are lattice oxygen ions, and the blue squares mark the oxygen vacancy sites. (B) Illustration of the oxygen exchange model. Applying a positive voltage leads to the extraction of oxygen *(light blue spheres)* from the oxide *(yellow spheres)* into the anode *(gray spheres)* as illustrated in the zoom box. The generated oxygen vacancy *(green spheres)* moves within the electric field toward the cathode *(dark blue spheres)*. Panel (A): Reprinted with permission from S. Bradley, A. Shluger, G. Bersuker, Electron-injection-assisted generation of oxygen vacancies in monoclinic HfO2, Phys. Rev. Appl. 4(6) (2015) 064008, https://doi.org/10.1103/PhysRevApplied.4.064008. Copyright 2015 by the American Physical Society. Panel (B): Reproduced with permission from D. Cooper, C. Baeumer, N. Bernier, A. Marchewka, C. La Torre, R.E. Dunin-Borkowski, S. Menzel, R. Waser, R. Dittmann, Anomalous resistance hysteresis in oxide ReRAM: oxygen evolution and reincorporation revealed by in situ TEM. Adv. Mater. 29 (2017), 1700212, https://doi.org/10.1002/adma.201700212. Copyright 2017, Wiley-VCH.

serve as trapping centers for a trap-assisted tunneling process. It should be mentioned that in fact the electronic d-states of the host cation are responsible for the conduction and not the oxygen vacancy itself. In the simulation models, anti-Frenkel pairs are generated according to a field-dependent generation rate and recombine with a field-independent static recombination rate. Both processes are related to an activation barrier. Besides the fact that it is questionable why the electric field should not affect the recombination barrier, it has been shown by means of atomistic simulations that the activation barrier for this process in HfO$_2$ is zero [22] and an immediate recombination should take place [23]. This observation is consistent with the findings of Bradley et al. [24], who showed that a $V_O^{\cdot\cdot}$-$O_i^{''}$ anti-Frenkel pair is not stable due to the strong Coulombic interaction between both species. The authors of this study proposed that the anti-Frenkel pair can be stabilized by a second process, the fast localization of two electrons ($2e^{''}$) at the newly formed $V_O^{\cdot\cdot}$ to form a neutral vacancy $V_O^{\times}$. Furthermore, the formation energy of the anti-Frenkel pair is proposed to lower in the direct vicinity of a doubly negatively charged oxygen vacancy [24, 25] (cf. Fig. 4.1A):

$$O_O + V_O^{''} \rightarrow 2V_O^{\times} + O_i^{''}. \tag{4.3}$$

This means that a bare oxygen vacancy would trap four electrons, of which two are released to neutralize the newly formed doubly positively charged oxygen vacancy $V_O^{\cdot\cdot}$.

Schie et al. critically reconsidered the anti-Frenkel pair generation model in monoclinic and cubic HfO$_2$ by means of a simple thermodynamic model as well as molecular static (MS) and molecular dynamic (MD) simulations [26]. The results of this study are discussed in the following.

The formation of anti-Frenkel pairs is often compared to the generation of electron–hole pairs in a semiconductor. The analogy to the band gap in the semiconductor is the Gibbs energy of anti-Frenkel disorder ΔG_{aF} in the ionic crystal. The corresponding energy band diagram is shown in Fig. 4.3A. The equilibrium chemical potential of the oxygen interstitial $\Delta\mu_{Oi}$ and the equilibrium chemical potential of the oxygen vacancy $\Delta\mu_{Vo}$ correspond to the conduction band edge and the valence band edge of a semiconductor, respectively. In thermal equilibrium, the product of the oxygen vacancy concentration and the concentration of oxygen interstitials can then be described by

$$\left[V_O^{..} \right]\left[O_i'' \right] = K_{iv}\left(T \right) = N_i N_v \exp\left(-\frac{\Delta G_{aF}\left(T \right)}{k_B T} \right) \tag{4.4}$$

where N_i and N_v are the temperature-independent densities of sites in the interstitial (i) and vacancy (v) sublattices, k_B is the Boltzmann constant, T is the temperature, and K_{iv} is the temperature-dependent equilibrium constant. If an electric field E is applied, the energy bands are tilted (cf. Fig. 4.3A). The gradient is zeE with $z=2$ being the charge number of the oxygen ions and e being the elementary charge. In this picture, the generation of an anti-Frenkel pair corresponds to an ion traversing the energy gap at a certain chemical potential by moving a distance d_{vi}. As the ions can only jump from

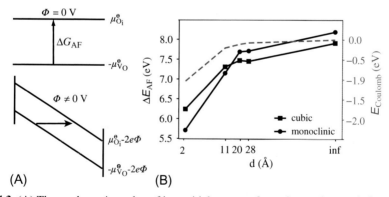

Fig. 4.3 (A) Thermodynamic analog of interstitial-vacancy formation to electron-hole-pair formation in a semiconductor without bias and under applied bias Φ. (B) Molecular static atomistic calculations of the formation energy ΔE_{AF} of an anti-Frenkel pair in the cubic and monoclinic phases of HfO$_2$ at different finite separations d. The *red dashed line* shows a simple calculation of the Coulomb contribution to the energies.

Reprinted with permission from M. Schie, S. Menzel, J. Robertson, R. Waser, R.A. De Souza, Field-enhanced route to generating anti-Frenkel pairs in HfO2, Phys. Rev. Mater. 2 (2018) 035002, https://doi.org/10.1103/PhysRevMaterials.2.035002. Copyright 2018 by the American Physical Society.

its regular lattice site to a neighboring interstitial site, this distance is about a few Å. Based on this consideration, a critical electric field strength

$$E_{aF}^{cr} = \frac{\Delta G_{aF}}{2ed_{vi}} \qquad (4.5)$$

for the generation of anti-Frenkel pairs can be defined. By determining ΔG_{aF} a critical field can be calculated. Typically, the formation energy of an oxygen vacancy is defined by the infinite separation of the oxygen vacancy and the removed oxygen ion. In reality, however, the oxygen ion will only move to an adjacent site, which will potentially give a lower forming energy. By means of MS atomistic simulation using a Mott-Littleton approach, it is shown that the formation energy of anti-Frenkel pairs in monoclinic and cubic HfO_2 reduces by about 2 eV when a separation of 2.2 Å is assumed compared to an infinite separation as shown in Fig. 4.3B. The obtained formation energy of about 6 eV is quite consistent with other studies, which reported values between 4.7 and 8 eV [27–30]. For a formation energy of 6 eV and $d_{vi}=2.5$ Å, a critical field of 12 GV/m = 12 V/nm results using Eq. (4.5). Assuming a typical oxide thickness of 5 nm, this means a voltage of 60 V needs to be applied to allow to form/set a cell based on the anti-Frenkel pair generation mechanism. This voltage is more than one order of magnitude higher than the forming/SET voltages observed in experiment, which are about a few V [7, 31–42]. Moreover, it exceeds the breakdown fields of the best insulators such as SiO_2 by a factor of 10.

The above discussion only considers a static picture, but forming and switching of VCM cells are dynamic processes. Thus, Schie et al. performed MD simulations at elevated temperatures and a varying electric field strength of $0 < E < 5$ GV/m. The calculations predict that for fields higher than 4 GV/m the entire crystal structure of both monoclinic and cubic HfO_2 became unstable, meaning that all ions start to move. Below 3 GV/m no anti-Frenkel pair formation takes place during the course of the simulations. Furthermore, V_o-O_i anti-Frenkel pairs were observed to form spontaneously in monoclinic HfO_2 at electric fields of about 3 GV/m but immediately recombined within a few ps. This result is consistent with ab-initio MD simulations by Clima et al. [23]. The latter MD simulations, however, do not include a possible stabilization of the anti-Frenkel pair due to current injection [24, 25].

To account for this possibility, Schie et al. investigated the recombination dynamics of a $[V_O^\times - O_i^{''}]$ pair using MD simulations at different electric fields. The neutral vacancy was obtained by reducing the next two nearest Hf^{4+} to Hf^{3+}. The simulation results show that even in this case the anti-Frenkel pair recombines within less than a picosecond. It is argued that there are still elastic interactions between vacancy and interstitial moieties. Moreover, the Coulomb interaction between vacancy and interstitial may only become negligible at large defect–defect separations, such that the vacancy together with its surroundings appear neutral [26]. To summarize, it is concluded that there is no viable route of anti-Frenkel pair formation in HfO_2.

The switching model based on the generation/recombination of anti-Frenkel pairs bears some more issues. First of all, if the continuous current flow really stabilizes the anti-Frenkel pair, it should be also stable independent of the applied voltage polarity. Thus, it is not clear how the anti-Frenkel pairs can recombine at all and the device can reset. Second, to reset the device oxygen interstitials are required, which are generated during the electroforming step. Consequently, a device employing an initially substoichiometric, electroforming-free oxide containing no oxygen interstitials should not switch at all. In fact, Alff and coworkers demonstrated that substoichiometric, electroforming-free HfO_x-based VCM cell can be switched reproducibly [43]. Based on all these considerations, one can conclude that a switching mechanism based on anti-Frenkel pair generation/recombination is not possible in HfO_2 and it is also not likely for other oxides.

4.2.2   Oxygen exchange and oxygen vacancy migration model

As pointed out by O'Hara et al., the generation of an oxygen vacancy in HfO_2 is more likely to occur by extracting the oxygen ion from the lattice and putting it into the anode [44]. This process was called an "extended Frenkel pair" by O'Hara, but is actually more precisely the description of a classical oxygen exchange reaction as known for solid-oxide fuel cells for decades. There are several reports in the literature supporting the idea of the occurrence of oxygen exchange reactions in VCM device stacks. While applying a positive bias to the top electrode, a bubble formation was observed for $SrTiO_3$-based [45], TiO_2 [39], and SiO_2-based [46] VCM devices. Sowinska et al. revealed by X-ray photoelectron spectroscopy that the Ti electrode in a $Ti/HfO_2/TiN$ VCM cell oxidizes during electroforming if a positive voltage is applied to it [47]. In this case, there is no evolution of oxygen gas, but TiO_x forms while the HfO_2 is reduced. Later the same group could show that a reversible interface reaction at the TiO_x/HfO_{2-x} interface takes place during SET/RESET cycling [48]. Moreover, it was demonstrated that the choice of the electrode material in a Ta_2O_5-based VCM device strongly influences the switching and electroforming properties [49]. The dynamics of the oxygen exchange reaction at the respective interface was proposed as the cause for this experimental observation. Using an in-situ TEM study, Cooper et al. revealed that the 8w switching in a $Pt/SrTiO_3/Nb:SrTiO_3$ VCM cell is based on oxygen evolution and reincorporation at the $Pt/SrTiO_3$ interface: an oxygen exchange reaction [50]. Further evidence for the occurrence of interfacial redox reactions was provided by Luebben et al., who studied the preforming state of Ta_2O_5- and HfO_2-based ReRAM stacks using cyclic voltammetry [51].

The electroforming mechanism based on oxygen exchange reactions and oxygen vacancy migration can be described as follows [1, 39]. During electroforming an oxygen exchange reaction occurs at the anode. This could be an oxygen evolution reaction at an inert electrode:

$$O_O^x \rightarrow \frac{1}{2}O_{2(g)} + V_O^{\cdot\cdot} + 2e''. \tag{4.6}$$

An oxygen ion on a regular lattice site close to the interface is extracted and gaseous oxygen forms. This oxygen extraction leads to the formation of a doubly positively charged oxygen vacancy, which is compensated by two electrons. The extracted oxygen could be stored in a cavity at the interface, which gives rise to the formation of bubbles as mentioned before [39, 45, 46]. Alternatively, oxygen could be stored in grain boundaries of the anode metal electrode. Instead of an oxygen evolution reaction, the extracted oxygen could react with the anode material forming a suboxide as reported for the Ti/HfO$_2$ interface by Sowinska [47]. In contrast to the anti-Frenkel pair generation model, the doubly positively charged oxygen vacancy is compensated by two electrons and not by one oxygen interstitial. Thus, the oxide material is donor-doped by the oxygen vacancy and the conductivity increases. No electron injection is required as in the case of the aFp model. The left behind oxygen vacancy migrates within the applied electric field toward the cathode. As the oxygen vacancy changes its place with the adjacent oxygen ion when it migrates, another oxygen ion is available for an oxygen exchange at the anode. The newly created oxygen vacancy will again migrate allowing for another oxygen exchange at the anode. In this way, the oxide material is reduced during the electroforming process. Due to local inhomogeneities and an additional thermal feedback at higher current densities, oxygen-vacancy rich filamentary regions can evolve. The inhomogeneities could be defects in the oxides, but also preferential sites for the oxygen exchange reaction, for example, at grain boundaries of the anode material. Any additional Joule heating, will lead to a thermal runaway, which favors the formation of filaments. For a quite homogenous interface and without any thermal feedback, homogenous switching might occur as well.

To simulate the electroforming process, a suitable model needs to take into account all relevant processes. The relevant ionic processes are the redox reaction at the metal/oxide interface and the migration of oxygen vacancies within the oxide. The relevant electronic processes are the electronic transport through the metal/oxide interfaces and the electron transport through the oxide. As we discuss filamentary switching, it is also necessary to include Joule heating in the simulation model. Marchewka et al. developed an axisymmetric 2D simulation model including all of these aspects [52]. This model is described by the following set of equation. The redox-reaction at the chemically active electrode is modeled using a Butler-Volmer type equation according to

$$J_{VO} = J_0 \left[\frac{N_{VO,M}N_{O,SC}}{N^*_{VO,M}N^*_{O,SC}} \exp\left(-\frac{\alpha z_{VO}e}{k_B T} \Delta\eta \right) - \frac{N_{O,M}N_{VO,SC}}{N^*_{O,M}N^*_{VO,SC}} \exp\left(\frac{(1-\alpha)z_{VO}e}{k_B T} \Delta\eta \right) \right] \quad (4.7)$$

with the exchange current density

$$J_0 = z_{VO}ek_{00} \exp\left(-\frac{\Delta G^\circ}{k_B T} \right) \left(N^*_{VO,M}N^*_{O,SC} \right)^{(1-\alpha)} \left(N^*_{O,M}N^*_{VO,SC} \right)^{\alpha}. \quad (4.8)$$

In Eqs. (4.7) and (4.8), $N_{O,SC}$ ($N_{VO,SC}$) denotes the oxygen ion (vacancy) concentration in the oxide layer, $N_{O,M}$ and $N_{VO,M}$ are the oxygen-defect concentrations in the metal electrode, $N_{O,SC}^*$, $N_{VO,SC}^*$, $N_{O,M}^*$, and $N_{VO,M}^*$ are the corresponding reference concentration values, α is the charge transfer coefficient and z_{VO} the number of involved electrons. The overpotential $\Delta\eta$ drives the reaction and is calculated from the difference between the Fermi level in the metal and the quasi-Fermi level of the electrons in the oxide. Further important parameters are the reaction rate constant

k_{00} and the Gibbs-free energy ΔG^0 of activation under standard conditions. Later, we show how the choice of the parameter values of k_{00} and ΔG^0 influences the filament evolution during the electroforming process.

To calculate the potential ψ within the oxide layer the Poisson equation

$$\nabla\left(\varepsilon\nabla\psi\right) = -e\left(n - \left[V_O^{\cdot}\right] - 2\left[V_O^{\cdot\cdot}\right]\right) \qquad (4.9)$$

is solved. Here, minority carries (holes) are neglected and it is assumed that the oxygen vacancies are twofold ionisable. This means the oxygen vacancies could be neutral or they are singly or doubly positively charged with respect to the lattice. For a doubly charged oxygen vacancies, the two compensating electrons are delocalized in the conduction band. For a neural/singly charged oxygen vacancy 2/1 electron(s) are/is localized at the neighboring cation. In Eq. (4.9), $[V_O^{\cdot}]([V_O^{\cdot\cdot}])$ denotes the singly (doubly) ionized oxygen vacancy concentration. This equation is solved along with the steady-state drift-diffusion equation for electrons

$$\nabla\left(\mu_n n\nabla\psi - D_n\nabla n\right) = \pm\partial j_{n,\text{tunnel}} / \partial x \qquad (4.10)$$

the time-dependent drift diffusion equation for the doubly ionized oxygen vacancies

$$\partial\left[V_O^{\cdot\cdot}\right] / \partial t - \nabla\left(\mu_{vo}\left[V_O^{\cdot\cdot}\right]\nabla\psi - D_{vo}\nabla\left[V_O^{\cdot\cdot}\right]\right) = -R_{VO,2} \qquad (4.11)$$

the rate equation for the singly ionized oxygen vacancies

$$\partial\left[V_O^{\cdot}\right] / \partial t = -R_{VO,1} \qquad (4.12)$$

and the rate equation for the neutral oxygen vacancies

$$\partial\left[V_O^{x}\right] / \partial t = -R_{VO,0}. \qquad (4.13)$$

In this model, it is assumed for simplicity that only the doubly ionized oxygen vacancies are mobile. For HfO_2, it has been shown that the migration barrier of an oxygen vacancy increases if the electrons are localized [53]. These barriers are so high (>3 eV) that the neutral and singly charged oxygen vacancies can be treated as immobile under the conditions of switching. In Eqs. (4.9)–(4.13), ε is the permittivity of the oxide, n is the electron concentration, μ_n (μ_{VO}) is the electron (oxygen-vacancy) mobility, D_n (D_{VO}) is the electron (oxygen-vacancy) diffusion coefficient, and N_{VO}^0 is the neutral oxygen vacancy concentration. $R_{VO,2}$, $R_{VO,1}$, and $R_{VO,0}$ represent the reaction rates that are derived from the donor ionization reactions along with oxygen-vacancy ionization statistics [54]. The local temperature in the device structure is calculated using the static heat equation

$$-\nabla k_{th}\nabla T = JF \qquad (4.14)$$

where k_{th} is the thermal conductivity, J is the current density, and F is the electric field. The temperature enters in the reaction rates [cf. Eq. (4.8)], the mobilities, and the diffusion coefficients. The latter are described according to

$$D_{VO} = D_0 \exp\left(-\frac{\Delta H_D}{k_B T}\right)\left(1 - \frac{\left[V_O^{..}\right]}{N_{VO,max}}\right). \tag{4.15}$$

The electron transport across the metal–insulator interfaces has two different contributions: Thermionic emission occurs for electrons with energy E_x higher than the conduction band edge maximum $E_{C,max}$ and tunneling occurs for energies between the conduction band minimum and maximum. The thermionic emission contribution is calculated by integration over the respective energies at position of the metal–insulator interfaces x_i according to

$$j_{TE}(x_i) = \frac{A^* T}{k_B} \int_{E_C(x_i)}^{\infty} N_{supply}(E_x)\, dE_x, \tag{4.16}$$

while the tunneling contribution is calculated as

$$j_{tunnel}(x) = \frac{A^* T}{k_B} \int_{E_{C,min}(x)}^{E_{C,max}(x)} T(E_x) N_{supply}(E_x)\, dE_x. \tag{4.17}$$

In Eqs. (4.16) and (4.17), A^* denotes the effective Richardson constant, $N_{supply}(E_x)$ denotes the supply function that describes the supply with carriers and is derived by integration of the occupancy function on both sides of the barrier, and $T(E_x)$ is the transmission coefficient that is calculated using the WKB approximation. The tunneling current enters as a local generation/recombination rate into the right-hand side of the drift-diffusion Eq. (4.10) for electrons. This set of equations is complemented by appropriate boundary conditions as outlined in Refs. [54–57].

The simulation model described by the set of Eqs. (4.7)–(4.17) has been applied to study the electroforming process of VCM cells [57]. In this study, the oxygen exchange reaction occurs at the high work function metal/oxide bottom interface, that is, the anode in this study. Furthermore, it is assumed that the oxygen exchange reaction takes place in a confined region at the center of the 2D axisymmetric simulation model. From a physical point of view this confinement can be related to the existence of preferred spots for the interface reaction, for example, at a grain boundary in the metal electrode. Fig. 4.4 shows the simulation results using the electroforming model for two different cases. In the first case a fast oxygen exchange rate of $k_{00} = 1 \times 10^{-22}\,cm^4/s$ was assumed, whereas a slow oxygen exchange rate of $k_{00} = 1 \times 10^{-34}\,cm^4/s$ was assumed in the second case. In both cases, the current increases over many orders of magnitudes while the device undergoes an electroforming process (Fig. 4.4A and B). The increase in current is associated with the increase of the amount of oxygen vacancies in a filamentary region. This reduction process already starts at low voltages (points marked with A, B, C) at which no significant Joule heating takes place. Starting with point C, Joule heating sets in, which leads to an acceleration of all involved ionic process. In contrast, to the case of a slow oxygen exchange rate, the Joule heating process leads to a thermal runaway for the case of a fast oxygen exchange reaction. The thermal

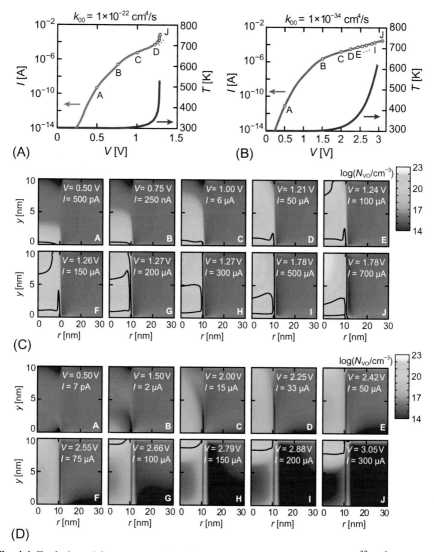

Fig. 4.4 Evolution of the current and maximum temperature for (A) $k_{00} = {}^\circ 10^{-22}$ cm^4/s and for (B) $k_{00} = {}^\circ 10^{-34}$ cm^4/s. (C) Maps of the oxygen-vacancy concentration at points A to J as marked in the *I-V* characteristics in (A). (D) Maps of the oxygen-vacancy concentration at points A to J as marked in the *I-V* characteristics in (B). The *black line* in the concentration maps marks the isoline of $N_{VO} = 1 \times 10^{20}$ cm^{-3} and illustrates the filament evolution.
Reprinted with permission from A. Marchewka, R. Waser, S. Menzel, Physical modeling of the electroforming process in resistive-switching devices (talk). 2017 International Conference on Simulation of Semiconductor Processes and Devices (SISPAD), September 7-9, Kamakura, Japan, 2017. © 2017 IEEE.

runaway manifests itself in the sharp increase in current and temperature in Fig. 4.4A. One reason for the different strength of the thermal feedback lies in the completely different filament evolution during the course of the electroforming process as illustrated in Fig. 4.4C and D. In the case of a fast oxygen exchange rate (Fig. 4.4C), there is always a high concentration of oxygen vacancies at the interface to the anode, that is, the bottom electrode. The conducting filament grows in this case from the anode to the cathode. In contrast, the oxygen vacancies drift relatively fast away from the anode interface in the case of the slow oxygen exchange reaction. This leads to an oxygen vacancy depletion at the anode interface and an accumulation of oxygen vacancies at the cathode. Eventually, a so-called virtual cathode evolves giving rise to a filament growth from the cathode to the anode. These different filament evolutions have an important impact on the electronic response of the system. Due to the asymmetry of the electrodes, that is, a high electrostatic barrier at the bottom electrode/oxide interface and a low electrostatic barrier at the top electrode/oxide interface, the oxygen vacancies affect the electronic transport in a different way if they are located close to the top or close to the bottom interface. If an oxygen vacancy is located close to the bottom interface with the high electrostatic barrier, it will lead to a bigger current increase than if located close to the top interface with the low electrostatic barrier. As the electrode with the higher electrostatic barrier (here the bottom electrode) has a bigger impact on the electronic transport properties, we call this electrode the electronically active electrode. In the case of the fast oxygen exchange reaction, the newly generated oxygen vacancies are closer to the bottom interface and a stronger electric feedback results. This will lead to strong thermal feedback. In the case of a slow oxygen exchange reaction, the temperature increase leads primarily to an accelerated drift of oxygen vacancies toward the cathode. Thus, the electronic feedback is less strong than in the case of a fast oxygen exchange reaction. The main result of this study, however, is the occurrence of the two different filament growth modes. If the oxygen exchange rate is fast (slow) compared to the ionic transport with the oxide, the filament grows from the anode (cathode) to the cathode (anode). This result is comparable with the growth modes discussed in ECM cells [58].

Experimentally, different growth modes have been reported for VCM systems, too. The "classical" growth mode from cathode to anode has been observed for example using electrocoloration experiments [1, 59, 60] and was proposed as the growth mode during the electroforming in VCM cells [1, 39]. The growth from anode to cathode was evidenced by TEM studies of electroformed Pt/Fe-doped $SrTiO_3$/Nb-doped $SrTiO_3$ VCM cells [61]. The TEM images showed the formation of multiple nanometer-sized filaments that extend from the anode (during electroforming) to the cathode. First indications that different growth mode can exist in one VCM cell were given by the fact an electroforming process can be triggered with opposing electroforming polarities [38]. The following switching cycling, however, was not affected by the forming polarity. Yalon et al. investigated the direction of the filament growth using a dedicated device structure composed of a bipolar transistor with an HfO_x-based VCM cell serving as emitter [62]. This structure enables the detection of a tunneling gap between the base of the bipolar transistor and a filament formed within the VCM cell emitter. If a tunneling gap is detected, the filament must have grown from the emitter electrode toward the transistor base. In this study, two different emitter electrodes were used: an inert electrode material and a material forming an oxygen exchange layer. The electroforming was performed with both polarities. In case of the inert emitter, a tunneling gap was only detected when the base of the transistor was used as anode. Thus, the filament grew from the cathode to the anode. In contrast, a tunneling gap was detected for both voltage polarities for the VCM cell including an oxygen exchange layer. Thus, the filament evolved from the cathode to the anode when the transistor base served as anode. If the oxygen exchange layer served as anode, the growth mode changed and the filament evolved from the anode to the cathode. This result is consistent with the simulation results in Fig. 4.4. The oxygen exchange layer allows for a fast exchange reaction rate compared to the ion migration rate in the oxide. This relation results in the filament evolution from anode to cathode (cf. Fig. 4.4A and C). In all other cases, the anode is more chemically inert and the redox reaction rate is slow compared to the ion migration. Thus, the filament grows from the cathode to the anode consistent with the results in Fig. 4.4B and D.

4.3   Conduction mechanism

The electronic transport through the VCM device stack is determined by the atomic configurations of the ionic defects. Oxygen vacancy defects can influence the electronic transport in different ways. On the one hand, they can modulate the conductivity of the oxide material as discussed in Section 4.3.1. On the other hand, they can modify the electrostatic barriers of the metal/oxide interfaces. These interface-related conduction mechanisms are discussed in Section 4.3.2. In principle different conduction mechanisms or a combination of them are conceivable as illustrated in Fig. 4.5. Electrons can be injected into the conduction band via electron tunneling or thermionic emission. They could travel via a trap-assisted-tunneling or a hopping process via defect states within the band gap of the oxide material. Moreover, trapped electrons could be released to the conduction band for high electronic fields according to the Poole-Frenkel mechanism.

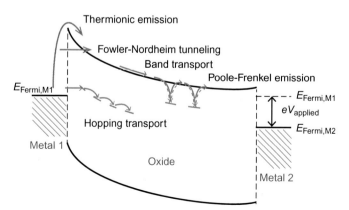

Fig. 4.5 Band diagram of a metal/oxide/metal structure under applied voltage illustrating the different electron transport mechanisms.

To distinguish between these different conduction mechanisms, a series of experiments is required. The conductivity should be measured at wide range of different temperatures at different electric fields with positive and negative polarity. It is conceivable that different conduction mechanisms will dominate in different voltage (including different voltage polarities) and temperature regimes. In the following an overview of the main conduction mechanisms discussed in the literature is given.

4.3.1 Bulk mechanism

Bulk conduction models discuss how the atomic configuration changes the overall resistance without taking into account the different metal/oxide interfaces. In these models, it is hard to explain a dependence of the conduction state on the choice of electrode materials or an asymmetric I-V characteristic with respect to the voltage polarity.

According to the oxygen exchange reaction (4.6), two electrons compensate the doubly positively charged oxygen vacancies, which thus act as mobile donors. More precisely, the electrons are released to the d^0-band of the transition metal. Consequently, the valence of the transition metal changes. Depending on the energetic level(s) of the oxygen-vacancy defect states, the electrons are free or localized. Assuming shallow donor levels, the electron concentration n in the conduction band is given by the oxygen vacancy concentration $[V_O^{..}]$ and the corresponding charge state z. Then, the oxide conductivity can be calculated by

$$\sigma = ne\mu_n = ze\left[V_O^{..}\right]\mu_n \tag{4.18}$$

where μ_n is the electron mobility. In this picture, the LRS would be represented by a homogenous oxygen vacancy concentration along the filament axis. In contrast, the HRS results for an inhomogeneous distribution with a depletion of oxygen vacancies in front of the electronically active electrode.

When the oxygen vacancy concentration and in turn the electron concentration increases, the atomic orbitals of the reduced transition metal cations will overlap. According to the Mottcriteria, the former semiconducting oxide will turn into a metal. From the perspective of the Mott criteria, it directly follows that on the atomic scale the LRS can be described by a chain of oxygen vacancies within the oxide matrix. Such a configuration has been calculated using density functional theory (DFT) for TiO_2, HfO_2, and Al_2O_3-based resistive switches [63]. The calculated partial density of states shows an electronically conducting channel along these oxygen vacancy chains. By disrupting this chain locally a noncontinuous partial charge density results and the HRS is obtained. Similar results have been obtained by Cartoixa et al. [64]. As the calculations were performed at 0 K, the proposed HRS is probably still well-conducting at operating temperatures. Moreover, the oxygen vacancy concentration in the LRS might be lower. It should be mentioned here that not the oxygen vacancies itself are conducting, but the electron conduction proceeds via the oxygen vacancy-mediated defect states within the oxide matrix, that is, the d^0 states of the cations.

In filamentary switching systems, the conductance depends highly on the local atomic configuration. Thus, it is quite reasonable that discrete conduction steps appear when the filament dimensions are as small as a few nanometers. Degraeve et al.

observed the occurrence of discrete conduction steps in a $TiN/HfO_2/Ti$ cell [65]. To describe these resistance states, the authors of this study introduced a quantum point conduction (QPC) model, which is similar to the one of Miranda et al. [66]. In this model, a narrow constriction is considered in which the electrons travel in a ballistic manner. Together with the electron confinement, quantized conduction appears. The number of defects in the constriction defines the number of available conductance modes. The QPC model shows a very good fit to the experimental data [65]. The QPC model, however, does not give any information about the position of the constriction within the oxide layer. Moreover, the occurrence of ballistic transport at the operating temperatures $>300\,K$ is doubtful from a theoretic point of view [67]. Nevertheless, the occurrence of discrete resistance steps can be attributed to the finite size of the filament and the limited number of possible atomic configurations.

Trap-assisted tunneling (TAT) is quite often reported as prevailing conduction mechanism in the HRS [14, 20]. In this case oxygen vacancy-induced defect states are considered as trap states and the electrons tunnel from one trap to the other. In the multiphonon TAT model proposed by Bersuker et al. [14, 20] the capture and emission of electrons leads to a heat dissipation at the position of the defect by energy relaxation. By assuming multiple current paths along trap states the authors could fit the temperature dependence of a HfO_x-based stack. This conduction model required the traps to lie deep in the band gap in order to prevent the excitation of electrons into the conduction band. As discussed earlier, the electrons provided by the oxygen vacancies appear in the d-states of the transition metal ion. Thus, the electron transport should occur by hopping between different reduced valent transition metal sites. Moreover, the assumption of pure TAT transport leads to a very high amount of oxygen vacancies in the LRS to explain the high currents [16, 17, 68] [69]. For high concentration of vacancies, and thus electrons, the Fermi-level will be closer to the conduction band. In this case, a transition from TAT to band conduction should be expected and a smaller oxygen vacancy concentration is possibly required to explain the LRS conduction.

In order to identify the dominant conduction mechanism, temperature-dependent measurements of the conductivity for different resistance states are inevitable. Graves et al. measured the temperature-dependent conductivity of a $Pt/TaO_x/Ta$ device for differently programmed resistance states (read at $0.2\,V$) from $300\,K$ down to $4\,K$ [70]. It was found that the T-dependence changes. At low temperatures the conductance state is almost temperature-independent and at higher temperatures a strong current increase with temperature was observed. This behavior is observed for different programmed resistance states between 0.2 and $42\,\mu S$. Based on the experimental data, the authors concluded that two parallel conduction paths exist. The first one describes the almost temperature-independent part and is modeled by Mott hoping via defect states. This conduction mechanism is supposed to describe the state of the device, where the number of defect states determines the conductance. The second parallel conduction path is considered to be a state-independent leakage current, which has been modeled by Schottky-emission. The slight asymmetry observed in the I-V characteristics has been attributed to the asymmetric injection of electrons from the electrodes into the defect states. The fitting resulted in a very low hopping distance in the region of interatomic distances, which puts the result into question as the wave functions of the defect states should already overlap.

4.3.2 Interface-related mechanism

The conduction mechanisms discussed in the previous chapter neglect the electron injection from the metal to the oxide material. The choice of the electrode material, however, plays already an important role during forming as discussed in Section 4.2. In addition, to enable a bipolar switching mechanism, the cell has to be asymmetric [1]. In fact, quite often asymmetric *I-V* characteristics with respect to the voltage polarity are reported and the modulation of the electron transport across one of the metal/oxide interfaces has been proposed as the dominating resistance switching mechanism [1, 45, 71–79].

Using a four-pad geometry with two Ti and two Pt electrodes on a TiO$_2$-based crystalline VCM device, Yang and coworkers could show that the TiO$_2$ forms a Schottky contact with the Pt electrode [80]. The two Ti/TiO$_2$ contacts in contrast show an ohmic behavior. By further switching experiments, the authors could identify that the switching took place at the Pt/TiO$_2$ interface by modifying its current blocking properties. In a further study, Yang et al. demonstrated that the chemical properties of the metal/oxide interface determine the barrier height rather than the work function difference [72]. For a non-noble electrode the electrode is oxidized and the oxide underneath is reduced. The resulting high concentration of positively charged oxygen vacancies (relative to the regular crystal lattice) lowers the Schottky barrier height and eventually an ohmic contact results. For noble metals, no redox reaction occurs and the Schottky barrier remains.

As the conduction mechanism depends on the local substoichiometry (e.g., the oxygen vacancy concentration), it is detrimental to know the exact distribution of the ionic defects in the LRS and HRS. Bäumer et al. used in-operando PEEM measurements to analyze the change in the donor concentration during resistive switching in a Graphene/SrTiO$_3$/Nb-doped SrTiO$_3$ VCM cell [76]. By acquiring spatially resolved O K-edge images a switching filament of about 500 nm in diameter can be identified. Moreover, the charge carrier concentration in the upper 2–3 nm of the filament can be quantified to be 1.5×10^{21} electrons per cm^{-3} in the LRS and 6.7×10^{20} electrons per cm^{-3} in the HRS. These parameters served as input for a simulation study using a parabolic band approximation. The simulation results reveal that the resistance change between LRS and HRS of more than two orders of magnitude by changing the electronic carrier concentration by only a factor of 2–3 can be explained by the dopant-induced barrier modulation at the graphene/SrTiO$_3$ interface. Due to the high dopant concentration, the electrostatic barrier is very thin and electron tunneling dominates over thermionic emission. This leads to very symmetric *I-V* characteristics with respect to the voltage polarity for both resistance states.

A thorough study of the current conduction mechanism of a Pt/SrTiO$_3$/Nb:SrTiO$_3$ VCM cell was presented by Funck et al. [81]. In this study, the temperature-dependence of the cell was measured under reverse bias of the Pt/SrTiO$_3$ interface as illustrated in Fig. 4.6A. For low reverse bias voltages, the current was found to increase with temperature (regime I) whereas for a higher voltage magnitude the temperature-dependence changed. In the latter voltage regime II, the current decreased with increasing temperature. The transition between these two regimes occurs at a so-called intersection voltage, at which the current is temperature-independent. An analytical fit assuming thermionic

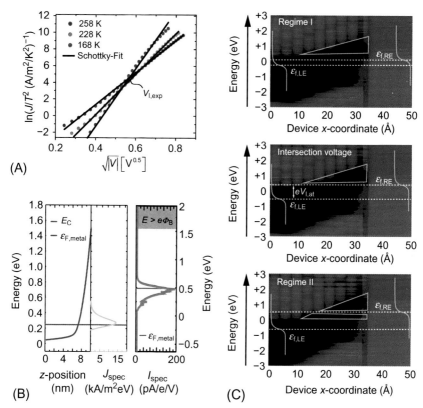

Fig. 4.6 (A) Analysis of the *I-V* relation in reverse direction of a Pt/SrTiO₃ Schottky interface using an analytical Schottky fit for three different temperatures. (B) The simulated spectral current density J_{spec} and conduction band profile at $V_{applied} = -0.25\,V$ using the continuum model is shown on the left. On the right, the spectral current I_{spec} is shown at $V = -0.96\,V$ calculated with the DFT-NEGF model. In both cases, the peak of the spectral current is well below the barrier height $e\Phi_B$. (C) Local density of states calculated with the DFT-NEGF model for regime I, regime II, and at the intersection voltage V_I. The Fermi distribution at the two electrodes is shown at the position of the left and right Fermi energy. The triangle indicates the conduction band edge at the interface and illustrates the tunneling barrier. Figures are reproduced with permission from C. Funck, A. Marchewka, C. Baeumer, P.C. Schmidt, P. Mueller, R. Dittmann, M. Martin, R. Waser, S. Menzel, A theoretical and experimental view on the temperature dependence of the electronic conduction through a schottky barrier in a resistively switching SrTiO3-based memory cell, Adv. Electron. Mater. 4 (2018), 1800062, Copyright 2018, Wiley-VCH.

emission and Schottky barrier lowering showed a reasonable agreement (solid line in Fig. 4.6A). The analytical fit, however, would predict a barrier lowering below zero, which cannot be physically motivated. The conduction mechanism was analyzed further using numerical simulations based on DFT with a nonequilibrium Green's function (NEGF) formalism and a continuum model based on a parabolic band approximation.

The continuum model solves the Eqs. (4.9)–(4.13), (4.16), and (4.17) in a one-dimensional (1D) domain. The simulation results of the DFT-NEGF model and the continuum model show the same temperature-dependency as observed in experiment and predicted by the analytical Schottky model. The analyses of the results, however, lead to a different interpretation of the conduction mechanism. Fig. 4.6B shows the calculated band diagram based on the continuum model and the calculated spectral current densities for the continuum and the DFT-NEGF model. Both models predict that the highest current contribution is obtained at an energy level way below the Schottky barrier height. Thus, the conduction mechanism is not a thermionic emission process, but an electron tunneling process. The temperature dependence of the conduction process follows from the analysis of the Fermi level of the injecting electrode in comparison to the conduction band edge of within the $SrTiO_3$. In regime I, the Fermi level of the injecting metal electrode is lower than the conduction band edge within the $SrTiO_3$ as shown in Fig. 4.6C. Thus, only thermally excited electrons in the Boltzmann tail of the metal Fermi distribution can tunnel through the interfacial barrier into the $SrTiO_3$ conduction band. As a consequence, the electronic transport increases with increasing temperature. At the intersection voltage, these two energy levels align and electrons at the Fermi edge can tunnel, too. The Boltzmann approximation is not valid anymore and the conduction mechanism becomes temperature-independent. In regime II, the Fermi level is above the conduction band edge and all electrons can tunnel. The observed current decrease with increasing temperature in this regime is likely due to phonon scattering in the oxide bulk.

As VCM cells, typically, employ a very thin oxide material, the application of a voltage leads to a severe band banding. Thus, electron tunneling into the conduction band should always be considered as possible conduction mechanism at least at higher voltages.

4.4 Switching mechanism

The basis of the resistive switching mechanism is the movement of the oxygen vacancies generated during the electroforming process within the oxide. According to the oxygen-exchange reaction (4.6) electrons compensate the oxygen vacancies. The electrons occupy the d^0 states of the host cation, leading to a valence change of the cation sublattice. As discussed in Section 4.3, the oxygen vacancies modulate the local conductivity as well as the interface-related transport. Thus, a redistribution of the oxygen vacancies will lead to a significant change of the total cell resistance. In the following discussion, we will focus on the redistribution of the ionic defects and exclude possible redox reactions. The SET/RESET switching voltages are typically lower than the electroforming voltage. One could argue that the high electroforming voltage is required to trigger any redox reaction and thus such interface reactions can be neglected for the SET/RESET switching. Despite this simplification the redistribution of the ionic defects can already explain the switching dynamics of c8w switching VCM cells, which are discussed in Section 4.4.1. In addition, this model can explain the transition from bipolar to CS (cf. Section 4.4.2).

4.4.1 c8w switching

The basic switching principle of filamentary c8w switching in an n-type oxide is illustrated in Fig. 4.7. The switching takes place in a region (disc) close to one of the electrodes, that is, the electronically active electrode. In the HRS, the concentration of the oxygen vacancies is low in the disc region. When applying a negative voltage to the electronically active electrode, the oxygen vacancies will migrate to this interface resulting in a more homogenous oxygen vacancy distribution. This leads to an increase of the local conductivity in the disc, but also modulates the electrostatic barrier. The latter would become thinner and lower enabling a high tunneling current. To reset the cell a positive voltage is applied to the electronically active electrode and the oxygen vacancies deplete close to this electrode restoring the HRS. According to the definition of the switching modes in the introduction, this switching mechanism describes a c8w switching mode. It should be mentioned that in p-type oxides such as manganites, the increase of the local oxygen vacancy concentration would lead to an increase of the local resistance. Thus, the switching mode would change to 8w as shown by the models of Rozenberg and coworkers [82, 83]. Here, focus is, however, on the switching in n-type oxides, which are more common.

If only a bulk conduction mechanism is considered, the minimal achievable resistance results for a homogenous defect distribution. A depletion of the ionic defects in one region (and a concurring accumulation in a different region) will effectively block the total current flow due to the high resistance of this region. In a filamentary picture, this depletion of ionic defects is often referred to as filament rupture. In fact, the current will typically still flow through this filamentary region only the conductance

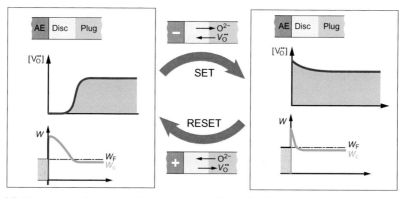

Fig. 4.7 Illustration of the switching mechanism. The switching takes place in a disc region close to the electronically active electrode (AE). The oxygen vacancies act as mobile donors. During SET a negative voltage is applied to the AE and the oxygen vacancies accumulate at the AE interface (right-hand side) resulting in a very thin a reduced electrostatic barrier at this interface. By applying a positive voltage to the AE, the oxygen vacancies are pushed away from the interface and the barrier is restored (RESET operation) as shown on the left. Reproduced with permission from R. Waser, R. Bruchhaus, S. Menzel, 'Redox-Based Resistive Random Access Memories'. Wiley. Copyright 2012, Wiley-VCH, 2012.

is worse. Such a bulk redistribution model was numerically implemented solving the current continuity equation, the drift-diffusion equation and the heat transfer equation to simulate the switching dynamics of HfO_2 [84] and Ta_2O_5-based [11] VCM cells. A similar model was applied to bulk switching GaO_x-based VCM cells [85]. Depending on the applied voltage range bipolar c8w switching properties can be obtained. The drawback of these models is the ambiguity of the RESET voltage polarity. Starting from a homogenous ionic defect distribution, the device would reset with both voltage polarities. For both voltage polarities, a depletion region will evolve leading to a higher cell resistance. This is consistent with a CS mode and certainly applies to VCM cells with two identical electrodes. The bulk model lacks the explanation for pure bipolar VCM switching.

Taking into account the electrostatic barriers at the metal/oxide interfaces solves the problem of the ambiguity of the RESET polarity. As mentioned before, the typical VCM device stack comprises one high electrostatic barrier and one ohmic contact. Marchewka et al. implemented a numerical model taking into account the contact potentials at the metal/oxide interfaces, the redistribution of the ionic defects, and Joule heating [54–56]. The simulation model solves the Eqs. (4.9)–(4.17) in a two-dimensional (2D) axisymmetric or in a 1D geometry. Whereas Eq. (4.14) is solved in the 2D model, the temperature is calculated according to

$$T = T_0 + \frac{L}{8k_{th}} j_{mean} V_{device} \qquad (4.19)$$

in the 1D model. In Eq. (4.19), T_0 is the ambient temperature, L is the oxide thickness, j_{mean} is the mean value of the current density in the insulating layer, and V_{device} is the voltage drop across the oxide.

Using the 2D axisymmetric model, c8w filamentary switching was successfully simulated [56]. In this model, a filamentary region with high concentration of oxygen vacancies is assumed at the center of the geometry. The bottom electrode forms a high electrostatic barrier, whereas a low electrostatic barrier exists at the top electrode/oxide interface. The bottom electrode is thus defined as the electronically active electrode. Fig. 4.8 shows the simulated *I-V* characteristic and the corresponding snap shots of the oxygen vacancy redistribution. For the simulation, a triangular voltage pulse with a voltage ramp rate of 10 V/s was used. In the HRS (A), the oxygen vacancies are slightly depleted close to the electronically active bottom electrode. When applying a negative voltage to the bottom electrode, the oxygen vacancies are attracted and accumulate at the bottom electrode (B-C), which sets the device to the LRS (D). When the polarity is reversed the oxygen vacancies deplete again at the bottom electrode resulting in the HRS (E-H). The simulated *I-V* characteristic shows an abrupt SET transition (B-C) and a very gradual RESET transition (E-G), which is typical for filamentary switching VCM devices. When considering only the redistribution close to the bottom interface, the model reproduces the cartoon of the switching mechanism in Fig. 4.7. The concentration maps, however, show that there is huge change in the oxygen vacancy concentration close to the ohmic top electrode. Nevertheless, the *I-V*

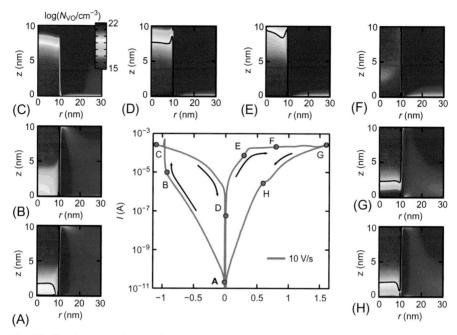

Fig. 4.8 Simulated *I-V* characteristics for a sweep rate of 10 V/s using the continuum model. A-H shows the maps of oxygen vacancy concentration during the sweep showing the redistribution of the oxygen vacancies. The black isoline of $N_{VO} = 3.2 \times 10^{19}$ cm^{-3} (equivalent to 100 oxygen vacancies inside the filament) serves as a guide to the eye to illustrate the evolution of the filament.

Reprinted with permission from A. Marchewka, R. Waser, S. Menzel, A 2D axisymmetric dynamic drift-diffusion model for numerical simulation of resistive switching phenomena in metal oxides. 2016 International Conference On Simulation of Semiconductor Processes and Devices (SISPAD), Nuremberg, Germany, September 6-8, 2016, vol., pp. 145-148, https://doi.org/10.1109/SISPAD.2016.7605168. © 2016 IEEE.

characteristic shows a bipolar c8w switching mode. The reason is the asymmetry of the two metal/oxide contact barriers. As the barrier height at the bottom electrode is a lot higher compared to the one at the top electrode, the total conduction is more sensitive to changes of the oxygen vacancy concentration close to the bottom electrode. The electron transport through the bottom interface defines the total cell resistance. In contrast to the bulk model, there will be no CS in this model as long as the barrier heights differ strongly (cf. Section 4.4.2 "Complementary switching").

Filamentary c8w switching VCM cells typically show a gradual RESET switching event, which can be exploited to program different resistance states [86–92]. The 2D axisymmetric simulation model was also applied to study the gradual RESET phenomenon of VCM cells [54]. In this study, the experimental RESET dynamics of a Pt/Ta$_2$O$_5$/Ta device were investigated by applying constant voltage stress with different

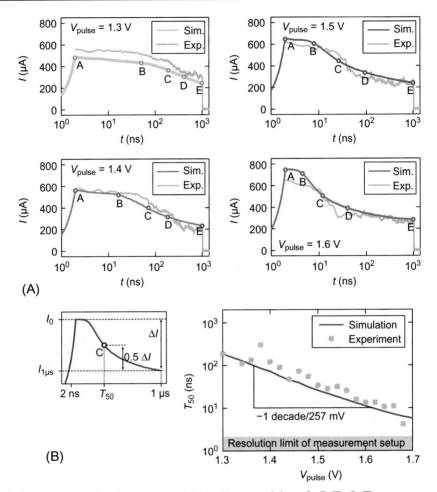

Fig. 4.9 (A) Comparison between simulated and measured data of a Pt/Ta$_2$O$_5$/Ta nano-crossbar device. Transient current behavior at different RESET voltages. Here, the voltages are defined with respect to the Pt electrode and not to the Ta electrode as in the original publication. (B) On the right, the 50% decay time τ_{50} is shown as a function of the pulse voltage. It is obtained from point C as illustrated in the schematic on the left.
Reproduced with permission from A. Marchewka, B. Roesgen, K. Skaja, H. Du, C.L. Jia, J. Mayer, V. Rana, R. Waser, S. Menzel, Nanoionic resistive switching memories: on the physical nature of the dynamic reset process. Adv. Electron. Mater., 2 (2016), 1500233/1-13, https://doi.org/10.1002/aelm.201500233. Copyright 2016, Wiley-VCH.

voltage amplitudes. The simulation results of the 2D axisymmetric model show a remarkable agreement with the experimental data as shown in Fig. 4.9. The simulation model could not only reproduce the trend of the RESET decay time τ_{50} as a function of the applied voltage (Fig. 4.9B), but also the shape of the current transient during the course of the RESET transition (Fig. 4.9A). A further analysis of the simulation results revealed the origin of the gradual RESET phenomenon. During the RESET process, the temperature-activated oxygen vacancy motion builds a concentration gradient, which counteracts the driving force of the applied electric field and decelerates the migration process. Eventually, drift and diffusion processes approach an equilibrium situation. Another important aspect is the moderate sensitivity of the current response to the change of the vacancy concentration. This prerequisite is fulfilled for the conduction mechanism dominated by the modulation of the contact potentials as described above, but also for the bulk conduction model according to Eq. (4.18).

In contrast to the RESET operation, the SET operation in c8w filamentary switching VCM cells is quite abrupt. It has been shown by a combination of experimental data and modeling that the temperature-enhanced ionic drift is the origin of the nonlinear SET switching kinetics in filamentary switching VCM systems [93]. Due to local Joule heating the temperature increases and accelerates the oxygen vacancy drift. Based on this model, the experimental set kinetics of a $SrTiO_3$-based VCM system could be modeled over >12 orders of magnitude in time [94, 95]. Moreover, it could be shown for a Ta_2O_5-based VCM cell that the switching time is indeed a function of the dissipated power [96]. Further, the role of Joule heating in determining the SET switching speed was further verified by decoupling the influence of temperature and electric field using an electric nanoheater device setup [97]. In a detailed SET kinetics study, Fleck et al. investigated the abrupt SET switching of $SrTiO_3$, ZrO_2, and Ta_2O_5-based VCM cells using constant voltage pulse with varying amplitude and analyzing the measured current transients [7]. In this study, it was observed that a slight gradual current increase precedes the abrupt SET switching event as shown in Fig. 4.10A. The gradual increase is characterized by the current increase per time called pre-SET slope.

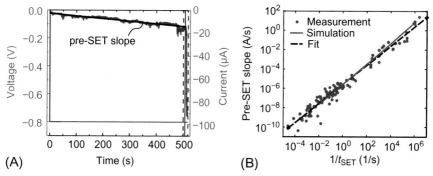

Fig. 4.10 (A) Transient current response *(red line)* of a $Pt/SrTiO_3/TiN$ device to a constant voltage of $-0.8\,V$ *(blue line)*. A gradual current increase, that is, the pre-SET slope precedes the abrupt switching event at about 511 s. (B) The pre-SET slope of the device is plotted against the inverse of the SET time. The experimental data obtained from SET kinetics measurements over 12 orders of magnitude is shown as blue circles. The simulation result using a compact model is shown as red solid line.

The pre-SET slope has a strong correlation to the switching time, which is defined as the point in time at which the abrupt current jump occurs. This correlation is valid over >10 orders of magnitude in time, suggesting that the two phenomena of gradual current increase and abrupt switching are closely linked to each other. To understand this behavior, the authors of this study developed a compact model, which is based on the change of the oxygen vacancy concentration close to the electronically active interface (disc region) due to ion migration. This concentration changes the resistance of this region as well as the contact barrier. The compact model, thus, represents a simplification of the model described by Eqs. (4.9)–(4.17). As shown in Fig. 4.10B, the compact models reproduces the experimental trend of the dependence of the pre-SET slope on the inverse switching time. By analyzing the simulation results, the origin of the two-step SET process could be revealed. The same physical process lies at the heart of both, the gradual current increase and the abrupt current jump. The pre-SET slope originates from an increasing oxygen vacancy concentration in the disc region, thus increasing the device conductivity in a nonvolatile manner. This gradual current increase coupled to local Joule heating leads to a positive feedback loop, which results in a thermal runaway. This runaway process is the origin of the sharp current increase in the current transient. The occurrence of the abrupt switching event is thus also an indicator for Joule heating taking place.

The occurrence of gradual SET switching and abrupt SET switching in one cell is reported for HfO$_2$-based VCM cells, too [98]. The two-step SET process, which is linked to a thermal runaway, can thus be seen as a very general behavior of c8w filamentary switching VCM systems.

4.4.2 Complementary switching

As mentioned before CS is observed in symmetric VCM device stacks [9, 99, 100]. Moreover, CS has also been observed in nonsymmetric devices with different metal electrodes [8, 10, 101–103]. For HfO$_2$- and Ta$_2$O$_5$-based devices with asymmetric electrodes, it was shown that the thickness of the ohmic electrode influences the switching mode [101]. For thin electrode layers CS was observed, whereas a pure bipolar switching was observed for thick electrode layers. For a medium electrode thickness, the Ta$_2$O$_5$-based devices showed an intermediate asymmetric CS *I-V* characteristic. The bulk simulation model by Nardi et al. [9] and Kim et al. [11] mentioned before would predict the occurrence of CS for all device stacks. These models do not account for any influence of the electrode materials. The simulation model described by Eqs. (4.9)–(4.17), however, takes into account the different electrode work functions in form of different contact barrier heights. As shown in the previous section, the higher barrier height will dominate the overall conduction. Thus, a pure bipolar switching is observed in Fig. 4.8. Using a 1D geometry, Marchewka et al. discussed the influence of the asymmetry of the two barrier heights [55]. Assuming very asymmetric barriers of 0.7 eV/0.1 eV, pure bipolar switching is obtained as shown in Fig. 4.11A. Identical barrier heights lead to the CS curves shown in Fig. 4.11B. Moreover, an asymmetric CS mode results for slightly asymmetric barriers of 0.7 and 0.5 eV indicating a gradual transition from bipolar to CS. These simulation results are consistent with the observed *I-V* characteristics for a Pt/Ta$_2$O$_5$/Ta/Pt device with varying Ta electrode thickness

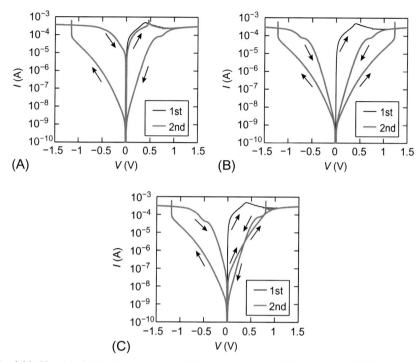

Fig. 4.11 Simulated *I-V* characteristics of (A) an asymmetric 1D structure exhibiting bipolar resistive switching, (B) a symmetric structure exhibiting complementary switching behavior, and (C) a slightly asymmetric structure indicating a transition between bipolar and complementary switching. The results are obtained using the numerical simulation model in a 1D domain [55].

Reprinted, with permission, from A. Marchewka, R. Waser, S. Menzel, Physical simulation of dynamic resistive switching in metal oxides using a Schottky contact barrier model. 2015 International Conference On Simulation of Semiconductor Processes and Devices (SISPAD), 9-11 September, Washington D.C, USA, 2015, https://doi.org/10.1109/SISPAD.2015.7292318. © 2015 IEEE.

[101]. The question is, however, why the Ta electrode thickness should influence the Ta work function. Schoenhals et al. argued that during electroforming the Ta electrode oxidizes partly [101]. In addition, O can be dissolved into the Ta electrode. The oxygen incorporation is reported to increase the work function of the electrode substantially [104]. In this regard, a higher concentration of oxygen in the electrode would lead to a higher work function and thus a higher barrier height. This means that by using a thicker electrode the oxygen concentration is lower than in a thinner electrode. Thus, for a thicker electrode a lower barrier height is expected as for a thinner electrode. This fits to the experimental observation: bipolar switching is observed for thick Ta electrodes, which would give the highest asymmetry in the barrier heights. Going to thinner Ta electrodes leads to a transition to CS, which is equivalent to a symmetrization of the barrier heights. For very thin Ta electrode thickness, it is also conceivable that the Ta

electrode oxidizes completely in a local filamentary region during the electroforming process. In this case, a symmetric $Pt/TaO_{5-x}/Pt$ structure results.

4.5   Conclusions and outlook

This chapter discussed the electroforming mechanism and the resistive switching mechanism of VCM devices. The electroforming mechanism is based on the oxygen exchange reaction at the anode and a migration of the generated oxygen vacancies within the oxide layer. Depending on the ratio between oxygen exchange rate and the oxygen-vacancy migration rate, different filament growth directions can be motivated. For a high oxygen exchange rate compared to the ion migration rate, the filament grows from the anode to the cathode. In contrast, for a VCM system with a fast ionic migration and a slow oxygen exchange reaction, the filament grows from the cathode to the anode.

Further, the switching mechanism of OxRAM is discussed. Assuming a redistribution of the ionic defects within the oxide material without considering possible oxygen exchange reaction leads to a filamentary c8w switching mode. This model can account for the abrupt SET switching as well as the gradual RESET switching. The physical origin of the SET and RESET dynamics is the temperature-assisted drift and diffusion of the mobile defects. During SET the combination of oxygen vacancy migration and Joule heating leads to a thermal runaway, which manifests in an abrupt current jump preceded by gradual current increase. The gradual RESET can be explained by drift and diffusion of oxygen vacancies approaching an equilibrium. Finally, the transition from bipolar to CS can be explained as a symmetrization of the electrostatic barrier heights at the two metal/oxide interfaces.

The models presented in Section 4.4 do not include possible redox reactions, which will be an important step forward especially for understanding endurance or retention failures. It was shown experimentally that the choice of the ohmic electrode material influences the switching dynamics of Ta_2O_5-based VCM cells [49]. This experimental observation was attributed to an oxygen exchange reaction at the ohmic electrode. This reaction would not change the switching polarity—it would be still c8w—but influences the SET and RESET dynamics. In contrast, an oxygen exchange reaction at the electronically active electrode would give rise to an 8w switching mode as demonstrated for a $Pt/SrTiO_3/Nb$-doped $SrTiO_3$ VCM cell [50, 76]. Moreover, it was shown that 8w switching and c8w switching can be obtained as competing mechanism in the same sample [6]. The investigation of the exact mechanism of the oxygen exchange reactions is an important task for the future to obtain reliable VCM devices. The challenge is how the redox reactions can be influenced by device design and the choice of materials.

The possible conduction mechanism of VCM cells is discussed in this chapter too. Up to now, a clear picture of the exact conduction mechanism is missing. The analysis of the conduction mechanism requires its characterization at different voltages, bias and at different temperatures. Most likely, the accurate description of the conduction mechanism in VCM cells involves different conduction mechanism occurring in parallel with different ones dominating the overall conduction in different voltage and temperature regimes.

References

[1]   R. Waser, R. Dittmann, G. Staikov, K. Szot, Redox-based resistive switching memories—nanoionic mechanisms, prospects, and challenges, Adv. Mater. 21 (2009) 2632–2663, https://doi.org/10.1002/adma.200900375.

[2]   A. Wedig, M. Luebben, D.-Y. Cho, M. Moors, K. Skaja, V. Rana, T. Hasegawa, K. Adepalli, B. Yildiz, R. Waser, I. Valov, Nanoscale cation motion in TaO_x, HfOx and TiOx memristive systems, Nat. Nanotechnol. 11 (2016) 67–74, https://doi.org/10.1038/nnano.2015.221.

[3]   A. Sawa, Resistive switching in transition metal oxides, Mater. Today 11 (2008) 28–36, https://doi.org/10.1016/S1369-7021(08)70119-6.

[4]   R. Muenstermann, T. Menke, R. Dittmann, R. Waser, Coexistence of filamentary and homogeneous resistive switching in Fe-doped $SrTiO_3$ thin-film memristive devices, Adv. Mater. 22 (2010) 4819–4822, https://doi.org/10.1002/adma.201001872.

[5]   J.S. Lee, S.B. Lee, B. Kahng, T.W. Noh, Two opposite hysteresis curves in semiconductors with mobile dopants, Appl. Phys. Lett. 102 (2013) 253503/1–4, https://doi.org/10.1063/1.4811556.

[6]   A. Schoenhals, C.M.M. Rosario, S. Hoffmann-Eifert, R. Waser, S. Menzel, D.J. Wouters, Role of the electrode material on the RESET limitation in oxide ReRAM devices, Adv. Electron. Mater. 4 (2018) 1700243/1–11, https://doi.org/10.1002/aelm.201700243.

[7]   K. Fleck, C. La Torre, N. Aslam, S. Hoffmann-Eifert, U. Böttger, S. Menzel, Uniting gradual and abrupt SET processes in resistive switching oxides, Phys. Rev. Appl. 6 (2016) 064015, https://doi.org/10.1103/PhysRevApplied.6.064015.

[8]   A. Schoenhals, S. Menzel, V. Rana, R. Waser, in: 3-bit Read Scheme for Single Layer Ta_2O_5 ReRAM, Non-Volatile Memory Technology Symposium (NVMTS), 2014, https://doi.org/10.1109/NVMTS.2014.7060845.

[9]   F. Nardi, S. Balatti, S. Larentis, D.C. Gilmer, D. Ielmini, Complementary switching in oxide-based bipolar resistive-switching random memory, IEEE Trans. Electron Devices 60 (2013) 70–77, https://doi.org/10.1109/TED.2012.2226728.

[10] A. Schoenhals, J. Mohr, D.J. Wouters, R. Waser, S. Menzel, 3-bit resistive RAM write-read scheme based on complementary switching mechanism, IEEE Electron Device Lett. 38 (2017) 449–452, https://doi.org/10.1109/LED.2017.2670642.

[11] S. Kim, S. Choi, W. Lu, Comprehensive physical model of dynamic resistive switching in an oxide memristor, ACS Nano 8 (2014) 2369–2376, https://doi.org/10.1021/nn405827t.

[12] E. Linn, R. Rosezin, C. Kügeler, R. Waser, Complementary resistive switches for passive nanocrossbar memories, Nat. Mater. 9 (2010) 403–406, https://doi.org/10.1038/nmat2748.

[13] B. Gao, S. Yu, N. Xu, L.F. Liu, B. Sun, X.Y. Liu, R.Q. Han, J.F. Kang, B. Yu, Y.Y. Wang, Oxide-based RRAM switching mechanism: a new ion-transport-recombination model, in: IEEE International Electron Devices Meeting 2008, 2008, pp. 563–566.

[14] G. Bersuker, D.C. Gilmer, D. Veksler, P. Kirsch, L. Vandelli, A. Padovani, L. Larcher, K. McKenna, A. Shluger, V. Iglesias, M. Porti, M. Nafria, Metal oxide resistive memory switching mechanism based on conductive filament properties, J. Appl. Phys. 110 (2011) 124518, https://doi.org/10.1063/1.3671565.

[15] N. Xu, L. Liu, X. Sun, X. Liu, D. Han, Y. Wang, R. Han, J. Kang, B. Yu, Characteristics and mechanism of conduction/set process in TiN/ZnO/Pt resistance switching random-access memories, Appl. Phys. Lett. 92 (2008) 232112/1–3, https://doi.org/10.1063/1.2945278.

[16] B. Butcher, G. Bersuker, D. Gilmer, L. Larcher, A. Padovani, L. Vandelli, R. Geer, P. Kirsch, Connecting the physical and electrical properties of Hafnia-based RRAM. in: Electron Devices Meeting (IEDM), 2013 IEEE International, 2013, pp. 22.2.1–22.2.4, https://doi.org/10.1109/IEDM.2013.6724682.

[17] X. Guan, S. Yu, H. Wong, On the switching parameter variation of metal-oxide RRAM-Part I: physical modeling and simulation methodology, IEEE Trans. Electron Devices 59 (2012) 1172–1182, https://doi.org/10.1109/TED.2012.2184545.

[18] A. Padovani, L. Larcher, O. Pirrotta, L. Vandelli, G. Bersuker, Microscopic modeling of HfOx RRAM operations: from forming to switching, IEEE Trans. Electron Devices 62 (2015) 1998–2006, https://doi.org/10.1109/TED.2015.2418114.

[19] Y. Zhao, P. Huang, Z. Chen, C. Liu, H. Li, B. Chen, W. Ma, F. Zhang, B. Gao, X. Liu, J. Kang, Modeling and optimization of bilayered TaOx RRAM based on defect evolution and phase transition effects, IEEE Trans. Electron Devices 63 (2016) 1524–1532, https://doi.org/10.1109/TED.2016.2532470.

[20] L. Vandelli, A. Padovani, L. Larcher, G. Bersuker, Microscopic modeling of electrical stress-induced breakdown in poly-crystalline hafnium oxide dielectrics, IEEE Trans. Electron Devices 60 (2013) 1754–1762, https://doi.org/10.1109/TED.2013.2255104.

[21] A. Padovani, D.Z. Gao, A.L. Shluger, L. Larcher, A microscopic mechanism of dielectric breakdown in SiO2 films: an insight from multi-scale modeling, J. Appl. Phys. 121 (2017) 155101, https://doi.org/10.1063/1.4979915.

[22] B. Traore, P. Blaise, E. Vianello, L. Perniola, B.D. Salvo, Y. Nishi, HfO_2-based RRAM: electrode effects Ti/HfO_2 interface, charge injection, and oxygen (O) defects diffusion through experiment and *Ab Initio* calculations, IEEE Trans. Electron Devices 63 (2016) 360–368.

[23] S. Clima, B. Govoreanu, M. Jurczak, G. Pourtois, HfOx as RRAM material—first principles insights on the working principles, Microelectron. Eng. 120 (2014) 13–18, https://doi.org/10.1016/j.mee.2013.08.002.

[24] S. Bradley, A. Shluger, G. Bersuker, Electron-injection-assisted generation of oxygen vacancies in monoclinic HfO_2, Phys. Rev. Appl. 4 (2015) 064008, https://doi.org/10.1103/PhysRevApplied.4.064008.

[25] S.R. Bradley, G. Bersuker, A.L. Shluger, Modelling of oxygen vacancy aggregates in monoclinic HfO_2: can they contribute to conductive filament formation? J. Phys. Condens. Matter 27 (2015) 415401, https://doi.org/10.1088/0953-8984/27/41/415401.

[26] M. Schie, S. Menzel, J. Robertson, R. Waser, R.A. De Souza, Field-enhanced route to generating anti-Frenkel pairs in HfO_2, Phys. Rev. Mater. 2 (2018) 035002, https://doi.org/10.1103/PhysRevMaterials.2.035002.

[27] B. Traore, P. Blaise, B. Sklenard, Reduction of monoclinic HfO2: a cascading migration of oxygen and its interplay with a high electric field, J. Phys. Chem. C 120 (2016) 25023–25029, https://doi.org/10.1021/acs.jpcc.6b06913.

[28] A. Foster, F. Gejo, A. Shluger, R. Nieminen, Vacancy and interstitial defects in hafnia, Phys. Rev. B Condens. Matter 65 (2002) 174117/1, https://doi.org/10.1103/PhysRevB.65.174117.

[29] J.X. Zheng, G. Ceder, T. Maxisch, W.K. Chim, W.K. Choi, First-principles study of native point defects in hafnia and zirconia, Phys. Rev. B 75 (2007) 104112.

[30] Y. Guo, J. Robertson, Materials selection for oxide-based resistive random access memories, Appl. Phys. Lett. 105 (2014) 223516/1–5, https://doi.org/10.1063/1.4903470.

[31] D. Ito, Y. Hamada, S. Otsuka, T. Shimizu, S. Shingubara, Oxide thickness dependence of resistive switching characteristics for Ni/HfOx/Pt resistive random access memory device, Jpn. J. Appl. Phys. 54 (2015) https://doi.org/10.7567/JJAP.54.06FH11. pp. 6FH11/1-4.

[32] Z. Fang, X.P. Wang, J. Sohn, B.B. Weng, Z.P. Zhang, Z.X. Chen, Y.Z. Tang, G. Lo, J. Provine, S.S. Wong, H.-P. Wong, D. Kwong, The role of Ti capping layer in HfOx-based RRAM devices, IEEE Electron Device Lett. 35 (2014) 912–914, https://doi.org/10.1109/LED.2014.2334311.

[33] P. Hua, N. Deng, A forming-free bipolar resistive switching in HfOx-based memory with a thin Ti cap, Chin. Phys. Lett. 31 (2014) https://doi.org/10.1088/0256-307X/31/10/107303. 107303/1-5.

[34] K.G. Young-Fisher, G. Bersuker, B. Butcher, A. Padovani, L. Larcher, D. Veksler, D.C. Gilmer, Leakage current-forming voltage relation and oxygen gettering in HfOx RRAM devices, IEEE Electron Device Lett. 34 (2013) 750–752, https://doi.org/10.1109/LED.2013.2256101.

[35] S. Yu, H.-Y. Chen, B. Gao, J. Kang, H.-S.P. Wong, HfOx-based vertical resistive switching random access memory suitable for bit-cost-effective three-dimensional cross-point architecture, ACS Nano 7 (2013) 2320–2325, https://doi.org/10.1021/nn305510u.

[36] Y.-S. Chen, H.-Y. Lee, P.-S. Chen, T.-Y. Wu, C.-C. Wang, P.-J. Tzeng, F. Chen, M.-J. Tsai, C. Lien, An ultrathin forming-free HfOx resistance memory with excellent electrical performance, IEEE Electron Device Lett. 31 (2010) 1473–1475, https://doi.org/10.1109/LED.2010.2081658.

[37] N. Raghavan, A. Fantini, R. Degraeve, P. Roussel, L. Goux, B. Govoreanu, D. Wouters, G. Groeseneken, M. Jurczak, Statistical insight into controlled forming and forming free stacks for HfOx RRAM, Microelectron. Eng. 109 (2013) 177–181. http://dx.doi.org/.

[38] C. Nauenheim, C. Kuegeler, A. Ruediger, R. Waser, Investigation of the electroforming process in resistively switching TiO$_2$ nanocrosspoint junctions, Appl. Phys. Lett. 96 (2010) 122902, https://doi.org/10.1063/1.3367752.

[39] J.J. Yang, F. Miao, M.D. Pickett, D.A.A. Ohlberg, D.R. Stewart, C.N. Lau, R.S. Williams, The mechanism of electroforming of metal oxide memristive switches, Nanotechnology 20 (2009) 215201, https://doi.org/10.1088/0957-4484/20/21/215201.

[40] S.B. Lee, H.K. Yoo, K. Kim, J.S. Lee, Y.S. Kim, S. Sinn, D. Lee, B.S. Kang, B. Kahng, T.W. Noh, Forming mechanism of the bipolar resistance switching in double-layer memristive nanodevices, Nanotechnology 23 (2012) https://doi.org/10.1088/0957-4484/23/31/315202. pp. 315202/1-9.

[41] M.-J. Lee, C.B. Lee, D. Lee, S.R. Lee, M. Chang, J.H. Hur, Y.-B. Kim, C.-J. Kim, D.H. Seo, S. Seo, U.-I. Chung, I.-K. Yoo, K. Kim, A fast, high-endurance and scalable non-volatile memory device made from asymmetric Ta$_2$O$_{5-x}$/TaO$_{2-x}$ bilayer structures, Nat. Mater. 10 (2011) 625–630.

[42] D.K. Gala, A.A. Sharma, D. Li, J.M. Goodwill, J.A. Bain, M. Skowronski, Low temperature electroformation of TaOx-based resistive switching devices, APL Mater. 4 (2016) 16101/1–6, https://doi.org/10.1063/1.4939181.

[43] S.U. Sharath, T. Bertaud, J. Kurian, E. Hildebrandt, C. Walczyk, P. Calka, P. Zaumseil, M. Sowinska, D. Walczyk, A. Gloskovskii, T. Schroeder, L. Alff, Towards forming-free resistive switching in oxygen engineered HfO2-x, Appl. Phys. Lett. 104 (2014) 063502, https://doi.org/10.1063/1.4864653.

[44] A. OHara, G. Bersuker, A.A. Demkov, Assessing hafnium on hafnia as an oxygen getter, J. Appl. Phys. 115 (2014) 183703, https://doi.org/10.1063/1.4876262.

[45] K. Szot, W. Speier, G. Bihlmayer, R. Waser, Switching the electrical resistance of individual dislocations in single-crystalline SrTiO$_3$, Nat. Mater. 5 (2006) 312–320, https://doi.org/10.1038/nmat1614.

[46] A. Mehonic, M. Buckwell, L. Montesi, M.S. Munde, D. Gao, S. Hudziak, R.J. Chater, S. Fearn, D. McPhail, M. Bosman, A.L. Shluger, A.J. Kenyon, Nanoscale transformations in metastable, amorphous, silicon-rich silica, Adv. Mater. 28 (2016) 7486–7493, https://doi.org/10.1002/adma.201601208.

[47] M. Sowinska, T. Bertaud, D. Walczyk, S. Thiess, M.A. Schubert, M. Lukosius, W. Drube, C. Walczyk, T. Schroeder, Hard x-ray photoelectron spectroscopy study of the electroforming in Ti/HfO2-based resistive switching structures, Appl. Phys. Lett. 100 (2012) 233509, https://doi.org/10.1063/1.4728118.

[48] T. Bertaud, M. Sowinska, D. Walczyk, S. Thiess, A. Gloskovskii, C. Walczyk, T. Schroeder, In-operando and non-destructive analysis of the resistive switching in the Ti/HfO2/TiN-based system by hard x-ray photoelectron spectroscopy, Appl. Phys. Lett. 101 (2012) https://doi.org/10.1063/1.4756897. 143501/1-5.

[49] W. Kim, S. Menzel, D.J. Wouters, Y. Guo, J. Robertson, B. Rösgen, R. Waser, V. Rana, Impact of oxygen exchange reaction at the ohmic interface in Ta_2O_5-based ReRAM devices, Nanoscale 8 (2016) 17774–17781, https://doi.org/10.1039/c6nr03810g.

[50] D. Cooper, C. Baeumer, N. Bernier, A. Marchewka, C. La Torre, R.E. Dunin-Borkowski, S. Menzel, R. Waser, R. Dittmann, Anomalous resistance hysteresis in oxide ReRAM: oxygen evolution and reincorporation revealed by in situ TEM, Adv. Mater. 29 (2017) 1700212, https://doi.org/10.1002/adma.201700212.

[51] M. Lübben, S. Wiefels, R. Waser, I. Valov, Processes and effects of oxygen and moisture in resistively switching TaOx and HfOx, Adv. Electron. Mater. 4 (2018) 1700458, https://doi.org/10.1002/aelm.201700458.

[52] A. Marchewka, R. Waser, S. Menzel, Physical modeling of the electroforming process in resistive-switching devices. in: 2017 International Conference on Simulation of Semiconductor Processes and Devices (SISPAD), September 7-9, Kamakura, Japan, 2017, pp. 133–136, https://doi.org/10.23919/SISPAD.2017.8085282.

[53] D. Duncan, B. Magyari-Koepe, Y. Nishi, Filament-induced anisotropic oxygen vacancy diffusion and charge trapping effects in hafnium oxide RRAM, IEEE Electron Device Lett. 37 (2016) 400–403, https://doi.org/10.1109/LED.2016.2524450.

[54] A. Marchewka, B. Roesgen, K. Skaja, H. Du, C.L. Jia, J. Mayer, V. Rana, R. Waser, S. Menzel, Nanoionic resistive switching memories: on the physical nature of the dynamic reset process, Adv. Electron. Mater. 2 (2016) https://doi.org/10.1002/aelm.201500233. 1500233/1-13.

[55] A. Marchewka, R. Waser, S. Menzel, Physical simulation of dynamic resistive switching in metal oxides using a Schottky contact barrier model. in: 2015 International Conference On Simulation of Semiconductor Processes and Devices (SISPAD), 9-11 September, Washington D.C, USA, 2015, pp. 297–300, https://doi.org/10.1109/SISPAD.2015.7292318.

[56] A. Marchewka, R. Waser, S. Menzel, A 2D axisymmetric dynamic drift-diffusion model for numerical simulation of resistive switching phenomena in metal oxides. in: 2016 International Conference On Simulation of Semiconductor Processes and Devices (SISPAD), Nuremberg, Germany, September 6-8, 2016, 2016, pp. 145–148, https://doi.org/10.1109/SISPAD.2016.7605168.

[57] A. Marchewka, R. Waser, S. Menzel, Physical modeling of the electroforming process in resistive-switching devices (talk), in: 2017 International Conference on Simulation of Semiconductor Processes and Devices (SISPAD), September 7-9, Kamakura, Japan, 2017.

[58] Y. Yang, P. Gao, L. Li, X. Pan, S. Tappertzhofen, S. Choi, R. Waser, I. Valov, W.D. Lu, Electrochemical dynamics of nanoscale metallic inclusions in dielectrics, Nat. Commun. 5 (2014) 4232/1-9.

[59] V. Havel, A. Marchewka, S. Menzel, S. Hoffmann-Eifert, G. Roth, R. Waser, Electroforming of Fe:STO samples for resistive switching made visible by electrocoloration observed by high resolution optical microscopy, MRS Online Proc. Lib. (2014) https://doi.org/10.1557/opl.2014.562.

[60] R. Waser, T. Baiatu, K.H. Hardtl, DC electrical degradation of perovskite-type titanates. II. Single crystals, J. Am. Ceram. Soc. 73 (1990) 1654–1662.

[61] H. Du, C. Jia, A. Koehl, J. Barthel, R. Dittmann, R. Waser, J. Mayer, Nanosized conducting filaments formed by atomic-scale defects in redox-based resistive switching memories, Chem. Mater. 29 (2017) 3164–3173, https://doi.org/10.1021/acs.chemmater.7b00220.

[62] E. Yalon, I. Karpov, V. Karpov, I. Riess, D. Kalaev, D. Ritter, Detection of the insulating gap and conductive filament growth direction in resistive memories, Nanoscale 7 (2015) 15434–15441, https://doi.org/10.1039/c5nr03314d.

[63] K. Kamiya, M.Y. Yang, B. Magyari-Kope, M. Niwa, Y. Nishi, K. Shiraishi, Physics in designing desirable ReRAM stack structure—atomistic recipes based on oxygen chemical potential control and charge injection/removal. in: 2012 IEEE International Electron Devices Meeting (IEDM 2012), 2012, https://doi.org/10.1109/IEDM.2012.6479078. 20.2 (4 pp.).

[64] X. Cartoixa, R. Rurali, J. Sune, Transport properties of oxygen vacancy filaments in metal/crystalline or amorphous HfO2/metal structures, Phys. Rev. B: Condens. Matter 86 (2012) 165445/1, https://doi.org/10.1103/PhysRevB.86.165445.

[65] R. Degraeve, L. Goux, S. Clima, B. Govoreanu, Y.Y. Chen, G.S. Kar, P. Rousse, G. Pourtois, D.J. Wouters, L. Altimime, M. Jurczak, G. Groeseneken, J.A. Kittl, Modeling and tuning the filament properties in RRAM metal oxide stacks for optimized stable cycling, in: Proceedings of the 2012 International Symposium On Vlsi Technology, Systems and Application (vlsi-Tsa), 2012, pp. 2. pp. 2 https://doi.org/10.1109/VLSI-TSA.2012.6210101.

[66] E. Miranda, D. Jimenez, J. Sune, The Quantum point-contact memristor, IEEE Electron Device Lett. 33 (2012) 1474–1476, https://doi.org/10.1109/LED.2012.2210185.

[67] D. Ielmini, R. Waser, Resistive Switching—From Fundamentals of Nanoionic Redox Processes to Memristive Device Applications, Wiley-VCH, 2016.

[68] S. Yu, X. Guan, H. Wong, Understanding metal oxide RRAM current overshoot and reliability using Kinetic Monte Carlo simulation. in: 2012 Ieee International Electron Devices Meeting (iedm 2012), 2012, https://doi.org/10.1109/IEDM.2012.6479105. 26.1 (4 pp.).

[69] R. Fang, W. Chen, L. Gao, W. Yu, S. Yu, Low-temperature characteristics of HfOx-based resistive random access memory, IEEE Electron Device Lett. 36 (2015) 567–569, https://doi.org/10.1109/LED.2015.2420665.

[70] C.E. Graves, N. Dávila, E.J. Merced-Grafals, S.-T. Lam, J. Paul Strachan, R.S. Williams, Temperature and field-dependent transport measurements in continuously tunable tantalum oxide memristors expose the dominant state variable, Appl. Phys. Lett. 110 (2017) 123501.

[71] J.J. Yang, J. Borghetti, D. Murphy, D.R. Stewart, R.S. Williams, A family of electronically reconfigurable nanodevices, Adv. Mater. 21 (2009) 3754–3758.

[72] J.J. Yang, J.P. Strachan, F. Miao, M. Zhang, M.D. Pickett, W. Yi, D.A.A. Ohlberg, G. Medeiros-Ribeiro, R.S. Williams, Metal/TiO2 interfaces for memristive switches, Appl. Phys. Mater. Sci. Process. 102 (2011) 785–789.

[73] C.W. Hsu, Y.F. Wang, C.C. Wan, I.T. Wang, C.T. Chou, W.L. Lai, Y.J. Lee, T.H. Hou, Homogeneous barrier modulation of TaOx/TiO2 bilayers for ultra-high endurance three-dimensional storage-class memory, Nanotechnology 25 (2014) 165202/1–7, https://doi.org/10.1088/0957-4484/25/16/165202.

[74] J.R. Jameson, Y. Fukuzumi, Z. Wang, P. Griffin, K. Tsunoda, G.I. Meijer, Y. Nishi, Field-programmable rectification in rutile TiO2 crystals, Appl. Phys. Lett. 91 (2007) 112101/1–3.

[75] S. Lee, J.S. Lee, J.-B. Park, Y.K. Kyoung, M.-J. Lee, T.W. Noh, Anomalous effect due to oxygen vacancy accumulation below the electrode in bipolar resistive switching Pt/Nb:SrTiO$_3$ cells, APL Mater. 2 (2014) 066103.

[76] C. Baeumer, C. Schmitz, A. Marchewka, D.N. Mueller, R. Valenta, J. Hackl, N. Raab, S.P. Rogers, M.I. Khan, S. Nemsak, M. Shim, S. Menzel, C.M. Schneider, R. Waser, R. Dittmann, Quantifying redox-induced Schottky barrier variations in memristive devices via in operando spectromicroscopy with graphene electrodes, Nat. Commun. 7 (2016) 12398, https://doi.org/10.1038/ncomms12398.

[77] T. You, Y. Shuai, W. Luo, N. Du, D. Bürger, I. Skorupa, R. Hübner, S. Henker, C. Mayr, R. Schüffny, T. Mikolajick, O.G. Schmidt, H. Schmidt, Exploiting memristive BiFeO$_3$ bilayer structures for compact sequential logics, Adv. Funct. Mater. 24 (2014) 3357–3365, https://doi.org/10.1002/adfm.201303365.

[78] E. Miranda, D. Jiménez, A. Tsurumaki-Fukuchi, J. Blasco, H. Yamada, J. Suñé, A. Sawa, Modeling of hysteretic Schottky diode-like conduction in Pt/BiFeO 3/SrRuO3 switches, Appl. Phys. Lett. 105 (2014) 082904.

[79] T. Fujii, M. Kawasaki, A. Sawa, Y. Kawazoe, H. Akoh, Y. Tokura, Electrical properties and colossal electroresistance of heteroepitaxial SrRuO3/SrTi1-xNbxO3 (0.0002 <= x <= 0.02) Schottky junctions, Phys. Rev. B 75 (2007) 165101.

[80] J.J. Yang, M.D. Pickett, X. Li, D.A.A. Ohlberg, D.R. Stewart, R.S. Williams, Memristive switching mechanism for metal/oxide/metal nanodevices, Nat. Nanotechnol. 3 (2008) 429–433.

[81] C. Funck, A. Marchewka, C. Baeumer, P.C. Schmidt, P. Mueller, R. Dittmann, M. Martin, R. Waser, S. Menzel, A theoretical and experimental view on the temperature dependence of the electronic conduction through a schottky barrier in a resistively switching SrTiO$_3$-based memory cell, Adv. Electron. Mater. 4 (2018) 1800062.

[82] M.J. Rozenberg, M.J. Sanchez, R. Weht, C. Acha, F. Gomez-Marlasca, P. Levy, Mechanism for bipolar resistive switching in transition-metal oxides, Phys. Rev. B 81 (2010) 115101.

[83] S. Tang, F. Tesler, F.G. Marlasca, P. Levy, V. Dobrosavljevic, M. Rozenberg, Shock waves and commutation speed of memristors, Phys. Rev. X 6 (2016) 11028/1–16, https://doi.org/10.1103/PhysRevX.6.011028.

[84] S. Larentis, F. Nardi, S. Balatti, D.C. Gilmer, D. Ielmini, Resistive switching by voltage-driven ion migration in Bipolar RRAM-part II: modeling, IEEE Trans. Electron Devices 59 (2012) 2468–2475, https://doi.org/10.1109/TED.2012.2202320.

[85] Y. Aoki, C. Wiemann, V. Feyer, H.-S. Kim, C.M. Schneider, H. Ill-Yoo, M. Martin, Bulk mixed ion electron conduction in amorphous gallium oxide causes memristive behaviour, Nat. Commun. 5 (2014) 3473/1–9, https://doi.org/10.1038/ncomms4473.

[86] L. Goux, Y. Chen, L. Pantisano, X. Wang, G. Groeseneken, M. Jurczak, D.J. Wouters, On the gradual unipolar and bipolar resistive switching of TiN\HfO$_2$\Pt memory systems, Electrochem. Solid State Lett. 13 (2010) G54–G56, https://doi.org/10.1149/1.3373529.

[87] J.H. Oh, K.C. Ryoo, S. Jung, Y. Park, B.G. Park, Effect of oxidation amount on gradual switching behavior in reset transition of Al/TiO2-based resistive switching memory and its mechanism for multilevel cell operation, Jpn. J. Appl. Phys. 51 (2012) 4DD16/1–5, https://doi.org/10.1143/JJAP.51.04DD16.

[88] F. Nardi, S. Larentis, S. Balatti, D. Gilmer, D. Ielmini, Resistive switching by voltage-driven ion migration in bipolar RRAMPart I: Experimental study, IEEE Trans. Electron Devices 59 (2012) 2461–2467.

[89]   S. Yu, Y. Wu, R. Jeyasingh, D. Kuzum, H.P. Wong, An electronic synapse device based on metal oxide resistive switching memory for neuromorphic computation, IEEE Trans. Electron Devices 58 (2011) 2729–2737, https://doi.org/10.1109/TED.2011.2147791.

[90]   J.H. Hur, K.M. Kim, M. Chang, S.R. Lee, D. Lee, C.B. Lee, M.J. Lee, Y.B. Kim, C.J. Kim, U.I. Chung, Modeling for multilevel switching in oxide-based bipolar resistive memory, Nanotechnology 23 (2012) 225702/1–5, https://doi.org/10.1088/0957-4484/23/22/225702.

[91]   L. Zhao, H. Chen, S. Wu, Z. Jiang, S. Yu, T. Hou, H.P. Wong, Y. Nishi, Multi-level control of conductive nano-filament evolution in HfO2 ReRAM by pulse-train operations, Nanoscale 6 (2014) 5698–5702, https://doi.org/10.1039/C4NR00500G.

[92]   S. Yu, Y. Wu, H. Wong, Investigating the switching dynamics and multilevel capability of bipolar metal oxide resistive switching memory, Appl. Phys. Lett. 98 (2011) 103514/1–3, https://doi.org/10.1063/1.3564883.

[93]   S. Menzel, M. Waters, A. Marchewka, U. Böttger, R. Dittmann, R. Waser, Origin of the ultra-nonlinear switching Kinetics in oxide-based resistive switches, Adv. Funct. Mater. 21 (2011) 4487–4492, https://doi.org/10.1002/adfm.201101117.

[94]   K. Fleck, N. Aslam, S. Hoffmann-Eifert, V. Longo, F. Roozeboom, W.M.M. Kessels, U. Böttger, R. Waser, S. Menzel, The influence of non-stoichiometry on the switching kinetics of strontium-titanate ReRAM devices, J. Appl. Phys. 120 (2016) 244502, https://doi.org/10.1063/1.4972833.

[95]   K. Fleck, U. Böttger, R. Waser, S. Menzel, Interrelation of sweep and pulse analysis of the SET process in SrTiO3 resistive switching memories, IEEE Electron Device Lett. 35 (2014) 924–926, https://doi.org/10.1109/LED.2014.2340016.

[96]   Y. Nishi, S. Menzel, K. Fleck, U. Boettger, R. Waser, Origin of the SET kinetics of the resistive switching in tantalum oxide thin films, IEEE Electron Device Lett. 35 (2013) 259–261, https://doi.org/10.1109/LED.2013.2294868.

[97]   M. Witzleben, K. Fleck, C. Funck, B. Baumkötter, M. Zuric, A. Idt, T. Breuer, R. Waser, U. Böttger, S. Menzel, Investigation of the impact of high temperatures on the switching kinetics of redox-based resistive switching cells using a high-speed nanoheater, Adv. Electron. Mater. 3 (2017) 1700294, https://doi.org/10.1002/aelm.201700294.

[98]   S. Brivio, E. Covi, A. Prodromakis, T. Serb, M. Fanciulli, S. Spiga, 'Gradual Set Dynamics in HfO2-Based Memristor Driven by Sub-Threshold Voltage Pulses, 2015.

[99]   Y. Yang, P. Sheridan, W. Lu, Complementary resistive switching in tantalum oxide-based resistive memory devices, Appl. Phys. Lett. 100 (2012) 203112/1–4.

[100] S.A. Mojarad, J.P. Goss, K.S.K. Kwa, P.K. Petrov, B. Zou, N. Alford, A. O'Neill, Anomalous resistive switching phenomenon, J. Appl. Phys. 112 (2012) 124516/1–7, https://doi.org/10.1063/1.4770489.

[101] A. Schoenhals, D.J. Wouters, A. Marchewka, T. Breuer, K. Skaja, V. Rana, S. Menzel, R. Waser, Critical ReRAM stack parameters controlling complementary versus bipolar resistive switching. in: Memory Workshop (IMW), 2015 IEEE International, 2015, pp. 73–76, https://doi.org/10.1109/IMW.2015.7150281.

[102] S.Z. Rahaman, Y.-D. Lin, H.-Y. Lee, Y.-S. Chen, P.-S. Chen, W.-S. Chen, C.-H. Hsu, K.-H. Tsai, M.-J. Tsai, P.-H. Wang, The role of Ti buffer layer thickness on the resistive switching properties of hafnium oxide-based resistive switching memories, Langmuir 33 (2017) 4654–4665, https://doi.org/10.1021/acs.langmuir.7b00479.

[103] S. Brivio, J. Frascaroli, S. Spiga, Role of metal-oxide interfaces in the multiple resistance switching regimes of Pt/HfO2/TiN devices, Appl. Phys. Lett. 107 (2015) 23504/1–5, https://doi.org/10.1063/1.4926340.

[104]   Z. Li, T. Schram, T. Witters, J. Tseng, S. De Gendt, K. De Meyer, Oxygen incorporation
        in TiN for metal gate work function tuning with a replacement gate integration approach,
        Microelectron. Eng. 87 (2010) 1805–1807, https://doi.org/10.1016/j.mee.2009.10.023.

Further reading

[105]   R. Waser, R. Bruchhaus, S. Menzel, Redox-Based Resistive Random Access Memories,
        Wiley, 2012.

Interface effects on memristive devices

Susanne Hoffmann-Eifert, Regina Dittmann
Peter Grünberg Institute (PGI-7), Forschungszentrum Jülich GmbH and JARA-FIT, Jülich, Germany

5.1 Introduction

Redox-based resistive switching memory (ReRAM) devices are considered as promising candidates for the next generation "beyond von Neumann" computing architectures, covering applications like computation in memory (CIM) devices, and machine learning (ML) employing artificial neural networks (ANN) where ReRAM fulfills the role of an artificial synapse [1, 2]. Challenges showing up with the implementation of the various approaches comprise the necessity for extreme scaling of the devices with the size of an individual memristive cell becoming smaller than $1000\,nm^3$ [3] and the requirement for high density, even three dimensional (3D), integration [4, 5].

The basic structure of a "standard" oxide-based ReRAM device comprises a transition metal oxide (MO) thin film sandwiched between an ohmic-type chemically active electrode, named oxygen exchange layer (OEL), and a Schottky-type chemically inert electrode [6]. The resistive switching behavior involves electrically induced nanoionic redox processes. Often accompanied by a change of the valence state of the metal ion these processes result in a change of the electronic conduction behavior in the ReRAM cell. So far, most of the resistive switching phenomena and models dealt with in the past, although considering a switching mechanism at the atomic scale, often introduce semiinfinite boundary conditions to reduce the complexity of the system for easier handling [7]. Taking filamentary-type oxide-based ReRAM cells, as an example: for standard laboratory [8] and industrial utilized [9] devices with a size in the range of 10^4–$10^6\,nm^2$ the approach of a conductive filament embedded in a cylindrical matrix that can be described in a semiinfinite manner [7] is a reasonable approximation. However, various studies [10–12] provide experimental proof on significant influences of interfacial reactions on the total switching performance. With further shrinkage of the device dimensions, ion exchange reactions across various interfaces will have to be considered in design and modeling (see Chapter 5) of the ReRAM cells in order to achieve optimized switching performance.

5.2 General considerations of ReRAM devices

5.2.1 Redox reactions

Following the established model of Waser et al. [6], resistive switching in transition MO-based ReRAM devices can be attributed to redox processes that occur at the

Advances in Non-volatile Memory and Storage Technology. https://doi.org/10.1016/B978-0-08-102584-0.00006-1

oxide-electrode interface accompanied by migration of mobile ions under applied electric field and elevated temperatures induced by Joule heating. This understanding builds the fundament of the switching model, which is described in detail in Chapter 5. Transition MOs may conduct oxygen anions and metal cations via vacancy or interstitial sites. The dominant defect species, the ion conduction mechanism and the ion mobility rely on the specific crystal structure and on the size and charge of the ionic species. Following the Kröger-Vink notation [13] defect species can be expressed by the type, the occupied lattice site and by their electronic charge with respect to this lattice site. For a divalent MO, characteristic species are (1) the host lattice sites, M_M^x and O_O^x, with "x" indicating charge neutrality with respect to the lattice cite, (2) metal dopants at lattice sites or interstitial sites ("i"), and (3) vacancies ("V"), acting as negatively charged acceptors or positive charged donors, indicated by "'" and "•", respectively. Typical defects to be considered in MOs are oxygen vacancies, oxygen interstitials, metal vacancies, and metal interstitials represented by $V_O^{••}$, $O_i^{''}$, $V_M^{''}$, and $M_i^{••}$, respectively. The defect formation and annihilation by means of exchange reactions can be described by the solid-state defect chemistry, which is based on equilibrium thermodynamics (see Ref. [14–16]). By this, the reduction of a MO accompanied with the formation of oxygen vacancies in the lattice and free oxygen in the gas phase, can be written as

$$O_O^x + 2\ M_M^x \rightleftarrows \frac{1}{2}O_2 + V_O^{••} + 2\ M'_M. \tag{5.1}$$

The oxygen evolution reaction leaves behind electrons ("e'") in the lattice, which remain in previously unoccupied states of the lowest energy level. In band insulator MOs, these are normally the empty states of the metal ion, which form the conduction band. Therefore, oxygen-vacancy formation due to reduction typically results in a valence change of the metal ions in vicinity of oxygen vacancies, as formulated in Eq. (5.1).

Defect chemistry provides a powerful tool for the understanding of defects and related redox reactions in solid-state compounds. However, special attention should be paid when applying the formalism to material of nano-size dimensions [17, 18], nanocrystalline and heterogeneous material [19], crystalline material with extended defects like, for example, dislocations and grain boundaries [20, 21], or nonstoichiometric material [22].

Resistive switching in transition MO-based ReRAM devices occurs by a valence change mechanism (VCM). Here the redox reactions, denoting a coupled reduction and oxidation reaction, are linked to the solid-state transport of ions at nanometer length scales. Due to a considerable ion mobility in the switching state, the memristive oxide can be described as a mixed ionic electronic conductor (MIEC). Fig. 5.1 provides an overview of different reactions that are conceivable in VCM-type ReRAM cells, with the MO switching layer sandwiched between the OEL named M* and the Pt Schottky electrode. The electrodes are considered to conduct electronic currents, whereas the MO layer may carry electronic and ionic currents. In addition to the redistribution of ionic charge carriers, formation and annihilation of the mobile defects can occur, typically at the electrode-oxide interfaces. The specific interface reaction

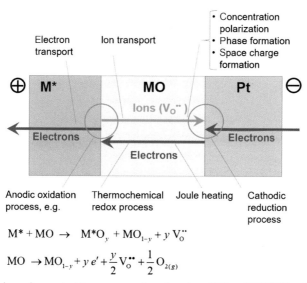

Fig. 5.1 Overview of conceivable processes (i.e., forming, SET and RESET), which may be relevant in VCM-type resistive switching. Here, Pt represents the noble metal, high work function electrode, and M* stands for the oxidizable ohmic electrode, while MO represents the mixed ionic-electronic conducting metal oxide.
Modified from R. Waser, R. Bruchhaus, S. Menzel, Redox-Based Resistive Random Access Memories. Wiley. © 2012 Wiley VCH, 2012. Credit: R. Waser, R. Bruchhaus, S. Menzel, Redox-Based Resistive switching Memories, in: R. Waser (Ed.) Nanoelectronics in Information Technology, 3rd ed., Wiley-VCH, 2012, pp. 683–710 (Chapter 30), Fig. 7.

is determined by the materials' combination. Several examples will be discussed in this chapter.

5.2.2   Area dependence of VCM-type resistive switching events

Considering the lateral dimensions of the ReRAM device, the redox processes can occur either in a spatially confined manner represented by one or multiple switching filaments, or in a spatially homogeneous, area-dependent manner. Fig. 5.2 gives a schematic representation of different scenarios.

From Fig. 5.2 important features of interface and bulk-controlled VCM-type memristive behavior become immediately apparent if focusing on the resistive switching area. (1) The concentration of mobile donor-type defects in the switching area is high. (2) The voltage- or current-driven change in the resistance is due to a redistribution of mobile defects close to the noble metal electrode while changing the barrier for electronic carrier injection at this Schottky contact. (3) Ionic charge carriers can be created or annihilated at the OEL-MO interface. (4) Thin film structures with an overall low concentration of mobile donors (Fig. 5.2A and B) tend to filamentary-type switching. This can show up either by an initially existing regime of higher defect concentration like, for example, a dislocation or a grain boundary, or by a filament

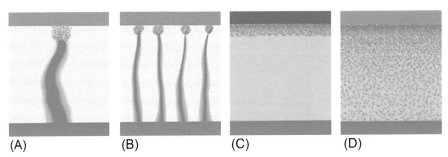

(A)   (B)   (C)   (D)

Fig. 5.2 Schematic representation of possible areas allocated to the occurrence of the nanoionic redox processes controlling the resistance state of the ReRAM device. Sandwiched between the OEL *(green)* and inert metal *(gray)* the MO is shown in yellow or green depending if the mobile donor concentration (mostly oxygen vacancies) is low or high, respectively. (A) Filamentary switching, (B) multifilamentary switching, (C) area-dependent interface-type switching, and (D) area-dependent bulk switching.
Reprinted with permission from C. Baeumer, R. Dittmann, Redox-Based Memristive Metal-Oxide Devices, Elsevier, Amsterdam, 2018, pp. 489–522, https://doi.org/10.1016/B978-0-12-811166-6.00020-0. © 2018 Elsevier Inc., Amsterdam. Credit: C. Bäumer, R. Dittmann, Redox-based memristive metal-oxide devices, in: N. Pryds, V. Esposito (Eds.) Metal Oxide-Based Thin Film Structures, Elsevier, Amsterdam, 2018, pp. 489–522 (Chapter 20), Fig. 20.2.

created by an electroforming step. This step typically requires a higher voltage to the cell compared to the SET voltage used during the resistive switching. (5) In contrast to filamentary-type devices, area-dependent interface-type switching devices reveal a moderately high, almost homogeneous distribution of mobile defects within the device. (6) In interface-related switching events typically, but not solely, one or both electrode metal-MO interface(s) are involved in the switching process whereas bulk switching (Fig. 5.2D) addresses changes of the dopant distribution inside the oxide bulk. Examples are GaO_x [23] and $(La,Sr)MnO_3$ [24].

This chapter focusses on interface effects on the resistive switching performance addressing structures comparable to the ones shown in Fig. 5.2A and C. The discussion will follow the switching model and nomenclature defined by Waser et al. [25], which is also in detail described in Chapter 5. The SET and RESET events define the switching from the high resistance state (HRS) to the low resistance state (LRS) and vice versa, respectively. In the "standard" VCM-type bipolar resistive switching, so-called "counter eightwise" (c8w) switching, the SET and RESET event are obtained applying a negative and positive voltage to the Schottky-type electrode, respectively.

5.2.3 Design of ReRAM devices

The next step in information processing will cover all types of cognitive applications requiring new computing architectures to cope with huge amount of data. The new concepts of neuromorphic computing such as ANN and CIM aim to overcome the limitations of classical von Neumann-based computers [2, 26]. ReRAM are intensively investigated due to their nonvolatility and energy efficiency, process compatibility with standard complementary metal-oxide-semiconductor (CMOS) technology, and the ability for device scaling and 3D integration [4, 27].

Technologically most relevant are therefore binary MOs already established in CMOS process flows by demands from various applications like, for example, high-k gate dielectric [28], ferroelectric field effect transistors [29], and tunneling diodes [30]. Thus CMOS compatibility makes materials like hafnium oxide [31–33], zirconium oxide [34], titanium oxide [3], tantalum oxide [35, 36], and aluminum oxide [37] most interesting for ReRAM devices. However, complex perovskite structures are studied as well for various reasons. $SrTiO_3$ is an excellent model material which can be obtained in single-crystalline form and which reveals n-type semiconducting behavior with donor doping [12]. $(Pr,Ca)MnO_3$ is an oxygen-ion conductor with p-type electronic conductivity, which enables area-dependent interfacial resistive switching [38].

Generally, the above-listed binary MOs with band gaps between about 3 eV for TiO_2 and 9 eV for Al_2O_3 are insulating in the stoichiometric case and doping is necessary to permit electrical conductance. In transition MOs, oxygen vacancies form intrinsic mobile donors, which can be generated intentionally by reduction [19, 22, 39]. Sometimes metal interstitials might show up as the dominating defect species [40]. However, this scenario is limited to an extreme nonequilibrium state [41]. In this chapter, we will therefore focus on oxygen vacancies as the dominating mobile defect in the MO layer.

Regarding the electrode configuration, a differentiation can be made between symmetric and asymmetric electrode configurations. Devices with symmetric electrodes can show resistive switching behavior if a gradient of oxygen deficiency built up creates a virtual electrode inside the switching oxide [42, 43]. Often such types of structures exhibit a limited reservoir of donor-type defects, rearranged during the switching process. This leads to complimentary switching (CS) behavior where both electrode/ MO interfaces almost equally act as the switching interface [44–47].

In this chapter, we focus on interface phenomena showing up in standard VCM-type ReRAM devices affecting (1) the pristine device state and the electroforming, and (2) the c8w switching behavior. Here, effects of ion transfer across interfaces often limit the reliability of the devices especially considering variability of the resistance states and retention issues. Furthermore, we describe (3) interface phenomena that can induce a reversed switching polarity resulting in the so-called 8w switching mode.

5.3 Pristine state of the ReRAM device and electroforming

The combination of the electrodes and the resistive switching MO layer defines the pristine state of the switching cell either electrical insulating or conducting, depending on the defect concentration in the MO layer. Electroforming processes locally modify the stack structure in order to enable stable and reversible resistive switching. Generally, both modifications come with ion transfer across interfaces. Highly insulating devices from MO layers with low oxygen-vacancy concentration mostly require an electro-forming step that equals a soft breakdown event. Local reduction of the MO might form one or more conductive filaments enabling stable resistive switching (see Fig. 5.2A and B). In contrast, electroforming of a stack containing a conducting oxygen-deficient MO_x layer typically comes with the formation of an insulating layer. Such devices might reveal homogeneous (see Fig. 5.2C) or multifilamentary like area-type switching behavior. In the following, the role of the two inherent interfaces of VCM-type ReRAM cells is described, in particular the interface of the MO to the noble metal and to the OEL. Here, also the films' microstructures, either amorphous or polycrystalline will be addressed, knowing that grain boundaries can be preferred leakage current paths.

5.3.1 Processes at the noble metal/oxide interface during forming

Standard VCM-type ReRAM cells built from MO layers deposited in their insulating stoichiometric phase require an electroforming step prior to a stable resistive switch-ing performance. Either being current or voltage driven the electroforming event can be understood as a soft breakdown event, which has to be controlled by means of an external circuitry. Sometimes a self-limitation by an internal resistive layer can also be observed. For most of the MOs under discussion, electroforming can be considered as an electronic current-driven thermal runaway process leading to the formation of ad-ditional oxygen vacancies in regimes of increased temperature and high electric field. Microscopically, the enthalpy for oxygen-vacancy formation can be considerable re-duced either under the influence of an electric field [48] or for the reason of an already existing local oxygen deficiency [39]. Oxygen deficient MO thin films are preferably grown via sputter deposition [49] or molecular beam epitaxy [50]. In contrast, atomic layer deposition (ALD) might be chosen for ReRAM fabrication at the technological level due to its ability of conformal coverage onto 3D structures especially for future massive integration of ReRAM cells. Thanks to their typically amorphous structure and highly insulating properties, ALD grown MO layers are favorable for use as high-k gate dielectrics and as storage capacitor in dynamic random access memories [51]. However, for the use in ReRAM devices this turns out disadvantageous while requiring high electroforming voltages. One method of compensation is by selecting a low work function electrode with a high activity for oxygen getter like, for example, Ti [52] and Hf [53]. Another possibility is crystallization anneal under reducing atmosphere, which results in the formation of grain boundaries known as sinks for defects [54, 55].

As discussed in Chapter 5 by Menzel and Waser the electroforming and resistive switching mechanism in transition MO VCM-type devices is most likely due to oxy-gen exchange reactions at the electrode-MO interfaces and oxygen-vacancy migration in the MO. During electroforming oxygen vacancies are generated in the MO thin film due to an oxygen exchange reaction occurring at the anode.

For the case that the inert electrode like, for example, Pt and TiN is used as the anode during electroforming, gaseous oxygen will form at the MO-electrode interface according to the oxygen evolution reaction:

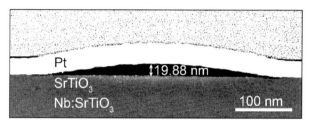

Fig. 5.3 Transmission electron micrograph of the cross section of a Nb:SrTiO$_3$/SrTiO$_3$/Pt ReRAM device taken after electroforming and resistive switching. The local delamination of the Pt electrode has been caused by oxygen gas release during the electroforming and switching process.

Reprinted with permission from Thomas Heisig, Christoph Baeumer, Ute N. Gries, Michael P. Mueller, Camilla La Torre, Michael Luebben, Nicolas Raab, Hongchu Du, Stephan Menzel, David N. Mueller, Chun-Lin Jia, Joachim Mayer, Rainer Waser, Ilia Valov, Roger A. De Souza, Regina Dittmann, Oxygen exchange processes between oxide memristive devices and water molecules, Adv. Mater. 30 (2018) 1800957. © 2018 Wiley VCH. Credit: T. Heisig, C. Baeumer, U.N. Gries, M.P. Mueller, C. La Torre, Mi. Luebben, N. Raab, H. Du, S. Menzel, D.N. Mueller, C.L. Jia, J. Mayer, R. Waser, I. Valov, R.A. De Souza, R. Dittmann, Oxygen exchange processes between oxide memristive devices and water molecules, Adv. Mater. 30(29) (2018) 18009570, 2018, Fig. 4e.

$$O_O^x \rightarrow \frac{1}{2} O_{2(g)} + V_O^{\bullet\bullet} + 2e'. \tag{5.2}$$

Molecular oxygen might diffuse through the electrode into the gas phase [56], or might be stored in a cavity at the interface. Formation of bubbles by the electroforming process is reported in several publications where relatively thick and insulating films undergoing an electroforming process with positive voltage applied to the chemically inert electrode have been studied [57, 58]. Fig. 5.3 shows a very intriguing transmission electron microscopy image of this type of cavity at the SrTiO$_3$/Pt interface after electroforming and switching of Nb:SrTiO$_3$/SrTiO$_3$/Pt devices.

In contrast, no such "bubbles" have been reported for technologically relevant nanostructured devices (see, e.g., Ref. [53]). A possible reason might be the smaller amount of oxygen that has to be removed in order to reduce the MO layer of a nanoscale device. This small amount of oxygen might diffuse through the electrode layer or could be stored either at the interface or at grain boundaries of the polycrystalline electrode material [59].

5.3.2 Processes at the OEL/oxide interface and their impact on the forming process

Bringing a MO in contact with a chemical reactive metal will immediately induce ion transfer across the interface. This is due to the differences in the free energy of MO formation of the two systems. In thermodynamic equilibrium, any mixture of a reactive metal and a MO will strive toward the MO for the equilibrium state. However, due to kinetic limitations the time for equilibration is strongly defined by the size of the ensemble, the temperature, the oxygen partial pressure, and additional counter-charges (see Refs. [60, 61]). By performing x-ray absorption spectroscopy (XAS) measurements, structural changes of the MO layer formed at the interface of the reactive metal toward the MO were identified by Cho et al. [62].

In this case, oxygen ions extracted from the MO can react with the chemically active material M* by formation of suboxides following the oxygen transfer reaction

$$M^{*}_{(bulk)} + MO \rightarrow M^{*}O_{y} + MO_{1-y} + y\ V_{O}^{\bullet\bullet} + 2ye'.$$ (5.3)

Solid-state reactions of MOs require ion migration that is limited by energy barriers of typically 1 eV. Large-scale classical molecular dynamics simulations run at 1000 K to calculate ion migration in amorphous and crystalline HfO_{2-x} provide activation enthalpies of migration between about 0.5 eV and close to 0.85 eV, while a subdiffusive behavior is predicted attributed to a nanoscale confinement of the migrating ions [63]. However, at dimensions of a ReRAM cell of about $100 \times 100 \times 10$ nm^3, the distances, oxygen vacancies have to migrate in order to induce a measurable effect on the current response, become close to a few nanometer or even a few unit cells. Therefore, possible ion exchange reactions across the interfaces of the ReRAM device should be considered even at room temperature.

Reactive metal electrodes deposited on insulating oxides can induce electrical conductivity through a redox-reaction involving oxygen transfer from the oxide to the metal. This phenomenon is exemplarily demonstrated in Fig. 5.4 for sputtered Ti layers deposited on top of epitaxial, insulating Fe-doped $SrTiO_3$ films of 20 nm thickness.

The initial resistance R_0, as read out by a low-voltage signal below the switching threshold, decreases abruptly between a Ti layer thickness of 5 and 10 nm, marking a transition from an insulating system that requires electroforming in order to enable RS properties to a conducting system without the need for electroforming [64].

Using hard x-ray photoelectron spectroscopy (HAXPES), it was possible to identify the underlying mechanism of the increased conductivity, which turned out to be a redox-reaction between electrode and insulator: $Ti + SrTiO_3 \leftrightarrow TiO_x + SrTiO_{3-x}$. Fig. 5.4 shows the photoemission spectra obtained using 4.2 keV X-rays for excitation. The escape depth of the photoelectrons in Ti increases to >10 nm when the electron kinetic energy is larger than 3.5 keV, enabling the examination of the buried Ti/SrTiO$_3$ interface. Through least-squares fitting techniques using Lorentzian-Gaussian functions, the shoulder at the low-binding energy side of the Ti 2p emission doublet can be attributed to the existence of multiple Ti oxidation states in the interface region of a 4 nm-thin Ti layer on SrTiO$_3$. Although there is no way to distinguish between Ti ions in the oxide and the electrode, the absence of a metallic Ti^0 is a firm indicator of oxygen removal from the oxide and transfer into the electrode. The deposition of a reactive Ti electrode can therefore be used to induce significant conductivity in the band-gap insulator SrTiO$_3$, presenting a viable way of controlling the redox state of the oxide through simple process engineering. This phenomenon is the basis for many state-of-the-art VCM systems in existence to date [65, 66].

The effect of the chemically active electrode material chosen for a certain MO-based ReRAM device is twofold. First, the ability of oxygen getter of the deposited

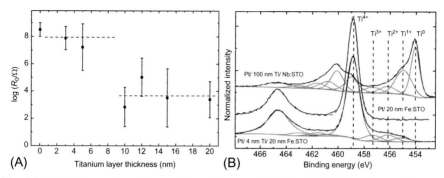

Fig. 5.4 (A) Thickness dependence of the resistance of Ti/ 20 nm Fe:STO interfaces with varying Ti thickness. The resistance drops by several orders of magnitude when 5 nm Ti are exceeded (*red lines* are a guide to the eye). (B) Hard X-ray photoemission spectroscopy (HAXPES) data of Ti layers of various thicknesses on SrTiO₃. All layers capped by a Pt layer to prevent surface oxidation.

Reproduced with permission from S. Stille, C. Lenser, R. Dittmann, A. Koehl, I. Krug, R. Muenstermann, J. Perlich, C.M. Schneider, U. Klemradt, R. Waser, Detection of filament formation in forming-free resistive switching SrTiO3 devices with Ti top electrodes, Appl. Phys. Lett. 100 (2012) 223503/1-4, https://doi.org/10.1063/1.4724108. Copyright 2012, AIP Publishing LLC. Credit: S. Stille, C. Lenser, R. Dittmann, A. Koehl, I. Krug, R. Muenstermann, J. Perlich, C.M. Schneider, U. Klemradt, R. Waser, Detection of filament formation in forming-free resistive switching SrTiO₃ devices with Ti top electrodes, Appl. Phys. Lett. 100(25) (2012) 223503, Figs. 1B and 2.

metal has to be considered, and with this, second, the additional oxygen transfer during the electrical-driven electroforming event. As shown, oxygen-ion transfer processes from the MO layer toward the chemically active MO already occur at the initial stage of the layer deposition. The driving force is defined by the differences in the standard free enthalpy of oxide formation for the involved metals. Experimental studies utilizing spectroscopic techniques combined with ab initio simulations provide a clear proof that the oxygen exchange reaction due to the chemical gradient creates oxidized (M^*O_y) and reduced (MO_{1-y}) interfacial regions of about 2–3 nm in the as-built device. By means of first-principle methods, Guo and Robertson [67] provided a guide for materials selection in oxide-based ReRAM. One of the results is shown in Fig. 5.5.

Analyzing the values of necessary electroforming voltages as a function of the thickness of the OEL reveals an indirect proof for the reduction of the MO layer by the oxygen getter effect. As an important result of the tailored oxygen deficiency in the resistive switching MO layer, the electroforming voltage can be strongly reduced by the use of OELs.

This is shown, for example, in the study of Chen et al. [68] for TiN/HfO₂/Hf/TiN stacks of a 5 nm-thick HfO₂ film grown by ALD and patterned into nano-crossbar structures of 35 nm×65 nm. Increasing the thickness of the Hf metal layer from 2 to 10nm enabled a reduction of the electroforming voltage from about 4.6 to 2.6 V (see Fig. 5.6).

Comparable results were published by Schoenhals et al. [46] for a device based on tantalum oxide. The authors studied the electroforming voltage of Pt/Ta₂O₅/Ta/Pt

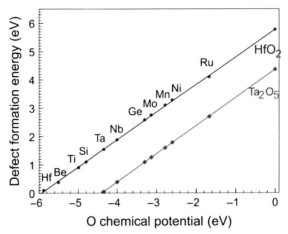

Fig. 5.5 Variation of O vacancy formation energy with local oxygen chemical potential for HfO_2 and Ta_2O_5 as examples. The scavenging metal (shown) can be used to set the oxygen chemical potential and thus control the oxygen-vacancy formation energy and oxygen vacancy concentration independently of the host oxide.

Reprinted with permission from Y. Guo, J. Robertson, Materials selection for oxide-based resistive random access memories, Appl. Phys. Lett. 105 (2014) 223516/1-5, https://doi. org/10.1063/1.4903470 © 2014 American Institute of Physics (AIP) Publishers.Credit: Y. Guo, J. Robertson, Materials selection for oxide-based resistive random access memories, Appl. Phys. Lett. 105(22) (2014) 223516, Fig. 5.

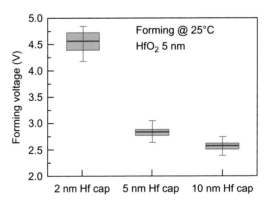

Fig. 5.6 Forming voltage distribution from the 25 cells statistics of TiN/5 nm ALD HfO_2/Hf/ TiN with Hf layer thickness of 2, 5, and 10 nm.

Reprinted with permission from Y.Y. Chen, G. Pourtois, S. Clima, L. Goux, B. Govoreanu, A. Fantini, R. Degraeve, G.S. Kar, G. Groeseneken, D.J. Wouters, M. Jurczak, 'Hf cap thickness dependence in bipolar-switching TiN\$HfO_2$\Hf\TiN RRAM device'. Nonvolatile Mem. 50 (2013) 3–9, https://doi.org/10.1149/05034.0003ecst © 2013 The Electrochemical Society. Credit: Y.Y. Chen, G. Pourtois, S. Clima, L. Goux, B. Govoreanu, A. Fantini, R. Degraeve, G.S. Kar, G. Groeseneken, D.J. Wouters, M. Jurczak, Hf cap thickness dependence in bipolar-switching TiN\$HfO_2$\Hf\TiN RRAM device, 2013 Nonvolatile Memories, ECS Transactions, 50(34) (2013) 3–9; Fig. 5A.

ReRAM devices built from a 5 nm-thick Ta_2O_5 with increasing thickness of the Ta OEL. Starting at an electroforming voltage of about 5.8 V for the symmetric $Pt/Ta_2O_5/Pt$ cell the electroforming voltage has been reduced to about 2.2 V for a Ta layer thicker than 3 nm. The transition from a thickness dependent to a thickness independent behavior shows the limitation of the oxygen exchange regime, which has to be considered for the electroforming step.

A correlation between the "Ta/O" substoichiometry in the MO layer and the electroforming voltage has recently been shown for thin films grown by sputtering and MBE [49] [69], respectively. Further analysis of interface reactions have been reported for many other VCM oxide/metal systems like, for example, Ta_2O_5/Ta [40], ZrO_2/Ti [70], and TiO_2/Al [71].

Interfacial ion transfer, especially the extraction of oxygen ions form the resistive switching layer during the process of electroforming is of fundamental importance for a stable operation of the device. Electroforming at voltages >5 V might easily lead to a hard breakdown. In addition, for technological applications, the electroforming voltage should reach the regime of the SET voltage necessary for resistive switching. Understanding the exchange reactions at the MO/M* interface during electroforming is of utmost importance for the design and control of devices which enable electroforming in the same voltage range as will be used for the resistive switching SET and RESET process.

5.3.3   Impact of the MO microstructure on the forming process

Mentioning an additional technologically interesting materials' combination $Ti/HfO_2/TiN$, the oxygen scavenging ability of Ti on HfO_2 has been proven by HAXPES analysis [72] and in addition, by oxygen tracer analysis using time of flight secondary ion mass spectroscopy [73]. Importantly, a significant influence of the HfO_2 thin film morphology on the Ti/HfO_2 interface reactivity could be revealed as shown in Fig. 5.7.

The electroforming process typically applied in standard VCM-type memristive devices is pursued utilizing the chemically active metal as the anode [72, 74]. The electroforming process that is driven by an electric field or current is, according to the established model described in Chapter 5, based on oxygen-vacancy formation due to ion transfer reaction and oxygen-vacancy migration toward the cathode electrode. This process rather results in the formation of oxygen-vacancy-rich conductive filaments than in a homogeneous reduction of the MO, probably originating from local inhomogeneity, which is enhanced by thermal feedback at higher current densities. Joule heating-induced thermal runaway further pushes the formation of filaments. Therefore, tailoring the formation of a single switching filament or, in contrast, a homogeneous switching device requires the control of sites preferential for enhanced oxygen exchange reactions under the application of an external electric field. Approaches toward this direction comprise the control of the microstructure like, for example, utilizing grain boundaries [75] or dislocations [76, 77] as preferential sites for filament formation. Besides changes of the bulk microstructure, even local modifications of the surface defect structure might promote filament formation by locally enhancing the oxygen exchange reaction [78, 79].

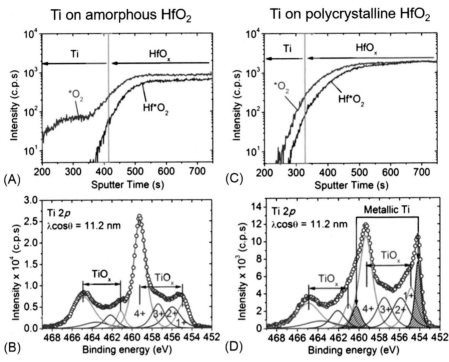

Fig. 5.7 Effect of the HfO₂ thin film morphology, either amorphous (A,B) or polycrystalline monoclinic (C,D), on the oxygen scavenging ability of the Ti electrode. The ToF-SIMS spectra (A,C) show the depth profile of the marked oxygen isotope introduced into the HfO₂ film during molecular beam epitaxy. The photoelectron spectra (HAXPES; B,D) of the Ti 2p electronic core levels are measured by in situ XPS after MBE deposition of Ti on HfO₂. Reprinted with permission from P. Calka, M. Sowinska, T. Bertaud, D. Walczyk, J. Dabrowski, P. Zaumseil, C. Walczyk, A. Gloskovskii, X. Cartoixa, J. Sune, T. Schroeder, Engineering of the chemical reactivity of the Ti/HfO2 Interface for RRAM: experiment and theory, ACS Appl. Mater. Interfaces 6 (2014), 5056–5060, https://doi.org/10.1021/am500137y © 2014 The American Chemical Society. Credit: P. Calka, M. Sowinska, T. Bertaud, D. Walczyk, J. Dabrowski, P. Zaumseil, C. Walczyk, A. Gloskovskii, X. Cartoixa, J. Sune, T. Schroeder, Engineering of the Chemical Reactivity of the Ti/HfO₂ Interface for RRAM: Experiment and Theory, ACS Appl. Mater. Interfaces, 6(7) (2014) 5056, Figs. 3 and 4.

Considering the role of grain boundaries on the change in mobility of certain species of charge carriers compared to the transport in the bulk phase, in general, it can be stated that the diffusivity of the oxygen ions and the ionic and electronic charge transport of the grain boundary region are significantly modified compared to the bulk material [41]. However, the direction of the change, either enhancement or decrease of the diffusivity, depends on the particular material's system. Typically, the oxygen diffusion coefficient in a distorted structure compared to the crystalline structure is enhanced for materials with a low diffusion coefficient in the bulk [e.g., (La,Sr)MnO₃₋$_\delta$] and reduced for good oxygen-ion conductors like yttria stabilized zirconium oxide

(YSZ) and $SrTiO_3$ [20, 41]. It is important to note that enhanced electronic conductivity in grain boundaries of acceptor-doped band-insulators [80] might result in locally enhanced Joule-heating thereby promoting filament formation in these regions.

5.3.4 Processes correlated with nonfilamentary forming

The interfacial oxidation of the reactive metal electrode in contact with the MO layer might even dominate the resistance properties of the device stack if the interfacial layer M^*O_y formed by oxidation of the metal electrode becomes higher insulating as the intentionally grown MO film. These types of devices might reveal either multifilamentary switching (see Fig. 5.2B) or interface-type switching (see Fig. 5.2C) through reversible oxidation and reduction of the interfacial oxide M^*O_y. Interesting for ReRAM device application, this mechanism of homogeneous interface-type switching enables a scaling of the device resistance states with the area. Model materials which

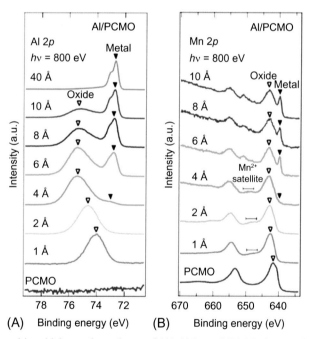

Fig. 5.8 Al-deposition thickness dependence of (A) Al 2p and (B) Mn 2p core-level PES spectra taken after the subsequent deposition of thin Al layers on a $(Pr,Ca)MnO_3$ film under vacuum conditions.

Reprinted with permission from R. Yasuhara, T. Yamamoto, I. Ohkuba, H. Kumigashira, M. Oshima, Interfacial chemical states of resistance switching metal/$Pr_{0.7}Ca_{0.3}MnO_3$ interfaces, Appl. Phys. Lett. 97 (2010), 132111, https://doi.org/10.1063/1.3496033 © 2010 American Institute of Physics (AIP) Publishing LLC. Credit: (A) R. Yasuhara, T. Yamamoto, I. Ohkuba, H. Kumigashira, M. Oshima, Interfacial chemical states of resistance switching metal/$Pr_{0.7}Ca_{0.3}MnO_3$ interfaces, Appl. Phys. Lett. 97 (2010) 132,111.

are most studied in the context of interface-type resistance switching are mixed valence manganites such as (Pr,Ca)MnO$_3$ in contact with for example Al and Y$_2$O$_3$:ZrO$_2$ [38, 81–83]. In-situ photoemission studies of the deposition of Al onto Pr$_{0.7}$Ca$_{0.3}$MnO$_3$ revealed the instantaneous oxidation of the growing Al layer [84]. Fig. 5.8A and B shows the Al thickness dependence of Al 2p and Mn 2p core level PES spectra for Al/PCMO structures, respectively. At the initial stage of Al deposition, a broad peak derived from Al oxide is observed. The peak position shifts toward higher binding energy and stays at 75.5 eV for an deposited Al film of about 0.4 nm. The chemical shift between the Al metal and the Al oxide of approximately 2.7 eV indicates the chemical state of the interfacial oxide to fully oxidized alumina. The complementary redox reaction was detected by means of the Mn 2p signal (Fig. 5.8B).

This interfacial oxygen exchange reaction is driven by the negative values of the Gibbs free energy change ΔG of the reaction: $Al + Pr_{0.7}Ca_{0.3}MnO_3 \rightarrow AlO_x + Pr_{0.7}Ca_{0.3}MnO_{3-x}$. The process is stopped due to the limited diffusion length of the oxygen ions in the (Pr,Ca)MnO$_3$ and in the grown Al$_2$O$_3$ at low temperatures.

Typically, area-switching devices are conductive in virgin state and an electroforming step is required to increase the resistance state of the cell by additional significant increase of the interfacial layer oxide thickness. The subsequent switching is by means of a thickness change of the interfacial tunnel barrier induced by redox-reactions under the applied electrical field.

5.4   Resistive switching in ReRAM devices

Standard VCM-type bipolar resistive switching cells composed of a thin MO layer sandwiched between a chemically inert, high work function electrode that forms a Schottky contact with the MO layer, and a chemically reactive, low work function metal OEL. This solely diffusion-driven oxygen-ion transfer, which took place during device fabrication, happens at the whole contact area, confirmed by spectroscopic analysis (see Section 5.4). Additional reduction of the MO by means of electroforming often leads to the formation of (one or more) conductive filaments originating from local inhomogeneity in the oxide or in the metal electrode causing inhomogeneous current flow and consecutive Joule heating. As shown in the previous section, the necessary electroforming voltage for a certain device stack can be considerable reduced by proper combinations of the thickness values of the insulating layer and the reactive metal [46, 53, 85], by the growth of oxygen deficient MOs [49, 86, 87], and by morphology control of the MO [73]. It is important to note that switching for "forming-free" devices still might take place via a local filament in contrast to real "area-dependent" switching (see Section 5.4.3). In the following, we will discuss interface phenomena during resistive switching of standard filamentary-type systems comprising two opposite polarity modes, namely 8w and c8w.

A large variety of experimentally observed phenomena can be nicely explained by assuming a redistribution of oxygen vacancies within the layer stacks. In that case, drift-diffusion models considering oxygen-blocking electrodes [88] can sufficiently describe the microscopic switching mechanism. However, electrode reactions during fabrication as well as during electrical biasing imply that they might play a role during the switching process as well. Indeed, experimental evidences for interfacial transfer reactions during switching affecting the ReRAM operation are frequently reported in the literature and some examples should be described here in more detail. These compile interface reactions understood as limiters for the programmable resistance ratio of a device stack, for the control of failure mechanisms like endurance failure during resistive switching cycling and read failure limiting retention of a programmed device state.

In standard filamentary VCM-type resistive switching of c8w polarity, a redistribution of oxygen vacancies affects the local conductivity as well as it changes interfacial barriers for electron transport. Here, the SET process initiated by a negative voltage applied to the inert Schottky electrode causes a drift of oxygen vacancies toward the insulating disc regime of the filament. The Schottky barrier for electron injection is reduced and the overall cell resistance of the device is decreased. For the RESET process a positive voltage is applied to the Schottky electrode pushing oxygen vacancies toward the chemically active electrode and increasing the insulation properties of the disc region (see Chapter 5 for details).

Although drift-diffusion models considering oxygen-blocking electrodes [88] can sufficiently describe the switching behavior, redox-type interfacial transfer reactions with either electrode become conceivable, especially if the thickness of the switching MO becomes below 10 nm.

5.4.1   Oxygen exchange at the MO/OEL interface during switching

Oxygen-ion transfer across the MO/OEL interface is accepted for being a necessary step in the electroforming of the device, generating oxygen vacancies in the insulating MO predominantly in the filament regime. However, there is no physical reason, which might hinder oxygen-ion interfacial transfer during resistive switching rather than kinetic limitations. Especially for devices that are considered as forming free, additional features in the resistive switching arising from this interfacial ion transfer might have to be considered. The group of Robertson [39] studied different MO/metal interface combinations. In particular, they could show that combinations, where the reactive metal exhibits a larger negative value for the oxidation enthalpy compared to the metal of the oxide, show a rather unstable switching due to the ongoing reduction of the MO during continuous resistive switching. Experimental evidence for this behavior was provided by Kim and coworkers for the system Pt/Ta_2O_5/M* [10]. The influence of the energy barrier of the oxygen transfer reaction between the MO and the reactive metal electrode was studied in nano-crossbar devices of 7 nm Ta_2O_{5-x} films sandwiched between a Pt Schottky electrode and different OE materials (W, Ta, Ti, Hf). The OE materials were chosen for the defect formation energy of oxygen vacancy with respect to the Ta_2O_5 layer. According to the theoretical calculations [39] the energy is negative for the Hf and the Ti, whereas it is positive for the Ta and the W. The difference is not only reflected in the decrease of the electroforming voltage along the series W-Ta-Ti-Hf, but also in the RESET behavior. Devices with the Ta and the W electrode show a stable switching behavior, whereas cells with the Hf and the Ti electrode suffer from early RESET failure. Further, $Pt/Ta_2O_5/W$ ReRAM devices in difference to the $Pt/Ta_2O_5/Ta$ cells show a faster RESET process and a higher resistance ratio under the same bias conditions. The authors present an advanced switching model where the oxygen exchange reaction at the MO/OEL interface plays a vital role in determining the resistance states. In the devices with Ti- and Hf-OEL there is a continuous thermodynamic driving force for oxygen-vacancy formation. The reduction of the kinetic barriers by Joule heating during the switching process can result in enhanced oxygen extraction during the SET event and reduced oxygen re-incorporation in the switching MO during the RESET process. Hence, during cycling the total amount of oxygen vacancies in the switching volume will steadily be increased along with the increase in device conductivity. This degradation process will finally result in a RESET failure. In contrast, for the Ta- and W-OEL stable switching can be obtained because at the Ta_2O_5/OEL interface the oxygen exchange during the SET and the RESET process is better balanced. Fig. 5.9 shows the change in the oxygen-vacancy profiles during the RESET process, characterized by a depletion at the Pt Schottky electrode and a pile up at the OEL interface by a redistribution of oxygen vacancies under the electric field (see Fig. 5.9A). The equilibrium shown in Fig. 5.9B is determined by equal contributions of diffusion of oxygen vacancies counteracting the concentration gradient build-up by the drift. However, the interfacial oxygen exchange at the MO/OEL interface will reduce the high oxygen-vacancy concentration lowering the back diffusion force, and result in a higher resistance in HRS. The authors concluded that an OE material with positively higher energy for oxygen-vacancy formation facilitates an increased resistance ratio as a deeper RESET process becomes feasible. This, however, comes at the expense of the SET speed.

The role of the OEL on the switching stability has been studied by Celano and coworkers [89] applying the technique of scalpel scanning probe method to the two popular systems, Hf/HfO_2 and Ta/Ta_2O_5, directly probing the volume of the oxygen exchange reservoir established between the filament and the OEL. The authors correlate the improved switching endurance and retention of the Ta_2O_5-based ReRAM cells to the wide lateral growth of the thermodynamic barrier protecting the filament at the interface to the OEL. They claim that stable switching by oxygen-ion exchange between the filament and its oxygen reservoir requires a minimum thickness of OEL/ oxide interface of ca. 3 nm.

A detailed study of oxygen-ion transfer across the MO/OEL interface of HfO_2-based ReRAM devices during switching has been reported by Bertaud et al. [90] Indeed, in operando X-ray photoelectron spectroscopy on $Ti/HfO_x/TiN$ devices gave experimental hints that the Ti electrode is directly involved in the switching process. Fig. 5.10 shows the Ti core levels of $Ti/HfO_x/TiN$ cells in different resistive states.

During electroforming, a significant decrease of the metal contribution as well as an increase in the Ti^{4+} contribution has been observed. In addition, although less pronounced, a small shift of the spectral weight from high valence Ti states (Ti^{3+}) to low valence Ti states (Ti^{1+}) has been observed during SET operation (see Fig. 5.10D).

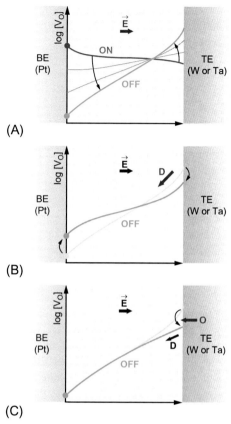

Fig. 5.9 (A) Redistribution of oxygen vacancies under the electric field E during the RESET process. The sketches are based on simulations of oxygen-vacancy concentration profiles during ReRAM switching (B) Diffusion of oxygen vacancies counteracting the concentration gradient by the drift process. (C) Interfacial oxygen exchange between metal oxide and the reactive metal electrode resulting in lower concentration of oxygen vacancy at the Pt interface. Reprinted with permission from W. Kim, S. Menzel, D. J. Wouters, Y. Guo, J. Robertson, B. Rösgen, R. Waser, V. Rana, Impact of oxygen exchange reaction at the ohmic interface in Ta2O5-based ReRAM devices, Nanoscale 8 (2016) 17774–17781, https://doi.org/10.1039/c6nr03810g © The Royal Society of Chemistry 2016. Credit: W. Kim, S. Menzel, D.J. Wouters, Y. Guo, J. Robertson, B. Rösgen, R. Wascr, V. Rana, Impact of oxygen exchange reaction at the ohmic interface in Ta2O5-based ReRAM devices, Nanoscale 8(41) (2016) 17,774, Fig. 8.

5.4.2 Oxygen exchange at the MO/noble metal interface during switching

While it is straightforward to discuss an interfacial oxygen transfer reaction between the reactive metal electrode and the MO, a variety of experiments also hint on a redox process taking place at the opposite interface built by the MO and the noble metal electrode. Since it has been clearly proved that oxygen is released during electroforming, resulting in a formation of bubbles underneath the Pt top electrode as shown in Fig. 5.3, one could also speculate of an oxygen release and reincorporation as possible switching mechanisms. In that case, the redox reaction might be restricted to the three-phase boundary or oxygen transport has to take place via grain boundaries or cracks in the Pt electrode. Since a large variety of publications show a dependence of the resistive state level (e.g., Refs. [90, 91]) or the RESET probability [92] on the ambient oxygen pressure, release and reincorporation as switching process has to be considered as possible scenario for noble metal electrodes. This effect as underlying mechanism for the 8w switching, obeying the opposite switching polarity compared to the standard c8w filamentary VCM-type switching has been recently demonstrated explicitly for the system Nb:$SrTiO_3$/$SrTiO_3$/Pt [58, 93].

In particular, in situ 8w switching of $SrTiO_3$ devices has been performed in an aberration corrected scanning transmission electron microscope (STEM) [93] in order to gain information about the changes of the oxygen-vacancy concentration during switching. Fig. 5.11A depicts a low magnification STEM images of the STO device. Fig. 5.11B and C shows that the Ti L-edge recorded at different distances from the Pt/$SrTiO_3$ interface exhibits significant changes between the LRS and the HRS. Fig. 5.11C shows the intensity of the L_3 edge which decreases with increasing Ti^{3+} concentration as a function of depth for the specimen in the LRS and HRS. In the HRS, the Ti^{3+} concentration profile within the switching STO thin film is very flat and comparable to the Nb-doped substrate. In contrast, the Ti^{3+} is strongly increased at the Pt interface in the LRS and decreases strongly towards the Nb-doped STO electrode. From this profile, it can clearly be excluded that the observed decrease of the resistance with positive biasing could be explained by an increase of the oxygen-vacancy concentration at the bottom Nb-doped STO electrode as expected for an internal redistribution of the oxygen vacancies by the applied field. The lower panel depicts the oxygen concentration profiles acquired by integrating the oxygen electron energy-loss spectroscopy (EELS) edge for the same region in the HRS and LRS. We find that an increase (decrease) in the overall O vacancy concentration within the device after positive (negative) biasing of the Schottky-type top electrode. Therefore, the observed 8w switching can be assigned to a release and reincorporation of oxygen at the top Pt electrode/oxide interface. This switching mechanism was explicitly confirmed by switching $SrTiO_3$ devices in $H_2^{18}O$ since subsequent secondary ion mass spectroscopy measurements reveal that ^{18}O is incorporated into the Pt/STO interface during the RESET process [58].

It is important to note that the oxygen exchange reaction at the MO/Pt interface is proposed as the process that limits the HRS achievable in the standard filamentary c8w switching of TiN/Ta_2O_{5-x}/Pt devices [94].

In TiO_2-based nano-sized ReRAM devices of the structure Pt/TiO_2/Ti/Pt built from 3- and 5-nm-thick TiO_2 layers sandwiched between Pt and Ti electrodes of $100 \times 100 \, nm^2$ size, Zhang et al. [95] reported two opposite switching polarities. The two coexisting switching polarities are identified as standard filamentary-type c8w and 8w switching. The either switching process can be selected by means of the negative voltage amplitude applied to the Pt Schottky electrode. If this voltage is lower than the threshold voltage necessary for c8w SET, the device resets into the HRS of

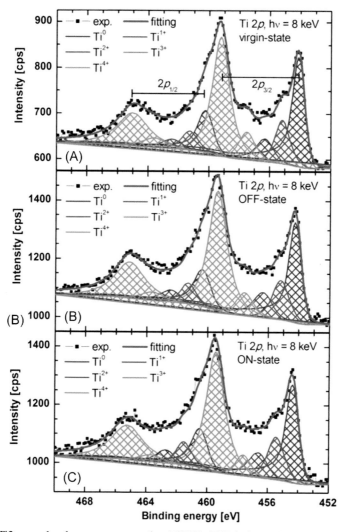

Fig. 5.10 Ti2p core-level spectra measured on Ti/HfO$_x$/TiN devices with a photon energy of 8 eV in different resistive states with fit curves considering five different Ti oxidation states. (A) virgin state, (B) OFF state (HRS), (C) ON state (LRS). The table in (D) summarizes the relative intensities of the oxidation levels varying in different resistive states.
Reprinted with permission from T. Bertaud, M. Sowinska, D. Walczyk, S. Thiess, A. Gloskovskii, C. Walczyk, T. Schroeder, In-operando and non-destructive analysis of the resistive switching in the Ti/HfO$_2$/TiN-based system by hard X-ray photoelectron spectroscopy, Appl. Phys. Lett. 101 (2012), 143501/1-5, https://doi.org/10.1063/1.4756897 © 2012 American Institute of Physics (AIP) Publishers. Credit: T. Bertaud, M. Sowinska, D. Walczyk, S. Thiess, A. Gloskovskii, C. Walczyk, T. Schroeder, In-operando and non-destructive analysis of the resistive switching in the Ti/HfO$_2$/TiN-based system by hard X-ray photoelectron spectroscopy, Appl. Phys. Lett. 101(14) (2012) 143501, Fig. 3.

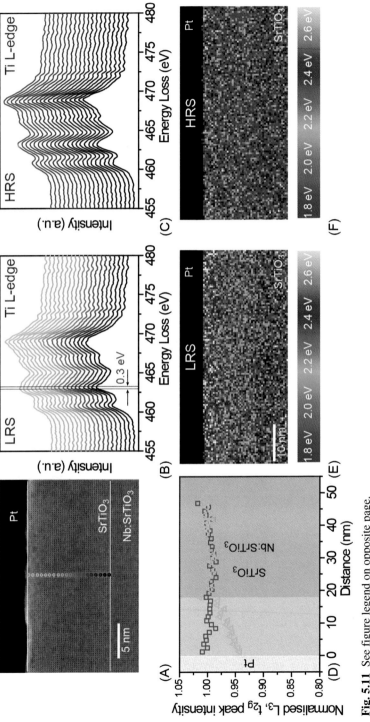

Fig. 5.11 See figure legend on opposite page.

the 8w hysteresis. If, in contrast, the negative voltage amplitude exceeds the necessary threshold, the device sets into the c8w LRS. Interestingly, a positive voltage applied to the device in c8w LRS or 8w HRS brings the device into the common state, which is the c8w HRS equaling the 8w LRS, respectively. The proposed switching model is shown in Fig. 5.12.

These coexisting switching polarities were attributed to two different switching phenomena based on oxygen-vacancy drift/diffusion and oxygen-ion transfer across the TiO_2/Pt interface. In consistency with the different phenomena discussed in this chapter, this extraordinary multifilamentary switching phenomenon in the nano-sized TiO_2 devices is observed for apparently forming-free devices. This type of structure reveals the lowest difference in driving forces for oxygen-vacancy drift/diffusion in the MO and oxygen transfer across interfaces, which become apparent in the electro-forming process.

The occurrence of both 8w and c8w switching polarity in one single device has been observed for many other systems such as Ta_2O_5 [94] and $SrTiO_3$ [96, 97] indicating a general relevance of the underlying oxygen exchange mechanisms for ReRAM materials.

5.4.3 Switching based on reversible changes of the interface barrier

For memristive systems based on conducting oxides (such as manganites or cobaltites) and reactive metal electrodes, electrical conduction is often controlled by an insulating layer built by interfacial oxidation of the OEL [98]. The electroforming treatment increases the thickness of the interfacial oxide layer significantly as has been shown using cross-sectional transmission electron microscopy (TEM) complemented by EELS and HAXPES [99, 100]. While TEM can observe the thickness increase directly, the

---

Fig. 5.11, Cont'd (A) High-resolution annular dark field STEM image of the region of interest in the LRS. (B) Ti L-edge EEL spectra for the device in the LRS. The spectra were acquired across the $SrTiO_3$ film in 1.2 nm steps from the bottom to the top, as indicated in (A). They show the evolution of the edges as a function of position up to a depth of 20 nm into the device. (C) Ti L-edge EEL spectra for the device in the HRS. Again, the spectra were acquired across the $SrTiO_3$ film in 1.2 nm steps from the bottom to the top, as indicated in a. (D) Upper panel: Intensity of the L_3 e_g edge as a function of depth into the specimen for the LRS and HRS to a depth of 55 nm into the device. Lower panel: Oxygen concentration profiles acquired by integrating the oxygen EELS edge for the same region in the HRS and LRS. (E) Map of the fitted Ti L_3 e_g peak width in the LRS. (F) Map of the fitted Ti L_3 e_g peak width in the HRS. Reprinted with permission from D. Cooper, C. Baeumer, N. Bernier, A. Marchewka, C. La Torre, R. E. Dunin-Borkowski, S. Menzel, R. Waser, R. Dittmann, Anomalous resistance hysteresis in oxide ReRAM: oxygen evolution and reincorporation revealed by in situ TEM, Adv. Mater. 29 (2017) 1700212, https://doi.org/10.1002/adma.201700212 © 2017 WILEY-VCH Verlag GmbH & Co. KGaA, Weinheim. Credit: D. Cooper, C. Baeumer, N. Bernier, A. Marchewka, C. La Torre, R.E. Dunin-Borkowski, S. Menzel, R. Waser, R. Dittmann, Anomalous resistance hysteresis in oxide ReRAM: oxygen evolution and reincorporation revealed by in situ TEM, Adv. Mater. (2017) 29(23) 1700212, Fig. 2.

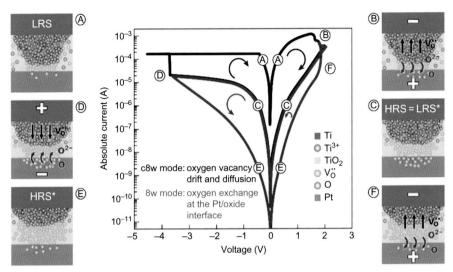

Fig. 5.12 Switching model for the two coexisting bipolar resistive switching modes with opposite polarities observed in nano-crossbar Pt/TiO$_2$/Ti/Pt devices. The occurrence can be understood from a competition of oxygen-vacancy drift and diffusion processes with an oxygen transfer reaction at the interface to the high-barrier Pt electrode. The corresponding ionic states and dominant processes marked in the I(V) plot are depicted as subfigures on the left and on the right. The color code of the subfigures is: Ti (TE), *violet square*; Pt (BE), *gray square*; TiO$_2$, *yellow square*; Ti^{3+}, *violet spheres*; double-charged oxygen vacancies V$_O^{\cdot\cdot}$, *green spheres*; and oxygen atoms, *blue spheres*.
Reprinted with permission from H. Zhang, S. Yoo, S. Menzel, C. Funck, F. Cüppers, D. J. Wouters, C. S. Hwang, R. Waser, S. Hoffmann-Eifert, Resistive switching modes with opposite polarity in Pt/TiO$_2$/Ti/Pt nano-sized ReRAM devices, ACS Appl. Mater. Interfaces, 10 (2018) 29766–29778, https://doi.org/10.1021/acsami.8b09068 © 2018 American Chemical Society. Credit: H. Zhang, S. Yoo, S. Menzel, C. Funck, F. Cüppers, D.J. Wouters, C.S. Hwang, R. Waser, S. Hoffmann-Eifert, Resistive Switching Modes with Opposite Polarity in Pt/TiO$_2$/Ti/Pt Nano-sized ReRAM Devices, ACS Appl. Mater. Interfaces, 10(35) (2018), 29766, Fig. 10.

nondestructive examination by HAXPES shows a spectral weight transfer from low to high oxidation states of the Ti. Due to the limited probing depth of photoemission, this amounts to an increase of the thickness of the TiO$_x$ layer. Fig. 5.13 demonstrates in detail how the application of an external electrical stimulus can reversibly change the valence state in a Ti/PCMO junction [83].

The pristine state (PS) of the junction was measured without any electrical treatment, while a junction in the electroformed state (FS) was examined after the application of a positive voltage sweep to the Ti top electrode that induced a nonvolatile change in resistance. LRS and HRS refer to states obtained by subsequent voltage treatments with negative and positive bias, respectively.

The valence change of the Ti top electrode is readily observed from the Ti 2p HAXPES spectra shown in Fig. 5.13A. The low binding energy component associated to metallic Ti (denoted Ti^{0+}) decreases with increasing resistance, while the high

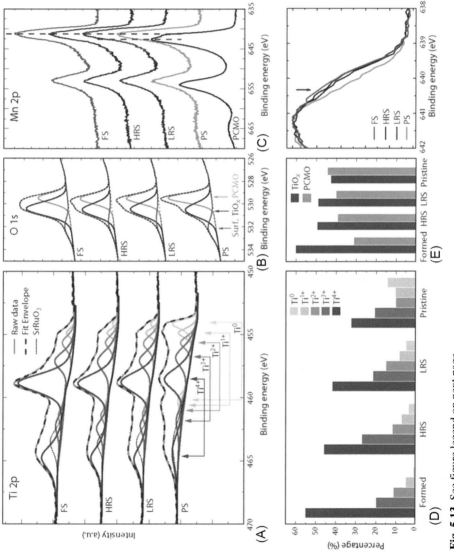

Fig. 5.13 See figure legend on next page.

binding energy component corresponding to fully oxidized Ti (marked Ti^{4+}) increases (PS → LRS → HRS → FS), see Fig. 5.13D for the area percentages. The O 1 s emission line in Fig. 5.13B directly demonstrates the oxygen transfer from the PCMO to the TiO_x layer [area percentages in Fig. 5.13E], while the Mn 2p line shows slight changes in the line shape associated with an increase in the ratio of Mn^{3+}/Mn^{4+} ions when going from LRS to HRS. The observation of oxidation of the Ti layer and reduction of the PCMO is direct evidence that the resistance change is caused by a redox-reaction at the interface.

5.5   Summary and outlook

In this chapter, we discussed interface effects in VCM-type bipolar switching devices with the focus of interfaces in the direction of the axis of the ReRAM cell. Focusing primarily on the most important defect species, which is based on (double) positively charged oxygen vacancies [101] it should be noted that motion of metal cations has been evidenced in certain cases and their role cannot be excluded [40]. Interfacial oxygen transfer reactions based on thermodynamic diffusion already happen when the different materials, that is, metal and MO, come into contact during device fabrication. The microstructure of the materials, in parts, affects the structure of the devices after the electroforming step controlled by the electric field and current-induced Joule heating. Electroforming of devices built form insulating oxide layers often results in the formation of locally confined conductive filaments. In contrast, electroforming of devices built from conducting oxides might lead to the oxidation of the reactive metal forming an insulating layer at the interface. In either cases, the subsequent resistive switching behavior is to a certain extend also determined by ion transfer processes across the electrode/switching MO-interface. Examples have been discussed in this chapter. Following the technological trend toward even smaller devices aiming at dimensions of only a few nanometer thickness of every layer and about $<100\,nm^2$ size in area [53], the ReRAM device behavior might be controlled by interface phenomena. In addition to transfer processes in the direction of the applied electric field, ion migration perpendicular to the axis of the filament needs to be discussed [102]. This

Fig. 5.13, Cont'd (A–C) Ti 2p/O 1 s/Mn 2p XPS spectra collected on the four regions, sorted from low resistance (pristine) to high resistance (formed). The spectrum named PCMO refers to the unpatterned reference sample. (D, E) Area percentages of the constituent components of the Ti 2p and O 1 s envelope, respectively. (F) Close-up of the Mn 2p 3/2 edge.
Reprinted with permission from A. Herpers, C. Lenser, C. Park, F. Offi, F. Borgatti, G. Panaccione, S. Menzel, R. Waser, R. Dittmann, Spectroscopic proof of the correlation between redox-state and charge-carrier transport at the interface of resistively switching Ti/PCMO devices, Adv. Mater. 26 (2014) 2730–2735, https://doi.org/10.1002/adma.201304054 © 2017 WILEY-VCH Verlag GmbH & Co. KGaA, Weinheim. Credit: A. Herpers, C. Lenser, C. Park, F. Offi, F. Borgatti, G. Panaccione, S. Menzel, R. Waser, R. Dittmann, Spectroscopic Proof of the Correlation between Redox-State and Charge-Carrier Transport at the Interface of Resistively Switching Ti/PCMO Devices, Adv. Mater. 26 (2014) 2730, Fig. 2.

effect can directly influence the performance of miniaturized pillar-shaped switching cells [36]. Shrinking the area for pillar-shaped switching cells to nano-sized dimensions will require proper sidewall encapsulation materials for the switching cells. Such materials will afford a careful selection, as they should not allow either ion transfer with the switching materials or ion migration of the involved species. Only with these tasks mastered, the realization of the shortest distance between two devices without significant cross talk between neighboring cells will become feasible.

References

[1] G.W. Burr, R.M. Shelby, A. Sebastian, S. Kim, S. Kim, S. Sidler, K. Virwani, M. Ishii, P. Narayanan, A. Fumarola, L.L. Sanches, I. Boybat, M. Le Gallo, K. Moon, J. Woo, H. Hwang, Y. Leblebici, Neuromorphic computing using non-volatile memory, Adv. Phys. X 2 (2017) 89–124, https://doi.org/10.1080/23746149.2016.1259585.

[2] Z. Wang, S. Joshi, S. Savel'ev, W. Song, R. Midya, Y. Li, M. Rao, P. Yan, S. Asapu, Y. Zhuo, H. Jiang, P. Lin, C. Li, J.H. Yoon, N.K. Upadhyay, J. Zhang, M. Hu, J.P. Strachan, M. Barnell, Q. Wu, H. Wu, R.S. Williams, Q. Xia, J.J. Yang, Fully memristive neural networks for pattern classification with unsupervised learning, Nat. Electron. 1 (2018) 137–145, https://doi.org/10.1038/s41928-018-0023-2.

[3] M.A. Zidan, J.P. Strachan, W.D. Lu, The future of electronics based on memristive systems, Nat. Electron. 1 (2018) 22–29, https://doi.org/10.1038/s41928-017-0006-8.

[4] B. Hudec, C.W. Hsu, I.T. Wang, W.L. Lai, C.C. Chang, T. Wang, K. Frohlich, C.H. Ho, C.H. Lin, T.H. Hou, 3D resistive RAM cell design for high-density storage class memory-a review, Sci. China Inform. Sci. 59 (2016) 61403/1–21, https://doi.org/10.1007/s11432-016-5566-0.

[5] J.A.J. Rupp, R. Waser, D.J. Wouters, Threshold Switching in Amorphous Cr-doped Vanadium Oxide for New Crossbar Selector, IEEE Xplore, 20164, https://doi.org/10.1109/IMW.2016.7495293.

[6] R. Waser, R. Dittmann, G. Staikov, K. Szot, Redox-based resistive switching memories—nanoionic mechanisms, prospects, and challenges, Adv. Mater. 21 (2009) 2632–2663, https://doi.org/10.1002/adma.200900375.

[7] S. Menzel, M. Waters, A. Marchewka, U. Böttger, R. Dittmann, R. Waser, Origin of the ultra-nonlinear switching Kinetics in oxide-based resistive switches, Adv. Funct. Mater. 21 (2011) 4487–4492, https://doi.org/10.1002/adfm.201101117.

[8] C. Wu, T.W. Kim, H.Y. Choi, D.B. Strukov, J.J. Yang, Flexible three-dimensional artificial synapsc networks with correlated learning and trainable memory capability, Nat. Commun. 8 (2017) 752/1, https://doi.org/10.1038/s41467-017-00803-1.

[9] G. Piccolboni, G. Molas, D. Garbin, E. Vianello, O. Cueto, C. Cagli, B. Traore, B. De Salvo, G. Ghibaudo, L. Perniola, Investigation of cycle-to-cycle variability in HfO$_2$-based OxRAM, IEEE Electron Device Lett. 37 (2016) 721–723, https://doi.org/10.1109/LED.2016.2553370.

[10] W. Kim, S. Menzel, D.J. Wouters, Y. Guo, J. Robertson, B. Rösgen, R. Waser, V. Rana, Impact of oxygen exchange reaction at the ohmic interface in Ta$_2$O$_5$-based ReRAM devices, Nanoscale 8 (2016) 17774–17781, https://doi.org/10.1039/c6nr03810g.

[11] C. La Torre, A. Kindsmueller, D.J. Wouters, C.E. Graves, G.A. Gibson, J.P. Strachan, R.S. Williams, R. Waser, S. Menzel, Volatile HRS asymmetry and subloops in resistive switching oxides, Nanoscale 9 (2017) 14414–14422, https://doi.org/10.1039/C7NR04896C.

[12] C. Baeumer, C. Schmitz, A. Marchewka, D.N. Mueller, R. Valenta, J. Hackl, N. Raab, S.P. Rogers, M.I. Khan, S. Nemsak, M. Shim, S. Menzel, C.M. Schneider, R. Waser, R. Dittmann, Quantifying redox-induced Schottky barrier variations in memristive devices via in operando spectromicroscopy with graphene electrodes, Nat. Commun. 7 (2016) 12398, https://doi.org/10.1038/ncomms12398.

[13] F.A. Kroeger, H.J. Vink, Relations between the concentrations of imperfections in crystalline solids, Solid State Phys. 3 (1956) 307–435, https://doi.org/10.1016/S0081-1947(08)60135-6.

[14] J. Maier, Defect chemistry: composition, transport, and reactions in the solid state; part I: thermodynamics, Angew. Chem. Int. Ed. Engl. 32 (1993) 313–335, https://doi.org/10.1002/anie.199303133.

[15] J. Maier, Defect chemistry and ion transport in nanostructured materials: part II. Aspects of nanoionics, Solid State Ion. 157 (2003) 327–334, https://doi.org/10.1016/S0167-2738(02)00229-1.

[16] D.M. Smyth, The Defect Chemistry of Metal Oxides, Oxford University Press, 2000. ISBN-13: 978-0195110142.

[17] F. Gunkel, R. Waser, A.H.H. Ramadan, R.A. De Souza, S. Hoffmann-Eifert, R. Dittmann, Space charges and defect concentration profiles at complex oxide interfaces, Phys. Rev. B 93 (2016) 245431.

[18] C. Ohly, S. Hoffmann-Eifert, X. Guo, R. Waser, Electrical conductivity of epitaxial $SrTiO_3$ thin films as a function of oxygen Partial pressure and temperature, J. Am. Ceram. Soc. 89 (2006) 2845–2852.

[19] K. Szot, M. Rogala, W. Speier, Z. Klusek, A. Besmehn, R. Waser, TiO_2—a prototypical memristive material, Nanotechnology 22 (2011) 254001/1–21, https://doi.org/10.1088/0957-4484/22/25/254001.

[20] V. Metlenko, A. Ramadan, F. Gunkel, H. Du, H. Schraknepper, S. Hoffmann-Eifert, R. Dittmann, R. Waser, R. De Souza, Do dislocations act as atomic autobahns for oxygen in the perovskite oxide $SrTiO_3$? Nanoscale 6 (2014) 12864–12876, https://doi.org/10.1039/C4NR04083J.

[21] K.K. Adepalli, M. Kelsch, R. Merkle, J. Maier, Influence of line defects on the electrical properties of single crystal TiO2, Adv. Funct. Mater. 23 (2013) 1798–1806, https://doi.org/10.1002/adfm.201202256.

[22] Y. Guo, J. Robertson, Oxygen vacancy defects in Ta2O5 showing long-range atomic re-arrangements, Appl. Phys. Lett. 104 (2014) https://doi.org/10.1063/1.4869553.

[23] Y. Aoki, C. Wiemann, V. Feyer, H.-S. Kim, C.M. Schneider, H. Ill-Yoo, M. Martin, Bulk mixed ion electron conduction in amorphous gallium oxide causes memristive behaviour, Nat. Commun. 5 (2014) 3473/1–9, https://doi.org/10.1038/ncomms4473.

[24] J. Carlos Gonzalez-Rosillo, R. Ortega-Hernandez, J. Jareno-Cerulla, E. Miranda, J. Sune, X. Granados, X. Obradors, A. Palau, T. Puig, Volume Resistive Switching in metallic perovskite oxides driven by the Metal-Insulator Transition, J. Electroceram. 39 (2017) 185–196, https://doi.org/10.1007/s10832-017-0101-2.

[25]. R. Waser, R. Bruchhaus, S. Menzel, Redox-based resistive random access memories. in: R. Waser (Ed.), Nanoelectronics and Information Technology, third ed., Wiley-VCH, Berlin, pp. 683–710, ISBN: 978-3-527-40927-3

[26] D. Ielmini, H.P. Wong, In-memory computing with resistive switching devices, Nat. Electron. 1 (2018) 333–343.

[27] S. Yu, H.-Y. Chen, B. Gao, J. Kang, H.-S.P. Wong, HfOx-based vertical resistive switching random access memory suitable for Bit-cost-effective three-dimensional cross-point architecture, ACS Nano 7 (2013) 2320–2325, https://doi.org/10.1021/nn305510u.

[28]   J. Robertson, High dielectric constant gate oxides for metal oxide Si transistors, Rep. Prog. Phys. 69 (2006) 327–396.

[29]   M.H. Park, Y.H. Lee, T. Mikolajick, U. Schroeder, C.S. Hwang, Review and perspective on ferroelectric HfO2-based thin films for, MRS Commun. (2018) 1–14, https://doi.org/10.1557/mrc.2018.175.

[30]   N. Alimardani, J.F. Conley, Enhancing metal-insulator-insulator-metal tunnel diodes via defect enhanced direct tunneling, Appl. Phys. Lett. 105 (2014) 82902/1–5, https://doi.org/10.1063/1.4893735.

[31]   L. Larcher, F.M. Puglisi, P. Pavan, A. Padovani, L. Vandelli, G. Bersuker, A compact model of program window in HfOx RRAM devices for conductive filament characteristics analysis, IEEE Trans. Electron Devices 61 (2014) 2668–2673, https://doi.org/10.1109/TED.2014.2329020.

[32]   S. Balatti, S. Ambrogio, Z. Wang, S. Sills, A. Calderoni, N. Ramaswamy, D. Ielmini, Voltage-controlled cycling endurance of HfOx-based resistive-switching memory, IEEE Trans. Electron Devices 62 (2015) 3365–3372, https://doi.org/10.1109/TED.2015.2463104.

[33]   L. Zhao, S. Clima, B. Magyari-Kope, M. Jurczak, Y. Nishi, Ab initio modeling of oxygen-vacancy formation in doped-HfOx RRAM: effects of oxide phases, stoichiometry, and dopant concentrations, Appl. Phys. Lett. 107 (2015) 13504/1, https://doi.org/10.1063/1.4926337.

[34]   M.-C. Wu, W.-Y. Jang, C.-H. Lin, T.-Y. Tseng, A study on low-power, nanosecond operation and multilevel bipolar resistance switching in Ti/ZrO 2/Pt nonvolatile memory with 1T1R architecture, Semicond. Sci. Technol. 27 (2012) 065010. (9 pp.)-06501. https://doi.org/10.1088/0268-1242/27/6/065010.

[35]   J.P. Strachan, G. Medeiros-Ribeiro, J.J. Yang, M.-X. Zhang, F. Miao, I. Goldfarb, M. Holt, V. Rose, R.S. Williams, Spectromicroscopy of tantalum oxide memristors, Appl. Phys. Lett. 98 (2011) 242114, https://doi.org/10.1063/1.3599589.

[36]   Y. Hayakawa, A. Himeno, R. Yasuhara, W. Boullart, E. Vecchio, T. Vandeweyer, T. Witters, D. Crotti, M. Jurczak, S. Fujii, S. Ito, Y. Kawashima, Y. Ikeda, A. Kawahara, K. Kawai, Z. Wei, S. Muraoka, K. Shimakawa, T. Mikawa, S. Yoneda, Highly reliable TaOx ReRAM with centralized filament for 28-nm embedded application. 2015 Symposium on VLSI Circuits (VLSI Circuits), Kyoto, Japan, 201514–15, https://doi.org/10.1109/VLSIC.2015.7231381.

[37]   B. Gao, B. Chen, R. Liu, F. Zhang, P. Huang, L. Liu, X. Liu, J. Kang, H. Chen, S. Yu, H. Wong, 3-D cross-point array operation on rmAlOy/rmHfOx -based vertical resistive switching memory, IEEE Trans. Electron Devices 61 (2014) 1377–1381, https://doi.org/10.1109/TED.2014.2311655.

[38]   B. Arndt, F. Borgatti, F. Offi, M. Phillips, P. Parreira, T. Meiners, S. Menzel, K. Skaja, G. Panaccione, D.A. MacLaren, R. Waser, R. Dittmann, Spectroscopic indications of tunnel barrier charging as the switching mechanism in memristive devices, Adv. Funct. Mater. (2017) 1702282. n/a, https://doi.org/10.1002/adfm.201702282.

[39]   Y. Guo, J. Robertson, Material selection for oxide-based resistive random access memories, Appl. Phys. Lett. 105 (2014) 223516, https://doi.org/10.1063/1.4903470.

[40]   A. Wedig, M. Luebben, D.-Y. Cho, M. Moors, K. Skaja, V. Rana, T. Hasegawa, K. Adepalli, B. Yildiz, R. Waser, I. Valov, Nanoscale cation motion in TaO_x, HfO_x and TiO_x memristive systems, Nat. Nanotechnol. 11 (2016) 67–74, https://doi.org/10.1038/nnano.2015.221.

[41]   R. De Souza, Limits to the rate of oxygen transport in mixed-conducting oxides, J. Mater. Chem. A 5 (2017) 20334–20350, https://doi.org/10.1039/C7TA04266C.

[42] J.J. Yang, M.D. Pickett, X. Li, D.A.A. Ohlberg, D.R. Stewart, R.S. Williams, Memristive switching mechanism for metal/oxide/metal nanodevices, Nat. Nanotechnol. 3 (2008) 429–433.

[43] E. Yalon, I. Karpov, V. Karpov, I. Riess, D. Kalaev, D. Ritter, Detection of the insulating gap and conductive filament growth direction in resistive memories, Nanoscale 7 (2015) 15434–15441, https://doi.org/10.1039/c5nr03314d.

[44] F. Nardi, S. Balatti, S. Larentis, D. Ielmini, Complementary switching in metal oxides: toward diode-less crossbar RRAMs. in: 2011 IEEE International Electron Devices Meeting (IEDM 2011), 2011, pp. 31.1/1–4, https://doi.org/10.1109/IEDM.2011.6131647.

[45] S. Balatti, S. Larentis, D.C. Gilmer, D. Ielmini, Multiple memory states in resistive switching devices through controlled size and Orientation of the conductive filament, Adv. Mater. 25 (2013) 1474–1478, https://doi.org/10.1002/adma.201204097.

[46] A. Schoenhals, D.J. Wouters, A. Marchewka, T. Breuer, K. Skaja, V. Rana, S. Menzel, R. Waser, Critical ReRAM stack parameters controlling complementary versus bipolar resistive switching. in: Memory Workshop (IMW), 2015 IEEE International, 2015, pp. 73–76, https://doi.org/10.1109/IMW.2015.7150281.

[47] T. Breuer, A. Siemon, E. Linn, S. Menzel, R. Waser, V. Rana, A HfO$_2$-based complementary switching crossbar adder, Adv. Electron. Mater. 1 (2015) 1500138, https://doi.org/10.1002/aelm.201500138.

[48] A.R. Genreith-Schriever, R.A. De Souza, Field-enhanced ion transport in solids: reexamination with molecular dynamics simulations, Phys. Rev. B: Condens. Matter 94 (2016) 224304, https://doi.org/10.1103/PhysRevB.94.224304.

[49] K. Skaja, M. Andrae, V. Rana, R. Waser, R. Dittmann, C. Baeumer, Reduction of the forming voltage through tailored oxygen non-stoichiometry in tantalum oxide ReRAM devices, Sci. Rep. 8 (2018) 10861/1–7, https://doi.org/10.1038/s41598-018-28992-9.

[50] S.U. Sharath, S. Vogel, L. Molina-Luna, E. Hildebrandt, C. Wenger, J. Kurian, M. Duerrschnabel, T. Niermann, G. Niu, P. Calka, M. Lehmann, H.J. Kleebe, T. Schroeder, L. Alff, Control of switching modes and conductance quantization in oxygen engineered HfO$_x$ based memristive devices, Adv. Funct. Mater. 27 (2017) 1700432/1–13, https://doi.org/10.1002/adfm.201700432.

[51] S.K. Kim, G.J. Choi, S.Y. Lee, M. Seo, S.W. Lee, J.H. Han, H.S. Ahn, S. Han, C.S. Hwang, Al-doped TiO$_2$ films with ultralow leakage currents for next generation DRAM capacitors, Adv. Mater. 20 (2008) 1429, https://doi.org/10.1002/adma.200701085.

[52] H. Kim, P. McIntyre, C. Chui, K. Saraswat, S. Stemmer, Engineering chemically abrupt high-k metal oxide/silicon interfaces using an oxygen-gettering metal overlayer, J. Appl. Phys. 96 (2004) 3467–3472, https://doi.org/10.1063/1.1776636.

[53] B. Govoreanu, G.S. Kar, Y.-Y. Chen, V. Paraschiv, S. Kubicek, A. Fantini, I.P. Radu, L. Goux, S. Clima, R. Degraeve, N. Jossart, O. Richard, T. Vandeweyer, K. Seo, P. Hendrickx, G. Pourtois, H. Bender, L. Altimime, D.J. Wouters, J.A. Kittl, M. Jurczak, 10x10 nm^2 Hf/HfO$_x$ Crossbar Resistive RAM with Excellent Performance, Reliability and Low-Energy Operation, IEDM Tech. Dig., 201131.6.1–31.6.4, https://doi.org/10.1109/IEDM.2011.6131652.

[54] B. Govoreanu, A. Redolfi, L. Zhang, C. Adelmann, M. Popovici, S. Clima, H. Hody, V. Paraschiv, I.P. Radu, A. Franquet, J.-C. Liu, J. Swerts, O. Richard, H. Bender, L. Altimime, M. Jurczak, Vacancy-modulated conductive oxide resistive RAM (VMCO-RRAM): an area-scalable switching current, self-compliant, highly nonlinear and wide On/Off-window resistive switching cell, in: Electron Devices Meeting (IEDM), 2013 IEEE International, 13, 2013, pp. 256–259.

[55] K.G. Young-Fisher, G. Bersuker, B. Butcher, A. Padovani, L. Larcher, D. Veksler, D.C. Gilmer, Leakage current-forming voltage relation and oxygen Gettering in HfO_x RRAM devices, IEEE Electron Device Lett. 34 (2013) 750–752, https://doi.org/10.1109/LED.2013.2256101.

[56] R. Merkle, J. Maier, How is oxygen incorporated into oxides? A comprehensive kinetic study of a simple solid-state reaction with $SrTiO_3$ as a model material, Angew. Chem. Int. Ed. 47 (2008) 3874–3894, https://doi.org/10.1002/anie.200700987.

[57] J.J. Yang, F. Miao, M.D. Pickett, D.A.A. Ohlberg, D.R. Stewart, C.N. Lau, R.S. Williams, The mechanism of electroforming of metal oxide memristive switches, Nanotechnology 20 (2009) 215201, https://doi.org/10.1088/0957-4484/20/21/215201.

[58] T. Heisig, C. Baeumer, U.N. Gries, M.P. Mueller, C. La Torre, M. Luebben, N. Raab, D. Hongchu, S. Menzel, D.N. Mueller, C.-L. Jia, J. Mayer, R. Waser, I. Valov, D. Souza, A. Roger, R. Dittmann, Oxygen Exchange Processes between Oxide Memristive Devices and Water Molecules, Adv. Mater. 30 (2018) 1800957.

[59] R. Stumpf, C. Liu, C. Tracy, Retardation of O diffusion through polycrystalline Pt by Be doping, Phys. Rev. B Condens. Matter 59 (1999) 16047–16052, https://doi.org/10.1103/PhysRevB.59.16047.

[60] J. Maier, Physical Chemistry of Ionic Materials. John Wiley & Sons, Ltd., 2004, https://doi.org/10.1002/0470020229.

[61] M. Lübben, S. Wiefels, R. Waser, I. Valov, Processes and effects of oxygen and moisture in resistively switching TaO_x and HfO_x, Adv. Electron. Mater. 4 (2018) 1700458, https://doi.org/10.1002/aelm.201700458.

[62] D. Cho, M. Lübben, S. Wiefels, K. Lee, I. Valov, Interfacial metal—oxide interactions in resistive switching memories, ACS Appl. Mater. Interfaces 9 (2017) 19287–19295, https://doi.org/10.1021/acsami.7b02921.

[63] M. Schie, M.P. Mueller, M. Salinga, R. Waser, R.A. De Souza, Ion migration in crystalline and amorphous HfO_x, J. Chem. Phys. 146 (2017) 94508/1–9, https://doi.org/10.1063/1.4977453.

[64] S. Stille, C. Lenser, R. Dittmann, A. Koehl, I. Krug, R. Muenstermann, J. Perlich, C.M. Schneider, U. Klemradt, R. Waser, Detection of filament formation in forming-free resistive switching $SrTiO_3$ devices with Ti top electrodes, Appl. Phys. Lett. 100 (2012) 223503/1–4, https://doi.org/10.1063/1.4724108.

[65] H.-S. Lee, T. Mizoguchi, J. Mistui, T. Yamamoto, S.-J.L. Kang, Y. Ikuhara, Defect energetics in $SrTiO_3$ symmetric tilt grain boundaries, Phys. Rev. B Condens. Matter 83 (2011) 104110/1, https://doi.org/10.1103/PhysRevB.83.104110.

[66] J.P. Strachan, J.J. Yang, L.A. Montoro, C.A. Ospina, A.J. Ramirez, A.L.D. Kilcoyne, G. Medeiros-Ribeiro, R.S. Williams, Characterization of electroforming-free titanium dioxide memristors, Beilstein J. Nanotechnol. 4 (2013) 467–473, https://doi.org/10.3762/bjnano.4.55.

[67] Y. Guo, J. Robertson, Materials selection for oxide-based resistive random access memories, Appl. Phys. Lett. 105 (2014) 223516/1–5, https://doi.org/10.1063/1.4903470.

[68] Y.Y. Chen, G. Pourtois, S. Clima, L. Goux, B. Govoreanu, A. Fantini, R. Degraeve, G.S. Kar, G. Groeseneken, D.J. Wouters, M. Jurczak, Hf cap thickness dependence in bipolar-switching $TiN\backslash HfO_2\backslash Hf\backslash TiN$ RRAM device, ECS Trans. 50 (2013) 3–9, https://doi.org/10.1149/05034.0003ecst.

[69] S.U. Sharath, M.J. Joseph, S. Vogel, E. Hildebrandt, P. Komissinskiy, J. Kurian, T. Schroeder, L. Alff, Impact of oxygen stoichiometry on electroforming and multiple switching modes in $TiN/TaO_x/Pt$ based ReRAM, Appl. Phys. Lett. 109 (2016) 173503/1–5, https://doi.org/10.1063/1.4965872.

[70] P. Parreira, G.W. Paterson, S. McVitie, D.A. MacLaren, Stability, bistability and instability of amorphous ZrO_2 resistive memory devices, J. Phys. D Appl. Phys. 49 (2016) 95111/1, https://doi.org/10.1088/0022-3727/49/9/095111.

[71] Y.H. Do, J.S. Kwak, Y.C. Bae, K. Jung, H. Im, J.P. Hong, Oxygen ion drifted bipolar resistive switching behaviors in TiO_2-Al electrode interfaces, Thin Solid Films 518 (2010) 4408–4411, https://doi.org/10.1016/j.tsf.2010.01.016.

[72] M. Sowinska, T. Bertaud, D. Walczyk, S. Thiess, M.A. Schubert, M. Lukosius, W. Drube, C. Walczyk, T. Schroeder, Hard X-ray photoelectron spectroscopy study of the electroforming in Ti/HfO_2-based resistive switching structures, Appl. Phys. Lett. 100 (2012) 233509, https://doi.org/10.1063/1.4728118.

[73] P. Calka, M. Sowinska, T. Bertaud, D. Walczyk, J. Dabrowski, P. Zaumseil, C. Walczyk, A. Gloskovskii, X. Cartoixa, J. Sune, T. Schroeder, Engineering of the chemical reactivity of the Ti/HfO_2 Interface for RRAM: experiment and theory, ACS Appl. Mater. Interfaces 6 (2014) 5056–5060, https://doi.org/10.1021/am500137y.

[74] C. Nauenheim, C. Kuegeler, A. Ruediger, R. Waser, Investigation of the electroforming process in resistively switching TiO_2 nanocrosspoint junctions, Appl. Phys. Lett. 96 (2010) 122902, https://doi.org/10.1063/1.3367752.

[75] S. Privitera, G. Bersuker, S. Lombardo, C. Bongiorno, D. Gilmer, Conductive filament structure in HfO_2 resistive switching memory devices, Solid State Electron. 111 (2015) 161–165, https://doi.org/10.1016/j.sse.2015.05.044.

[76] C. Lenser, Z. Connell, A. Kovacs, R. Dunin-Borkowski, A. Koehl, R. Waser, R. Dittmann, Identification of screw dislocations as fast-forming sites in Fe-doped $SrTiO_3$, Appl. Phys. Lett. 102 (2013) 183504.

[77] A. Koehl, H. Wasmund, A. Herpers, P. Guttmann, S. Werner, K. Henzler, H. Du, J. Mayer, R. Waser, R. Dittmann, Evidence for multifilamentary valence changes in resistive switching $SrTiO_3$ devices detected by transmission X-ray microscopy, APL Mater. 1 (2013) 042102https://doi.org/10.1063/1.4822438.

[78] N. Raab, D.O. Schmidt, H. Du, M. Kruth, U. Simon, R. Dittmann, Au nanoparticles as template for defect formation in memristive $SrTiO_3$ thin films, Nanomaterials 8 (2018) 869.

[79] P. Calka, E. Martinez, V. Delaye, D. Lafond, G. Audoit, D. Mariolle, N. Chevalier, H. Grampeix, C. Cagli, V. Jousseaume, C. Guedj, Chemical and structural properties of conducting nanofilaments in TiN/HfO_2-based resistive switching structures, Nanotechnology 24 (2013) 85706/1–9, https://doi.org/10.1088/0957-4484/24/8/085706.

[80] R. Hagenbeck, R. Waser, Influence of temperature and interface charge on the grain-boundary conductivity in acceptor-doped $SrTiO_3$ ceramics, J. Appl. Phys. 83 (1998) 2083–2092. USA.

[81] A. Sawa, Resistive switching in transition metal oxides, Mater. Today 11 (2008) 28–36, https://doi.org/10.1016/S1369-7021(08)70119-6.

[82] A. Sawa, R. Meyer, Interface type switching. in: D. Ielmini, R. Waser (Eds.), Resistive Switching: From Fumdamentals of Nanoionic Redox Processes to Memristive Device Applications, Wiley-VCH Verlag GmbH & Co. KGaA, 2016, pp. 457–482, https://doi.org/10.1002/9783527680870 (Chapter 16).

[83] A. Herpers, C. Lenser, C. Park, F. Offi, F. Borgatti, G. Panaccione, S. Menzel, R. Waser, R. Dittmann, Spectroscopic proof of the correlation between redox-state and charge-carrier transport at the interface of resistively switching Ti/PCMO devices, Adv. Mater. 26 (2014) 2730–2735, https://doi.org/10.1002/adma.201304054.

[84] R. Yasuhara, T. Yamamoto, I. Ohkuba, H. Kumigashira, M. Oshima, Interfacial chemical states of resistance switching metal/Pr0.7Ca0.3MnO3 interfaces, Appl. Phys. Lett. 97 (2010) 132111, https://doi.org/10.1063/1.3496033.

[85] R. Bruchhaus, C.R. Hermes, R. Waser, Memristive switches with two switching polarities in a forming free device structure, MRS Online Proc. Libr. 1337 (2011) 73–78, https://doi.org/10.1557/opl.2011.858.

[86] S.U. Sharath, T. Bertaud, J. Kurian, E. Hildebrandt, C. Walczyk, P. Calka, P. Zaumseil, M. Sowinska, D. Walczyk, A. Gloskovskii, T. Schroeder, L. Alff, Towards forming-free resistive switching in oxygen engineered HfO_{2-x}, Appl. Phys. Lett. 104 (2014) 063502https://doi.org/10.1063/1.4864653.

[87] S.U. Sharath, J. Kurian, P. Komissinskiy, E. Hildebrandt, T. Bertaud, C. Walczyk, P. Calka, T. Schroeder, L. Alff, Thickness independent reduced forming voltage in oxygen engineered HfO_2 based resistive switching memories, Appl. Phys. Lett. 105 (2014) 73505/1–4, https://doi.org/10.1063/1.4893605.

[88] A. Marchewka, B. Roesgen, K. Skaja, H. Du, C.L. Jia, J. Mayer, V. Rana, R. Waser, S. Menzel, Nanoionic resistive switching memories: on the physical nature of the dynamic reset process, Adv. Electron. Mater. 2 (2016) 1500233/1–13, https://doi.org/10.1002/aelm.201500233.

[89] U. Celano, J. Op de Beeck, S. Clima, M. Luebben, P.M. Koenraad, L. Goux, I. Valov, W. Vandervorst, Direct probing of the dielectric scavenging-layer interface in oxide filamentary-based valence change memory, ACS Appl. Mater. Interfaces 9 (2017) 10820–10824, https://doi.org/10.1021/acsami.6b16268.

[90] T. Bertaud, M. Sowinska, D. Walczyk, S. Thiess, A. Gloskovskii, C. Walczyk, T. Schroeder, In-operando and non-destructive analysis of the resistive switching in the $Ti/HfO_2/TiN$-based system by hard x-ray photoelectron spectroscopy, Appl. Phys. Lett. 101 (2012) 143501/1–5, https://doi.org/10.1063/1.4756897.

[91] T. Bertaud, D. Walczyk, C. Walczyk, S. Kubotsch, M. Sowinska, T. Schroeder, C. Wenger, C. Vallee, P. Gonon, C. Mannequin, V. Jousseaume, H. Grampeix, Resistive switching of HfO_2-based metal-insulator-metal diodes: impact of the top electrode material, Thin Solid Films 520 (2012) 4551–4555, https://doi.org/10.1016/j.tsf.2011.10.183.

[92] L. Goux, P. Czarnecki, Y.Y. Chen, L. Pantisano, X.P. Wang, R. Degraeve, B. Govoreanu, M. Jurczak, D.J. Wouters, L. Altimime, Evidence of oxygen-mediated resistive-switching mechanism in $TiN/HfO_2/Pt$ cells, Appl. Phys. Lett. 97 (2010) 243509.

[93] D. Cooper, C. Baeumer, N. Bernier, A. Marchewka, C. La Torre, R.E. Dunin-Borkowski, S. Menzel, R. Waser, R. Dittmann, Anomalous resistance hysteresis in oxide ReRAM: oxygen evolution and reincorporation revealed by in situ TEM, Adv. Mater. 29 (2017) 1700212, https://doi.org/10.1002/adma.201700212.

[94] A. Schönhals, C.M.M. Rosario, S. Hoffmann-Eifert, R. Waser, S. Menzel, D.J. Wouters, Role of the electrode material on the RESET limitation in oxide ReRAM devices, Adv. Electron. Mater. 4 (2017) 1700243/1–11, https://doi.org/10.1002/aelm.201700243.

[95] H. Zhang, S. Yoo, S. Menzel, C. Funck, F. Cüppers, D.J. Wouters, C.S. Hwang, R. Waser, S. Hoffmann-Eifert, Resistive switching modes with opposite polarity in $Pt/TiO_2/Ti/Pt$ nano-sized ReRAM devices, ACS Appl. Mater. Interfaces 10 (2018) 29766–29778, https://doi.org/10.1021/acsami.8b09068.

[96] R. Muenstermann, T. Menke, R. Dittmann, R. Waser, Coexistence of filamentary and homogeneous resistive switching in Fe-doped $SrTiO_3$ thin-film memristive devices, Adv. Mater. 22 (2010) 4819–4822, https://doi.org/10.1002/adma.201001872.

[97] M. Kubicek, R. Schmitt, F. Messerschmitt, J.L.M. Rupp, Uncovering two competing switching mechanisms for Epitaxial and ultrathin strontium titanate-based resistive switching bits. ACS Nano 9 (2015) 10737–10748, https://doi.org/10.1021/acsnano.5b02752.

[98] M. Hasan, R. Dong, H.J. Choe, D.S. Lee, D.J. Seong, M.B. Pyun, H. Hwang, Uniform resistive switching with a thin reactive metal interface layer in metal-$La_{0.7}Ca_{0.3}MnO_3$-metal heterostructures, Appl. Phys. Lett. 92 (2008) 202102.

[99]   F. Borgatti, C. Park, A. Herpers, F. Offi, R. Egoavil, Y. Yamashita, A. Yang, M. Kobata, K. Kobayashi, J. Verbeeck, G. Panaccione, R. Dittmann, Chemical insight into electro-forming of resistive switching manganite heterostructures, Nanoscale 5 (2013) 3954–3960, https://doi.org/10.1039/c3nr00106g.

[100]   S. Asanuma, H. Akoh, H. Yamada, A. Sawa, Relationship between resistive switching characteristics and band diagrams of $Ti/Pr_{1-x}Ca_xMnO_3$ junctions, Phys. Rev. B 80 (2009) 235113/1–8, https://doi.org/10.1103/PhysRevB.80.235113.

[101]   R. Waser, M. Aono, Nanoionics-based resistive switching memories, Nat. Mater. 6 (2007) 833–840.

[102]   S. Kumar, Z. Wang, X. Huang, N. Kumari, N. Davila, J.P. Strachan, D. Vine, A.L. David Kilcoyne, Y. Nishi, R.S. Williams, Oxygen migration during resistance switching and failure of hafnium oxide memristors, Appl. Phys. Lett. 110 (2017) 103503, https://doi.org/10.1063/1.4974535.

Further reading

[103]   C. Baeumer, R. Dittmann, Redox-Based Memristive Metal-Oxide Devices. Elsevier, Amsterdam, 2018 489–522, https://doi.org/10.1016/B978-0-12-811166-6.00020-0.

Spin-orbit torque magnetoresistive random-access memory (SOT-MRAM)

Chong Bi, Noriyuki Sato, Shan X. Wang
Department of Materials Science and Engineering, Stanford University, Stanford, CA,
United States, Department of Electrical Engineering, Stanford University, Stanford, CA,
United States

6.1 Introduction

Magnetic tunnel junction (MTJ), basic element of magnetoresistive random-access memory (MRAM), consists of two ferromagnetic electrodes separated by a tunnel barrier layer. When the magnetized directions between the two ferromagnetic electrodes are in parallel configuration, the resistance of MTJ is low; when they are in antiparallel configuration, the resistance of MTJ is high. The low and high-resistance states can be used for recording information like other types of memory. The alternation between parallel and antiparallel configuration is a typical write operation in MRAM, which is implemented by reversing magnetization direction of one ferromagnetic electrode (free layer or storage layer) while keeping magnetization of the other ferromagnetic electrode (reference layer) fixed. In MRAM, a current rather than a magnetic field must be used for switching the free layer from the viewpoint of scalability. Conventionally, the applied current passes through the tunnel barrier layer of a MTJ to generate spin-transfer torque (STT) for switching free layer but the write current shares the same path as read current, as shown in Fig. 6.1. The large write current passing through the barrier layer will cause aging issues of tunnel barrier. In STT-MRAM, an ultrathin tunnel barrier, around 1 nm, must be used to obtain a low resistance-area (RA) product that guarantees enough write current density under the voltage capability of access transistors. However, the ultrathin tunnel barrier layer may reduce the perpendicular magnetic anisotropy (PMA) and tunneling magnetoresistance (TMR) of MTJs. The former directly determines the thermal stability of MTJs and thus the minimum dimension of a single MTJ in MRAM; the latter is related to the resistance ratio between low and high-resistance states and decides read margins of MRAM. Moreover, the ultrathin barrier layer further limits thickness tolerance of the tunnel barrier in manufacturing production since any small variation in the barrier thickness can dramatically change MTJ resistance due to its exponential dependence on the tunnel barrier thickness.

For spin-orbit torque (SOT)-MRAM device, the general method for switching the free layer is by using an in-plane current-induced SOT generated at the interface between the free layer and the SOT layer, as shown in Fig. 6.1. The SOT layer consists

Advances in Non-volatile Memory and Storage Technology. https://doi.org/10.1016/B978-0-08-102584-0.00007-3

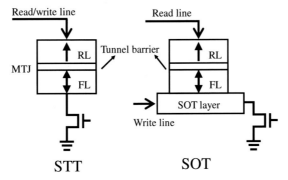

Fig. 6.1 Schematic representation of STT and SOT memory cell. *FL*, free layer; *RL*, reference layer.

Table 6.1 Representative parameters for STT-MRAM and SOT-MRAM

	RA ($\Omega\,\mu m^2$)	Switching current density (A/cm^2)	Switching time (ns)	External magnetic field (Oe)	References
STT-MRAM	<9	3×10^6	2–10	0	[1]
SOT-MRAM	Depends on three-terminal or two-terminal	5.4×10^6	0.18–0.4	50–1000 or 0	[2–4]

of materials showing strong spin Hall effects (SHE), such as heavy metals, which can generate strong SOT under an applied in-plane current. In the SOT-induced magnetization switching, the large write current does not pass through the tunnel barrier and thus the reliability and endurance are highly improved compared to STT switching. The read operation utilizes an independent current path with much smaller sensing current to detect MTJ resistance states. The separated read and write paths also reduce the read error ratio, and more importantly, it does not need RA product as low as STT-MRAM. This is because the write current passes through highly conducting heavy metal layers, not tunnel barriers, and the critical switching current can be easily satisfied by access transistors. Currently, one main obstacle for SOT-MRAM in application is that an in-plane magnetic field is required for SOT deterministic switching, which places undue burden in realizing MRAM products. The comparison of selected metrics of STT-MRAM and SOT-MRAM are given in Table 6.1. Notably, sub-ns switching speed has been achieved in SOT-MRAM, making it attractive for applications such as lower-level cache, traditionally the domain of static random access memory (SRAM).

Improving access speed of MRAM, which is determined by the switching time of the free layer, is another critical issue. In STT-MRAM, the magnetic switching is a precessional process as shown in Fig. 6.2, in which the switching time is inversely proportional to the applied current density and limited by the damping constant of

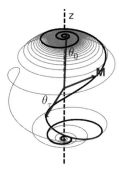

Fig. 6.2 Simulated STT induced precessional switching. **M** is the magnetization of the free layer. The initial state of **M** is at θ_0 [5].

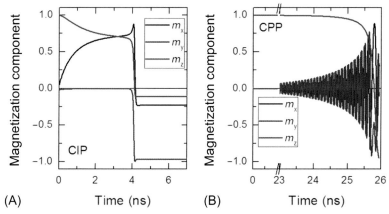

(A)   Time (ns)   (B)   Time (ns)

Fig. 6.3 Simulation results of (A) SOT (current in plane (CIP)) and (B) STT (current perpendicular to plane (CPP)) switching [7]. The easy axis is along the **z** direction; magnetization settles to ±**z** eventually.

the free layer, usually in the time scale of several nanoseconds. In the initial stage of STT switching, the magnetization of the reference and free layer is collinear and STT is zero, and thermal agitation of the free layer magnetization is necessary to yield nonzero torque. This leads to an additional incubation delay time as well as a distribution of STT switching time [5, 6]. For fast switching, a sufficiently large write current is required. As mentioned above, the large current is usually limited by the RA product of MTJs and power capability of access transistors. In SOT-MRAM, SOT is generated in adjacent heavy metal layers and orthogonal to the magnetization of the free layer. There is no incubation time during SOT switching and thus it promises an ultrafast switching process [7, 8]. Fig. 6.3 shows the comparison between SOT and STT switching. Unlike STT switching in which magnetization shows an oscillation behavior before reaching its final stable state, SOT can quickly switch the magnetization to a stable state, and the switching time can be as low as several hundred picoseconds [8]. The modeling parameters in Fig. 6.3 are: damping parameter $\alpha = 0.1$, PMA field $H_{K,eff}=0.4$ T, applied magnetic field along **x**-axis $H_x=200$ Oe, diameter of the magnetic free layer $D=30$ nm, and thickness of magnetic free layer $t_F=1$ nm.

6.2 Spin-orbit torque and SOT switching

6.2.1 Discovery of SOT

Spin torques enable manipulation of magnetic devices through an applied current. Usually, the spin torques are generated by a spin-polarized current when the applied current passes through a ferromagnetic reference layer in MTJ or spin valve structures (CPP in Fig. 6.3B). An alternative way to generate spin torques is by applying an in-plane current in the heavy-metal/ferromagnet bilayers (current in plane (CIP) in Fig. 6.3A). SHE in the heavy-metal layer and the Rashba effect at the heavy-metal/ferromagnet interface are suggested to cause the spin torques in this system. Rashba effect exists in an asymmetric crystal structure lacking inversion symmetry. It has been predicted in nonmagnetic semiconductors [9], two-dimensional (2D) systems [10, 11], and the surface of a metal [12–14]. SHE arises from asymmetric electron scattering with opposite spin directions in a nonmagnetic material which has an extrinsic [15, 16] or intrinsic scattering [17] origin. This asymmetric scattering generates a lateral spin current when applying a longitudinal electron current. As a result, nonequilibrium spin accumulation with opposite spin polarization is created at the two surfaces of a nonmagnetic material. Since both effects originate from spin-orbit coupling, the in-plane current-generated spin torque is usually referred as SOT. Similar to STT in MTJ, SOT can also induce magnetization precession [18, 19] and switching [20–22] as well as damping constant change of an adjacent ferromagnet [23]. SOT effective field was first predicted in ferromagnets by theory [24–26] and then measured in an ultrathin ferromagnetic layer [27]. The SOT effective field was initially attributed to Rashba effect after its discovery, but it was suggested later that SHE in the adjacent heavy-metal layer could also induce similar effects [21]. So far, pronounced SOT effects can only be observed in the systems with both Rashba effect and SHE, such as heavy-metal/ferromagnet system, which leads to many debates on whether Rashba effect or SHE dominates SOT behaviors.

The first MRAM-compatible metallic system showing SOT effects is Pt/Co/AlO$_x$ structures with strong PMA [27]. In this structure, the domain nucleation probability induced by an in-plane current strongly depends on an applied in-plane magnetic field (Fig. 6.4) [27]. More importantly, the in-plane field-modulated domain nucleation only happens when the applied field is orthogonal to current, that is, along the predicted Rashba field direction in theory. When the sample structure becomes symmetric, for example, in Pt/Co/Pt structure, the domain nucleation becomes insensitive with applied in-plane fields. This control experiment confirms the importance of structural inversion asymmetry, a critical condition for generating Rashba effect, in the in-plane magnetic field-modulated domain nucleation. By measuring the current-induced domain nucleation rate under different in-plane fields, a large Rashba-like effective field of 1 T per 10^8 A/cm^2 was estimated. Since the involved Pt layer has a large spin Hall angle, the

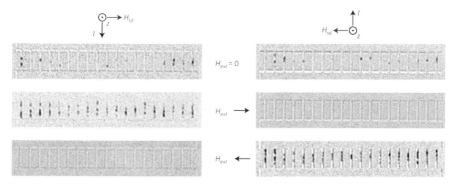

Fig. 6.4 Current-induced domain nucleation along an array of patterned Pt/Co/AlO$_x$ wires (parallel to current I, each wire is 0.5 μm wide and 5 μm long) under different in-plane magnetic fields [27]. In-plane current (I), external magnetic field (H_{ext}), and Rashba effective field (H_{sd}) are also illustrated. When H_{ext} is parallel with H_{sd}, the domain nucleation rate increases; when they are antiparallel, the domain nucleation rate decreases.

spin accumulation at the Pt/Co interface will modify the damping constant of the Co layer [23] that can also result in the change of domain nucleation rate [21]. Therefore, the observed Rashba-like effective field may also have a SHE origin. The Rashba-like effective field is also observed in Ta/CoFeB/MgO system, but with an opposite field direction compared to the Pt-based system [28], indicating opposite signs of the Rashba coefficients or spin Hall angles in the two systems. Another typical feature of SOT systems is the fast domain wall (DW) motion driven by an applied current, up to 400 m/s in Pt/Co/AlO$_x$ system [29], much faster than that in a conventional structure (<100 m/s) without SOT. Although this fast DW motion was attributed to Rashba effect [29], it turns out that SHE can also be a governing factor for this DW motion [30].

The striking development of SOT is the discovery of SOT-induced magnetic switching [20, 22] as discussed below. This SOT switching not only shows promising prospects for improving write process of emerging MRAM products, but also reveals plentiful underlying physics in the field of nanoscale and ultrafast spintronics. Since the initial experiments many efforts have been focused on the SOT material/ferromagnet system and SOT effects have been evaluated by various methods [31–34]. For example, a widely used quantitative method is to detect second harmonic signal generated by magnetization oscillation resulting from current-generated SOT (Fig. 6.5) [31]. This harmonic measurement shows that the current-generated transverse and longitudinal effective fields strongly depend on the thickness of Ta layer in a Ta/CoFeB/MgO structure. When the Ta layer is thicker, the transverse field is about three times larger than the longitudinal field [31]. These effective fields even change sign when reducing Ta thickness, suggesting competing contributions from two distinct sources. Since both Rashba effects and SHE can dramatically change when the Ta thickness is less than the spin diffusion length of Ta, these results are reasonable and suggest that these effective fields can originate from both effects. These effective fields still exist when a thin metal layer is inserted between the heavy-metal and ferromagnetic layer in a heavy-metal/ferromagnet structure [32].

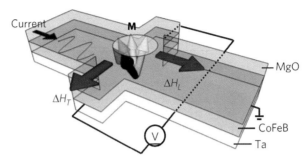

Fig. 6.5 Schematic representation of SOT measurement through second harmonic signals generated by magnetization oscillation [31]. ΔH_T and ΔH_L are transverse and longitudinal SOT effective fields, respectively.

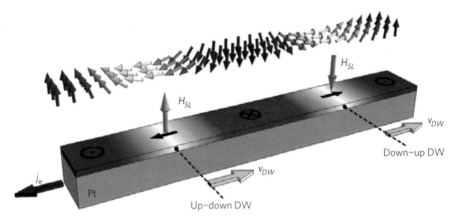

Fig. 6.6 Illustration of left-handed chiral Néel DWs in Pt-based system [35]. v_{DW} is the DW velocity, which is against j_e, the electron current. H_{SL} indicates an effective field associated with a Slonczewski-like torque generated by an applied in-plane current in the Pt layer. H_{SL} reverses in the center of up-down and down-up DWs.

Combining SOT and Dzyaloshinskii-Moriya interaction (DMI) at the heavy-metal/ferromagnet interface, unique DW motion modulated by an in-plane magnetic field has also been investigated [35, 36]. DMI, an antisymmetric exchange interaction that favors a spin canting of otherwise (anti)parallel aligned magnetic moments, is predicted to promote chiral Néel DWs in theory [37, 38] and the combination of SOT and DMI can lead to fast DW motion [39, 40]. The internal magnetization of a Néel DW in a nanowire aligns with the nanowire axis as shown in Fig. 6.6. For a left-handed chiral texture stabilized by DMI, the internal magnetization of up-down and down-up DWs are opposite. The distinctive feature of these chiral DWs is that an in-plane magnetic field can modulate current-driven DW motion separately for up-down and down-up DWs. When an in-plane magnetic field is applied along the nanowire axis, the current-driven DW motion can be promoted or suppressed depending on the relative direction between the applied in-plane field and internal magnetization of DWs. When they are in parallel configuration, the DW motion is promoted; in antiparallel configuration, the DW motion is suppressed. Experimentally, without applied in-plane magnetic field, opposite DW motion is observed in Ta- and Pt-based ferromagnetic structures [35], indicating that SHE with opposite signs in the two systems mainly contributes to the DW motion. Under an in-plane field along the nanowire axis, the current-driven DW velocity can be increased (decreased) for up-down (down-up) DWs, which can be explained only through a Néel type of chiral DW texture. The current-driven DW dynamics directly confirms DW chirality and rigidity in this system, and suggests that the underlying mechanisms are due to SOT and DMI.

6.2.2   SOT-induced magnetization switching

SOT switching was first observed in Pt/Co/AlO$_x$ structures [20], the same structure where the current-induced effective fields and unique DW motion had been widely investigated [27, 29]. As shown in Fig. 6.7, when applying an in-plane current in the Pt layer, the magnetization of the Co layer can be switched to up or down states under an in-plane magnetic field along to the current direction. The switching direction is determined by both the current and in-plane field direction. This switching behavior was unexpected because, whether for Rashba effect or SHE, the current-induced effective fields are always in-plane and should not switch a perpendicular magnetization

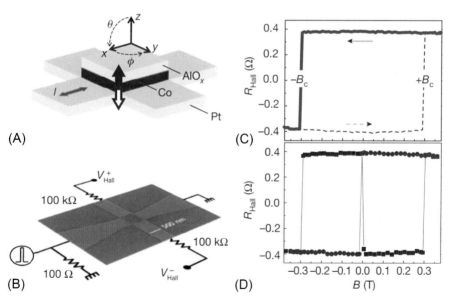

Fig. 6.7 (A) Schematic representation of the in-plane current-induced magnetization switching in Pt/Co/AlO$_x$ structure. (B) Scanning electron micrograph of the Pt/Co/AlO$_x$ device. (C) Anomalous Hall resistance (R_H) as a function of applied in-plane magnetic field. (D) R_H as a function of in-plane field after positive *(squares)* and negative *(circles)* current-induced switching [20]. $\boldsymbol{B}$ is applied nearly parallel to the current direction, but with 2 degrees offset with respect to the ideal in-plane direction, which is used to define the residual component B_z unambiguously.

according to the conventional STT theory. On the other hand, the switching only occurs when the applied in-plane field is collinear with current direction, indicating that the symmetry in the current direction must be broken during the switching process. The critical switching current density due to SOT increases with decreasing current pulse width, similar to STT-induced magnetic switching [41]. To explain the SOT switching, an effective perpendicular magnetic field due to Rashba effect was proposed [20]:

$$\boldsymbol{B}_{Sz} \approx \left(\hat{z} \times \boldsymbol{j}\right) \times \boldsymbol{B} \tag{6.1}$$

where $\boldsymbol{B}$ is the external (or applied) magnetic field, $\boldsymbol{j}$ is the applied current, and $\hat{z}$ points out-of-plane direction.

At almost the same time, the in-plane current-induced switching was demonstrated in a Ta/CoFeB/MgO system [22]. Because of the opposite spin Hall angle between Ta and Pt, the switching direction reverses for Ta/CoFeB/MgO and Pt/Co/AlO$_x$ structures, strongly indicating a SHE origin in this switching. Moreover, the in-plane current switching of an in-plane magnetized CoFeB was also demonstrated [22] and the switching direction is consistent with SHE-induced spin polarization at the Ta/CoFeB interface. This is the first work to demonstrate this type of switching in MRAM-compatible ferromagnet CoFeB that allows a high TMR. Although an in-plane field is required during the switching, it demonstrates an alternative mechanism to switch MTJ, in addition to widely used STT. A macrospin model was also proposed to explain the switching behavior [21]. As shown in Fig. 6.8, this model includes the current-induced damping-like torque (τ_{ST}), magnetic field-induced torque (τ_{ext}), and anisotropy field-induced torque (τ_{an}). The switching direction is determined by the equilibrium condition under all these torques:

$$\tau_{tot} \equiv \tau_{ST} + \tau_{ext} + \tau_{an} = 0. \tag{6.2}$$

According to this equation, reversing current or in-plane field directions will lead to opposite switching, as shown in Fig. 6.8. Although the field-like torque and possible

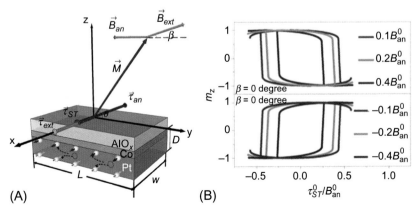

(A)   (B)

Fig. 6.8 (A) Illustration of spin torques in the macrospin model. (B) The simulated magnetization switching by using the macrospin model [21].

DMI are not considered in this model, it can qualitatively explain SOT switching behaviors, especially opposite switching directions in Pt- and Ta-based structures. For a micro-sized sample, it is reasonable that the switching arises from a domain nucleation and then domain expansion until full switching is reached. This switching process was directly imaged by using magneto-optical Kerr effect [42] and time-resolved X-ray images [43].

However, some works suggest that the DW motion may dominate the switching process regardless of initial domain nucleation. This indicates that certain unusual switching behavior may occur in SOT-induced switching [44–46], for example, the switching direction is not determined by external in-plane magnetic field direction. One proposed unusual switching model is based on in-plane field-modulated chiral DW motion. As discussed above, the chiral DW velocity can be modulated separately for up-down and down-up DWs, which can induce an asymmetric domain expansion. Specifically, if the applied in-plane field promotes the up-down DW motion and suppresses down-up DW motion as shown in Fig. 6.9C, a down domain will be shrunken and an up domain will be expanded. Therefore, the relative DW motion between two types of DWs can also induce magnetization switching [44, 47]. This chiral DW

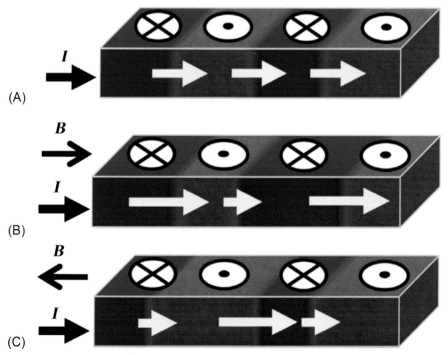

Fig. 6.9 Illustration of chiral DW motion dominated magnetization switching [47]. (A) When the DW velocity keeps the same for all DWs, the magnetization state can also be kept and there is no magnetization switching. (B, C) When the relative velocity between up-down and down-up DWs is not zero, one directional domains are expanded and the switching happens. The switching direction is determined by the sign of relative velocity.

motion-induced magnetization switching direction can be the same or opposite with the switching direction determined by the macrospin model. In a single ferromagnetic layer, the chiral DW motion-induced switching direction is the same as that predicted by the macrospin model.

An exceptional example is SOT switching in synthetic antiferromagnets (SAFs), in which the relative velocity between up-down and down-up DWs can be modulated by the strength and the direction of an applied in-plane field [44]. In SAFs, when the applied in-plane field is larger than the interlayer coupling field and interfacial DMI field, the relative internal magnetization of a DW between top and bottom ferromagnetic layers becomes parallel from an antiparallel configuration. Correspondingly, the relative velocity between up-down and down-up DWs changes sign around this field. Therefore, it is expected that the switching direction reveres with the increasing field strength according to the chiral DW motion model [44] even when the magnetic field direction remains unchanged. In contrary, macrospin model predicts the same switching directions for the same in-plane field direction [21]. Experimental demonstration of the SOT switching of SAF was performed by using a thicker top magnetic layer (TML) in Pt/BML/Ru/TML structures [44] (BML=bottom magnetic layer). Fig. 6.10A shows the velocity of two types of DW as a function of an applied in-plane field, in which there are four regions with opposite sign of relative DW

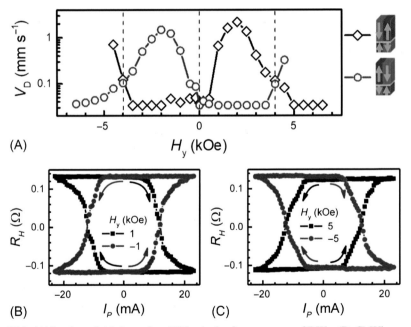

Fig. 6.10 (A) In-plane field-dependent DW velocity for two types of DWs. (B, C) When the relative velocity between two types of DWs changes sign, the magnetization switching direction reverses even with the same field direction. For example, for +1 kOe and +5 kOe fields, the switching directions are opposite, consistent with opposite sign of relative velocity as shown in (A) [44].

velocity. The corresponding SOT switching direction reverses between each region, completely consistent with the sign change of relative DW velocity as claimed in chiral DW motion model. The DW motion dominated switching was also confirmed through magneto-optical Kerr imaging and could completely explain a competing spin current-induced SOT switching by using two heavy metals with opposite spin Hall angles as the SOT layer [48].

In MRAM application, SOT switching usually requires another independent terminal for reading the resistance states of MTJ cells, in addition to two terminals for writing information. Although the three-terminal devices have many advantages as discussed above and may also facilitate to integrate SOT devices into present logic circuits or cache memory like a three-terminal field-effect transistor, two-terminal SOT devices are also attractive for high-density memory application because of a much smaller cell size. Fig. 6.11 shows a schematic representation of two-terminal SOT devices. Similar to STT-MRAM as shown in Fig. 6.1, the two-terminal SOT device has a very thin and narrow heavy-metal underlayer so that an in-plane current can also be generated when applying an out-of-plane current. The generated in-plane current creates SOT and further switches the free layer of a MTJ. The in-plane current density (J_{in}) can be calculated from this equation [4]:

$$J_{in}\left(\text{max.}\right) = \frac{\pi d_{MTJ}^2}{4w\left(t_{HM} + t_{FM}\right)} J_{out} \tag{6.3}$$

where w is the width of the underlayer, d_{MTJ} is the MTJ diameter, J_{out} is the out-of-plane current density, and t_{HM} and t_{FM} are the thickness of the underlayer and ferromagnetic layer, respectively. To achieve a larger J_{in}, a thinner underlayer is preferred, but on the other side, the thinner underlayer will reduce SOT efficiency when the thickness is comparable with the spin-diffusion length of the underlayer (typically several nm). Therefore, choosing a proper thickness of the underlayer is very important to get a high switching efficiency.

The demonstration of two-terminal SOT switching was performed in a MTJ structure by using Ta as the underlayer [4]. As shown in Fig. 6.12, a MTJ pillar

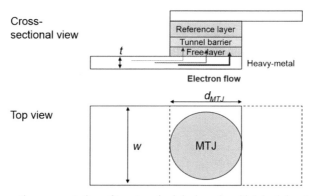

Fig. 6.11 Schematic representation of two terminal devices, in which both in-plane and out-of-plane current can be generated.

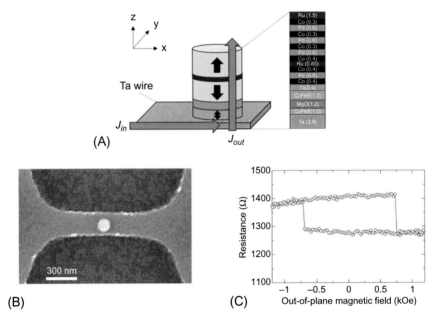

Fig. 6.12 (A) Structure and layout of two-terminal SOT devices. (B) Scanning electron micrograph of a MTJ pillar on the Ta underlayer. (C) Perpendicular resistive switching curve under an applied magnetic field.

with the diameter of 110 nm sitting on a 220-nm-wide Ta underlayer wire was fabricated by using standard electron beam lithography processes. The top electrode was then deposited for measurement. During the measurement, a 10 μs switching current pulse was applied first and then a smaller sensing current was applied to detect the resistance of MTJ. Fig. 6.13 shows the out-of-plane current-induced MTJ switching under an in-plane field. Like a three-terminal SOT switching device, the opposite switching direction for a positive and a negative in-plane field was observed, indicating a SOT-dominated MTJ switching since STT switching direction does not relate to the in-plane field direction. Without applied in-plane magnetic fields, the MTJ can also be switched due to STT as shown in Fig. 6.13C. The switching direction is consistent with that under a negative external field, but the critical switching current is about three times larger than that of the SOT-dominated switching. These results indicate that the writing current density of a conventional STT-MRAM can be reduced by about 70% through the two-terminal SOT switching. This two-terminal switching was also confirmed by using a wider underlayer. According to Eq. (6.3), the J_{in}, and thus the J_{in}-induced SOT, is inversely proportional to the width of the underlayer. Therefore, for a wider underlayer, a weaker SOT contribution is expected, and thus a much larger switching current is required. This is completely consistent with experimental results as shown in Fig. 6.14, in which the critical switching current gradually approaches to the STT-switching current with increasing the width of the underlayer.

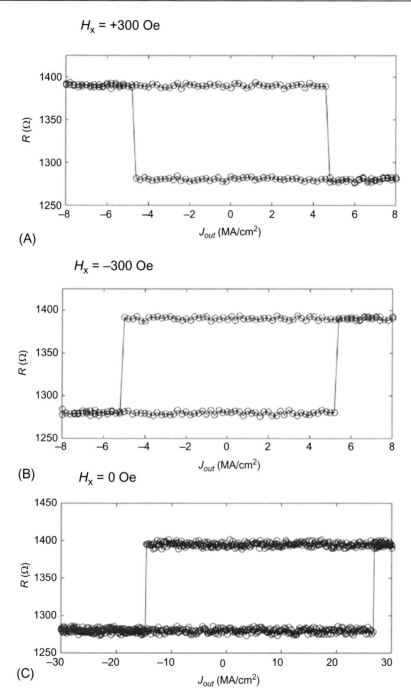

Fig. 6.13 Out-of-plane current-induced MTJ switching in a two-terminal SOT device under a (A) positive, (B) negative, and (C) zero magnetic field.

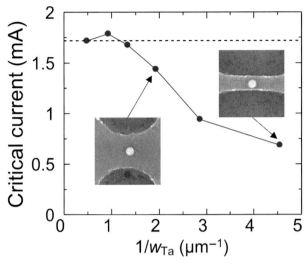

Fig. 6.14 The critical switching current of the two-terminal SOT devices as a function of the width of Ta underlayer. The MTJ diameters are 110 ± 5 nm for all devices. Insets show two typical scanning electron micrographs of the devices. The applied in-plane magnetic field is 100 Oe during the SOT switching.

6.2.3   Ultrafast SOT switching

One advantage of SOT switching is its ultrafast switching speed. As discussed above, SOT switching can be achieved in several hundreds of picoseconds, much faster than STT switching. Fig. 6.15 shows a schematic representation of an ultrafast switching measurement setup for SOT devices. A $100 \, \Omega$ resistor is connected in parallel

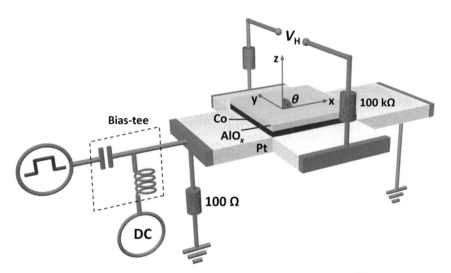

Fig. 6.15 Schematic representation of ultrafast SOT device measurements [2].

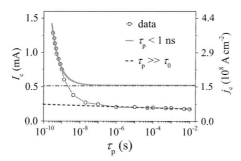

Fig. 6.16 Critical SOT switching current as a function of switching time [2].

with the sample to reduce significant reflection. The switching direction is detected through anomalous Hall effects by applying an independent sensing current. The dc sensing current and pulse current are separated by a bias tee. In a Pt/Co/AlO$_x$ structure, the switching time as low as 180 ps was demonstrated [2]. Under a constant in-plane field, the critical switching current shows two different regimes as a function of switching time, as shown in Fig. 6.16. In a short-time scale (<1 ns), the switching current sharply increases as reducing the length of applied current pulse, whereas in a longer time scale (>1 μs), the switching current shows a weak dependence on the switching time. The time-dependent switching current supports the DW motion dominated switching mechanism. Since the DW velocity is proportional to the applied current, for a sample with fixed size, the switching time is inversely proportional to the critical switching current, consistent with the observations in the short-time scale. The estimated DW velocity from the fast SOT switching is also reasonable, with a magnitude of several hundred meters per second, agreeing with current-driven DW velocity in other independent DW motion measurements [29]. The switching current in a longer time scale is a typical thermally assisted regime like STT switching, in which the thermal fluctuations help magnetization to overcome reversal energy barrier. The fast SOT switching directly confirms a negligibly small incubation time in SOT switching and demonstrates a promising candidate for ultrafast MRAM and cache memories.

The dynamic switching process during the fast SOT switching was also directly observed by time-resolved X-ray [43]. In a circular-shaped Co dots with the diameter of 500 nm, a 2-ns current pulse with the rise time of 150 ps was applied. X-ray images were recorded at time intervals of 100 ps. Fig. 6.17 shows four series of consecutive images corresponding to four possible combinations of current and field polarity. All the magnetization states were switched by domain nucleation and propagation and no appreciable incubation delay was observed. A clear DW front moving (as indicated by arrows in the center images) from fixed nucleation points (solid dots in center-left images) indicates the switching process is reproducible and deterministic. The domain nucleation favors to take place at the edge of the sample where the DMI and in-plane field concur to tilt the magnetization according to the micromagnetic simulations [49, 50]. The X-ray microscopy data provide a consistent picture of the SOT switching process as well as the influences of various effects such as field- and damping-like torque and

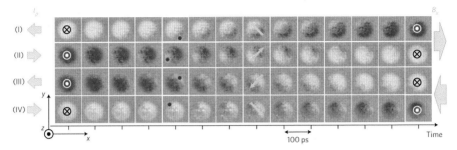

Fig. 6.17 Evolution of magnetization during the switching process imaged by X-ray microscopy. I_p and B_x indicate the applied current and in-plane field directions, respectively. The *solid dots* in the center-left images are domain nucleation points and the *arrows* in the center images indicate DW propagation direction [43].

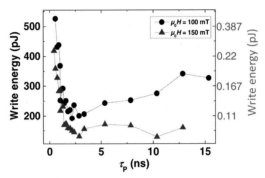

Fig. 6.18 Typical write energy of SOT-MRAM as a function of switching time [51]. *Left scale* shows the write energy for a 635 nm wide Ta underlayer and *right scale* shows the write energy extrapolated for a 50 nm wide Ta underlayer.

DMI on the domain nucleation and propagation during the switching process. The switching is robust with respect to repeated cycling events up to 10^{12} [43].

Finally, SOT switching of CoFeB-based MTJ with the switching time down to 400 ps was also demonstrated for SOT-MRAM application [51]. In addition to a dramatically increase of switching current when the switching time is down to 1 ns, it is interesting that a minimum write energy can be achieved by using current pulses with the length between 1 and 3 ns (Fig. 6.18). This minimum write energy happens in the transition region between the thermally activated regime for long switching current pulse and the short switching pulse regime. The minimum write energy of 95 fJ at 1.5 ns can be reached, which is the lowest switching energy so far in perpendicular MRAM technology [51].

6.2.4 Field-free SOT switching

One major obstacle in the development of SOT-MRAM is the required in-plane field that leads to the scaling problems in application. To realize the field-free SOT switching, there are several ways to avoid the applied in-plane field. The first demonstrated field-free switching is by using a nonuniform top oxide layer [52]. The experiment was carried out on a sputtered Ta/CoFeB/TaO$_x$ structure with a wedged TaO$_x$ layer. The sample was patterned into a Hall bar structure with the transverse direction along the wedge direction, as shown in Fig. 6.19. The wedge TaO$_x$ layer is expected to break the symmetry by creating a lateral structural asymmetry. Since the PMA of the CoFeB layer is very sensitive to the thickness of the adjacent thin oxide layer, the wedged TaO$_x$ layer actually creates a nonuniform PMA in the entire film. The nonuniform PMA inevitably induces a slight tilt of perpendicular magnetization, similar to the effects induced by an applied in-plane field during SOT switching. In fact, the tilted magnetization can also be induced by gradually altering the thickness of CoFeB through carefully controlled ion milling [53] because PMA is also very sensitive to the thickness of the ferromagnetic layer. The tilted PMA-induced lateral asymmetry is an effective configuration to achieve field-free SOT switching. In this method of field-free SOT switching, it is reasonable that the field-free switching directions are opposite for those spatial positions with an increasing and decreasing PMA along the wedge direction [52]. This is because the increasing PMA and decreasing PMA lead to opposite magnetization tilt directions and create reverse asymmetry. Although the nonuniform oxide layer or ferromagnetic layer is not practical for commercial manufacturing processes, it motivates researchers worldwide to further develop more ingenious field-free SOT switching devices for practical uses.

After the first demonstration of field-free SOT switching, many reports focus on the replacement of the applied in-plane field by using an interlayer coupling field from another ferromagnetic layer [54, 55] or exchange bias field from an antiferromagnetic layer [56, 57]. Fig. 6.20A shows a schematic representation of field-free SOT switching structures with the interlayer coupling generated from a top in-plane magnetized ferromagnetic layer. The antiferromagnetic layer on the top of the entire structure is

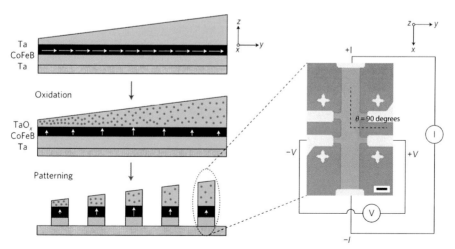

Fig. 6.19 Illustration of growth and patterning of SOT devices by using a nonuniform TaO$_x$ layer in Ta/CoFeB/TaO$_x$ structure. The measurement configuration is also shown at the right.

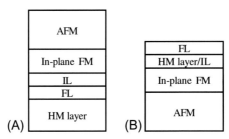

Fig. 6.20 Schematic illustration of two types of field-free SOT switching (IL (A) on the top or (B) at the bottom of FL) by using interlayer coupling structure. *AFM*, antiferromagnet; *FL*, (ferromagnetic) free layer; *FM*, ferromagnet; *HM*, heavy metal; *IL*, (nonmagnetic) interlayer coupling layer.

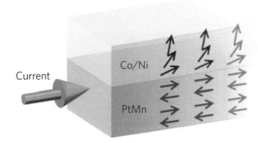

Fig. 6.21 Schematic representation of a SOT device by using antiferromagnet as the SOT layer [56].

used for pinning the adjacent in-plane ferromagnetic layer to the current direction. This structure usually can get a relatively strong PMA and SOT because the heavy-metal layer directly neighbors the free layer. However, it is not compatible with a MTJ structure since a tunnel layer and reference layer on the top of the free layer are necessary to generate TMR. Another structure for the field-free SOT switching is shown in Fig. 6.20B, in which the heavy-metal layer is served as both the SOT and coupling space layer. To get a considerable interlayer coupling, the heavy-metal layer must be very thin, usually less than 2 nm and comparable with its spin diffusion length, which will reduce both PMA and SOT efficiency and cannot show a clear switching behavior [55]. The preferred switching direction in these devices is determined by both the current and interlayer coupling directions. For the ferromagnetic and antiferromagnetic coupling with the same pinning direction, the field-free switching direction reverses.

Antiferromagnets have also been demonstrated to show strong SOT [58], and thus it is possible to replace the heavy-metal layer by using an antiferromagnet which can provide both SOT and exchange bias field simultaneously, as shown in Fig. 6.21. In this structure, SHE-induced spin accumulation at the antiferromagnet/ferromagnet interface provides SOT to switch the ferromagnet and the in-plane exchange bias field generated by the antiferromagnet breaks the symmetry. However, PMA of the ferromagnet in this structure is very weak by using an antiferromagnetic layer as the buffer layer, especially for CoFeB free layer promising high TMR in MTJs. To get a perpendicularly magnetized ferromagnet, Co/Ni multilayers [56] in which PMA has a weak dependence on the buffer layer are often adopted. Another approach to get PMA is to insert the antiferromagnetic layer between the heavy-metal and ferromagnetic layer in a heavy-metal/ferromagnet/oxide structure, but the antiferromagnet layer must be very thin to maintain PMA of the heavy-metal/ferromagnet structure [57]. With an in-plane pinned direction along the current direction, field-free SOT switching was demonstrated in these antiferromagnet-based structures [56, 57]. Due to the weak PMA, the SOT switching in this structure shows a gradually switching behavior like a memristor [56]. According to the DW motion model, the weak PMA cannot provide a sufficiently large DW velocity to enable sharp switching [44].

In the heavy metals and antiferromagnets, SOT generated at their surface is orthogonal to the perpendicular magnetization of the adjacent ferromagnet and an in-plane field is required to assist SOT-induced deterministic switching. In ferromagnets, it has been demonstrated that a spin current with an out-of-plane polarization, can be also generated by an applied in-plane current. The spin polarization depends on the magnetization direction of the ferromagnet [59]. Together with in-plane SOT, if the out-of-plane polarized spin current can be used for switching an adjacent perpendicularly magnetized ferromagnet, there will be no in-plane field required during the switching process. As shown in Fig. 6.22, by considering an in-plane magnetized ferromagnet adjacent to a normal metal, the spin current resulting from the in-plane current, flowed through the normal metal, and had an out-of-plane component of the spin polarization in addition to in-plane components. There are two processes during the generation of the spin current [60]. First, the applied in-plane electric field creates a net spin propagation normal to the ferromagnet/normal metal interface. Second, the polarization of the net spin current injected into the normal metal is established through interfacial spin-orbit scattering. Two distinct mechanisms are involved in the interfacial spin-orbit scattering: spin-orbit filtering and spin-orbit procession [60]. The former gives a net spin polarization along $y = z \times E$ direction, identical to that of SHE-induced spin current. Here, z is the normal direction of the ferromagnet/normal metal interface and E is the electric field. The latter gives a net spin polarization in the $m \times y$ direction, where m is the magnetization vector of the in-plane magnetized ferromagnet. Therefore, for a ferromagnet magnetized along the x-direction, this mechanism can generate a spin current flowing into the normal metal with an out-of-plane polarization. The experimental demonstration of the out-of-plane SOT was performed

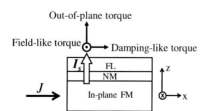

Fig. 6.22 Illustration of SOT devices by using an in-plane magnetized ferromagnet as the SOT layer and the schematic representation of current-induced torques in a perpendicularly magnetized ferromagnet (FL). I_s, spin current; *NM*, normal metal.

in CoFeB/Ti and NiFe/Ti structures and a considerable SOT was detected [60]. The SOT was large enough to switch an adjacent perpendicularly magnetized ferromagnet without an applied in-plane field. Interestingly, the signs of the SOT and thus field-free SOT switching directions in both structures were opposite (Fig. 6.23). The opposite SOT seems to relate with the anomalous Hall effects that also show opposite signs in both structures, but on the other hand, it is determined by the ferromagnet/normal metal interface. Since spin polarization due to current in this system is collinear with the perpendicular magnetization of the free layer like STT switching, it is unclear if its switching speed can be as high as the orthogonal SOT switching in a heavy-metal/ferromagnet system.

Although the field-free SOT switching has been widely investigated, it is only demonstrated to switch a single ferromagnetic layer, not a MTJ structure used in SOT-MRAM. These field-free switching approaches either are not compatible with MTJ structure or cannot provide strong PMA in practical applications. For the emerging SOT-MRAM applications, the demonstration of field-free switching of MTJs with strong PMA and high TMR ratios remain to realized experimentally.

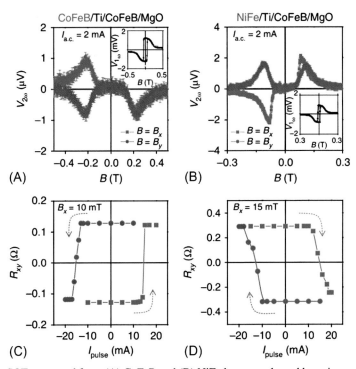

Fig. 6.23 SOT generated from (A) CoFeB and (B) NiFe layers evaluated by using second harmonic measurements. *Reverse peaks* in these two figures indicate opposite SOT. (C, D) Current-induced magnetization switching in (C) CoFeB and (D) NiFe-based systems under a positive in-plane field. The switching directions are also opposite in these two systems.

6.2.5 Further optimization of SOT switching

To achieve low-power consumption of SOT-MRAM, further improving SOT switching, especially reducing the critical switching current, is required. The critical switching current is directly determined by the effective SOT applied in the free layer. The effective SOT originates from SHE of adjacent heavy metals, but strongly depends on the transparence of the heavy-metal/ferromagnet interface. To get strong SOT, various materials with large spin Hall angle are explored. Table 6.2 shows the spin Hall angle of several typical materials. W with β phase shows a largest spin Hall angle in all MTJ-compatible heavy metals, which probably is the most promising heavy metal material for current SOT-MRAM application. Although topological insulator and WTe_2 are reported to have considerable spin Hall angles, [67] their conductivities are too low and fabrication processes are not compatible with current semiconductor process. Several reports show that the spin Hall angle can be improved by introducing oxygen during film growth [68, 69], even for the light metals with a negligible spin Hall angle when pure [68]. With an oxygen doped level of 12.1%, the spin Hall angle of W is improved to −0.49 (from −0.3 when pure). [69] The large spin Hall angle remains even at high oxygen concentrations, although the electrical resistivity, microstructure, and thickness of the W film are further changed by the high oxygen doping, indicating that the enhancement of spin Hall angle may originate from the oxygen-modulated W/ferromagnet interface. Similarly, in a Cu layer which has a very small spin Hall angle when pure, such oxidation process can enhance its spin Hall angle by two orders of magnitude, nearly reaching the very large value of the spin Hall angle of a pure Pt layer [69].

Controlling interfacial transparency of the heavy-metal/ferromagnet layer is another approach to improve SOT. By using both the spin-torque ferromagnetic resonance (ST-FMR) [70] and second harmonic measurements [71], it has been demonstrated that the spin Hall angle of a given SOT layer strongly depends on the composition, thickness, and post-deposition processing protocols of the adjacent ferromagnet layer. For Pt/Co structures with a highly transparent Pt/Co interface, the measured spin Hall

Table 6.2 Spin Hall angle of typical materials at room temperature

Materials	Spin Hall angle (%)	Measurements	References
Pt	8.5 ± 0.9	STT-FMR	[61]
Au	0.25 ± 0.1	SP	[62]
β-Ta	-12 ± 4	STT-FMR	[22]
β-W	-30 ± 2	STT-FMR	[63]
Ru	3.3	STT-FMR	[64]
IrMn	18 ± 1	SP	[65]
PtMn	31 ± 1	SP	[65]
Bi_2Se_3	200–350	STT-FMR	[66]
WTe_2	1.3	STT-FMR	[67]

SP, spin pumping; *STT-FMR*, spin torque ferromagnetic resonance.

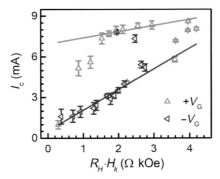

Fig. 6.24 Critical SOT switching current in Pt/Co/GdO$_x$ structure as a function of $R_H \cdot H_k$ under a clean (+V_G)/contaminated (−V_G) Pt/Co interface. R_H is the anomalous Hall resistance; H_k is the perpendicular anisotropy field. *Solid lines* are linear fitting results and their slopes are inversely proportional to the effective spin Hall angle [72].

angle is about two times larger than that in Pt/NiFe structures with low transparency [70]. Since the spin Hall angle is an intrinsic property of SOT materials, it should not depend on the adjacent ferromagnet if the SHE-generated spin torque can be completely transferred to the adjacent ferromagnet. This contradiction indicates that the interfacial transparency plays a central role in generating high-efficiency SOT. The importance of interfacial transparency is also confirmed by altering the heavy-metal/ferromagnet interface through insertion of an atomically thin magnetic layer. Direct confirmation of the effects of interfacial transparency on SOT switching efficiency was performed in the same structure through gate voltage control [72]. In a Pt/Co/oxide structure, the transparency of the Pt/Co interface can be controlled by voltage-driven oxygen motion. The critical SOT switching current strongly depends on the applied gate voltage. As shown in Fig. 6.24, after applying a positive gate voltage (corresponding to a cleaner or more transparent Pt/Co interface), the estimated spin Hall angle is about 3.5 times larger than that after applying a negative gate voltage.

Interestingly, the SOT efficiency can also be enhanced by modulating another interface of the adjacent ferromagnet. Specifically, in a Pt/ferromagnet/capping-layer structure, when the thickness of the ferromagnet layer is smaller than spin dephasing length, a Ru capping layer can largely boost the absorption of spin current in the ferromagnet layer and thus substantially enhances the strength of SOT acting on the ferromagnet layer. Compared to a MgO capping layer, the switching current can be reduced by about 75% and SOT efficiency can be enhanced by about 1.7 times by inserting a 0.6 nm Ru capping layer between the ferromagnet and MgO layer [64]. Fig. 6.25 illustrates the possible mechanism of SOT enchantment with a Ru capping layer. In Pt/ferromagnet/MgO structures, the transverse spin currents from the Pt layer cannot be fully absorbed in the ferromagnet and then are reflected at the ferromagnet/MgO interface; in the Pt/ferromagnet/Ru structures, the Ru layer may absorb the incoming spin current as a spin sink that could enhance the spin current absorption in the ferromagnet. Although the insertion of Ru layer between the ferromagnet and MgO layer may reduce TMR of a MTJ [73, 74], this result highlights that modulating the

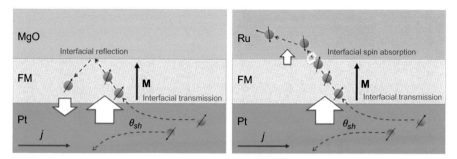

Fig. 6.25 Schematic representation of enhanced SOT by modulating capping layers [64].

top ferromagnet/capping-layer interface can also enhance SOT efficiency, in addition to controlling the heavy-metal/ferromagnet interface where the SOT originates.

6.3   SOT-MRAM and future trends

The complementary metal-oxide-semiconductor (CMOS) technology is now approaching its fundamental limitations in areal density, power, and performance as the size of a single device reaches the physical limits of silicon. Spintronic devices provide a low-power alternative to the CMOS technology to stretch Moore's law both in logic and memory applications. Here, several potential applications of SOT memory in the field of conventional semiconductor memories are prospected.

Multiport memories are widely adopted in a microprocessor as shared memory allowing simultaneous array accesses, but they are facing critical challenges in bit-cell leakage, scalability, and reliability issues. SOT memory technology shows a potential to overcome these challenges. For example, one read and one write (1R1W transistor) bit-cell architecture of SOT-MRAM is proposed [75] to avoid sneaky current path during operational periods. It consists of an MTJ cell and two access transistors separating write and read paths. During the read and write operations, only the corresponding access transistor is on, controlled by its own word line. The independent write and read paths indicate that they can be optimized separately and operated simultaneously with a negligible influence on each other. This unique read/write path design is very important for realizing multiport MRAM [76, 77]. In multiport memories, the read and write contention can be inherently resolved at a bit-cell level, delivering a dual-port characteristic. Fig. 6.26 shows a block diagram of a typical dual-port SOT memory. Two decoders are introduced to facilitate two different addresses to access the bit-cell array and read-write periphery circuitry to perform memory operations. The read and write circuits are the same as those in standard SOT-MRAM design. The multiport design is implemented on the top of the single-port SOT-MRAM with just an additional modification of accessibility to the bit-cell by introducing another transistor in the write terminal as shown in Fig. 6.27. This is because the bit-cell inherently provides the characteristics for both write and read operations that simplify the overall architecture design. The introduced transistor

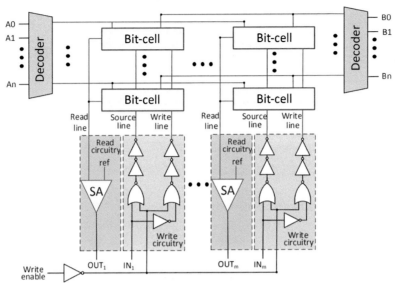

Fig. 6.26 Block diagram of 1R1W multiport SOT-MRAM [76].

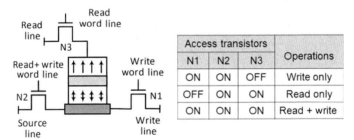

Access transistors			Operations
N1	N2	N3	
ON	ON	OFF	Write only
OFF	ON	ON	Read only
ON	ON	ON	Read + write

Fig. 6.27 Modified SOT bit-cell architecture and its operations with respect to their access transistors [76].

eliminates asymmetric current distribution in the write current path due to the write access transistor in a single-port SOT-MRAM, which in turn balances write latency. The
extra access transistor is required to be on for both read and write operations. The bit-cell
operations with respect to their access transistors are also shown in Fig. 6.27. Fig. 6.28
demonstrates the write and read current flow path for different operations. When only
write is performed, a bidirectional current path is established using write circuitry and
the direction of the write current is shown in Fig. 6.28A. During the write operation,
the two access transistors connected to the write terminals are always on and the read
access transistor is off. For read only operations, the write access transistor is off and
the read access transistor and extra transistor in the write terminal are on. A read only
current path is shown in Fig. 6.28B. When both read and write operations are performed
simultaneously, all the three transistors are on. In this case, the write current shows the
same behaviors as Fig. 6.28A, but the read current path must be altered based on the

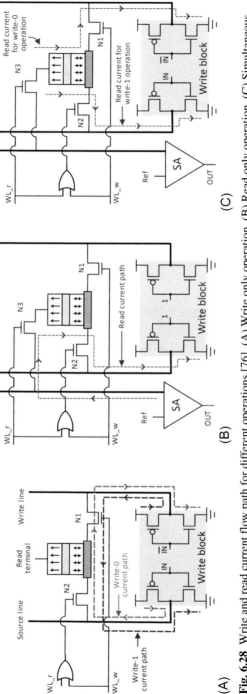

Fig. 6.28 Write and read current flow path for different operations [76]. (A) Write only operation. (B) Read only operation. (C) Simultaneous read-write on same bit-cell.

write current flow directions, as shown in Fig. 6.28C. Simulation results show that this multiport memory architecture is highly beneficial for read operations and can increase the write energy efficiency by up to 68% [76].

In-memory computing architecture and nonvolatile memory technology have been proposed to integrate memory and logic together, with promise to realize a memory-oriented processing for large datasets at exabyte scale (10^{18} bytes/s or flops). SOT-MRAM array is suggested to work as the nonvolatile memory in the in-memory computing [78]. Such SOT-MRAM array design can implement a reconfigurable in-memory logic without additional logic circuits like conventional logic-in-memory designs. The designed in-memory device can be used to process data locally, being much less power hungry and providing much shorter distance data communication than conventional Von Neumann computing systems [78–80]. Fig. 6.29 shows the details on circuit realization of SOT-MRAM in-memory computing platform. The architecture of memory sub-array is depicted in Fig. 6.29A, where the read/write path of a specific bit-cell is enabled by the row/column decoders. Fig. 6.29D shows the modified row/column decoders that can enable either single line (memory operation) or double lines (logic computation), depending on the selected addresses. For write operation in memory, the writing voltage on write bit-line is generated by the voltage drivers [81], which produces sufficient current for fast memory switching. For read operations and logic computation, a small sense current is injected into the read path and then a sense voltage is generated. As shown in Fig. 6.29E, the modified sense circuit can realize memory read, AND/NAND, OR/NOR, and XOR/XNOR functions, through the combination of two sense amplifiers, external CMOS logic gates and control units. The simulated results show that the SOT-MRAM can reproduce the desired functions well.

SOT-MRAM cells are also suggested to design graphics processing unit (GPU) register file which provides higher energy efficiency than SRAM and STT-MRAM [82]. The energy saving can be as much as 44.3% compared to SRAM register file, without harming performance. A SOT-MRAM-based register file architecture is shown in Fig. 6.30, in which the SOT-MRAM replaces conventional SRAM with the same read/write latency. SOT-MRAM may also replace SRAM embedded in the microcontroller used in intermittently powered systems with higher energy-efficiency [83–85]. Simulation results indicate that this SOT-MRAM configuration leads to significant memory energy benefits of 2.6× on average, compared to SRAM+STT-MRAM memory configurations [85]. To match semiconductor memories in both speed and density, multilevel SOT memories with a higher capacity are also proposed [86, 87].

6.4 Conclusions

In this chapter, we have presented an overview of SOT-MRAM. Although SOT-MRAM is still in its infancy, it has been shown promising in various memory applications with high-speed and low-energy consumption, such as cache memory, nonvolatile register [88], and stand-alone memory. However, there are many remaining challenges in its development, including optimizing materials and structures, clarifying SOT switching mechanisms, and achieving practical field-free

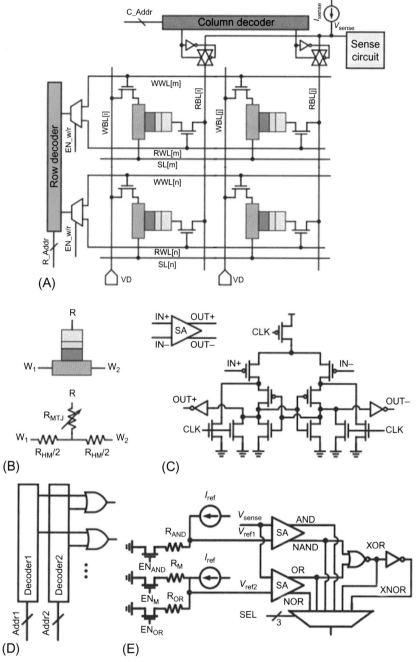

Fig. 6.29 (A) The modified memory architecture. (B) Equivalent resistive model of the SOT-MRAM cell. (C) The schematic representation of sense amplifier (SA). (D) Modified decoder providing single/multiple lines enable function. (E) Modified sense circuit for both memory and in-memory computing operations [78].

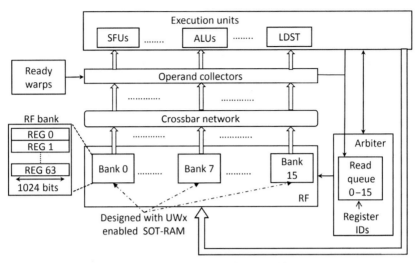

Fig. 6.30 SOT-MRAM-based GPU register file architecture [82].

switching. New SOT materials such as topological insulators and 2D materials may be utilized to reduce switching current in the future. Further improvements of field-free SOT switching of MTJ with strong PMA and high TMR ratios are also expected. On the other hand, skyrmions with smaller sizes than patterned bits and high current-driven velocities are observed in the SOT-compatible structures [89], making SOT memory with higher densities and ultrafast speeds tantalizing. In addition, further improving SOT-MRAM design, such as one transistor one Schottky diode SOT-MRAM, is proposed to reduce the cell size of SOT-MRAM [90]. Compared to state-of-the-art STT-MRAM [91], SOT-MRAM directly addresses several critical issues in MRAM development, such as high write currents (especially at fast writing), the trade-off between RA and TMR of MTJ bits, and the conflicting requirements of data retention and fast programming. With ingenious and global efforts in industry and academia [92, 93], practical SOT-MRAM products for cache memory or ultrafast neural networks may become realistic in a few years.

Acknowledgments

The authors wish to thank Dr. Wilman Tsai for a critical reading and revision of this book chapter, and Drs Daniel Worledge and Ian Young for valuable input. S.X.W. wishes to thank TSMC, Stanford SystemX Alliance, Stanford Center for Magnetic Nanotechnology, and the NSF Center for Energy Efficient Electronics Science (E^3S) for financial support. This work was supported in part by ASCENT, one of six centers in JUMP, a Semiconductor Research Corporation (SRC) program sponsored by Defense Advanced Research Projects Agency. The experimental work has benefited from the equipment and tools at Stanford Nanofabrication Facility, Stanford Nano Shared Facilities, and Michigan Lurie Nanofabrication Facility (LNF) which are supported by the National Science Foundation (NSF).

References

[1]   M. Wang, et al., Current-induced magnetization switching in atom-thick tungsten engineered perpendicular magnetic tunnel junctions with large tunnel magnetoresistance, Nat. Commun. 9 (2018) 671.

[2]   K. Garello, et al., Ultrafast magnetization switching by spin-orbit torques, Appl. Phys. Lett. 105 (2014) 212402.

[3]   S. Shi, Y. Ou, S.V. Aradhya, D.C. Ralph, R.A. Buhrman, Fast low-current spin-orbit-torque switching of magnetic tunnel junctions through atomic modifications of the free-layer interfaces, Phys. Rev. Appl. 9 (2018) 011002.

[4]   N. Sato, F. Xue, R.M. White, C. Bi, S.X. Wang, Two-terminal spin–orbit torque magnetoresistive random access memory, Nat. Electron. 1 (2018) 508–511.

[5]   H. Liu, et al., Dynamics of spin torque switching in all-perpendicular spin valve nanopillars, J. Magn. Magn. Mater. 358–359 (2014) 233–258.

[6]   T. Devolder, C. Chappert, J.A. Katine, M.J. Carey, K. Ito, Distribution of the magnetization reversal duration in subnanosecond spin-transfer switching, Phys. Rev. B 75 (2007) 64402.

[7]   K.-S. Lee, S.-W. Lee, B.-C. Min, K.-J. Lee, Threshold current for switching of a perpendicular magnetic layer induced by spin Hall effect, Appl. Phys. Lett. 102 (2013) 112410.

[8]   W. Legrand, R. Ramaswamy, R. Mishra, H. Yang, Coherent subnanosecond switching of perpendicular magnetization by the fieldlike spin-orbit torque without an external magnetic field, Phys. Rev. Appl. 3 (2015) 64012.

[9]   D. Awschalom, N. Samarth, Spintronics without magnetism, Physics (College Park, Md) 2 (2009) 50.

[10] J. Inoue, G.E.W. Bauer, L.W. Molenkamp, Diffuse transport and spin accumulation in a Rashba two-dimensional electron gas, Phys. Rev. B 67 (2003) 33104.

[11] V.M. Edelstein, Spin polarization of conduction electrons induced by electric current in two-dimensional asymmetric electron systems, Solid State Commun. 73 (1990) 233–235.

[12] O. Krupin, et al., Rashba effect at magnetic metal surfaces, Phys. Rev. B 71 (2005) 201403.

[13] J. Henk, M. Hoesch, J. Osterwalder, A. Ernst, P. Bruno, Spin–orbit coupling in the L-gap surface states of Au(111): spin-resolved photoemission experiments and first-principles calculations, J. Phys. Condens. Matter 16 (2004) 7581–7597.

[14] S. LaShell, B.A. McDougall, E. Jensen, Spin splitting of an Au(111) surface state band observed with angle resolved photoelectron spectroscopy, Phys. Rev. Lett. 77 (1996) 3419–3422.

[15] S. Zhang, Spin Hall effect in the presence of spin diffusion, Phys. Rev. Lett. 85 (2000) 393–396.

[16] J.E. Hirsch, Spin Hall effect, Phys. Rev. Lett. 83 (1999) 1834–1837.

[17] J. Sinova, et al., Universal intrinsic spin Hall effect, Phys. Rev. Lett. 92 (2004) 126603.

[18] L. Liu, T. Moriyama, D.C. Ralph, R.A. Buhrman, Spin-torque ferromagnetic resonance induced by the spin Hall effect, Phys. Rev. Lett. 106 (2011) 36601.

[19] L. Liu, C.-F. Pai, D.C. Ralph, R.A. Buhrman, Magnetic oscillations driven by the spin Hall effect in 3-terminal magnetic tunnel junction devices, Phys. Rev. Lett. 109 (2012) 186602.

[20] I.M. Miron, et al., Perpendicular switching of a single ferromagnetic layer induced by in-plane current injection, Nature 476 (2011) 189–193.

[21] L. Liu, O.J. Lee, T.J. Gudmundsen, D.C. Ralph, R.A. Buhrman, Current-induced switching of perpendicularly magnetized magnetic layers using spin torque from the spin Hall effect, Phys. Rev. Lett. 109 (2012) 96602.
[22] L. Liu, et al., Spin-torque switching with the giant spin Hall effect of tantalum, Science 336 (2012) 555–558.
[23] K. Ando, et al., Electric manipulation of spin relaxation using the spin Hall effect, Phys. Rev. Lett. 101 (2008) 36601.
[24] A. Manchon, S. Zhang, Theory of nonequilibrium intrinsic spin torque in a single nanomagnet, Phys. Rev. B 78 (2008) 212405.
[25] A. Manchon, S. Zhang, Theory of spin torque due to spin-orbit coupling, Phys. Rev. B 79 (2009) 94422.
[26] A. Matos-Abiague, R.L. Rodríguez-Suárez, Spin-orbit coupling mediated spin torque in a single ferromagnetic layer, Phys. Rev. B 80 (2009) 94424.
[27] I.M. Miron, et al., Current-driven spin torque induced by the Rashba effect in a ferromagnetic metal layer, Nat. Mater. 9 (2010) 230–234.
[28] T. Suzuki, et al., Current-induced effective field in perpendicularly magnetized Ta/CoFeB/MgO wire, Appl. Phys. Lett. 98 (2011) 142505.
[29] I.M. Miron, et al., Fast current-induced domain-wall motion controlled by the Rashba effect, Nat. Mater. 10 (2011) 419–423.
[30] P.P.J. Haazen, et al., Domain wall depinning governed by the spin Hall effect, Nat. Mater. 12 (2013) 299–303.
[31] J. Kim, et al., Layer thickness dependence of the current-induced effective field vector in Ta|CoFeB|MgO, Nat. Mater. 12 (2013) 240–245.
[32] X. Fan, et al., Observation of the nonlocal spin-orbital effective field, Nat. Commun. 4 (2013) 1799.
[33] K. Garello, et al., Symmetry and magnitude of spin–orbit torques in ferromagnetic heterostructures, Nat. Nanotechnol. 8 (2013) 587–593.
[34] X. Fan, et al., Quantifying interface and bulk contributions to spin–orbit torque in magnetic bilayers, Nat. Commun. 5 (2014) 3042.
[35] S. Emori, U. Bauer, S.-M. Ahn, E. Martinez, G.S.D. Beach, Current-driven dynamics of chiral ferromagnetic domain walls, Nat. Mater. 12 (2013) 611–616.
[36] K.-S. Ryu, L. Thomas, S.-H. Yang, S. Parkin, Chiral spin torque at magnetic domain walls, Nat. Nanotechnol. 8 (2013) 527–533.
[37] M. Heide, G. Bihlmayer, S. Blügel, Dzyaloshinskii-Moriya interaction accounting for the orientation of magnetic domains in ultrathin films: Fe/W(110), Phys. Rev. B 78 (2008) 140403.
[38] X.Z. Yu, et al., Real-space observation of a two-dimensional skyrmion crystal, Nature 465 (2010) 901–904.
[39] A. Thiaville, S. Rohart, É. Jué, V. Cros, A. Fert, Dynamics of Dzyaloshinskii domain walls in ultrathin magnetic films, EPL (Europhys. Lett.) 100 (2012) 57002.
[40] A.V. Khvalkovskiy, et al., Matching domain-wall configuration and spin-orbit torques for efficient domain-wall motion, Phys. Rev. B 87 (2013) 20402.
[41] R.H. Koch, J.A. Katine, J.Z. Sun, Time-resolved reversal of spin-transfer switching in a nanomagnet, Phys. Rev. Lett. 92 (2004) 88302.
[42] G. Yu, et al., Magnetization switching through spin-Hall-effect-induced chiral domain wall propagation, Phys. Rev. B 89 (2014) 104421.
[43] M. Baumgartner, et al., Spatially and time-resolved magnetization dynamics driven by spin–orbit torques, Nat. Nanotechnol. 12 (2017) 980–986.

[44] C. Bi, et al., Anomalous spin-orbit torque switching in synthetic antiferromagnets, Phys. Rev. B 95 (2017) 104434.

[45] J. Yoon, et al., Anomalous spin-orbit torque switching due to field-like torque–assisted domain wall reflection, Sci. Adv. 3 (2017) e1603099.

[46] J.M. Lee, et al., Oscillatory spin-orbit torque switching induced by field-like torques, Commun. Phys. 1 (2018) 2.

[47] C. Bi, M. Liu, Reversal-mechanism of perpendicular switching induced by an in-plane current, J. Magn. Magn. Mater. 381 (2015) 258–262.

[48] Q. Ma, et al., Switching a perpendicular ferromagnetic layer by competing spin currents, Phys. Rev. Lett. 120 (2018) 117703.

[49] E. Martinez, et al., Universal chiral-triggered magnetization switching in confined nanodots, Sci. Rep. 5 (2015) 10156.

[50] N. Mikuszeit, et al., Spin-orbit torque driven chiral magnetization reversal in ultrathin nanostructures, Phys. Rev. B 92 (2015) 144424.

[51] M. Cubukcu, et al., Ultra-fast perpendicular spin–orbit torque MRAM, IEEE Trans. Magn. 54 (2018) 9300204.

[52] G. Yu, et al., Switching of perpendicular magnetization by spin–orbit torques in the absence of external magnetic fields, Nat. Nanotechnol. 9 (2014) 548–554.

[53] L. You, et al., Switching of perpendicularly polarized nanomagnets with spin orbit torque without an external magnetic field by engineering a tilted anisotropy, Proc. Natl. Acad. Sci. U. S. A. 112 (2015) 10310–10315.

[54] A. van den Brink, et al., Field-free magnetization reversal by spin-Hall effect and exchange bias, Nat. Commun. 7 (2016) 10854.

[55] Y.-C. Lau, D. Betto, K. Rode, J.M.D. Coey, P. Stamenov, Spin–orbit torque switching without an external field using interlayer exchange coupling, Nat. Nanotechnol. 11 (2016) 758–762.

[56] S. Fukami, C. Zhang, S. DuttaGupta, A. Kurenkov, H. Ohno, Magnetization switching by spin–orbit torque in an antiferromagnet–ferromagnet bilayer system, Nat. Mater. 15 (2016) 535–541.

[57] Y.-W. Oh, et al., Field-free switching of perpendicular magnetization through spin–orbit torque in antiferromagnet/ferromagnet/oxide structures, Nat. Nanotechnol. 11 (2016) 878–884.

[58] J. Železný, P. Wadley, K. Olejník, A. Hoffmann, H. Ohno, Spin transport and spin torque in antiferromagnetic devices, Nat. Phys. 14 (2018) 220–228.

[59] A.M. Humphries, et al., Observation of spin-orbit effects with spin rotation symmetry, Nat. Commun. 8 (2017) 911.

[60] S.C. Baek, et al., Spin currents and spin–orbit torques in ferromagnetic trilayers, Nat. Mater. 17 (2018) 509–513.

[61] A. Ganguly, et al., Thickness dependence of spin torque ferromagnetic resonance in $Co_{75}Fe_{25}$/Pt bilayer films, Appl. Phys. Lett. 104 (2014) 72405.

[62] V. Vlaminck, J.E. Pearson, S.D. Bader, A. Hoffmann, Dependence of spin-pumping spin Hall effect measurements on layer thicknesses and stacking order, Phys. Rev. B 88 (2013) 64414.

[63] C.-F. Pai, et al., Spin transfer torque devices utilizing the giant spin Hall effect of tungsten, Appl. Phys. Lett. 101 (2012) 122404.

[64] X. Qiu, et al., Enhanced spin-orbit torque via modulation of spin current absorption, Phys. Rev. Lett. 117 (2016) 217206.

[65] W. Zhang, et al., Spin Hall effects in metallic antiferromagnets, Phys. Rev. Lett. 113 (2014) 196602.

[66] A.R. Mellnik, et al., Spin-transfer torque generated by a topological insulator, Nature 511 (2014) 449–451.

[67] D. MacNeill, et al., Control of spin–orbit torques through crystal symmetry in WTe2/ferromagnet bilayers, Nat. Phys. 13 (2016) 300–305.

[68] H. An, Y. Kageyama, Y. Kanno, N. Enishi, K. Ando, Spin–torque generator engineered by natural oxidation of Cu, Nat. Commun. 7 (2016) 13069.

[69] K.-U. Demasius, et al., Enhanced spin–orbit torques by oxygen incorporation in tungsten films, Nat. Commun. 7 (2016) 10644.

[70] W. Zhang, W. Han, X. Jiang, S.-H. Yang, S.P. Parkin, S. Role of transparency of platinum–ferromagnet interfaces in determining the intrinsic magnitude of the spin Hall effect, Nat. Phys. 11 (2015) 496–502.

[71] C.-F. Pai, Y. Ou, L.H. Vilela-Leão, D.C. Ralph, R.A. Buhrman, Dependence of the efficiency of spin Hall torque on the transparency of Pt/ferromagnetic layer interfaces, Phys. Rev. B 92 (2015) 64426.

[72] C. Bi, et al., Electrical control of metallic heavy-metal–ferromagnet interfacial states, Phys. Rev. Appl. 8 (2017) 34003.

[73] T. Nozaki, et al., Spin-dependent quantum oscillations in magnetic tunnel junctions with Ru quantum wells, Phys. Rev. B 70 (2004) 172401.

[74] P. LeClair, et al., Sign reversal of spin polarization in Co/Ru/Al2O3/Co magnetic tunnel junctions, Phys. Rev. B 64 (2001) 100406.

[75] Y. Seo, K.-W. Kwon, X. Fong, K. Roy, High performance and energy-efficient on-chip cache using dual port (1R/1W) spin-orbit torque MRAM, IEEE J. Emerg. Sel. Top. Circuits Syst. 6 (2016) 293–304.

[76] R. Bishnoi, F. Oboril, M.B. Tahoori, Low-power multi-port memory architecture based on spin orbit torque magnetic devices. in: Proceedings of the 26th edition on Great Lakes Symposium on VLSI—GLSVLSI '16, ACM Press, 2016, pp. 409–414, https://doi.org/10.1145/2902961.2903022.

[77] D. Fan, S. Angizi, Z. He, In-memory computing with spintronic devices. in: 2017 IEEE Computer Society Annual Symposium on VLSI (ISVLSI), IEEE, 2017, pp. 683–688, https://doi.org/10.1109/ISVLSI.2017.116.

[78] Z. He, S. Angizi, F. Parveen, D. Fan, High performance and energy-efficient in-memory computing architecture based on SOT-MRAM, in: 2017 IEEE/ACM International Symposium on Nanoscale Architectures (NANOARCH), IEEE, 2017, pp. 97–102, https://doi.org/10.1109/NANOARCH.2017.8053725.

[79] L. Chang, Z. Wang, Y. Zhang, W. Zhao, Reconfigurable processing in memory architecture based on spin orbit torque, in: 2017 IEEE/ACM International Symposium on Nanoscale Architectures (NANOARCH), IEEE, 2017, pp. 95–96, https://doi.org/10.1109/NANOARCH.2017.8053713.

[80] D. Fan, Z. He, S. Angizi, Leveraging spintronic devices for ultra-low power in-memory computing: logic and neural network, in: 2017 IEEE 60th International Midwest Symposium on Circuits and Systems (MWSCAS), IEEE, 2017, pp. 1109–1112, https://doi.org/10.1109/MWSCAS.2017.8053122.

[81] R. Bishnoi, M. Ebrahimi, F. Oboril, M.B. Tahoori, Architectural aspects in design and analysis of SOT-based memories, in: 2014 19th Asia and South Pacific Design Automation Conference (ASP-DAC), IEEE, 2014, pp. 700–707, https://doi.org/10.1109/ASPDAC.2014.6742972.

[82] S. Mittal, et al., Architecting SOT-RAM based GPU register file, in: 2017 IEEE Computer Society Annual Symposium on VLSI (ISVLSI), IEEE, 2017, pp. 38–44, https://doi.org/10.1109/ISVLSI.2017.17.

[83] D. Fan, S. Angizi, Energy efficient in-memory binary deep neural network accelerator with dual-mode SOT-MRAM, in: 2017 IEEE International Conference on Computer Design (ICCD), IEEE, 2017, pp. 609–612, https://doi.org/10.1109/ICCD.2017.107.

[84] R. Zand, A. Roohi, R.F. DeMara, Energy-efficient and process-variation-resilient write circuit schemes for spin Hall effect MRAM device, IEEE Trans. Very Large Scale Integr. Syst. 25 (2017) 2394–2401.

[85] A. Raha, et al., Designing energy-efficient intermittently powered systems using spin-Hall-effect-based nonvolatile SRAM, IEEE Trans. Very Large Scale Integr. Syst. 26 (2018) 294–307.

[86] B. Zeinali, M. Esmaeili, J.K. Madsen, F. Moradi, Multilevel SOT-MRAM cell with a novel sensing scheme for high-density memory applications, in: 2017 47th European Solid-State Device Research Conference (ESSDERC), IEEE, 2017, pp. 172–175, https://doi.org/10.1109/ESSDERC.2017.8066619.

[87] Y. Kim, X. Fong, K.-W. Kwon, M.-C. Chen, K. Roy, Multilevel spin-orbit torque MRAMs, IEEE Trans. Electron Devices 62 (2015) 561–568.

[88] G. Di Pendina, K. Jabeur, G. Prenat, Hybrid CMOS/Magnetic process design kit and SOT-based non-volatile standard cell architectures, in: 2014 19th Asia and South Pacific Design Automation Conference (ASP-DAC), IEEE, 2014, pp. 692–699, https://doi.org/10.1109/ASPDAC.2014.6742971.

[89] A. Fert, N. Reyren, V. Cros, Magnetic skyrmions: advances in physics and potential applications, Nat. Rev. Mater. 2 (2017) 17031.

[90] Y. Seo, K.-W. Kwon, K. Roy, Area-efficient SOT-MRAM with a Schottky diode, IEEE Electron Device Lett. 37 (2016) 982–985.

[91] S. Yuasa, K. Hono, G. Hu, D.C. Worledge, Materials for spin-transfer-torque magnetoresistive random-access memory, MRS Bull. 43 (2018) 352–357.

[92] T.A. Gosavi, S. Manipatruni, S.V. Aradhya, G.E. Rowlands, D. Nikonov, I.A. Young, S.A. Bhave, Experimental demonstration of efficient spin–orbit torque switching of an MTJ with sub-100 ns pulses, IEEE Trans. Magn. 53 (2017) 1–7.

[93] R. Ramaswamy, J.M. Lee, K. Cai, H. Yang, Recent advances in spin-orbit torques: moving towards device applications, Appl. Phys. Rev. 5 (2018) 031107.

Spin-transfer-torque magnetoresistive random-access memory (STT-MRAM) technology

T. Hanyu, T. Endoh, Y. Ando, S. Ikeda, S. Fukami, H. Sato, H. Koike, Y. Ma, D. Suzuki, H. Ohno
Tohoku University, Sendai, Japan

7.1 Introduction

This chapter discusses nonvolatile magnetoresistive random-access memory (MRAM) technology based on spin-transfer torque (STT). It reviews new directions in very large-scale integrated circuits (VLSIs) made possible by the technology. The increase in both power consumption and interconnection delay are two challenges which need to be addressed in order to enhance the performance of the next generation of VLSIs. The use of high-performance nonvolatile memory in the back-end-of-line process (BEOL) of VLSI is expected to result in reductions of both power consumption and interconnection delay. The nonvolatile memory suitable for VLSI applications needs to have features of access time of less than 10 ns, virtually unlimited endurance, operation voltage compatible with CMOS circuits, and scalability. STT-MRAM technology is being developed to meet these requirements. In this section, current materials and device technology using STT are reviewed with particular emphasis on magnetic tunnel junction (MTJ) having perpendicular easy axis for magnetization utilizing interface perpendicular anisotropy at CoFeB-MgO. The development of STT-MRAM devices based on MTJ technology is discussed in the following:

- a high-density device characterized by nonvolatile main memory and high-speed storage
- a high-speed device characterized by nonvolatile cache memory

Nonvolatile MTJ-based logic-in-memory architecture is shown to be capable of realizing both compactness, resulting in reduced interconnection delay, and low power consumption in a number of applications. Finally, as examples of STT-MRAM applications, recent development results of STT-MRAM-based pattern recognition processors and microprocessors are briefly added. Future trends as well as further reading for interested readers are presented in the last part of this section.

7.2 Materials and devices

Nonvolatile memory suitable for VLSI memory applications needs to have an access time of less than 10 ns, virtually unlimited endurance, and scalable write scheme, all at a small dimension compatible with the associated CMOS technology node. MTJs

Advances in Non-volatile Memory and Storage Technology. https://doi.org/10.1016/B978-0-08-102584-0.00008-5

with STT and switching [1] are being developed to satisfy these requirements. This subsection reviews the materials and device technology for MTJs used in:

- prototype MRAMs [2–10]
- logic circuits with embedded MTJs [1, 11–20].

Fig. 7.1A shows a schematic diagram of a typical memory cell consisting of a MTJ-one transistor in STT-MRAM. An MTJ consists of two ferromagnetic electrode layers separated by a tunnel barrier. One of the two magnetic layers is called a recording layer in which information is stored as the direction of magnetization. The other layer called a reference layer is engineered in such a way to maintain a fixed magnetization direction. The resistance of a MTJ is different from the parallel (P) and antiparallel (AP) configurations of the magnetization directions of the two magnetic layers.

MTJs are classified into two types depending on the direction of magnetic easy axes of the two magnetic layers. (A magnetic easy axis is the direction in which magnetization assumes the lowest energy.) These two types are as follows:

- in-plane easy axis (i-MTJ)
- perpendicular easy axis (p-MTJ)

These types are shown in Fig. 7.1B. Unless an electrode material has crystalline perpendicular anisotropy which is strong enough to overcome in-plane shape

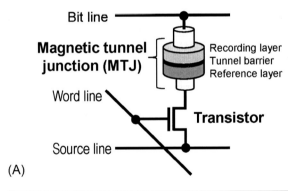

(A)

MTJ type	Aspect ratio $k = L/F$	Parallel	Antiparallel
In-plane anisotropy MTJ (i-MTJ)	> 1		
Perpendicular anisotropy MTJ (p-MTJ)	1		

(B)

Fig. 7.1 Schematic diagram of (A) one MTJ-one transistor cell and (B) magnetization configuration of MTJ with in-plane easy axis (i-MTJ) and with perpendicular easy axis (p-MTJ). $k = L/F$ is aspect ratio.

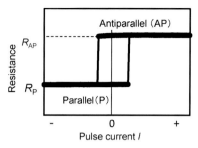

Fig. 7.2 Schematic diagram of STT writing.

anisotropy, MTJs exhibit in-plane easy axis. Read and write operations are performed by applying low current to interrogate the resistance of the MTJ and by applying high current above the threshold current for switching, respectively. The MTJ is selected by the bit line and word line (WL) connected to a transistor, as shown in Fig. 7.1A. Magnetization switching of recording layer by STT writing results from transfer of spin angular momentum from conduction electrons to the magnetization of the recording layer [21–26]. P or AP configuration can be selected by the polarity of the write current through the target MTJ (see Fig. 7.2). The resistance difference between the two magnetization configurations (A and AP) is a result of spin-dependent tunneling. Tunneling conserves spin direction. Consequently, the conduction depends on the density of states available at the Fermi energy for the majority and minority spin states and hence density-of-states spin polarization. This resistance change is defined as tunnel magnetoresistance (TMR) ratio $= \Delta R/R = (R_{AP} - R_P)/R_P$, where R_{AP} and R_P are the resistance of AP and P magnetization configurations (see Figs. 7.1B and 7.2), respectively.

In 1975, Julliere found the TMR ratio was 14% at 4.2 K in Fe/Ge(-O)/Co junction [27]. Maekawa and Gafvert showed correspondence between the magnetization configuration and the resistance of the device at low temperatures in Ni/NiO/(Ni, Fe, or Co) junctions in 1982 [28]. In 1995 TMR ratios of more than 10% at room temperature (RT) were reported independently in amorphous Al-O barrier MTJs by Miyazaki and Tezuka at Tohoku University [29] and Moodera and co-workers at MIT [30]. The TMR ratio in Al-O barrier MTJs has increased year by year and reached 70% which is close to the limit of the TMR ratio expected from the spin polarization at the Fermi energy. In 2001, Butler et al. [31] and Mathon and Umerski [32] showed theoretically that TMR ratios can reach 100% to even 1000% based on first-principles electronic structure calculations on fully ordered (001)-oriented Fe/MgO/Fe MTJs. These predictions were later confirmed experimentally for fully (001) epitaxial Fe/MgO/Fe MTJs prepared using molecular beam epitaxy in 2004 by Yuasa et al. [33, 34] and for CoFe/MgO/CoFe MTJs prepared using a combination of ion-beam and magnetron sputtering in 2004 by Parkin et al. [35]. In 2005, Djavaprawira et al. [36] and Hayakawa et al. [37] demonstrated a TMR ratio of over 200% for MgO barrier i-MTJs with sputtered $Co_{60}Fe_{20}B_{20}$ and $Co_{40}Fe_{40}B_{20}$, respectively. The CoFeB/MgO MTJs are especially important from a mass-production point of view because they are deposited on thermally oxidized Si using a conventional sputtering method and are then annealed to attain

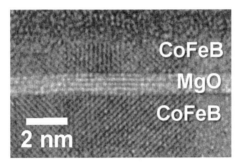

Fig. 7.3 Cross-sectional TEM image of CoFeB-MgO barrier i-MTJ with high TMR ratio.

high TMR. TMR ratios up to 604% at RT, approaching the theoretically predicted value [31, 32], have been observed in CoFeB/MgO barrier MTJs [38–41]. Fig. 7.3 is a cross-sectional transmission electron microscope (TEM) image of high-TMR CoFeB-MgO barrier MTJ structure with bcc (001)-oriented CoFeB electrodes and MgO(001) barrier for the coherent tunneling of $\Delta 1$ band electrons. It was reported in Ref. [42] that Co or Fe atom sits on the O atom in MgO layer from atomic-resolution imaging using high-angle annular dark field scanning TEM at CoFeB/MgO interface from [110] and [100] directions.

MTJs for VLSI application need to satisfy the following requirements:

(1) a small size (F nm, where F represents the feature size),
(2) a low current for STT write (the rule of thumb is critical switching current $I_C < F \mu A$, because of the transistor drivability),
(3) a high TMR ratio for fast sensing (>100%),
(4) a high thermal stability factor of the recording layer $\Delta = E/k_B T$ (>60, where E denotes the energy barrier between P and AP configurations, k_B the Boltzmann constant, and T the absolute temperature), and
(5) capability to withstand annealing at 350–410°C for BEOL post annealing without losing its high TMR ratio.

The switching current I_C for coherent magnetization reversal in i-MTJ and p-MTJ can be expressed as

$$\text{for i-MTJ}, I_C = J_C A = \alpha \frac{2e}{\hbar} \frac{M_S A t}{g(\theta)}\left(H_K + \frac{M_S}{\mu_0} \pm H_{\text{ext}} \right), \tag{1}$$

$$\text{for p-MTJ}, I_C = J_C A = \alpha \frac{2e}{\hbar} \frac{M_S A t}{g(\theta)}\left(H_K - \frac{M_S}{\mu_0} \pm H_{\text{ext}} \right) = \alpha \frac{2e}{\hbar} \frac{M_S H_K^{\text{eff}} V}{g(\theta)}, \tag{2}$$

where J_C is the critical switching current density, A the junction area, α the magnetic damping constant, e the elementary charge, M_S the saturation magnetization, t the recording layer thickness, $\hbar$ the reduced Planck constant, and $g(\theta)$ a function of the spin polarization of the tunnel current and the relative magnetization angle between the recording and the reference layers, H_K the anisotropy field that represents crystalline anisotropy, H_{ext} the external field, H_K^{eff} the effective anisotropy field that includes both crystalline and shape anisotropies, and V $(=At)$ the volume of recording layer [31, 32, 43–45]. p-MTJs can achieve lower I_C and higher Δ than i-MTJs, because I_C is reduced by the demagnetizing field component and reduced junction area $(A \propto kF^2$, see Fig. 7.1B) while Δ can be increased using high magnetic anisotropy materials. For this reason, STT-MRAMs with p-MTJs are being developed [4, 6–8, 10].

To attain perpendicular anisotropy, a number of material systems were explored as a recording layer material including rare-earth/transition metal alloys [46–49], $L1_0$-ordered (Co, Fe)-Pt alloys [44, 50, 51], as well as multilayers [52–56]. For example, STT switching with $4.7\,MA/cm^2$ in 130-nm-diameter TbFeCo/CoFeB/MgO/CoFeB/TbFeCo p-MTJ with Δ of 107 was reported although the TMR ratio showed a value of 15%[49] mainly due to the lack of annealing tolerance. It was also found that 50-nmϕ p-MTJs with Fe alloy doped with Pt and/or other elements resulted in a switching current of $10\,\mu A$ and Δ of 32 although the TMR ratio showed a low value of 23% [44]. It turned out that conventional perpendicular materials examined do not satisfy the requirements for VLSI applications at the same time. This is understood in the following way; to realize a high TMR ratio a (100) CoFeB-MgO-CoFeB structure is needed, but adding the structure to an electrode material with perpendicular anisotropy which is incompatible with the (001) structure results in either destruction of the (001) structure after annealing (hence low TMR) or the $M_s t$ value (see Eq. 2) being too large to observe switching.

In 2010, a perpendicular magnetic easy axis in Ta/CoFeB/MgO stacks with an ultrathin CoFeB layer was developed, where CoFeB-MgO interface anisotropy was responsible for the anisotropy [45, 57–59]. According to the first-principles calculation, the interfacial perpendicular anisotropy between oxide and ferromagnetic metal has its origin in the hybridization of Fe 3d and O 2p orbitals [60]. The prediction by the first-principle calculations has been supported by the experiment using x-ray magnetic circular dichroism on CoFeB/MgO stack with varied CoFeB thickness [61]. Good TMR and STT properties of CoFeB-MgO p-MTJs were reported soon after the initial report. This approach is particularly suitable for p-MTJ because no additional material is required other than CoFeB-MgO, a crucial materials combination for realizing a high TMR ratio [45]. Fig. 7.4A shows a schematic diagram of typical CoFeB-MgO p-MTJ. The junction diameter in the CoFeB-MgO p-MTJ is identified from a top view (Fig. 7.4B) taken by a scanning electron microscope and is 40 nm. The MTJ structures consist of, from the substrate side, Ta(5)/Ru(10)/Ta(5)/$Co_{20}Fe_{60}B_{20}(1.0)$/MgO(0.85)/$Co_{20}Fe_{60}B_{20}(1.7)$/Ta(5)/Ru(5) (numbers in parenthesis are nominal thicknesses in nanometers). Fig. 7.4C is a cross-sectional TEM image of the p-MTJ annealed at 300°C. This confirms that the top and bottom CoFeB layers and MgO barrier are also continuous in the thickness of nm order. Fig. 7.5A shows major and minor loops of the out-of-plane resistance versus field $(R\text{-}H)$ curve in the CoFeB-MgO p-MTJ with 40 nm in diameter annealed at 300°C. The TMR ratio is 124% with resistance area product $RA = 18\,\Omega\mu m^2$. In addition, no degradations of the TMR ratio and RA were observed after annealing at 350°C. Fig. 7.5B shows the resistance versus current density $(R\text{-}J)$ curves measured at a current-pulse duration $\tau_P = 300\,\mu s$, without an external magnetic field. Clear switching is observed at a current density of a few MA/cm^2.

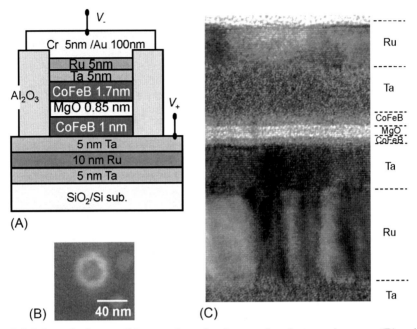

Fig. 7.4 Schematic diagram (A), a top view taken by scanning electron microscope (B) and a cross-sectional TEM image (C) in typical CoFeB-MgO p-MTJ annealed at 300°C.

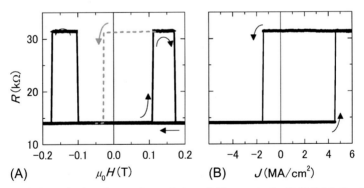

Fig. 7.5 (A) Major and minor loops of out-of-plane R-H curve in the CoFeB-MgO p-MTJ annealed at 300°C. The *solid and dashed lines* correspond to major and minor loops, respectively. (B) R-J curve at current-pulse durations of 300 μs.

For p-MTJs with a single CoFeB-MgO interface, Δ shows a nearly constant value down to a critical junction size below which Δ starts to reduce. This constant value of Δ is a result of nucleation-type reversal [62], where a nucleation embryo having a size and length scale determined by material parameters governs the magnetization reversal, and sets the highest attainable Δ of a stack. It was also found that Δ increases linearly as the thickness of the recording layer increases [63], until the junction size is larger than the critical size of the nucleation embryo. It is therefore necessary to increase the thickness of the recording layer, increasing the contribution of in-plane anisotropy, while maintaining the perpendicular easy axis. To realize this, a double CoFeB-MgO interface structure was adopted to increase the interface anisotropy [64]. It has been shown that the double-interface MgO/CoFeB/Ta/CoFeB/MgO recording structure in p-MTJs having a synthetic ferrimagnetic (SyF) reference layer (Fig. 7.6) increases Δ by approximately 60 at a dimension of 29 nm in diameter, while keeping comparable intrinsic critical current density compared to the single interface CoFeB-MgO recording structure [65]. This can be understood by a decrease of α increasing the total CoFeB recoding layer thickness [45]. A fast switching speed by STT [66], low write error rate [67], and high endurance [68] have also been demonstrated in p-MTJ with a single interface CoFeB-MgO recording structure. Consequently, CoFeB/MgO interfacial anisotropy technology is becoming an indispensable building block in developing nonvolatile STT-MRAM for VLSIs.

It should be noted that a trade-off exists between write current I_C and thermal stability Δ as shown in Eq. (2). Current CoFeB-MgO interface anisotropy technology appears to take us to the beginning of a 1X-nm generation. In order to scale down the device dimension further, the switching current and device dimension need to be reduced while maintaining a required E, that is, Δ. This requires development of perpendicular anisotropy electrode materials with high anisotropy and low α. Based on recent experiments, there is an empirical correlation between saturation magnetization and the damping parameter α. The saturation magnetization of $L1_0$-MnAl is a half value of that of CoFeB. It has been shown that high-quality MnAn films with a surface roughness of 0.4 nm, an important factor for obtaining a high TMR ratio, exhibit a high anisotropy of $1 \times 10^6 \, \text{J/m}^3$ with low saturation magnetization of 0.67 T and a damping parameter α of 0.003; it is 1/10 in comparison with that of CoFeB (see Fig. 7.7) [69].

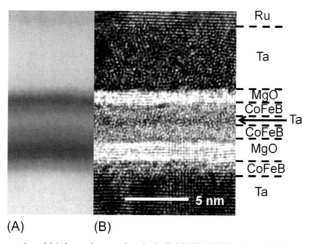

Fig. 7.6 Cross-sectional high angle annular dark field (HAADF) image (A) and TEM image (B) in p-MTJ with double CoFeB-MgO interfaces.

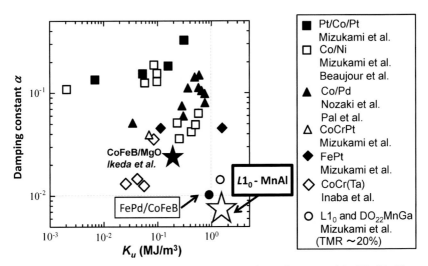

Fig. 7.7 Damping constants vs magnetic anisotropy K_u for various materials. $L1_0$-MnAl achieve the very large K_u and low magnetic damping.

7.3   Improving memory storage

A lot of power is consumed in facilities like data centers not only by the computers themselves but also by the air conditioning systems whose temperature is raised by the heat generated from the computers. The main reason attributed to the increase in computer operating power is leakage of current from semiconductor integrated circuits due to the device scaling. Fig. 7.8A represents the state-of-the-art computer memory hierarchy in which working memories such as SRAM cache and DRAM main memory are volatile. By scaling the feature sizes of MOSFETs consisting of the memories the subthreshold leakage increases exponentially causing a large increase in static current leakage in SRAMs. The speed gap between the SRAM cache and the DRAM main memory and that between the DRAM main memory and the storage (SSD, HDD) are another bottleneck in the computer's performance gain by scaling.

Fig. 7.8B suggests a possible new memory hierarchy where the last level (LL) cache is constructed by nonvolatile STT-MRAMs and the two speed gaps are filled with nonvolatile (NV)-main memory and high-speed nonvolatile cache to boost performance both being based on STT-MRAMs [70–72]. This structure can solve the power problem because the large static current leakage can be eliminated by shutting down the power supply of the nonvolatile cache memory when it is idle. The use of STT-MRAM as nonvolatile (NV)-main memory replacing DRAM cache and as high-speed storage and storage cache is also effective in resolving the speed-gap problem. Fig. 7.8C depicts the next step, an extreme nonvolatile computer system. In this second generation, logic is also made nonvolatile by using spintronics.

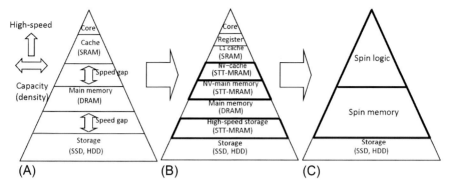

Fig. 7.8 The problems of large power consumption and speed gap can be solved by adopting STT-MRAMs. (A) The conventional memory hierarchy, (B) nonvolatile memory hierarchy of first generation using STT-MRAMs, (C) second generation in which logic is also made nonvolatile using spintronics.

7.3.1 Nonvolatile MRAM for main memories

A memory cell consisting of an MTJ and a metal-oxide semiconductor field-effect transistor (MOSFET) as the MTJ's selective device can achieve a high-density memory aimed at a nonvolatile main memory [73, 74]. Fig. 7.9A illustrates a conventional one transistor and one MTJ (1T1MTJ) cell in which the MOSFET is n-type (NFET). The bottom layer (closer to the silicon surface) of an MTJ is usually the pinned layer. In this configuration the current I_{UP} that flows from the bottom pin to the top pin to switch the MTJ from parallel state to antiparallel state is reduced because the NFET operates in the saturation region. The current I_{DOWN} that flows from the top pin to the bottom pin to switch the MTJ from antiparallel state to parallel state can be larger than I_{UP}. However, it is also usual that the critical current needed to switch an MTJ from

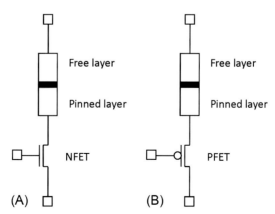

Fig. 7.9 One transistor and one MTJ (1T1MTJ) memory cell equivalent circuits. (A) The conventional 1T1MTJ cell using NFET as the selective device, (B) a proposed 1T1MTJ cell using PFET as the selective device.

parallel to antiparallel I_{C0} is larger than the critical current required to switch the MTJ from antiparallel state to parallel state I_{C1}. Therefore, the asymmetry of the MOSFET current drivability in I_{UP} and I_{DOWN} is not consistent with the asymmetry of the MTJ critical currents I_{C0} and I_{C1}, making the switching difficult. The inconsistency can be resolved either by reversing the connection by wiring [75] or by making the MTJ top pinned [76]. However, the former method increases the cell size and the latter introduces a difficulty into the process.

A 1T1MTJ memory cell using a p-type MOSFET (PFET) as the selective device has also been shown to be a solution resolving the asymmetry, as shown in Fig. 7.9B [77–79]. By using a PFET as the MTJ's selective device, the asymmetry of the MOSFET current drivability can be consistent with the asymmetry of the MTJ critical current without changing the bottom-pin process making both the switching and the process easier. Fig. 7.10 shows the relationship between the currents NFET and PFET can flow and the MTJ's critical currents in switching for both directions. This shows that the NFET cannot flow a large enough current to switch the MTJ from parallel to antiparallel while the PFET can flow large enough currents to switch the MTJs for both directions.

The 1T1MTJ cell of present day is larger than the 1T1C DRAM cell used in main memories because the critical currents required in switching MTJs are still large because of the relatively large junction size. However, since it has been shown that the critical currents for MTJ switching are reduced according to D^2 (D = MTJ's diameter), the currents can be reduced drastically by scaling the MTJ's size. This current reduction contributes to the reduction in the channel width of the selective transistor in the 1T1MTJ cell making the 1T1MTJ cell much smaller. It is roughly estimated that the channel width of the selective transistor of 1T1MTJ cell can be as small as the minimum feature size F if the diameter of MTJ is reduced to $D = 25$ nm in $F = 22$-nm technology node and that the 1T1MTJ cell size can match the 1T1C DRAM cell. Furthermore, given that scaling F to 20 nm or smaller becomes difficult in DRAM due to its capacitor scaling issue the 1T1MTJ cell may replace the 1T1C DRAM cell in 20 nm node and beyond. Another advantage of the 1T1MTJ cell over the 1T1C DRAM cell in terms of integration density is that the constraint on the switching transistor in the 1T1MTJ cell is much more relaxed than that on the DRAM switching transistor. DRAM's switching transistor must satisfy both the large current drivability and the

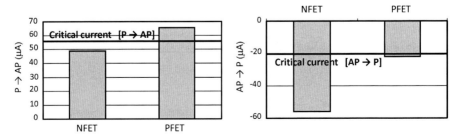

Fig. 7.10 Comparison between the current MOSFETs can flow and the MTJ's critical currents in switching where P and AP stand for parallel state and antiparallel state, respectively.

very small cut-off current to retain the charge stored in the capacitor for more than a few seconds at high temperature for an average cell. The switching transistor of the 1T1MTJ cell, on the other hand, has no severe constraint on the cut-off feature. This difference also means that the channel length of 1T1MTJ cell can be smaller than the 1T1C DRAM cell in 20-nm technology node and beyond thereby achieving a smaller cell size.

7.3.2 Nonvolatile MRAM for cache memories

Although the 1T1MTJ cell can provide a high-density nonvolatile memory suitable for a main memory its access time is not fast enough to be applied to cache memories [5]. One major reason is that a reference current or voltage is required to sense the data that are read from the 1T1MTJ cell. In general, generating an accurate reference current or voltage on a chip is very difficult [80]. MTJ-based differential pair type cells have been proposed in high-speed applications [81, 82]. They need no reference current or voltage for the data to be read by a sense amplifier because the read is double-ended in the sense that a pair of true and complementary signals are read from the cells. Therefore, the access times are expected to be much faster than that of the 1T1MTJ cell. However, they have six or more MOSFETs in addition to two MTJs, making the cell size larger than the conventional SRAM.

To overcome the large cell size, a 4T2MTJ cell has been proposed that has the potential to become smaller than the conventional SRAM cell [83, 84] as shown in Fig. 7.11A. Two PFETs in the conventional SRAM cell are replaced by a pair of MTJs. The pinned layers must be connected to the power line (PL) for the cell to be read nondestructively. The operation of the cell is very similar to the SRAM, the only difference is that PL is clocked low and high to switch the two MTJs to the opposite states. Fig. 7.11B shows the cell's physical layout. Only four NFETs are drawn in Fig. 7.11B. Two MTJs in the cell can be put on the drain areas between the two gates. Since the cell's footprint is determined by the four NFETs the cell can be smaller

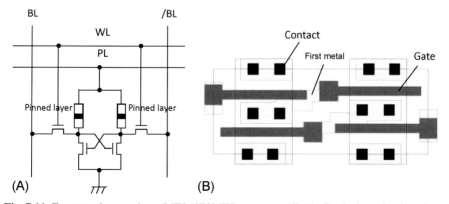

Fig. 7.11 Four transistor and two MTJ (4T2MTJ) memory cell. (A) Equivalent circuit and (B) physical layout in which only four MOSFETs are drawn with two MTJs put on the drain areas in the middle.

than the SRAM cell as the channel widths of the MOSFETs are as small as those in the SRAM. It was predicted that the cell size can become smaller than the SRAM in 45-nm technology node and beyond by scaling the thickness of the tunnel dielectrics (MgO) of the MTJs along with their areas [83, 84].

To reduce the power consumption in memories using the 4T2MTJ cell it is important to control PL according to the cell's operational modes because the cell consumes a static current between PL and GND through the MTJ in the antiparallel MTJ. A fine-grained power-gating scheme has been proposed to achieve a low-power and high-performance memory [17, 85]. In this scheme the PL of 32 cells along a WL is connected together to be controlled by a PL driver as shown in Fig. 7.12. The PL driver is CMOS AND-type one that drives PL high and low in correspondence to WL from a row decoder and grain select signal (GSL) from a column decoder. When cells are to be written or read the corresponding 32b-cell group (grain) is activated by raising PL high. It is worth noting that almost all other grains remain inactive with PL low, limiting the static current to that by the activated grain. Thus the fine-grained power-gating scheme can reduce the power drastically in contrast to SRAM in which all cells consume static current leakage. The fine-grained power-gating scheme also has the merit of achieving fast wake-up and power-off times realizing fast access and cycle times of memory. The number of cells controlled by a PL driver (grain size) needs to be determined in the trade-off between the operation power and the array size. The operation power increases as the grain size increases because the static current that flows in the activated grain increases as the grain size increases. On the other hand, the array size increases as the grain size decreases because the area overhead of the PL drivers increases as the grain size decreases. The size 32 was an optimized number in the trade-off for the CMOS AND-type PL driver. Fig. 7.13 shows the microphotograph of

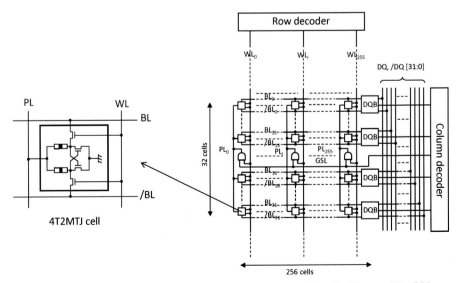

Fig. 7.12 32b fine-grained power-gating scheme applied to a 4T2MTJ cell array. PL of 32 cells along a WL are connected together to be controlled by a CMOS AND-type PL driver.

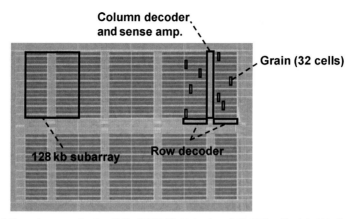

Fig. 7.13 Microphotograph of a 1 Mb STT-MRAM using 4T2MTJ cell with 32b fine-grained power-gating scheme that was fabricated in 90-nm CMOS and 100-nm perpendicular MTJ processes.

a 1 Mb STT-MRAM using the 4T2MTJ cell with the 32b fine-grained power-gating scheme that was fabricated in 90-nm CMOS and 100-nm perpendicular MTJ processes. The memory cell size is $2.19\,\mu m^2$ and the 1 Mb macro size is $3.54\,mm^2$. The functionality of the cells has been confirmed by using a simple write/read test pattern. The cell size is about twice as big as the SRAM cell counterpart in 90-nm technology node. This is due to the large MTJ's critical currents in switching in 100-nm MTJ size. The channel width of the MOSFETs in the 4T2MTJ cell must be large enough to supply the critical currents. It was shown that the channel width divided by the channel length can be reduced by scaling the MTJ size. Therefore, the 4T2MTJ cell size can be smaller than the SRAM counterpart in 45-nm technology node and beyond by scaling together with the MgO thinning as mentioned previously.

The CMOS AND-type PL driver that was used in the 32b grain is not compact enough to be applied to a much smaller grain. A single MOSFET is not suitable for the PL driver for the 4T2MTJ cell because PL voltage must be driven both high and low (PL needs to be low when one of the two MTJs is to be switched from parallel to antiparallel). To overcome this, a two-transistor bootstrap-type PL driver consisting of only two NFETs [13] has been proposed. Fig. 7.14 illustrates an 8b grain with 4T2MTJ cells controlled by the two-transistor bootstrap-type PL driver. The PL driver has only two NFETs, one working as a driver to drive the PL line and the other working as a barrier transistor that is cut off to boost the voltage in the isolated gate of the driver NFET beyond the power supply voltage V_{dd} through the gate-channel capacitance of the driver NFET to avoid the driver NFET to operate in saturation region as shown in Fig. 7.15. This PL driver is very compact and can drive PL high and low efficiently, being suitable to control a smaller grain. Another merit of using this two-transistor bootstrap-type PL driver is that the driver itself does not consume a static current due to subthreshold leakage at all unlike the CMOS AND-type PL driver. As a circuit unit the number of PL drivers is the largest next to the memory cells especially when the grain size is small. Therefore, this static power is not negligible. Fig. 7.16 compares the 128 kb subarray area and the 1 Mb write

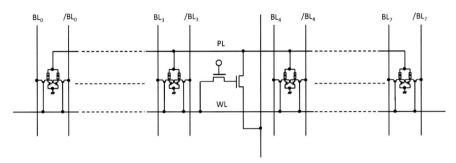

Fig. 7.14 A two-transistor bootstrap-type PL driver applied to a 8b grain using 4T2MTJ cells.

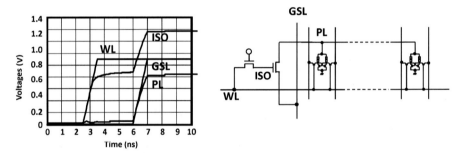

Fig. 7.15 Operation waveforms of the two-transistor bootstrap-type PL driver. At first, WL is raised, making the node ISO precharged to V_{dd}-V_{th} where V_{th} is the threshold voltage of the barrier NFET. Then, GSL is raised to boost the ISO beyond V_{dd} due to the gate-channel capacitance coupling. It is worth noting that the barrier NFET is cut off in this boosting operation.

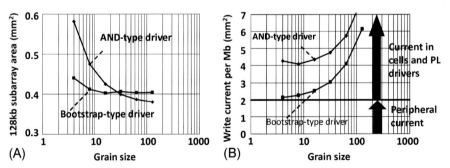

Fig. 7.16 (A) Comparison of 128 kb subarray area between using CMOS AND-type PL driver and two-transistor bootstrap-type PL driver. (B) Comparison of averaged write current per Mb in 45 ns cycles between using CMOS AND-type PL driver and two-transistor bootstrap-type PL driver.

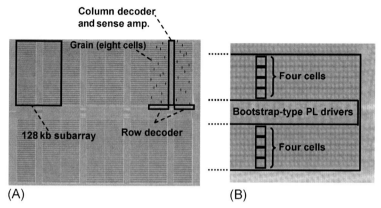

Fig. 7.17 (A) Microphotograph of a 1 Mb STT-MRAM using 4T2MTJ cell with 8b fine-grained power-gating scheme and two-transistor bootstrap-type PL driver that was fabricated in 90-nm CMOS and 100-nm perpendicular MTJ processes. (B) A magnified view of a part of 2 kb unit that includes 256 8b grains that are controlled by two-transistor bootstrap-type PL drivers (4 cells in both sides of a PL driver).

current for the CMOS AND-type PL driver and the two-transistor bootstrap-type PL driver as a function of grain size. Fig. 7.17A shows the microphotograph of a 1 Mb STT-MRAM using 4T2MTJ cell with 8b super-fine-grained power-gating scheme in which two-transistor bootstrap-type PL drivers are used. The magnified view for a 2 kb unit that includes 8b grains is shown in Fig. 7.17B, each of the grains having four cells in both sides of the PL driver (eight cells in total). The cell efficiency remains high (59%) even though the macro adopts a super-fine-grained power-gating scheme of 8b grain thanks to the compact PL driver with only two NFETs.

Though MTJ has strong potential to be used in high-performance nonvolatile working memories, one concern is that it requires a large current to switch it very quickly, namely in a few nanoseconds or faster [86]. The switching is also probabilistic when MTJs are to be switched by a moderate current [87]. To work around these issues, an auto-data backup STT-MRAM with background write scheme has been proposed [88]. The general concept of the background write scheme for an auto-data backup cell is illustrated in Fig. 7.18A. There are two memories in a bit cell. Memories 1 and 2 have write times t_{w1} and t_{w2} and retention times t_{r1} and t_{r2}, respectively. We assume that $t_{w1} < t_{w2}$ and $t_{r1} \ll t_{r2}$. When a write cycle from outside through the switching FETs gated by WL is performed in a period t so that $t_{w1} < t < t_{w2}$, the data can be written to the memory 1 but memory 2 is not written in the write period. However, if memory 1 can switch memory 2 by using the data that are written to memory 1 for more than t_{w2} without control from outside or even autonomously after WL is lowered, memory 2 is also written without any help from outside. This auto-data backup can be performed without interrupting the write operation specification in the memory chip because the fast write cycles can be maintained with the relatively slow auto-data backup performed in the background. This background write can achieve a memory cell with the write time t_{w1} and the retention time t_{r2}. In Fig. 7.18B, we adopted a CMOS latch as

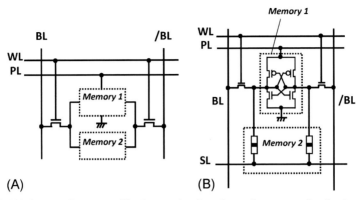

Fig. 7.18 (A) A general concept of background write scheme for an auto-data backup cell. (B) An actual embodiment of the auto-data backup cell. Memory 1 is a CMOS latch and memory 2 is a pair of MTJs.

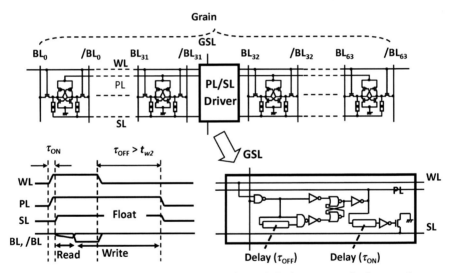

Fig. 7.19 A grain and a PL driver that eliminate the static leakage current in the auto-data backup cell for the background write scheme. Sixty-four cells are controlled by the PL driver. The PL driver includes two delays. One is to measure a period longer than t_{w2}, while the other is for a very short period needed in power on sequence to load data from a pair of MTJs to CMOS latch. The timing chart is also shown.

memory 1 and a pair of MTJ to achieve a nonvolatile working memory cell having a fast write performance equivalent to SRAM.

However, again, it is very important to shut PL down when the auto-data backup operation is completed to eliminate the static current that flows in the cell. Therefore, a timer which measures t_{w2} has been implemented into a PL driver used for a power-gating scheme. Fig. 7.19 shows the PL driver that controls a grain along a WL with its operational waveforms. The timing chart shows that PL activation is expanded to a longer

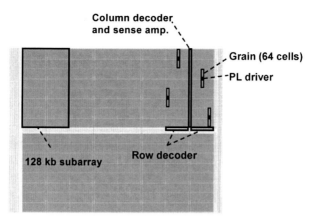

Fig. 7.20 Microphotograph of a 1 Mb STT-MRAM using the auto-data backup cell with 64b fine-grained power-gating scheme that was fabricated in 90-nm CMOS and 100-nm perpendicular MTJ processes.

period than t_{w2} to ensure that a pair of MTJs can be switched by using data written to a CMOS latch after WL is shut down. The PL driver includes two times. One is for τ_{OFF} that measure t_{w2}. The other is for τ_{ON} that is a nanosecond or shorter necessary for the data to be correctly loaded from a pair of MTJs to a CMOS latch. Fig. 7.20 shows the microphotograph of a 1 Mb STT-MRAM using the auto-data backup cell with 64b fine-grained power-gating scheme in which the PL driver in Fig. 7.19 are used to achieve the background write. It was fabricated in 90-nm CMOS and 100-nm perpendicular MTJ processes. The read and write cycle times are shown to be 1.5 and 2.1 ns by simulation, respectively. Moreover, sensing delay of 0.5-ns STT-MRAM with 2T2MTJ type cell was also reported [89]. STT-MRAMs with differential-type memory cell are reaching the fast access time enough to be applied to the LL cache.

7.4 Improving logic-in-memory architecture

Demand for high-performance and low-power logic LSIs has continued to increase, reflecting recent highly intelligent electronics applications. However, the increase in power dissipation, especially the standby power dissipation, limits the performance in recent LSIs with nanometer-scaled transistors. This is caused by constant current leakage due to on-chip volatile memory modules such as static RAM and flip-flops. In addition, interconnection delay as well as power dissipation has to be managed to ensure LSIs are high in performance and low in power consumption. In order to counter such challenges in logic LSI, there are primarily two approaches;

(1) utilizing a novel LSI architecture where intra-chip data transfer between memory and logic modules is localized as much as possible and global wire counts are minimized,

(2) realizing a compact logic-LSI with a power-gating capability where the power supply to each fine-grained functional logic-circuit component can be shut down quickly (or turned on quickly) whenever its operation status changes to a standby mode (or comes back to an active mode again),

Approach (1) is achieved by utilizing a novel logic-LSI-oriented hardware structure, called "logic-in-memory (LIM)" architecture [90]. In the LIM architecture storage elements are distributed over the logic-circuit to achieve low power dissipation for data transmission and short interconnection delay greatly reducing global interconnection wires.

In order to apply a power-gating technique (Approach (2)) in reducing the power dissipation of logic LSIs as efficiently as possible every storage element must be nonvolatile because stored data must remain even when the power supply is shut down by the power gating. MTJ with a spin-injection write capability has all the following features for such an application; a reasonably large resistance ratio, virtually unlimited endurance, fast read/write accessibility, scalability, complementary MOS (CMOS)-process compatibility, and nonvolatility. MTJ is thus suited to implement such an MOS/MTJ-hybrid logic circuit with logic-in-memory architecture. Moreover, MTJ not only acts as a "storage element," but can also be designed to act as a "logic element," because its resistance is programmable in accordance with its stored status. This use of MTJ-based logic-circuit design allows the storage function to be merged into logic-circuit plane resulting in a compact logic LSI [72].

7.4.1 MTJ-based nonvolatile ternary content-addressable memory

As a typical example of nonvolatile logic-in-memory architecture, MTJ-based nonvolatile ternary content-addressable memories (TCAM) have been designed and fabricated [91–94]. TCAM is a functional memory for high-speed data retrieval that performs a fully parallel search and fully parallel comparison between an input key and stored words. Currently, its high bit cost and high power dissipation, higher than those of standard semiconductor memories such as static random access memory limits the fields to which TCAM can be applied. Table 7.1 shows the truth table of a TCAM cell function. Its rich functionality makes data search powerful and flexible but with conventional CMOS realization there is an associated cost of a complicated logic circuit with two-bit storage elements. Fig. 7.21 shows the design philosophy of realizing the TCAM cell circuit compactly with a nonvolatile storage capability. In the case of both conventional volatile TCAM cell structure and conventional nonvolatile

Table 7.1 Truth table of the TCAM cell function

Stored data		Search input	Current	Match result
B	(b1, b2)	S	comparison	ML
0	(0, 1)	0	$I_Z < I_Z'$	1 (*Match*)
		1	$I_Z > I_Z'$	0 (*Mismatch*)
1	(1, 0)	0	$I_Z > I_Z'$	0 (*Mismatch*)
		1	$I_Z < I_Z'$	1 (*Match*)
X (don't care)	(0, 0)	0	$I_Z < I_Z'$	1 (*Match*)
		1	$I_Z < I_Z'$	1 (*Match*)

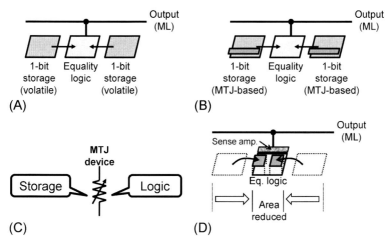

Fig. 7.21 Design philosophy of making a compact TCAM cell circuit: (A) conventional TCAM cell structure, (B) conventional NV-TCAM cell structure, (C) MTJ device merging storage and logic functions, and (D) proposed NV-TCAM cell structure.

TCAM cell structure without using LIM architecture shown in Fig. 7.21A and B, respectively, the bit cost is high. In contrast, when two-bit storage elements are merged into a logic-circuit part by using the LIM architecture as shown in Fig. 7.21C, the proposed TCAM cell structure becomes compact and nonvolatile as in Fig. 7.21D. Fig. 7.22A and B compare a conventional volatile TCAM cell circuit and the proposed nonvolatile one. The conventional CMOS-based volatile TCAM cell circuit consumes 12 MOS transistors (12T-TCAM circuit structure) while the proposed one takes just 4 MOS transistors with 2 MTJ devices (4T-2MTJ circuit structure). Note that MTJs do not affect the total TCAM cell-circuit one, because MTJs are fabricated onto the CMOS plane. Compact realization due to NV-LIM architecture has the advantage of improving the performance of the circuit by inserting a driver as shown in Fig. 7.22C. Fig. 7.23 summarizes the comparison of TCAM word circuits with 144 cells. By the appropriate division of the TCAM word circuit, the activation ratio of the TCAM can be minimized. Fig. 7.24 shows the variety of the segment-based TCAM structures. In the case of the three-segment-based NV-TCAM structure, where the first segment, the second segment and the rest consists of 3-bit, 7-bit, and 134-bit cells, respectively, its average activation ratio becomes as low as 2.8%, which indicates that 97% or more TCAM cells can be in standby mode on average by the fine-grained power gating. Fig. 7.25 shows the fabricated nonvolatile TCAM test chip under 90-nm CMOS and 100-nm MTJ technologies.

7.4.2 MTJ-based nonvolatile field-programmable gate array

MTJ-based nonvolatile look-up table (LUT) circuit for an instant power-ON/OFF field-programmable gate array (FPGA) is another example that shows the clear advantage of nonvolatile logic-in-memory architecture [13, 95, 96]. Fig. 7.26 shows the

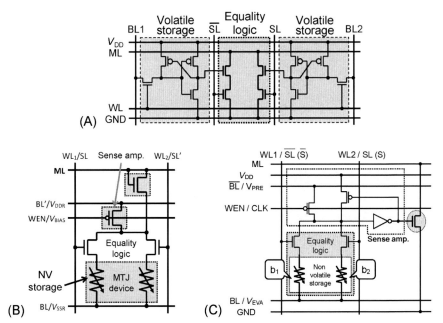

Fig. 7.22 TCAM cell-circuit design: (A) conventional volatile TCAM cell circuit, (B) proposed 4T-2MTJ NV-TCAM cell circuit, and (C) proposed 7T-2MTJ NV-TCAM cell circuit.

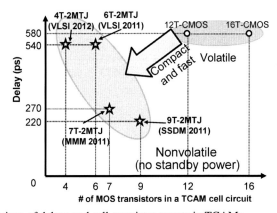

Fig. 7.23 Comparison of delays and cell-transistor counts in TCAMs.

overall structure of a nonvolatile FPGA where each LUT circuit in the configuration logic block (CLB) stores logical configuration data into nonvolatile storage elements in the present case MTJs. Therefore, whenever an LUT circuit is in a standby mode its power supply can be shut down completely eliminating the wasted standby power dissipation. Although the use of MTJs makes the LUT circuit nonvolatile, its hardware cost is increased when volatile storage elements are replaced with nonvolatile ones as shown in Fig. 7.27A. To circumvent such hardware overhead in Fig. 7.27B, MTJs are merged into the combinational logic circuit in the LUT circuit by using NV-LIM

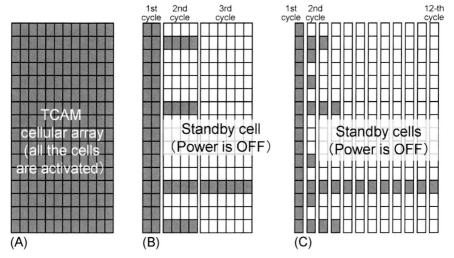

Fig. 7.24 Power-gating-oriented TCAM search schemes: (A) fully parallel search scheme, (B) series-parallel search scheme, and (C) bit-serial search scheme.

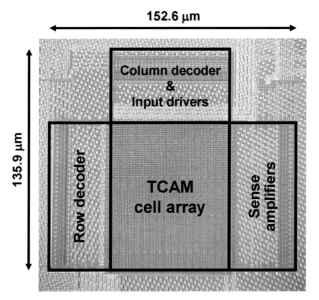

Fig. 7.25 Fabricated nonvolatile TCAM test chip.

architecture, resulting in a highly compact LUT circuit (only a single sense amplifier (SA) is required in the proposed structure). Fig. 7.28 shows a fabricated two-input nonvolatile LUT-circuit test chip. In practical FPGA the LUT circuit generally has four inputs or more. Since a multi-input LUT circuit requires many MTJs the variation of the resistance values of MTJs becomes critical. For stable LUT operation of multi-input LUT circuits additional MTJs can be added to adjust the operating point

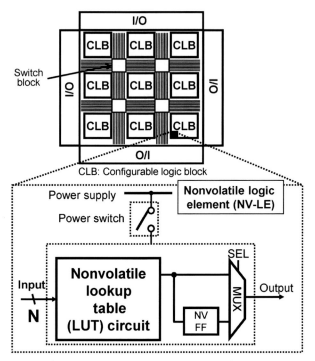

Fig. 7.26 Overall structure of a nonvolatile FPGA.

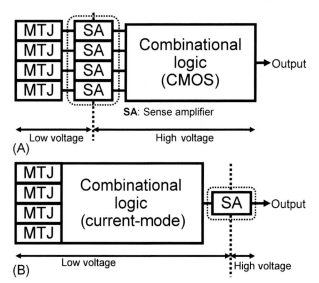

Fig. 7.27 Design philosophy of a compact nonvolatile LUT circuit: (A) Conventional approach and (B) proposed NV-LIM architecture-based approach.

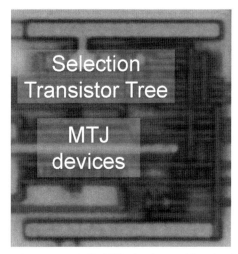

Fig. 7.28 Fabricated two-input nonvolatile LUT-circuit test chip using MTJ devices.

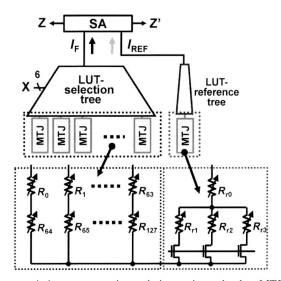

Fig. 7.29 Resistance-variation compensation technique using redundant MTJ devices.

of the LUT function. Fig. 7.29 shows an example where both two and three additional MTJs are inserted into the LUT-selection tree and the LUT-reference tree, respectively. Table 7.2 summarizes the comparison of six-input LUT circuits. This clearly demonstrates that the proposed NV-LIM-based NV-LUT circuit is implemented compactly with shorter delay and less active power dissipation in comparison with those using the conventional approach [95, 97]. Consequently, NV-FPGA chip embedded 3000 6-input NV-LUT circuits has been successfully fabricated under 90 nm-CMOS/100 nm-MTJ technologies as shown in Fig. 7.30 [98].

Table 7.2   Comparison of six-input nonvolatile LUT circuits

	CMOS-based[a]	**Proposed**
Number of devices	590 Tr. + 128 MTJ	222 Tr. + 132 MTJ
Delay[b]	194 ps	150 ps
Active power @ 1 GHz[b]	7.43 µW	5.72 µW
Standby power	0 µW	0 µW
$R_P = 1.27\,k\Omega$, $R_{AP} = 2.70\,k\Omega$, $I_{W1} = 112\,\mu A$, $I_{W0} = -180\,\mu A$		

[a] It is composed of 64-bit MTJ-Based Nonvolatile SRAM (448 Tr.), nMOS selector tree (126 Tr.), inverters (12 Tr.), and output buffer (4 Tr.).
[b] SPICE simulation under a 90 nm-CMOS technology.

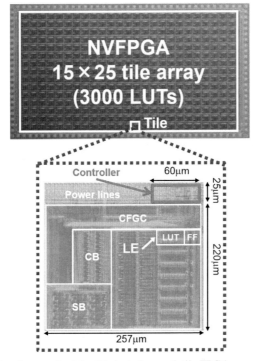

Fig. 7.30 Chip photomicrograph of the fabricated nonvolatile FPGA.

A nonvolatile flip-flop (NV-FF) is an essential component not only for an NV-FPGA, but also for a general nonvolatile logic LSI since temporal data of each function block must be clock-synchronized, backed up before power-off, and recalled just after power-on. While various types of MTJ-based NV-FFs have been reported, there are important issues for the MTJ-based NV-FF design. The first issue is the stochastic nature in MTJ switching. Even in the same MTJ device, the actual time to complete the write operation varies dramatically. Therefore, a longer-time write current pulse is required for the reliable write operation than that of average one, which causes a large amount of energy consumption. The large amount write energy consumption is critical for the backup operation because temporal data in each NV-FF must be stored into the MTJ device before power-off. Thus, the time during which the energy saved by the power-gating technique equals to the energy lost by the backup/recall operations (referred as break-even time) becomes long. As a result, a fine-grained power-gating technique cannot be applicable and the potential of nonvolatile logic LSI is not fully utilized. To overcome this issue, a self-terminated NV-FF, which makes it possible to minimize write energy for the MTJ device by monitoring the voltage change in MTJ switching and terminating write current, has been reported [99–104]. In particular, a three-terminal MTJ device is focused on to realize high-speed nonvolatile memories and high-performance logic gates. Since the write current path is separated from the read current path in the three-terminal MTJ device, both a sense circuit for reading MTJ resistance and a write driver for applying write current can be individually optimized.

Fig. 7.31A shows the schematic diagram of the self-terminated NV-FF, which is composed of a master latch, a slave latch, a write driver, a nonvolatile storage cell, and a self-termination circuit, and Fig. 7.31B shows the comparison of nonvolatile storage cells using the two-terminal MTJ device and the three-terminal MTJ device. In case of 3T-MTJ-based nonvolatile storage cell, the read-operation path is separated from the write operation path, which makes it possible to relax design space exploration.

Fig. 7.32 shows the comparison of 20 cases of backup energy consumption in conventional MTJ-based NV-FF using the worst-cased-oriented method and those of the self-terminated NV-FF. By utilizing the self-terminated mechanism, the backup energy consumption is greatly reduced. In fact, the average backup energy is reduced by 69% compared with that of a conventional nonself-terminated method.

7.4.3   MTJ-based nonvolatile random-logic LSI

Finally, an MTJ-based nonvolatile random-logic LSI with LIM architecture is discussed [18, 105, 106]. In designing such a large-scaled logic circuit with NV-LIM architecture it is important to establish a semiautomatic LSI CAD tool. Fig. 7.33 shows a circuit simulator, SPICE, where the MTJ device model with spin-transfer-torque write is built-in as one of the basic circuit components. A subcircuit macro-model emulating MTJ characteristics were discussed previously [107–109] and its behavior was simulated by importing it to the circuit simulator. However, it could not achieve sufficient simulation speed and accuracy since the use of the complex subcircuit model resulted in a high CPU time penalty. The proposed SPICE simulator incorporates MTJ model parameters and can perform accurate simulation of a complex circuit operation including MTJ switching behavior with a small computational cost. Actually, the proposed simulator achieves MTJ behavior up to 50 times faster than with a conventional subcircuit-based one. Fig. 7.34 shows a layout-design example of an NV-LIM-based random logic circuit. All the NV-LIM-based circuit primitives are automatically generated in this design. Fig. 7.35 shows a chip photomicrograph of a hardware accelerator for motion-vector extraction. The number of PEs arranged in a 5×5 grid is 25. The number of MOS transistors is about 0.5 million and that of MTJ devices is about 13,000.

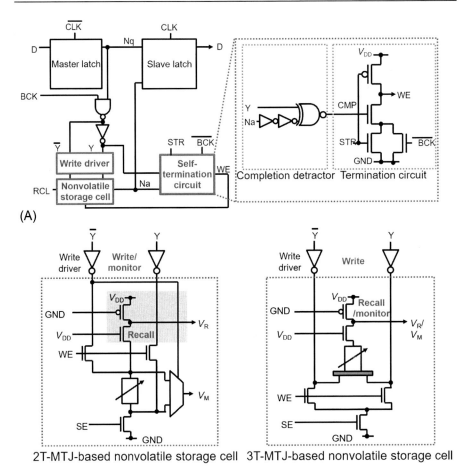

(A)

(B)

2T-MTJ-based nonvolatile storage cell   3T-MTJ-based nonvolatile storage cell

Fig. 7.31 Design of a self-terminated NV-FF: (A) overall schematic and (B) Nonvolatile storage cell configurations.

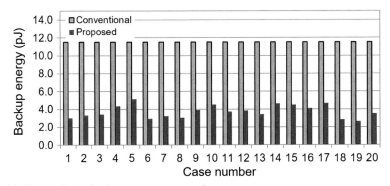

Fig. 7.32 Comparison of write energy consumptions.

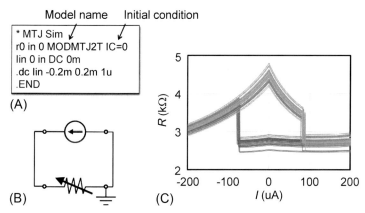

Fig. 7.33 STT-MTJ device model built in SPICE simulator: (A) example of a netlist, (B) corresponding equivalent circuit, and (C) simulated waveforms.

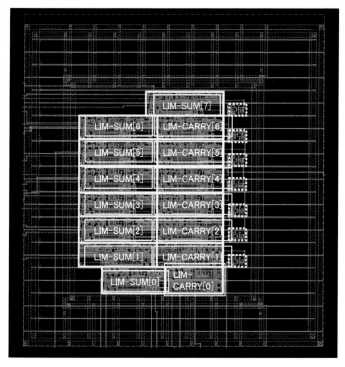

Fig. 7.34 Layout-design example of an NV-LIM-based random logic circuit.

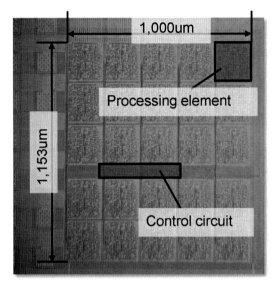

Fig. 7.35 Fabricated NV-LIM-based hardware-accelerator LSI chip for motion-vector extraction.

7.5 Application of STT-MRAM technology

Various technologies using the developed STT-MRAMs explained above have been growing to take advantages of their low-power and high-speed features. This section introduces two typical cases of such applications, pattern recognition processor, and microprocessor.

7.5.1 Pattern recognition processor

The nearest neighbor search function, which is one of the most crucial core technologies for pattern classification, requires ultralow-power operation especially for battery powered systems like smartphones and sensor networks. STT-MRAM can contribute to achieve both the ultralow-power operation and high performance [110–113]. A high-density nonvolatile associative memory based on the 4T2MTJ STT-MRAM [17, 85] for such applications was developed [114, 115].

In order to adopt STT-MRAM to the associative memory chip with maximum performance, two novel circuits were proposed. The first circuit, the truly compact current-mode circuitry, can realize flexibly controllable and high-parallel similarity evaluation. This makes the nonvolatile associative memory adaptable to any dimensionality and component-bit of template data. The second circuit, the compact dual-stage time-domain minimum searching circuit, can freely extend the system for more template data by connecting multiple nonvolatile associative memory cores without additional circuits for integrated processing. The STT-MRAM module and the computing circuit modules in this chip are synchronously power gated to completely

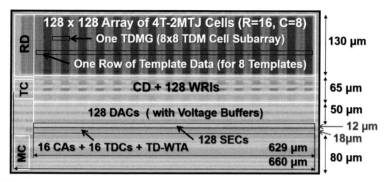

Fig. 7.36 Microphotograph of the developed nonvolatile associative memory core chip fabricated using a 90-nm CMOS/70-nm p-MTJ five-metal hybrid process on a 300 mm wafer.

eliminate standby power and maximally reduce operation power by only activating the currently accessed circuit blocks.

Fig. 7.36 shows the microphotograph of nonvolatile associative memory chip fabricated using a 90-nm CMOS/70-nm perpendicular-magnetic-tunnel-junction hybrid process on a 300 mm wafer, as a proof-of-concept chip designed with a 128×128 4T2MTJ STT-MRAM array. The chip operation demonstration for 16-D 8-bit texture pattern recognition is shown in Fig. 7.37. The chip operates at 40 MHz, and its operation power is only 130 μW, that is, 99.2% power improvements compared with the latest conventional works in both volatile and nonvolatile approaches [113, 116–119]. This result makes the proposed nonvolatile associative memory particularly superior for a large majority of better-powered recognition VLSI systems for nearest neighbor search with a large-scale template database.

7.5.2 Microprocessor

Enabling sudden power-down by introducing nonvolatile memories to a microprocessor is very effective for its power reduction by a more frequent entry in the power-down mode. Since NV-FFs and STT-MRAMs using MTJ have excellent features such as fast store/recall operation and high endurance [120, 121], a microprocessor with NV-FFs and/or STT-MRAMs can achieve power-gating operation with minimum performance degradation by a very small delay time to enter into/exit out of the power-down mode [122].

For practical use of power-gated microprocessors, however, it is necessary to prepare a method of controlling the power-on/off procedure, in order to ensure correct power-gating operation. In order to address this issue, the essential functions required for such power-gated microprocessors were proposed [123, 124]. It includes two functions: (a) adding a power-off (poff) instruction to realize power-gating control by software, and (b) an on-chip power control circuit activated by the poff instruction.

Fig. 7.38 displays the microphotograph of the test chip that employed the above two functions, fabricated using 90-nm CMOS and an additional 100-nm MTJ process. These functions enable the power-off procedure for the MPU to be executed

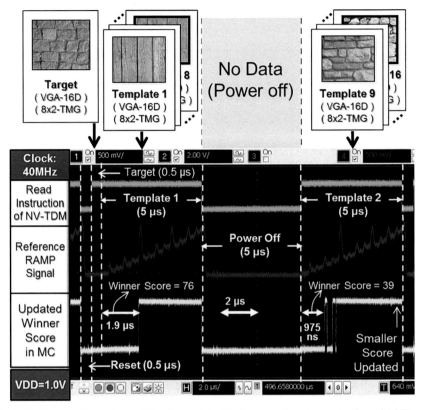

Fig. 7.37 Measured waveform of the fabricated chip for operation demonstration of 16-D 8-bit texture pattern recognition at 40 MHz using 1.0 V power supply.

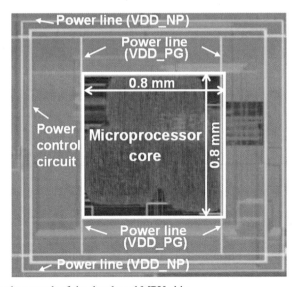

Fig. 7.38 Microphotograph of the developed MPU chip.

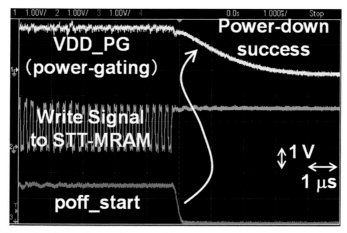

Fig. 7.39 Operational waveforms of successful power gating.

appropriately. Successful operation of power-gating was observed as shown in Fig. 7.39. As a result of measurement, the energy reduction effects for this microprocessor were estimated, for example, an operation energy reduction of 1/28 can be achieved when the operation duty is 10%, in the condition of a sufficient number of idle clock cycles.

7.6   Future trends

This section discusses future directions on various fronts involved in STT-MRAM development.

7.6.1   Materials and devices

In order to scale the switching current for STT in an MTJ a low α is required while maintaining high thermal stability. As shown in Table 7.3, K_{eff} and α in double CoFeB-MgO MTJ with 20 nmϕ and 2.6 nm-thick recording layer were calculated from Eq. (2) using the experimental values of I_C, and Δ. When it assumed that 10 nmϕ and 15 nmϕ

Table 7.3 K_{eff} and α for scaling of MTJ

F (nm)	I_C (μA)	J_C (10^{10}A/m^2)	Δ	K_{eff} (MJ/m^3)	α
20	27$_{(P \rightarrow AP)}$	8.59	60	0.304	0.0039
15	15	8.49	80 (100)	0.721 (0.902)	0.00164 (0.00131)
10	10	12.73	80 (100)	1.623 (2.029)	0.00109 (0.00088)

MTJs can realize the I_C of F μA and the heat stability of 80–100, the K_{eff} and α were estimated from Eq. (2). The realization of both $I_C = F$ μA and $\Delta = 80$–100 in the 10 (15) nmϕ MTJs needs to increase K_{eff} to 5.3–6.7 (2.4–3.0) times and decrease α to 1/4–1/5 (1/2–1/3) as compared with those of 20 nmϕ MTJs.

Mn-based-ordered alloys (i.e., MnGa and MnAl) which satisfy high magnetic anisotropy and low α at the same time are capable of realizing such a recording layer [69, 125–128]. In p-MTJs with MnGa/MgO/CoFe(B) stack a TMR ratio of 5% at RT was observed. Insertion of Fe or Co between Mn-based alloy and MgO helps increase the TMR ratio to 40% [126–128]. However, this is not high enough. In this system a further increase of the TMR ratio and demonstration of STT switching are the following steps. It is also necessary to perform an in parallel search of other material systems with higher thermal stability and lower α for further dimension reduction of MTJ.

There are two approaches to construct a memory cell using MTJ devices. The two-terminal cell such as STT-MTJ has the advantage of scalability for high-density memory embedded in VLSIs. The three-terminal cell utilizes domain wall motion by the spin-polarized current [129, 130] or magnetization switching by spin-orbit torque [131–135] for write operation. The three-terminal cell can increase the speed of operation over the two-terminal cell because the write and read paths are separated to avoid disturbance and is a particularly strong candidate for substitution of embedded SRAMs. Cell structures strongly affect performance, area efficiency, and reliability of nonvolatile memory and CMOS logic applications consequently an optimized and perhaps combined approach is needed to satisfy the requirements of specific applications. The spin-orbit torque-induced magnetization switching also allows analog-control of magnetization state, paving a pathway toward neuromorphic computing. In addition, using heterostructure with antiferromagnet-ferromagnet bilayer, an analog evolution of resistance state is observed [136]. A proof-of-concept demonstration of neuromorphic computing has been demonstrated using an artificial neural network where the analog spin-orbit torque devices are utilized as artificial synapses [137].

Write energy of current-driven write operation for the two- and three-terminal cells is still a few orders larger than switching CMOS. Therefore, along with developing schemes to reduce write energy, it is necessary to explore other write (magnetization switching) schemes to reduce the dynamic power of these devices. One possibility is to utilize an electric field E to control the magnetic anisotropy of magnetic layer in an MTJ. Electric field-control of magnetism through carrier density modulation which was first demonstrated for III–V ferromagnetic semiconductors [138–142] has been applied to 3d transition-metal and its alloy ferromagnets [143–148]. Recently, electric-field control of magnetism such as the change of coercivity and magnetization reversal has been experimentally shown on CoFeB-MgO p-MTJs paving the way to an entirely new and energy efficient write scheme for MRAM and associated devices [57, 58, 149, 150].

7.6.2 Memory

It is predicted that a memory static power component especially for LSIs used in portable devices will increase year on year and will become a large portion of the total power consumption [151]. The main part of the static power consumption comes from SRAM caches (the state-of-the-art memory hierarchy in computers is shown in Fig. 7.8A) whose capacities are increasing for improved performance. This is due to the subthreshold leakage increase by MOSFET's channel length's scaling in obtaining high-speed operation as well as high-density integration. Another problem is speed gaps in the memory hierarchy. One is between the SRAM cache and the DRAM main memory. The other is between the DRAM main memory and the storage. To solve the above issues, a new memory hierarchy based on nonvolatile memories using STT-MRAMs was proposed as shown in Fig. 7.8B. In Section 7.3 we presented circuit technologies realizing the STT-MRAMs that can be used in the new memory hierarchy. Furthermore, we proposed an ultimate memory hierarchy as shown in Fig. 7.8C in which all working segments in the computers are made nonvolatile by using spintronics. One of the key technologies in the spin logic is an MTJ-based nonvolatile logic-in-memory [120]. In the architecture, MTJ-based nonvolatile memory elements are distributed over the logic circuit to eliminate long wirings between memory and logic. The resultant effects expected are twofold:

(1) performance gain and power reduction by eliminating the long wiring and making a fine-grained power gating and
(2) chip size reduction by placing MTJs above logic circuits without area overhead and by decreased device count.

7.6.3 Logic

Further advancement of the nonvolatile MTJ-based logic-gate families described in this chapter represent a rich future direction of achieving lower power yet higher performance logic VLSI. In addition, establishing design methods to automatically minimize the power-delay-area product in both general-purpose and special-purpose logic VLSIs is needed. A wide variety of possibilities have just begun to surface.

7.7 Conclusion

This chapter has given an overview of the STT-MRAM technology. STT-MRAM has two advantages that have not been explicitly discussed. One is that the physics involved in switching and TMR is reasonably well established. The other is that there is a large research community of magnetism and magnetics as well as an industrial community of hard disk drives. So it is already known, for example, that a nonvolatile bit of less than 10 nm in diameter is possible thanks to the hard disk industry that has established it in the recording media. In fact, MTJs with its feature size (diameter) of less than 10 nm has been successfully fabricated in accordance with great contribution of the recent materials science. Moreover, it is time to call for circuit-device collaboration as the performance of an MTJ has become reasonably high. Several examples are shown in this chapter, which emphasize the advantage of nonvolatility of MTJs and mask other shortcomings of current MTJ technology. This chapter also proposes that with virtually unlimited endurance and with its back-end-of-line compatibility, MTJs can be implemented as part of logic on a CMOS logic plane that can store information without the need for power. This demonstration of nonvolatile logic-in-memory architecture opens up a number of new possibilities that could not be realized by CMOS alone. Several examples of logic-in-memory realization not just design but in real chips are included to emphasize this new possibility.

7.8   Sources of further information and advice

Ikeda, S., Hayakawa, J., Lee, Y. M., Matsukura, F., Ohno, Y., Hanyu, T. and Ohno, H. (2007) 'Magnetic tunnel junctions for spintronic memories and beyond', *IEEE Trans. Electron Devices*, 54, 991–1002.

Ikeda, S. Sato, H., Yamanouchi, M., Gan, H. D., Miura, K., Mizumuma, K., Kanai, S., Fukami, S., Matsukura, F., Kasai, N. and Ohno, H. (2012) 'Recent progress of perpendicular anisotropy magnetic tunnel junctions for nonvolatile VLSI', *SPIN*, 3, 1240003-(1)-(12).

Fukami, S. and Ohno, H. (2017) 'Magnetization switching schemes for nanoscale three-terminal spintronics devices', Japanese Journal of Applied Physics (JJAP), Vol.56, No.8, pp.0802A1.

Oogane, M., Watanabe, K., Saruyama, H., Hosoda, M., Shahnaz, P., Kurimoto, Y., Kubota, M. and Ando, Y. (2017) 'L1$_0$-ordered MnAl thin films with high perpendicular magnetic anisotropy', Japanese Journal of Applied Physics (JJAP), Vol.56, No.8, pp.0802A2.

Kanai, S., Matsukura, F. and Ohno, H. (2017) 'Electric-field-induced magnetization switching in CoFeB/MgO magnetic tunnel junctions', Japanese Journal of Applied Physics (JJAP), Vol.56, No.8, pp.0802A3.

Kobayashi, D., Hirose, K., Makino, T., Onoda, S., Ohshima, T., Ikeda, S., Sato, H., Enobio, E. C. I., Endoh, T. and Ohno, H. (2017) 'Soft errors in 10-nm-scale magnetic tunnel junctions exposed to high-energy heavy-ion radiation', Japanese Journal of Applied Physics (JJAP), Vol.56, No.8, pp.0802B4.

Onizawa, N., Tamakoshi, A. and Hanyu, T. (2017) 'Evaluation of reinitialization-free nonvolatile computer systems for energy-harvesting Internet of things applications', Japanese Journal of Applied Physics (JJAP), Vol.56, No.8, pp.0802B7.

Acknowledgment

Part of the authors' work described here have been supported by the project "Research and Development of Ultra-Low Power Spintronics-Based VLSIs" through the "Funding Program for World-Leading Innovative R&D on Science and Technology (FIRST Program)" by the Japan Society for the Promotion of Science (JSPS) initiated by the Council for Science and Technology Policy (CSTP) and "Research and Development of Spintronics Material and Device Science and Technology for a Disaster-Resistant Safe and Secure Society" program under Research and Development for Next-Generation Information Technology of Ministry of Education, Culture, Sports, Science, and Technology (MEXT). The authors thank Professors Takashi Ohsawa and Naoki Kasai for their support.

References

[1] H. Ohno, T. Endoh, T. Hanyu, i.N. Kasa, S. Ikeda, Magnetic tunnel junction for nonvolatile CMOS logic, in: IEDM Tech. Dig, 2010, pp. 218–221.

[2] M. Hosomi, H. Yamagishi, T. Yamamoto, K. Bessho, Y. Higo, K. Yamane, H. Yamada, M. Shoji, H. Hachino, C. Fukumoto, H. Nagao, H. Kano, A novel nonvolatile memory with spin torque transfer magnetization switching: spin-RAM, in: IEDM Tech. Dig, 2005, pp. 459–462.

[3] T. Kawahara, R. Takemura, K. Miura, J. Hayakawa, S. Ikeda, Y.M. Lee, R. Sasaki, Y. Goto, K. Ito, T. Meguro, F. Matsukura, H. Takahashi, H. Matsuoka, H. Ohno, 2 Mb SPRAM (SPin-Transfer Torque RAM) with bit-by-bit bi-directional current write and parallelizing-direction current read, IEEE J. Solid State Circuits 43 (2008) 109–120.

[4] T. Kishi, H. Yoda, T. Kai, T. Nagase, E. Kitagawa, M. Yoshikawa, K. Nishiyama, T. Daibou, M. Nagamine, M. Amano, S. Takahashi, M. Nakayama, N. Shimomura, H. Aikawa, S. Ikegawa, S. Yuasa, K. Yakushiji, H. Kubota, A. Fukushima, M. Oogane, T. Miyazaki, K. Ando, Lower-current and fast switching of a perpendicular TMR for high speed and high density spin-transfer-torque MRAM, in: IEDM Tech. Dig, 2008, pp. 1–4.

[5] R. Takemura, T. Kawahara, K. Miura, H. Yamamoto, J. Hayakawa, N. Matsuzaki, K. Ono, M. Yamanouchi, K. Ito, H. Takahashi, S. Ikeda, H. Hasegawa, H. Matsuoka, H. Ohno, A 32-Mb SPRAM with 2T1R memory cell, localized bi-directional write driver and '1'/'0' dual-array equalized reference scheme, IEEE J. Solid-State Circuits 45 (2010) 869–879.

[6] K. Tsuchida, T. Inaba, K. Fujita, Y. Ueda, T. Shimizu, Y. Asao, T. Kajiyama, M. Iwayama, K. Sugiura, S. Ikegawa, T. Kishi, T. Kai, M. Amano, N. Shimomura, H. Yoda, Y. Watanabe, A 64Mb MRAM with clamped-reference and adequate-reference schemes, in: ISSCC Dig. Tech. Pap, 2010, pp. 258–259.

[7] D.C. Worledge, G. Hu, P.L. Trouilloud, D.W. Abraham, S. Brown, M.C. Gaidis, J. Nowak, E.J. O'Sullivan, R.P. Robertazzi, J.Z. Sun, W.J. Gallagher, Switching distributions and write reliability of perpendicular spin torque MRAM, in: IEDM Tech. Dig, 2010, pp. 296–299.

[8] S. Chung, K.M. Rho, S.D. Kim, H.J. Suh, D.J. Kim, H.J. Kim, S.H. Lee, J.H. Park, H.M. Hwang, S.M. Hwang, J.Y. Lee, Y.B. An, J.U. Yi, Y.H. Seo, D.H. Jung, M.S. Lee, S.H. Cho, J.N. Kim, G.J. Park, G. Jin, A.D. Smith, V. Nikitin, A. Ong, X. Tang, Y. Kim, J.S. Rho, S.K. Park, S.W. Chung, J.G. Jeong, S.J. Hong, Fully integrated 54nm STT-RAM with the smallest bit cell dimension for high density memory application, in: IEDM Tech. Dig, 2010, pp. 304–307.

[9] J.M. Slaughter, N.D. Rizzo, J. Janesky, R. Whig, F.B. Mancoff, D. Houssameddine, J.J. Sun, S. Aggarwal, K. Nagel, S. Deshpande, S.M. Alam, T. Andre, P. LoPresti, High density ST-MRAM technology, in: IEDM Tech. Dig, 2012, pp. 673–676.

[10] E. Kitagawa, S. Fujita, K. Nomura, H. Noguchi, K. Abe, K. Ikegami, T. Daibou, Y. Kato, C. Kamata, S. Kashiwada, N. Shimomura, J. Ito, H. Yoda, Impact of ultra low power and fast write operation of advanced perpendicular MTJ on power reduction for high-performance mobile CPU, in: IEDM Tech. Dig, 2012, pp. 677–680.

[11] A. Mochizuki, H. Kimura, M. Ibuki, T. Hanyu, TMR-based logic-in-memory circuit for low-power VLSI, IEICE Trans. Fundam. Electron. Commun. Comput. Sci. E88-A (2005) 1408–1415.

[12] S. Matsunaga, J. Hayakawa, S. Ikeda, K. Miura, H. Hasegawa, T. Endoh, H. Ohno, T. Hanyu, Fabrication of a nonvolatile full adder based on logic-in-memory architecture using magnetic tunnel junctions, Appl. Phys. Express 1 (2008). 091301(1)-(3).

[13] D. Suzuki, M. Natsui, S. Ikeda, H. Hasegawa, K. Miura, J. Hayakawa, T. Endoh, H. Ohno, T. Hanyu, Fabrication of a nonvolatile lookup-table circuit chip using magneto/semiconductor-hybrid structure for an immediate-power-up field programmable gate array, in: IEEE 2009 Symposia on VLSI Circuits, Dig. Tech. Papers, 2009, pp. 80–81.

[14] W. Zhao, C. Chappert, V. Javerliac, J.-P. Noziere, High speed, high stability and low power sensing amplifier for MTJ/CMOS hybrid logic circuits, IEEE Trans. Magn. 45 (2009) 3784–3787.

[15] S. Matsunaga, A. Katsumata, M. Natsui, S. Fukami, T. Endoh, H. Ohno, T. Hanyu, Fully parallel 6T-2MTJ nonvolatile TCAM with single-transistor-based self match-line discharge control, in: IEEE Symp. VLSI Circuits, 2011, pp. 298–299.

[16] S. Matsunaga, S. Miura, H. Honjou, K. Kinoshita, S. Ikeda, T. Endoh, H. Ohno, T. Hanyu, A 3.14 um2 4T-2MTJ-cell fully parallel TCAM based on nonvolatile logic-in-memory architecture, in: IEEE Symp. VLSI Circuits, 2012, pp. 44–45.

[17] T. Ohsawa, H. Koike, S. Miura, H. Honjo, K. Tokutome, S. Ikeda, T. Hanyu, H. Ohno, T. Endoh, 1Mb 4T-2MTJ nonvolatile STT-RAM for embedded memories using 32b fine-grained power gating technique with 1.0ns/200ps wake-up/power-off times, in: 2012 Symposium on VLSI Circuits Digest of Technical Papers, 2012, pp. 46–47.

[18] M. Natsui, D. Suzuki, N. Sakimura, R. Nebashi, Y. Tsuji, A. Morioka, T. Sugibayashi, S. Miura, H. Honjo, K. Kinoshita, S. Ikeda, T. Endoh, H. Ohno, T. Hanyu, Nonvolatile logic-in-memory array processor in 90nm MTJ/MOS achieving 75% leakage reduction using cycle-based power gating, in: 2013 IEEE International Solid-State Circuits Conference (ISSCC) Dig. Tech. Papers, 2013, pp. 194–195.

[19] S. Matsunaga, N. Sakimura, R. Nebashi, Y. Tsuji, A. Morioka, T. Sugibayashi, S. Miura, H. Honjo, K. Kinoshita, H. Sato, S. Fukami, M. Natsui, A. Mochizuki, S. Ikeda, T. Endoh, H. Ohno, T. Hanyu, Fabrication of a 99%-energy-less nonvolatile multi-functional CAM chip using hierarchical power gating for a massively-parallel full-text-search engine, in: IEEE Symp. VLSI Circuits, 2013, pp. C106–C107.

[20] T. Ohsawa, S. Miura, K. Kinoshita, H. Honjo, S. Ikeda, T. Hanyu, H. Ohno, T. Endoh, A 1.5nsec/2.1nsec random read/write cycle 1Mb STT-RAM using 6T2MTJ cell with background write for nonvolatile-memories, in: IEEE Symp. VLSI Circuits, 2013, pp. C110–C111.

[21] J.C. Slonczewski, Current-driven excitation of magnetic multilayers, J. Magn. Magn. Mater. 159 (1996) L1–L7.

[22] L. Berger, Emission of spin waves by a magnetic multilayer traversed by a current, Phys. Rev. B 54 (1996) 9353–9358.

[23] M. Tsoi, A.G.M. Jansen, J. Bass, W.-C. Chiang, M. Seck, V. Tsoi, P. Wyder, Excitation of a magnetic multilayer by an electric current, Phys. Rev. Lett. 80 (1998) 4281–4284.

[24] E.B. Mayer, D.C. Ralph, J.A. Katine, R.N. Louie, R.A. Buhrman, Current-induced switching of domains in magnetic multilayer devices, Science 285 (1999) 867–870.

[25] J.A. Katine, F.J. Albert, R.A. Buhrman, E.B. Myers, D.C. Ralph, Current-driven magnetization reversal and spin-wave excitations in Co/Cu/Co pillars, Phys. Rev. Lett. 84 (2000) 3149–3152.

[26] F.J. Albert, J.A. Katine, R.A. Buhrman, D.C. Ralph, Spin-polarized current switching of a Co thin film nanomagnet, Appl. Phys. Lett. 77 (2000) 3809–3811.

[27] M. Julliere, Tunneling between ferromagnetic films, Phys. Lett. A 54 (1975) 225–226.

[28] S. Maekawa, U. Gafvert, Electron-tunneling between ferromagnetic films, IEEE Trans. Magn. 18 (1982) 707–708.

[29] T. Miyazaki, N. Tezuka, Giant magnetic tunneling effect in Fe/Al2O3/Fe junction, J. Magn. Magn. Mater. 139 (1995) L231–L234.

[30] J.S. Moodera, L.R. Kinder, T.M. Wong, R. Meservey, Large magnetoresistance at room-temperature in ferromagnetic thin-film tunnel junctions, Phys. Rev. Lett. 74 (1995) 3273–3276.

[31] W.H. Butler, X.-G. Zhang, T.C. Schulthess, J.M. MacLaren, Spin dependent tunneling conductance of Fe/MgO/Fe sandwiches, Phys. Rev. B 63 (2001). 054416(1)-(12).

[32] J. Mathon, A. Umerski, Theory of tunneling magnetoresistance of an epitaxial Fe/MgO/Fe(001) junction, Phys. Rev. B 63 (2001). 220403(1)-(4).

[33] S. Yuasa, A. Fukushima, T. Nagahama, K. Ando, Y. Suzuki, High tunnel magnetoresistance at room temperature in fully epitaxial Fe/MgO/Fe tunnel junctions due to coherent spin-polarized tunneling, Jpn. J. Appl. Phys. 43 (2004) L588–L590.

[34] S. Yuasa, T. Nagahama, A. Fukushima, Y. Suzuki, K. Ando, Giant room-temperature magnetoresistance in single-crystal Fe/MgO/Fe magnetic tunnel junctions, Nat. Mater. 3 (2004) 868–871.

[35] S.S.P. Parkin, C. Kaiser, A. Panchula, P.M. Rice, B. Hughes, M. Samant, S.-H. Yang, Giant tunneling magnetoresistance at room temperature with MgO (100) tunnel barriers, Nat. Mater. 3 (2004) 662–867.

[36] D.D. Djayaprawira, K. Tsunekawa, M. Nagai, H. Maehara, S. Yamagata, N. Watanabe, S. Yuasa, Y. Suziki, K. Ando, 230% room-temperature magnetoresistance in CoFeB/MgO/CoFeB magnetic tunnel junctions, Appl. Phys. Lett. 86 (2005). 092502(1)-(3).

[37] J. Hayakawa, S. Ikeda, F. Matsukura, H. Takahashi, H. Ohno, Dependence of giant tunnel magnetoresistance of sputtered CoFeB/MgO/CoFeB magnetic tunnel junctions on MgO barrier thickness and annealing temperature, Jpn. J. Appl. Phys. 44 (2005) L587–L589.

[38] S. Yuasa, A. Fukushima, H. Kubota, Y. Suzuki, K. Ando, Giant tunneling magnetoresistance up to 410% at room temperature in fully epitaxial Co/MgO/Co magnetic tunnel junctions with bcc Co(001) electrodes, Appl. Phys. Lett. 89 (2006). 042505(1)-(3).

[39] J. Hayakawa, S. Ikeda, Y.M. Lee, F. Matsukura, H. Ohno, Effect of high annealing temperature on giant tunnel magnetoresistance ratio of CoFeB/MgO/CoFeB magnetic tunnel junctions, Appl. Phys. Lett. 89 (2006). 232510-(1)-(3).

[40] Y.M. Lee, J. Hayakawa, S. Ikeda, F. Matsukura, H. Ohno, Effect of electrode composition on the tunnel magnetoresistance of pseudo-spin-valve magnetic tunnel junction with a MgO tunnel barrier, Appl. Phys. Lett. 90 (2007). 212507-(1)-(3).

[41] S. Ikeda, J. Hayakawa, Y. Ashizawa, Y.M. Lee, K. Miura, H. Hasegawa, M. Tsunoda, F. Matsukura, H. Ohno, Tunnel magnetoresistance of 604% at 300 K by suppression of Ta diffusion in CoFeB/MgO/CoFeB pseudo-spin-valves annealed at high temperature, Appl. Phys. Lett. 93 (2008). 082508-(1)-(3).

[42] Z. Wang, M. Saito, P.K. McKenna, S. Fukami, H. Sato, S. Ikeda, H. Ohno, Y. Ikuhara, Atomic-scale structure and local chemistry of CoFeB−MgO magnetic tunnel junctions, Nano Lett. 16 (1510) (2016).

[43] S. Mangin, D. Ravelosona, J.A. Katine, M.J. Carey, B.D. Terris, E.E. Fullerton, Current-induced magnetization reversal in nanopillars with perpendicular anisotropy, Nat. Mater. 5 (2006) 210–215.

[44] H. Yoda, T. Kishi, T. Nagase, M. Yoshikawa, K. Nishiyama, E. Kitagawa, T. Daibou, M. Amano, N. Shimomura, S. Takahashi, T. Kai, M. Nakayama, H. Aikawa, S. Ikegawa, M. Nagamine, J. Ozeki, S. Mizukami, M. Oogane, Y. Ando, S. Yuasa, K. Yakushiji,

H. Kubota, Y. Suzuki, Y. Nakatani, T. Miyazaki, K. Ando, High efficient spin transfer torque writing on perpendicular magnetic tunnel junctions for high density MRAMs, Curr. Appl. Phys. 10 (2010) e87–e89.

[45] S. Ikeda, K. Miura, H. Yamamoto, K. Mizunuma, H.D. Gan, M. Endo, S. Kanai, J. Hayakawa, F. Matsukura, H. Ohno, A perpendicular-anisotropy CoFeB–MgO magnetic tunnel junction, Nat. Mater. 9 (2010) 721–724.

[46] N. Nishimura, T. Hirai, A. Koganei, T. Ikeda, K. Okano, Y. Sekiguchi, Y. Osada, Magnetic tunnel junction device with perpendicular magnetization films for high-density magnetic random access memory, J. Appl. Phys. 91 (2002) 5246–5249.

[47] A.C. Cabrera, C.H. Chang, C.C. Hsu, M.C. Weng, C.C. Chen, C.T. Chao, J.C. Wu, Y.H. Chang, T.H. Wu, Perpendicular magnetic tunneling junction with double barrier layers for MRAM application, IEEE Trans. Magn. 43 (2007) 914–916.

[48] H. Ohmori, T. Hatori, S. Nakagawa, Perpendicular magnetic tunnel junction with tunneling magnetoresistance ratio of 64% using MgO (100) barrier layer prepared at room temperature, J. Appl. Phys. 103 (2008). 07A911-(1)-(3).

[49] M. Nakayama, T. Kai, N. Shimomura, M. Amano, E. Kitagawa, T. Nagase, M. Yoshikawa, T. Kishi, S. Ikegawa, H. Yoda, Spin transfer switching in TbCoFe/CoFeB/MgO/CoFeB/TbCoFe magnetic tunnel junctions with perpendicular magnetic anisotropy, J. Appl. Phys. 103 (2008). 07A710-(1)-(3).

[50] G. Kim, Y. Sakuraba, M. Oogane, Y. Ando, T. Miyazaki, Tunneling magnetoresistance of magnetic tunnel junctions using perpendicular magnetization L10-CoPt electrodes, Appl. Phys. Lett. 92 (2008). 172502-(1)-(2).

[51] M. Yoshikawa, E. Kitagawa, T. Nagase, T. Daibou, M. Nagamine, K. Nishiyama, T. Kishi, H. Yoda, Tunnel magnetoresistance over 100% in MgO-based magnetic tunnel junction films with perpendicular magnetic L10–FePt electrodes, IEEE Trans. Magn. 44 (2008) 2573–2576.

[52] K. Mizunuma, S. Ikeda, J.H. Park, H. Yamamoto, H.D. Gan, K. Miura, H. Hasegawa, J. Hayakawa, F. Matsukura, H. Ohno, MgO barrier-perpendicular magnetic tunnel junctions with CoFe/Pd multilayers and ferromagnetic insertion layers, Appl. Phys. Lett. 95 (2009). 232516-(1)-(3).

[53] K. Yakushiji, K. Noma, T. Saruya, H. Kubota, A. Fukushima, T. Nagahama, S. Yuasa, K. Ando, High magnetoresistance ratio and low resistance–area product in magnetic tunnel junctions with perpendicularly magnetized electrodes, Appl. Phys. Express 3 (2010). 053003-(1)-(3).

[54] K. Yakushiji, T. Saruya, H. Kubota, A. Fukushima, T. Nagahama, S. Yuasa, K. Ando, Ultrathin Co/Pt and Co/Pd superlattice films for MgO-based perpendicular magnetic tunnel junctions, Appl. Phys. Lett. 97 (2010). 232508-(1)-(3).

[55] K. Mizunuma, M. Yamanouchi, S. Ikeda, H. Yamamoto, H.D. Gan, K. Miura, J. Hayakawa, F. Matsukura, H. Ohno, Pd layer thickness dependence of tunnel magnetoresistance properties in CoFeB/MgO-based magnetic tunnel junctions with perpendicular anisotropy CoFe/Pd multilayers, Appl. Phys. Express 4 (2011). 023002-(1)-(3).

[56] S. Ishikawa, H. Sato, M. Yamanouchi, S. Ikeda, S. Fukami, F. Matsukura, H. Ohno, Magnetic properties of MgO-[Co/Pt] multilayers with a CoFeB insertion layer, J. Appl. Phys. 113 (2013). 17C721-(1)-(3).

[57] M. Endo, S. Kanai, S. Ikeda, F. Matsukura, H. Ohno, Electric-field effects on thickness dependent magnetic anisotropy of sputtered MgO/Co40Fe40B20/Ta structures, Appl. Phys. Lett. 96 (2010). 212503-(1)-(3).

[58] S. Kanai, M. Endo, S. Ikeda, F. Matsukura, H. Ohno, Magnetic anisotropy modulation in Ta/CoFeB/MgO structure by electric fields, J. Phys. Conf. Ser. 266 (2011) 012092.

[59] M. Yamanouchi, R. Koizumi, S. Ikeda, H. Sato, K. Mizunuma, K. Miura, H.D. Gan, F. Matsukura, H. Ohno, Dependence of magnetic anisotropy on MgO thickness and buffer layer in Co20Fe60B20-MgO structure, J. Appl. Phys. 109 (2011) 07C712.

[60] R. Shimabukuro, K. Nakamura, T. Akiyama, T. Ito, Electric field effects on magnetocrystalline anisotropy in ferromagnetic Fe monolayers, Phys. E. 42 (2010) 1014–1017.

[61] S. Kanai, M. Tsujikawa, Y. Miura, M. Shirai, F. Matsukura, H. Ohno, Magnetic anisotropy in Ta/CoFeB/MgO investigated by x-ray magnetic circular dichroism and first-principles calculation, Appl. Phys. Lett. 105 (2014) 222409.

[62] H. Sato, M. Yamanouchi, K. Miura, S. Ikeda, H.D. Gan, K. Mizunuma, R. Koizumi, F. Matsukura, H. Ohno, Junction size effect on switching current and thermal stability in CoFeB/MgO perpendicular magnetic tunnel junctions, Appl. Phys. Lett. 99 (2011). 042501-(1)-(3).

[63] H. Sato, M. Yamanouchi, K. Miura, S. Ikeda, R. Koizumi, F. Matsukura, H. Ohno, CoFeB thickness dependence of thermal stability factor in CoFeB/MgO perpendicular magnetic tunnel junctions, IEEE Magn. Lett. 3 (2012). 3000204(1)-(4).

[64] H. Sato, M. Yamanouchi, S. Ikeda, S. Fukami, F. Matsukura, H. Ohno, Perpendicular-anisotropy CoFeB-MgO magnetic tunnel junctions with a MgO/CoFeB/Ta/CoFeB/MgO recording structure, Appl. Phys. Lett. 101 (2012). 022414-(1)-(4).

[65] H. Sato, M. Yamanouchi, S. Ikeda, S. Fukami, F. Matsukura, H. Ohno, MgO/CoFeB/Ta/CoFeB/MgO recording structure in magnetic tunnel junctions with perpendicular easy axis, IEEE Trans. Magn. (2013) 4437–4440.

[66] D.C. Worledge, G. Hu, D.W. Abraham, J.Z. Sun, P.L. Trouilloud, J. Nowak, S. Brown, M.C. Gaidis, J. O'Sullivan, R.P. Robertazzi, Spin torque switching of perpendicular Ta|CoFeB|MgO-based magnetic tunnel junctions, Appl. Phys. Lett. 98 (2011). 022501-(1)-(3).

[67] J.J. Nowak, R.P. Robertazzi, J.Z. Sun, G. Hu, D.W. Abraham, P.L. Trouilloud, S. Brown, M.C. Gaidis, E.J. O'Sullizan, W.J. Gallagher, D.C. Worledge, Demonstration of ultralow bit error rates for spin-torque magnetic random-access memory with perpendicular magnetic anisotropy, IEEE Magn. Lett. 2 (2011). 3000204-(1)-(4).

[68] C. Yoshida, T. Sugii, Reliability study of magnetic tunnel junction with naturally oxidized MgO barrier, in: IEEE Inter. Reliability Physics Symp, 2012. 2A.3.1–2A3.

[69] M. Hosoda, M. Oogane, M. Kubota, T. Kubota, H. Saruyama, S. Iihama, H. Naganuma, Y. Ando, Fabrication of L10-MnAl perpendicularly magnetized thin films for perpendicular magnetic tunnel junctions, J. Appl. Phys. 111 (2012). 07A324-(1)-(3).

[70] T. Endoh, T. Ohsawa, H. Koike, T. Hanyu, H. Ohno, Restructuring of memory hierarchy in computing system with spintronics-based technologies, in: 2012 Symposium on VLSI Technology Digest of Technical Papers, 2012, pp. 89–90.

[71] T. Endoh, H. Koike, S. Ikeda, T. Hanyu, H. Ohno, An overview of nonvolatile emerging memories—spintronics for working memories, IEEE J. Emerging Sel. Top. Circuits Syst. 6 (2016) 109–119.

[72] T. Hanyu, T. Endoh, D. Suzuki, H. Koike, Y. Ma, N. Onizawa, M. Natsui, S. Ikeda, H. Ohno, Standby-power-free integrated circuits using MTJ-based VLSI computing, Proc. IEEE 104 (10) (2016) 1844–1863.

[73] H. Koike, S. Miura, H. Honjo, T. Watanabe, H. Sato, S. Sato, T. Nasuno, Y. Noguchi, M. Yasuhira, T. Tanigawa, M. Muraguchi, M. Niwa, K. Ito, S. Ikeda, H. Ohno, T. Endoh, 1T1MTJ STT-MRAM cell array design with an adaptive reference voltage generator for improving device variation tolerance, in: 2015 IEEE International Memory Workshop (IMW), 2015, pp. 141–144.

[74] H. Koike, S. Miura, H. Honjo, T. Watanabe, H. Sato, S. Sato, T. Nasuno, Y. Noguchi, M. Yasuhira, T. Tanigawa, M. Muraguchi, M. Niwa, K. Ito, S. Ikeda, H. Ohno, T. Endoh, Demonstration of yield improvement for on-via MTJ using a 2-Mbit 1T-1MTJ STT-MRAM test chip, in: 2016 IEEE International Memory Workshop (IMW), 2016, pp. 56–59.

[75] C.J. Lin, S.H. Kang, Y.J. Wang, K. Lee, X. Zhu, W.C. Chen, X. Li, W.N. Hsu, Y.C. Kao, M.T. Liu, W.C. Chen, Y.C. Lin, M. Nowak, N. Yu, L. Tran, 45nm low power CMOS logic compatible embedded STT MRAM utilizing a reverse-connection 1T/1MTJ cell, in: 2011 International Electron Device Meeting (IEDM), 2009, pp. 279–282.

[76] Y.M. Lee, C. Yoshida, K. Tsunoda, S. Umehara, M. Aoki, T. Sugii, High scalable STT-MRAM with MTJs of top-pinned structure in 1T/1MTJ cell, in: 2010 Symposium on VLSI Technology Digest of Technical Papers, 2010, pp. 49–50.

[77] H. Koike, T. Endoh, A study for adopting PMOS memory cell for 1T1R STT-RAM with asymmetric switching current MTJ, in: 2011 International Conference on Solid State Devices and Materials (SSDM), 2011, pp. 961–962.

[78] H. Koike, T. Ohsawa, S. Miura, H. Honjo, S. Ikeda, T. Hanyu, H. Ohno, T. Endoh, Wide operational margin capability of 1kbit STT-MRAM array chip with 1-PMOS and 1-bottom-pin-MTJ type cell, in: 2013 International Conference on Solid State Devices and Materials (SSDM), 2013, pp. 1094–1095.

[79] H. Koike, T. Ohsawa, S. Miura, H. Honjo, S. Ikeda, T. Hanyu, H. Ohno, T. Endoh, Wide operational margin capability of 1 kbit spin-transfer-torque memory array chip with 1-PMOS and 1-bottom-pin-magnetic-tunnel-junction type cell, Jpn. J. Appl. Phys. 53 (2014) 04ED13.

[80] T. Ohsawa, K. Hatsuda, K. Fujita, F. Matsuoka, T. Higashi, Generation of accurate reference current for data sensing in high-density memories by averaging multiple pairs of dummy cells, IEEE J. Solid State Circuits 46 (2011) 2148–2157.

[81] S. Yamamoto, S. Sugahara, Nonvolatile static random access memory using magnetic tunnel junctions with current-induced magnetization switching architecture, Jpn. J. Appl. Phys. 48 (2009) 043001.

[82] K. Abe, K. Nomura, S. Ikegawa, T. Kishi, H. Yoda, S. Fujita, Hierarchical cache memory based on MRAM and nonvolatile SRAM with perpendicular magnetic tunnel junctions for ultra low power system, in: 2010 International Conference on Solid State Devices and Materials (SSDM), 2010, pp. 1144–1145.

[83] T. Ohsawa, F. Iga, S. Ikeda, T. Hanyu, H. Ohno, T. Endoh, Studies on static noise margin and scalability for low-power and high-density nonvolatile SRAM using spin-transfer-torque (STT) MTJs, in: 2011 International Conference on Solid State Devices and Materials (SSDM), 2011, pp. 959–960.

[84] T. Ohsawa, F. Iga, S. Ikeda, T. Hanyu, H. Ohno, T. Endoh, High-density and low-power nonvolatile static random access memory using spin-transfer-torque magnetic tunnel junction, Jpn J. Appl. Phys. 51 (2012) 02BD01.

[85] T. Ohsawa, H. Koike, S. Miura, H. Honjo, K. Kinoshita, S. Ikeda, T. Hanyu, H. Ohno, T. Endoh, A 1Mb nonvolatile embedded memory using 4T2MTJ cell with 32b fine-grained power gating scheme, IEEE J. Solid State Circuits 48 (2013) 1511–1520.

[86] T. Aoki, Y. Ando, D. Watanabe, M. Oogane, T. Miyazaki, Spin transfer switching in the nanosecond regime for CoFeB/MgO/CoFeB ferromagnetic tunnel junctions, J. Appl. Phys. 103 (2008) 103911.

[87] Z. Diao, Z. Li, S. Wang, Y. Ding, A. Panchula, E. Chen, L.-C. Wang, Y. Huai, Spin-transfer torque switching in magnetic tunnel junctions and spin-transfer torque random access memory, J. Phys. Condens. Matter 19 (2007) 165209.

[88] T. Ohsawa, S. Miura, K. Kinoshita, H. Honjo, S. Ikeda, T. Hanyu, H. Ohno, T. Endoh, A 1.5nsec/2.1nsec random read/write cycle 1Mb STT-RAM using 6T2MTJ cell with background write for nonvolatile e-memories, in: 2013 Symposium on VLSI Circuits Digest of Technical Papers, 2013, pp. C110–C111.

[89] T. Ohsawa, S. Miura, H. Honjo, S. Ikeda, T. Hanyu, H. Ohno, T. Endoh, A 500ps/8.5ns array read/write latency 1Mb twin 1T1MTJ STT-MRAM designed in 90nm CMOS/40nm MTJ process with novel positive feedback S/A circuit, in: 2014 International Conference on Solid State Devices and Materials (SSDM), 2014, pp. 458–459.

[90] W.H. Kautz, Cellular logic-in-memory arrays, IEEE Trans. Computer C-18 (1969) 719–727.

[91] S. Matsunaga, N. Sakimura, R. Nebashi, Y. Tsuji, A. Morioka, T. Sugibayashi, S. Miura, H. Honjo, K. Kinoshita, H. Sato, S. Fukami, M. Natsui, A. Mochizuki, S. Ikeda, T. Endoh, H. Ohno, T. Hanyu, Fabrication of a 99%-energy-less nonvolatile multi-functional CAM chip using hierarchical power gating for a massively-parallel full-text-search engine, in: 2013 Symposia on VLSI Circuits, Digest of Technical Papers, C9-2, 2013, pp. 106–107.

[92] S. Matsunaga, S. Miura, H. Honjo, K. Kinoshita, S. Ikeda, T. Endoh, H. Ohno, T. Hanyu, A 3.14 um2 4T-2MTJ-cell fully parallel TCAM based on nonvolatile logic-in-memory architecture, in: 2012 Symposia on VLSI Circuits, Digest of Technical Papers, 2012, pp. 44–45.

[93] S. Matsunaga, A. Katsumata, M. Natsui, T. Endoh, H. Ohno, T. Hanyu, Design of a Nine-Transistor/two-magnetic-tunnel-junction-cell-based low-energy nonvolatile ternary content-addressable memory, Jpn. J. Appl. Phys. (JJAP) 51 (2012) 02BM06-1–02BM06-6.

[94] S. Matsunaga, A. Katsumata, M. Natsui, T. Endoh, H. Ohno, T. Hanyu, Design of a 270ps-access 7T-2MTJ-cell nonvolatile ternary content-addressable memory, J. Appl. Phys. (JAP) 111 (7) (2012) 07E336.

[95] D. Suzuki, M. Natsui, T. Endoh, H. Ohno, T. Hanyu, Six-input lookup table circuit with 62% fewer transistors using nonvolatile logic-in-memory architecture with series/parallel-connected magnetic tunnel junctions, J. Appl. Phys. (JAP) 111 (7) (2012) 07E318.

[96] D. Suzuki, Y. Lin, M. Natsui, T. Hanyu, A 71%-area-reduced six-input nonvolatile lookup-table circuit using a three-terminal magnetic-tunnel-junction-based single-ended structure, Jpn. J. Appl. Phys. (JJAP) 52 (2013) 04CM04-1–04CM04-6.

[97] D. Suzuki, M. Natsui, S. Ikeda, T. Endoh, H. Ohno, T. Hanyu, Design of a variation-resilient single-ended nonvolatile 6-input lookup table circuit with a redundant-MTJ-based active load for smart IoT applications, Institute of Engineering Technology (IET) Electron. Lett. 53 (7) (Mar. 2017) 456–458.

[98] D. Suzuki, M. Natsui, A. Mochizuki, S. Miura, H. Honjo, H. Sato, S. Fukami, S. Ikeda, T. Endoh, H. Ohno, T. Hanyu, Fabrication of a 3000-6-input-LUTs embedded and block-level power-gated nonvolatile FPGA chip using p-MTJ-based logic-in-memory structure, in: Symp. VLSI Circuits Dig. Tech. Papers, Jun. 2015, pp. 172–173.

[99] D. Chabi, W. Zhao, E. Deng, Y. Zhang, N.B. Romdhane, J.-O. Klein, C. Chappert, Ultra low power magnetic flip-flop based on check pointing/power gating and self-enable mechanisms, IEEE Trans. Circuits Syst. I 61 (6) (2014) 1755–1765.

[100] D. Suzuki, M. Natsui, A. Mochizuki, T. Hanyu, Cost-efficient self-terminated write driver for spin-transfer-torque RAM and logic, IEEE Trans. Magn. 50 (11) (Nov. 2014) 3402104.

[101] D. Suzuki, T. Hanyu, Magnetic-tunnel-junction based low-energy nonvolatile flip-flop using an area-efficient self-terminated write driver, J. Appl. Phys. (JAP) 117 (Jan. 2015) 17B504-1–17B504-3.

[102] D. Suzuki, T. Hanyu, A greedy power saving of MTJ-based nonvolatile FPGA with self-terminated logic-in-memory structure, in: Proc. of Int. Conf. on Field-Programmable Logic and Applications (FPL), Aug. 2016, pp. 1–4.

[103] D. Suzuki, T. Hanyu, Design of a low-power nonvolatile flip-flop using 3-terminal magnetic-tunnel-junction-based self-terminated mechanism, Jpn. J. Appl. Phys. (JJAP) 56 (4S) (Mar. 2017) 04CN06.

[104] T. Hanyu, D. Suzuki, N. Onizawa, M. Natsui, Three-terminal MTJ-based nonvolatile logic circuits with self-terminated writing mechanism for ultra-low-power VLSI processor, in: Design, Automation & Test in Europe (DATE), Mar. 2017, pp. 548–553.

[105] N. Sakimura, R. Nebashi, Y. Tsuji, H. Honjo, T. Sugibayashi, H. Koike, T. Ohsawa, S. Fukami, T. Hanyu, H. Ohno, T. Endoh, High-speed simulator including accurate MTJ models for spintronics integrated circuit design, in: Proc. IEEE International Symposium on Circuits and Systems (ISCAS), 2012, pp. 1971–1974.

[106] M. Natsui, A. Tamakoshi, A. Mochizuki, H. Koike, H. Ohno, T. Endoh, T. Hanyu, Stochastic behavior-considered VLSI CAD environment for MTJ/MOS-hybrid microprocessor design, in: Proc. IEEE International Symposium on Circuits and Systems (ISCAS), 2016, pp. 1878–1881.

[107] W. Zhao, et al., Macro-model of spin-transfer torque based magnetic tunnel junction device for hybrid magnetic-CMOS design, in: IEEE International Behavioral Modeling and Simulation Workshop, 2006, pp. 40–43.

[108] A. Kostrov, SPICE macro-model of magnetic tunnel nanostructure for digital applications and memory cells, in: MEMSTECH, 2010, pp. 20–23.

[109] J.D. Harms, et al., SPICE macromodel of spin-torque-transfer-operated magnetic tunnel junctions, IEEE Trans. Electron Devices 57 (6) (2010) 1425–1430.

[110] Y. Ma, T. Endoh, A novel neuron circuit with nonvolatile synapses based on magnetic-tunnel-junction for high-speed pattern learning and recognition, in: 2015 Asia-Pacific Workshop on Fundamentals and Applications of Advanced Semiconductor Devices, 2015, pp. 273–278.

[111] Y. Ma, S. Miura, H. Honjo, S. Ikeda, T. Hanyu, H. Ohno, T. Shibata, T. Endoh, A 600-μW ultra-low-power associative processor for image pattern recognition employing magnetic tunnel junction (MTJ) based nonvolatile memories with novel intelligent power-gating (IPG) scheme, in: 2015 International Conference on Solid State Devices and Materials (SSDM), 2015, pp. 1172–1173.

[112] Y. Ma, T. Endoh, Effect of MTJ resistance fluctuations on synapse stability of MTJ-based nonvolatile neuron circuit for high-speed object recognition, in: 2016 Asia-Pacific Workshop on Fundamentals and Applications of Advanced Semiconductor Devices, 2016, pp. 258–262.

[113] Y. Ma, S. Miura, H. Honjo, S. Ikeda, T. Hanyu, H. Ohno, T. Endoh, A 600-μW ultra-low-power associative processor for image pattern recognition employing magnetic tunnel junction-based nonvolatile memories with autonomic intelligent power-gating scheme, Jpn. J. Appl. Phys. 55 (2016) 04EF15.

[114] Y. Ma, S. Miura, H. Honjo, S. Ikeda, T. Hanyu, H. Ohno, T. Endoh, A compact and ultra-low-power STT-MRAM-based associative memory for nearest neighbor search with full adaptivity of template data format employing current-mode similarity evaluation and time-domain minimum searching, in: 2016 International Conference on Solid State Devices and Materials (SSDM), 2016, pp. 83–84.

[115] Y. Ma, S. Miura, H. Honjo, S. Ikeda, T. Hanyu, H. Ohno, T. Endoh, A spin transfer torque magnetoresistance random access memory-based high-density and ultralow-power associative memory for fully data-adaptive nearest neighbor search with current-mode similarity evaluation and time-domain minimum searching, Jpn. J. Appl. Phys. 56 (2017) 04CF08.

[116] T. Bui, T. Shibata, A scalable architecture of associative processors employing nano functional devices, in: 2009 International Conference on Ultimate Integration of Silicon (ULIS), 2016, pp. 213–216.
[117] F. An, T. Akazawa, S. Yamazaki, L. Chen, H.J. Mattausch, A coprocessor for clock-mapping-based nearest Euclidean distance search with feature vector dimension adaptability, in: 2014 Custom Integrated Circuits Conference (CICC), 2014, pp. 1–4.
[118] F. An, K. Mihara, S. Yamasaki, L. Chen, H.J. Mattausch, Associative memory for nearest neighbor search with high flexibility of reference-vector number due to configurable dual-storage space, in: 2015 International Conference on Solid State Devices and Materials (SSDM), 2015, pp. 144–145.
[119] T. Akazawa, S. Sasaki, H.J. Mattausch, Word-parallel coprocessor architecture for digital nearest Euclidean distance search, in: European Solid-State Circuits Conf. (ESSCIRC), 2013, pp. 267–270.
[120] H. Ohno, T. Endoh, T. Hanyu, N. Kasai, S. Ikeda, Magnetic tunnel junction for nonvolatile CMOS logic, in: IEDM Tech. Dig, 2010, pp. 218–221.
[121] T. Endoh, S. Togashi, F. Iga, Y. Yoshida, T. Ohsawa, H. Koike, S. Fukami, S. Ikeda, N. Kasai, N. Sakimura, T. Hanyu, H. Ohno, A 600MHz MTJ-based nonvolatile latch making use of incubation time in MTJ switching, in: IEDM Tech. Dig, 2011, pp. 75–78.
[122] H. Koike, T. Ohsawa, N. Sakimura, R. Nebashi, Y. Tsuji, A. Morioka, S. Miura, H. Honjo, T. Sugibayashi, S. Ikeda, T. Hanyu, H. Ohno, T. Endoh, A power-gated MPU with 3-microsecond entry/exit delay using MTJ-based nonvolatile flip-flop, in: 2013 Asian Solid-State Circuits Conference (A-SSCC), 2013, pp. 317–320.
[123] H. Koike, T. Ohsawa, S. Miura, H. Honjo, K. Kinoshita, S. Ikeda, T. Hanyu, H. Ohno, T. Endoh, A power-gated 32bit MPU with a power controller circuit activated by deep-sleep-mode instruction achieving ultra-low power operation, in: 2014 International Conference on Solid State Devices and Materials (SSDM), 2014, pp. 448–449.
[124] H. Koike, T. Ohsawa, S. Miura, H. Honjo, S. Ikeda, T. Hanyu, H. Ohno, T. Endoh, Power-gated 32 bit microprocessor with a power controller circuit activated by deep-sleep-mode instruction achieving ultra-low power operation, Jpn. J. Appl. Phys. 54 (2015) 04DE08.
[125] S. Mizukami, F. Wu, A. Sakuma, J. Walowski, D. Watanabe, T. Kubota, X. Zhang, H. Naganuma, M. Oogane, Y. Ando, T. Miyazaki, Long-lived ultrafast spin precession in manganese alloys films with a large perpendicular magnetic anisotropy, Phys. Rev. Lett. 106 (2011) 117201.
[126] T. Kubota, Q.L. Ma, S. Mizukami, X.M. Zhang, H. Naganuma, M. Oogane, Y. Ando, T. Miyazaki, Dependence of tunnel magnetoresistance effect on Fe thickness of perpendicularly magnetized L10-Mn62Ga38/Fe/MgO/CoFe junctions, Appl. Phys. Express 5 (2012). 043003-(1)-(3).
[127] Q.L. Ma, T. Kubota, S. Mizukami, X.M. Zhang, H. Naganuma, M. Oogane, Y. Ando, T. Miyazaki, Magnetoresistance effect in L10-MnGa/MgO/CoFeB perpendicular magnetic tunnel junctions with Co interlayer, Appl. Phys. Lett. 101 (2012). 032402-(1)-(4).
[128] H. Saruyama, M. Oogane, Y. Kurimoto, H. Naganuma, Y. Ando, Fabrication of L10-ordered MnAl films for observation of tunnel magnetoresistance effect, Jpn. J. Appl. Phys. 52 (2013). 063003-(1)-(4).
[129] S. Fukami, T. Suzuki, K. Nagahara, N. Ohshima, Y. Ozaki, S. Saito, R. Nebashi, N. Sakimura, H. Honjo, K. Mori, C. Igarashi, S. Miura, N. Ishiwata, T. Sugibayashi, Low-current perpendicular domain wall motion cell for scalable high-speed MRAM, in: IEEE Symp. VLSI Technology, 2009, pp. 230–231.

[130] S. Fukami, M. Yamanouchi, H. Honjo, K. Kinoshita, K. Tokutome, S. Miura, S. Ikeda, N. Kasai, H. Ohno, Electrical endurance of Co/Ni wire for magnetic domain wall motion device, Appl. Phys. Lett. 89 (2013). 222410(1)-(4).

[131] I.M. Miron, K. Garello, G. Gaudin, P.-J. Zermatten, M.V. Costache, S. Auffret, S. Bandiera, B. Rodmacq, A. Schuhl, P. Gambardella, Perpendicular switching of a single ferromagnetic layer induced by in-plane current injection, Nature 476 (2011) 189–193.

[132] L. Liu, C. Pai, Y. Li, H.M. Tseng, D.C. Ralph, R.A. Buhrman, Spin-torque switching with the giant spin Hall effect of tantalum, Science 336 (2012) 555–558.

[133] C. Pai, L. Liu, Y. Li, H.M. Tseng, D.C. Ralph, R.A. Buhrman, Spin transfer torque devices utilizing the giant spin Hall effect of tungsten, Appl. Phys. Lett. 101 (2012). 122404-(1)-(4).

[134] M. Yamanouchi, L. Chen, J.Y. Kim, M. Hayashi, H. Sato, S. Fukami, S. Ikeda, F. Matsukura, H. Ohno, Three terminal magnetic tunnel junction utilizing the spin Hall effect of iridium-doped copper, Appl. Phys. Lett. 102 (2012). 212408-(1)-(4).

[135] S. Fukami, T. Anekawa, C. Zhang, H. Ohno, A spin–orbit torque switching scheme with collinear magnetic easy axis and current configuration, Nat. Nanotechnol. 11 (2016) 621–625.

[136] S. Fukami, C. Zhang, S. DuttaGupta, A. Kurenkov, H. Ohno, Magnetization switching by spin–orbit torque in an antiferromagnet–ferromagnet bilayer system, Nat. Mater. 15 (2016) 535–540.

[137] W.A. Borders, H. Akima, S. Fukami, S. Moriya, S. Kurihara, Y. Horio, S. Sato, H. Ohno, Analogue spin–orbit torque device for artificial-neural-network-based associative memory operation, Appl. Phys. Express 10 (2017) 013007.

[138] H. Ohno, D. Chiba, F. Matsukura, T. Omiya, E. Abe, T. Dietl, Y. Ohno, K. Ohtani, Electric-field control of ferromagnetism, Nature 408 (2000) 944–946.

[139] D. Chiba, M. Yamanouchi, F. Matsukura, H. Ohno, Electrical manipulation of magnetization reversal in a ferromagnetic semiconductor, Science 301 (2003) 943–945.

[140] D. Chiba, F. Matsukura, H. Ohno, Electric-field control of ferromagnetism in (Ga,Mn) As, Appl. Phys. Lett. 89 (2006). 162505(1)-(3).

[141] D. Chiba, M. Sawicki, Y. Nishitani, Y. Nakatani, F. Matsukura, H. Ohno, Magnetization vector manipulation by electric fields, Nature 455 (2008) 515–518.

[142] D. Chiba, Y. Nakatani, F. Matsukura, H. Ohno, Simulation of magnetization switching by electric-field manipulation of magnetic anisotropy, Appl. Phys. Lett. 96 (2010). 192506-(1)-(3).

[143] M. Weisheit, S. Fähler, A. Marty, Y. Souche, C. Poinsignon, D. Givord, Electric field-induced modification of magnetism in thin-film ferromagnets, Science 315 (2007) 349–351.

[144] T. Maruyama, Y. Shiota, T. Nozaki, K. Ohta, N. Toda, M. Mizuguchi, A.A. Tulapurkar, T. Shinjo, M. Shiraishi, S. Mizukami, Y. Ando, T. Suzuki, Large voltage-induced magnetic anisotropy change in a few atomic layers of iron, Nat. Nanotechnol. 4 (2009) 158–161.

[145] D. Chiba, S. Fukami, K. Shimamura, N. Ishiwata, K. Kobayashi, T. Ono, Electrical control of the ferromagnetic phase transition in cobalt at room temperature, Nat. Mater. 10 (2011) 853–856.

[146] Y. Shiota, T. Nozaki, F. Bonell, S. Murakami, T. Shinjo, Y. Suzuki, Induction of coherent magnetization switching in a few atomic layers of FeCo using voltage pulses, Nat. Mater. 11 (2012) 39–43.

[147] W.-G. Wang, M. Li, S. Hageman, C.L. Chien, Electric-field-assisted switching in magnetic tunnel junctions, Nat. Mater. 11 (2012) 64–68.

[148] H. Meng, R. Sbiaa, M.A.K. Akhtar, R.S. Liu, V.B. Naik, C.C. Wang, Electric field effects in low resistance CoFeB-MgO magnetic tunnel junctions with perpendicular anisotropy, Appl. Phys. Lett. 100 (2012). 122405-(1)-(3).
[149] S. Kanai, M. Yamanouchi, S. Ikeda, Y. Nakatani, F. Matsukura, H. Ohno, Electric field-induced magnetization reversal in a perpendicular-anisotropy CoFeB-MgO magnetic tunnel junction, Appl. Phys. Lett. 101 (2012). 122403-(1)-(3).
[150] S. Kanai, Y. Nakatani, M. Yamanouchi, S. Ikeda, F. Matsukura, H. Ohno, In-plane magnetic field dependence of electric field-induced magnetization switching, Appl. Phys. Lett. 101 (2013). 072408-(1)-(4).
[151] ITRS, System drivers, in: International Technology Roadmap for Semiconductors, 2011, pp. 12. 2011 edition.

Further reading

[152] T. Ohsawa, S. Ikeda, T. Hanyu, H. Ohno, T. Endoh, A 1-Mb STT-MRAM with zero-array standby power and 1.5-ns quick wake-up by 8-b fine-grained power gating, in: 2013 5th IEEE International Memory Workshop (IMW), 2013, pp. 80–83.

3D-NAND Flash memory and technology

8

R. Shirota
National Chiao Tung University, Hsinchu, Taiwan

8.1 Introduction

In the early stages of 3D-NAND Flash development, there were two basic ideas of the memory array organization, as shown in Fig. 8.1C and D [1]. First, architecture was formed by the vertical channels and horizontally stacked gates (Fig. 8.1C), whereas the second one utilized vertically positioned gate array combined with the horizontal channel stack (Fig. 8.1D). However, until now, 3D-NAND products, which had entered mass production and therefore were available on the market, all belong to the first category. Thus the following review will focus on the vertical channel topology NAND memory IC.

The first high-density, multistacked vertical channel 3D-NAND was introduced by Toshiba at the 2007 Symposium on VLSI Technology, and it went by the name of BiCS (Bit Cost Scalable) flash, which attracted much attention as a new 3D architecture for memory [2]. Two years later, Samsung presented its own 3D memory, the so-called TCAT (Terabit Cell Array Transistor) Flash [3]. In both cases, vertical channels were formed by cutting deep trenches through the multiple gate-stacked layers, with the silicon-nitride insulator coating used for the charge storage. Later, Samsung was the first to successfully launch TCAT 3D-NAND Flash, entering mass production in 2013. Nowadays, Samsung products are using the V-NAND (3D-Vertical NAND) cell, which in fact is the refined version of the original TCAT cell made in 2014.

The first generation of V-NAND was built on the 24-layer stack, with a memory capacity of 128 Gbit, featuring 2 bit/cell operation [4]. A few years later, in 2016, the Toshiba/Sandisk group began joint production of the 3D-NAND with 3 bit/cell capability [5]. This was anticipated by the Micron/Intel group, who entered the 3 bit/cell 3D-NAND market in 2015. Unlike Samsung or Toshiba/WD, Micron/Intel's 3D-cell [6] was using FG (floating gate) to store the charge. Until now, Samsung, Toshiba along with WD (Western Digital; previously known as "Sandisk"), and Micron strive for the leadership, continuing their heated competition for the shares in the 3D-NAND Flash memory market.

This chapter provides an insight into the state of the art in current 3D-NAND Flash technology. It will review 3D-NAND cell structures, memory array architectures, operation methods, and major reliability issues.

Advances in Non-volatile Memory and Storage Technology. https://doi.org/10.1016/B978-0-08-102584-0.00009-7

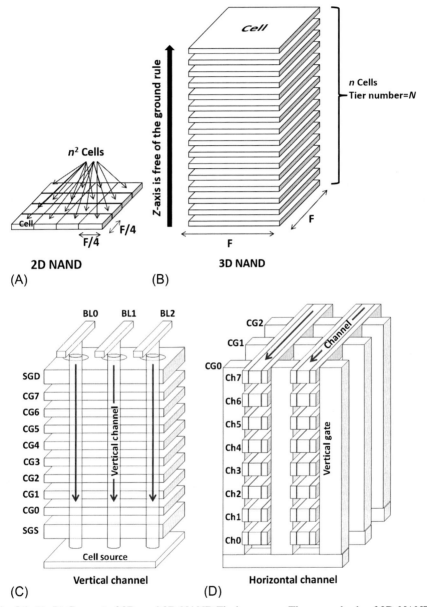

Fig. 8.1 (A, B) Concept of 2D- and 3D-NAND Flash memory. The ground rule of 3D-NAND can be relaxed to $\times \sqrt{N}$ (N=No. of tiers). (C, D) Structure of two kinds of 3D-NAND cell array. (C) 3D-cell array with vertical channel. (D) 3D-cell array with vertical gate.

8.2   Current status of the 3D-NAND market and production

At the 2018 ISSCC, the Toshiba/WD group presented the trend of memory density for TLC (3 bit/cell), as shown in Table 8.1 [10]. Fig. 8.2 illustrates that the memory capacity of 2D-NAND has already saturated the market since 2015. However, instead of 2D-NAND, 3D-NAND started to have a role in extending the memory capacity at the steady pace of about 1.5 times per year. Moreover, Samsung presented novel QLC (4 bit/cell) 3D-NAND technology on the same occasion [11]. Due to this rapid increase in memory capacity and continuing advances in memory technology, 3D-NAND-equipped SSD has been quickly replacing HHD in the storage market. Furthermore, these changes in market trend caused the portion of SSD in the NAND Flash market to rise from 23% in 2013 to 43% in 2017, as shown in Fig. 8.3.

8.3   V-NAND

8.3.1   Cell array structure and basic operation

First, we explain Samsung's 3D-NAND technology (V-NAND) in detail. A cross-sectional view of the cell array is shown in Fig. 8.4A [9]. A single memory string is formed by the circular, cylinder-shaped vertical channel made by poly-Si and is surrounded with vertically stacked insulators. The top of the vertical channel is connected to the bit line (BL) through the metal shunt layer. At the bottom of the string, a vertical channel is tied to the epitaxially grown p-Si pillar (black area), whereas the epitaxial p-Si is connected to p-Si substrate. The gate stack consists of 55 layers in total, which includes 48 word lines (WL), and seven other gates. The breakdown of seven gates is not clear. These probably include two select gates (SG), four dummy WL, and one bottom gate near the substrate. Here, the bottom gate is named "Sub"-gate, and its function is the electrical connection of the vertical channel to SL during reading and programming by forming the inversion layer at the Si surface below the Sub-gate. The bottom stack gate on the Sub-gate and two top stack gates act as the source side (GSL) and drain side (SSL) SGs, respectively. The nail-shaped black line is the source line (SL) made of tungsten. During manufacturing, source n^+-Si area is formed by the n-type impurity implantation at the bottom of SL through the gate space, before the SL metal fill. Therefore, SL is electrically connected to the source-n^+ region. SG, WL stacks are isolated from each other by the SL split, which extends perpendicularly to BL. Thus, SG, WL, and SL are orthogonal to BL, which is the same as in the 2D-NAND.

As the number of WL layers increases, the total mold stack grows higher. This leads to the enlargement of the difference in channel-hole size (L) and gate-stack thickness (T) between the upper and lower WL, as shown in Fig. 8.4B. Corresponding structural discrepancies can cause a difference in program-erase speed, cell-to-cell interference, and

Table 8.1 Comparison of several companies' 3D-NAND with 3 bit-per-cell technology, presented at ISSCC 2018.

ISSCC	J.W. Im et al. [7]	D. Kang et al. [8]	T. Tanaka et al. [6]	R. Yamashita et al. [5]	C. Kim et al. [9]	H. Maejima et al. [10]
			3D-NAND (3 bit/cell) comparison			
Company	Samsung	Samsung	Micron/Intel	Toshiba/WD	Samsung	Toshiba/WD
Technology	32WL layer, V-NAND	48WL layer, V-NAND	Floating gate, CMOS under array	64WL layer, BiCS	64WL layer, V-NAND	96WL layer, BiCS
Memory capacity	128 Gb	256 Gb	768 Gb	512 Gb	512 Gb	512 Gb
# of plane	2	2	4	2	2	2
Die size (mm^2)	68.8	97.6	179.2	132	128.5	86.1
Bit density (Gb/mm^2)	1.86	2.62	4.29	3.88	3.98	5.95

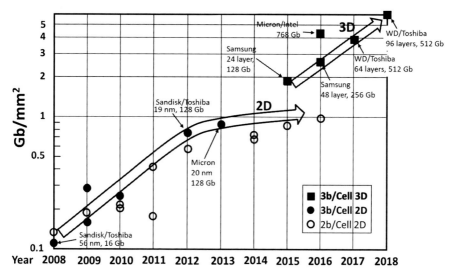

Fig. 8.2 Trend of the memory density.

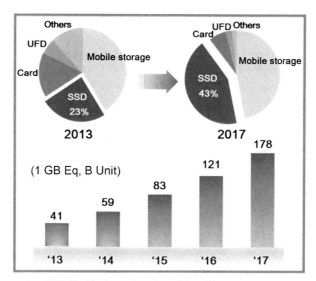

Fig. 8.3 The growth of NAND Flash market (from ISuppli).

retention, for the different WLs. The top view of the cell array is graphically illustrated in Fig. 8.5. The TEM photograph in Fig. 8.4A corresponds to the cross-sectional view indicated by the dotted line (B-B'). The 60 degrees tilt angle-arrayed vertical channels, sandwiched by the two neighboring SLs, along with the cells inside one area, are forming a so-called subblock. Four subblocks compose a "Block." Cells in one block are erased simultaneously, which is the same as for the 2D-NAND.

Next, the details of the cell array structure are explained, based on the original patent by Samsung [12]. Fig. 8.6 is the cross-sectional view of the cell array. The hole

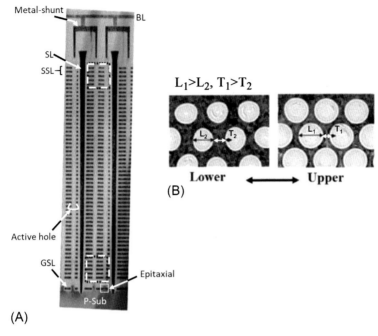

Fig. 8.4 (A) Cross-sectional view of NAND cell array. (B) Top view of the *dotted area* on the left-hand side. Figures to the left and right show the active channel-hole's area at the bottom and top of the string, respectively.

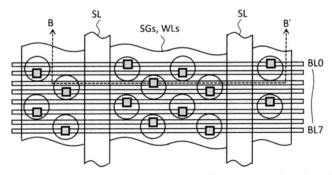

Fig. 8.5 Schematic top view of the NAND cell array, which corresponds to the TEM photograph in Fig. 8.4A and B.

region (1) consists of a poly-Si vertical channel (2), stacked insulators (3), and the central insulator (4). Thus, the vertical channel has a macaroni shape, which is considered to have better reliability characteristics [13]. The insulator layer (3) is made from the stack of three insulator coats whose middle layer is made of Si-nitride and which can store the charge delivered by program/erase operation. The bottom oxide sheet (nearby the channel) is made from the Si-oxide-like complex insulators for band-gap-engineering. Charges tunnel between the channel and the Si-nitride layer

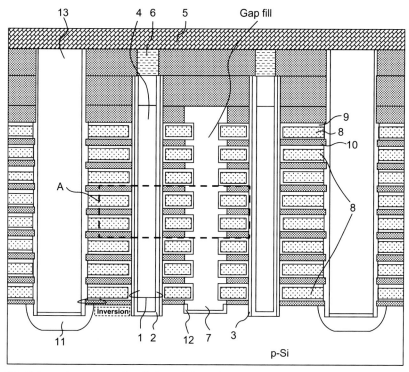

Fig. 8.6 Schematic cross-sectional view of the cell array.

through the bottom oxide. The vertical channel (2) is connected to the BL (5) through the BL plug (6). Insulator (10) and gate layers (8) are alternately stacked. The stacked gates (8) are surrounded by the insulator layer (9), which becomes the buffered layer between the gate and the vertically stacked insulator layers (3). SG and CG are isolated by the space (7, 13). Source n^{+}-layer (9) is made by the implantation of n-type impurity through the space (13), which extends perpendicularly to BL (5). Gate-cut space (10), which is filled by the insulator, is perpendicularly extended to BL (5) as well. Thus, SG and WL (CG line) become the long stripe line orthogonal to BL (5). The enlarged view of the dotted area (A) is shown in Fig. 8.7. The top view of channel (2) has a midair shape and is surrounded by the insulator (3).

The top view of the cell array is shown in Fig. 8.8. There is no gate cut in Fig. 8.5. Thus, whether there is a gate cut or not depends on the V-NAND generation. The tilt shape holes (1) can minimize the distance between neighboring channels.

An equivalent circuit of the cell array, which forms one block, is shown in Fig. 8.9. This figure was presented at the 2014 ISSCC featuring the 24 WL stack [4]. BL and SL side SGs are named SSL(1–n) and GSL(1–n), respectively. A total of 26 WLs include 2 dummy and 24 operational WLs. Eight SSLs are independently decoded. However, WL and GSL are electrically connected inside one block. Thus, each vertical NAND string is selected by decoding one BL and one SSL. The Sub-gate is not exhibited in Fig. 8.9, because the Sub-gate is common at least in one block.

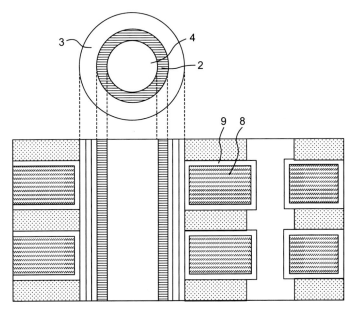

Fig. 8.7 Schematic top and cross-sectional view of the cell array.

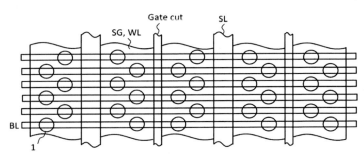

Fig. 8.8 Schematic top view of Fig. 8.6.

This paragraph explains the basics of the read operation using Figs. 8.6 and 8.9. For reading, read bias is applied to selected WL, higher voltage (Vpassr) is applied to the other WLs and SG, and selected BL is applied as a positive bias with the SL grounded. By applying high voltage to the Sub-gate, the inversion layer (a) is formed at the surface of p-Si below the Sub-gate, which connects the vertical channel to SL, as shown in Fig. 8.6. The data are judged based on whether or not there is the cell current flow from SL to BL, and this is the same as for 2D-NAND.

Bias setting for the program operation is depicted in Fig. 8.10. When the string connected to SSL0 is selected to program, other strings connected to SSL1–7 are facing the program-inhibit mode. In this case, high voltage (Vpgm) is applied to the selected WL, and some positive bias (Vpassw) between 0 V to Vpgm is applied to the rest of the WLs. When BL is grounded, electrons are injected

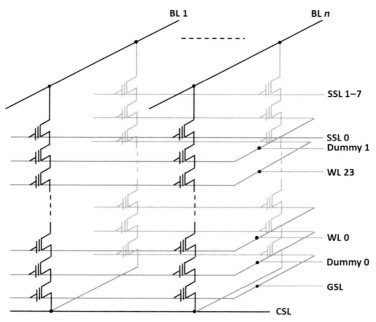

Fig. 8.9 Equivalent circuit inside one block.

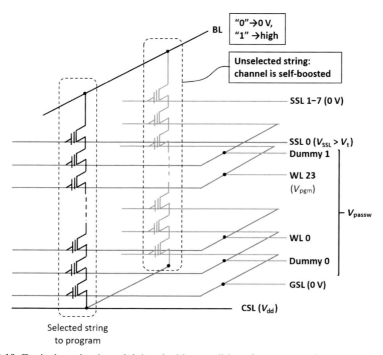

Fig. 8.10 Equivalent circuit explaining the bias conditions for programming.

to the Si-nitride layer at the selected cell. At this time, program-inhibit strings must be self-boosted by cutting off the leakage from grounded BL. In the case of 2D-NAND, BL connected to the program-inhibit string can be easily isolated by applying Vdd to the BL itself. This is one of the differences of 3D-NAND programming, and it depends on the number of SSLs in one block. When Gate thickness becomes thinner, more than two SSLs will be necessary to cut off the leakage from BL. At the selected string, with BL biased at a voltage higher than Vt of SSL, the channel becomes floating and disconnected from SL and BL. Thus, the channel potential is self-boosted and there is no electron injection into the Si-nitride layer.

The difference in cell characteristics for each WL is another concern, as shown in Fig. 8.11 [9]. As the number of WL layers increases, the total stack length becomes longer, so that the difference in channel-hole size (L) and gate-stack thickness (T) between the upper and lower WL stage becomes enlarged. These structural differences between WLs can cause a difference in program-erase speed, cell-to-cell interference, and retention. The program-erase speed of the lower WLs is normally faster than that of the upper WL because of the higher coupling ratio originated by the smaller channel-hole CD, and the retention characteristics are poor because of the thinner gate stack by insufficient step coverage when depositing gate-stack layers.

For the block erase, high voltage is applied to p-Si with all WLs in the selected block grounded, which is the same as in the 2D-NAND erase operation. Electron emission or hole injection from channel to Si-nitride layer then occurs, which in turn reduces the Vt of all cells in a selected block. In the case of the BiCS cell proposed by Toshiba [2], the vertical channel is directly connected to SL. Thus, to apply high voltage to the vertical channel, bias is applied to SL with SSL potential lower than SL. Then, hole generation by GIDL (gate-induced drain leakage) will happen at the edge of the SL adjacent to SSL, and, consequently, holes will be injected into the vertical channel. This is one of the big differences between V-NAND and the previous BiCS Flash technologies.

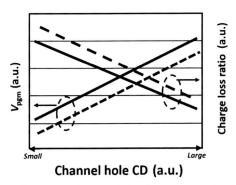

Fig. 8.11 The program and retention characteristics depending on the channel-hole size.

8.3.2 Process flow

The process to fabricate V-NAND is reviewed in the following text. First, Si-oxide (10) and Si-nitride layers (14) are sequentially deposited as shown in Fig. 8.12. Next, the oxide/nitride stack is etched, which forms the rectangle space called gate cut (7). Next, the bottom of the space (7) is thermally oxidized (12). After applying insulator fill to space (7), the cap oxide layer (15) is deposited. Next, a cylindrical hole (1) is formed, which finally becomes the active channel and charge-storage layer. After hole etching (1), the epitaxy p-Si layer is grown through the Si substrate into the hole area (1). Later, three insulator layers are sequentially deposited (3).

Next, thinner poly-Si is deposited and the bottom of the poly-Si and stacked insulator layers (3) are etched simultaneously. Then the surface of the p-Si is exposed at the bottom of the hole. Next, second poly-Si is deposited and the first and second poly-Si pillar inside the cylindrical hole are therefore electrically connected to p-Si. In this way, the first thin poly-Si layer acts as a buffer layer to protect the side insulator coat (3) from etching ambient. After the removal of them, the second poly-Si layer is deposited. Thus, the first and second poly-Si layers become the vertical channel. The purpose in forming the epitaxy layer is twofold. First, it makes the buffer layer to prevent the over-etching of the Si surface during the removal of the first poly-Si and stacked insulators (3). This is possible since the etching rate of crystal Si is lower than that of poly-Si. Second, this kind

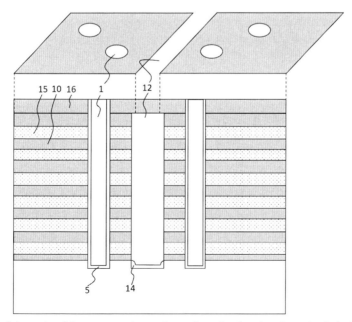

Fig. 8.12 Schematic of the cross-section and top view after forming the active hole (1).

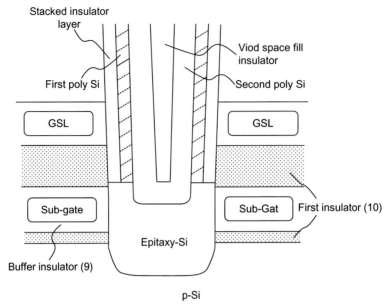

Fig. 8.13 The enlarged drawing of active area near to the Si surface.

of structure improves the *I-V* characteristics of the Sub-gate transistor, which is used to regulate the current flowing from SL to the vertical channel (2). There is a midair space at the center of the cylindrical hole (macaroni-shaped channel), and this space is filled by an insulator (4). The expected shape of the vertical channel near the Si substrate is shown in Fig. 8.13. The boundary of the epitaxy Si and the poly-Si channel must be located between the Sub-gate and GSL. Thus, wider spacing is required between the Sub-gate and GSL than is required between the two neighboring WLs.

After forming a vertical channel inside the active hole, the insulator layer is deposited (16), followed by the creation of the rectangular space parallel to the gate cut (7), as shown in Fig. 8.14. Next, the Si-nitride layer (14) lying next to oxide layer (10) is etched by phosphoric acid wet etching, which forms the void space (17), as shown in Fig. 8.15. Then, a thin buffer insulator (9) is deposited, as shown in Fig. 8.16.

Next, a tungsten metal layer (18) is deposited as shown in Fig. 8.17A. Then, tungsten is partially etched, except for the part that filled the stack voids, and thus it becomes gates (8), as shown in Fig. 8.17B.

After the gate formation, an insulator in the space for SL is etched, and the second insulator is deposited. Then, n^+ impurities are subsequently implanted, as shown in Fig. 8.18.

After forming the source n^+ layer, tungsten SL metal deposition and subsequent BL formation continue. Finally, NAND cell array formation is completed, as shown in Fig. 8.4A and B.

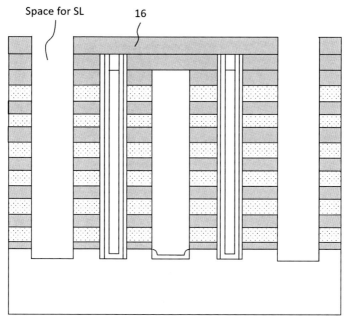

Fig. 8.14 Drawing of the cross-sectional view after forming the SL space.

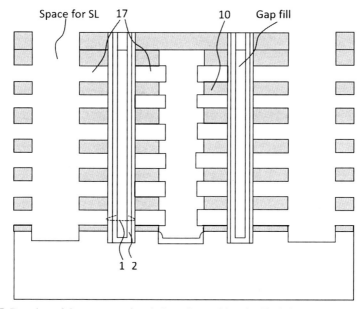

Fig. 8.15 Drawing of the cross-sectional view after etching the Si-nitride layer.

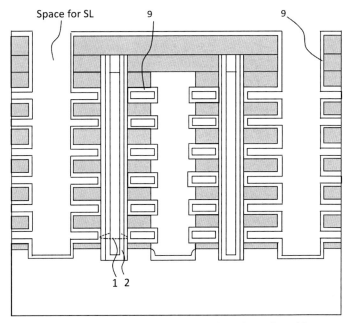

Fig. 8.16 Drawing of the cross-sectional view after insulator-layer deposition.

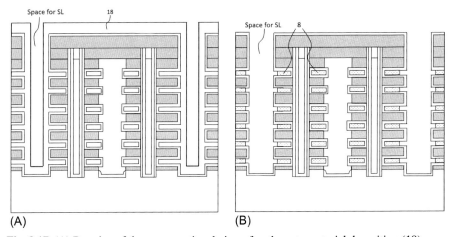

Fig. 8.17 (A) Drawing of the cross-sectional view after the gate-material deposition (19). (B) The cross-section after partially etching gate material. Remaining ones become SG, WL, and Sub-gates.

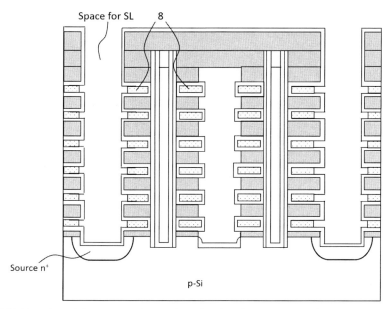

Fig. 8.18 Drawing of the cross-sectional view after forming the source n$^+$ layer.

8.3.3 Characteristics and performance of V-NAND

Samsung presented 256 Gb and 512 Gb V-NAND chips at 2016 [8] and 2017 ISSCC [14], respectively. The die photograph and the table of key parameters of those are shown in Fig. 8.19. The 48.WL stack was used for 256 Gb Flash, which matched the NAND structure presented at IMW in 2017 [9]. The single-string area can be roughly estimated from the cell-area occupancy. Hereby, the string area is around 3.6×10^4 nm^2 and 3.4×10^4 nm for the 256 Gb and 512 Gb chip, respectively. Thus, one string area is almost the same for each generation. Since it is hard to scale down the active hole area, the string area cannot be scaled either. However, effective cell size (= 1 string area/number of WL stack) can be scaled, and it is inversely proportional to the number of the WL stack. Thus, the effective cell size has already become smaller than the size of 2D-NAND's minimal rule ($>5 \times F^2$ nm^2; $F \sim 14$ nm).

Next, memory organization (page, block size) is compared for both cases. In the 256 Gb chip, one block consists of 576 pages, which in turns is composed of $567 = 3 \times 48 \times 4$; (3 bit/cell) × 48 (WL stack) × (4SSL). Thus, 256 and 512 Gb ICs follow four SSL structures that are different from that of 128 Gb's (8 SSL). As such, one block's bit size can be kept almost the same, even though the number of WL stacks increases.

The string current tends to decrease due to the increased channel-hole height, as shown in Fig. 8.20 [9]. Furthermore, the read voltage applied to the unselected WL (Vpassr) needs to be lowered to prevent the degradation of read disturbance due to the increased read count by generation. This will accelerate the decrease of the string current, and the current limit for sensing will be soon reached. Thus, Vpassr must be optimized according to the number of the WL stack.

	2016	2017
Bits per cell	3	→
Density	256 Gb	512 Gb
Chip size	97.6 mm² (2.62 Gb/mm²)	128.5 mm² (3.98 Gb/mm²)
Technology	48 WL stacked layer	64 WL stacked layer
Organization	16 KB/page,576 pages/BLK,3776BLK/die	16KB/page,768pages/BLK,5748BLK/die
I/O bandwidth	Max. 1 Gb/s	→
VCCQ	1.8 V	1.8 V (Legacy), 1.2 V (Low power)
t_{BERS}	3.5 ms (Typ)	→
t_{PROG}	660µs	700µs
t_R	45µs	60µs
Gb/mm²	2.62	3.98
1 string area	~3.6 × 10⁴ nm²	~3.4 × 10⁴ nm²
Effective cell size	~750 nm²	~530 nm²

2016 ISSCC, Samsung

2017 ISSCC, Samsung

Fig. 8.19 Two die photographs and key features are listed.

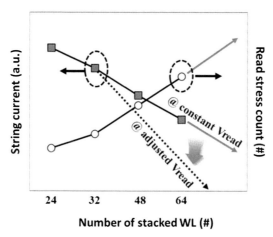

Fig. 8.20 String current trend by increasing the number of stacked WL.

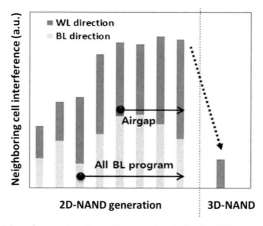

Fig. 8.21 Cell-to-cell interference trend over the past decade for the 2D planar cell and the 3D cell.

Fig. 8.21 shows the trends of cell-to-cell interference over the past decade for the 2D planar cell [9]. The gate feature size has been reduced to 15 nm by going beyond the limit of photolithography with the introduction of DPT (Double Patterning Technology) and QPT (Quadruple Patterning Technology). Because of the reduction of the space between neighboring cells, cell-to-cell interference tends to increase over the generations even after adding air-gap structures and improving the operating conditions such as utilizing the all bit-line program scheme. In the case of 3D-NAND, WL-to-WL space is equal to the cell-to-cell distance, which also slightly impacts the string-area size. Thus, the cell-to-cell interference can be reduced by optimizing the WL-WL space.

By reducing the cell-to-cell interference, the Vt distribution shift of 3D-NAND after programming neighboring cells is smaller than that of the 2D-NAND. Therefore, the multibit programming sequence of 3D-NAND can be simplified. As shown on

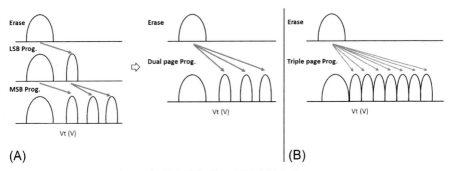

(A)   (B)

Fig. 8.22 Program algorithm of (A) 2 bit/cell and (B) 3 bit/cell.

the left-hand side of Fig. 8.22A, 2D-NAND uses two-step programming for 2 bit/cell (lower and upper bit sequential programming). Using a two-step programming sequence with either the "shadow" or the "reprogram" algorithm, Vt shift caused by the cell-to-cell interference can be mitigated. Meanwhile, 3D-NAND can program multibit data simultaneously due to the reduced effect of the cell-to-cell interference, as shown on the right side of Fig. 8.22A. A 3 bit/cell programming operation can be performed in a single step as well, as illustrated in Fig. 8.22B. Therefore, the programming time can be reduced up to 50% of that of a typical planar 3b/cell NAND Flash device [7].

Fig. 8.23 depicts initial Vt distribution together with Vt distribution after 5K P/E cycles. P/E cycling was performed at an ambient temperature of 55°C. After 5K cycles, Vt distribution is still similar to the initial one. For this particular case, high-speed programming time (700 μs) was achieved, as well as excellent endurance, as shown in Fig. 8.24 [9]. Their recorded Vt distribution is available to use even after 100K cycles. Such significant reliability improvement was achieved by means of the so-called tunnel barrier engineering, that is, customization of the tunneling oxide manufacturing process.

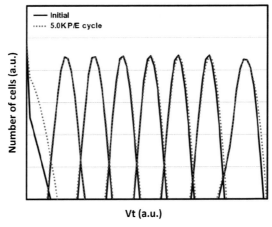

Fig. 8.23 Vt distribution after 3 bit/cell programming.

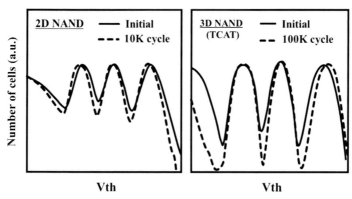

Fig. 8.24 The comparison of Vt distribution of 2D and 3D-NAND [9].

Thus, the insulator layer between the SiN charge-storage layer and the poly-Si channel must not be the single Si-dioxide layer, but rather the combination of multiple insulator grades with different gap barrier heights.

8.4 BiCS NAND

Fig. 8.25 shows a TEM photograph of Toshiba's NAND cell array, disclosed in December 2016 and posted to Toshiba's official website [15].

The picture shows that there is an epitaxial layer at the bottom of the vertical channel, and there is a bottom gate near the Si substrate similar to the Sub-gate in the V-NAND. Moreover, the SL metal line is revealed as well. Thus, the configuration is different from previous BiSC [2, 16], but is somehow similar to that of V-NAND. Details of the cell structure are unknown, but it seems that the BiCS3 array is very different from the old version. SL is composed of two materials. It seems that SL's top layer is made by tungsten, which is the same as in the case of V-NAND. However, the inner layer is made by some poly-Si–like material, which is different from V-NAND.

Toshiba/WD presented the 512 Gb NAND with a 96 WL layer stack at the 2018 ISSCC [10]. The key feature is shown in Fig. 8.26.

From the cell-area occupancy, one string area can be roughly estimated. Here, the string area is around 3.4×10^4 nm, which is similar to Samsung's 512 Gb. The effective cell size is around $350 \, \text{nm}^2$.

8.5 Micron/intel 3D-NAND

8.5.1 Cell array structure and its process

The first generation of microns 3D-NAND has 32 tiers of active WLs plus additional tiers for dummy WL and source/drain select gates. Fig. 8.27 shows an SEM cross-section of the NAND string, which is formed above the silicon substrate. There

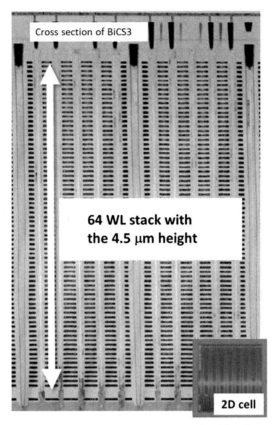

Fig. 8.25 The cross-sectional view of the third generation of BiCS cell array.

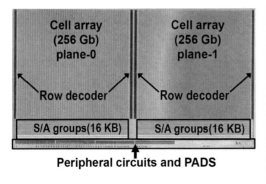

Capacity	512 Gb (3 bit/cell)
Technology	96-WL-Layer, BiCS
Die size (bit density)	86.1 mm² (5.95 Gb/mm²)
Organization	(16 KB + ECC) / Page, 1152 Pages / Block →18 MB/Block, (1822 + EXT) Blocks / Plane, 2 Planes
Throughput	Read(tR): 58 µs (ABL) Prog: 57 MB/s
I/O	533 Mbps DDR, X8
Power Supply	Vcc : 2.3 V to 3.6 V Vccq: 1.8 V

Fig. 8.26 The 512 Gb die photographs and key features are listed.

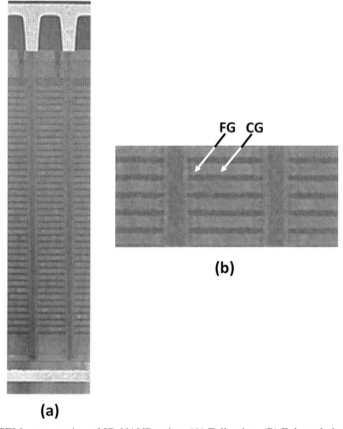

Fig. 8.27 SEM cross-section of 3D-NAND string. (A) Full string. (B) Enlarged view of the NAND cell.

are two big structural differences from the V-NAND topology. First, the charge storage layer is implemented with the floating gate (FG), not the Si-nitride layer. Second, the source of the string is made by the n^+ polysilicon source plate lying at the bottom of the vertical channel. On the other hand, the source of V-NAND has an SL metal line connected to the n^+ source on the p-Si substrate, as shown in Fig. 8.4.

The basic process flow used for separating FG of one cell from that of adjacent cells above and below is shown schematically in Fig. 8.28. After the cell hole etch through the WL tiers, CG is recessed back and the interpoly dielectric is formed. Following this, FG is deposited and is etched back to form an isolated FG for each cell. This is followed by the tunnel-oxide and channel formation. While the NAND cells are FG-based cells, the source and drain select devices are single-gate oxide transistors.

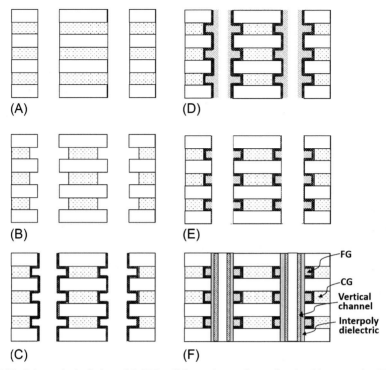

Fig. 8.28 Schematic depiction of the FG cell formation and steps involved in separating FG between cells. (A) Tier deposition and 3D cell hole formation. (B) CG recess. (C) IPD formation. (D) FG deposition. (E) FG isolation etch. (F) Tunnel-oxide and channel formation.

8.5.2   CMOS under cell array

V-NAND uses a metal gate by replacing the insulator layer with the tungsten gate. However, FG-based 3D-NAND has no process to replace one insulator layer to the metal fill for WL. Thus, WL must be made with an impurity-doped poly-Si, which results in the higher resistance of WL than that in the case of the metal WL. To keep the RC delay of the WL down at a certain acceptable level, the length of WL must be shorter compared to the length of metal WL. Consequently, the cell array must be divided into smaller unit areas to reduce the WL length, which causes an extension of the row decoders area. Moreover, a source plate has to be divided into several parts as well to reduce resistance. To cope with NAND's footprint increase, CMOS under cell array (CUA) architecture was introduced [17]. According to CUA, core circuits are buried under the cell array to reduce the chip area. The 3D-NAND architecture with CMOS under array (CUA) achieves a minimal cell footprint and die size by placing the CMOS logic below the NAND array and by keeping SG devices (or SGS, SGD for source/drain side) along the strings. The strings are landing on an n^+ polysilicon source plate fabricated after the CMOS, as shown in Fig. 8.29.

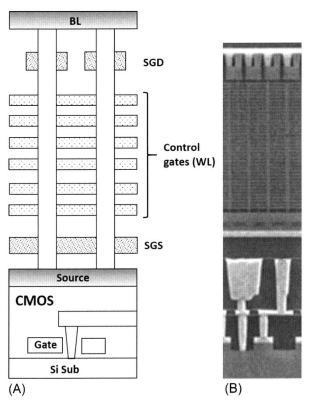

Fig. 8.29 (A) Schematic cross-section of 3D-NAND strings in a CUA architecture. (B) Array SEM cross-section.

Presented at the 2016 ISSCC was 768 Gb 3D-NAND, which was fabricated with CUA technology, the buried part including not only a row decoder but also a page buffer and other circuits [6]. Fig. 8.30 shows a memory tile that is 2 kB × 8-BL long and 256 blocks wide. Each NAND block is divided into 16-block segments, and each tile includes 256-block segments. CMOS circuits are formed underneath the array. A total of 2 kB page buffers and 256 string drivers are deployed, along with their associated drivers and column/block redundancy circuits, which are necessary to drive the memory array above them. WLs descend in the middle of the tile and are connected to their string drivers. BLs descend at the edges of the tile and are connected to their page buffers. WL and BL are divided into 16 and 4 segments without any die-size penalty. The segmented BL and WL result in faster access times. Since defective BLs and blocks are replaced with redundant elements within each tile, a greater defect tolerance is realized. The common source plate (CSP) is also divided into 64 plates, and the CSP resistance is low enough to allow all-bit-line sensing with minimal CSP bounce.

Fig. 8.31 shows the layout and how to connect WL to the string driver, and BL to the bottom page buffer under the cell array. Because there is a large area underneath the cell array, several core circuits are well arranged to resolve the territorial problem.

Architecture to connect CMOS core circuits under cell array

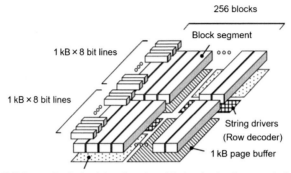

Fig. 8.30 A memory tile with CMOS under array.

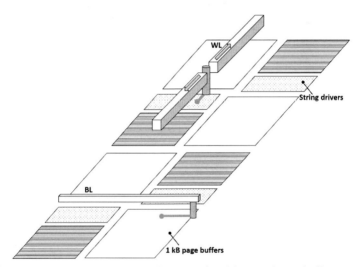

Fig. 8.31 The structure to connect WL and BL to string drivers and page buffers, respectively.

As shown in Fig. 8.32, a 4×8 arrangement of tiles forms the whole array that is capable of storing 768 Gbits. The array itself is logically divided into four planes and allows simultaneous 16 kB×4 plane operation. Charge pumps, controller logic circuits, and global drivers are placed at the bottom of the die. All of the bonding pads and DQ buffers are located on one side. A total of four metal layers are used in this device to efficiently minimize the periphery area. One metal layer above the array is used for the BLs and the remaining one is used for power and global routing that interconnects tiles and the periphery. Two metal layers are used under the array for routing in x- and y-directions.

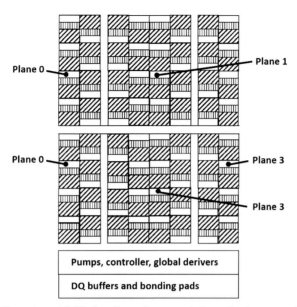

Fig. 8.32 Chip floor plan with 32 tiles. Four planes are independently operated.

8.5.3   Cell characteristics

The basic cell characteristics of FG-based 3D-NAND were discussed at the 2015 IEDM [18]. Cell-to-cell interference has been a key challenge for 2D-NAND scaling. The gate all-around structure of 3D-NAND makes it less susceptible to interference from neighboring cells, as shown in Fig. 8.33. The vertical cell-to-cell space for 3D-NAND can be longer than the 20 nm 2D-NAND, which can also reduce the interference. Therefore, the interference of 3D-NAND is only 20% of that of the 20 nm 2D-NAND.

Next, consider the program disturb of 3D-NAND. The 3D-NAND array architecture is sensitive to the increased occurrence of a program disturb (Fig. 8.34) due to the fact that WL is being shared by multiple NAND strings on the same bit line, which is the same as V-NAND in Fig. 8.10.

The gate-all-around structure can allow a 100% boost ratio, but provides the various boost leakage mechanisms as well, as shown in Fig. 8.35A. Five principles of undesired electron injection in the channel of the inhibit string are listed: (a) the electron/hole pair generation in the channel, (b) band-to-band tunneling, (c) electron injection from S/D, (d) trap-assisted tunneling, (e) junction leakage avalanche multiplication. The published program inhibit characteristics are quite good, as shown in Fig. 8.35B. However, the temperature dependence about the increase of the electron hole pair generation at high temperatures is not known.

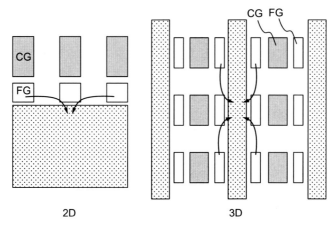

Fig. 8.33 Gate all-around structure of the 3D-NAND makes it less susceptible to the interference from the neighboring cells.

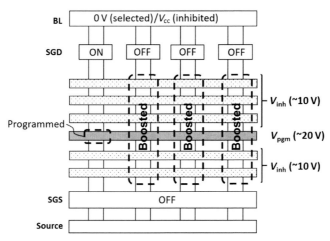

Fig. 8.34 3D-NAND programming-selected/inhibit cell biases and disturb conditions.

Erase operation using n^+ CSP is shown in Fig. 8.36 [17]. GIDL is generated at both the SGs at the channel edges, and holes are moved into the channel.

Fig. 8.37 shows the GIDL data measured. Even though V_{DS} is less than 1 V, there is a reasonable amount of GIDL current generated. This comes from the gate all-around structure. As shown in Fig. 8.38, the floating channel potential becomes negative due to its capacitive coupling to SG. Thus, the potential difference between the drain and the channel is larger than V_{DS}, and GIDL current can be generated.

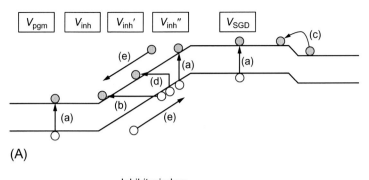

(A)

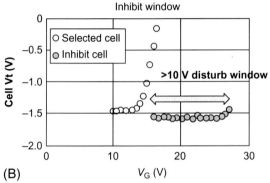

(B)

Fig. 8.35 (A) Various possible mechanisms loosing boosted potential in the inhibit string. (B) Program inhibit window characterization.

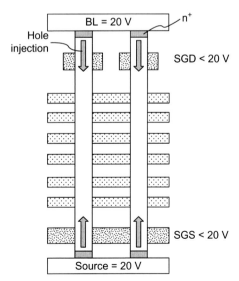

Fig. 8.36 Erase bias condition to generate GIDL current.

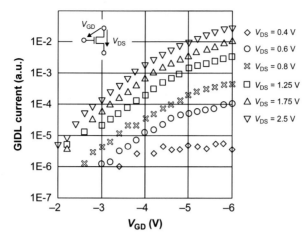

Fig. 8.37 Experimental SG GIDL vs. gate to drain potential (V_{GD}) for various drain-to-source biases (V_{DS}), 30°C. Inset: bias-naming convention for SG GIDL measurements.

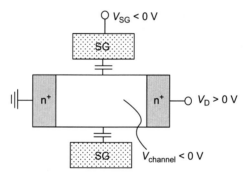

Fig. 8.38 The structure explaining GIDL for the cell-erase operation.

8.6   3D-NAND reliability

8.6.1   Retention characteristics

Samsung presented one method to improve the retention characteristics [14]. By the erase operation, holes are trapped in the trap layer (Si-nitride). Some of them are in the shallow trapped states, which can easily move to channel poly-Si or shift inside the trap layer. These holes have a high probability of moving during baking time, which shifts Vt distribution. This effect is especially severe for the shallow trapped hope positioned next to an electron, which will cause charge neutralization by the lateral movement of the hole to the electron side, as shown in Fig. 8.39.

To improve the retention characteristics, a two-step annealing pulse is introduced, as shown in Fig. 8.40A, which also demonstrates the flow diagram used for this annealing process. First, a high-voltage pulse of Vbias for electrical annealing is applied to even WLs while biasing odd WLs and the poly-Si channel to ground after the erase. Second, Vbias is applied to odd WLs while biasing even WLs and the channel to ground. Vbias should be carefully selected so as not to program erased cells. The timing overhead for annealing is reduced to less than 3% of $t_{block-erase}$. Fig. 8.40B illustrates the conceptual behavior of holes during the annealing process. Some holes in shallow traps can be detrapped by the strong electric field. The measurement result of Vth distribution after retention, shown in Fig. 8.41, indicates that the two-step annealing pulse scheme slightly helps to improve the reliability characteristics.

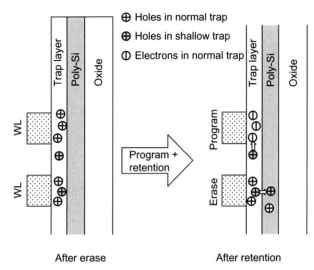

Fig. 8.39 Degradation of the retention characteristics with the scaling down of the WL space.

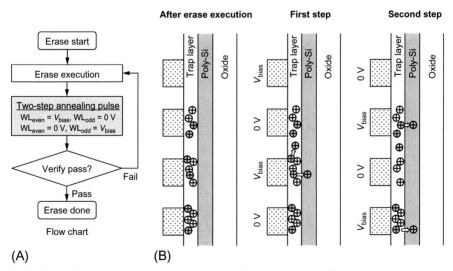

Fig. 8.40 (A) Flow chart of erase operation using the two-step annealing pulse scheme, (B) conceptual behavioral model of annealing pulse scheme.

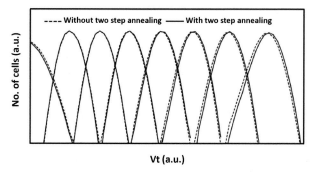

Fig. 8.41 Measured Vth distribution after retention.

8.6.2 Temperature dependence of cell I-V curve and Vt instability by random telegraph noise

The Politecnico di Milano group investigated the temperature dependence of the I-V curve of memory cells. Fig. 8.42 shows the string current characteristics of a cell in (A) 16 nm 2D and (B) 3D-NAND arrays with a temperature between $-10°C$ and $125°C$ [18]. In the case of 3D-NAND, saturation current increases as temperature increases. This is the reverse trend of 2D-NAND. This difference comes from the presence of grain boundaries in the poly-Si channel of 3D-strings. The high density of acceptor-like traps at these boundaries creates energy barriers that electrons have to cross as they travel through the channel of the memory cells. These barriers represent the bottleneck for electron transport in 3D devices, and the increase of their conductivity when temperature is increased typically overwhelms the reduction of electron mobility coming from the higher phonon scattering. As a result, cell current in 3D-arrays increases with temperature. To investigate in more detail, the temperature effects in two NAND technologies, Vt at the three I_{cell} values (I^{high}, I^{medium}, and I^{low}) in Fig. 8.42 are shown in Fig. 8.43. Results reveal a comparable decrease of Vt with temperature for both 2D and 3D cells. However, 3D-NAND appears rather insensitive to the Vt extraction level.

Fig. 8.44 shows the temperature dependence of the saturation current (I_{sat}) with three background data patterns (other cells' data in one string). These follow an Arrhenius temperature activation both in (A) 2D and (B) 3D-NAND strings. In the former case, I_{sat} decreases when the temperature increases, resulting in a positive Ea of 7 meV, which is almost independent of the strings back-pattern.

The I_{sat} in 3D-string increase with temperature is owing to the conductivity increase at the grain boundaries. This growth leads to a negative E_A, which, moreover, displays a nonnegligible dependence on the strings back-pattern. The E_A of 3D-strings, in particular, grows in magnitude when increasing the Vt level of the unselected cells in the string.

Random telegraph noise (RTN) of 3D-NAND was also studied by the same group [19]. Fig. 8.45 shows the cumulative distribution (F) and its complement ($1 - F$) of the Vt shift experienced by a cell between two read operations separated by an idle

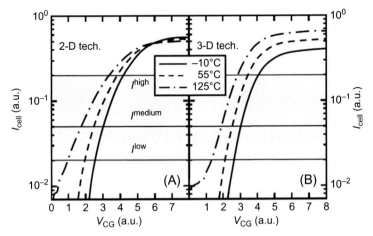

Fig. 8.42 Measured I_{cell}-V_{CG} characteristics of the NAND cell. (A) 16 nm 2D and (B) 3D-NAND array. A random back pattern was presented in either case.

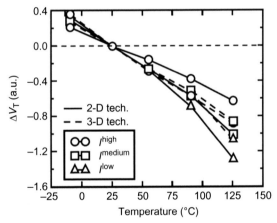

Fig. 8.43 Vt shift with temperature for the cells in Fig. 8.42. Vt is extracted at the three levels (high, medium, low current) in Fig. 8.42.

time (around 10 min), in the case of (A) 2D and (B) 3D-NAND arrays. The distribution of 2D-NAND displays clear exponential tails branching off a Gaussian function. The average value λ of these latter statistics, which equals the slope of the exponential tails of the RTN distribution in Fig. 8.45A, has been constantly increasing with the scaling of the NAND technology, on account of the miniaturization of cell dimensions. In turn, the possibility of largely reducing λ thanks to the increase of the cell area has been one of the most relevant benefits triggering the development of the first 3D-arrays. The reduction of λ in 3D-arrays is clearly visible when comparing parts (A) and (B) of Fig. 8.45.

Moving now to the temperature sensitivity of the results, Fig. 8.45A shows that the RTN distribution of 2D-arrays is almost independent of the experimental temperature.

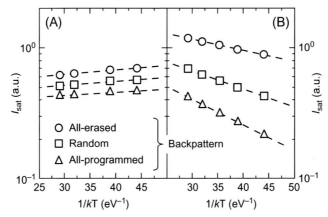

Fig. 8.44 Saturation current (I_{sat}) as a function of $1/kT$ for different back-patterns. (A) 2D and (B) 3D-NAND, respectively.

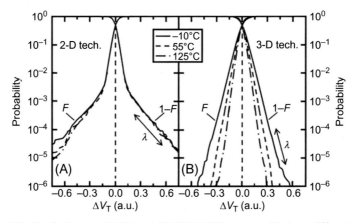

Fig. 8.45 RTN distribution on (A) 2D and (B) 3D-NAND arrays with three different temperatures.

In the case of 3D-arrays, instead, the RTN distribution in Fig. 8.45B shows a nonnegligible enlargement when the temperature is decreased. In particular, the reduction of the temperature gives rise to an increase in the slope of the distribution tails. This increase can be explained in terms of more percolative conduction in the polysilicon channel of 3D devices when the temperature is decreased, due to the stronger impact of the barrier heights at the polysilicon grain boundaries on electron transport.

To understand the temperature dependence in Fig. 8.45, refer to the electron density profile shown in Fig. 8.46. Nonuniformity becomes stronger when reducing the temperature from (c) 400 K to (b) 300 K to (a) 250 K, along the longitudinal direction with one electron trap. The enhancement of nonuniformities with the reduction of temperature can be explained by considering that the increase of V_{WL} needed to reach the same I_{cell}, when the temperature is reduced, leads to a stronger inversion of the

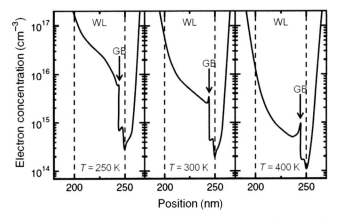

Fig. 8.46 The electron concentration profiles along the longitudinal direction with a trap at the grain boundary (GB).

inner parts of the polysilicon grains and to a higher occupancy of the acceptor states at the grain boundaries.

Heretofore, reliability issues such as retention and RTN were reviewed. These two aspects are closely related to the 3D-NAND's structure and to each other. One characterizes Si-nitride charge storage layer of V-NAND. The other depends on features of the poly-Si channel with grain boundaries. Comparing to the advanced 2D-NAND array with miniaturized cell size, the reliability of the 3D-NAND seems to be improved owing to the extension of gate length and width.

However, there are more aspects of the 3D-NAND to be considered. That is, the vertical channel has no p-substrate and is isolated from SL and BL when both SGs cut off. Therefore, after electrons are confined in the channel after reading, these electrons remain in the channel until the next operation. During the time interval before the next operation, some of the electrons in the poly-Si channel may be trapped in the grain boundary or Si-oxide surface. These trapped electrons will cause a Vt shift comparable to that from the previous study. This is one of the issues to consider. Thus we have to carefully examine the 3D-NAND's inherent reliability issues to achieve a high-quality memory array.

8.7 Future 3D-NAND technology trend

8.7.1 Number of WL stack

Because a single string's footprint is hard to scale any further, WL stack is growing to increase memory capacity. The main issue is that the stacking process is finite, thus, the focus point here would be *how many numbers of WL stack (N) is possible?* The first item to consider is the degradation of electrical characteristics. First, cell current reduction impacts the read margin. Second, read disturb has to be considered as well. The number of read stress is proportional to Read cycles $\times (N-1) \times t_{read}$ (mainly discharge time of BL for sensing). As the number of the WL stack (N) increases, t_{read} is also increasing because of the reduction of cell current. Thus, to avoid a buildup of the read disturb, some improvement of the process for poly-Si channel formation will be required to increase the magnitude of the cell current.

Another issue posed by the increase of the WL number will be the difficulty of the process to stack many WLs. Vertical hole etching will be especially hard as N increases. One kind of workaround solution is to reduce the WL thickness and the WL-to-WL spacing. A second method is that the stacking process may be divided into several parts to avoid high-aspect ratio etching, as shown in Fig 8.47 [20]. However, yet another concern is that process cost will be increased by the multihole etching and the subsequent deposition of insulators and channel poly-Si.

The other approach to reduce the chip size will be CMOS under the cell array (CUA). The Micron/Intel group has already used CUA because cell array topology in use is divided into more than 32 tiles, as shown in Fig. 8.32. However, V-NAND can also reduce the chip size by using CUA, even though the number of the plane is as small as two. The concern with using CUA for V-NAND will be the source-plate process. In the case of V-NAND, there is no source plate, but the vertical channel is directly connected to the p-Si substrate. However, to fabricate CMOS under cell array, this structure must be altered.

The alternative approach is to use the vertical gate cell structure [21] (Fig. 8.1D), which has not yet entered production. However, it could be another candidate for the future higher-density NAND, due to its capability to reduce the string area. As shown in Fig. 8.48, the length across the channel width in the gate vertical cell is almost the same as the unit length of the channel vertical cell. However, the length across WL can

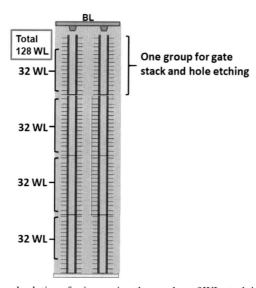

Fig. 8.47 Problems and solutions for increasing the number of WL stack in 3D-NAND.

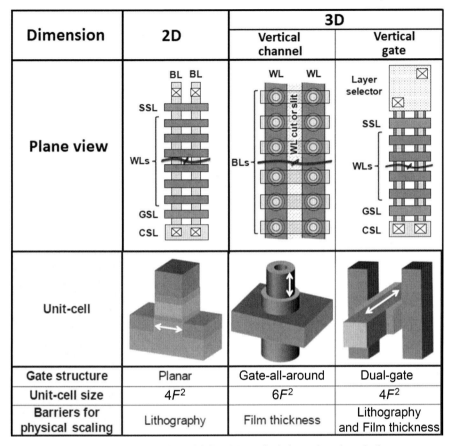

Dimension	2D	3D	
		Vertical channel	Vertical gate
Plane view			
Unit-cell			
Gate structure	Planar	Gate-all-around	Dual-gate
Unit-cell size	$4F^2$	$6F^2$	$4F^2$
Barriers for physical scaling	Lithography	Film thickness	Lithography and Film thickness

Fig. 8.48 Comparison of the cell size with 2D, 3D vertical channel, and vertical gate.

be smaller than the unit length of the channel vertical cell. Thus, the string area can be lessened up to 66% of the string area of the vertical channel [22].

Several new proposals for the next generation 3D-NAND are expected for the further extension of the application of mass storage.

8.8 Conclusion

In 2013, 3D-NAND was launched into production and it thoroughly penetrated the mass-storage market. This chapter reviews several major, feasible 3D-NAND technologies already in production. Concerning electrical characteristics, 3D-NAND has certain advantages compared to the 2D-NAND, mainly due to the larger channel dimensions (physical length and width) and sufficient spacing between neighboring cells.

However, process integration and multiple steps associated with 3D-NAND gate stacking require a quite complex and sophisticated production sequence. Thus, it is difficult to make an accurate prediction of the future technology trend. It is clear, though, that several major issues will be posed by an increase in the number of WL levels, which is directly linked to the memory capacity, as long as the horizontal string area scaling remains difficult. Furthermore, it is seriously expected that new difficulties will arise with each of the next memory generations, demanding novel technologies and fueling new research directions to ensure the survival of the 3D-NAND Flash in future nonvolatile memory markets.

Acknowledgment

Part of this work was supported by A. Suzuki in Macronix and Nina Mitiukhina in Lite-ON Tech.

References

[1]   K. Sakui, Next generation memory technology, in: Short Course of VLSI Symposium on Technology, 2013.
[2]   H. Tanaka, M. Kido, K. Yahashi, M. Oomura, R. Katsumata, M. Kito, Y. Fukuzumi, M. Sato, Y. Nagata, Y. Matsuoka, Y. Iwata, H. Aochi, A. Nitayama, Bit cost scalable technology with punch and plug process for ultra high density flash memory, in: VLSI Symposium on Technology, 2007, pp. 14–15.
[3]   J. Jang, H.-S. Kim, et al., Vertical Cell Array Using TCAT(Terabit Cell Array Transistor) Technology for Ultra High Density NAND Flash Memory, 2009,pp.192–193.
[4]   K.-T. Park, J.-M. Han, et al., Three-dimensional 128Gb MLC vertical NAND flash memory with 24-WL stacked layers and 50MB/s high-speed programming, IEEE J. Solid State Circuits (2014) 334–335. Dig. Tech Papers.
[5]   R. Yamashita, S. Magia, et al., A 512Gb 3b/cell flash memory on 64-word-line-layer BiCS technology, in: IEEE ISSCC Dig. Tech Papers, 2017, pp. 196–198.
[6]   T. Tanaka, M. Helm, et al., A 768Gb 3b/cell 3D-floating-gate NAND flash memory, in: IEEE ISSCC Dig. Tech Papers, 2016, pp. 142–144.
[7]   J.-W. Im, W.-P. Jeong, et al., A 128Gb 3b/cell V-NAND flash memory with 1Gb/s I/O rate, in: IEEE ISSCC Dig. Tech Papers, 2015, pp. 130–132.
[8]   D. Kang, W. Jeong, et al., 256Gb 3b/cell V-NAND flash memory with 48 stacked WL layers, in: IEEE ISSCC Dig. Tech Papers, 2016, p. 1301.
[9]   H. Kim, S.-J. Ahn, Y.-G. Shin, K. Lee, E. Jung, Evolution of NAND flash memory: from 2D to 3D as a storage market leader, in: IEEE IMW Dig, 2017, pp. 1–4.
[10] H. Maejima, K. Kanda, et al., A 512Gb 3b/cell 3D flash memory on a 96-word-line-layer technology, in: IEEE ISSCC Dig. Tech Papers, 2018, pp. 336–337.
[11] S. Lee, C. Kim, et al., A 1Tb 4b/cell 64-stacked-WL 3D-NAND flash memory with 12MB/s program throughput, in: IEEE ISSCC Dig. Tech Papers, 2018, pp. 340–342.
[12] J.-J. Shim, H.-S. Kim, W.-K. Lee, J.-Y. Lim, S.-M. Hwang, Semiconductor device with vertical structures that penetrate conductive patterns and interlayer insulating patterns, 2012, USP No. 9.209.244 B2.

[13] Y. Fukuzumi, R. Katsumata, et al., Optimal integration and characteristics of vertical array devices for ultra-high density, bit-cost scalable flash memory, in: IEDM Tech. Dig, 2007, pp. 449–452.

[14] C. Kim, J.-H. Cho, et al., A 512Gb 3b/cell 64-stacked WL 3d V-NAND flash memory, in: IEEE ISSCC Dig. Tech Papers, 2017, pp. 202–203.

[15] T. Maruyama, https://www.toshiba.co.jp/about/ir/jp/pr/pdf/opr20161207_3.pdf, 2016.

[16] R. Katsumata, M. Kito, et al., Pipe-shaped BiCS flash memory with 16 stacked layers and multi-level-cell operation for ultra high density storage devices, in: VLSI Sympo. on Tech, 2009, pp. 176–177.

[17] C. Caillat, S. Gowda, et al., 3D-NAND GIDL-assisted body biasing for erase enabling CMOS under array (CUA) architecture, in: IEEE IMW Dig, 2017, pp. 1–4.

[18] K. Parat, C. Dennison, A floating gate based 3D-NAND technology with CMOS under array, in: IEDM Tech. Dig, 2015, pp. 48–51.

[19] G. Nicosia1, A. Mannara1, et al., Impact of temperature on the amplitude of RTN fluctuations in 3D-NAND flash cells, in: IEDM Tech. Dig, 2017. pp. 48–51, 521–524.

[20] S. Aritome, NAND flash revolution, in: IEEE IMW Dig, 2016, pp. 1–4.

[21] M. Kiyotoshi, A. Yamamoto, Y. Ozawa, F. Arai, R. Shirota, Semiconductor memory and method of manufacturing the same, 2006, USP No. 8,008,732 B2.

[22] J. Choi, K.-S. Seol, 3D approaches for non-volatile memory, in: VLSI Sympo. on Tech, 2011, pp. 178–179.

Further reading

[23] D. Resnati, A. Goda, G. Nicosia, C. Miccoli, A.S. Spinelli, C.M. Compagnoni, Temperature effects in NAND flash memories: a comparison between 2D and 3D-arrays, IEEE Electron Device Lett. 38 (2017) 461–464.

Advances in oxide-based conductive bridge memory (CBRAM) technology for computing systems

9

Gabriel Molas, Michel Harrand, Cécile Nail, Philippe Blaise
CEA LETI Minatec Campus, Grenoble, France

9.1   Introduction

Resistive random access memory (RRAM) is based on the reversible formation/disruption of a conductive filament in a resistive layer providing a low- and a high-resistance states. Various classifications exist. Depending on the filament composition, two types of RRAM can be distinguished: OXRAM, where the filament is based on oxygen vacancies, and conductive bridging RAM (CBRAM), where the filament results from the dissolution of an active electrode, that can be made of silver (Ag) or copper (Cu) in most cases.

RRAM technology has many advantages. It is a low-cost two-terminal device and the number of integration steps is lower than in standard flash. It also shows a low-voltage operation: typical RRAM operating voltages are 1–3 V, which is much lower than ~20 V of flash NAND memories. Then, RRAM is a fast memory, with typical programming time of 100 ns. Less than 10-ns programming times were also reported in the literature. Very good endurance was also demonstrated in the literature, with window margin (WM) maintained after 10^{12} cycles [1]. Finally, good scalability was put in evidence: scaled memory devices were presented, demonstrating the scalability of the technology: 10-nm crossbar OXRAM [2] and 5-nm liner CBRAM [3]; further density improvements could be obtained later with vertical random access memory (VRRAM) architecture [4].

Despite some RRAM-based products start to enter the market [5–7], RRAM is still an emerging memory; maturity and reliability still need to be improved before mass production could be envisaged.

In this chapter, we propose to discuss how emerging resistive memories can break the memory bottleneck in computing applications, both in design and technology perspectives. A specific focus is made on oxide-based CBRAM with Cu-based active electrode.

This chapter is structured as follows: we first present the main memory design architectures that can be used to implement a RRAM memory array. Then, it is quickly recalled how and why current computer memory hierarchy is organized, then showed where this technology could be used in this memory hierarchy, what they can bring,

Copyright © 2019 Elsevier Ltd. All rights reserved.

and what should their main characteristics be for that purpose. We then discuss the technology aspects. We start from the material properties of the layers constituting the memory stack. Based on ab initio calculations, we provide insights on the conductive filament composition, and the competition between Cu and V_O diffusion mechanisms in various metal-oxide resistive layers. Then we present an in-depth electrical characterization of RRAM. The performances of emerging memories are deeply discussed in terms of WM, speed, endurance, retention, and consumption. Trade off and correlation existing among these features are analyzed, clarifying the potential offered by each RRAM stack. The limits and roadblocks of emerging memories (among them the critical point of variability) are discussed, and potential solutions are proposed. Finally, we evaluate how the device performances can match the product specifications for computing applications. The different levels where each emerging memory concepts could be integrated and would be the most efficient are discussed.

9.2   RRAM design and system aspects

9.2.1   RRAM memory architecture

There are three main types of RRAM memory architectures so far in the literature: 1T-1R, Crossbar, and VRRAM, each having advantages and drawbacks and suited for different applications.

9.2.9.1   1T-1R

A 1T-1R cell consists in a resistance serially connected with a MOS transistor. These cells can be assembled in a memory array by connecting all the transistor gates of cells belonging to the same line to metal word lines, all the transistor drains of cells belonging to the same column to bit lines, and all transistors sources of either a line or a column of cells to a power supply line. A cell can be read by applying a voltage on the selected word line and reading the current on the selected bit line (Figs. 9.1 and 9.2).

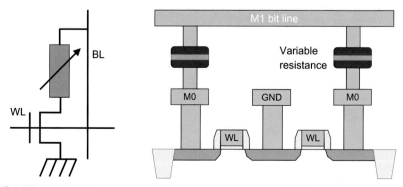

Fig. 9.1 Electrical and cross-section schematic representation of 1T1R cell.

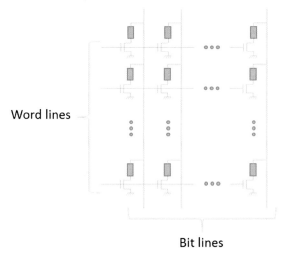

Fig. 9.2 Schematic representations of a 1T1R cell array.

This organization has the merit of simplicity and of reliable read-and-write operations; the drawback is the area due to the cell transistor. Actually, as RRAM rely on a higher voltage than those supported by modern CMOS technologies, especially for forming operations often required just after fabrication and require nonnegligible currents, the transistor cannot be of minimum size and will dictate the area of the memory array; the area of a PCM 1T-1R memory array requiring a rather high programing current is then close to SRAM area. Global Foundries STT-MRAM cell is announced to have an area of $45F^2$, while an SRAM cell is generally assumed to be around $145F^2$, which means that an STT-MRAM 1T-1R memory array should be close to a third of an SRAM array. RRAM usually require much less current than PCM but a bit higher than STT-MRAM, and a 1T-1R RRAM-based memory cell should have an in-between area.

9.2.9.2   Crossbar memory array

A crossbar memory cell is a variable resistance serially connected to a selector (Fig. 9.3). This selector is a nonlinear resistance, that is to say, a resistance whose value is much higher when the voltage applied to it is low. This cell occupies the place of a via between perpendicular metal lines.

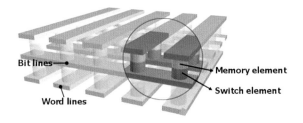

Fig. 9.3 Crossbar memory array 3D view from Ref. [7a].

A cell is programmed by applying a voltage between the upper and the lower metal lines to which it is connected; without selectors, a current could flow through cells connected to the upper selected line, then cells connected to these half-selected cells, then cells connected to the lower selected line (sneak paths). The selector is designed in such a way that the current is practically cut off when the voltage applied to a cell is half the programming voltage, in order to reduce to a minimum the current flowing through such memory arrays are extremely dense, as the memory cell area is reduced to a single via. Moreover, several layers can be deposited one above the other, even increasing the density. At least part of the peripheral circuits can be put below the memory array.

However, a large enough ratio between off and on resistances and moderate programming current values are required for a correct operation of such a memory array.

Actually, in a cross-point memory, the array size is limited by the voltage drop on the word lines and bit lines due to the resistance of the metal lines, the programming current, and the leakage current of the selectors of the unselected cells. But on the same time, an advantage of density versus the 1T-1R architecture can only be obtained if the memory arrays are large enough to locate the peripheral circuits below the memory array built on the back-end of line levels of the circuit. In practice, the minimum viable memory array size is around $2K \times 2K$, that is, 4 Mb. However, only 1 bit can be programmed at a time in such an array in order to limit the current flowing though the metal line and therefore the voltage drop along these lines. For these reasons, the minimum capacity of such a memory is around $4M \times 1b$ to be area efficient. An $\times 32b$ memory requires a capacity of at least 128 Mb, as composed of 32 arrays.

9.2.9.3   Vertical resistive random access memory

VRRAM is analog to crossbar memories, but the bit lines are made with a vertical pillar.

These memories are made using a process comparable to vertical NAND flash, suitable for very low cost/very high-density stand-alone memories, but not for embedded memories, as there process is not compatible with logic technologies.

To fabricate such a memory, uniform silicon oxide and metal layers are successively deposited; then holes are ditched and the different materials composing the RRAM cell and its selector are deposited along the internal circumference of the ditched hole, that is, afterwards filled with metal to realize the vertical pillar (Fig. 9.4).

9.2.2   Error correction

Memory yield and reliability can be improved by using error correction code (ECC). The principle is to add several bits to each word, page, or block to insert a code enabling detection of one or several errors, location of these errors, and therefore correction of them. The most well-known codes are the Hamming ones that are rather simple and fast to encode and decode; for these reasons, they are often used for SRAM and DRAM protection. However, more efficient codes like BCH and LDPC enabling the correction of more errors with a lower amount of redundancy, but requiring a more complex and slower encoding and decoding processes are preferred for denser,

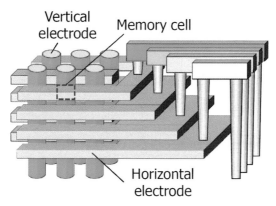

Fig. 9.4 3D view of a RRAM integrated in vertical architecture (VRRAM) from Ref. [4].

cheaper, but slower memories like flash. Obviously, adding a few microseconds to correct errors is not a big deal for 100-μs latency memories like flash, but would be a serious problem for fast memories like DRAMs.

Regarding the application to RRAM, as for flash, there will be some cells belonging to the tail distribution where it will not be possible to correctly recover their state, either because they are stuck to a fixed state, or because their R_{ON} or R_{OFF} state is too far from the main values. Depending of the probability of apparition of such defective cells and the capacity of the memory chip, a more or less efficient code will have to be applied.

To give an order of magnitude, it can be calculated that for a 128 GB memory, a BCH code correcting 8b on a 512B (i.e., 4096b) codeword enables a yield higher than 90% (i.e., more than 90% of the dies appear as without any defect) if the uncorrected error rate (i.e., the proportion of erroneous bits in a single die) is less than 10^{-4}, that is, the WM is ensured at 4σ [8]. Such a code requires a small increase of the chip capacity of 8/4096 = 0.2%.

However, such a code is complex to decode and can require more than 1000 cycles for this decoding operation, that might require a few microseconds. This might impact the latency of RRAM that are likely to be much faster than flash devices. With the increase of density of flash devices, SSD manufacturers now tend to use LDPC codes that are more efficient than BCH, but even more complex and long to decode. Also, BCH as well as LDPC codes protect very long words, what is not a problem for a block-based memory as flash, but may be a constraint for byte addressable memories like RRAM.

In a tentative to solve these problems, some simpler correction codes have been proposed in the literature, by Ref. [9] (Fig. 9.5).

Such a code provides as many fields as there are errors that can be decoded: each field indicates the erroneous bit position and the value of this bit. As in more conventional codes, errors on these redundancy fields can also be corrected. However, at the contrary of Hamming, BCH, Read-Solomon, or LDPC codes, this code is specific to yield enhancement and not adapted to transient errors correction, as they require a special operation to encode the additional field.

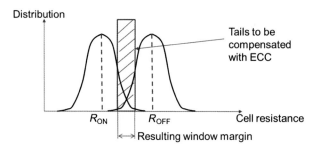

Fig. 9.5 Cell resistance distribution and window margin.

9.2.3   Applications of RRAM in computers architecture

9.2.9.1   Computer memory hierarchy

Modern computers exhibit a deep memory hierarchy.

The reason is that today nonvolatile memories used to permanently store programs and data are far too slow to be used directly by the computer cores. These cores, running at several GHz, need to retrieve their instructions and data in memories with subnanosecond access times that are only achievable by small capacities SRAM. These memories, being too small to store the integrality of programs and data, are organized as caches. When the searched data or instruction is not present in this first-level cache (L1), the system searches them in a larger—but slower—second-level (L2) cache, generally shared by several cores. Sometimes, there is even a third-level cache—even bigger and slower—generally shared by all the processor chip cores. Some chips implement a last-level cache (LLC)—either L2 or L3—in embedded DRAM (eDRAM) technology, all other use SRAM. When the searched data are not in either of these caches, the data are looked for in the main memory, much bigger (1–128 GB). This main memory is today realized using DRAM technology, much denser and then cheaper than SRAM but difficult in being compatible with the logic technology used by the processor chip, and for this reason located outside the processor chip. This DRAM technology, which exhibits access times around 30–40 ns, that is, much faster than today nonvolatile technology—hard disk drive or flash—is however volatile and even need to be frequently refreshed to keep its data. Then, at the end of the chain, stands the storage memory, with access times of hundreds of microseconds (flash) to tenths of milliseconds (hard disk drives) (Fig. 9.6).

9.2.9.2   Usage of RRAM technologies in the computer memory hierarchy

We can think of using RRAM technologies in several places in this hierarchy, each having its own advantages with respect to the current implementation:

- We can use it in the storage memory area, either as a replacement of current flash technology, or, at least in the short term as an intermediate step between the main memory and the flash or disks: storage class memory (SCM); in this case, the RRAM will be interfaced via an I/O like interface, as SATA, Ethernet, PCIe, or whatever interface that could emerge in

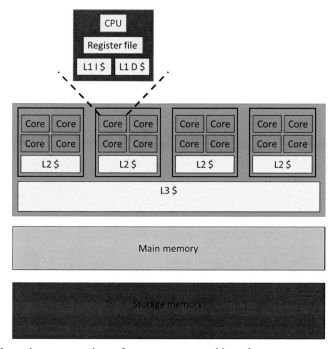

Fig. 9.6 Schematic representations of computer memory hierarchy.

this class of interfaces, and the transfer of data with the main memory could be managed by the OS of the chip as it is today for storage memory.
- We can use it in the main memory area, either as a DRAM companion chip, located on the same memory bus, or even as a replacement of the DRAM.
- We can use it as LLC replacement or complement; it is unlikely that RRAM technology could become fast enough to be used as first-level cache.

These usages are summarized in Fig. 9.7.

9.2.9.3 Storage

Filling the gap
The storage memory market being essentially cost driven, it is likely that RRAM will only replace the current well-established flash technology when it will be competitive in cost with it; as flash density continues to increase thanks to monolithic 3D integration, this will take some times to happen. In the meantime, RRAM could be used as an intermediate step between flash and the main memory, as for instance an ultrafast SSD used to store data most often accessed. The large difference on latency is likely to make acceptable a higher cost.

Today, the gap in latency between the main memory, that is, in the order of 30–50 ns and the one of flash-based storage, which is about 100 µs (much more in writing) is tremendous, and having a technology like RRAM enabling microsecond scale latencies would undoubtedly be a big improvement for data centric applications as big data and database ones.

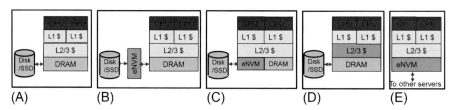

Fig. 9.7 Possible usage of RRAM in computer memory hierarchy. (A) Current system. (B) Storage: Filling the gap between SSD and DRAM, ... or SSD replacement. (C) Main memory: Filling the gap between SSD and DRAM, ... or DRAM replacement. (D) Embedded memory: LLC cache replacement. (E) Universal memory: Main/storage.

However, such usage may impose an improvement on the network and software through which data are accessed. The faster network interface used for SSD today is PCIe, which can present a latency in the order of the microsecond; this is not a big deal for a flash-based SSD where the memory itself has a latency of around $100\,\mu s$, but would be for a technology where the latency is itself around $1\,\mu s$. Lower latency I/O networks and hardware accelerated page replacement management should then be envisaged.

Flash replacement

Even if not for the short term, it is likely that RRAM will be one day cost competitive with flash as we do not see any reason for 3D integration could not be applied at the same scale as for flash, and it is also likely that RRAM will scale much further than flash. It is then foreseeable that we could one day completely replace the current storage technologies by RRAM.

It is what the Hewlett-Packard research team had in mind with their prototype called "The Machine," where they use a pool of processors connected to a 4 PB RRAM-based storage memory. For the reasons invoked upper, the interconnect was a photonic one, allowing a much smaller latency than current interconnect technologies (Fig. 9.8).

In this case, the storage management will be greatly facilitated compared to the "filling the gap" scenario, as only one layer of storage is used, making not useful to differentiate high-speed from low-speed access data.

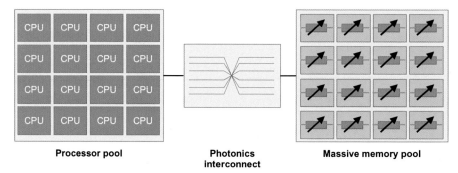

Fig. 9.8 HP's "The Machine" architecture principle inspired from Ref. [10].

RRAM specifications for storage application

The RRAM requirements for storage application can be summarized in Fig. 9.9.

Latency A latency of a few microseconds seems to be easily achievable for RRAM technology and would represent a ×100 improvement w.r.t. current flash technology.

Endurance As data are generated and updated in the main memory before being sent to the storage memory, data are not extremely often written and a moderate endurance is acceptable. Also, wear leveling can be used for RRAM, as well as for flash. However, endurance should be made much better than flash—that is, around 10^4–10^5 cycles—as the RRAM being much faster, they will be written more often. As a conclusion, an endurance of 10^6–10^7 cycles should be acceptable.

Cost and power consumption Cost is the most stringent parameter for a storage memory. If we want to replace flash with RRAM technology, the cost per bit should be at least as good as flash. However, as an intermediate memory between main memory and storage memory, an intermediate cost, around two times the flash cost/bit can be acceptable, as they will be used in a much lower quantity than the storage memory and their performances are so much higher.

In both case, the power consumption should be made of the same order of magnitude.

9.2.9.4   Main memory applications

Filling the gap

Inserting RRAM components on the main memory bus, beside DRAM components, will present significant system benefits.

It will make the data integrity management more performant; data can be quickly secured locally, on a word-by-word basis, instead of using current cumbersome journaling or check-pointing schemes to protect data from events as a loss of power supply [11, 12]. This can significantly reduce data traffic in a data center, as it is estimated that data integrity management could represent up to 80% of the file system usage in PetaFlop data centers. As RRAM should become much denser and cheaper than DRAM, it will enable much greater capacity main memories; this is especially important for big data applications, where storing big multidimensional tables in the main memory enables a tremendous performance advantage because these tables often needs to be accessed in a different order than the one it has been stored in the storage memory that can only be accessed sequentially. It will also simplify atomic operations in transactional data bases, as RRAM are byte addressable and nonvolatile.

Storage	Latency	Endurance	Retention	Cost/b	Power consumption
Filling the gap	1 – 5 µs	> 10^6	10 years @ 85°C	< 2 × flash cost	< flash
Replacement	1 – 5 µs	> 10^6	10 years @ 85°C	~flash cost	< flash

Fig. 9.9 RRAM requirements for storage applications.

They will anyway propose a much more efficient solution than current NVDIMMs composed of a mix of DRAM and flashes powered by a bulky supercapacitor during the data transfer from DRAM to flash when the main power supply is lost.

DRAM replacement

If RRAM can be made fast and enduring enough, they can even completely replace DRAM. This presents the same system benefits as above, but with an improved simplicity—only one type of components in the main memory, therefore simpler to manage—and a cost advantage.

Even if RRAM cannot reach the DRAM 30–40 ns latency, a bandwidth improvement can be obtained because of the larger capacity enabled by their increased density: this improved bandwidth can be got by splitting the memory in many banks accessed in parallel. The access control and data transmission do not need to be realized in an additional logic technology chip as it is currently the case in hybrid memory cubes (HMC) and high-bandwidth memories (HBM) DRAM stack arrays because as the memory resistors are located in the back-end of line, the RRAM components could use the very same fast transistors that are used in logic technologies.

RRAM specifications for main memory application

The RRAM requirements for main memory application can be summarized in Fig. 9.10.

Latency To replace DRAM, we need a comparable latency, that is, at least less than 50 ns. If RRAM are used beside DRAM components, a somewhat higher latency can be acceptable.

Endurance The memory being here much faster than for storage applications, a higher endurance is needed. If the RRAM is used beside DRAM, it will not be written every cycle and 10^9 cycles should be sufficient. If the RRAM is used as a replacement of DRAM, the same 10^9 cycles can still be acceptable assuming a smart wear-leveling is used, ensuring all bits are equally used, and the capacity of the memory is high enough to prevent the whole memory to be written so much times due to the speed limitation of the technology.

Retention In this application, the RRAM is not required to store the data infinitely, assuming a storage memory is used for that. However, to benefit from the system advantages listed upper, the retention should be high enough to be sure that in the

Storage	Latency	Endurance	Retention	Cost/b	Power consumption
Filling the gap	< 200 ns	> 10^9	> 5 days @ 85°C	< ½ DRAM cost	< DRAM
Replacement	< 50 ns	> 10^9*	> 5 days @ 85°C	< ½ DRAM cost	< DRAM

* Assuming smart wear-leveling is used

Fig. 9.10 RRAM requirements for main memory applications.

event of a power supply loss, it can be restored before the data vanishes; also, a too low retention would require refresh operations that would negatively impact the reliability of the memory. A few days retention time should then be an acceptable order of magnitude.

Cost and power consumption In this application, the RRAM is to be compared to DRAM, not to flash. A lower cost per bit—let say half the price—than DRAM should be easily achievable and would make the technology attractive. Power consumption should not be higher than the DRAM ones to be competitive.

9.2.9.5   Embedded memory application

The RRAM could also be used as LLC; other levels would definitely require a too high speed. In this case, the cache capacity could be made much higher, diminishing the external bandwidth requirements. If jointly used with an RRAM-based main memory, interesting savings can also be obtained in data recovery operations.

The main requirement for a cache application is summarized in Fig. 9.11.

Here, the most stringent requirements are speed and endurance. We cannot rely on wear-leveling in this case as the cache capacity will not be sufficient in regard of its bandwidth, making each bit being written frequently whatever we do. Such a high endurance of 10^{16} cycles is likely to reserve this application to STT-MRAM. However, some papers propose hybrid caches where the most frequently written cache line is stored in SRAM while the other ones are stored in RRAM [13].

As far as cost and power consumption are concerned, the reference is eDRAM; the RRAM has to be cheaper and less consuming than eDRAM to be competitive.

9.2.9.6   Universal memory

Let us now dream about a kind of "universal memory" that could be used for both the current main memory and the current storage memory of a computer. This memory could be connected to the processor main memory bus (assuming there would not be any electrical issue as connection length...); there would not be any storage memory as we know it today, as the main memory would replace it as well.

Such technology would have several advantages. First, there would be no need to store data sequentially as it is the case today, as RRAM are byte addressable: data could be updated on the flow. File systems may also disappear and be replaced by something randomly accessible. Then, check-pointing mechanisms would be still necessary to make sure data are maintained in a coherent state even in the case of an unexpected event as a power supply loss, but it could be done in other ways that it is done today.

Latency	Endurance	Retention	Cost/b	Power consumption
< 30 ns	> 10^{16}	> 5 days	< eDRAM cost	< eDRAM

Fig. 9.11 RRAM requirements for embedded cache applications.

Latency	Endurance	Retention	Cost/b	Power consumption
< 50 ns	> 10^9 if smart wear-leveling is applied	10 years @ 85°C	~flash cost	< flash

Fig. 9.12 RRAM requirements for a universal memory.

Moreover, no boot would be required; the operating system could be maintained resident in the main memory, as it is the same as the storage one; when the user would switch off his computer, a rapid back-up of the processor internal volatile caches would be sufficient before effectively completely switch off the power supply, as the main memory itself would not need to be backed-up; when switching on again, the computer would wake up instantaneously in the exact state it was before switch off. This drastically solves the bandwidth issues between main memory and storage memory; they are the same! New operating systems are required to take benefit of all these points [14].

However, such a memory would merge current DRAM and flash requirements; a tentative specification would be as indicated in Fig. 9.12.

As for main memory application, a latency below 50 ns and an endurance larger than 10^9 cycles would be required, and as for storage memory application, a 10-year retention and a cost per bit comparable to flash are required.

Unfortunately, getting all these characteristics in the same technology seems very difficult in the foreseeable future.

9.2.4 Conclusions

9.2.9.1 Memory design aspects

As discussed in the first part of this chapter, cross-point and VRRAM architectures are the best suited for large, high-density memories, while 1T-1R, due to its simplicity, can be a good solution for small- and medium-sized embedded memories.

VRRAM are promising because of the reduced cost of providing a large number of layers, but might be slower than conventional cross-point memories. For this reason, VRRAM would be a good candidate for flash replacement, while cross-point would be better suited for DRAM replacement.

In any case, the main technology factors to increase density are a low programming current and a low leakage current for half-selected cells in order to reduce voltage drop in word lines and bit lines.

9.2.9.2 System aspects

As discussed in the second part of this chapter, RRAM can be used at the following places of computer architectures, with different specification requirements:

- Storage: between SSD and main memory or replace SSD

    The main advantage with respect to flash technology is the speed that is expected to be multiplied by a factor 100–1000. However, the network bandwidth must be increased to take benefit of it. The main requirements for this usage are density, cost, and retention, and the best memory architecture is likely to be the VRRAM one.

- Main memory: on the memory bus, together with DRAM or replace DRAM

    Big benefits for big data/data base applications and on check-pointing traffic are expected in using RRAM for this application. The main requirement for this usage is speed, and the best memory architecture is likely to be the cross-point one.

- Last-level cache

    A RRAM LLC would increase cache density and thus enable larger cache sizes that would reduce memory wall issues. The main requirement for this usage is speed and endurance. As the cache is generally too small to make wear-leveling efficient, it is likely that an STT-MRAM 1T-1R memory would be the best suited one, despite some hybrid—PCM and SRAM—caches have been proposed in the literature.

- Universal memory

    Finally, it could be dreamed of using RRAM technology for building a main memory that would also serve as storage memory. This would make a revolution in computer architecture and operating systems, as we could benefit of features like instantaneous boot, and randomly access to data with no need for a file system. However, such an universal memory would require the same cost and retention characteristics of current flash together with the speed and endurance of DRAM; unfortunately, gathering all these characteristics in the same technology appears to be a too big constraints in the foreseeable future.

9.3 RRAM technology and performances

9.3.1 Introduction

In the technology section, we provide insights on the conductive filament composition of various RRAM based on first-principle calculations. Then the electrical characteristics of RRAM are presented. We highlight the tradeoffs and correlations existing between various memory features. We show that it is possible to boost some performances at the price of the degradation of some others, playing with technology tuning (integrated materials…) or operating conditions (SET current, programming time and voltage…). Then we discuss how variability affects the overall performances of an RRAM-based memory array. Finally, based on these inputs and on recommendations of Section 9.2 (Design), we evaluate how RRAM can enter the memory hierarchy and for which function of the computing system that they would be the most promising.

9.3.2 Filament composition

In this section we investigate the filament composition in conductive-bridging resistive memories (CBRAM). To this aim we use first-principle calculations to identify the most favorable species composing the conductive filament (among them: Cu from the top electrode (TE), metal atoms in the metal-oxide resistive layer, oxygen vacancies):

- Calculation of defect enthalpy formation energy ΔH using DFT allows us to identify the most favorable defects and energy cost to insert or generate a specific species in the resistive layer. Impact of electrodes and dielectric stoichiometry can be taken into account.

- Calculation of energy migration barrier $E_{migration}$ by molecular dynamics quantifies the energy cost of the diffusion of species (V_O, Cu...) between two equivalent sites. Lowest energies allow us to identify the most favorable events and to give insights on filament composition.

9.3.9.1   Al₂O₃ resistive layer with Cu-based electrode

For alumina, a 160-atoms supercell consisting of four γ-Al_2O_3 unit cells is calculated [15]. This structure is the closest crystalline form of our amorphous ALD material with a density of $3.45\,\mathrm{g\,cm^{-3}}$ vs the measurement value of $3.1\,\mathrm{g\,cm^{-3}}$. Different positions are tested for each defect, especially for interstitials, up to four inequivalent sites are compared (see Fig. 9.13).

The band gap obtained at $3.43\,\mathrm{eV}$ is corrected in G_0W_0 at $6.5\,\mathrm{eV}$ which corresponds to the one measured by ellipsometry [16]. As expected, this result of $3.43\,\mathrm{eV}$ is lower than 7–$8\,\mathrm{eV}$ reported in the literature due to GGA approximations, well known for underestimating dielectric bandgap. Corrections can be done to extract exact values and estimate error on other results [17]. For TE, we used a 96-atom supercell consisting of eight Cu_2Te_3Ge unit cells [18]. This structure is the closest crystalline form of our amorphous PVD material with a density of 6.0 vs $6.1\,\mathrm{g\,cm^{-3}}$ (measured).

All results are based on a band alignment between band gap of alumina and Cu_2Te_3Ge which is done considering:

- G_0W_0 gap correction with an opening of more than $2\,\mathrm{eV}$ at conduction band and $-0.5\,\mathrm{eV}$ at valence band
- Internal photo-emission measurement of the barrier of Al_2O_3/Cu_2Te_3Ge at $3.6\,\mathrm{eV}$ [19, 20].

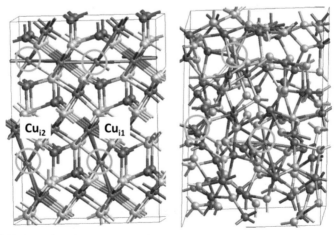

Fig. 9.13 γ-Al_2O_3 and a-Al_2O_3 structure with different copper interstitial positions circled in *orange*. O in *purple*, Al in *green black*, Cu in *orange*. Cu_{i1} and Cu_{i2} are mentioned and used for migration barrier calculation.

Thermodynamic of defects

Al$_2$O$_3$ post process defects In this section, we first study Al$_2$O$_3$'s native point defects that may appear after deposition. Calculations are based on the following equation:

$$\Delta H = U_{\text{Defect}} - U_{\text{Pristine}} + \Sigma n_i \cdot \mu_i + q \cdot \left(E_{\text{Fermi}} + \Delta V + E_{\text{VB}} \right) \tag{9.1}$$

With ΔH the formation enthalpy of the defect, U_{Defect} the defect total energy, U_{Pristine} the initial state total energy, n_i and μ_i the number of chemical formula and the reference chemical potential for the exchange, q the charges exchanged in the system, E_{Fermi} the electrode Fermi level considered in contact, ΔV the electrostatic energy shift [17], and E_{VB} the highest valence band energy level.

Results are reported in Fig. 9.14. It shows the formation enthalpy of intrinsic Al$_2$O$_3$ defects depending on oxygen concentration during deposition process (from $\mu_O = \mu_{O2}/2$ to metal-rich condition) and considering a bottom electrode (BE) with a work function close to W. It points out that deposited Al$_2$O$_3$ may be rich in aluminum Frenkel pair (Al$_{\text{FP}}$) and aluminum vacancies (V$_{\text{Al}}$) charged -3 for O-rich condition.

To summarize these results: deposited Al$_2$O$_3$ may be rich in aluminum Frenkel pair (Al$_{\text{FP}}$) and aluminum vacancies (V$_{\text{Al}}$) charged -3 for O-rich condition.

Al$_2$O$_3$ defects after CuTe$_x$Ge$_y$ deposition Then, we study aluminum point defects after CuTe$_x$Ge$_y$ deposition. As we consider exchanges between Al$_2$O$_3$ and CuTe$_2$Ge$_3$, formula for a defect charged $+q$, D^{+q}, generating a complementary defect D in TE becomes:

$$\Delta H = U_{\text{Al}_2\text{O}_3:D^{+q}} - U_{\text{Al}_2\text{O}_3} + U_{\text{CuTe}_2\text{Ge}_3:D} - U_{\text{CuTe}_2\text{Ge}_3} + q \cdot \left(E_{\text{Fermi}} + \Delta V + E_{\text{VB}} \right)$$

With D, the defect introduced in Al$_2$O$_3$ (e.g., Cu$_i^{+1}$, Te$_i^{-2}$, V$_O^0$, V$_{\text{Al}}^{-3}$), D the complementary defect introduced in TE (e.g., V$_{\text{Cu}}$, V$_{\text{Te}}$, O$_i$, Al$_i$, respectively), and E_{Fermi} the electrode Fermi level in contact (e.g., W, Cu, Pt, or Ti).

Results are presented in Fig. 9.15 and the lowest defects are charged. After CuTe$_x$Ge$_y$ deposition, considering charge of species and a Fermi level close to Cu,

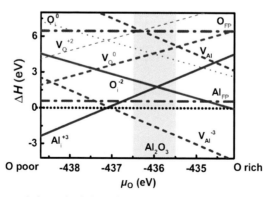

Fig. 9.14 Formation enthalpy calculation of intrinsic Al$_2$O$_3$ defects. W Fermi level is considered. *Yellow zone* highlights stoichiometric Al$_2$O$_3$ process conditions.

the most favorable interstitials and vacancies we could obtain are respectively, copper interstitial (Cu_i^{+1}), aluminum vacancy (V_{Al}^{-3}), germanium interstitial (Ge_i^{+2}), and oxygen vacancy (V_O). Tellurium interstitial (Te_i^{-2}) appears as unfavorable. It can be noted that Ge is used to stabilize CuTe layer and to keep it amorphous during integration process, to prevent any segregation and phase separation. Typically, 10%–20% Ge is used [21, 22] in CuTe alloys. As a consequence, Ge is not expected to have any impact in the electrical behavior of CuTe. Moreover, CuTe bound is polar and can easily be broken in $Cu^{\delta+}$ and $Te^{\gamma-}$ ions. They will be mobile under the electric field. Ge calculations are just used to confirm this affirmation. We thus focus on Cu and Te elements in the last section.

To summarize these results: the most favorable defects that can appear in Al_2O_3 after $CuTe_xGe_y$ deposition are: copper interstitial (Cu_i^{+1}), aluminum vacancy (V_{Al}^{-3}), and oxygen vacancy (V_O). Tellurium interstitial (Te_i^{-2}) appears as unfavorable. Germanium interstitial (Ge_i^{+2}) is less favorable than Cu_i and is just used to stabilize CuTe layer.

Al_2O_3 defects during RRAM operation Finally, for the CBRAM working principle, we study the case of copper in defective alumina. The reference for oxide becomes either a V_O-rich Al_2O_3 (Al_2O_3 with one V_O defect) or a Al_{FP}-rich Al_2O_3 (Al_2O_3 with one Al_{FP} defect) or a V_{Al}^{-3}-rich Al_2O_3 (Al_2O_3 with one V_{Al}^{-3} defect) instead of stoichiometric Al_2O_3. V_O in Al_2O_3 can come from:

- Al_2O_3 process condition in O-poor atmosphere
- O exchange with TE during forming process requiring 5.3 eV (Fig. 9.15)

Al_{FP} and V_{Al}^{-3} come almost spontaneously from process (Fig. 9.14). As TE is positively polarized during forming, Al_i^{+3} insertion in TE is not allowed. Therefore, V_{Al}^{-3} creation does not seem favorable during forming process. Results are based on the following equation:

$$\Delta H = U_{Al_2O_3:D_0^{+q_0}+D^{+q}} - U_{Al_2O_3:D_0^{+q_0}} + U_{CuTe_2Ge_3:D} - U_{CuTe_2Ge_3} + \left(q - q_0\right)$$
$$\left(E_{Fermi} + \Delta V + E_{VB}\right)$$

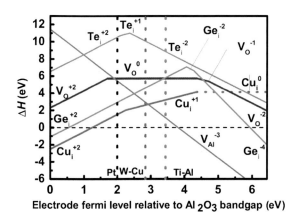

Fig. 9.15 Formation enthalpy calculation of defects in $Al_2O_3/CuTe_xGe_y$ system. Fermi level of three types of metal (Pt, W-Cu, Ti-Al) are represented *(dashed vertical lines)*.

With $D_0^{+q_0}$, the defect initially present in defective Al_2O_3 due to process or forming (e.g., Al_{FP}, V_O or V_{Al}^{-3}), D the studied defect (Cu_i, Te_i...) and $\bar{D}$ its complementary defect in TE (V_{Cu}, V_{Te}...).

The most favorable exchanges are highlighted in Fig. 9.16. Only defects that can lead to a conductive path through alumina are taken into account. Defects are then considered when they introduce energy level in alumina band gap. Cu introduction is favorable in interstitial site or replacing an Al site. If Al_2O_3 is V_O rich, O substitution is the most favorable. In the case of Al_{FP}-rich and V_{Al}^{-3}-rich, Al substitution is the most favorable. We can note that there is no favorable state found for a tellurium insertion and that Ge insertion is very expensive compared to Cu.

To summarize these results: the most favorable defect in a V_O-rich Al_2O_3 structure is Cu by O substitution. In the case of Al_{FP}-rich and V_{Al}^{-3}-rich, Cu by Al substitution is the most favorable. Tellurium and germanium insertions are very expensive compared to Cu [23].

Diffusion in Al_2O_3 While thermodynamic approach only considers initial and final defect energy states, more insights are required to quantify the energetic barrier to diffuse Cu ion in Al_2O_3. For this purpose, energy migration barriers are calculated using NEB-based (nudged elastic band) approaches. Results correspond to migrations from Cu_{i1} site to Cu_{i2} (see Fig. 9.13). It can be noted that in this work, defect formation energy barrier is not considered.

This would require an interface model between Al_2O_3 and $CuTe_xGe_y$ with charge injection effect [24] which is well beyond the scope of this study. However, we are in the case of nonreversible dielectric breakdown during forming with a high local electric field. This particular event is happening at nonequilibrium. Therefore, we expect

Defect D	$\Delta H\, Al_2O_3$ (eV)	$\Delta H\, Al_{FP}$-rich (eV)	$\Delta H\, V_O$-rich (eV)	$\Delta H\, V_{Al}^{-3}$-rich (eV)
Cu_i^{+1}	2.5	2	3^1	x
$(Cu_{Al}+Al_i)^{+1}$	$1.1\,^2$	$0.6\,^3$	3.2	x
Cu_{Al}^{-2}	5.6	x	x	3.2
Cu_O^{+1}	5.8	8	$0.6\,^4$	x
V_O	5.3	x	x	x
V_{Al}^{-3}	2.8	x	x	x
Al_{FP}	0.7	x	x	x
Te_i^{-2}	10	8.5	7	x
Te_{Al}^{-3}	7.4	x	x	7.3
Ge_O^{+2}	9.6	x	3.3	x

[1] Cu interstitial defect next to V_O already present in Al_2O_3 with V_O
[2] Al_{FP} is created and Cu substitute Al site
[3] Cu substitute Al site already present in Al_2O_3 with Al_{FP}
[4] Cu substitute O site already present in Al_2O_3 with V_O

Fig. 9.16 Exchange energy ΔH (eV) between TE and γ-Al_2O_3. Al_2O_3 is either stoichiometric or includes oxygen vacancies or aluminum Frenkel pairs or aluminum vacancies.

the first limiting event to be the sensitivity of species to the electric field. With high diffusion barrier, it is indeed impossible to move in the resistive layer. In the case of Ag/GeS$_2$ CBRAM behavior with very low migration barrier and low local electric field, this defect formation energy barrier could be the first limiting event and would need to be calculated. Cu_i^+ diffusion barrier in Al$_2$O$_3$ structure is found close to 4 eV (Fig. 9.17).

Cu migration barriers in defective alumina structures are calculated as well with a V_O or Al_{FP} next to Cu migration path. In particular, Cu_i^+ diffusion barriers in Al$_2$O$_3$-V_O and in Al$_2$O$_3$-Al_{FP} are evaluated at 1.4 eV and 0.7 eV, respectively. Cu can pass through Cu_O or Cu_{Al}-Al_i state. Consequently, oxygen vacancies and aluminum vacancies in Al$_2$O$_3$ would be very likely to facilitate Cu diffusion. It can be noted that Te_i^{-2} migration in Al_{FP}-rich Al$_2$O$_3$ is evaluated at more than 7 eV (not shown) while Te_{Al}^{-3} migration in V_{Al}^{-3}-rich is evaluated at 3.1 eV (Fig. 9.17B).

To summarize these results: Te diffusion in V_{Al}^{-3} is evaluated at 3.1 eV while Cu_i^{+1} diffusion is the most favorable next to V_O or Al_{FP} with a migration barrier of 1.4 and 0.7 eV, respectively. O and Al diffusions in order to find out if oxygen or aluminum can be favorable in Al$_2$O$_3$ to help Cu diffusion, O_i^{2-}, V_O, Al_i^{3+}, and V_{Al}^{3-} migration barriers are evaluated.

Results are shown in Fig. 9.18 with associated path screen shots. It reveals that oxygen interstitial and aluminum vacancy have the lowest barriers. Indeed, we find oxygen vacancy diffusion barrier at 3.4 and 4.7 eV for threefold and fourfold coordinated oxygen, respectively. This really high value indicates that V_O is immobile. Aluminum interstitial charged +3 is quite hard to diffuse directly from one site to another (>5.5 eV not shown). A more favorable path is found by concerted movement and barrier is evaluated around 2.3 eV. Meanwhile, oxygen interstitial and aluminum vacancy have their highest barrier around 0.54 and 1.5 eV, respectively (by concerted movement).

To summarize these results: V_O and Al_i^{+3} are hardly mobile in Al$_2$O$_3$ whereas O_i^{-2} and V_{Al}^{-3} have the lowest energy migration barriers at 0.54 and 1.5 eV, respectively.

To conclude, Fig. 9.19 summarizes energy costs to introduce various species in alumina. Based on these results, a new filament model is proposed. In the case of a

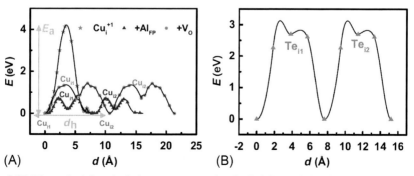

Fig. 9.17 First-principle calculations to compute barrier height to diffuse from Cu_{i1} to Cu_{i2} sites (see Fig. 9.13): (A) Cu_i^{+1} in stoichiometric or defective Al$_2$O$_3$ systems and (B) Te$_i$ in V_{Al}^{-3}-rich Al$_2$O$_3$.

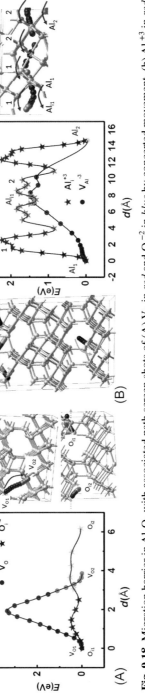

Fig. 9.18 Migration barrier in Al_2O_3 with associated path screen shots of (A) V_O in *red* and O_i^{-2} in *blue* by concerted movement, (b) Al_i^{+3} in *red* by concerted movements and V_{Al}^{-3} in *blue*.

Defect	ΔH(eV)	E_d(eV)	d_h(Å)
Cu_i^{+1}	2.5	4	6
V_O	3.8	3.5	4
V_{Al}^{-3}	3.1	1.5	15
Al_i^{+3}	-0.7	> 2.3	> 8
O_i^{-2}	0.25	0.54	6
Cu_i^{+1} in V_O	0.6	1.4	10
Cu_i^{+1} in Al_{FP}	1.1	0.7	4
Cu_{Al}^{+1} in V_{Al}^{-3}	3.2	x	x
Te_i^0 in V_{Al}^{-3}	7.3	3.1	8
Te_i^{-2} in Al_{FP}	8.5	7	10

(A)

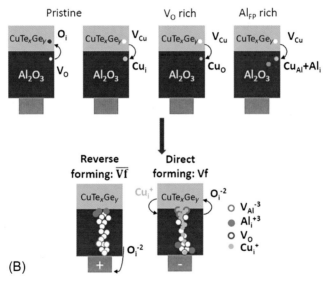

(B)

Fig. 9.19 A: Summary of DFT calculations, indicating enthalpy formation energy ΔH, migration energy barrier height E_d, and hopping distance d_h. B: Illustration of the most energetically favorable exchanges between Al_2O_3 and $CuTe_xGe_y$ during forming and schematic movements happening during forming.

reverse forming (F) oxygen diffuse toward BE, while aluminum vacancies accumulate near it. When a positive voltage is applied to the TE, copper is inserted in the oxide while oxygen is going in the TE and aluminum vacancies accumulate near it. As we have seen previously, SET voltage is much smaller than forming voltage. Moreover, due to the fact that V_O are immobile in Al_2O_3, SET process may mostly imply Cu migration in a V_O-rich structure: a hybrid Cu/V_O filament is proposed [23].

9.3.9.2 HfO₂ resistive layer with Cu-based electrode

For hafnia, a 96-atoms supercell consisting of 8 monoclinic-HfO_2 unit cells was calculated. This structure is the closest crystalline form of our amorphous PVD material with a density of $9.8 \, \mathrm{g\,cm^{-3}}$ vs the measurement value of $10 \, \mathrm{g\,cm^{-3}}$. Different positions were tested for each defect, especially for interstitials, up to four inequivalent sites were compared (see Fig. 9.20).

The band gap obtained at 3.73 eV is corrected [25] in G_0W_0 at 5.2 vs 5.5 eV measured by ellipsometry. For the TE we used a 96-atom supercell consisting of eight Cu_2Te_3Ge unit cells [18]. This structure is the closest crystalline form of our amorphous PVD material with a density of $6.0 \, \mathrm{g\,cm^{-3}}$ vs $6.1 \, \mathrm{g\,cm^{-3}}$ (measured). All results are based on a band alignment between the band-gap of hafnia and Cu_2Te_3Ge which was done considering:

- G_0W_0 gap correction with an opening of 1 eV at the conduction band and -0.5 eV at the valence band
- Electron affinity of HfO_2 at 2 eV [26]

To complete the study, an amorphous a-HfO_2 model with a density of $9.4 \, \mathrm{g\,cm^{-3}}$ was implemented. This model was generated using the melt and quench method: a crystalline monoclinic-HfO_2 with Schottky defects was used for the initial structure to lower density with the eight HfO_2 unit cells considered above. It was melted at 2800 K for 10 ps and cooled to 300 K for 10 ps with a time step of 1 fs. 1 k-point was used during the molecular dynamics while 2 * 2 * 2 k-points were used for the defect calculations. The band gap is obtained at 3 eV free of defect against 3.73 eV in the crystalline structure.

Thermodynamic of defects

We studied HfO_2's point defects in contact with a $CuTe_x$-based TE. As we consider exchanges between HfO_2 and Cu_2Te_3Ge, calculations are based on the following equation for an interstitial of copper charged $+q$ (Cu^{+q_i}):

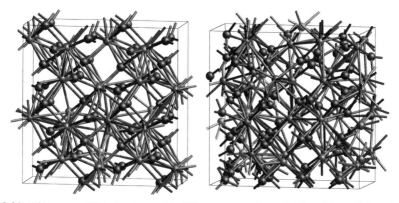

Fig. 9.20 HfO_2 and a-HfO_2 structure with different copper interstitial positions. O in *red*, Hf in *blue*, Cu in *orange*. Cu_{i1} and Cu_{i2} are used for migration barrier calculation.

$$\Delta H = U_{\text{HfO}_2 : \text{Cu}^{+qi}} - U_{\text{HfO}_2} + U_{\text{CuTe}_2\text{Ge}_3 : \text{VCu}} - U_{\text{CuTe}_2\text{Ge}_3} + q \cdot \left(E_{\text{Fermi}} + \Delta V + E_{\text{VB}} \right)$$

With ΔH the formation enthalpy of the defect, U_{Defect} the total energy of the defect, U_{Pristine} the total energy of the initial state, n_i and μ_i the number of chemical formula and the reference chemical potential for the exchange, q the charges exchanged in the system, E_{Fermi} the Fermi level of the electrode in considered contact, ΔV the electrostatic energy shift [17], and E_{VB} the highest valence band energy level. Results are presented in Figs. 9.21 and 9.22 with the lowest charged defects. Considering the charge of the species and a TE in contact based on Cu (vertical orange line), the most favorable interstitials and vacancies we could obtain are, respectively, copper interstitial (Cu_i^{+1}), and oxygen Vacancy (V_O^{+2}).

Tellurium interstitial and hafnium vacancies are not considered here because of their high formation enthalpy energy. As explained in the previous chapter, Ge is used to stabilize CuTe layer and keep it amorphous during integration process, to prevent any segregation and phase separation. Typically, 10%–20% Ge is used [21, 22] in CuTe alloys. As a consequence, Ge is not expected to have any impact in the electrical

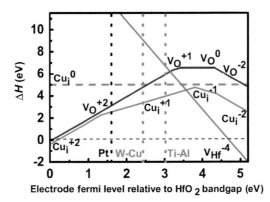

Fig. 9.21 Formation enthalpy calculation of defects in $\text{HfO}_2/\text{CuTe}_x\text{Ge}_y$ system. Fermi level of three types of metal (Pt, W-Cu, Ti-Al) is represented *(dashed vertical lines)*.

Defect	$\Delta H \text{HfO}_2$ (eV)	$\Delta H \text{V}_O$-rich (eV)
Cu_i^{+1}	3.4	2.6
V_O^{+2}	4.8	x
$\text{V}_{\text{Hf}}^{-4}$	9.1	x
O_{FP}	4.5	x
Te_i^{+2}	9.1	x
Ge_i^{+2}	5.6	x

Fig. 9.22 Formation enthalpy ΔH (eV) between the TE and HfO_2 either stoichiometric or including oxygen vacancies.

behavior of CuTe. Moreover, CuTe bound is polar and can easily be broken in $Cu^{\delta+}$ and $Te^{\gamma-}$ ions. They will be mobile under the electric field. Ge calculations are just used to confirm this affirmation. We thus focus on Cu ions in the following sections.

Based on these two candidates, we studied the case of copper introduction in defective hafnia. The reference for the oxide becomes then a V_O-rich HfO_2 (HfO_2 with one V_O defect) instead of stoichiometric HfO_2. V_O in HfO_2 can come from

- HfO_2 process condition in O-poor atmosphere
- O exchange with the TE during forming process requiring 4.8 eV

The results are summarized in Fig. 9.22 and based on the following equation:

$$\Delta H = U_{HfO_2:V_O+Cu^{+q}} - U_{HfO_2:V_O} + U_{CuTe_2Ge_3:VCu} - U_{CuTe_2Ge_3} + q \cdot (E_{Fermi} + \Delta V + E_{VB})$$

Cu introduction is favorable in interstitial site while if HfO_2 is V_O rich, Cu introduction is the easiest in the interstitial site next to V_O [27].

HfO_2 defect diffusion

The most favorable defects that can be inserted in HfO_2 were extracted. However, thermodynamic approach only considers initial and final energy states of the defects. Therefore, this section focuses on Cu diffusion to quantify the energetic barrier height to move it between two equivalent sites. Energy migration barriers are calculated using NEB-based approaches. The presented results correspond to Cu migration from Cu_{i1} site to Cu_{i2} (see Fig. 9.20). The diffusion barrier of Cu_i^+ in HfO_2 structure was found close to 2.5 eV (Fig. 9.23). We also calculated Cu migration barriers in defective hafnia structures with a V_O next to the Cu migration path. In particular, the diffusion barriers of Cu_i^+ in HfO_2-V_O is evaluated at 1.28 eV. Cu can pass through a Cu_O state. To complete this result, calculations were performed on the amorphous model. The diffusion barrier of Cu_i^+ in a-HfO_2 is found close to 3 and 2.5 eV with Cu_i^+ next to V_O [27].

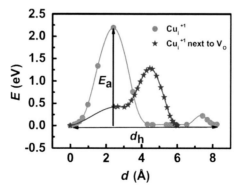

Fig. 9.23 First-principle calculations to compute the barrier height to diffuse from Cu_{i1} to Cu_{i2} sites (see Fig. 9.20): Cu_i^{+1} diffusion barrier in stoichiometric or defective HfO_2 systems.

9.3.9.3 Summary

This study is done for three different materials: Al_2O_3, HfO_2, and Gd_2O_3. Fig. 9.24 summarizes the main results. If thermodynamic and kinetic energies are taken into account, Al_2O_3 and Gd_2O_3 show $(Cu_iV_O)^{+1}$ defect as the most favorable while HfO_2 shows favorable energies for $(Cu_iV_O)^{+1}$ and $V^{+2}{}_O$.

9.3.3 Tradeoff and correlation existing among memory features

9.3.9.1 Time voltage dilemma

Fig. 9.25 shows the forming time as a function of voltage applied on memory anode, on CBRAM with Al_2O_3 resistive layers (3.5 and 5 nm thick), with $CuTe_x$ ion source layer. Time voltage dependence is illustrated: increasing the applied voltage leads to an exponential decrease of the switching time. It is thus possible to adjust the switching time playing with programming voltage. This exponential dependence was put in evidence on all oxide-based CBRAM with various resistive layers, top and BEs [28], and same behavior was also found on OXRAM devices [29]. Moreover, similar laws are measured for both forming, SET and RESET operations [3]. Ideally, an abrupt characteristic is beneficial. Indeed, high slope results in a good tradeoff in terms of reading disturb immunity (with very high switching time at low reading voltage) and

Defect	ΔH_{HfO_2} (eV)	$E_{a\,HfO_2}$ (eV)	$\Delta H_{Al_2O_3}$ (eV)	$E_{a\,Al_2O_3}$ (eV)	ΔH_{GdO_x} (eV)	$E_{a\,GdO_x}$ (eV)
Cu_i^{+1}	3.4	2.5	2.5	4	3.5	2
V_O^x	2.5 (+2)	0.7 (+2)	3.8 (0)	3.5 (0)	4.1 (0)	x
$(Cu_iV_O)^{+1}$	2.6	1.3	0.6	1.4	1.2	3.44
Filament	Hybrid: Cu_i and V_O moving		Cu_i next to V_O		Cu_i next to V_O	

Fig. 9.24 Migration energy barrier, E_a, and formation enthalpy, ΔH, between the TE and different oxides either stoichiometric or including oxygen vacancies (Cu_iV_O defect).

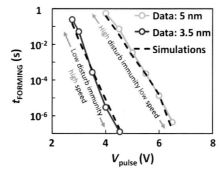

Fig. 9.25 Time voltage dilemma illustrated on CBRAM with Al_2O_3 resistive layers (3.5 and 5 nm thick), with $CuTe_x$ ion source layer. Simulations come from Kinetic Monte Carlo calculations from Ref. [28].

fast programming behavior (short programming time at high programming voltage). Playing with the RRAM stack it is possible to adjust the dependence: tuning top and BEs (thus electrodes work functions) leads to a shift of the characteristics, while playing with resistive layer thickness and dielectric constant allows to change the slope of the curve, improving the memory tradeoff between programming speed and immunity to reading disturb.

9.3.9.2   Retention consumption tradeoff

Programming current strongly affects RRAM stability. Indeed, when a low programming current is used, conductive filament is thinner. During bake, Cu atoms composing the filament diffuse in the resistive layer, gradually disrupting the conductive path.

When $R_{ON} < 10^5 \Omega$, the filament is sufficiently thick to be stable over time. For higher R_{ON}, the percolation becomes dominant and the filament is more sensitive to the Cu diffusion in the electrolyte during bake, leading to low-resistance state (LRS) instability. Concerning OFF state, a higher R_{OFF} leads to a more stable high-resistance state (HRS). The height of the remaining OFF state filament in the electrolyte is reduced as the RESET conditions are strengthened. Thus, this leads to a larger gap between the remaining filament and the Cu-based TE, which prevents the reconstruction of a conductive path during bake. Moreover, it was demonstrated that for a given initial R_{OFF}, the HRS stability is degraded when the memory has been previously SET to a low R_{ON} [30]. This can be explained as follows: a lower R_{ON} leads to a thicker remaining filament, yielding to more Cu atoms in the filament able to reconnect to the TE. ON and OFF states retention characteristics were measured at various temperatures (130–300°C) with various I_{SET} (Fig. 9.26). The failure time criterion was fixed to a critical variation R_{OFF}/R_{ON} of a factor 2 (Arrhenius graphs in Fig. 9.26). Using a low programming current, LRS is controlling the lifetime while for high I_{SET} HRS plays the critical role. Moreover, we can notice that the maximum operating temperature for a 10-year retention is drastically increased for higher SET currents (250°C for $I_{SET} = 80 \mu A$).

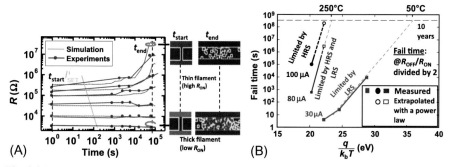

Fig. 9.26 (A) ON state retention characteristics at 130°C of Cu-based CBRAM [30] for various programming currents from 20 to 200 µA. High I_{SET} corresponds to better stability [28]. (B) CBRAM failure time as a function of baking temperature (Arrhenius graphs) for various programming currents.

Fig. 9.27 summarizes the impact of SET and RESET conditions on LRS and HRS retention behavior.

9.3.9.3   Endurance/speed tradeoff

Endurance characteristics of $HfO_2/CuTe_x$ CBRAM were measured for various programming pulse widths (see Fig. 9.28). As programming time is reduced, WM gets lower. However, 10^8 cycles endurance can be achieved with programming SET and RESET times down to 50 ns, with clear separation between high-resistive state (HRS) and low-resistive state (LRS).

Consequently, high endurance and fast speed can be obtained with specific RRAM stacks, promising for main memory applications.

9.3.9.4   WM/retention/endurance tradeoff

Following the understanding of filament properties, this section focuses on material stack comparison to highlight a link between RRAM performances and microscopic

	SET and RESET conditions	Filament morphology	Retention performances
LRS	Strong SET (low R_{ON})	Thick filament	Stable LRS
HRS	Strong SET (low R_{ON})	Thick filament basis	Unstable HRS
	Strong RESET (high R_{OFF})	Large gap	Stable HRS

Fig. 9.27 Impact of SET and RESET conditions on LRS and HRS retention behavior.

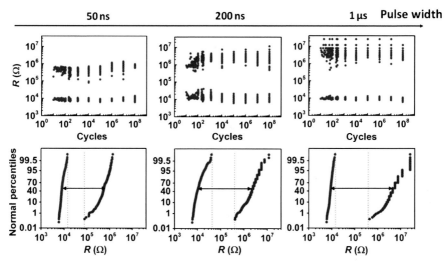

Fig. 9.28 Endurance characteristics of $HfO_2/CuTe_x$ CBRAM devices, for various programming pulse widths (50 ns to 1 µs). Resistance distributions for low- and high-resistive states are also reported for each condition [1].

features. We first study three memory characteristics on different RRAM technologies such as endurance, WM and retention, and expose their interdependence. Focusing on one technology, programming conditions such as current, voltage, and time need to be optimized and a trade-off between endurance and WM is established. Comparison between memory stacks is then realized, highlighting a correlation between endurance plus WM improvement and retention degradation. Studying this last feature from a material point of view, we compare different oxides based on the density functional theory (DFT) results presented before, providing insights on conductive filament composition in different stacks. Combining previous experiments and simulations, a link between memory characteristics and material microscopic parameters is established, through the ion-energy migration barrier. A correlation between RRAM performances, material properties and electrical parameters is exposed and is used to choose the suitable material for a defined application using RRAM technology.

As an RRAM industrial benchmark is not yet well defined, literature on their performances is collected. Recently, good characteristics were reported, making RRAM potential candidates to replace flash for both stand alone, storage class applications [31] and embedded products [32, 33]. Endurance over 10^8 cycles is targeted for storage class compare to 10^5–10^6 cycles for stand alone applications [34] while for automotive applications temperature stability over 175°C is required [35]. For example, in C.Y. Chen et al. [36] endurance up to 10^{12} cycles was reported with TiN/GdO/Hf/TiN stack with a retention below 150°C, promising for storage class memory applications. Excellent stability was demonstrated in M. Barci et al. [30] where a retention up to 350°C with an endurance up to 10^3 cycles was shown. However, it is very challenging to combine all features such as having good cycling, stable retention at high temperature as well as high enough WM depending on the targeted density. Indeed, intrinsic RRAM variability is a challenge to avoid WM closure [37]. In S. Sills et al. [31] a 10 MB array integration is demonstrated with a median memory window of three orders of read current which leads to an overlap at the −4.6¾ level (>99.9% of the devices). Fig. 9.29 shows memory performances data based on experimental results

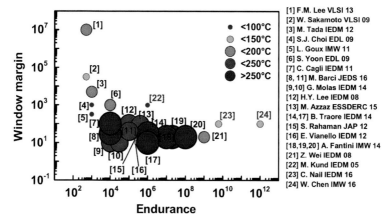

Fig. 9.29 Window margin (mean value) as a function of RRAM maximum achieved cycles reported in the literature depending on maximum temperature stability for retention after 24 h.

reported in the literature. It illustrates a general trend where high WM corresponds to low endurance.

Meanwhile, there are some outliers showing more promising characteristics combining high WM and endurance (blue references). However, these technologies exhibit lower temperature stability for 24 h retention. In order to gather a maximum of data, temperature stability criterion is taken as the maximum demonstrated temperature with no resistance variation during 24 h. Following this observation, the interrelationship between these three memory features requires more investigation.

In order to study the trade-off between WM and endurance, we extract the maximum number of cycles a cell can reach for different RESET voltages (V_{RESET}), giving different WM, setting time pulse at 1 µs. The maximum number of cycles (N_{cmax}) is achieved when RESET operation becomes impossible, meaning oxide layer was led to an irreversible breakdown, leading to WM closure. Fig. 9.30B and C show V_{RESET} influence on R_{OFF} and endurance, respectively. Fig. 9.30A shows typical curves to extract these data. R_{OFF} increases with V_{RESET} until a threshold value is reached where WM is then degraded. WM degradation after 2 V can be explained by a negative SET following the RESET due to defect injection from the BE [38]. On the other hand, the maximum number of cycles decreases as V_{RESET} increases. Results reveal a trade-off between WM and endurance: staying out of an early degradation with V_{RESET} below a threshold value, improving WM leads to a degraded endurance.

In order to compare different materials, this study is extended to four RRAM technologies. We chose to fix the programming current to 100 µA, which is close to the

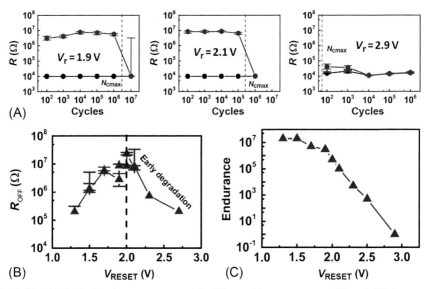

Fig. 9.30 (A) Typical endurance curves with different V_{RESET} to extract (B) and (C). Impact of V_{RESET} on TiN/Al$_2$O$_3$/CuTe$_x$ sample on (B) R_{OFF}, error bars corresponds to standard deviation on 10 cells and (C) endurance (maximum number of achieved cycles).

optimum programming current insuring large WM and high number of cycles. Taking the RESET parameter giving the best endurance for each technology, we compared their endurance (N_{cmax}) and their best retention (highest stability temperature T_{max} for 24-h baking time). Results are shown in Fig. 9.31 for three of them.

W/HfO$_2$/CuTe$_x$ shows up to $6 \cdot 10^9$ cycles with a constant WM while its maximum temperature stability is around 100°C. TiN/Al$_2$O$_3$/CuTe$_x$ shows an endurance up to 10^7 cycles and a temperature stability up to 200°C while TiN/GdO$_x$/CuTe$_x$ shows an endurance up to $3 \cdot 10^3$ cycles and a stable retention up to 300°C. These three RRAM are a good example to illustrate the trade-off between endurance and retention: good cycling goes with poor retention (W/HfO$_2$/CuTe$_x$ technology), while stable retention is reached at the cost of limited endurance (TiN/GdO$_x$/CuTe$_x$ technology). Fig. 9.32A represents reached endurance for each RRAM technology for different RESET Voltage conditions, implying different WM. Focusing on one material, WM can be increased at the expense of degraded endurance. Having compared different materials (for a given WM), it can be concluded that endurance is improved as retention is degraded (Fig. 9.32B). Through these results, trade-off between endurance, WM and retention is revealed. It should be noted that similar behavior was observed and put in evidence in OXRAM [27].

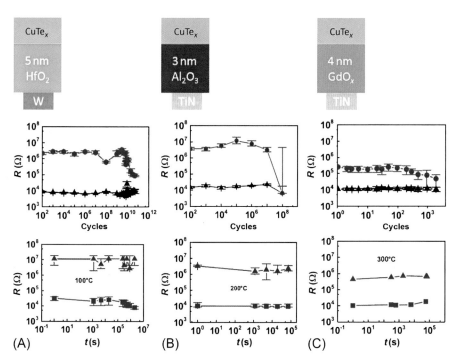

Fig. 9.31 Endurance and retention performances for (A) W/HfO$_2$/CuTe$_x$, (B) TiN/Al$_2$O$_3$/CuTe$_x$, and (C) TiN/GdO$_x$/CuTe$_x$ memory technologies.

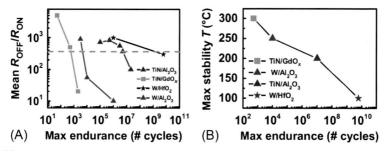

Fig. 9.32 (A) Window margin as a function of maximum endurance. For each RRAM technology, various WM were achieved changing the RESET conditions. *Gray line* is used to extract (B) for a constant WM. (B) Maximum stable retention temperature after 24-h baking time as a function of maximum endurance for a given WM ($R_{OFF}/R_{ON} \sim 400$).

Combining electrical characterization and DFT results, we can put in evidence a clear correlation between memory characteristics and the ion migration barrier. Fig. 9.33 represents memory endurance and retention temperature as a function of calculated $E_{migration}$.

The number of cycles reachable by a cell decreases with $E_{migration}$ (Fig. 9.33A) while temperature stability increases (Fig. 9.33B). This phenomenon can be understood with a simple description of ion migration in the resistive layer. Fig. 9.33 illustrates how endurance is facilitated by a low migration barrier; a low energy will lead to an easy formation and disruption of the filament. A high migration energy barrier will lead to stable retention; making the filament more difficult to be dissolved. In HfO_2, as Cu_iV_O and V_O calculated migration energies are of the same order of magnitude (see Fig. 9.24), we considered a hybrid filament resulting from both Cu and V_O diffusion.

Thus, $E_{migration}$ is calculated as the average value between Cu_iV_O and V_O in Fig. 9.33. In conclusions, it clearly appears that the ion migration energy is linked to the trade-off between memory endurance and retention.

To discuss in more details the previous observation, we propose the following description: a RRAM device can be characterized by a maximum energy, E_{max}, that can be sustained before oxide degradation and breakdown. This maximum energy is reached when a maximum number of cycles, N_{cmax}, is achieved. RRAM is then led to irreversible breakdown. For each cycle, provided energy, E_{cycle}, depends on the programming conditions. The maximum energy can thus be expressed with N_{cmax} and E_{cycle}. A fraction of E_{cycle} is transferred to ions and serves to switch the RRAM (E_{switch}) while another is transferred to the network and leads to material degradation such as defect generation. In our case, endurance failure is detected when RRAM cannot be erased anymore. The cell is then stuck in R_{ON} state meaning there are too much defects in the oxide which leads to a permanent conductive path. Creation of defects occurs during cycling and can be accelerated by the increase of V_{RESET} [38]. As it was previously demonstrated, increasing R_{OFF}/R_{ON} ratio leads to a drop in the number of cycles a cell can sustain. Thus, the maximum energy is reached faster if WM increases. In our description, we link the WM to the number of atoms involved during filament formation and disruption ($N_{moved\ atoms}$). Increasing SET programming condition leads to bigger conductive path, then composed of more atoms and resulting in lower R_{ON} and

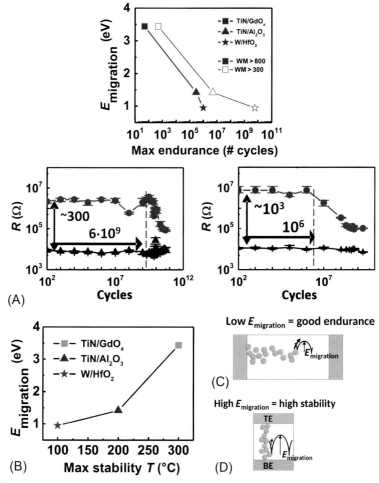

Fig. 9.33 $E_{migration}$ impact study on (A) endurance with typical curves to extract endurance at different WM and (B) retention. Schema of RRAM mechanism facilitating (C) endurance and (D) retention.

higher WM. However, increasing R_{OFF} means a bigger free defect area in resistive layer meaning that increasing RESET condition (time and voltage) is equivalent to increase the number of moving atoms during RESET. On the other hand, we also demonstrated that ion migration barrier impacts memory endurance. If $E_{migration}$ increases, the number of achieved cycle decreases. In our description, we thus assume that $E_{migration}$ is the required energy to move and drift each ion in the oxide, during filament formation and disruption. So, switching energy can be expressed as $E_{switch} = E_{migration} * N_{moved\ atoms}$. The maximum energy of the dielectric can thus be expressed with the maximum number of cycles, the WM (through $N_{moved\ atoms}$) and $E_{migration}$. In S. Balatti et al. [38] a similar description of this phenomenon is given. Degradation of the oxide is controlled by activation energy and V_{RESET}. Irreversible endurance failure is associated to a high defect injection in conductive filament. As explained previously, the switching energy (transferred to filament ions) only represents a fraction of the total supplied energy. Energy surplus is transferred to the network and degrades the oxide. We can express the total provided energy with E_{switch} and a degradation parameter α with $E_{switch} = \alpha * E_{cycle}$, $0 < \alpha < 1$. α evolves with oxide degradation: the more the RRAM is cycled, the more the oxide is degraded, leading to a higher energy amount transferred to the network. We thus assume $\alpha = 1$ for a fresh cell (without defect) and gradually decreases with cycling. Finally, we can express the maximum energy that the dielectric can sustain depending on $N_{cyclemax}$, E_{switch}, and α:

$$E_{max} \sim N_{cyclemax} * E_{cycle}$$

$$E_{cycle} \sim E_{switch} / \alpha \sim E_{migration} * N_{moved\ atoms} / \alpha$$

$$E_{max} \sim N_{cmax} * E_{migration} * N_{moved\ atoms}$$

$$E_{max} = f\left(\text{Endurance, Retention, WM}\right)$$

where E_{max} the maximum energy a dielectric can sustain before breakdown, $N_{cyclemax}$ the maximum number of achieved cycles, E_{cycle} the total energy supplied per cycle, E_{switch} the energy needed to move the atoms to create the filament, α the degradation parameter such as $E_{switch} = \alpha * E_{cycle}$ and $0 < \alpha < 1$, $E_{migration}$ the activation energy to move one atom in the oxide and $N_{moved\ atoms}$ the number of involved atoms in the filament formation and disruption (Fig. 9.34).

$E_{migration}$ can be associated to the activation energy experimentally extracted from retention measurements. To this aim, activation energy is extracted from failure time evolution with baking temperature assuming an Arrhenius law. Fig. 9.35 shows correlations between simulated (migration energy) and experimental (activation energy) $E_{migration}$ and memory endurance.

Very good agreement is observed between simulated and experimental $E_{migration}$. Temperature stability failure can then be well associated with movement of Cu and V_O atoms which will disrupt or form again the filament [22]. R_{OFF} failure, on Fig. 9.35, reveals a filament formation. In conclusion, the maximum dielectric energy is linked to the memory performances and microscopic features of the memory technology: WM (through the number of atoms involved in the filament formation and disruption), memory endurance (maximum number of cycles), and memory retention (through ion migration barrier). Consequently, stronger programming conditions (current, time or voltage) enlarge the WM, decreasing N_{cmax} (and endurance) for a given memory technology (fixed E_{max}), as observed in Fig. 9.32A. Moreover, changing memory stack, it makes possible to tune the migration energy and favor either endurance or retention (Figs. 9.32 and 9.33), for a fixed WM.

Following observations must also be made:

- Dielectric optimization is of great importance; improving the material quality can lead to E_{max} increase and general improvement of the WM/endurance/retention trade-off. In the literature, J. Woo et al. demonstrated that breakdown strength of a material can be improved by thermal treatment [39].

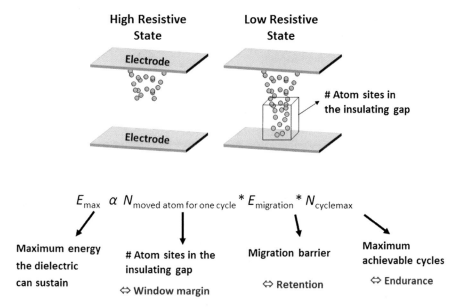

Fig. 9.34 Summary of the correlation between endurance, retention, and window margin in RRAM. Maximum energy the device can sustain during endurance before dielectric breakdown depends on the dielectric properties (migration barrier) and program conditions (window margin).

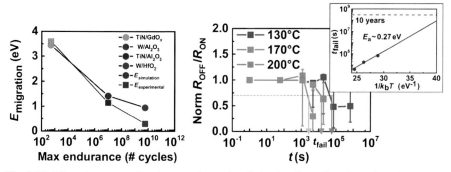

Fig. 9.35 Migration energy barrier experimental and simulated as a function of maximum endurance. Extraction example of $E_{experimental}$ for W/HfO$_2$. Experimental energy is calculated as the activation energy of failure time t_{fail}, extracted from Arrhenius graphs, from retention characteristics at various baking temperatures.

- TE/oxide combination needs to be optimized to tune migration barrier. Unfortunately, this study needs to be done for each combination. However, some guidance can be given to avoid high diffusion in oxides such as: controlling oxide grain boundaries, material structure, and material density. It can be added that small metal atoms are an advantage for interstitial diffusion which is usually the most favorable.
- Understanding the defect generation during memory endurance is also critical. For a given WM, different sets of programming parameters (current, time, voltage…) and memory stack

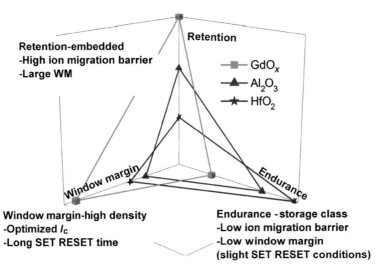

Fig. 9.36 Radar plot of the performance trade-off for the various RRAM classes reported in this work, and critical key parameters (material microscopic properties and operating conditions).

features (top, BEs…) should minimize endurance failure, preventing early oxide degradation (Fig. 9.30). Moreover, smart programming procedures is a way to adjust the energy provided to the system, reducing the energy surplus transferred to the network and degrading the memory dielectric.

The link between memory technology microscopic features and memory performances can be used to design an optimized memory stack. However, operating conditions need to be adjusted to reach the specifications of a given memory application. Fig. 9.36 summarizes optimization for three applications. Low migration barrier and optimized WM favor endurance for storage class memory (proposed stack: W/HfO$_2$/CuTe$_x$). Optimized SET current and long-time programming conditions enlarge WM for data storage (proposed stack: TiN/Al$_2$O$_3$/CuTe$_x$). High migration barrier and large WM are favorable for embedded applications requiring high stability (proposed stack: TiN/GdO$_x$/CuTe$_x$).

9.3.4 Variability of resistive memories

Variability understanding is still the major technical roadblock for RRAM widespread adoption. The fundamental variability limits of filamentary-based RRAM were identified, showing that the main limiting components in 1T-1R architectures are: the select transistor when the conductive filament (CF) is formed or set to a LRS, and the MIM stack properties when CF is brought to a HRS [40, 41]. Variability results from the stochastic character of filament formation and disruption, leading to distribution of RRAM resistance with a log-normal law. Both HRS and LRS resistances are distributed. A general correlation between resistance standard deviation and median value was demonstrated (Fig. 9.37). This dependence is confirmed on various OXRAM and CBRAM stacks, with various integrated materials and thicknesses. In other words, high RRAM resistance is associated to high variability and poor control of the resistance value [42].

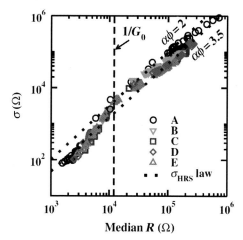

Fig. 9.37 Resistance standard deviation dependence with resistance median value after forming for TiN/HfO$_2$ 10 nm/Ti 10 nm OXRAM.
From Alessandro Grossi, Cristian Zambelli, Piero Olivo, Etienne Nowak, Gabriel Molas, Jean Francois Nodin, Luca Perniola, Cell-to-cell fundamental variability limits, IEEE Electron Device Lett. 39 (1) (2018), 27–30.

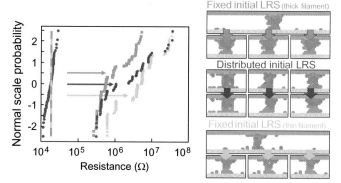

Fig. 9.38 Simulated HRS-resistance distributions starting from two initial HRS (*green*: same R_{LRS} but different filament shapes) or spread R_{LRS} distributions (*purple*) [43].

In the case of low atoms count in the filament or residual filament, slight atom movements can lead to a strong modification of the filament morphology and thus to a high variation of current, temperature, and energy. This low and very much dispersed energy along with low programming current I_C induces high ON state resistance variability. Finally, a thin filament (low I_C) is more sensitive to Cu (or V$_O$, or Ag…) positions, which is also increasing the R_{LRS} spread.

R_{HRS} distributions were measured and simulated as seen in Fig. 9.38, using the KMC model described in Ref. [43]. RESET simulations were performed, starting either from LRS, assuming a distributed resistance to study cycle to cycle variability, or from two unique initial filaments to study intrinsic variability. In opposition to LRS, it appears that HRS dispersion has two contributions: one depending on the initial filament morphology and the other on the stochastic RESET operation itself. HRS distribution obtained starting from fixed LRS demonstrates the intrinsic variability character of RESET operation (a unique filament can lead to various HRS).

Filament dissolution during RESET is driven by electric field and temperature. Depending on the starting disruption point of the filament, the morphology of the residual filament can be greatly modified as temperature is strongly reduced. This reduction happens once current flow and Joule heating are cut. Stochastic ion migration also increases the intrinsic variability of RESET operation. Moreover, depending on the initial morphology of the LRS filament, the mean value of HRS distribution is modified. Considering a given initial R_{LRS}, a thick filament leads to a lower R_{HRS} than a thinner one due to more Cu atoms to remove. This strong cycle to cycle HRS variability is not directly tied to R_{LRS} but to the filament shape. However, by increasing I_C and thus decreasing R_{LRS}, the conductive filament globally gets thicker and R_{HRS} decreases, leading to a dependence of R_{HRS} with R_{LRS}.

9.3.5 Summary—Oxide CBRAM vs OXRAM

In summary, both OXRAM and oxide CBRAM integrate an oxide layer (HfO_2, Ta_2O_5, $Al_2O_3...$) as switching material. While in OXRAM the conductive filament is made of oxygen vacancies, filament formation in oxide CBRAM relies on the diffusion of a metal ion coming from an active electrode (generally made of Cu).

In an integration point of view, oxide CBRAM can require some additional developments with respect to standard CMOS process. Cu integration can lead to some integration and contamination precautions. Moreover, etching Cu-based alloys and may not be possible with standard reactive ion etching, and may require ion-beam etching.

Oxide-based CBRAM and OXRAM show many similarities regarding the electrical behavior. They require a forming operation; forming voltage being in general higher than subsequent SET voltage. They also have similar programming voltages and times; sub-10-ns programming speed was demonstrated for both technologies. Moreover, they both show an exponential dependence between switching time and programming voltage [28, 29]. In the case of chalcogenide CBRAM, at least two slopes are measured in the time voltage characteristics, due to the competition between ion generation (oxidation and reduction) and diffusion [44]. Concerning variability, oxide-based CBRAM and OXRAM show resistance intrinsic variability due to the stochastic character of filament formation and dissolution. Variability increases with resistance median value, and is of the same order of magnitude for OXRAM and oxide CBRAM. Finally, these two technologies show good scalability; demonstrations of small active areas down to 10 nm were presented in both cases.

Depending on the memory stack, they show various electrical properties and device characteristics. In general, looking at data reported in the literature, OXRAM exhibit better endurance and retention but smaller WM than CBRAM [1]. However, both OXRAM or CBRAM can exhibit excellent endurance [27] or very stable retention [45] depending on the integrated materials. Indeed, for both of them, we could correlate memory characteristics and performances tradeoff with ion migration energy. Thus, depending on the TE, filament composition (V_O, Cu and V_O, Cu...) and resistive material

(HfO_2, Ta_2O_5, Al_2O_3...), the memory stack can be designed to favor one memory feature or the other, to target a specific memory application. The second-order optimization can then be achieved playing with technological adjustments of OXRAM and CBRAM device, such as doping and interfaces [46, 47]. In that sense, oxide CBRAM can be considered along with OXRAM as a flavor of RRAM, giving the opportunity to play with memory performances tradeoff depending on the integrated materials.

9.4   Which emerging memory for which new system?

In the previous sections, we presented the performances of emerging RRAM, and we highlighted the correlation existing between the various memory features. According to this, it is very challenging to envisage a universal memory fulfilling all the requirements of a computing system. In a more conservative approach, a specific emerging RRAM with adapted performances could replace a specific computing segment.

In the first section of this chapter, we proposed various computing system architectures, implementing an emerging memory. Configurations are summarized in Fig. 9.39.

In this section, we investigate how emerging RRAM can enter computing system hierarchy, analyzing the best possible configurations.

To this aim, we discuss latency, endurance, retention, cost, and consumption.

9.4.1   Storage memory

Two options can be envisaged for RRAM as storage memory: RRAM can fill the gap between Disk/SSD and DRAM main memory thanks to its lower latency time than flash and higher density than DRAM, or replace SSD memory to insure storage functions. These two potions are summarized in Fig. 9.40.

Required latency time of 1–5 μs is in line with RRAM performances, which allow <1 μs programming and reading times with good reliability. 10^6 cycles endurance is also possible with RRAM as illustrated in previous paragraphs of this chapter. 10 years retention at 85°C could also be envisaged as stable retention behavior was demonstrated with several optimized RRAM technologies.

The first challenge for an emerging memory to replace a storage memory is to be competitive in terms of cost per bit. Indeed, NAND flash memory in 3D architecture

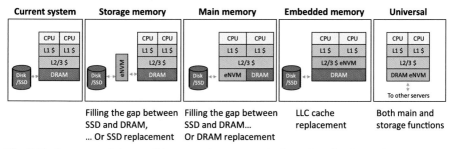

Fig. 9.39 Summary of the possible computing system configurations implementing an emerging memory (eNVM) in various segments.

Storage	Latency time	Endurance	Retention	Cost/b	Consumption
Filling the gap	$1-5\,\mu s$	$> 10^6$	10 years	< 2 × flash cost	< flash
RRAM (OXRAM)	Ok	Ok	Ok >85°C	Vertical	10 pJ
Replacement	$1-5\,\mu s$	$> 10^6$	10 years	~flash cost	< flash
RRAM (OXRAM)	Ok	Ok	Ok >85°C	Vertical	10 pJ

Fig. 9.40 Summary of RRAM requirements to be integrated as a storage memory in computing systems.

has a very low cost per bit, below 1$ per GB, even in single-cell configuration [48]. Consequently, the only opportunity for RRAM to achieve low cost per bit will be to integrate in vertical arrays, saving expensive critical lithographic steps [49].

However, vertical RRAM process implies severe integration challenges, among them high aspect ratio multilayer etching, controlled conformal deposition, back-end selector development...

The second challenge concerns RRAM variability as previously described. Indeed, as high densities (>128 GB) are required for storage applications, good control of resistance tails is essential, for both retention and endurance performances. Moreover, low RRAM programming currents (<100 μA) have to be considered to reduce voltage drop issues in the memory array. As previously discussed in the previous section, reducing programming current results in higher variability due to lower number of involved atoms in the conductive filament. Two paths of improvement can be envisaged. First, variability can be directly addressed, investigating new RRAM technologies (stacks, materials, tuning the filament composition) to reduce intrinsic resistance dispersion. Second approach is to take advantage of the fast programming time with respect to storage memory bitrate requirements. Consequently, smart programming schemes could be used to improve resistance distribution around the targeted value, repeating several programming programming/reading pulses. For instance, several 100-ns programming pulses with verification readings could be performed in 1–5 μs leading to improved control of RRAM states. In Ref. [50], smart procedure allowed us to keep a WM even at 4σ on RRAM programmed at 100 μA, repeating 50 ns pulses in 1 μs total time, what is consistent with storage class memory programming time specifications. In addition, as discussed in Section 9.2.2, an efficient error correction scheme will be required.

Finally, last challenge concerns programming consumption. It can be estimated at the first order by $V_{prog} * t_{prog} * I_{prog}$, leading to ~10 pJ assuming a programming current of 100 μA and programming time of 100 ns for a programming voltage of ~1 V. However, we saw in the previous paragraph that longer pulses or smart algorithms may have to be used to improve resistance distributions. In case several successive pulses are used to program the RRAM, consumption can significantly increase due to intrinsic variability of RRAM switching time. Optimized patterns can be envisaged in this case to limit extra consumption in smart procedures. In particular, ramp voltage programming was demonstrated to be much more energy saving than constant voltage programming, leading in principle to a programming energy of 10 pJ for 4σ distributions with in-die smart algorithm [50].

In conclusions, RRAM performances should be in line with storage memory specifications. Main performance challenge will be related to variability that becomes critical at high densities. On the other hand, reaching lower cost than flash with vertical RRAM appears today hardly achievable [48]. Thus, filling the gap option should be more realistic rather than SSD replacement, at least in first step.

9.4.2 Main memory

Again, RRAM is envisaged to fill the gap between Disk/SSD and DRAM main memory, or replace DRAM memory to insure main memory functions. These two options are summarized in Fig. 9.41:

With respect to storage requirements, working specifications are more stringent in terms of speed and endurance, and less aggressive in terms of retention and density (cost).

Two hundred nanoseconds latency time is consistent with RRAM speed. A programming time lower than 50 ns was also demonstrated but in this latter case, variability will start to become an issue. Indeed, multiple pulses with program verify will hardly be achieved in 50 ns.

Retention specifications will not be the main issue. Moreover, as only a few days would be required, $<100\,\mu A$ programming current could be envisaged, what would be beneficial for voltage drop issues and consumption considerations.

Relaxing retention will lead to endurance improvement (Fig. 9.32). 10^9 cycles can thus be achieved with RRAM assuming that stability is not key.

Concerning costs, RRAM will have to be competitive with DRAM. In order to achieve typical main memory densities (several GB), crossbar configurations can be adopted [51]. Crossbar integration is much less critical in terms of integration process, and several layers could be stacked to increase the density with reasonable costs.

In conclusions, for main memory applications, the main challenge will be to insure both fast speed and long endurance with high reliability.

9.5 Conclusions

Potential of RRAM, and in particular oxide-based CBRAM was discussed to enter the memory hierarchy in computing systems. Thanks to their fast speed, good endurance, and retention, RRAM appear as good candidates for both data storage and main memory applications.

Physical insights on filament composition were proposed. The strong role of oxygen vacancies was stressed on the diffusion of Cu in the resistive layer. Electrical performances of various RRAM technologies were presented. In particular, correlation

Storage	Latency time	Endurance	Retention	Cost/b	Consumption
Filling the gap	< 200 ns	> 10^9	> 5 days	< ½ DRAM cost	< DRAM
RRAM [1]	Ok	Variability	Ok	Ok?	10 pJ
Replacement	< 50 ns	> 10^9 *	> 5 days	< ½ DRAM cost	< DRAM
RRAM [1]	Variability	Variability	> 5 days	Ok?	>1 pJ

Fig. 9.41 Summary of RRAM requirements to be integrated as a main memory in computing systems.

between WM, endurance and retention was demonstrated, proving that technology and operating conditions can be adjusted in order to reach the requirements for a given application. Variability was discussed as one of the main roadblock for RRAM integration, especially for high densities. Optimized programming schemes and ECC were proposed as potential solutions.

Finally, based on these theoretical and experimental inputs, RRAM potential for computing system applications was discussed. RRAM offer very good opportunities for both main and data storage memories. In both cases, two options are possible: one module replacement (keeping the current system configuration) or system modification, if emerging RRAM is able to combine short latency, good endurance, and low cost/high density. Combination of both good endurance and large WM (required for high densities) will be challenging, or only at the cost of degraded retention. Moreover, it will be hard to reduce programming current (~10 µA) and programming time (below 10 ns), keeping good endurance and WM (for main memory applications). Variability issue may lead to correction algorithm requirement to insure readability for high densities, or redundancy for DRAM-type memories. Vertical integration will be required to insure high densities (storage) at low cost. Finally, next step challenge will be to develop high-quality back-end selector (with excellent endurance, high ON/OFF ratio and low leakage current) to enable crossbar and vertical integrations.

References

[1] C. Nail, G. Molas, P. Blaise, G. Piccolboni, B. Sklenard, C. Cagli, M. Bernard, A. Roule, M. Azzaz, E. Vianello, C. Carabasse, R. Berthier, D. Cooper, C. Pelissier, T. Magis, G. Ghibaudo, C. Vallée, D. Bedau, O. Mosendz, B. De Salvo, L. Perniola, Understanding RRAM endurance, retention and window margin trade-off using experimental results and simulations, in: IEDM 2016 Tech. Dig, 2016, pp. 95–98.

[2] B. Govoreanu, G.S. Kar, Y.-Y. Chen, V. Paraschiv, S. Kubicek, A. Fantini, I.P. Radu, L. Goux, S. Clima, R. Degraeve, N. Jossart, O. Richard, T. Vandeweyer, K. Seo, P. Hendrickx, G. Pourtois, H. Bender, L. Altimime, D.J. Wouters, J.A. Kittl, M. Jurczak, 10x10nm2 Hf/HfOx crossbar resistive RAM with excellent performance, reliability and low-energy operation, in: IEDM 2011 Tech. Dig., 31.6, 2011, pp. 729–732.

[3] J. Guy, G. Molas, P. Blaise, C. Carabasse, M. Bernard, A. Roule, G. Le Carval, V. Sousa, H. Grampeix, V. Delaye, A. Toffoli, J. Cluzel, P. Brianceau, O. Pollet, V. Balan, S. Barraud, O. Cueto, G. Ghibaudo, F. Clermidy, B. De Salvo, L. Perniola, Experimental and theoretical understanding of forming, SET and RESET operations in conductive bridge RAM (CBRAM) for memory stack optimization, in: IEDM 2014 Tech. Dig, 2014, pp. 152–155.

[4] I.G. Baek, et al., IEDM 2011 Tech. Dig, 2011, pp. 737–740.

[5] Panasonic, https://www.rram-info.com/panasonic-and-umc-co-develop-and-produce-rram-chips-2019.

[6] Adesto, http://www.adestotech.com/products/mavriq/.

[7] Crossbar, https://www.crossbar-inc.com/en/products/t-series/.

[7a] M.-J. Lee, Y. Park, B.-S. Kang, S.-E. Ahn, C. Lee, Kihwan Kim, W. Xianyu, G. Stefanovich, J.-H. Lee, S.-J. Chung, Y.-H. Kim, C.-S. Lee, J.-B. Park, I.-G. Baek, I.-K. Yoo, 2-Stack 1D-1R cross-point structure with oxide diodes as switch elements for high density resistance RAM applications, IEDM 2007, Tech. Dig. pp. 771–774.

[8] N. Mielke, T. Marquart, N. Wu, J. Kessenich, H. Belgal, E. Schares, F. Trivedi, E. Goodness, L.R. Nevill, Bit error rate in NAND Flash memories, in: Proc. of 2008 Int. Rel. Phys. Symp, 2008, pp. 9–19.

[9] S. Schechter, et al., Use ECP, not ECC, for hard failures in resistive memories, in: ISCAS10, 2010.

[10] Milojicic, http://fr.slideshare.net/linaroorg/hkg15the-machine-a-new-kind-of-computer-keynote-by-dejan-milojicic.

[11] J. Condit, E.B. Nightingale, C. Frost, E. Ipek, B. Lee, D. Burger, D. Coetzee, Better I/O through byte-addressable, persistent memory, in: Proc. of the 2009 ACM SIGOPS Symposium on Operating Systems Principles, 2009, pp. 133–146.

[12] J. Ren, J. Zhao, S. Khan, J. Choi, Y. Wu, O. Mutlu, ThyNVM: enabling software-transparent crash consistency in persistent memory systems, in: Proc. of 2015 Annual IEEE/ACM International Symposium on Microarchitecture (MICRO), 2015, pp. 672–685.

[13] X. Wu, J. Li, L. Zhang, E. Speight, R. Rajamony, Y. Xie, Hybrid cache architecture with disparate memory technologies, in: Proc. of 2009 Int. Symp. on Computer Architecture, 2009, pp. 34–45.

[14] K. Bailey, L. Ceze, S.D. Gribble, H.M. Levy, Operating system implications of fast, cheap, non-volatile memory, in: Proc. of 2011 USENIX Conference on Hot Topics in Operating Systems, 2011.

[15] E. Menéndez-Proupin, G. Gutiérrez, Electronic properties of bulk gamma Al2O3, Phys. Rev. B 72 (3) (July 2005) 035116, https://doi.org/10.1103/PhysRevB.72.035116.

[16] L. Masoero, P. Blaise, G. Molas, J.P. Colonna, M. Gély, J.P. Barnes, G. Ghibaudo, B. De Salvo, Defects-induced gap states in hydrogenated γ-alumina used as blocking layer for non-volatile memories, Microelectron. Eng. 88 (7) (July 2011) 1448–1451, https://doi.org/10.1016/j.mee.2011.03.029.

[17] C. Freysoldt, B. Grabowski, T. Hickel, J. Neugebauer, G. Kresse, A. Janotti, C.G. Van de Walle, First-principles calculations for point defects in solids. Rev. Mod. Phys. 86 (1) (March 2014) 253–305, https://doi.org/10.1103/RevModPhys.86.253.

[18] G.E. Delgado, A.J. Mora, M. Pirela, A. Velásquez-Velásquez, M. Villarreal, B.J. Fernández, Structural refinement of the ternary chalcogenide compound Cu2gete3 by X-ray powder diffraction. Phys. Status Solidi A 201 (13) (October 2004) 2900–2904, https://doi.org/10.1002/pssa.200406850.

[19] K. Sankaran, L. Goux, S. Clima, M. Mees, J.A. Kittl, M. Jurczak, L. Altimime, G.-M. Rignanese, G. Pourtois, Modeling of copper diffusion in amorphous aluminum oxide in CBRAM memory stack. ECS Trans. 45 (3) (April 2012) 317–330, https://doi.org/10.1149/1.3700896.

[20] F. De Stefano, V.V. Afanas'ev, M. Houssa, L. Goux, K. Opsomer, M. Jurczak, A. Stesmans, Influence of metal electrode stoichiometry on the electron barrier height at CuxTe1-x/Al2o3 interfaces for CBRAM applications, Microelectron. Eng. 120 (May 2014) 9–12, https://doi.org/10.1016/j.mee.2013.08.016.

[21] W. Devulder, K. Opsomer, M. Jurczak, L. Goux, C. Detavernier, Influence of alloying the copper supply layer on the retention of CBRAM, in: DeepDyve, May 2015.

[22] K. Ohba, S. Yasuda, T. Mizuguchi, K. Aratani, M. Shimuta, A. Kouchiyama, M. Ogasawara, Memory Component, Memory Device, and Method of Operating Memory Device, August 2011. U.S. Classification 365/148, 257/2, 257/4, 257/E45.003; International Classification H01L45/00, G11C11/00; Cooperative Classification G11C11/161, H01L45/145, H01L45/1253, H01L45/1233, H01L45/1633, H01L45/146, H01L45/1266, H01L45/085, H01L27/2472, H01L27/2436, G11C13/0011, G11C11/16; European Classification H01L45/12E4, H01L45/14C2, H01L45/14C, H01L27/24F, H01L45/08M, H01L27/24H2, H01L45/16D6, G11C13/00R5B.

[23] C. Nail, P. Blaise, G. Molas, M. Bernard, A. Roule, A. Toffoli, L. Perniola, C. Vallee, Atomistic mechanisms of copper filament formation and composition in Al2O3-based conductive bridge random access memory, J. Appl. Phys. 122 (2017) 024503.

[24] B. Traoré, P. Blaise, E. Vianello, L. Perniola, B.D. Salvo, Y. Nishi, HfO2-based RRAM: electrode effects, Ti/HfO2 interface, charge injection, and oxygen (O) defects diffusion through experiment and ab initio calculations, IEEE Trans. Electron Devices 63 (1) (January 2016) 360–368, https://doi.org/10.1109/TED.2015.2503145.

[25] L.G. Ferreira, M. Marques, L.K. Teles, Approximation to density functional theory for the calculation of band gaps of semiconductors, Phys. Rev. B 78 (12) (September 2008) 125116, https://doi.org/10.1103/PhysRevB.78.125116.

[26] W. Zheng, K.H. Bowen, J. Li, I. Dąbkowska, M. Gutowski, Electronic structure differences in ZrO2 vs HfO2, J. Phys. Chem. A 109 (50) (December 2005) 11521–11525, https://doi.org/10.1021/jp053593e.

[27] C. Nail, G. Molas, P. Blaise, B. Sklenard, R. Berthier, M. Bernard, L. Perniola, G. Ghibaudo, C. Vallée, A link between CBRAM performances and material microscopic properties based on electrical characterization and atomistic simulations, IEEE Trans. Electron Devices 64 (11) (2017) 4479–4485.

[28] J. Guy, G. Molas, P. Blaise, M. Bernard, A. Roule, G. Le Carval, V. Delaye, A. Toffoli, G. Ghibaudo, F. Clermidy, B. De Salvo, L. Perniola, Investigation of forming, SET and data retention of conductive-bridge random-access memory for stack optimization, IEEE Trans. Electron Devices 62 (11) (2015) 3482–3489.

[29] E. Vianello, O. Thomas, G. Molas, O. Turkyilmaz, N. Jovanović, D. Garbin, G. Palma, M. Alayan, C. Nguyen, J. Coignus, B. Giraud, T. Benoist, M. Reyboz, A. Toffoli, C. Charpin, F. Clermidy, L. Perniola, Resistive memories for ultra-low-power embedded computing design, in: IEDM 2014 Tech. Dig, 2014, pp. 144–147.

[30] M. Barci, J. Guy, G. Molas, E. Vianello, A. Toffoli, J. Cluzel, A. Roule, M. Bernard, C. Sabbione, L. Perniola, B. De Salvo, Impact of SET and RESET conditions on CBRAM high temperature data retention, in: Proc. of IRPS 2014, 2014, pp. 5E.3.1–5E.3.4.

[31] S. Sills, S. Yasuda, J. Strand, A. Calderoni, K. Aratani, A. Johnson, N. Ramaswamy, A copper ReRAM cell for storage class memory applications. in: 2014 Symposium on VLSI Technology (VLSI-Technology): Digest of Technical Papers, June 2014, pp. 1–2, https://doi.org/10.1109/VLSIT.2014.6894368.

[32] J.R. Jameson, P. Blanchard, C. Cheng, J. Dinh, A. Gallo, V. Gopalakrishnan, C. Gopalan, B. Guichet, S. Hsu, D. Kamalanathan, D. Kim, F. Koushan, M. Kwan, K. Law, D. Lewis, Y. Ma, V. McCaffrey, S. Park, S. Puthenthermadam, E. Runnion, J. Sanchez, J. Shields, K. Tsai, A. Tysdal, D. Wang, R. Williams, M. Kozicki, J. Wang, V. Gopinath, S. Hollmer, M. Van Buskirk, Technical digest, in: International Electron Devices Meeting, IEDM, 2013, https://doi.org/10.1109/IEDM.2013.6724721.

[33] Z. Wei, K. Eriguchi, S. Muraoka, K. Katayama, R. Yasuhara, K. Kawai, Y. Ikeda, M. Yoshimura, Y. Hayakawa, K. Shimakawa, T. Mikawa, S. Yoneda, Distribution projecting the reliability for 40 nm ReRAM and beyond based on stochastic differential equation, in: 2015 IEEE International Electron Devices Meeting (IEDM), December 2015, pp. 7.7.1–7.7.4, https://doi.org/10.1109/IEDM.2015.7409650.

[34] ITRS, Executive report, 2015.

[35] R. Strenz, Embedded Flash technologies and their applications: status and outlook. in: 2011 International Electron Devices Meeting, December 2011, pp. 9.4.1–9.4.4, https://doi.org/10.1109/IEDM.2011.6131521.

[36] C.Y. Chen, L. Goux, A. Fantini, A. Redolfi, G. Groeseneken, M. Jurczak, Doped Gd-O based RRAM for embedded application, in: 2016 IEEE 8th International Memory Workshop (IMW), May 2016, pp. 1–4, https://doi.org/10.1109/IMW.2016.7495266.

[37] A. Fantini, L. Goux, R. Degraeve, D.J. Wouters, N. Raghavan, G. Kar, A. Belmonte, Y.Y. Chen, B. Govoreanu, M. Jurczak, Intrinsic switching variability in HfO2 RRAM, in: 2013 5th IEEE International Memory Workshop, May 2013, pp. 30–33, https://doi.org/10.1109/IMW.2013.6582090.

[38] S. Balatti, S. Ambrogio, Z.Q. Wang, S. Sills, A. Calderoni, N. Ramaswamy, D. Ielmini, Pulsed cycling operation and endurance failure of metal-oxide resistive (RRAM), in: 2014 IEEE International Electron Devices Meeting, December 2014, pp. 14.3.1–14.3.4, https://doi.org/10.1109/IEDM.2014.7047050.

[39] J. Woo, A. Belmonte, A. Redolfi, H. Hwang, M. Jurczak, L. Goux, Introduction of WO3 layer in a Cu-based Al2o3 conductive bridge RAM system for robust cycling and large memory window, IEEE J. Electron Devices Soc. 4 (3) (May 2016) 163–166, https://doi.org/10.1109/JEDS.2016.2526632.

[40] S. Ambrogio, S. Balatti, A. Cubeta, A. Calderoni, N. Ramaswamy, D. Ielmini, Statistical fluctuations in hfox resistive-switching memory: Part I—set/reset variability, IEEE Trans. Electron Devices 61 (8) (Aug 2014) 2912–2919.

[41] A. Grossi, E. Nowak, C. Zambelli, C. Pellissier, S. Bernasconi, G. Cibrario, K.E. Hajjam, R. Crochemore, J. Nodin, P. Olivo, L. Perniola, Fundamental variability limits of filament-based RRAM, in: IEEE Int. Electron Devices Meeting (IEDM), Dec 2016, pp. 4.7.1–4.7.4.

[42] A. Grossi, C. Zambelli, P. Olivo, E. Nowak, G. Molas, J. Francois Nodin, L. Perniola, Cell-to-cell fundamental variability limits, IEEE Electron Device Lett. 39 (1) (2018) 27–30.

[43] J. Guy, G. Molas, C. Cagli, M. Bernard, A. Roule, C. Carabasse, A. Toffoli, F. Clermidy, B. De Salvo, L. Perniola, Guidance to reliability improvement in CBRAM using advanced KMC modelling, in: Proc. of 2017 IRPS, 2017.

[44] F. Longnos, E. Vianello, G. Molas, G. Palma, E. Souchier, C. Carabasse, M. Bernard, B. De Salvo, D. Bretegnier, J. Liebault, On disturb immunity and P/E kinetics of Sb-doped GeS2/Ag conductive bridge memories, in: Proc. of Int. Mem. Workshop 2013, 2013, pp. 96–99.

[45] M. Barci, J. Guy, G. Molas, E. Vianello, A. Toffoli, J. Cluzel, A. Roule, M. Bernard, C. Sabbione, L. Perniola, B.D. Salvo, Impact of SET and RESET conditions on CBRAM high temperature data retention. in: 2014 IEEE International Reliability Physics Symposium, June 2014, pp. 5E.3.1–5E.3.4, https://doi.org/10.1109/IRPS.2014.6860677.

[46] L. Zhao, S. Clima, B. Magyari-Köpe, M. Jurczak, Y. Nishi, Ab initio modeling of oxygen-vacancy formation in doped-HfOx RRAM: effects of oxide phases, stoichiometry, and dopant concentrations, Appl. Phys. Lett. 107 (1) (2015) 013504.

[47] G. Molas, E. Vianello, F. Dahmani, M. Barci, P. Blaise, J. Guy, A. Toffoli, M. Bernard, A. Roule, F. Pierre, C. Licitra, B. De Salvo, L. Perniola, Controlling oxygen vacancies in doped oxide based CBRAM for improved memory performances, in: IEDM 2014 Tech. Dig, 2014, pp. 136–139.

[48] P.-Y. Chen, C. Xu, Y. Xie, S. Yu, 3D RRAM design and benchmark with 3D NAND Flash, in: Proc. of ICITIC 2014, 2014.

[49] I.G. Baek, C.J. Park, H. Ju, D.J. Seong, H.S. Ahn, J.H. Kim, M.K. Yang, S.H. Song, E.M. Kim, S.O. Park, C.H. Park, C.W. Song, G.T. Jeong, S. Choi, H.K. Kang, C. Chung, Realization of vertical resistive memory (VRRAM) using cost effective 3D process, in: IEDM 2011 Tech. Dig., 31.8, 2011, pp. 737–740.

[50] G. Sassine, C. Nail, L. Tillie, D. Alfaro Robayo, A. Levisse, C. Cagli, K. El, J.F. Hajjam, E. Nodin, M. Vianello, G. Bernard, E.N. Molas, Sub-pJ consumption and short latency time in RRAM arrays for high endurance applications, in: Proc. of IRPS 2018, 2018.

[51] T.-Y. Liu, et al., A 130.7mm^2 2-layer 32Gb ReRAM memory device in 24nm technology, in: Proc. of 2013 ISSCC, 12.1, 2013, pp. 210–212.

Further reading

[52] Daolin_Cai, https://www.researchgate.net/profile/Daolin_Cai/publication/272366718/figure/download/fig 4/AS:267505344184346@1440789652609/Figure-1-Schematic-of-the-1T1R-structure-memory-cell-b-Schematic-cross-section-of-the.png.

[53] J. Guy, G. Molas, E. Vianello, F. Longnos, S. Blanc, C. Carabasse, M. Bernard, J.F. Nodin, A. Toffoli, J. Cluzel, P. Blaise, P. Dorion, O. Cueto, H. Grampeix, E. Souchier, T. Cabout, P. Brianceau, V. Balan, A. Roule, S. Maitrejean, L. Perniola, B. De Salvo, Investigation of the physical mechanisms governing data-retention in down to 10nm nano-trench Al2O3/CuTeGe conductive bridge RAM (CBRAM), in: IEDM 2013 Tech. Dig, 2013, pp. 742–745.

Selector devices for x-point memory

Jeonghwan Song, Yunmo Koo, Jaehyuk Park, Seokjae Lim, Hyunsang Hwang
Department of Materials Science and Engineering, Pohang University of Science and Technology, Pohang, South Korea

10.1   Introduction

Crossbar arrays capable of implementing device cross sections of ~$4F^2$ have been studied for ultrahigh-density storage architectures, where F is the minimum feature size of the array [1]. For this purpose, two-terminal emerging memory devices such as phase change memory (PCM) [2] and resistive RAM (RRAM) [3] have been proposed in storage devices. To operate the storage device in the crossbar array, different voltages are applied to the line edges of the row and column where the *selected cell* is located, which causes a *net voltage drop* in the selected cell. When the net voltage drop is greater than the threshold voltage of the storage device, the data states change through the program/erase process. On the other hand, under low-voltage state, the stored data states can be read without change of data.

However, the crossbar array architecture has inevitable problems such as undesired current flow and data disturbance due to a nonzero net voltage drop in *half-selected cells* located in the same column or row line as the selected cells and *unselected cells* near the selected cell. Undesired current flow such as leakage and sneak path currents that can occur in both the unselected and the half-selected cells during the read operation degrades the sense margin of the crossbar array. In addition, the data disturbance is a critical problem that can cause data loss due to the high net voltage in the half-selected cells during the program/erase process.

Therefore, an access device called a *selector device* is required to cause the desired net voltage drop only in the selected cells in crossbar array [4]. The selector device, which acts as a switch, is turned on when a voltage above a threshold value is applied that leads to increase in the net voltage drop to the storage device. However, in the low-voltage regime, the off state is maintained, which suppresses in undesired current flow and data disturbance.

To achieve this goal, the selector device has lots of requirements in the electrical characteristics and device fabrication process. Among them, the extremely low OFF current (I_{OFF}) characteristic is the most important parameter that can increase in the sensing margin of crossbar array. On the other hand, the ON-current density must be high enough so that there is no limit on the required current during programming/erasing of the storage device. Thus, the selector device should have highly nonlinear *I-V* characteristics. Other requirements include fast switching speeds, infinite cycling endurance, excellent device uniformity, large voltage margins (voltage window

Advances in Non-volatile Memory and Storage Technology. https://doi.org/10.1016/B978-0-08-102584-0.00011-5

separating the ON and OFF states), compatible operating voltage conditions with storage devices, low-temperature fabrication processes, high thermal stability, 3D stackable, and two-terminal device structure. CMOS transistors are the best selector devices with the required electrical characteristics, but the three-terminal device architecture is not suitable for ultrahigh-density crossbar array architectures. In addition, the P-N diodes rectifying the current flow are limited in use due to the high-temperature process conditions. Therefore, various selector devices have been suggested based on a new operating mechanisms to meet the criteria for the selector device such as mixed ionic electronic conduction (MIEC) device, tunnel barrier type device, insulator-metal transition (IMT), Ovonic threshold switching (OTS), and conductive-bridging RAM (CBRAM)-type selector device.

Among them, this chapter introduces *threshold-type* selector devices, such as IMT, OTS, and CBRAM-type, which have the potential for atomic scale scalability, and discusses about the guidelines for future development of selector devices.

10.2   Insulator-metal-transition selector

Certain transition metal oxide such as NiO, VO_2 [5], and NbO_2 [6] are highlighted thanks to the special properties that change from an insulating state to a metallic state under certain conditions such as a specific temperature [7], optical excitation [8], or an electric field [9]. These materials change conductivity by rearranging their crystal structure at critical temperature called Mott-Peierls transition, which is shown in Fig. 10.1 [10, 11].

Recently, electrically driven IMT (E-IMT) devices suggested as a selector device which has bidirectional threshold characteristics for ReRAM x-point applications thanks to their simple metal/insulator/metal structure and fast transition speed. M. Son et al. suggested the VO_2-based threshold switch device as a selector [12] as shown in Fig. 10.2.

When the applied voltage on VO_2 device exceeds a critical voltage called threshold voltage (V_{th}), the device changes its state from insulating state to metallic state. The transition mechanism of VO_2 is usually interpreted as the result of joule heating by electric field. As shown in Fig. 10.1A, the VO_2 changes its state at about 340 K. When a voltage is applied to the VO_2 device, joule heating occurs due to the current flowing through the material. When the temperature rising due to joule-heating exceeds 340 K on the VO_2 device, the VO_2 material changes to a metallic state. Actually, S. Kumar showed VO_2 changes from an insulating state to a metallic state when the internal temperature exceeds 340 K due to joule heating (Fig. 10.3A) [13]. However, VO_2-based selector has limitations as a selector because their transition temperature (~340 K, Fig. 10.3B) is too close to room temperature (RT) [14], reducing control precision in circuit. For replacement, some groups suggested the NbO_2-based selector device which has high transition temperature (1080 K, Fig. 10.3C) [15, 16].

In their research, ultrathin NbO_2 film was fabricated by reactive sputtering method and patterned to nanoscale device. Similar to VO_2 device, the NbO_2 selector

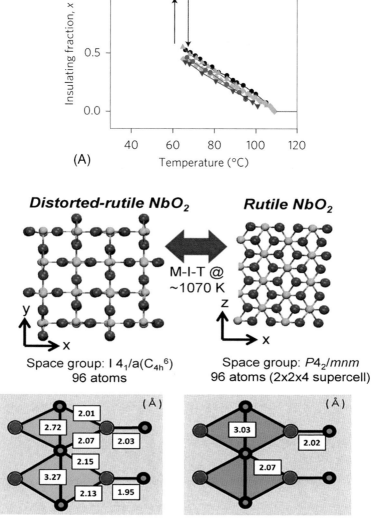

Fig. 10.1 (A) IMT characteristic according to temperature in VO_2 [10]. (B) IMT characteristics in NbO_2 [11].

exhibits repeatable threshold characteristics over 1000 switching cycles (Fig. 10.3C and D) [15]. Moreover, NbO_2 device has additional unique characteristics as a selector such as fast transition speed (<50 ns) and high thermal stability (>430 K). E. Cha et al. also announced that the device characteristics such as forming voltage (V_F), V_{th}, and I_{OFF} can be controlled by thickness and device area engineering by using TiN/NbO_2/W device [17]. According to their research, the V_{th} can be reduced by the thickness of NbO_2 layer and the I_{OFF} can be reduced by reducing the device size as shown in Fig. 10.4. Thanks to controllability of the NbO_2 device, it is easy to

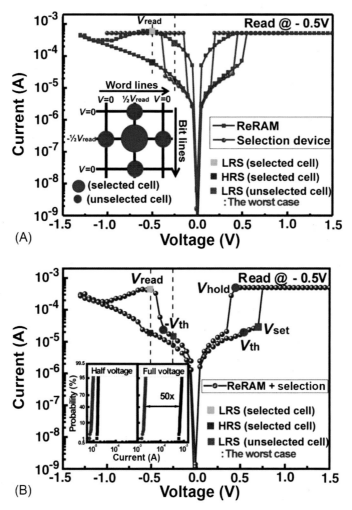

Fig. 10.2 (A) *I-V* characteristics of VO$_2$ selector device and ReRAM device and (B) the *I-V* characteristics of 1S1R structure [12].

fabricate NbO$_2$ device which has proper device characteristics with variety ReRAM devices for 1S1R structure.

Later, in order to develop the NbO$_2$ device through a more accurate analysis, many groups studied the mechanism of the E-IMT phenomenon of NbO$_2$ device like VO$_2$ device. Many groups initially attempted to interpret the E-IMT of NbO$_2$ as a result of joule heating just as VO$_2$ [18]. However, the joule-heating model conflicts with the fact that transition temperature of NbO$_2$ (1080 K) is much higher than the temperature that can be induced by Joule heating of the insulating state of NbO$_2$ [19]. For more accurate mechanism analysis, a number of groups tried to show that NbO$_2$ can change its resistance far below the IMT temperature of NbO$_2$

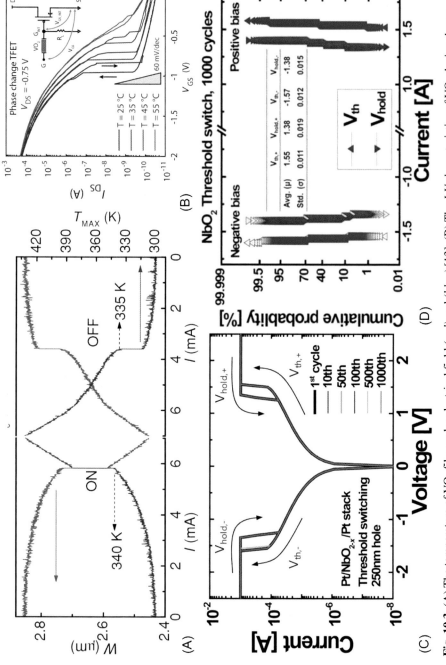

Fig. 10.3 (A) The temperature of VO₂ film under electrical field (*red*) and (*blue*) [13]. (B) The *I-V* characteristics of VO₂ device shows that it cannot act as selector at over 340 K [14]. (C) The *I-V* characteristics of IMT behavior of Pt/NbO₂/Pt devices. (D) Distribution of V_{th} and V_{hold} of Pt/NbO₂/Pt device during 1000 cycles [15].

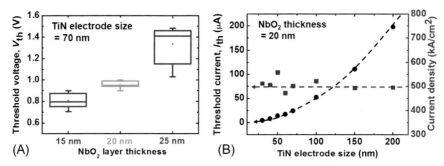

Fig. 10.4 (A) Distribution of V_{th} and (B) I_{th} (I_{OFF}) of TiN/NbO$_2$/W devices according to thickness and size of NbO$_2$ layer. V_{th} decreases as thickness of NbO$_2$ layer decreases, and I_{th} decreases as device size of NbO$_2$ decrease [17].

(1080 K) based on thermal runaway model with Poole-Frenkel simulation [20–22]. According to Funck C. et al.'s description, they claim that IMT can be started at much lower temperatures than the original transition temperature of NbO$_2$ (1080 K) because of combination of barrier lowering by electrical field and joule-heating effect as shown in Fig. 10.5A [20]. Actually, the simulation result (Fig. 10.5) shows that the temperature of starting point of transition is almost 320 K which is much lower temperature than original transition temperature of NbO$_2$, and the temperature and conductivity can self-accelerated until reach to the metallic state by their thermal-runaway model.

Similarly, J. Park et al. adopted the field-induced nucleation theory to clarify the mechanism of NbO$_2$ E-IMT by measuring the delay time of the film at various temperature and voltage [23]. Despite mechanism analysis and various attempts to improve the performance of IMT selectors, the I_{OFF} of the IMT selector is relatively high compare with other selector candidates such as CBRAM-type selectors and OTS selectors. To reveal the origin of leakage current of IMT devices, some groups try to C-AFM analysis to find the leakage point of the film [24, 25].

Fig. 10.6 shows that the that the local amorphous state in polycrystalline thin-film and the grain boundaries can cause leakage current, especially with large defect density [24, 25]. These defects generate the conduction subbands between conduction band and valance band of IMT materials (NbO$_2$, VO$_2$) and these generate the leakage current. Moreover, the interface defects between electrode and IMT materials also can generate the leakage current because the defects can pin the Schottky barrier height and it can not be formed perfectly [26]. To passivate the defects and reduce the leakage current, many groups tried to insert the barrier structure which can eliminate both of interfacial and bulk defects [24, 26, 27].

X. Liu et al. insert a thin HfO$_x$ layer between bottom electrode and NbO$_2$ IMT layer which forms the narrow conductive filament during forming process. During their research, the most of defects of the NbO$_2$ film are suppressed and a low I_{OFF} can be obtained compared with the single NbO$_2$ device since the other HfO$_x$ portions without a conductive filament are in a very high resistance state, as shown in Fig. 10.7A and B. Similarly, J. Park et al. also inserted the NiO$_y$ barrier layer in both

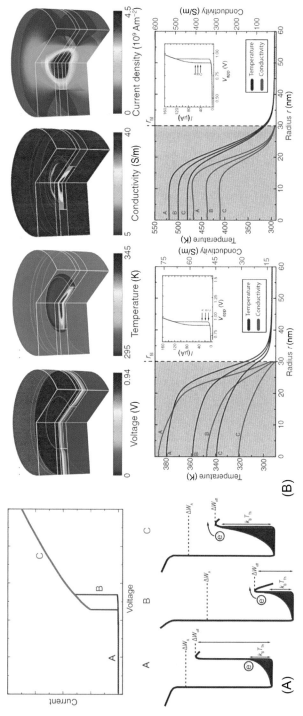

Fig. 10.5 (A) Schematic diagram of the energy barrier to transit to a metallic state from an insulating state of the NbO$_2$ device. In process A, the device cannot change because of insufficient energy to transition. In process B and C, the device can change its state due to a combination of joule-heating induced thermal energy and energy barrier lowering by an electrical field. (B) A thermal simulation of a NbO$_2$ device during operation shows that the transition can occur much lower temperature (~320 K) than the transition temperature of the NbO$_2$ device (~1070 K) under an electrical field [20].

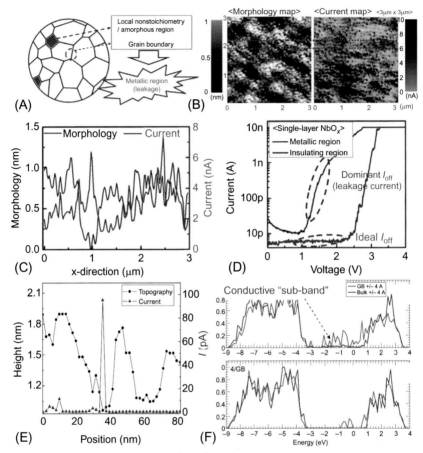

Fig. 10.6 (A) The schematic diagram of leakage source of NbO$_2$ film. (B) The morphology and current map of the NbO$_2$ film, and (C-D) line and point profiles *(red box in (B))* show that the morphologically shrink position which corresponding to defects generate the leakage current [24]. (E) Also the current map of HfO$_2$ device shows similar result of NbO$_2$ device. (F) The defects in IMT materials generate the conductive subband between valance band and conduction band [25].

side of between electrode and NbO$_2$ IMT layer (W/NiO$_y$/NbO$_2$/NiO$_y$/W) to form the perfect Schottky barrier height which can reduce the I_{OFF} of NbO$_2$-based IMT device (Fig. 10.7C).

Additionally, suggested NbO$_2$-based selector by J. Park et al. also shows excellent characteristics as a selector devices such as high I_{ON}/I_{OFF} ratio (>5400), high operating temperature (>453 K), fast transition speed (<2 ns), and drift-free operation as shown in Fig. 10.8. Due to the excellent selector characteristics of suggested NiO$_y$ inserted NbO$_2$ IMT device, they also can show a significantly improved readout margin (up to 2^9 word lines) is possible in a large x-point memory array [26].

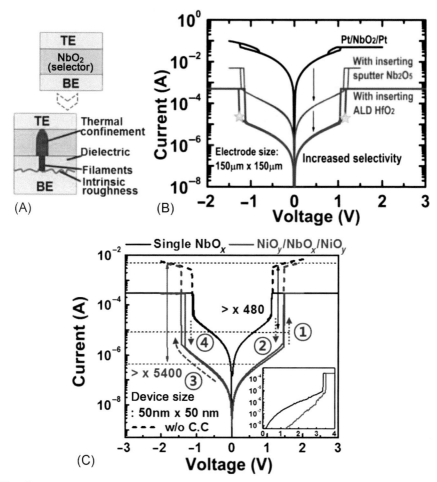

Fig. 10.7 (A–B) The *I-V* characteristics of barrier inserted Pt/HfO₂/NbO₂/Pt device [27]. (C) W/NiO_y/NbO₂/NiO_y/W device. I_{OFF} of both devices shows reduced I_{OFF} compare with single layer NbO₂ device [26].

10.3 Ovonic threshold switching

OTS phenomenon is a field-dependent volatile threshold switching (TS) occurring in amorphous chalcogenide alloys, first reported by S.R. Ovshinsky in 1968 [28]. Recently, various OTS selector devices fabricated on wafers with patterned nanoscale electrodes have been reported whose performances are illustrated in Fig. 10.9. The first distinguishing factor of OTS among various types of TS phenomena is that it has comprehensively superior properties satisfying the various requirements of selector requirements. OTS material is highly insulating in its "OFF" state (~ 20 GΩ at 0.2 V (Fig. 10.9A) [29]), effectively minimizing current

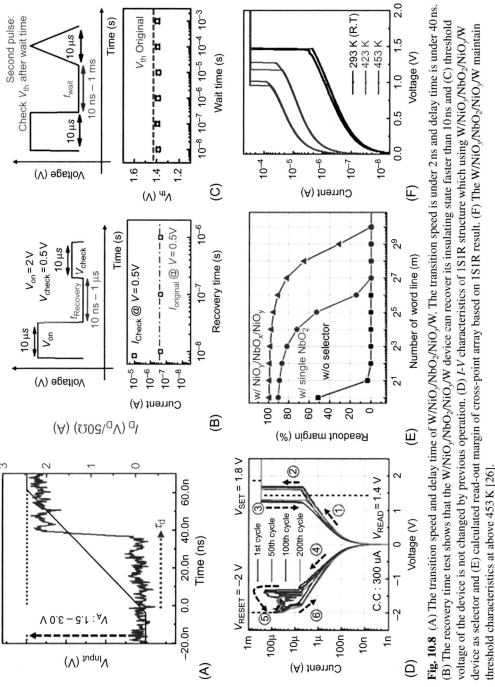

Fig. 10.8 (A) The transition speed and delay time of W/NiO$_y$/NbO$_2$/NiO$_y$/W. The transition speed is under 2ns and delay time is under 40ns. (B) The recovery time test shows that the W/NiO$_y$/NbO$_2$/NiO$_y$/W device can recover its insulating state faster than 10ns and (C) threshold voltage of the device is not changed by previous operation. (D) I-V characteristics of 1S1R structure which using W/NiO$_y$/NbO$_2$/NiO$_y$/W device as selector and (E) calculated read-out margin of cross-point array based on 1S1R result. (F) The W/NiO$_y$/NbO$_2$/NiO$_y$/W maintain threshold characteristics at above 453 K [26].

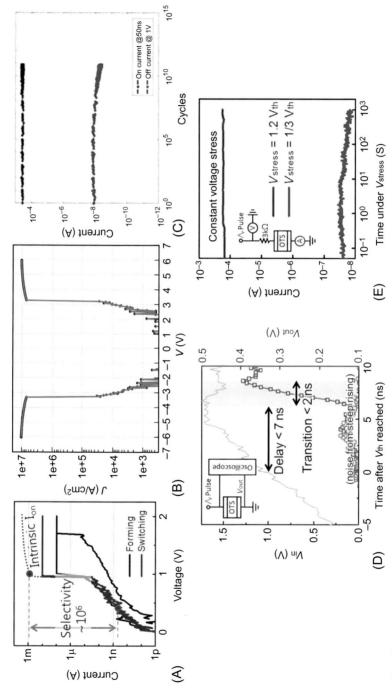

Fig. 10.9 Reported performance of OTS selector devices.

flow through the device (30 MA/cm^2 (Fig. 10.9B) [30]). Once the external electric field exceeds the threshold voltage (V_{th}), OTS material exhibits a rapid and volatile resistance decrease and turns into its low resistance "ON" state (< 1 kΩ [29]) through "S"-shaped negative differential resistance change, allowing high current supply (> 1.1 × 10^7 A/cm^2 [31]) for memory operation. This resistance change in OTS material is nondestructive and repeatable (> 8 × 10^{12} cycles (Fig. 10.9C) [32] and > 10^{11} cycles [33]), instant (short delay time < 2 ns (Fig. 10.9D) [34]), fast (transition time < 5 ns [35]), abrupt (switching slope < 1 mV/dec [29]), thus suitable for selector device application. Electrical stress test results show that both ON and OFF states are electrically stable. Under more than 10^3 s of either constant current (300 μA [33]) or constant voltage (1.2 V_{th} (Fig. 10.9E) [34]) stress, performances of OTS devices were securely maintained.

To explain the OTS phenomenon, various theoretical studies have been conducted. Representative schematics of the studies including thermally induced instability by A.C. Warren [36], Shockley-Read-Hall recombination with impact

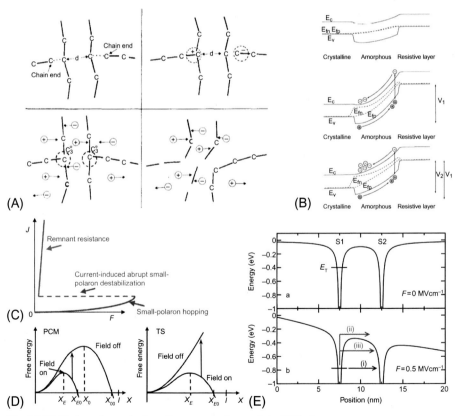

Fig. 10.10 Various theoretical studies proposed to explain OTS phenomenon.

ionization by D. Adler (Fig. 10.10A) [37] and A. Pirovano (Fig. 10.10B) [38], polaron destabilization by D. Emin (Fig. 10.10C) [39], nucleation theory by M. Nardone (Fig. 10.10D) [40], and thermally assisted hopping model by D. Ielmini (Fig. 10.10E) [41]. Many of the above models explain the OTS as an electronic phenomenon with possible secondary thermal effects, suggesting the second distinguishing factor of OTS among various types of TS phenomena that it does not involve any atomic arrangements while switching [32]. Due to this advantage, OTS has a high potential in extremely fast operation speed [30, 34, 35], and superb cycling endurance ($> 8 \times 10^{12}$ cycles [32] and $> 10^{11}$ cycles [33]) among TS devices.

One drawback of OTS selector devices was its complex material (four to five elements for plausible device performance in usual [28, 31, 33, 36, 37]). However, recent reports showed that simple binary OTS materials such as GeTe, ZnTe, and SiTe could be showed even with improved selector device performances such as higher OFF resistance, lower ON resistance, faster operation speed, higher cycling endurance, and higher thermal stability (Fig. 10.11A) [29, 34]. However, to improve performances of the binary OTS devices further, many groups started to seek for new OTS materials and reported various OTS devices whose electrical properties are illustrated in Fig. 10.11. As a result, OTS has been reported in a number of chalcogenide materials (Fig. 10.11B–F) [28–37, 42–46]. While having Te or Se as a core element, the other components can be chosen from various species such as Ge, Si, P, As, Sb, Bi, Zn, or even N, B, or C. The freedom of choice of material combination greatly enhances the tenability of the OTS material [29, 32, 33, 42–46].

In recent OTS studies, it has become an important topic to improve thermal stability without sacrificing device performance. The thermal stability of various OTS devices reported in recent papers is illustrated in Fig. 10.12. To satisfy thermal budget of CMOS process, selector devices have to endurance temperature higher than 400°C but many of the reported OTS devices have shown performance degradation near 300°C (Fig. 10.12A and B) [33, 34, 42]. Consequently, approaches to investigate the performance degradation mechanism to improve the thermal stability have been actively carried out by field leading researches groups over the world (Fig. 10.12C) [33, 34, 42, 43]. Since, OTS material with high thermal stability withstanding 500°C annealing has already been reported (Fig. 10.12D and E) [31, 32], device scale high thermal stability would be hopefully achieved in the near future.

OTS is clearly one of the closest candidates to the commercialization of selector devices, but there are some important issues that are obviously need to be improved in advance. Analysis on device reliability issues such as cell-to-cell or cycle-to-cycle uniformity is not sufficient and needs further research. In addition, despite the fact that various different mechanisms have been suggested to explain the OTS behavior [32, 36–41, 43], a unified model is still missed. Therefore, a detailed study of the physics behind the OTS phenomena is needed to fully understand and tailor different electrical properties.

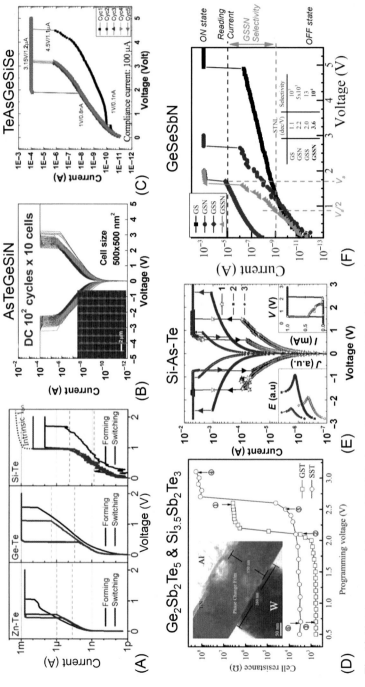

Fig. 10.11 Diverse material pool of OTS devices enhances the tunability of their device properties.

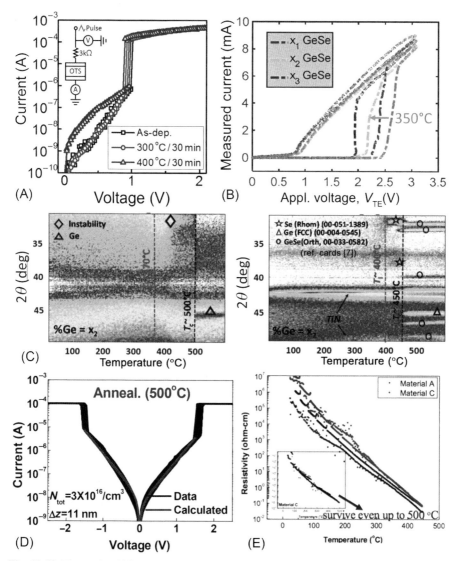

Fig. 10.12 Thermal stability of OTS selector devices and studies to explain the device performance degradation in high temperature.

10.4 CBRAM-type selector

As a new approach, recently, there is growing interest in using instability of the filament in conventional conductive bridge random access memory (CBRAM) devices for selector applications. When the CBRAM device is programmed under a lower current compliance, the small and unstable Cu or Ag filament is formed. Accordingly, the spontaneous self-rupturing of the unstable filament is occurred when the applied bias is removed which can be used as a TS devices. Various materials have been reported for the CBRAM-type TS device as a selector device [47–60].

Song et al. suggested the use of a Ag/Titanium oxide-based threshold switch (Ag/TiO$_2$/Pt) as a selector device in a crossbar array [55]. Fig. 10.13A shows the measured *I-V* characteristics. The Ag/TiO$_2$/Pt device showed high selectivity (~10^7) and steep slope (<5 mV/dec). The exponential relation between time and voltage was also verified (Fig. 10.13B).

The Ag/amorphous Si-based devices (Ag/a-Si/Pt) were reported by the same group in Ref. [56]. A faster dissolution of the metal filament and lower down the I_{OFF} were achieved by enhancing diffusivity of the Ag atoms and removing leakage paths in the a-Si film of the device through hydrogen doping of the a-Si layer.

Midya et al. reported Ag/Hafnium oxide-based (Pd/Ag/HfO$_x$/Ag/Pd) selector device [58]. The selector device exhibits a high selectivity of 10^{10}, steep turn on slope of <1 mV/dec, and high endurance beyond 10^8 cycles. They also demonstrated the vertical integration of Pd/Ag/HfO$_x$/Ag/Pd threshold switch on top of a Pd/Ta$_2$O$_5$/TaO$_x$/Pd memristor device. Fig. 10.14A shows a TEM of the threshold switch device and memristor device, while Fig. 10.14B shows the measured *I-V* characteristics. The presence of the select device successfully suppresses the leakage current at the half read region.

Bricalli et al. reported Ag/Silicon oxide-based (Ag/SiO$_x$/C) selector device which shows stable switching voltage ($+V_{th} \approx 2$ and $-V_{th} \approx -0.5$ V) and $I_{OFF} \approx 1$ pA under cycling, as shown in Fig. 10.15A [60]. They also characterized 1S1R structures by serially connecting the Ag/SiO$_x$/C threshold switch with a discrete Ti/SiO$_x$/C RRAM device. Fig. 10.15B shows the measured *I-V* characteristics of 1S1R device. The leakage current was significantly suppressed at the half read region.

As shown above, the CBRAM-type selector devices have shown excellent properties such as ultralow leakage current and sharp transition. Since the switching (formation and rupture of an Ag or Cu filament) occurs without causing electrical break-down of the dielectric, the low I_{OFF} can be maintained under cycling. Also, the abrupt transition is attributed to the filamentary mechanism.

However, the CBRAM-type selector device lost TS characteristics and exhibit memory characteristics after electrical pulse with certain current compliance, due

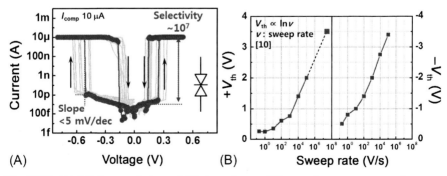

Fig. 10.13 (A) *I-V* characteristics and (B) dependence of V_{th} on voltage sweep rate of Ag/TiO$_2$/Pt selector device.

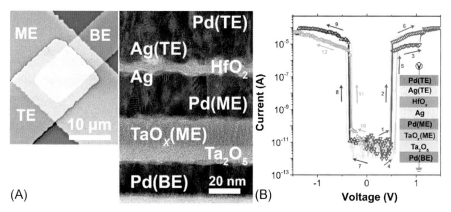

Fig. 10.14 (A) Cross-sectional TEM image and (B) I-V characteristics of an integrated 1S-1R device consisting of a Pd/Ag/HfO$_x$/Ag/Pd selector and a Pd/Ta$_2$O$_5$/TaO$_x$/Pd memristor.

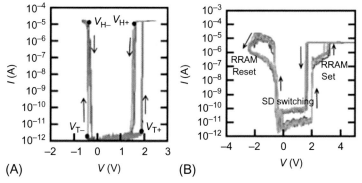

Fig. 10.15 (A) I-V characteristics of an Ag/SiO$_x$-based (Ag/SiO$_x$/C) selector device and (B) 1S-1R device consisting of a Ag/SiO$_x$/C selector and a Ti/SiO$_x$/C RRAM.

to relatively thick filament formation. The maximum on-current of Ag/TiO$_2$/Pt [55], Pd/Ag/HfO$_x$/Ag/Pd [58], and Ag/SiO$_x$/C [60] selector devices are 10, 100, and 80 μA, respectively.

Also, the operation of the CBRAM-type selector device is based on the ionic motion [61], the switching speed is slower than the electronic-based systems such as IMT and OTS devices. The turn-on operation is performed by applying a voltage. Therefore, the turn-on (delay and transition) speed can be boosted by increasing applied voltage due to the exponential relation between time and voltage (Fig. 10.13B) and a delay speed of less than 75 ns [58] and a transition speed of less than 10 ns [55] have been reported. However, the turn-off (relaxation) speed is slower than the turn-on speed because it has a self-rupturing mechanism without an applied voltage. The Ag/TiO$_2$/Pt selector device can switch to the off-state when time delay is more than 1 μs after turn-on operation but it remains on on-state when time delay is less than 500 ns [62]. The Pd/Ag/HfO$_x$/Ag/Pd selector device shows turn off (relaxation) speed of 250 ns [58].

Therefore, the further research efforts are needed to improve the maximum on-current and relaxation speed of CBRAM-type selector device.

Song et al. proposed the AgTe/TiO$_2$/Pt stack to maintain selector characteristics with fast filament dissolution under high on-current stress [63]. The Te allows an extraction of Ag out of the switching layer because Ag tends to form Ag-Te phases which are more favorable [64, 65]. Therefore, the moderate Te content (35% Te) can acts as an additional driving force for fast filament dissolution which shows the TS behavior even at high operation current (Fig. 10.16A). In addition, the selector device with AgTe TE (AgTe/TiO$_2$/Pt) shows turn off (relaxation) speed of 100 ns which is 10 times faster than that of selector device with Ag TE (Ag/TiO$_2$/Pt). The enhanced on-current and relaxation speed were attributed to the formation of Ag-Te phase which accelerates the filament dissolution.

Zhao et al. obtained selector (Ag/defective-grain/SiO$_2$/Pt) with very high on-current (500 μA) and fast on/off switching speed (<0.1/1 μs) by introducing defective graphene (DG) layer between active electrode and switching layer, as shown in Fig. 10.17 [66]. They fabricated discrete graphene defects with various concentrations/sizes by irradiation of accelerated Si$^+$ ions. The cation injecting path to the switching layer is modulated by the discrete atomic-scale graphene defects and it results in the formation of discrete tiny conductive filaments which can be easily spontaneous ruptured.

In order to improve the performance of CBRAM-type selector device, various attempts have been made such as utilizing AgTe alloy electrode or DG layer. However, as described earlier, the different selector performances are exhibited in various materials. Therefore, in-depth study of the switching mechanism of the selector device is essential to provide guidelines for improving selector performance. The turn-on operation is the same as the conventional CBRAM device. The turn-on process occurs when a sufficient positive voltage is applied to the active electrode. The overall process involves the following steps: (i) anodic dissolution of the active electrode; (ii) migration of metal cations in the switching layer electrode under the external electric field; and (iii) reduction of metal cations and growth of the metal filaments from the counter electrode [61]. In the case of a turn-off operation in which the filament

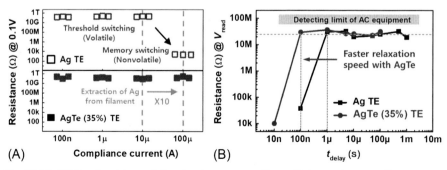

Fig. 10.16 (A) Read resistance following turn-on operation with increasing compliance current and (B) measured relaxation speed of selector devices with Ag TE and AgTe (35%) TE.

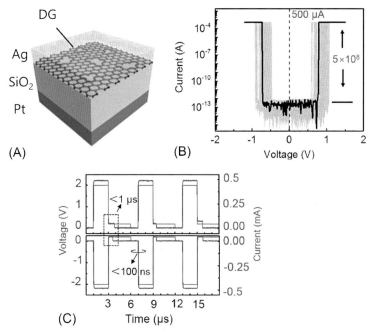

Fig. 10.17 (A) Schematic illustrations and (B) *I-V* characteristics and switching speed of an Ag/defective-grain/SiO₂/Pt selector device.

ruptures when the applied voltage is removed, the mechanism has not yet been established, but various attempts have been made to understand it.

Hsiung et al. reported the volatile characteristics of Ag/TiO₂/Pt cell and observed the filament breaks up into a chain of nanospheres [54]. They explained the phenomena by using Rayleigh instability theorem, or so-called Thomas-Gibbs theorem. Wang et al. also observed spontaneous formation of Ag particles to minimize interfacial energy between Ag and the dielectric in a planar Au/SiOₓNᵧ:Ag/Au device when removing the electrical biasing after forming a continuous Ag filament [53]. Ambrogio et al. reported that the ion migration-induced mechanical stress affects ion hopping in two migration directions, therefore leading to switching asymmetry and even spontaneous CF dissolution in Ag/GeS₂/W system [48].

In addition, various mechanisms such as steric repulsion [55], electromotive force [47], and tunneling barrier modulation [59] have been proposed. However, this mechanism only qualitatively explains why the filaments are broken in each material. Therefore, quantitative studies based on physical evidence are needed to find out which parameters of the material determine the selector properties.

Recently, Shukla et al. conducted a quantitative analysis based on the first principle calculations on the effect of the energy difference (between the cluster configuration in the high-resistance state, and the filament configuration in the low-resistance state) on the selector characteristics, as shown in Fig. 10.18 [67]. The larger the energy difference value, the stronger the force to break the filament.

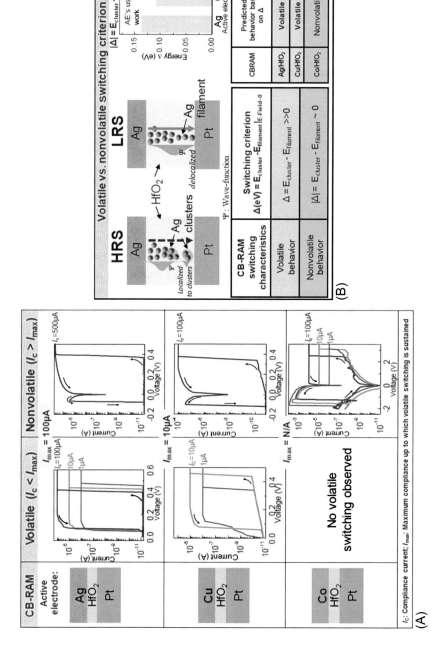

Fig. 10.18 (A) Experimental switching characteristics for various active electrodes and (B) criterion for volatile vs nonvolatile switching.

The calculated energy difference values of Ag and Cu are 0.1354 and 0.1076 eV, respectively. This is consistent with the fact that most of the reported CBRAM-type selectors are Ag-based and Ag-based selectors show a higher maximum current than Cu-based selectors.

10.5 Conclusion

In this chapter, we demonstrate a selector device to compensate for the limitations of the two-terminal emerging memories to implement an ultrahigh-density crossbar arrays. The crossbar array consists of emerging memories suffers from undesired data disturbance and current flow. To minimize undesirable characteristics without changing the intrinsic operating characteristics of the memory device, there are various requirements of the selector device related with electrical characteristics (low OFF current, high ON-current density, fast switching speeds, infinite cycling endurance, large voltage margins, and compatible operating voltage conditions with storage devices) and fabrication process (low-temperature fabrication processes, high thermal stability, 3D stackable, and two-terminal device structure).

Though various candidates for the selector devices have been reported to replace conventional switch elements (CMOS transistor and P-N diode, etc.), we tried to demonstrate the researches about threshold-type selector devices such as IMT, OTS, and CBRAM-type devices due to their potential for atomic scale scalability.

The selector device with IMT behaviors has been widely researched based on excellent electrical characteristics and fabrication process compatibility. However, IMT device suffers from high OFF current problem because the metal and insulator phase are separated by very small electronic energy difference. Nevertheless, various studies have been conducted to reduce OFF current by suppressing the leakage path in insulators using additional barrier layers.

Recently, the OTS selector devices have shown the ideal selector characteristics of the crossbar array architecture. Especially, the simple binary-composition OTS have overcome the fabrication complexity which was critical bottleneck for the OTS selector. However, further enhancement of reliability issues such as cell-to-cell or cycle-to-cycle uniformity and thermal stability must be implemented based on understanding the switching mechanisms.

The CBRAM-type selector devices exhibit excellent nonlinear I-V characteristics. However, it exhibits slow turn OFF speed and even loss of selector characteristics at high operating current conditions. Therefore, to improve the switching speed, many research groups have studied the origin of filament instability based on their hypothesis and provided guidelines for ideal selector characteristics.

Nonidealities of the crossbar array architecture hinder the implementation of ultradense storage systems. However, the development of an ideal selector with insight into the switch mechanism will lead to overcoming the problems of crossbar arrays and the scaling limit of conventional memory technologies.

References

[1] E.P.G. Wright, Electric connecting device, U.S. patent 2 667 542, Sept. 25, 1954.
[2] D. Kau, T. Stephen, I.V. Karpov, R. Dodge, B. Klehn, J.A. Kalb, J. Strand, A. Diaz, N. Leung, J. Wu, S. Lee, T. Langtry, K. Chang, C. Papagianni, J. Lee, J. Hirst, S. Erra, E. Flores, N. Righos, H. Castro, G. Spadini, A stackable cross point phase change memory, in: Proc. IEEE Int. Electron Devices Meeting, Dec. 2009, pp. 27.1.1–27.1.4.
[3] R. Waser, R. Dittmann, G. Staikov, K. Szot, Redox-based resistive switching memories— nanoionic mechanisms, prospects, and challenges, Adv. Mater. 21 (Jul. 2009) 2632–2663.
[4] G.W. Burr, R.S. Shenoy, K. Virwani, P. Narayanan, A. Padilla, B. Kurdi, H. Hwang, Access devices for 3D crosspoint memory, J. Vac. Sci. Technol. 32 (4) (Jun. 2014) 040802.
[5] A. Zylbersztejn, N.F. Mott, Metal-insulator transition in vanadium dioxide, Phys. Rev. B 11 (11) (Jun. 1975) 4383.
[6] V. Eyert, The metal-insulator transition of NbO_2: an embedded Peierls instability, Europhys. Lett. 58 (6) (Apr. 2002) 851.
[7] F.J. Morin, Oxides which show a metal-to-insulator transition at the Neel temperature, Phys. Rev. Lett. 3 (1) (Jul. 1959) 34.
[8] A. Cavalleri, M. Rini, H.H.W. Chong, S. Fourmaux, T.E. Glover, P.A. Heimann, J.C. Kieffer, R.W. Schoenlein, Band-selective measurements of electron dynamics in VO2 using femto-second near-edge X-ray absorption, Phys. Rev. Lett. 95 (6) (Aug. 2005) 067405.
[9] M. Son, J. Lee, J. Park, J. Shin, G. Chio, S. Jung, W. Lee, S. Kim, S. Park, H. Hwang, Excellent selector characteristics of nanoscale VO_2 for high-density bipolar ReRAM applications, IEEE Electron Device Lett. 32 (11) (Nov. 2011) 1579–1581.
[10] J. Wei, Z. Wang, W. Chen, D.H. Cobden, New aspects of the metal-insulator transition in single-domain vanadium dioxide nanobeams, Nat. Nanotechnol. 4 (May 2009) 420–424.
[11] E. Cha, J. Woo, D. Lee, et al., Nanoscale (∼10nm) 3D vertical ReRAM and NbO_2 threshold selector with TiN electrode, in: Proc. IEEE Int. Electron Devices Meeting, Dec. 2013, pp. 10.5.1–10.5.4.
[12] M. Son, J. Lee, J. Park, J. Shin, G. Choi, S. Jung, W. Lee, S. Kim, S. Park, H. Hwang, Excellent selector characteristics of nanoscale VO_2 for high-density bipolar ReRAM applications, IEEE Electron Device Lett. 32 (11) (Aug. 2011) 1579–1581.
[13] S. Kumar, M.D. Pickett, J.P. Strachan, G. Gibson, Y. Nishi, R.S. Williams, Local temperature redistribution and structural transition during joule-heating-driven conductance switching in VO_2, Adv. Mater. 25 (42) (Nov. 2013) 6128–6132.
[14] W.A. Vitale, E.A. Casu, A. Biswas, T. Rosca, C. Alper, A. Krammer, G.V. Luong, Q.T. Zhao, S. Mantl, A. Schuler, A.M. Ionescu, A steep-slope transistor combining phase-change and band-to-band-tunneling to achieve a sub-unity body factor, Sci. Rep. 7 (355) (Feb. 2017) 1–10.
[15] S. Kim, X. Liu, J. Park, S. Jung, W. Lee, J. Woo, J. Shin, G. Choi, C. Cho, S. Park, D. Lee, E. Cha, B. Lee, H. Lee, S. Kim, S. Jung, H. Hwang, Ultrathin (<10nm) Nb_2O_5/NbO_2 hybrid memory with both memory and selector characteristics for high density 3D vertically stackable RRAM applications, in: Symposium on VLSI Technology, 2012, pp. T18.3.
[16] X. Liu, S.M. Sadaf, M. Son, J. Shin, J. Park, J. Lee, S. Park, H. Hwang, Diode-less bilayer oxide (WO_x-NbO_x) device for cross-point resistive memory applications, Nanotechnology 22 (47) (Nov. 2011) 175702.
[17] E. Cha, J. Park, J. Woo, D. Lee, A. Prakash, H. Hwang, Comprehensive scaling study of NbO_2 insulator-metal-transition selector for cross point array application, Appl. Phys. Lett. 108 (15) (Apr. 2016) 153502.

[18] S.K. Nandi, X. Liu, D.K. Venkatachalam, R.G. Eliman, Threshold current reduction for the metal-insulator transition in NbO_{2-x}-selector devices: the effect of ReRAM integration, J. Phys. D 48 (19) (Apr. 2015) 195105.

[19] X. Liu, S. Md, M. Son, J. Park, J. Shin, W. Lee, K. Seo, D. Lee, H. Hwang, Co-occurrence of threshold switching and memory switching in $Pt/NbO_x/Pt$ cells for crosspoint memory applications, IEEE Electron Device Lett. 33 (2) (Feb. 2012) 236–238.

[20] C. Funck, S. Menzel, N. Aslam, H. Zhang, A. Hardtdegen, R. Waser, S.H. Eifert, Multidimensional simulation of threshold switching in NbO_2 based on an electric field triggered thermal runaway model, Adv. Electron. Mater. 2 (7) (Jun. 2016) 201600169.

[21] S. Slesazeck, H. Mähne, H. Wylezich, A. Wachowiak, J. Radhakrishnan, A. Ascoli, R. Tetzlaff, T. Mikolajick, Physical model of threshold switching in NbO_2 based memristors, RSC Adv. 5 (124) (Nov. 2015) 102318–102322.

[22] G.A. Gibson, S. Musunuru, J. Zhang, K. Vandenberghe, J. Lee, C.C. Hsieh, W. Jackson, Y. Jeon, D. Henze, Z. Li, S. Williams, An accurate locally active memristor model for S-type negative differential resistance in NbO_x, Appl. Phys. Lett. 108 (2) (Jan. 2016) 023505.

[23] J. Park, E. Cha, I. Karpov, H. Hwang, Dynamics of electroforming and electrically driven insulator-metal transition in NbO_2 selector, Appl. Phys. Lett. 108 (23) (Jun. 2016) 232101.

[24] J. Park, E. Cha, D. Lee, S. Lee, J. Song, J. Park, H. Hwang, Improved threshold switching characteristics of multi-layer NbO_x for 3-D selector application, Microelectron. Eng. 147 (Apr. 2015) 318–320.

[25] G. Bersukera, J. Yum, L. Vandelli, A. Padovani, L. Larcher, V. Iglesias, M. Porti, M. Nafría, K. McKenna, A. Shluger, P. Kirsch, R. Jammy, Grain boundary-driven leakage path formation in HfO_2 dielectrics, Solid State Electron. 65–66 (Jul. 2011) 146–150.

[26] J. Park, T. Hadamek, A.B. Posadas, E. Cha, A.A. Demkov, H. Hwang, Multi-layered $NiO_y/NbO_x/NiO_y$ fast drift-free threshold switch with high Ion/Ioff ratio for selector application, Sci. Rep. 7 (Jun. 2017) 4068.

[27] X. Liu, S.K. Nandi, D.K. Venkatachalam, K. Belay, S. Song, R.G. Elliman, Reduced threshold current in NbO_2 selector by engineering device structure, IEEE Electron Device Lett. 35 (10) (Aug. 2014) 1055–1057.

[28] S.R. Ovshinsky, Reversible electrical switching phenomena in disordered structures, Phys. Rev. Lett. 21 (20) (Nov. 1968) 1450–1453.

[29] Y. Koo, K. Baek, H. Hwang, Te-based amorphous binary OTS device with excellent selector characteristics for X-point memory applications, in: Symp. VLSI Tech. Dig, Jun. 2016, pp. T86–T87.

[30] S. Yasuda, K. Ohba, T. Mizuguchi, H. Sei, M. Shimuta, K. Aratani, T. Shiimoto, T. Yamamoto, T. Sone, S. Nonoguchi, J. Okuno, A. Kouchiyama, W. Otsuka, K. Tzutsui, A cross point Cu-ReRAM with a novel OTS selector for storage class memory applications, in: Symp. VLSI Tech. Dig, Jun. 2017, pp. T30–T31.

[31] M. Lee, D. Lee, H. Kim, H. Choi, J. Park, H. Kim, Y. Cha, U. Chung, I. Yoo, K. Kim, Highly-scalable threshold switching select device based on chalcogenide glasses for 3D nanoscaled memory arrays, in: Proc. IEEE Int. Electron Devices Meeting, Dec. 2012, pp. 2.6.1–2.6.3.

[32] S.R. Ovshinsky, An introduction to ovonic research, J. Non-Cryst. Solids 2 (1970) 99–106.

[33] H.Y. Cheng, W.C. Chien, I.T. Kuo, E.K. Lai, Y. Zhu, J.L. Jordan-Sweet, A. Ray, F. Carta, F.M. Lee, P.H. Tseng, M.H. Lee, Y.Y. Lin, W. Kim, R. Bruce, C.W. Yeh, C.H. Yang, M. BrightSky, H.L. Lung, An ultra high endurance and thermally stable selector based on TeAsGeSiSe chalcogenides compatible with BEOL IC integration for cross-point PCM, in: Proc. IEEE Int. Electron Devices Meeting, Dec. 2017, pp. 2.2.1–2.2.4.

[34] Y. Koo, S. Lee, S. Park, M. Yang, H. Hwang, Simple binary ovonic threshold switching material SiTe and its excellent selector performance for high-density memory array application, IEEE Electron Device Lett. 38 (5) (May 2017) 568–571.

[35] S. Kim, H. Kim, S. Choi, Intrinsic threshold switching responses in AsTeSi thin film, J. Alloys Compd. 667 (Jan. 2016) 91–95.

[36] A.C. Warren, Reversible thermal breakdown as a switching mechanism in chalcogenide glasses, IEEE Trans. Electron Devices 20 (2) (1973) 123–131.

[37] D. Adler, M.S. Shur, M. Silver, S.R. Ovshinsky, Threshold switching in chalcogenide-glass thin films, J. Appl. Phys. 51 (6) (Jun. 1980) 3289–3309.

[38] A. Pirovano, A.L. Lacaita, A. Benvenuti, F. Pellizzer, R. Bez, Electronic switching in phase-change memories, IEEE Trans. Electron Devices 51 (3) (Mar. 2004) 452–459.

[39] D. Emin, Current-driven threshold switching of a small polaron semiconductor to a metastable conductor, Phys. Rev. B 74 (3) (2006) 035206.

[40] M. Nardone, V.G. Karpov, D.C.S. Jackson, I.V. Karpov, A unified model of nucleation switching, Appl. Phys. Lett. 94 (2009) 103509.

[41] D. Ielmini, Threshold switching mechanism by high-field energy gain in the hopping transport of chalcogenide glasses, Phys. Rev. B 78 (3) (2008) 035308.

[42] B. Govoreanu, G.L. Donadio, K. Opsomer, W. Devulder, V.V. Afanas'ev, T. Witters, S. Clima, N.S. Avasarala, A. Redolfi, S. Kundu, O. Richard, D. Tsvetanova, G. Pourtois, C. Detavernier, L. Goux, G.S. Kar, Thermally stable integrated Se-based OTS selectors with >20 MA/cm^2 current drive, >3.10^3 half-bias nonlinearity, tunable threshold voltage and excellent endurance, in: Symp. VLSI Tech. Dig, Jun. 2017, pp. T92–T93.

[43] S. Clima, B. Govoreanu, K. Opsomer, A. Velea, N.S. Avasarala, W. Devulder, I. Shlyakhov, G.L. Donadio, T. Witters, S. Kundu, L. Goux, V. Afanasiev, G.S. Kar, G. Pourtois, Atomistic investigation of the electronic structure, thermal properties and conduction defects in Ge-rich Ge$_x$Se$_{1-x}$ materials for selector applications, in: Proc. IEEE Int. Electron Devices Meeting, Dec. 2017, pp. 4.1.1–4.1.4.

[44] A. Verdy, G. Navarro, V. Sousa, P. Noé, M. Bernard, F. Fillot, G. Bourgeois, J. Garrione, L. Perniola, Improved electrical performance thanks to Sb and N doping in Se-rich GeSe-based OTS selector devices, in: IEEE International Memory Workshop (IMW), May 2017.

[45] F. Rao, Z. Song, Y. Cheng, M. Xia, K. Ren, L. Wu, B. Liu, S. Feng, Investigation of changes in band gap and density of localized states on phase transition for Ge$_2$Sb$_2$Te$_5$ and Si$_{3.5}$Sb$_2$Te$_3$ materials, Acta Mater. 60 (2012) 323–328.

[46] J. Lee, G. Kim, Y. Ahn, J. Park, S. Ryu, C. Hwang, H. Kim, Threshold switching in Si-As-Te thin film for the selector device of crossbar resistive memory, Appl. Phys. Lett. 100 (Mar. 2012) 123505.

[47] J.V.D. Hurk, E. Linn, H. Zhang, R. Waser, I. Valov, Volatile resistance states in electrochemical metallization cells enabling non-destructive readout of complementary resistive switches, Nanotechnology 25 (42) (Sep. 2014) 425202.

[48] S. Ambrogio, S. Balatti, S. Choi, D. Ielmini, Impact of the mechanical stress on switching characteristics of electrochemical resistive memory, Adv. Mater. 26 (23) (2014) 3885–3892.

[49] J. Song, A. Prakash, D. Lee, J. Woo, E. Cha, S. Lee, H. Hwang, Bidirectional threshold switching in engineered multilayer (Cu$_2$O/Ag:Cu$_2$O/Cu$_2$O) stack for cross-point selector application, Appl. Phys. Lett. 107 (11) (Sep. 2015) 113504.

[50] Q. Luo, X. Xu, H. Liu, H. Lv, T. Gong, S. Long, Q. Liu, H. Sun, W. Banerjee, L. Li, N. Lu, M. Liu, Cu BEOL compatible selector with high selectivity (>10^7), extremely low off-current (~pA) and high endurance (>10^{10}), in: Proc. IEEE IEDM, Dec. 2015, pp. 10.4.1–10.4.4.

[51] W. Chen, H.J. Barnaby, M.N. Kozicki, Volatile and non-volatile switching in Cu-SiO$_2$ programmable metallization cells, IEEE Electron Device Lett. 37 (5) (2016) 580–583.

[52] J. Woo, D. Lee, E. Cha, S. Lee, S. Park, H. Hwang, Control of Cu conductive filament in complementary atom switch for cross-point selector device application, IEEE Electron Device Lett. 35 (1) (Jan. 2014) 60–62.

[53] Z. Wang, S. Joshi, S.E. Saveliev, H. Jiang, R. Midya, P. Lin, M. Hu, N. Ge, J.P. Strachan, Z. Li, Q. Wu, M. Barnell, G.-L. Li, H.L. Xin, R.S. Williams, Q. Xia, J.J. Yang, Memristors with diffusive dynamics as synaptic emulators for neuromorphic computing, Nat. Mater. 16 (1) (2017) 101–108.

[54] C.-P. Hsiung, H.-W. Liao, J.-Y. Gan, T.-B. Wu, J.-C. Hwang, F. Chen, M.-J. Tsai, Formation and instability of silver nanofilament in Ag-based programmable metallization cells, ACS Nano 4 (9) (2010) 5414–5420.

[55] J. Song, J. Woo, A. Prakash, D. Lee, H. Hwang, Threshold selector with high selectivity and steep slope for cross-point memory array, IEEE Electron Device Lett. 36 (7) (Jul. 2015) 681–683.

[56] J. Yoo, J. Woo, J. Song, H. Hwang, Threshold switching behavior of Ag-Si based selector device and hydrogen doping effect on its characteristics, AIP Adv. 5 (12) (Dec. 2015) 127221.

[57] N. Shukla, B. Grisafe, R.K. Ghosh, N. Jao, A. Aziz, J. Frougier, M. Jerry, S. Sonde, S. Rouvimov, T. Orlova, S. Gupta, S. Datta, Ag/HfO$_2$ based threshold switch with extreme non-linearity for unipolar cross-point memory and steep-slope Phase-FETs, in: Proc. IEEE IEDM, Dec. 2016, pp. 34.6.1–34.6.4.

[58] R. Midya, Z. Wang, J. Zhang, S.E. Savel'ev, C. Li, M. Rao, M.H. Jang, S. Joshi, H. Jiang, P. Lin, K. Norris, N. Ge, Q. Wu, M. Barnell, Z. Li, H.L. Xin, R.S. Williams, Q. Xia, J.J. Yang, Anatomy of Ag/hafnia-based selectors with 10^{10} nonlinearity, Adv. Mater. 29 (12) (Jan. 2017) 1604457.

[59] H. Sun, Q. Liu, C. Li, S. Long, H. Lv, C. Bi, Z. Huo, L. Li, M. Liu, Direct observation of conversion between threshold switching and memory switching induced by conductive filament morphology, Adv. Funct. Mater. 24 (36) (2014) 5679–5686.

[60] A. Bricalli, E. Ambrosi, M. Laudato, M. Maestro, R. Rodriguez, D. Ielmini, SiO$_x$-based resistive switching memory (RRAM) for crossbar storage/select elements with high on/off ratio, in: Proc. IEEE IEDM, Dec. 2016, pp. 4.3.1–4.3.4.

[61] I. Valov, R. Waser, J.R. Jameson, M.N. Kozicki, Electrochemical metallization memories–fundamentals, applications, prospects, Nanotechnology 22 (25) (Jun. 2011) 254003.

[62] J. Song, J. Woo, S. Lee, A. Prakash, J. Yoo, K. Moon, H. Hwang, Steep slope field-effect transistors with Ag/TiO2-based threshold switching device, IEEE Electron Device Lett. 37 (7) (Jul. 2016) 932–934.

[63] J. Song, J. Park, K. Moon, J. Woo, S. Lim, J. Yoo, D. Lee, H. Hwang, Monolithic integration of AgTe/TiO$_2$ based threshold switching device with TiN liner for steep slope field-effect transistors, in: Proc. IEEE IEDM, Dec. 2016, pp. 25.3.1–25.3.4.

[64] W. Devulder, K. Opsomer, J. Meersschaut, M. Deduytsche, M. Jurczak, L. Goux, C. Detavernier, Combinatorial study of Ag–Te thin films and their application as cation supply layer in CBRAM cells, ACS Comb. Sci. 17 (5) (2015) 334–430.

[65] L. Goux, K. Opsomer, R. Degraeve, R. Müller, C. Detavernier, D.J. Wouters, M. Jurczak, L. Altimime, J.A. Kittl, Influence of the Cu-Te composition and microstructure on the resistive switching of Cu-Te/Al$_2$O$_3$/Si cells, Appl. Phys. Lett. 99 (5) (Aug. 2011) 053502.

[66] X. Zhao, J. Ma, X. Xiao, Q. Liu, L. Shao, D. Chen, S. Liu, J. Niu, X. Zhang, Y. Wang, R. Cao, W. Wang, Z. Di, H. Lv, S. Long, M. Liu, Breaking the current-retention dilemma in cation-based resistive switching devices utilizing graphene with controlled defects, Adv. Mater. (Feb. 2018) 1–9. 1705193.

[67] N. Shukla, R.K. Ghosh, B. Grisafe, S. Datta, Fundamental mechanism behind volatile and non-volatile switching in metallic conducting bridge RAM, in: Proc. IEEE IEDM, Dec. 2017, pp. 4.3.1–4.3.4.

Part Two

Emerging opportunities

Ferroelectric memories

11

Cheol Seong Hwang, Thomas Mikolajick[†,‡]*
*Department of Materials Science and Engineering, and Inter-University Semiconductor Research Center, Seoul National University, Seoul, South Korea, [†]NaMLab gGmbH, Dresden, Germany, [‡]Institute of Semiconductors and Microsystems, TU Dresden, Dresden, Germany

11.1 Introduction

In a typical computing system, different types of memories are required based on their access speed versus density and cost per bit positioning. Close to the processor, random access memories (RAMs) are required to enable data exchange between the processing unit and the memory that is fast enough not to slow down the system speed and not limited to cycling. Traditionally, this was the role of the semiconductor memory in the flavor of either a static random access memory (SRAM) or a dynamic random access memory (DRAM) [1]. Note that the term RAM is therefore sometimes used as a synonym for a semiconductor memory, for example, in PCRAM (for phase change RAM) or RRAM (for resistive RAM), although these concepts are far away from fulfilling a RAM-type specification regarding write endurance in many cases. When computing systems were developed to be more distributed and mobile, more and more requirements for realizing nonvolatile memories also came up. However, due to the voltage-time dilemma [2], it is virtually impossible to realize a RAM, having infinite endurance, symmetric read/write that is nonvolatile based on charge storage. With the invention of Flash memories, the semiconductor memories, therefore, extended beyond RAMs with NOR Flash still maintaining random read access. In NAND Flash, another step away from random access was taken, and semiconductor memories even entered the arena of data storage, which had been the domain of magnetic and optical storage systems for a few decades. With faster and faster computing speed, however, the memory access becomes the bottleneck for the performance and a major driver for energy consumption. Therefore, it can be foreseen that even more memories may enter the memory hierarchy. Additionally, the trend toward the internet of things calls for nonvolatile memories with small to medium density in essentially every system that needs to be very effectively integrated into existing high-performance CMOS technology. Under these boundary conditions, four driving forces for establishing new memory technologies can be summarized:

1. Realization of memories that go beyond the scaling capabilities of DRAM or NAND Flash. DRAM has already come very close to its scaling limits, and NAND Flash is using a three-dimensional (3D) structure to overcome the scaling limits of the individual memory cells.
2. Realization of nonvolatile RAM technology. This has been the holy grail of semiconductor memories for a long time.

Advances in Non-volatile Memory and Storage Technology. https://doi.org/10.1016/B978-0-08-102584-0.00012-7

3. Establishment of new technologies that can bridge the large performance gap between DRAM and NAND Flash. These solutions are often called storage class memories [3].

4. Realization of very cheap nonvolatile memories that can be easily integrated into a CMOS baseline compared to the conventional floating gate or charge trapping-based Flash memories.

Therefore, in the last two decades, a number of technologies that promise to either replace or supplement the existing memory technologies namely ferroelectric memories, magnetic RAMs, phase change memories as well as resistive switching memories in different flavors have emerged [4]. The first products are available in the market for each of them. Among them, ferroelectric memories have the unique selling point of nonvolatility combined with lower write energy than all the other candidates. Therefore, this chapter deals with the different flavors of ferroelectric memories and shows how they can fit into the requirements of the future memory hierarchy.

After this introduction, the second section introduces ferroelectricity as an important and useful material property of certain dielectrics. The third section gives an overview of the most important and technologically relevant ferroelectric material. In the fourth section, a historical view on ferroelectric memories is provided. Reliability topics will then be covered in Section 11.5 and the following Sections 11.6–11.8 introduce the different possibilities of realizing memories based on ferroelectrics namely a capacitor-based 1T-1C ferroelectric RAM (FeRAM), an antiferroelectric RAM (AFRAM), a ferroelectric field-effect transistor (FeFET), and finally ferroelectric resistive switching memories (FeRRAM). In Section 11.9, applications going beyond pure memories are shortly reviewed, and Section 11.10 summarizes the chapter and gives some prospects for the future development.

11.2 Ferroelectricity

11.2.1 Origin of ferroelectricity from crystal point groups perspective

11.2.1.1 Symmetry breaking distortion

Ferroelectricity is known to arise from two stable polarization states that can be sustained in the absence of an applied electric field, and such spontaneous polarization can be produced by various arrangements of ions in the crystal structure of the material. The characteristic of symmetry-breaking distortion enables ferroelectric materials to ensure the presence of discrete states and, therefore, enhance the possibility of polarization switching between the states by applied electric field [5]. Temperature is an important variable that determines the transition between a polar ferroelectric phase and a nonpolar paraelectric phase via structural changes in the dimension of a crystal unit cell and the associated strain leads to distinct anomalies in the dielectric, elastic, thermal, and other properties of the material.

11.2.1.2   Structural classification of ferroelectrics

Particular crystal structure of ferroelectrics can be experimentally verified by X-ray scattering analysis whereby atoms and solids scatter X-rays to produce the interaction between electromagnetic radiation field and the electrons of the atom. Atomic arrangements determined by X-rays are classified into three different classes based on the predominant nature of the atomic displacements required to reverse the polarization. The simplest "one-dimensional" class refers to atomic arrangements, which are all parallel to the polar axis. In the "two-dimensional" class, the polarity is oriented by atomic displacements in a plane containing the polar axis and often rotated about an axis perpendicular to that plane. The "three-dimensional" class involves polarity orientation with similar magnitude displacements in all three dimensions. The most typical type of one-dimensional class is the perovskite family of ABO_3 such as $BaTiO_3$, $LiNbO_3$, and $LiTaO_3$ in which all oxygen octahedra have a threefold axis parallel to the polar axis. Polarization reversal by the rotation about axes normal to mirror planes is a major characteristic of two-dimensional ferroelectrics, and their best-known example is the barium cobalt fluoride family of $BaMF_4$ (M = Mg, Mn, Fe, Co, Ni, Zn) in which infinite sheets of MF_6 octahedra are normal to an orthorhombic b-axis. Three-dimensional ferroelectric materials generally contain one or more structural features of discrete tetrahedral or molecular ions, hydrogen bonds, or boron-oxygen frameworks, and their well-known examples are the families of the gadolinium molybdate, guanidinium aluminum sulfate hexahydrate (GASH), and the potassium dihydrogen phosphate (KDP) [6]. This classification determines the static crystallographic configuration of ferroelectrics and other studies of diffuse scattering and local probes, such as pair-density correlation function (PDF) analysis, and extended X-ray absorption fine-structure spectroscopy (EXAFS) can be utilized to examine the local distortions and fluctuations of the crystal structure [5].

11.2.2   Landau theory

Landau theory is a symmetry-based formalism to analyze an equilibrium behavior near a phase transition [7, 8]. In theory, a transition from one phase with a symmetry to another phase with a different symmetry is described by an order parameter of the phases. The order parameter is a physical entity, which is zero at a high symmetry (disordered) phase and has a finite value at a low symmetry (ordered) phase. This symmetry-based transition theory was first applied to the paraelectric-ferroelectric transition of a homogeneous bulk system by Devonshire, and the theory earned a name of Landau-Devonshire theory (LD theory) [9–11]. In LD theory, a spontaneous polarization induced by symmetry breaking of ionic displacement corresponds to the order parameter. In the vicinity of paraelectric-ferroelectric transition, the free energy of the system can be approximated by a power series concerning the order parameter, and each coefficient of the expansion term should be symmetry compatible. For example, in the transition of a perovskite-type $BaTiO_3$ where a high symmetry of cubic phase transits to a low symmetry of tetragonal phase, the free energy has two global minimum points of the order parameters (polarization vectors) along one axis, where from the viewpoint of thermodynamics, these two states are equivalent. Therefore, the odd power terms in the free energy expansion, which conserves the sign of the order parameter, must be zero. In a homogeneous bulk system, the free energy in the vicinity of the paraelectric-ferroelectric transition can be expressed as a power series with only even power terms. When truncated by the 6th order term, the free energy density f_P can be written as follows:

$$f_P = \frac{1}{2}aP^2 + \frac{1}{4}bP^2 + \frac{1}{6}cP^6 - EP. \tag{11.1}$$

In this uniaxial system, where E is defined as an electric field, a strain field energy can be ignored, and the origin of the free energy for the nonpolarized, unstrained state is assumed to be zero. Here, f_P refers to the free energy density and the total free energy of the system. This simple polynomial can be used as a tool to directly explain a singular behavior of the order parameter around the transition. More specifically, determining an equilibrium state of the system is equivalent to finding a minimum state with respect to the order parameter, P, and the specific thermodynamic variables can be obtained by differentiating the free energy. Therefore, each coefficient in the polynomial determines the behavior of the transition. The only temperature-dependent coefficient in the polynomial is a, which has a positive value above a characteristic temperature T_0, called Curie-Weiss temperature, and a negative value below T_0:

$$a = a_0\left(T - T_0\right). \tag{11.2}$$

For all the known ferroelectric materials, the signs of a_0 and c are always positive, while b can take both signs. Above T_0, the free energy landscape transitions to a single well potential configuration, which corresponds to the typical quadratic energy form of the paraelectric phase, while below T_0, the free energy function preserves the double well potential shape whose minimum points correspond to the spontaneous polarizations of the ferroelectric phase:

$$\frac{df_p}{dP} = 0, \tag{11.3}$$

$$E = aP + bP^3 + cP^5. \tag{11.4}$$

The solutions of Eq. (11.3) correspond to the equilibrium states of polarization defined as P_o, and Eq. (11.4) gives E as a function of the polarization. By differentiating this equation concerning P again, a linear dielectric susceptibility, χ can be obtained in a reciprocal form:

$$\kappa = \frac{1}{\chi} = a_0\left(T - T_0\right) + 3bP_0^2 + 5cP_0^4 \tag{11.5}$$

where κ is the reciprocal susceptibility. In this equation, the order parameter P should be substituted for the equilibrium point P_0. For paraelectric phase above T_0, the equilibrium P_0 becomes 0, and κ becomes as follows:

$$\kappa^{T>T_0} = \frac{1}{\chi} = a_0\left(T - T_0\right). \tag{11.6}$$

This equation shows an inverse proportionality to $T - T_0$ of the linear susceptibility of paraelectrics, which is experimentally confirmed as Curie-Weiss behavior in most ferroelectric materials for $T > T_0$. The free energy formulas of this system far above $(T \gg T_0)$ and far below $(T \ll T_0)$ are universal for most ferroelectrics. However, a pathway between these two phases is dependent on the coefficient b. In other words, the sign of b determines whether the order parameter P develops continuously or discontinuously during the transition as discussed further.

11.2.2.1 Second-order transition

A transition where the order parameter develops continuously is called a second-order transition, which occurs when the sign of the coefficient b is positive. Under this circumstance, the equilibrium P_0 can be readily obtained by ignoring the coefficient c in the vicinity of the transition ($P_0 \sim 0$ near T_0, then $P_0^6 \ll 1$) as follows for $T < T_0$ and $E = 0$ from Eq. (11.4):

$$a_0 \left(T - T_0\right) + bP_0^2 = 0. \tag{11.7}$$

Therefore, P_0 is obtained as:

$$P_0 = \sqrt{\frac{a_0 \left(T_0 - T\right)}{b}} \text{ as } T \to T_0. \tag{11.8}$$

From Eq. (11.8), it is obvious that the spontaneous polarization of ferroelectrics becomes zero at $T = T_0$, and above T_0 the material becomes nonpolarized paraelectric. When P_0 in Eq. (11.5) is substituted for the expression in (11.8), the reciprocal susceptibility of the system below T_0 can be obtained as follows:

$$\kappa^{T < T_0} = 2a_0 \left(T_0 - T\right). \tag{11.9}$$

Eqs. (11.6) and (11.9) show that the linear susceptibility χ during the second-order transition diverges. This is another characteristic behavior of the second-order transition. The features of the transition are summarized in Fig. 11.1. In many thin film ferroelectrics deposited on a substrate, the effects of clamping can be large enough to shift the transition temperature or the order of the transition itself by altering the sign of the coefficient b [12]. While most ferroelectrics materials except a few examples (e.g., triglycine sulfate, or TGS) fall into the first-order ferroelectrics, the first-order transition can become the second order in the clamped system, which is usually observed in $BaTiO_3$ [13].

11.2.2.2 First-order transition

A transition at which the order parameter develops discontinuously is called a first-order transition, which occurs when the sign of the coefficient b is negative. In the case of $b < 0$, the last term in Eq. (11.4) cannot be ignored when finding the equilibrium below T_0. In addition, the first-order transition point is different from T_0. The first-order transition occurs when Eq. (11.3) and the ferroelectric energy (Eq. 11.1) itself also becomes zero (equal to that of the origin), and the associated transition temperature T_c, which is called Curie temperature, can be obtained as follows:

$$T = T_c = T_0 + \frac{3}{16} \frac{b^2}{a_0 c}. \tag{11.10}$$

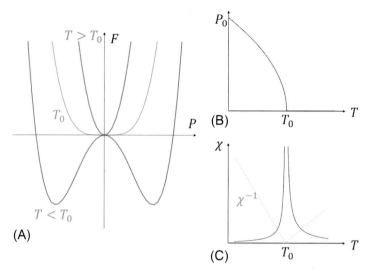

Fig. 11.1 (A) Free energy diagram of second-order ferroelectric. (B) Spontaneous polarization $P_0(T)$ as a function of temperature. (C) Linear susceptibility $\chi(T)$ as a function of temperature.

As shown in Eq. (11.10), T_c is higher than T_0, which means that a first-order ferro-electrics can maintain a nonzero spontaneous polarization above T_0 up to T_c. When Eq. (11.10) is solved for T_0 and substituted back into Eq. (11.4), the nonzero equilibrium polarization P_0 below T_c can be obtained:

$$P_0 = \sqrt{\dfrac{-b + \sqrt{\left(\dfrac{b}{2}\right)^2 + 4a_0c\left(T_c - T\right)}}{2c}} \quad \text{as } T \to T_c. \tag{11.11}$$

Note that P_0 does not converge to zero but a finite value at $T=T_c$, while above T_c, the global minimum of the free energy becomes a nonpolarized state ($P_0=0$) as shown in Fig. 11.2(A). In other words, the order parameter develops discontinuously at $T=T_c$ in the first-order transition, which is critically different from the case of a second-order phase transition. The linear susceptibility of the system below T_c can be obtained by substituting Eq. (11.11) into (11.5). With appropriate approximations in the vicinity of the transition point, κ becomes:

$$\kappa^{T<T_c} = 8a_0\left(T_c - T\right) + \dfrac{3b^2}{4c}. \tag{11.12}$$

Above T_c, the reciprocal susceptibility given as Eq. (11.6) can be expressed as follows:

$$\kappa^{T>T_c} = a_0\left(T - T_c\right) + \dfrac{3b^2}{16c}. \tag{11.13}$$

It is noteworthy that unlike the second-order transition, the linear susceptibility does not diverge at the transition. Since the first-order transition does not reflect a singularity

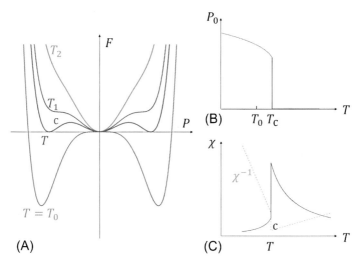

Fig. 11.2 (A) Free energy diagram of first-order ferroelectrics. (B) Spontaneous polarization $P_0(T)$ as a function of temperature. (C) Linear susceptibility $\chi(T)$ as a function of temperature.

in thermodynamic potential, a low-temperature phase can persist above T_c as a metastable state, and a high-temperature phase can also persist below T_c as a metastable state as shown in Fig. 11.2(A). These metastable states can generate hysteresis in the first-order transition curve. The feature of the transition is summarized in Fig. 11.2. Most ferroelectrics are identified as the first-order ferroelectrics. Typical examples are perovskite-type ferroelectrics, such as $BaTiO_3$ or $Pb(Zr,Ti)O_3$. Recently found doped-HfO_2 ferroelectrics, of which crystal structure is fluorite type, also fall into this category [14].

11.3 Ferroelectric materials

11.3.1 Conventional ferroelectrics

The $PbZr_xTi_{1-x}O_3$ $(0 < x < 1)$ (PZT) material is the most extensively studied material system for the development of ferroelectric memory devices, such as ferroelectric RAM (FRAM) or ferroelectric FET (FeFET). This material is composed of the solid solution of $PbTiO_3$ and $PbZrO_3$ perovskite materials, where the spontaneous polarization is exhibited by the displacement of Zr^{4+} or Ti^{4+} ions located in the oxygen ion octahedron in the crystal structure as shown in Fig. 11.3. Below the Curie temperature T_c, the stable cation positions appear to slightly deviate from the center of the oxygen ion octahedron forming a spontaneous electric dipole (P_0). The direction of P_0 can be switched by the electric field making the material ferroelectric. In the Ti-rich $(x < 0.52)$ region, PZT yields a tetragonal crystal structure, and the spontaneous polarization is formed toward six face-center $\langle 001 \rangle$ directions. On the other hand, PZT in the Zr-rich $(x > 0.52)$ region yields a rhombohedral crystal structure, and the spontaneous polarization can be formed toward eight $\langle 111 \rangle$ directions of the pseudo-cubic perovskite structure. However, this composition material is usually antiferroelectric wherein the

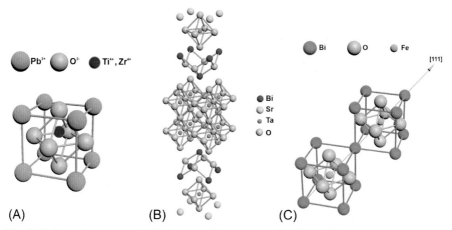

Fig. 11.3 Crystal structure of (A) the perovskite structure for $Pb(Zr,Ti)O_3$, (B) Aurivillius structure for $SrBi_2Ta_2O_9$, and (C) the perovskite structure for $BiFeO_3$.

direction of P_0 in the crystal structure is opposite in every other unit cell. In particular, $PbZr_{0.52}Ti_{0.48}O_3$, the phase boundary composition of the rhombohedral and tetragonal crystal structures, is called the morphotropic phase boundary (MPB), where the transition between two phases can easily occur with the application of the electric field. This phase transition by the electric field can be accompanied by large lattice strain response [15]. Therefore, PZT materials of the MPB composition are frequently used in micro-electro-mechanical system applications due to their high piezoelectric response [15]. On the other hand, when the Ti composition ratio is further increased, the c/a ratio is increased accordingly, and the P_s value appears to be high. These compositions are used in FRAM or pyroelectric applications [15]. Polycrystalline PZT materials have a remanent polarization (P_r) value of ~10–40 $\mu C/cm^2$, a permittivity of ~300–400, and a coercive field (E_c) of ~50–70 kV/cm. It shows a high T_c of ~400°C in the Ti-rich composition, which is considered to be a suitable property for the memory applications [16]. The PZT thin film can usually be crystallized at ~450–650°C. On the other hand, when the PZT material is used with the metal electrode, such as Pt, a severe fatigue phenomenon generally can occur in ~10^6 cycles [17]. It was reported that the fatigue problem could be avoided when PZT is used in contact with the oxide electrode (described in detail in a later section) [17, 18].

The $SrBi_2Ta_2O_9$ (SBT) material exhibits the bi-layered structure (Aurivillius phase), where the Bi_2O_2 layers are interposed between the pseudo-perovskite layers as shown in Fig. 11.3. The P_r value reaches approximately 5–10 $\mu C/cm^2$, which is slightly smaller than that of PZT. E_c is also slightly smaller than that of PZT as ~30–50 kV/cm. However, unlike PZT, SBT materials are not only lead free but have the advantage that fatigue phenomenon does not appear on Pt or other metal electrodes [19]. Therefore, it does not require the complex deposition processes to use the oxide electrodes. The crystallization temperature reaches ~650–800°C, which is higher than that of PZT thin films by ~150°C.

$BiFeO_3$ (BFO) is a multiferroic material with the rhombohedral crystal structure, where both ferroelectric and antiferromagnetic properties can appear simultaneously in a single phase. T_c and the Neel temperature (T_N) of BFO are ~1370°C and ~637°C, respectively [20, 21]. Therefore, it is theoretically possible to conceive such devices that control the magnetization of the material by the application of the electric field (or vice versa) at room temperature. BFO is a lead-free material and has a large theoretical P_s of 90–95 $\mu C/cm^2$, which is much larger than that of PZT [22, 23]. However, E_c appears to be between ~300 and ~550 kV/cm, which is a much larger value than that of PZT. The spontaneous polarization appears in the $\langle 111 \rangle$ direction on the pseudo-cubic perovskite structure, and the bandgap is reported to range from 2.0 to 2.7 eV, which is lower than that of PZT or SBT (3.0–3.5 eV). Therefore, it is almost close to a wide-bandgap semiconductor rather than an insulator. This is considered as a disadvantage for using BFO in major memory applications, such as FRAM and FeFET, where the ferroelectric thin film needs to be insulating to minimize the leakage current. Meanwhile, this small bandgap enables it to be used as optoelectronics, such as the switchable photovoltaic devices [24]. In the photovoltaic device, the light energy can be converted into the electrical energy by separating the hole and electron in the heterojunction region, and in the ferroelectric semiconductor material such as BFO, the direction of the photocurrent can be switched according to the poling direction of the spontaneous ferroelectric polarization [24]. Recently, BFO has been extensively examined as a material for the domain-wall-type resistance switching memory, which is described in more detail in Section 11.8. This is another example of utilizing the relatively smaller bandgap of BFO.

11.3.2   Fluorite-based ferroelectrics

11.3.2.1   Overview

Since ferroelectricity and antiferroelectricity (field-induced ferroelectricity to be exact) were first reported by Boeske et al. in 2011 for Si-doped HfO_2 thin films [25], the ferroelectric properties were extensively studied in the doped-HfO_2 thin films [25–32], HfO_2-ZrO_2 solid solution thin films [33–36], and the undoped HfO_2 thin films [37, 38]. Unlike the conventional ferroelectrics based on the perovskite-type crystal structure, of which ferroelectric properties degrade at the thickness smaller than ~100 nm, the FE-HfO_2 thin films can exhibit excellent ferroelectric properties at extremely low thicknesses of ~10 nm owing to the high bandgap of ~5.5 eV [39]. Moreover, the coercive field (generally 1–2 MV/cm), which is higher than that of perovskite-type ferroelectrics by ~1 order, enables a sufficient memory window of FeFETs with such a small film thickness (see Section 11.7). These properties of HfO_2 brought the renewed interest in the ferroelectric memories. In addition, the HfO_2 and ZrO_2 thin films have already been used in the current semiconductor industries, such as high-k gate dielectrics of logic transistors and dielectric materials of DRAM capacitors [40, 41]. Therefore, there is no compatibility issue of these materials with the Si-based complementary metal oxide semiconductor field-effect transistor (CMOSFET) technology, which is not the case for the abovementioned PZT, SBT, and BFO materials.

It is well known that HfO_2 materials exhibit a monoclinic phase (space group $P2_1/c$; m-phase) at room temperature and atmospheric pressure. The m-phase exhibits a low dielectric constant (15–20) similar to that of the amorphous state and a simple linear dielectric response. In the bulk state, phase transitions are made from the m-phase to the tetragonal phase (space group $P4_2/nmc$; t-phase) at ~1800°C, and to the cubic phase (space group $Pm3m$; c-phase) at ~2600°C [39]. In the high-pressure region, o1 phase (space group $Pbca$) begins to appear around ~4 GPa, and o2 phase (space group $Pnma$) appears around ~14.5 GPa at room temperature [42]. ZrO_2 material also shows similar crystal structures, and all the phases present in the phase diagram of the two materials are centrosymmetric with no evidence of ferroelectricity. Therefore, it was surprising that ferroelectricity was found in HfO_2 and its solid solution system with ZrO_2. Up to now, this ferroelectric phase is known to be formed only in the thin film conditions, and the ferroelectricity in the bulk state has not yet been reported.

The crystal structure of the ferroelectric phase has been identified as the orthorhombic phase (o-phase) with the space group $Pca2_1$ [43]. It is known that dopant type, encapsulation effect, interface/grain boundary energy, and epitaxial effect contribute to the stabilization of the o-phase. However, the quantitative relation with the free energy of the phase and the interrelations between them are not fully understood until now (described in detail later). Compared to the conventional perovskite-type ferroelectric material, it is classified as the fluorite-type ferroelectric material considering the overall crystal structure of the o-phase. As shown in Fig. 11.4, the spontaneous polarization appears due to the displacement of the four threefold coordinated O^{2-} ions among the eight O^{2-} ions (four threefold and four fourfold coordinated) present in eight tetrahedral sites surrounded by Hf^{4+} ions. Four O^{2-} anions can be displaced in the c-axis direction by a high enough external electric field, and the magnitude of the spontaneous polarization is estimated as ~50 $\mu C/cm^2$ according to first principle calculations [44]. However, ferroelectric HfO_2 thin films are

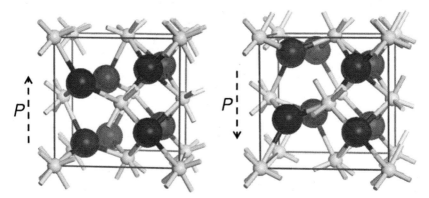

Fig. 11.4 Polar orthorhombic structure of the HfO_2 thin film. The spontaneous polarization direction is indicated by the black arrows. The red spheres are oxygen ions, and the light blue spheres are hafnium ions. The left four oxygen ions are threefold coordinated while the right four oxygen ions are fourfold coordinated.

Reproduced from T.S. Böscke, J. Müller, D. Bräuhaus, U. Schröder, U. Böttger, Ferroelectricity in hafnium oxide thin film, Appl. Phys. Lett. 99 (2011) 102903, with permission from AIP Publishing.

deposited as polycrystalline thin films with almost random orientations. Therefore, its P_r value decreases to ~10–25 µC/cm^2 according to phase composition and film quality. Especially, a significantly large P_r of ~40 µC/cm^2 was reported in La-doped HfO$_2$ thin films [45].

It is known that field-induced ferroelectricity appears when dopants smaller than Hf, such as Si, Al, and Zr, are doped in the HfO$_2$ thin films at higher concentrations than those exhibiting the optimum FE characteristics [46]. This appears to be a double hysteresis that is electrically similar to that of the antiferroelectricity. However, it is not explained by the action of the actual antiparallel dipole in the crystal structure; instead, this electrical phenomenon originates from the field-induced phase transformation between the structurally similar t-phase and the o-phase [47]. When the external electric field is applied to the thin film, the o-phase, which is a polar phase, can be reversibly induced by the -EP term of the free energy of the ferroelectric material, resulting in the induction of spontaneous polarization [48]. This phase transition has the nature and characteristic of first-order transition with a small activation energy [48].

11.3.2.2 Wake-up effect

Conventional perovskite-type ferroelectric materials exhibit deterioration of ferroelectric properties after a number of electric field cyclings. Although materials like PZT also show a "wake-up" phenomenon, this characteristic behavior is predominant in the HfO$_2$ materials in which P_r increases and the P–E hysteresis is sharpened as the electric field-cycling progresses. This phenomenon has been intensively studied in the HfO$_2$ thin film area [39, 48, 49]. The pinched hysteresis observed in doped HfO$_2$ thin films in the pristine state is explained due to the existence of defects, such as Vo, domain pinning, and internal bias field [39]. Especially near the TiN electrode, the oxygen-deficient layer is known to be easily formed by the slight oxidation of the TiN electrode, which can increase the depolarization field and make the polarization switching nonuniform [39]. Wake-up in doped HfO$_2$ thin films has been attributed to the redistribution of Vo due to the applied external electric field [39]. It has been verified that such redistribution of Vo is based on the thermally activated process [50, 51]. In the study on HfO$_2$-ZrO$_2$ solid solution thin film, it was especially emphasized that the t-phase near the TiN electrode could be destabilized and permanently transformed into the o-phase after the application of the external field [52]. Meanwhile, transmission electron microscopy study has shown that the phase transition from the m-phase to the o-phase occurs during the wake-up in the Gd-doped HfO$_2$ thin film [53]. In memory applications, this wake-up effect can be undesirable because the memory operations need the constant FE characteristics even after repetitive external field cycling. In this regard, the change of the Zr composition to Hf$_{0.6}$Zr$_{0.4}$O$_2$ in the HfO$_2$-ZrO$_2$ solid solution or the simultaneous use of the Si and Al dopants has been suggested (described in more detail in a later section) [52, 54]. Additionally, this material shows the typical fatigue behavior known from optimized PZT between Pt electrodes. Indeed, the La-doped Hf$_{0.5}$Zr$_{0.5}$O$_2$ film, which withstands the electrical cycling up to ~10^{11} cycles due to the lower leakage current induced by the La-doping, showed rather significant fatigues [32].

11.3.2.3 Dopants

It is known that electrical properties can be greatly changed depending on the dopant types although the cause of the change is not elucidated in-depth yet. However, several studies have revealed dopant-dependent changes in view of the valence number and the ionic size of the dopant [31, 49, 55]. The effect of the valence number has been studied in doped HfO_2 thin film deposited by an RF-sputtering method for several dopants (Sc^{3+}, Zr^{4+}, and Nb^{5+}), which have a similar ionic radius to that of Hf^{4+} ions [31]. The induced phase change depending on the type of dopant could be explained as the incorporation and/or elimination of Vos to maintain the charge neutrality in the HfO_2 thin film [31, 55]. Vo was reported to decrease the free energy of the high-symmetry (ferroelectric) phases [55]. Meanwhile, the FE response is characteristically found only with dopants such as Zr, Si, and Al, of which the ionic size is similar or smaller than that of the Hf^{4+} ion. Such behavior has been explained by the stabilized high-symmetry phases, t-phase or c-phase, at high doping concentrations whose concentration varies depending on the dopant size [56]. When the ionic size is smaller than that of Hf^{4+} ion, the dopant stabilizes the tetragonal phase by forming a local bonding configuration similar to that in the tetragonal structure, while it stabilizes the cubic structure when the ionic size is larger than Hf ion [57]. In this regard, the field-induced ferroelectric response can only be found with dopants which are smaller than (or similar to) the Hf^{4+} ion since it is based on the field-induced phase transition between the t-phase and the o-phase [39].

A comprehensive study on the effect of dopant species on the stabilization of the o-phase was conducted using computational simulation by Batra and coworkers [58]. They reported that lanthanide series metals, the lower half of alkaline earth metals (Ca, Sr, and Ba) and Y, are the most suitable dopants to promote ferroelectricity in HfO_2, and the result was consistent with the experimental results from the chemical solution deposited HfO_2 thin films doped with various dopants [59]. From the overall trend, divalent or trivalent ions with larger radii and smaller electronegativity could decrease the free energy of the o-phase more strongly.

11.3.2.4 HfO₂-ZrO₂ solid solution thin film

Most of the dopants exhibit the maximum P_r within the doping concentration of around or below 10 at.%, whereas the maximum P_r of the $Hf_{1-x}Zr_xO_2$ thin film is generally observed at ~50 at.% doping concentration. $Hf_{1-x}Zr_xO_2$ shows a solid solution in the entire concentration range due to the structural similarity of the HfO_2 and ZrO_2. The value of the P_r is around ~15–25 $\mu C/cm^2$ when it is deposited by atomic layer deposition [60]. The crystallization temperature is ~400–600°C, which is significantly lower than that of the reported doped HfO_2 thin films, such as Si-doped HfO_2, which needs 800–1000°C to be fully crystallized. This property can be a great advantage for the integration into memory technologies.

The change of stability between the t-phase and the o-phase with the Zr content in the $Hf_{1-x}Zr_xO_2$ thin film has been intensively analyzed based on the Landau-Ginzburg-Devonshire (LGD) model [60, 61]. In the LGD model, the T_0 value determines the stability of the nonpolar t-phase, which has higher-symmetry compared to the polar o-phase, and T_0 can be expressed as a function of Zr content. As the Zr concentration increases, the T_0 value can decrease, and the t-phase becomes more stabilized [61]. At the boundary between these two phases, the phase transition appears as of first-order nature. A broken hysteresis loop can be observed in the $Hf_{0.4}Zr_{0.6}O_2$ thin film [14], which corresponds to the boundary composition between the FE and the FFE characteristics, and of which behavior generally can be observed in the ferroelectric materials with a first-order transition at temperatures between T_c and T_1 [62]. The 0 K first principle calculations for the o-phase HfO_2 crystal structure also shows qualitatively similar energy landscape with the first-order transition of the LGD model [44].

11.3.2.5 Top electrode encapsulation effect

Most of the doped HfO_2 thin films are in an amorphous state in their as-deposited state because the thermal energy provided during the deposition process is not sufficient to fully crystallize the thin film. Therefore, they are subjected to an additional crystallization heat treatment step after deposition. In this case, two conditions are possible: postmetallization annealing (PMA) condition, in which the crystallization annealing is performed after the top electrode deposition, and postdeposition annealing (PDA), in which crystallization annealing is performed before the top electrode is deposited. The PMA condition is reported to exhibit better ferroelectricity than that of the PDA condition in most doped HfO_2 thin films, and this points to the fact that the top encapsulation effect acts as a factor that suppresses the formation of the m-phase during the crystallization heat treatment [25, 27, 34, 39]. A certain grain crystallized during the high-temperature annealing process may act as a driving force for the phase transition from the t/o-phase to a stable m-phase when it is cooled down to room temperature. In this condition, the upper electrode can physically limit the volume expansion and/or shear deformation required during the phase transition from t/o-phase to m-phase, increasing the activation energy of the phase transition [39]. As another explanation, the top electrode can suppress the formation of the m-phase by inhibiting the ion diffusion on the surface during high-temperature heat treatment, thereby preventing the grain growth [39].

11.3.2.6 Thickness effect

It is known that the stabilization of the FE o-phase in doped HfO_2 thin films is influenced by various factors such as doping, interface/grain boundary energy, top encapsulation, impurities, and thermal treatment conditions. However, the quantitative influences of the individual components on the thermodynamic free energy of the o-phase are not fully elucidated to date. Meanwhile, the maximum P_r of the doped HfO_2 thin film appears at a thickness near ~10 nm, and it gradually decreases as the thickness of the thin film is increased or decreased [40, 51, 63]. This behavior can be explained based on the loss of the surface (exactly interface/grain boundary) energy effect with the increase in grain size [57]. First principles calculations conducted by Materlik et al. showed that the FE o-phase could not be easily stabilized over various phases, such as m-, high-symmetry (t/o/c-) or high-pressure (o1- and o2-) phases, by simply increasing the temperature or applying specific strain in certain directions onto the lattices in $Hf_{1-x}Zr_xO_2$ [57]. On the other hand, the FE o-phase could be stabilized by only assuming the surface energy generated in grains in the $Hf_{0.5}Zr_{0.5}O_2$ and HfO_2 thin films in the small thickness range of 7–15 and 3–5 nm, respectively. [57]. This can be possible because the grains at small thicknesses have a large surface-to-volume ratio, and in this environment, the high-symmetry phase, which has lower surface energy than the m-phase, can be preferred over the m-phase. Also, it was shown that the FE o-phase could be stabilized in the HfO_2 thin films without any dopant simply due to the surface energy effect, and it was confirmed later experimentally [37]. For $Hf_{0.5}Zr_{0.5}O_2$ thin films, the electrical properties were reported to change from ferroelectric to antiferroelectric with decreasing film thickness below 8 nm [51].

In this calculation, however, the interface/grain boundary energy is assumed to be identical to the surface energy. In reality, the interface/grain boundary energy is considered to be much smaller due to a smaller fraction of dangling bonds. In this regard, the effect of surface energy becomes much smaller than the value assumed in the calculation. In recent studies, it was reported that the thermodynamic free energy of the o-phase could not be stabilized in a large Zr composition range when considering the various factors in the HfO_2-ZrO_2 solid solution thin film, such as grain size distribution, film thickness, Zr content, and top encapsulation effect [57, 64]. In the wide composition range, the m-phase is still found to be the most stable phase. Therefore, the kinetic path during the evolution of the o-phase was emphasized instead [57, 64]. The t-phase nuclei observed in the as-deposited thin film grows to be one of the coarse grains during the thermal treatment. Large activation energy for the phase transition inhibits the t-phase to m-phase transition (~300 meV/f.u.), whereas it only requires a small activation energy of ~30 meV/f.u. for the o-phase transition. [64]. Therefore, it appears that the fundamental reason for the evolution of the FE o-phase is not based on thermodynamics alone and kinetic effects (t → o is favored, but t → m is disfavored) have an important contribution.

11.4   Reliability

Any nonvolatile memory needs to deal with three basic reliability effects. Specific to ferroelectrics is another phenomenon called imprint. Therefore, we need to consider the following aspects when looking at reliability:

- Retention is the ability of the memory cell to maintain its state over a longer period. In a nonvolatile memory, 10 years at an elevated temperature are typically required.
- Endurance is the ability of the memory cell to be continuously cycled between the different memory states.
- Disturbs are caused by signals that are applied to the word line (WL) and bit line (BL) and also affect cells that are not subject to the selected read or write operation.
- Imprint is the tendency of stabilizing the polarization state where the ferroelectric cell is stored in. Consequently, after retention bake, the stored state is even enhanced, but the opposite logic state may show a read error after reprogramming.

Disturbs are strongly dependent on the array architecture. In the 1T-1C architecture covered in the following section, disturb is a minor issue as the MOS sect device protects the unselected cells from disturbs. For 1T FeFETs, the array architecture needs to be considered with the concept of inhibiting the unselected cells. Some comments on that can be found in Section 11.7. In this section, the retention, endurance, and imprint in ferroelectric devices are described. Fig. 11.5 shows how these three effects can be understood in terms of the polarization-electric field hysteresis curves. The classical retention loss corresponds to a depolarization of the ferroelectrics (Fig. 11.5(A)). Most ferroelectrics show a more or less strong fatigue effect when subjected to extensive bipolar cycling. It should be mentioned here that in the early phase of the life cycle the opposite can happen as well. During the so-called wake-up phase, the polarization can increase with cycling. Finally, the imprint is specific to ferroelectrics. They tend to stabilize the state they have been stored in for a longer time [65]. This may seem innocent at first glance since retention gain will be observed rather than retention loss. However, as it becomes clear from looking at Fig. 11.5(C),

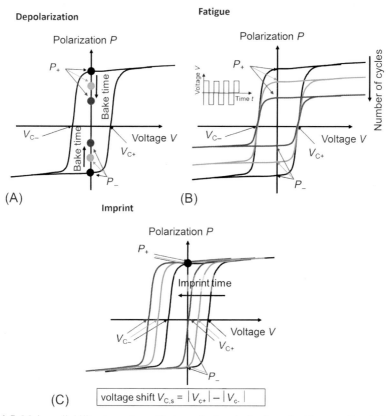

Fig. 11.5 Major reliability degrading effects on the level of a ferroelectric capacitor. (A) Depolarization is the loss of remanent polarization with time manifests itself as retention loss. (B) The reduction of the remanent polarization with field cycling is called fatigue and limits the endurance of the ferroelectric. (C) The imprint effect tends to stabilize the state the capacitor is stored in reflected by a parallel shift of the hysteresis and may result in a read fail of the opposite polarization state after reprogramming.

the opposite polarization state will significantly suffer a polarization loss, and a read error can occur after reprogramming.

Since imprint complicates the retention assessment, retention measurements in ferroelectric devices are usually more complex than in other nonvolatile memories. As a minimum, the other state needs to be checked after retention measurement. However, in practically important cases several combinations of stored and new states are typically reported [66] to achieve a better view of the practically most critical case.

For PZT-based FeRAM, the most challenging reliability aspects that need to be solved were fatigue and imprint. Both could be solved by using oxide electrodes such as IrO_2 or $SrRuO_3$ along with additional dopants in PZT. This points to the direction that the degradation has a common physical origin which is connected to oxygen vacancy-related defects. In HfO_2-based ferroelectrics, the filed cycling degradation is stronger than in optimized PZT. However, comparing $Hf_{0.5}Zr_{0.5}O_2$ films to non-optimized PZT films, the similar behavior can be seen in both cases [67]. Besides the more general aspects discussed in this section, many reliability aspects need to be viewed in the context of the concrete memory and material concept, and more detailed information can be found in the following sections.

11.5 Ferroelectric RAM

11.5.1 General concept of a 1T-1C ferroelectric memory (writing, reading)

FeRAM has a similar structure to the DRAM but is different to each other due to the presence of the nonvolatile P_r to show a nonvolatile RAM performance. The binary digital data are represented by $+P_r$ and $-P_r$ states of the ferroelectric capacitor, which is accessed through a serially connected cell-select MOSFET. Unlike DRAM, where the charge sharing between the cell capacitor and BL is naturally achieved when the access MOSFET opens, the stored charge ($+P_r$ or $-P_r$) must be driven by a read voltage, which is applied to the plate electrode. The sense amplifier detects the voltage variation of the BL by comparing it with the voltage of a reference BL when the stored charge in the ferroelectric capacitor is shared with the operating BL. Therefore, FeRAM has a similar performance as the DRAM, and thus, shares similar scaling limit: stored charge density. Due to the involvement of fatigue and several other reliability issues, the reference voltage generation scheme is not as simple as that of DRAM. Also, the limited fatigue-endurance often limits the endurance cycles below 10^{16}, which is the general assumption for a RAM. These factors have deterred the commercialization of FeRAM along with the difficulties related to the fabrication of three-dimensional capacitor composed of high-performance ferroelectric layers. These trends, however, are now under significant reformulation due to the recent development of the doped-HfO_2-based ferroelectric thin films. Next, the writing and reading methods are shortly explained. The subsequent sections give more details about the cell configuration and material properties necessary for the high-performance FeRAM.

During the process of writing, $+V_{DD}$ is applied to the ferroelectric capacitor via the BL or the PL to obtain "1" or "0" data, respectively. When the WL is selected for the selected cell, which implies that the cell-select MOSFET is turned on, voltage is applied either to the BL or the plate line (PL). To write "0" data, BL = 0V and PL = V_{DD} are applied, whereas BL = V_{DD} and PL = 0V to write "1" data. After writing, the data are retained and stays nonvolatile, which can be represented as $+P_r$ and $-P_r$ in the hysteresis curve.

During the process of reading, BL must be pre-charged to 0V before selecting the WL. This is in contrast to DRAM where the BL needs to be precharged to $V_{DD}/2$ in order to allow a symmetrical access. Then, WL is selected with V_{DD} applied to PL. If the cell possesses "0" data, relatively small amount of electric charges is collected to boost BL to small voltage (ΔSV) without any polarization reversal. If the cell possesses "1" data, polarization is reversed, causing large electric charge difference, and BL is charged up by large voltage (ΔLV) by the sense amplifier connected to the BL. Compared to the reference voltage (V_{ref}) obtained from the reference cell, ΔSV is reduced to 0V, and ΔLV is raised to V_{DD}.

Reading "1" data, however, causes the data destruction to "0." Therefore, rewriting after reading "1" data is needed. After reading "1" data, the BL voltage level remains at V_{DD}, whereas the PL voltage is 0V, causing "1" data to be restored.

11.5.2   Cell concepts (1T-1C vs 2T-2C; stacked vs. offset vs. chain)

FeRAMs are conventionally classified into two categories: 1T-1C and 2T-2C. Similar to DRAM, each type consists of a transistor and a capacitor, two of each for 2T-2C, which are responsible for access control and data storage, respectively. It is notable that the capacitors of 1T-1C and 2T-2C FeRAM require a plate line which applies different voltage pulses for the read and write operation while the plate line of the DRAM is grounded. 1T-1C FeRAM reads data by comparing the states of the BL voltage with the reference voltage whereas 2T-2C compares the states of the paired capacitors [68]. When reading data "1," in which the polarization directions of ferroelectrics and applied voltage are opposite, because the polarization reversal charge is proportional to the cell area, 2T-2C is intrinsically doubling the sensing margin for the read operation compared to 1T-1C. Moreover, both cells are cycled identically and therefore degradation mechanisms partially cancel out as well. Nonetheless, the V_{dd} of 2T-2C may not be fully applied to each cell capacitor as it is reduced by the floating BL, and a large number of capacitors for 2T-2C is disadvantageous for device integration and scaling (Fig. 11.6).

The offset cell structure, where the transistors and capacitors are aligned in a planar manner and connected to each other using the exiting metallization layers, has a relatively large cell size that limits the development of device integration and scaling. This structure was initially developed due to the absence of integration technology for vertical connection of capacitor onto the MOSFET. The more recent integrated structure of a capacitor vertically stacked on a transistor was devised, but its efficiency regarding the cell size and fabrication was still not comparable with those of DRAM cells (Fig. 11.7). The chain cell structure of transistors sharing the source and

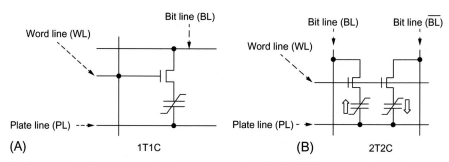

Fig. 11.6 Schematic representations of (A) 1T-1C and (B) 2T-2C FeRAM circuit [68].

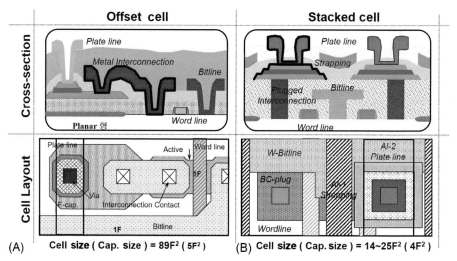

Fig. 11.7 Cross section and top view of FeRAM (A) offset and (B) stacked-cell.

drain can theoretically reduce the cell size down to $4F^2$ [69], and its configuration of short-circuited capacitors enhanced the problem of refresh operation by achieving fast nondriven half-V_{dd} cell-plate operation [70] In this chain cell structure, the voltage can be applied to the targeted ferroelectric capacitor by turning off the corresponding MOSFET when the voltage is applied between the two endpoints of the cell string. Nonetheless, the inherent noise that may arise from such dense structure of chain FeRAM has been an issue that impedes the performance of each device (Fig. 11.8).

11.5.3 Problems with PZT integration

Conventional materials utilized for FeRAM, such as PZT, $BaTiO_3$, and SBT, generally have a Schottky barrier height of only ~1 eV at the electrode interface, causing the leakage current and the electrical breakdown. Consequently, the comparable larger thickness is required to prevent such problems. Moreover, a number of specific issues need to be solved [71] when integrating such complex materials into a CMOS process. First, the

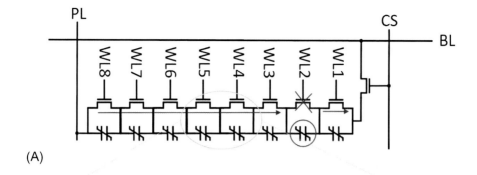

(A)

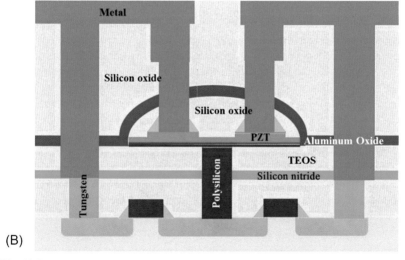

(B)

Fig. 11.8 (A) Circuit diagram of one chain of a chain FeRAM. WL2 is low, and all the other WLs are high. Therefore, the current will flow through the transistors for WL1, and WL3–WL8 (indicated by the red arrows) but through the capacitor for the cell connected to WL2. (B) Cross-section of two neighboring chain FeRAM cell blocks with stacked capacitor configuration.

contamination issue by the used metals needs to be solved. This is done naturally in the 1T-1C FeRAM structure by separating the capacitor from the base CMOS. Second, in the stacked capacitor cell a temperature stable plug connection between the transistor and the bottom electrode of the capacitor needs to be established. This was a major reason why offset cells were used for many years. However, for PZT it was possible to reduce the crystallization temperature low enough to make the barriers, such as TaSiN, remain intact, which cannot be the case for SBT due to its high crystallization temperature [72]. The lowering of the crystallization temperature also solved issues of doping deactivation that was specifically pronounced in SBT-based technologies [73]. Finally, the complex oxide structures of perovskites and layered perovskites are prone to degradation in reducing ambient, requiring complicated barrier structures [74]. All these issues could be solved in state of the art devices fabricated in 0.13 μm technology [75]. Further reduction in feature size creates a sensing problem for PZT [76], and therefore, there could be the two options for further scaling: a material with significantly higher remanent polarization or a three-dimensional capacitor as in DRAM. Attempts to use BFO, which has significantly higher P_R, have failed due to the leakage issues [76]. Making PZT ferroelectric in three-dimensional structures has proven to be nontrivial. Although larger structures can show promising results, the sidewalls typically turn into a nonferroelectric pyrochlore phase in smaller structures [77], which are to be solved. It seems, therefore, that PZT-based FeRAMs will stay limited to applications that can live with low densities and large feature sizes, resulting in a high cost per bit.

11.5.4   *Prospects of using ferroelectric hafnium oxide in FeRAM*

In contrast to PZT, ferroelectric hafnium oxide can be straightforwardly integrated into 3D structures [78]. However, the high coercive field of the ferroelectric hafnium oxide imposes two challenges that need to be solved for widespread commercial success. First, a higher switching voltage compared to the PZT-based counterparts is necessary. Current switching voltage may be acceptable as a single device but should become larger for an extension of the technology described in Ref. [75]. Even more problematic is the fact that the high field cycling leads to limited endurance combined with the effects of wake-up and fatigue. Since a RAM will need virtually unlimited endurance, which has been achieved in PZT-based devices [78], significant improvement is required. The AFRAM described in the following section was recently proposed as an alternative solution to this challenge, but more research on the cycling degradation should be conducted. It needs to be mentioned that most of the studies that dealt with the field cycling stability of metal/ferroelectric/metal capacitors have been conducted using large area capacitors. In an integrated device, the stress on the capacitor is expected to be reduced, and enhanced data are expected, although this issue remains speculative. It is nevertheless clear that further improvement is required for doped hafnium oxide to match the cycling performance of PZT since optimized PZT structures show more qualified performance than doped hafnium oxide even on the large capacitor level [79].

11.6   Antiferroelectric RAM

The high coercive field and the limited cycling endurance are the top challenges to solve for realizing FeRAM using doped hafnium oxide. As was shown in Section 11.3, a stable antiferroelectric hysteresis with two hysteresis loops can be generated when hafnium oxide is doped with dopants having an atomic radius smaller than that of hafnium. For the sake of simplicity, this will be called antiferroelectric in the following. Note that the physical origin is not a classic antiferroelectric crystal but a field-induced ferroelectric effect as explained in Section 11.3. Even pure ZrO_2 shows this behavior that can, therefore, be observed in state of the art DRAM capacitors operated at higher voltages [80, 81]. Since capacitors based on these antiferroelectric materials show a significantly better cycling endurance compared to their ferroelectric counterparts, it would be desirable to utilize this material. However, the missing remanent polarization makes the antiferroelectric useless for a nonvolatile memory at first glance. The idea of the antiferroelectric random access memory (AFERAM) is to shift the hysteresis loop in a way that one of the two hysteresis loops will be centered around an applied voltage of 0 V. For this, a fixed and stable bias field needs to be generated. In principle two approaches are possible to establish such a fixed bias in the capacitor. Electrodes having a significant difference in work function can be used [80], and second, another thin insulator can be placed on top or below the ferroelectric hafnium oxide in order to either create interface dipoles [82] or fixed charges [83] at the insulator/insulator interface. Since the additional dielectric required to form the fixed charges or interface dipole layers imposes an additional depolarization field, optimization of the latter is difficult and first attempts showed only very short data retention [84]. Therefore, work function approach will be focused in the following discussion. Fig. 11.9 illustrates the principle idea and the experimental results with respect to a ZrO_2 capacitor between one TiN and one RuO_x electrode [85]. When the internal bias field is applied, the antiferroelectric capacitor clearly behaves like a ferroelectric capacitor. Note that the coercive field observed is lowered compared to the ferroelectric counterpart where for example, $Hf_{0.5}Zr_{0.5}O_2$ is used [33].

For making this concept viable as a nonvolatile memory, the endurance and retention properties needed to be further verified. Fig. 11.10 shows the results of all relevant retention tests for FeRAM devices. As can be seen, not only the same state retention but also the critical "other state" (see Section 11.3 for details) retention test is passed. In Fig. 11.10(B), the endurance test is shown, and it can be seen that the promise of much more stable cycling indeed holds true.

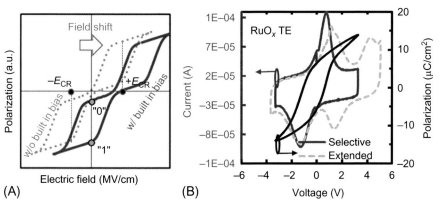

Fig. 11.9 Antiferroelectric RAM: (A) principle idea of shifting the *P–E* characteristics of an antiferroelectric material by implementing a built-in bias field. (B) Experimental verification using an antiferroelectric ZrO_2. The bias field is generated by using a RuO_x top electrode and a TiN bottom electrode. The hysteresis curve (black; right axis) is nearly centered. The corresponding *I–V* curve is shown for the region the hysteresis curve has been extracted (red curve) as well as for an extended region (gray curve) demonstrating that the other part of the antiferroelectric hysteresis is still observable [85].

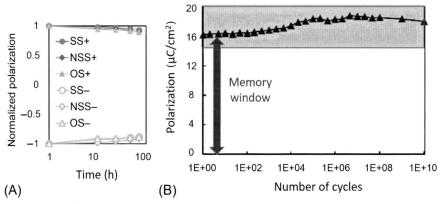

Fig. 11.10 Reliability assessment of ZrO_2 antiferroelectric capacitors with built-in bias created by the work function difference of a RuO_x top electrode and a TiN bottom electrode. (A) Retention measured at 100°C for all possible states (SS: same state; NSS: new same state; OS: opposite sate. The + and − signs indicate the two polarization states). For a description of the testing sequence see [85]. (B) Endurance test for 10^{10} cycles [85].

11.7 Ferroelectric field-effect transistor

The idea of a FeFET is very simple. If the gate dielectric of a MOSFET is replaced by a ferroelectric material, the characteristics of the transistor are expected to change based on the polarization charge in the ferroelectric material. However, since the ferroelectric can switch quite fast while the current-voltage characteristics are recorded, unlike a charge trapping or a floating-gate transistor, the characteristics will not simply reflect the transistor characteristics shifted by the dipole charge, but the switching itself will have two important implications on the transfer characteristics of the resulting ferroelectric FET. For simplicity, it will be assumed the switched charge ($2P_R$) of the ferroelectric layer is large enough to switch the transistor between strong inversion and accumulation. The first consequence of the ferroelectric switching is the fact that under the assumption that the memory window is defined by the distance between the two points of the ferroelectric switching and is not a strong function of the polarization of the ferroelectric itself [86, 87]. The second consequence is that the on-switching of the transistor is not defined by the transistor physics but by the ferroelectric switching. Consequently, the subthreshold slope can be much better compared to a transistor having a pure dielectric stack with the same capacitor equivalent thickness [86]. Under certain assumptions for the ferroelectric switching process, the transistor can even have a subthreshold slope below the Boltzmann limit [88]. Note that this can happen even without considering a stabilized negative capacitance (NC) effect [89], but it will show the typical ferroelectric hysteresis which would be absent in the case of a stabilized NC (see Section 11.9 for more discussions). Fig. 11.11 shows how the transfer characteristics of a FeFET result from the transfer characteristic of the MOSFET combined with the hysteresis of the ferroelectric layer.

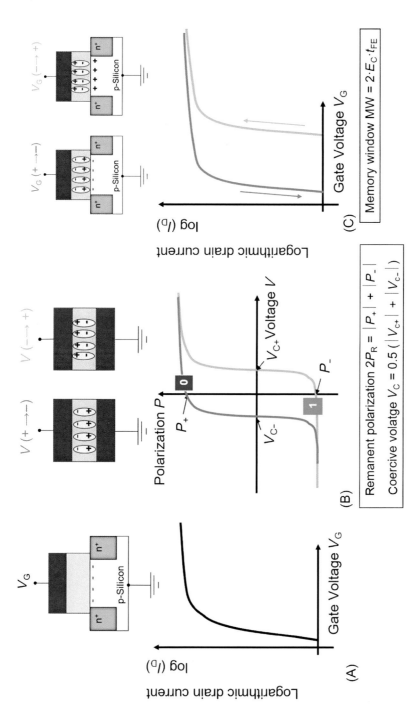

Fig. 11.11 Ferroelectric field-effect transistor as a combination of a MOSFET and a ferroelectric capacitor. (A) Schematic cross section of a MOSFET (top) together with the logarithmic transfer characteristics (bottom). (B) Schematic cross section of a ferroelectric capacitor in the two different polarization states (top) together with the polarization–voltage hysteresis (bottom). (C) Schematic cross section of the ferroelectric field-effect transistor in the two possible polarization states (top) together with the resulting logarithmic transfer characteristics of the device.

The concept of the FeFET comes with two distinct advantages compared to the ca-pacitor based FeRAM. Namely, the nondestructive read and the fact that the effect of the polarization switching of the transistor will stay constant during scaling and, there-fore, the device is much easier to scale down compared to a capacitor-based FeRAM were a certain minimum switched charge needs to be obtained. Therefore, this concept has been studied since the 1970s (see Section 11.4 of this section). However, in a tran-sistor configuration, a semiconductor will take the place of one electrode, and, there-fore, possesses inherent second capacitance in series to the ferroelectric capacitor. The fact that normally at least a thin interface layer is necessary to realize a good interface between ferroelectric and semiconductor adds to this capacitance. Consequently, a depolarization field exists during the retention period which tends to weaken the po-larization of the ferroelectric [90]. When using perovskites or layered perovskites that have very high permittivity in a few hundred range, this depolarization field becomes very strong. For about three decades, therefore, it was impossible to fabricate FeFETs with real nonvolatile retention although considerable effort was put into both material research [91] and device architecture [92]. Finally, in 2004 a device based on a rather thick SBT ferroelectric and a HfO_2-based interface layer could show 10 years of reten-tion [93]. However, the very thick SBT is an essential element of achieving this goal together with a reasonable memory window and therefore this result, although being a breakthrough, could not change the practical application perspective of FeFETs yet.

When ferroelectricity in hafnium oxide was discovered in 2011, it became immedi-ately clear that this would be a natural choice to use the material in a FeFET. Hafnium oxide has been the mainstream gate dielectric in high-*k*/metal gate technologies since 2007 [94]. The combination of having a much lower permittivity together with a sig-nificantly higher coercive field can solve the retention and the scaling issue of tradi-tional perovskite-based FeFETs. Therefore, scaling to the 28 nm [95] and 22 nm [96] node is straightforward, and already first fully integrated arrays have been demon-strated [96, 97]. Fig. 11.12 shows the cross section, and the transfer characteristic of a FeFET integrated into a 28 nm CMOS process together with the threshold voltage (V_T) distribution of a 64 kB array and the respective retention and cycling data. It becomes clear that a typical flash like specification at faster and more flexible overwrites at lower cost seems to be within reach.

Going significantly toward a RAM type of specification then mainly cycling en-durance needs to be improved. In case of the FeFET during reading, the device will be highly sensitive to any trapped charge. Therefore, the memory window will be narrowed significantly earlier compared to a capacitor-based structure [98]. Based on this observation, charge trapping needs to be reduced significantly. The difference in permittivity between the currently unavoidable interface layer of the silicon channel and the ferroelectric hafnium oxide layer results in a situation where a larger fraction of the voltage drop arises on the interface oxide. The polarization of the ferroelectric worsens this situation [99]. Therefore, the endurance optimization may not only focus on improving the material stack but can also be done by changing the device architec-ture [99]. Promising data using a gate last process give first evidence that significantly higher endurance may be achieved at the single cell level [100]. An essential element of the overall memory optimization of any transistor-based memory concepts is array

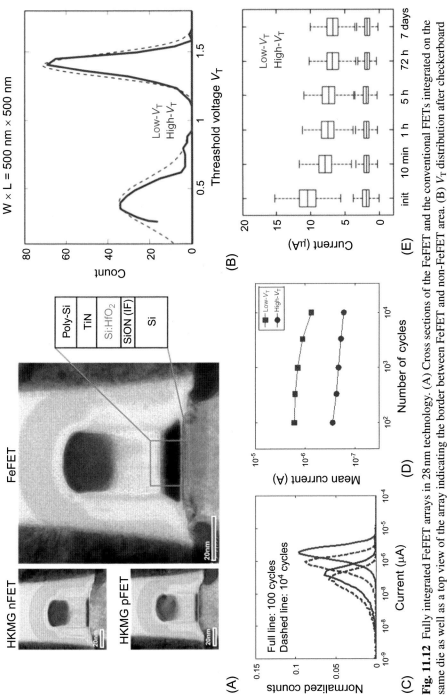

Fig. 11.12 Fully integrated FeFET arrays in 28 nm technology. (A) Cross sections of the FeFET and the conventional FETs integrated on the same die as well as a top view of the array indicating the border between FeFET and non-FeFET area. (B) V_T distribution after checkerboard programming for 500 nm × 500 nm cells. (C) V_T distribution of the high and low V_T state before and after 10^4 write cycles. (D) Maximum of the data presented in (C) as a function of cycle count. (E) Retention behavior measured for 7 days at 250°C.

effects. Because both programming and erase operations need to apply a homogenous field across the ferroelectric, the architecture needs to allow symmetrical access also from the semiconductor electrode. Therefore, architectures like the common ground NOR that has asymmetric access on the side of the transistor terminals are not among the most promising options. As a consequence, either the AND architecture, a variant of the NOR array where an individual drain and source lines are available for cell access [101] or the NAND architecture [102] seem to be appropriate. The later had already been proposed using SBT-based memory cells in the past [103], and recently the first demonstration for the applicability of Al-doped hafnium oxide as the ferroelectric in 3D NAND architectures has been made [104]. The NAND-based developments could pave the way toward realizing storage class-type memories using FeFETs. However, this issue is still in a very early stage.

Coming back to the more general array effects, by using a $V_{DD}/3$ inhibit scheme, disturbs in an AND architecture can be controlled for a cycling specification comparable to Flash memories [105]. In case of much higher endurance on the single cell level, the inhibit concepts need to be revisited as well. It needs to be mentioned here that up to now transistor-based concepts are only used for typical nonvolatile memories with cycling specs up to 10^6. Once the cycling endurance is drastically increased, the array operation also needs to be considerably reworked, or a select device needs to be added to control the BL and WL disturbs.

An alternative approach to increase the cycling endurance of the FeFET was proposed in Ref. [106]. Here, the device is operated in a subcycle at significantly reduced fields that will also reduce the charge injection into the dielectric stack. As a tradeoff, the retention is significantly reduced, and a regular refresh operation is required. This is an interesting approach, and the data presented in Ref. [106] are very promising. However, it remains to be demonstrated that the concept can work at small feature sizes where only a few grains are involved in the ferroelectric switching. In such case, a very abrupt switching can be observed [107], which could prohibit the proposed subloop operation. This abrupt switching events could only be observed as scaled down FeFETs became available. They are interpreted as the switching of the single domains [108] and, therefore, allow us to study the dynamics of single domains much more precisely as has been possible before. Therefore, the possibility to scale down FeFETs to the state-of-the-art feature sizes does not only pave the way toward a practical and economical viable FeFET memory but can also shed new light on the physics of nanoscaled ferroelectrics.

11.8   Ferroelectric resistance switching RAM

This chapter describes the methods to use the ferroelectric polarization to induce reversible resistance switching of the ferroelectric thin films. It covers two topics; ferroelectric tunnel junction (FTJ), which has been discussed within the community for more than two decades and more recently emerged ferroelectric domain wall (DW) memory. FTJ is based on the fundamental idea that the incomplete screening of the ferroelectric polarization by the involvement of finite screening length of the metal electrodes or interfacial nonferroelectric layers can induce different internal (depolarization) field across the ferroelectric film depending on the polarization direction. Since the current flow is mostly dependent on the tunneling mechanism of carriers, which involves a defect-mediated mechanism that is prone to the resistance switching mechanism in oxide-based resistance switching RAM, has hindered its development toward commercialization. In contrast, the DW memory is dependent on the local current flow along the mostly charge-uncompensated DWs, such as head-to-head DW s. Therefore, DW memory operation is dependent on the creation and erasure of the DWs, of which configuration should be appropriately adjusted to conduct sufficient electricity. Due to these limitations, the exploration of both FTJ and DW memory has been limited to the epitaxial thin films grown on single crystal oxide substrates, which are the critical drawback of these device concepts. Nonetheless, these devices are more compatible with the passive array structure, such as cross-bar array, compared with the transistor-based integration architecture discussed in Section 11.5.

11.8.1 Ferroelectric tunnel junction

The concept of FTJ memory was first proposed by Esaki in the 1970s [109]. FTJ has a simple capacitor structure consisting of top and bottom electrodes with different materials and ultrathin ($<\sim5$ nm) ferroelectric film. The device is a type of resistance change memory that uses the tunneling electroresistance (TER) phenomenon which indicates that the amount of tunneling current across ferroelectric capacitors varies depending on the polarization direction [110, 111]. Since the read operation of conventional FeRAM switches the polarization, the destructive readout is inevitable. On the other hand, during a read operation, the nondestructive readout is possible in FTJ, as in FeFET, because the tunneling current is sensed with a voltage lower than the coercive voltage. In addition, since capacitive readout is not performed, it is possible to miniaturize the device. Also, it includes ultrathin ferroelectric material, which makes it advantageous in terms of scaling compared to conventional FeRAM devices. However, it had been demonstrated that retaining the ferroelectricity at such thickness is challenging due to the almost inevitable involvement of the depolarization field effect [112, 113]. In contrast, recently developed first principles calculations show that ferroelectricity can occur even at a thickness of several nanometers [114, 115]. Subsequently, ferroelectricity was observed in a ferroelectric thin film of 10 nm or less due to the development of high-quality thin film manufacturing technology [116–121].

Several theories have been proposed to explain the cause of the TER phenomenon in FTJ. The most widely accepted theory is the direct tunneling current change due to the distorted potential caused by the asymmetric charge screening length of the top and bottom electrodes [110, 111]. The ideal metal has no charge screening length because it completely compensates the polarization. However, in reality, a finite screening length is present (~0.05 nm). This screening length generates a depolarization field, which produces a tilted potential barrier as shown in Fig. 11.13. According to Thomas-Fermi theory, the screening length depends on the electronic density of state at the Fermi level, so different screening lengths appear at the top and bottom electrodes ($\delta1$, $\delta2$). Then, as shown in Fig. 11.13, potential barriers with different average barrier height

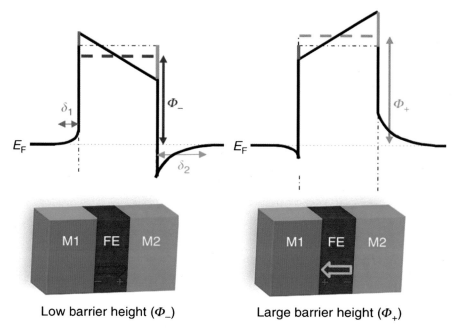

Low barrier height (Φ_-)   Large barrier height (Φ_+)

Fig. 11.13 Polarization-induced variation of the tunnel barrier height and the potential profile in FTJ [122].

are generated along the polarization direction ($\Phi-$, $\Phi+$). Direct tunneling current decreases exponentially when the average barrier height increases, so high resistance and low resistance state occur depending on the polarization direction [110, 111].

Gruverman et al. observed the TER phenomenon experimentally as a method of measuring the tunneling current in the domain pattern using piezoresponse force microscopy (PFM) and conducting atomic force microscopy (C-AFM). Fig. 11.14(A) and (B) shows the domain pattern generated by the applied voltage using PFM and the tunneling current flowing through it. The larger the poling voltage, the more tunneling current is generated because there is more conducting domain [123].

The write and read operations of the FTJ memory are performed as follows. Electric field larger than the coercive field is applied to the device and switching the polarization direction to write 0 (insulating state) or 1 (conducting state). In the read operation, electric field smaller than the coercive field is applied, and the polarization state is read by sensing the magnitude of the tunneling current without switching the polarization. Fig. 11.14(C) shows the measurement of the tunneling current of epitaxial $BaTiO_3$ films with different polarization states. The results show that the tunneling currents vary significantly depending on the polarization direction (red and green dots), and this behavior is well explained by the Wentzel-Kramers-Brillouin (WKB) approximation (solid black lines) describing the direct tunneling behavior [123].

The disadvantages of FTJ to be overcome are relatively low on-off current ratio and complicated device fabrication process. Several studies have been conducted to

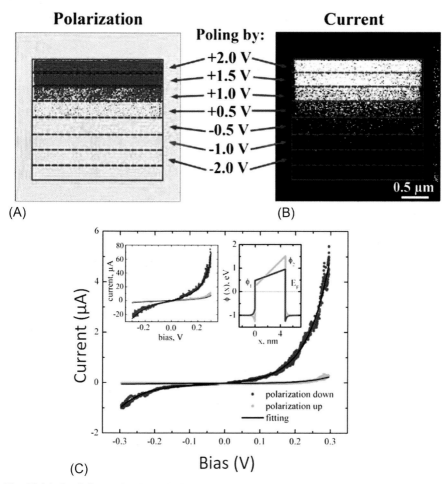

Fig. 11.14 Spatially resolved correlation between the onset of (A) polarization reversal and (B) a change in electrical conductance. (C) *I–V* curves for two opposite polarization directions in the 4.8 nm-thick (2.4 nm for inset) BaTiO₃ film measured by C-AFM. Solid lines fitting of the experimental data by the WKB model [123].

improve these drawbacks. To improve on-off current ratio, Wen et al. studied the use of highly doped semiconductors as one electrode [124]. In the semiconductor electrode, carrier depletion and accumulation occur along the polarization direction. This increases the on-off ratio by significantly changing the screening length [124].

The use of ultrathin perovskite epitaxial film which exhibits ferroelectricity at 5 nm or less leads to a difficulty in device processing. As ferroelectricity is observed in the polycrystalline-doped HfO₂ of <10 nm, which is deposited by atomic layer deposition technic, continuous research is being conducted to apply this material to FTJ. This material has advantages over conventional perovskite materials in that it has simple fabrication process and good Si compatibility. However, the on-off current ratio can deteriorate due to leakage current through defects such as grain boundaries. In addition, since the ferroelectricity of doped HfO_2 varies greatly depending on the thickness it is necessary to confirm whether the ferroelectricity is maintained at a thickness of sub-5 nm. Nonetheless, this material also provides a new horizon for the research in this field.

An alternative to using such ultrathin films has been proposed in Refs. [125, 126]. In this realization, the polarization reversal is achieved by a significantly thicker ferroelectric hafnium oxide layer, and the actual tunneling happens through a dedicated thin tunneling layer in series to the ferroelectric. The main drawback of such an implementation is the rather low current through the device. However, its exploitation is just in a very infant stage and it seems to be fitting very well with the properties of doped hafnium oxide.

11.8.2 Domain wall memory

Ferroelectric DWs are the boundary between the two nearby ferroelectric domains, which can be explored to store binary data. They can be used as a path for the fluent electrical current flow within an insulating matrix under the appropriate condition and geometry [127–131]. DW is similar to a twin-boundary in crystallographic terms, so it does not contain many defects which can produce electrical carriers, as in grain boundaries in many insulating ceramics. However, due to the polar nature of the ferroelectric materials, a DW can be either polar or nonpolar, where the former has a high chance of involving free carriers. A typical example of nonpolar DW is a 180° wall in perovskite ferroelectric material, such as $PbTiO_3$, where the antiparallel polarization directions of the two nearby domains have a parallel direction to the 180° DW. Therefore, the 180° DW has no polarization charge discontinuity (in fact no polarization charge itself) across the wall along the direction normal to the DW, and thus, the DW does not have any driving force to induce free carriers. In contrast, the head-to-head or tail-to-tail configuration of 180° domains inevitably involves polar DWs between the two nearby domains due to the polarization discontinuity, of which (electrostatic) energy cost is very high making them highly unfavorable under usual (short circuit) conditions. However, for example, during the ferroelectric polarization switching in epitaxial thin films, mediated by the reverse domain nucleation and growth, polar DW must be temporally formed within the film thickness until the wall passes through the entire film thickness. Another method to form such polar DW is to make a lateral device geometry, where the lateral electric field along the surface direction can form such domain and DW configurations. Due to the involvement of the very high electrostatic energy at these DWs, carriers injected from the electrodes or thermally generated within the bulk region of the ferroelectric materials are drawn to them to mitigate the serious electrostatic energy. For example, when the head-to-head domains are formed, electrons injected from the nearby metal electrode or generated from the oxygen vacancy within the ferroelectric bulk oxide can be located along the DW to cancel the positive-to-positive bound charge at the DW [132]. In bulk single crystals, certain DWs have shown an enhanced electrical conductivity compared with the insulating bulk region, which has been proven by adopting conductive atomic force microscopy [133].

When an electric field is applied along the DW direction under such a circumstance, the gathered carriers at the DW can move rather freely along the applied (reading) field direction making the ferroelectric layer conducting electrically. On the other hand, when the reading field is applied to the ferroelectric memory device, which does not contain just-mentioned polar DWs, the electrical conductivity must be much lower. Therefore, these mechanisms can be used to store binary digital data, that is, involving conductive DW within the memory cell corresponds to data 1, whereas the other case does data 0. Nonetheless, there have been several problems in realizing such effect for high-density memory. First, the electric field for the data writing and reading are generally perpendicular, as can be readily understood from the fact that when the formation of the polar DW is made along one crystallographic direction, reading of the domain state should be performed along the perpendicular direction. This operation needs four electrodes (two for writing and two for reading) which are very unfavorable for high-density integration. Second, the charged DW can draw other ionic defects, such as oxygen vacancies or other impurity ions, which adversely interferes with the fluent and clean erasure and regeneration of the DWs. Third, the wall current is generally too low to be detected by a high-speed amplifier. The typical wall current has been in the orders of pA to nA, but at least μA is necessary to overcome the Johnson-Nyquist limit [134, 135] due to the inevitable presence of thermal noise.

Jiang et al. have reported the fundamental feasibilities of using the wall current along the DWs formed within the epitaxial $BiFeO_3$ (BFO) film, which is of rhombohedral distortion of the (pseudo) cubic perovskite structure of the BFO [136]. BFO is an especially appropriate material for this purpose because it contains 71° and 109° DWs in addition to the typical 180° DW so an appropriate arrangement of the DWs could be used for both data writing and reading, meaning that the above-mentioned first problem might be solved. In addition, it could be either n-type or p-type wide-bandgap oxide semiconductor due to the involvement of oxygen vacancies or cation (typically Bi) vacancies, respectively. It also has a lower bandgap (2.1–2.8 eV) compared with other standard ferroelectrics, such as $Pb(Zr,Ti)O_3$ (3.0–3.5 eV), making it possible to solve the third problem mentioned above.

Therefore, Sharma et al. recently reported a prototypical DW memory device using a persistent conductive DWs in BFO epitaxial film [137]. They have proven that a pair of lateral electrodes can be used to write (form or erase) DWs within the memory cell by applying a voltage > coercive voltage, and the same set of electrodes are also used as a reading electrode by applying a reading voltage ≪ coercive voltage. The detected currents were only 5–20 pA, which, however, are too low to operate sense amplifiers at sufficiently high speeds. It also suffers from a fundamental reliability issue related to instability of the charge-uncompensated persistent domain boundaries. Here, the "persistent" means that the DWs remained un-erased even when the write voltage was removed, so the formed DWs have driving force to attract several (undesirable) ionic defects to compensate for the uncompensated ferroelectric polarization charges, in addition to the desired electronic carriers. The stabilization of the created DWs was accomplished by the charge-compensation by the carriers injected from the electrodes.

Jiang et al. presented another step-forward work in the DW memory field, triggered by the work of Sharma et al., where the DWs are temporarily formed only during the application of the reading voltage [138]. In this case, the writing corresponds to polarize the ferroelectric epitaxial BFO film between the two lateral electrodes uniformly either parallel or antiparallel to the subsequent reading field direction. When the writing (or polarizing) field direction is identical to the reading field direction, the DWs are not formed, whereas the opposite writing field direction induces the formation of DW during the reading field application. The former indicates an off-state whereas the latter does an on-state, so the binary digital data can be recorded as a direction of domains of which direction can be read out subsequently by a current flow, not by a charge integration method, as in a conventional FeRAM. The formed DWs can be temporal or persistent depending on the degree of charge compensation of the formed walls, which can be controlled by either the electrode geometry or wiring pulse width/time. It was found that the temporal mode guarantees stable switching cycles up to 10^{7-8}, which is a favorable performance for the stable storage class memory. It was revealed that the involvement of the 71° and 109° domains during the eventual formation of charge-uncompensated 180° DWs was the critical step to induce the high wall current (14 nA), which could be further improved to 300 nA in the more recent report [138] only with a single pair of electrodes. The configuration of field-induced reverse domains usually has a triangular shape within the gap between the planar two electrodes, due to the reverse domain nucleation and growth mechanism. The involvement of fringing field (evolution of electric field along the directions other than the nominal applied field direction) at the edge of the nanoscale electrodes also helped to configure the DW shape, which simultaneously allows both domain formation and high current flow. Therefore, this is a highly feasible solution to all the problems mentioned above for the ferroelectric DW memory.

Nonetheless, there is still the most critical problem remained for such a DW memory to be used in the mainstream semiconductor memory; Si-compatibility. Up to now, all these DW-related memory devices are fabricated on epitaxial ferroelectric thin films grown on single crystal oxide substrate such as $SrTiO_3$ with epitaxial oxide electrode, typically $SrRuO_3$, by mostly pulse laser deposition method. These are not Si-compatible at all. Therefore, it is impending to evolve these proven technologies into poly-crystalline ferroelectric thin films grown on Si wafer, or epitaxial ferroelectric oxide platform on Si wafer. Some of the recent progress in the bonding of $LiNbO_3$ single crystal to Si wafer and cutting it to make a FOS (ferroelectric on Si) wafer using smart-cut technique is notable in this regard. Additional problems of implementing such technology to the resistance-switching platform, such as the cross-bar architecture, are another critical field for the further study, which is over the coverage of this chapter. These devices are more compatible with the passive array structure, such as cross-bar array, compared with the transistor-based integration architecture discussed in Section 11.5. It is recommended for the readers to read review articles describing the integration strategies of resistive switching memories [139, 140].

11.9 Other applications of ferroelectrics integrated into CMOS

Besides using the nonvolatile polarization of a ferroelectric to store binary information, the features of a ferroelectric can be used in integrated circuits in various ways. The first and very straightforward approach is to integrate the memory function itself into the logic operation. Kimura and co-workers showed how logic functions could be implemented in complementary ferroelectric capacitor (CFC) logic circuits [141]. When using FeFETs instead of normal transistors, a CMOS inverter can be operated as both a normal logic gate as well as a memory cell [142]. Finally, when an NMOS-type gate is realized with a FeFET as a switching device and an additional load device a logical operation between the value stored in the ferroelectric and the external input that can mimic bot a NAND and a NOR function is possible [143]. Moreover, in the later realization, the logic function can be reconfigured between NAND and NOR by either using a source voltage on the FeFET or implementing the FeFET in a fully depleted silicon-on-insulator (FDSOI) technology and makes use of the back bias effect to shift the V_T of the FeFET device. All these realizations can be categorized as a memory-in-logic type of applications where the separation of memory and signal processing is abolished to ease the so-called "von Neumann bottleneck" [144].

A more radical way to overcome the limitations of the von-Neumann architecture is the switch to neuromorphic computing solutions. Two basic circuit elements will be required namely neurons that generate a signal based on a threshold and synapses that can realize certain learning rules like spike-time-dependent plasticity [145]. Owing to their inherent memory function, FeFETs can mimic the function of a synapse [146]. Recently, also doped hafnium oxide integrated into a scaled down 28 nm technology has been shown to be able to realize spike-time-dependent plasticity [147]. However, due to the abrupt switching of aggressively scaled FeFET devices [108], a somewhat larger device with multiple domains needs to be used to achieve a more analog type of switching. Moreover, the natural property of ferroelectric hafnium oxide thin films to accumulate excitation arising from the progressive polarization reversal through ferroelectric nucleation was recently used to experimentally demonstrate the simple but yet important all-or-nothing functionality of integrate-and-fire neurons using a single nanoscale ferroelectric transistor [148]. This offers unprecedented possibilities to mimic the biological brain by building all-ferroelectric computing systems using ferroelectric neutrons as well as ferroelectric synapses.

Another interesting use of a FeFET was recently proposed. The mentioned feature of abrupt switching is stochastic in a certain voltage range [107]. This property can be used to generate a true random number generator by operating the device exactly in the range where the probability of obtaining a "0" or "1" is about equal [149]. Another appealing application of the ferroelectric thin film is to utilize its possible NC effect as discussed in more detail below (Section 11.9.1). Yet another application is to explore them to the energy storage and conversion fields (Section 11.9.2).

11.9.1   NCFET

11.9.1.1   Background of NC in ferroelectrics

NC in ferroelectrics had been expected from Landau theory. Landau theory expressed free energy of ferroelectrics as a double well potential function of polarization. Noting that a reciprocal susceptibility is equivalent to the curvature of the free energy, the region near the maximum point of the free energy—polarization curve of which $P \sim 0$, which corresponds to the Landau barrier between two equilibrium states, has NC. However, the experimental demonstration of such intriguing effect is challenging because the material avoids this state as it corresponds to maximum energy state, and rather tends to be polarized into positive or negative polarization state. As demand for overcoming performance degradation of modern electronic devices as relentless scaling arises, NC in ferroelectrics recaptures the interests of researchers as a potential rescue to this issue [150–152]. Salahuddin et al. suggested a novel idea that NC on Landau barrier can be stabilized by a serial connection of a dielectric layer to the ferroelectrics [89]. The idea has been experimentally confirmed by follow-up studies [153–158]. In addition, experimental evidence that a single ferroelectric layer can show a NC has been proposed by Khan et al. [159].

In theoretical respects, however, suggested NC models so far mainly have assumed a homogeneous polarization of ferroelectrics (single domain state) so that they have a fundamental limitation to analyze NC behavior of multidomain ferroelectrics. The multi-domain state in ferroelectrics comprises the fundamental understanding of the ferroelectric switching and almost all the related phenomena. The multidomain state is a highly unfavorable circumstance that hampers the emergence of the NC effect originally suggested. On the purpose of understanding NC in multi-domain states, various models have been suggested [160–162]. The basic idea on NC in ferroelectrics has been originated from simple Landau theory, but due to the complexity of physics in ferroelectrics, the NC models have not been established firmly yet, and as a result, it is still challenging to implement NC experimentally and interpret the experimental results.

11.9.1.2   Theoretical models and experimental results of NCFET

A conceptual idea that negative capacitance FET (NCFET) which is a steep-slope FET with a subthreshold swing (SS) lower than a theoretical Boltzmann limit (SS ~60 mV/dec), as well as an increased on current, can be implemented by applying ferroelectrics as an NC insulator has been suggested by assuming a constant capacitance C_s of a semiconductor substrate [136]. However, as for a real NCFET operation, the capacitance of semiconductors is highly nonlinear to the applied voltage so that new models to understand NC operation in a ferroelectric-semiconductor bilayer structure should be proposed. Theoretical studies combining the well-established semiconductor models with Landau theory of ferroelectrics has been firstly performed by Jimenez and Chen for each structure of double gate FET and bulk FET [163, 164]. In their models, a voltage applied on the ferroelectric insulator (V_{ins}) is written as an odd polynomial of P by the LD theory, as follows:

$$V_{ins} = 2\alpha t_{ins} P + 4\beta t_{ins} P^3 + 6\gamma t_{ins} P^5. \tag{11.14}$$

By considering the approximation that surface charge density, Q at the semiconductor interface is equal to P in ferroelectrics, $Q = P + \epsilon_0 E \approx P$, a load line analysis for serial capacitors is performed based on Kirchhoff's voltage law, as follows:

$$V_g - V_{FB} = V_{ins} + \psi_s, \tag{11.15}$$

where V_{FB} is a flat band voltage, and ψ_s is the surface potential of the semiconductor. As shown in Fig. 11.15, the load line analysis can be performed by finding a cross point between two nonlinear capacitance functions for the given gate voltage. Fig. 11.15 shows that as the V_g increases, an initial cross-point A develops to point B linearly. However, when V_{ins} increases higher than a coercive voltage of the ferroelectrics, the cross point becomes snapped back to point C abruptly, and the snapback occurs once again at the transition from C to D. In other words, NC phenomenon emerges, in which the voltage across the ferroelectrics decreases while the gate voltage increases, and in turn, the surface potential increases abruptly with respect to the gate voltage.

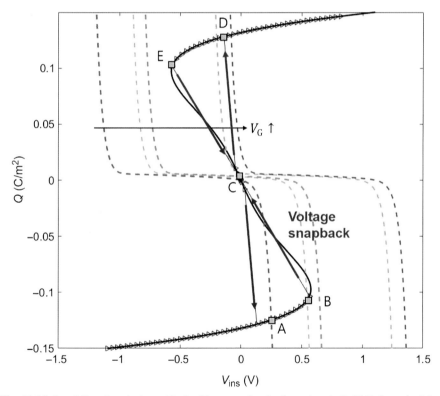

Fig. 11.15 Load-line description with $Q - V_{ins}$ curve for the ferroelectric BaTiO$_3$ layer (solid line) with 200 nm thickness, and $Q - (V_g - V_{FB} - \psi_s)$ curve for p-type semiconductor (dotted line) of 10^{18} cm^{-3} dopant concentration. As the gate voltage increases, the dotted line shifts in parallel from left to right. The voltage snapback occurs at the transition B → C and C → D for a positive gate sweep. The triangular marks correspond to the cross points of two curves for positive and negative gate sweep.

This is an obvious contrast to a switching behavior of a conventional MOSFET where a body factor ($dV_g/d\psi_s$) cannot be lower than 1, and it is a direct origin of the steep-sloped switching of NCFET.

Nonetheless, in this scenario, the NC is not fully stabilized and the surface potential suffers a typical hysteresis of ferroelectrics depending on the previous state of P_s. Therefore, in the viewpoints of engineering, it is crucial to reduce this hysteresis for ideal steep-slope NCFET operation. The theoretical model reveals that a thickness of ferroelectrics is one of the critical factors to be engineered. Starting from the NCFET theories, various groups have been reporting the experimental results of NCFET device operation with sub-60 mV/dec SS values in HfO_2-based transistor [165–167] with various configurations of a device structure including Ge substrate, MoS_2 2D material, and FinFET. Still, the most significant conceptual and practical challenge is to stabilize the NC state and decrease or eliminate the hysteresis voltage. As previously mentioned in Section 11.7, this hysteresis is the fundamental asset for the normal FeFET, operation that should be avoided for NCFET is by stabilizing the NC region. However, as described above it is still not proven that this can be achieved in a reasonable operation range and more research is necessary on the microscopic switching mechanism. Given the fact that an integration in scaled transistors is necessary, only ferroelectric HfO_2 can be used to realize a NCFET from today's point of view. However, the microscopic details of domain switching in ferroelectric HfO_2 are not as clear as in conventional ferroelectrics and so far only indirect data-based either on fitting simulations [162] or on switching dynamics in scaled devices [107] exist. Therefore, more research will be necessary to finally conclude if or if not NCFET can be a practical viable concept.

11.9.2   Energy storage and energy harvesting

The electrostatic supercapacitors using dielectric capacitors in energy storage technology are promising because they can exhibit much higher power densities than the existing conventional Li-ion batteries or the newly developed electrochemical supercapacitors due to the fast charging and discharging speeds of several ~μs [168, 169]. On the other hand, the electrostatic supercapacitors generally have lower energy storage densities (ESD) than other methods. Therefore, several 3D nanostructures to increase the ESD per projected area were previously reported, such as Si nanotrench or the anodized aluminum oxide [170]. It was reported that the significantly increased recoverable energy density could be achieved when the antiferroelectric polarization switching is exploited owing to its large polarizability and low P_R. [170]. The ESD in the ferroelectric thin film can be expressed by the following equation.

$$W = \int E dP \tag{11.16}$$

Eq. (11.16) indicates that the estimated work done on a capacitor by a power source is complicated when the electric field is a function of P, as in the case of the ferroelectric or antiferroelectric capacitor. Fig. 11.16(A) depicts the energy storage behavior in the (+) voltage region of the antiferroelectric P–E curve. Various perovskite-type ferroelectric bulk ceramics or thin films have been studied as an electrostatic supercapacitor

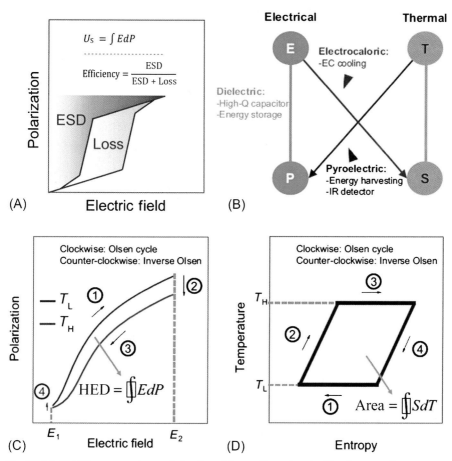

Fig. 11.16 (A) The energy storage behavior of the antiferroelectric-like electrostatic supercapacitor. (B) Schematic diagram of the coupling between the electrical and thermal properties of the polar material. (C) The Olsen cycle for the energy harvesting between the two different temperature states, and (D) the corresponding temperature-entropy change.

application, and the ESD value of about a few of or tens of J/cm^3 is currently reported, respectively. In the case of a fluorite-based ferroelectric material, ~46–61 J/cm^3 has been reported [61, 171–173]. Common endurance tests of the electrostatic supercapacitor are thermal stability and cyclic electric field switching because the potential environments usually require the high repetitive operation over a wide temperature range (~−100–~200°C).

Pyroelectric energy harvesting and electrocaloric cooling of the ferroelectric materials exploit the pyroelectricity and the electrocaloric effect of polarizable materials. In the case of the pyroelectric energy harvesting, the temperature change of the material over time is converted into the electrical energy. Conversely, electrocaloric cooling consumes the electrical energy to perform refrigeration operation like a heat pump, where the electric field induces entropy changes inside the material to absorb and release heat.

Fig. 11.16(B) shows how these applications are related to the coupling of the electrical and the thermal properties of the polar material. The specific thermodynamic cycle is generally exploited to increase the efficiency of the cycle, and the Olsen cycle was reported to exhibit the highest efficiency to date [174]. Fig. 11.16(C) shows the Olsen cycle process in the P–E hysteresis of the polar material, and Fig. 11.16(D) shows the corresponding temperature-entropy change. When the cycle proceeds clockwise (Olsen-cycle), the harvestable energy density (HED) is identical to the internal area of the hysteresis in Fig. 11.16(C). For the polycrystalline Si-doped HfO_2 and $Hf_{0.3}Zr_{0.7}O_2$ thin film, the high HED values of 20.27 and 11.5 J/cm^3 have been observed, respectively, owing to the larger applicable electric field than other perovskite-type ferroelectric materials by approximately one order of magnitude [175, 176]. In a typical application environment, however, the power density is generally low due to the very slow (<1 Hz) external temperature change, and in this regard, the nanoscale pyroelectric systems or the nanogenerators can be a viable solution [177, 178]. On the contrary, when the Olsen cycle is performed counterclockwise, the cooling operation is performed by absorbing the heat corresponding to $T_L\Delta S$ at the low temperature and releasing the heat of $T_L\Delta S$ (which is larger than the previous one) at the high temperature by consuming the energy corresponding to the internal area of the P–E hysteresis theoretically. In general, it is known that the large adiabatic temperature change (ΔT) or the adiabatic entropy change (ΔS) can be obtained in the vicinity of the Curie temperature. The promising properties of $\Delta T = {\sim}45.3$ K and $\Delta S = {\sim}361$ mJ/(K cm^3) were reported in the $Pb_{0.8}Ba_{0.2}ZrO_3$ thin film [179]. Recent reports on $Hf_xZr_{1-x}O_2$ are also intriguing to show both positive and negative ECE properties. Positive ECE values were $\Delta T = {\sim}13.4$ K and $\Delta S = {\sim}16.7$ J/(K kg), whereas negative ECE values were $\Delta T = {\sim}-10.8$ K and $\Delta S = {\sim}-10.9$ J/(K kg). Further improvements could be achieved through film optimization, and the combination of both positive and negative ECEs from a single material system shows high potential to be used in the next-generation chip-cooling system.

11.10 Summary and outlook

Ferroelectric materials with their two stable remanent polarization states are ideally suited for low write-power nonvolatile memories. Many ferroelectric materials with excellent properties exist, but most of them commonly have rather complicated crystal structures, which make it hard to be achieved and maintain when integrated into a semiconductor process.

There are several possibilities to integrate a ferroelectric material into a memory cell. The straightforward solution is to replace the dielectric in a transistor—one capacitor DRAM cell by a ferroelectric material. This option called ferroelectric random access memory (FeRAM) has been successfully developed using PZT as the ferroelectric material. However, since the switched charges during polarization reversal are sensed, the capacitor still needs a certain minimum area for reliable sensing, and below, a 3D capacitor would be required ~100-nm feature size. The complexity of integrating both the ferroelectric material and the required electrodes into a 3D structure has limited the success of the FeRAM, therefore, to technologies of 130 nm and above.

The newly discovered ferroelectricity in hafnium oxide could change this picture since the material can be deposited by the well-established atomic layer deposition techniques, and thus, can easily realize three-dimensional integrated structures. However, the high coercive field currently limits the voltage scalability and the endurance of the ferroelectric hafnium oxide leaving research with some remaining challenges.

The second mainstream option of integrating a ferroelectric memory cell is to place the ferroelectric into the gate stack of a transistor to realize a FeFET. This option has been limited by the very high permittivity and the low coercive field of conventional ferroelectric material like PZT, SBT, and the likes. The inherent de-polarization field of such a structure also limits the nonvolatile retention, and the low coercive field dictates very thick layers to achieve a reasonable memory window. Again, doped hafnium oxide can solve these issues, and the scaling gap between FeFETs and conventional logic was overcome very rapidly. This development places FeFETs as a very promising option for future embedded nonvolatile memories on the research and development roadmap. Moreover, the programming and erase mechanism allows the integration into a NAND type architecture opening a path toward high-density storage or storage class memories.

Besides these options, two further possibilities exist that allow a resistive readout—the FTJ and the DW memory. Both options are still in the early development stage and significant innovations are required to realize real memory arrays. However, in the successful case the resistance sensing may have some scaling advantages over the other two were the polarization charge is read out directly (FeRAM) or indirectly (FeFET).

Finally, there exists a number of possible applications beyond the field of semiconductor memories. Memory-in-Logic and neuromorphic computing being two fields that still are related to the traditional memory application but use a specific way of operating the device. The inherent variability of the switching in very small FeFETs can be used to realize a true random number generator. Finally, the switching of the ferroelectric may provide a path to realize a differential NC effect that could boost the internal voltage in a FeFET device and by this get a subthreshold behavior with <60 mV/dec slope as well as higher on-current. This would be one of the most important contributors to further scaling. However, the domain structure of a ferroelectric is making the stabilization of the NC in a way that would allow logic operation still doubtful. Again ferroelectric hafnium oxide would be the material of choice to supply the necessary CMOS compatibility, and too little is known so far about the details of the domain structure and its reversal to give a fair assessment of the realization potential of this approach.

Finally, ferroelectrics come with the piezo and pyroelectric effect, which have a potential for energy storage, energy harvesting as well as sensors and actuators. Specifically, for the CMOS compatible hafnium oxide, this research field is still in its infancy, and many possibilities remain to be explored.

In summary, ferroelectric materials have a great potential for nonvolatile memories but also some other device applications. With complex perovskite (like PZT) or layered perovskite materials (like SBT) the integration into CMOS processes limits the practical application in electronic devices. The discovery of the ferroelectricity in doped hafnium oxide has widened the possible application fields in integrated devices again that were limited by the integration difficulties of perovskites or layered perovskite until a few years ago.

Acknowledgment

CSH acknowledges the support from Global Research Laboratory (GRL) program of National Research Foundation of Korea (2012K1A1A2040157). He also thanks his group members at Seoul National University for preparing the manuscript.

References

[1] C.H. Lam, Storage class memory, in: 10th IEEE International Conference on Solid-State and Integrated Circuit Technology, Shanghai, 2010, 2010, pp. 1080–1083.
[2] H. Schroeder, V. Zhirnov, R.K. Cavin, R. Waser, Voltage-time dilemma of pure electronic mechanisms in resistive switching memory cells, J. Appl. Phys. 107 (2010) 054517.
[3] R.F. Freitas, W.W. Wilcke, Storage-class memory: the next storage system technology, IBM J. Res. Dev. 52 (4.5) (2008) 439–447.
[4] A. Chen, A review of emerging non-volatile memory (NVM) technologies and applications, Solid State Electron. 125 (2016) 25–38.
[5] K. Rabe, et al., Physics of Ferroelectrics a Modern Perspective, Springer, 2007.
[6] M.E. Lines, et al., Principles and Applications of Ferroelectrics and Related Materials, Clarendon Press, 2009.
[7] L.D. Landau, On the theory of phase transitions II, Phys. Z. Sowjun 11 (1937) 545.
[8] L.D. Landau, On the theory of phase transitions II, Zh. Eksp. Teor. Fiz. 7 (1937) 627.
[9] A.F. Devonshire, XCVI. Theory of barium titanate. London, Edinburgh, and Dublin Philos. Mag. J. Sci. 40 (309) (1949) 1040–1063, https://doi.org/10.1080/14786444908561372.
[10] A.F. Devonshire, CIX. Theory of barium titanate—part II. London, Edinburgh, and Dublin Philos. Mag. J. Sci. 42 (333) (1951) 1065–1079, https://doi.org/10.1080/14786445108561354.
[11] A.F. Devonshire, Theory of ferroelectrics. Adv. Phys. 3 (10) (1954) 85–130, https://doi.org/10.1080/00018735400101173.
[12] E. Fatuzzo, et al., Ferroelectricity in SbSI. Phys. Rev. 127 (6) (1962) 2036–2037, https://doi.org/10.1103/physrev.127.2036.
[13] M.J. Haun, E. Furman, S.J. Jang, H.A. McKinstry, L.E. Cross, Thermodynamic theory of $PbTiO_3$, J. Appl. Phys. 62 (1987) 3331.
[14] M.H. Park, H.J. Kim, Y.H. Lee, Y.J. Kim, T. Moon, K.D. Kim, S.D. Hyun, C.S. Hwang, Two-step polarization switching mediated by a nonpolar intermediate phase in $Hf_{0.4}Zr_{0.6}O_2$ thin films, Nanoscale 8 (2016) 13898–13907.
[15] B. Noheda, D.E. Cox, Bridging phases at the morphotropic boundaries of lead oxide solid solutions, Phase Transit. 79 (5) (2006) 13.
[16] B. Noheda, D.E. Cox, G. Shirane, J.A. Gonzalo, L.E. Cross, S.E. Park, A monoclinic ferroelectric phase in the $Pb(Zr_{1-x}Ti_x)O_3$ solid solution, Appl. Phys. Lett. 74 (1999) 2059.
[17] C.B. Eom, et al., Fabrication and properties of epitaxial ferroelectric heterostructures with ($SrRuO_3$) isotropic metallic oxide electrodes, Appl. Phys. Lett. 63 (1993) 2570.

[18] J.F. Scott, M. Dawber, Oxygen-vacancy ordering as a fatigue mechanism in perovskite ferroelectrics, Appl. Phys. Lett. 76 (2000) 3801.

[19] N. Setter, D. Damjanovic, L. Eng, G. Fox, S. Gevorgian, S. Hong, Ferroelectric thin films: review of materials, properties, and applications, J. Appl. Phys. 100 (2006) 051606.

[20] J.R. Teague, R. Gerson, W.J. James, Dielectric hysteresis in single crystal $BiFeO_3$, Solid State Commun. 8 (1970) 1073–1074.

[21] S.V. Kiselev, R.P. Ozerov, G.S. Zhdanov, Detection of magnetic order in ferroelectric $BiFeO_3$ by neutron diffraction, Sov. Phys. Dokl. 7 (1963) 742–744.

[22] J.B. Neaton, C. Ederer, U.V. Waghmare, N.A. Spaldin, K.M. Rabe, First-principles study of spontaneous polarization in multiferroic $BiFeO_3$, Phys. Rev. B 71 (2005) 014113.

[23] H.J. Lee, M.H. Park, Y.J. Kim, C.S. Hwang, J.H. Kim, H. Funakubo, H. Ishiwara, Improved ferroelectric property of very thin Mn-doped $BiFeO_3$ films by an inlaid Al_2O_3 tunnel switch, J. Appl. Phys. 110 (2011) 074111.

[24] T. Choi, S. Lee, Y.J. Choi, V. Kiryukhin, S.-W. Cheong, Switchable ferroelectric diode and photovoltaic effect in $BiFeO_3$, Science 324 (2009) 63.

[25] T.S. Böscke, J. Müller, D. Bräuhaus, U. Schröder, U. Böttger, Ferroelectricity in hafnium oxide thin film, Appl. Phys. Lett. 99 (2011) 102903.

[26] J. Müller, U. Schröder, T.S. Böscke, I. Müller, U. Böttger, et al., Ferroelectricity in yttrium-doped hafnium oxide, J. Appl. Phys. 110 (2011) 114113.

[27] S. Mueller, J. Mueller, A. Singh, S. Riedel, J. Sundqvist, et al., Incipient ferroelectricity in Al-doped HfO_2 thin films, Adv. Funct. Mater. 22 (2012) 2412.

[28] S. Mueller, C. Adelmann, A. Singh, S. Van Elshocht, U. Schroeder, et al., Ferroelectricity in Gd-doped HfO_2 thin films, ECS J. Solid State Sci. Technol. 1 (2012) N123.

[29] T. Schenk, S. Mueller, U. Schroeder, R. Materlik, A. Kersch, et al., in: Strontium doped hafnium oxide thin films: wide process window for ferroelectric memories, 2013 Proceedings of the European Solid-State Device Research Conference (ESSDERC), IEEE, Bucharest, 2013, pp. 260–263. Available from: https://doi.org/10.1109/ESSDERC.2013.6818868.

[30] S. Starschich, U. Boettger, An extensive study of the influence of dopants on the ferroelectric properties of HfO_2, J. Mater. Chem. C 5 (2017) 333.

[31] L. Xu, T. Nishimura, S. Shibayama, T. Yajima, S. Migita, A. Toriumi, Kinetic pathway of the ferroelectric phase formation in doped HfO_2 films, J. Appl. Phys. 122 (2017) 124104.

[32] A.G. Chernikova, D.S. Kuzmichev, D.V. Negrov, M.G. Kozodaev, S.N. Polyakov, A.M. Markeev, Ferroelectric properties of lightly doped $La{:}HfO_2$ thin films grown by plasma-assisted atomic layer deposition, Appl. Phys. Lett. 111 (2016) 132903.

[33] J. Müller, T.S. Böscke, D. Bräuhaus, U. Schröder, U. Böttger, Ferroelectric $Zr_{0.5}Hf_{0.5}O_2$ thin films for nonvolatile memory applications, Appl. Phys. Lett. 99 (2011) 112901.

[34] J. Müller, T.S. Böscke, U. Schröder, S. Mueller, D. Bräuhaus, U. Böttger, L. Frey, T. Mikolajick, Ferroelectricity in Simple Binary ZrO_2 and HfO_2, Nano Lett. 12 (2012) 4318.

[35] M.H. Park, H.J. Kim, Y.J. Kim, W. Lee, T. Moon, C.S. Hwang, Evolution of phases and ferroelectric properties of thin $Hf_{0.5}Zr_{0.5}O_2$ films according to the thickness and annealing temperature, Appl. Phys. Lett. 102 (2013) 242905.

[36] M.H. Park, H.J. Kim, Y.J. Kim, T. Moon, C.S. Hwang, Study on the degradation mechanism of the ferroelectric properties of thin $Hf_{0.5}Zr_{0.5}O_2$ films on TiN and Ir electrodes, Appl. Phys. Lett. 105 (2014) 072902.

[37] P. Polakowski, J. Müller, Ferroelectricity in undoped hafnium oxide, Appl. Phys. Lett. 106 (2015) 232905.

[38] K.D. Kim, M.H. Park, H.J. Kim, Y.J. Kim, T. Moon, Y.H. Lee, S.D. Hyun, T. Gwon, Ferroelectricity in undoped-HfO$_2$ thin films induced by deposition temperature control during atomic layer deposition, J. Mater. Chem. C 4 (2016) 6484.

[39] M.H. Park, Y.H. Lee, H.J. Kim, Y.J. Kim, T. Moon, K.D. Kim, J. Müller, A. Kersch, U. Schroeder, T. Mikolajick, C.S. Hwang, Ferroelectricity and antiferroelectricity of doped thin HfO$_2$-based films, Adv. Mater. 27 (2015) 1811–1831.

[40] V. George, S. Jahagirdar, C. Tong, K. Smits, S. Damaraju, S. Siers, V. Naydenov, T. Khondker, S. Sarkar, P. Singh, in: Penryn: 45-nm Next Generation Intel CoreTM 2 Processor, Proc. of IEEE Asian Solid-State Circuits Conference, 2007, pp. 14–17.

[41] C.S. Hwang, Atomic Layer Deposition for Semiconductors, Springer, New York, USA, 2013.

[42] O. Ohtaka, H. Fukui, T. Kunisada, T. Fujisawa, K. Funakoshi, W. Utsumi, T. Irifune, K. Kuroda, T. Kikegawa, Phase relations and volume changes of hafnia under high pressure and high temperature, J. Am. Ceram. Soc. 84 (2001) 1369.

[43] X. Sang, E.D. Grimley, T. Schenk, U. Schroeder, J.M. LeBeau, On the structural origins of ferroelectricity in HfO$_2$ thin films, Appl. Phys. Lett. 106 (2015) 162905.

[44] T.D. Huan, V. Sharma, G.A. Rossetti Jr., R. Ramprasad, Pathways towards ferroelectricity in hafnia, Phys. Rev. B 90 (2014) 064111.

[45] J. Müller, T.S. Böscke, Y. Yurchuk, P. Polakowski, J. Paul, D. Martin, T. Schenk, K. Khullar, A. Kersch, W. Weinreich, S. Riedel, K. Seidel, A. Kumar, T.M. Arruda, S.V. Kalinin, T. Schlosser, R. Boschke, V.R. Bentum, U. Schröder, T. Mikolajick, Ferroelectric hafnium oxide: a CMOS-compatible and highly scalable approach to future ferroelectric memories, in: IEEE International Electron Devices Meetings 10.8.1, 2013.

[46] U. Schroeder, et al., Impact of different dopants on the switching properties of ferroelectric hafniumoxide. Jpn. J. Appl. Phys. 53 (8S1) (2014) https://doi.org/10.7567/jjap.53.08le02.

[47] S.E. Reyes-Lillo, K.F. Garrity, K.M. Rabe, Antiferroelectricity in thin film ZrO$_2$ from first principles, Phys. Rev. B 90 (2014) 140103.

[48] D. Zhou, J. Xu, Q. Lu, Y. Guan, F. Cao, X. Dong, J. Mueller, T. Schenk, U. Schroeder, Wake-up effects in Si-doped hafnium oxide ferroelectric thin films, Appl. Phys. Lett. 103 (2013) 192904.

[49] D. Martin, J. Mueller, T. Schenk, T.M. Arruda, A. Kumar, E. Strelcov, E. Yurchuk, S. Mueller, D. Pohl, U. Schroeder, S.V. Kalinin, T. Mikolajick, Ferroelectricity in Si-doped HfO$_2$ revealed: a binary lead-free ferroelectric, Adv. Mater. 26 (2014) 8198.

[50] T. Schenk, U. Schroeder, M. Pesic, M. Popovici, Y.V. Pershin, T. Mikolajick, Electric field cycling behavior of ferroelectric hafnium oxide, ACS Appl. Mater. Interfaces 6 (2014) 19744.

[51] M.H. Park, H.J. Kim, Y.J. Kim, Y.H. Lee, T. Moon, K.D. Kim, S.D. Hyun, C.S. Hwang, Study on the size effect in Hf$_{0.5}$Zr$_{0.5}$O$_2$ films thinner than 8 nm before and after wakeup field cycling, Appl. Phys. Lett. 107 (2015) 192907.

[52] M.H. Park, H.J. Kim, Y.J. Kim, Y.H. Lee, T. Moon, K.D. Kim, S.D. Hyun, F. Fengler, U. Schroeder, C.S. Hwang, Effect of Zr content on the wake-up effect in Hf$_{1-x}$Zr$_x$O$_2$ films, ACS Appl. Mater. Interfaces 8 (24) (2016) 15466.

[53] E.D. Grimley, T. Schenk, X. Sang, M. Pešić, U. Schroeder, T. Mikolajick, J.M. LeBeau, Structural changes underlying field-cycling phenomena in ferroelectric HfO$_2$ thin films, Adv. Electron. Mater. 2 (2016) 1600173.

[54] P.D. Lomenzo, Q. Takmeel, C. Zhou, C.-C. Chung, S. Moghaddam, J.L. Jones, T. Nishida, Mixed Al and Si doping in ferroelectric HfO$_2$ thin films, Appl. Phys. Lett. 107 (2015) 242903.

[55] M.H. Park, T. Schenka, M. Hoffmann, S. Knebela, J. Gärtnera, T. Mikolajicka, U. Schroeder, Effect of acceptor doping on phase transitions of HfO_2 thin films for energy related applications, Nano Energy 36 (2017) 381–389.

[56] C.-K. Lee, E. Cho, H.-S. Lee, C.S. Hwang, S. Han, First-principles study on doping and phase stability of HfO_2, Phys. Rev. B 78 (2008) 012102.

[57] R. Materlik, C. Künneth, A. Kersch, The origin of ferroelectricity in $Hf_{1-x}Zr_xO_2$: a computational investigation and a surface energy model, J. Appl. Phys. 117 (2015) 134109.

[58] Batra, et al., Chem. Mater. 29 (2017) 9102.

[59] S. Starschich, U. Boettger, An extensive study of the influence of dopants on the ferroelectric properties of HfO_2. J. Mater. Chem. C 5 (2) (2017) 333–338, https://doi.org/10.1039/c6tc04807b.

[60] S.J. Kim, et al., Large ferroelectric polarization of $TiN/Hf_{0.5}Zr_{0.5}O_2/TiN$ capacitors due to stress-induced crystallization at low thermal budget. Appl. Phys. Lett. 111 (24) (2017) 242901, https://doi.org/10.1063/1.4995619.

[61] K.D. Kim, Y.H. Lee, T. Gwon, Y.J. Kim, H.J. Kim, T. Moon, S.D. Hyun, H.W. Park, M.H. Park, C.S. Hwang, Scale-up and optimization of HfO_2-ZrO_2 solid solution thin films for the electrostatic supercapacitors, Nano Energy 39 (2017) 390–399.

[62] W.J. Merz, Double Hysteresis Loop of $BaTiO_3$ at the Curie Point, Phys. Rev. 91 (1953) 513.

[63] M.H. Park, Y.H. Lee, H.J. Kim, T. Schenk, W. Lee, K.D. Kim, F.P.G. Fengler, T. Mikolajick, U. Schroeder, C.S. Hwang, Surface and grain boundary energy as the key enabler of ferroelectricity in nanoscale hafnia-zirconia: a comparison of model and experiment, Nanoscale 9 (2017) 9973–9986.

[64] M.H. Park, Y.H. Lee, H.J. Kim, Y.J. Kim, T. Moon, K.D. Kim, S.D. Hyun, T. Mikolajick, U. Schroeder, C.S. Hwang, Understanding the formation of the metastable ferroelectric phase in hafnia–zirconia solid solution thin films, Nanoscale 10 (2018) 716.

[65] G.H. Kim, H.J. Lee, A.Q. Jiang, M.H. Park, C.S. Hwang, An analysis of improved hysteresis loops for ferroelectric $Pb(Zr,Ti)O_3$ thin film capacitor using the switching transient current measurements, J. Appl. Phys. 105 (2009) 044106.

[66] J. Rodriguez, et al., in: Reliability of ferroelectric random access memory embedded within 130 nm CMOS, 2010 IEEE International Reliability Physics Symposium, Anaheim, CA, 2010, pp. 750–758.

[67] F.P. Fengler, et al., Domain pinning: comparison of Hafnia and PZT based ferroelectrics, Adv. Electron. Mater. 3 (4) (2017).

[68] Y. Nishi, Advances in Non-Volatile Memory and Storage Technology, Woodhead Publishing, 2014.

[69] Nagel, et al., in: New highly scalable 3 dimensional chain FeRAM cell with vertical capacitor, Symposium on VLSI Technology, Digest of Technical Papers, 2004, pp. 146–147.

[70] D. Takashima, et al., High-density chain ferroelectric random access memory, IEEE J. Solid-State Circuits 33 (5) (1998).

[71] C.-U. Pinnow, T. Mikolajick, Material aspects in emerging nonvolatile memories, J. Electrochem. Soc. 151 (2004) K13–K19.

[72] R.F. Schnabel, et al., in: Stack capacitor integration with buried oxygen barrier using chemical mechanical polishing of noble metals, 2001 International Symposium on VLSI Technology, Systems, and Applications, Proceedings of Technical Papers (Cat. No. 01TH8517), Hsinchu, 2001, pp. 264–266.

[73] M. Röhner, T. Mikolajick, R. Hagenbeck, N. Nagel, Integration of FeRAM devices into a standard CMOS process—impact of ferroelectric anneals on CMOS characteristics, Integr. Ferroelectr. 47 (2002) 61–70.

[74] W. Hartner, G. Schindler, P. Bosk, Z. Gabric, M. Kastner, G. Beitel, T. Mikolajick, C. Dehm, C. Mazuré, Integration of H2 barriers for ferroelectric memories based on SrBi$_2$Ta$_2$O$_9$ (SBT), Integr. Ferroelectr. 31 (1–4) (2006) 273–284.

[75] H.P. McAdams, R. Acklin, T. Blake, X.-H. Du, J. Eliason, J. Fong, W.F. Kraus, D. Liu, S. Madan, T. Moise, S. Natarajan, N. Qian, Y. Qiu, K.A. Remack, J. Rodriguez, J. Roscher, A. Seshadri, S.R. Summerfelt, A 64-Mb embedded FRAM utilizing a 130-nm 5LM Cu/FSG logic process, IEEE J. Solid State Circuits 39 (2004) 667–677.

[76] K. Maruyama, M. Kondo, S.K. Singh, H. Ishiwara, New ferroelectric material for embedded FRAM LSIs, Fujitsu Sci. Tech. J. 43 (2007) 502–507.

[77] J.-M. Koo, B.-S. Seo, S. Kim, S. Shin, J.-H. Lee, H. Baik, J.-H. Lee, J.H. Lee, B.-J. Bae, J.-E. Lim, D.-C. Yoo, S.-O. Park, H.-S. Kim, H. Han, S. Baik, J.-Y. Choi, Y.J. Park, Y. Park, Fabrication of 3D trench PZT capacitors for 256Mbit FRAM device application, IEDM Techn. Digest. (2005) 340–343.

[78] F. Chu, E. Kim, D. Kim, S. Emley, Enhanced Endurance Performance of 0.13 µm Nonvolatile F-RAM Products, Cypress White Paper, 2016.

[79] H. Itokawa, K. Natori, S. Yamazaki, G. Beitel, K. Yamakawa, in: High performance PZT capacitor using highly crystalline SRO bottom electrode for Mbit FeRAM devices, Extended Abstracts of the 2003 International Conference on Solid State Devices and Materials, Tokyo, 2003, pp. 40–41. 40 – B-2-2.

[80] M. Pešić, M. Hoffmann, C. Richter, T. Mikolajick, U. Schroeder, Nonvolatile random access memory and energy storage based on antiferroelectric like hysteresis in ZrO$_2$, Adv. Funct. Mater. 26 (2016) 7486–7494.

[81] M. Pesic, S. Knebel, M. Hoffmann, C. Richter, T. Mikolajick, U. Schroeder, How to make DRAM non-volatile? Anti-ferroelectrics: a new paradigm for univcrsal memories, in: IEEE International Electron Devices Meeting (IEDM), 2016, pp. 11.6.1–11.6.4.

[82] K. Kita, A. Toriumi, Origin of electric dipoles formed at high-k/SiO2 interface, Appl. Phys. Lett. 94 (2009) 132902.

[83] D.K. Simon, et al., On the control of the fixed charge densities in Al$_2$O$_3$-Based silicon surface passivation schemes, ACS Appl. Mater. Interfaces 7 (2015) 51.

[84] M. Pesic, et al., Built-in bias generation in anti-ferroelectric stacks: methods and device applications. IEEE J. Electron Devices Soc. 6 (2018) 1019–1025, https://doi.org/10.1109/jeds.2018.2825360.

[85] M. Pesic, et al., Anti-ferroelectric ZrO$_2$, an enabler for low power non-volatile 1T-1C and 1T random access memories, in: 2017 47th European Solid-State Device Research Conference (ESSDERC), Leuven, 2017, pp. 160–163.

[86] S.L. Miller, P.J. McWhorter, Physics of the ferroelectric nonvolatile memory field effect transistor, J. Appl. Phys. 72 (1992) 5999.

[87] J. Müller, et al., Integration challenges of ferroelectric hafnium oxide based embedded memory, ECS Trans. 69 (3) (2015) 85–95.

[88] J. Van Houdt, P. Roussel, Alternative Explanation for the Steep Subthreshold Slope in Ferroelectric FETs, 2018arXiv180203590V, 2018.

[89] S. Salahuddin, S. Datta, Use of negative capacitance to provide voltage amplification for low power nanoscale devices, Nano Lett. 8 (2) (2008) 405–410.

[90] T.P. Ma, J.-P. Han, Why is nonvolatile ferroelectric memory field-effect transistor still elusive? IEEE Electron Device Lett. 23 (2002) 386–388.

[91] D.R. Lampe, S. Sinharoy, E. Stepke, H. Buhay, Integration of UHV-grown ferroelectric films into nonvolatile memories, in: IEEE 7th International Symposium on Applications of Ferroelectrics, Urbana-Champaign, IL, 1990, pp. 177–180.

[92] E. Tokumitsu, G. Fujii, H. Ishiwara, Electrical properties of metal-ferroelectric-insulator-semiconductor (MFIS)- and metal-ferroelectric-metal-insulator-semiconductor (MFMIS)-FETs using ferroelectric $SrBi_2Ta_2O_9$ film and $SrTa_2O_6$/SiON buffer layer, Jpn. J. Appl. Phys. 39 (Part 1) (2000) 2125–2130. No. 4B (2000).

[93] S. Sakai, R. Ilangovan, Metal–ferroelectric–insulator–semiconductor memory FET with long retention and high endurance, IEEE Electron Device Lett. 25 (2004) 369–371.

[94] M.T. Bohr, R.S. Chau, T. Ghani, K. Mistry, The high-k solution: microprocessors entering production this year are the result of the biggest transistor redesign in 40 years, IEEE Spectr. 44 (2007) 29–35.

[95] J. Müller, J. Müller, E. Yurchuk, T. Schlösser, J. Paul, R. Hoffmann, S. Müller, D. Martin, S. Slesazeck, P. Polakowski, J. Sundqvist, M. Czernohorsky, P. Kücher, R. Boschke, M. Trentzsch, K. Gebauer, U. Schröder, T. Mikolajick, Ferroelectricity in HfO_2 enables nonvolatile data storage in 28 nm HKMG, Proceeding of IEEE Symposia on VLSI Technology, 2012, pp. 25–26.

[96] S. Dünkel, et al., A FeFET based super-low-power ultra-fast embedded NVM technology for 22 nm FDSOI and beyond, in: 2017 IEEE International Electron Devices Meeting (IEDM), San Francisco, CA, 2017, pp. 19.7.1–19.7.4.

[97] M. Trentzsch, et al., A 28 nm HKMG super low power embedded NVM technology based on ferroelectric FETs, in: 2016 IEEE International Electron Devices Meeting (IEDM), San Francisco, CA, 2016, pp. 11.5.1–11.5.4.

[98] E. Yurchuk, J. Müller, J. Paul, R. Hoffmann, S. Müller, D. Martin, S. Slesazeck, U. Schröder, J. Sundqvist, T. Schlösser, R. Boschke, V.R. Bentum, M. Trentzsch, T. Mikolajick, Origin of the endurance degradation in the Novel HfO_2-based 1T ferroelectric non-volatile memories, Proceedings of the IRPS, 2014.

[99] J. Muller, et al., High endurance strategies for hafnium oxide based ferroelectric field effect transistor, in: 2016 16th Non-Volatile Memory Technology Symposium (NVMTS), Pittsburgh, PA, 2016, pp. 1–7.

[100] K. Chatterjee, et al., Self-aligned, gate last, FDSOI, ferroelectric gate memory device with 5.5-nm $Hf_{0.8}Zr_{0.2}O_2$, high endurance and breakdown recovery, IEEE Electron Device Lett. 38 (10) (2017) 1379–1382.

[101] V. Zhirnov, T. Mikolajick, Flash memories, in: R. Waser (Ed.), Nanoelectronics and Information Technology, Wiley-VCH, 2012, pp. 623–634.

[102] S. Mueller, S. Slesazeck, T. Mikolajick, J. Müller, P. Polakowski, S. Flachowsky, in: Next-generation ferroelectric memories based on FE-HfO_2, 2015 Joint IEEE International Symposium on the Applications of Ferroelectric (ISAF), International Symposium on Integrated Functionalities (ISIF), and Piezoelectric Force Microscopy Workshop (PFM), Singapore, 2015, pp. 233–236.

[103] S. Sakai, et al., Highly scalable Fe(ferroelectric)-NAND cell with MFIS(metal-ferroelectric-insulator-semiconductor) structure for sub-10nm tera-bit capacity NAND flash memories, in: 2008 Joint Non-Volatile Semiconductor Memory Workshop and International Conference on Memory Technology and Design, Opio, 2008, pp. 103–105.

[104] K. Florent, et al., First demonstration of vertically stacked ferroelectric Al doped HfO_2 devices for NAND applications, in: 2017 Symposium on VLSI Technology, Kyoto, 2017, pp. T158–T159.

[105] S. Mueller, S. Slesazeck, S. Henker, S. Flachowsky, P. Polakowski, J. Paul, E. Smith, J. Müller, T. Mikolajick, Correlation between the macroscopic ferroelectric material properties of Si:HfO_2 and the statistics of 28 nm FeFET memory arrays, Ferroelectrics 497 (1) (2016) 42–51.

[106] C.H. Cheng, A. Chin, Low-leakage-current DRAM-like memory using a one-transistor ferroelectric MOSFET with a Hf-based gate dielectric, IEEE Electron Device Lett. 35 (1) (2014) 138–140.

[107] H. Mulaosmanovic, et al., Switching kinetics in nanoscale hafnium oxide based ferroelectric field-effect transistors, ACS Appl. Mater. Interfaces 9 (4) (Jan. 2017) 3792–3798.

[108] H. Mulaosmanovic, et al., Evidence of single domain switching in hafnium oxide based FeFETs: enabler for multi-level FeFET memory cells, Proc. IEEE Int. Electron Devices Meeting (IEDM), 201526.8.1–26.8.3.

[109] L. Esaki, R.B. Laibowitz, P.J. Stiles, Electron transport in Nb-Nb oxide-Bi tunnel junctions, Phys. Lett. A 36 (5) (1971) 429–430.

[110] H. Kohlstedt, N.A. Pertsev, J. Rodríguez Contreras, R. Waser, Theoretical current-voltage characteristics of ferroelectric tunnel junctions, Phys. Rev. B 72 (2005) 125341.

[111] M.Y. Zhuravlev, R.F. Sabirianov, S.S. Jaswal, E.Y. Tsymbal, Giant electroresistance in ferroelectric tunnel junctions, Phys. Rev. Lett. 94 (246802) (2005).

[112] I.P. Batra, B.D. Silverman, Thermodynamic stability of thin ferroelectric films, Solid State Commun. 11 (1972) 291.

[113] I.P. Batra, P. Würfel, B.D. Silverman, New type of first-order phase transition in ferroelectric thin films, Phys. Rev. Lett. 30 (1973) 384.

[114] P. Ghosez, K.M. Rabe, Microscopic model of ferroelectricity in stress-free ultrathin films, Appl. Phys. Lett. 76 (2000) 2767.

[115] B. Meyer, D. Vanderbilt, Ab initio study of $BaTiO_3$ and $PbTiO_3$ surfaces in external electric fields, Phys. Rev. B 63 (2001) 205426.

[116] A.G. Zembilgotov, N.A. Pertsev, H. Kohlstedt, R. Waser, Ultrathin epitaxial ferroelectric films grown on compressive substrates: compctition bctween the surface and strain effects, J. Appl. Phys. 91 (2002) 2247.

[117] J. Junquera, P. Ghosez, Critical thickness for ferroelectricity in perovskite ultrathin films, Nature 422 (2003) 506.

[118] C.H. Ahn, K.M. Rabe, J.-M. Triscone, Ferroelectricity at the nanoscale: local polarization in oxide thin films and heterostructures, Science 303 (2004) 488.

[119] T. Tybell, C.H. Ahn, J.M. Triscone, Ferroelectricity in thin perovskite films, Appl. Phys. Lett. 75 (1999) 856.

[120] S.K. Streiffer, J.A. Eastman, D.D. Fong, C. Thompson, A. Munkholm, M.V. Ramana Murty, O. Auciello, G.R. Bai, G.B. Stephenson, Observation of nanoscale stripe domains in ferroelectric thin films, Phys. Rev. Lett. 89 (2002) 067601.

[121] D.D. Fong, G.B. Stephenson, S.K. Streiffer, J.A. Eastman, O. Auciello, P.H. Fuoss, C. Thompson, Ferroelectricity in ultrathin perovskite films, Science 304 (2004) 1650.

[122] V. Garcia, M. Bibes, Ferroelectric tunnel junctions for information storage and processing, Nat. Commun. 5 (2014) 4289.

[123] A. Gruverman, D. Wu, H. Lu, Y. Wang, H.W. Jang, C.M. Folkman, M.Y. Zhuravlev, D. Felker, M. Rzchowski, C.-B. Eom, E.Y. Tsymbal, Tunneling electroresistance effect in ferroelectric tunnel junctions at the nanoscale, Nano Lett. 9 (10) (2009) 3539–3543.

[124] Z. Wen, C. Li, D. Wu, A. Li, N. Ming, Ferroelectric-field-effect-enhanced electroresistance in metal/ferroelectric/semiconductor tunnel junctions, Nat. Mater. 12 (2013) 617–621.

[125] S. Fujii, Y. Kamimuta, T. Ino, Y. Nakasaki, R. Takaishi, M. Saitoh, First demonstration and performance improvement of ferroelectric HfO_2-based resistive switch with low operation current and intrinsic diode property, in: 2016 IEEE Symposium on VLSI Technology, Honolulu, HI, 2016, pp. 1–2.

[126] M. Pešić, et al., Physical and circuit modeling of HfO_2 based ferroelectric memories and devices, in: 2017 IEEE SOI-3D-Subthreshold Microelectronics Technology Unified Conference (S3S), Burlingame, CA, USA, 2017, pp. 1–4.

[127] A. Crassous, T. Sluka, A.K. Tagantsev, N. Setter, Polarization charge as a reconfigurable quasi-dopant in ferroelectric thin films, Nat. Nanotechnol. 10 (2015) 614–618.

[128] S. Farokhipoor, B. Noheda, Conduction through 71o domain walls in $BiFeO_3$ thin films, Phys. Rev. Lett. 107 (2011) 127601.

[129] J. Seidel, L.W. Martin, Q. He, Q. Zhan, Y.-H. Chu, A. Rother, M.E. Hawkridge, P. Maksymovych, P. Yu, M. Gajek, N. Balke, S.V. Kalinin, S. Gemming, F. Wang, G. Catalan, J.F. Scott, N.A. Spaldin, J. Orenstein, R. Ramesh, Conduction at domain walls in oxide multiferroics, Nat. Mater. 8 (2009) 229–234.

[130] A. Lubk, M.D. Rossell, J. Seidel, Q. He, S.Y. Yang, Y.H. Chu, R. Ramesh, M.J. Hÿtch, E. Snoeck, Evidence of sharp and diffuse domain walls in $BiFeO_3$ by means of unit-cell-wise strain and polarization maps obtained with high resolution scanning transmission electron microscopy, Phys. Rev. Lett. 109 (2012) 047601.

[131] J.A. Mundy, J. Schaab, Y. Kumagai, A. Cano, M. Stengel, I.P. Krug, D.M. Gottlob, H. Doğanay, M.E. Holtz, R. Held, Z. Yan, E. Bourret, C.M. Schneider, D.G. Schlom, D.A. Muller, R. Ramesh, N.A. Spaldin, D. Meier, Functional electronic inversion layers at ferroelectric domain walls, Nat. Mater. 16 (2017) 622–627.

[132] A.Q. Jiang, H.J. Lee, C.S. Hwang, T.A. Tang, Resolving the Landauer paradox in ferroelectric switching by high-field charge injection, Phys. Rev. B 80 (2009) 024119.

[133] M. Schröder, A. Haußmann, A. Thiessen, E. Soergel, T. Woike, L.M. Eng, Conducting domain walls in lithium niobate single crystals, Adv. Funct. Mater. 22 (2012) 3936–3944.

[134] J.B. Johnson, Thermal agitation of electricity in conductors, Phys. Rev. 32 (1928) 97–109.

[135] H. Nyquist, Thermal agitation of electric charge in conductors, Phys. Rev. 32 (1928) 110–113.

[136] A.Q. Jiang, C. Wang, K.J. Jin, X.B. Liu, J.F. Scott, C.S. Hwang, T.A. Tang, H.B. Lu, G.Z. Yang, A resistive memory in semiconducting $BiFeO_3$ thin-film capacitors, Adv. Mater. 23 (10) (2011) 1277–1281.

[137] P. Sharma, Q. Zhang, D. Sando, C.H. Lei, Y. Liu, J. Li, V. Nagarajan, J. Seidel, Nonvolatile ferroelectric domain wall memory, Sci. Adv. 3 (e1700512) (2017).

[138] J. Jiang, Z.L. Bai, Z.H. Chen, L. He, D.W. Zhang, Q.H. Zhang, J.A. Shi, M.H. Park, J.F. Scott, C.S. Hwang, A.Q. Jiang, Temporary formation of highly conducting domain walls for non-destructive read-out of ferroelectric domain-wall resistance switching memories, Nat. Mater. 17 (2018) 49–56.

[139] J.Y. Seok, S.J. Song, J.H. Yoon, K.J. Yoon, T.H. Park, D.E. Kwon, H. Lim, G.H. Kim, D.S. Jeong, C.S. Hwang, A review of three-dimensional resistive switching Cross-Bar array memories from the integration and materials property points of view, Adv. Funct. Mater. 34 (24) (2014) 5316–5339.

[140] C.S. Hwang, Prospective of semiconductor memory devices: from memory system to materials, Adv. Electr. Mater. 6 (1) (2015) 1400056.

[141] H. Kimura, T. Hanyu, M. Kameyama, Y. Fujimori, T. Nakamura, H. Takasu, Complementary ferroelectric-capacitor logic for low-power logic-in-memory VLSI, IEEE J. Solid-State Circuits 39 (6) (2004) 919–926.

[142] S. Sakai, M. Takahashi, Recent progress of ferroelectric-gate field-effect transistors and applications to nonvolatile logic and FeNAND flash memory, Materials 3 (11) (2010) 4950–4964.

[143] E.T. Breyer, H. Mulaosmanovic, T. Mikolajick, S. Slesazeck, Reconfigurable NAND/ NOR logic gates in 28 nm HKMG and 22 nm FD-SOI FeFET Technology, in: 2017 IEEE International Electron Devices Meeting (IEDM), San Francisco, CA, 2017, pp. 28.5.1–28.5.4.

[144] W. Haensch, Scaling is over — what now? in: 75th Annual Device Research Conference (DRC), South Bend, IN, 2017, 2017, pp. 1–2.

[145] H.Z. Shouval, S.S.-H. Wang, G.M. Wittenberg, Spike timing dependent plasticity: a consequence of more fundamental learning rules, Front. Comput. Neurosci. 01 (2010).

[146] Y. Nishitani, Y. Kaneko, M. Ueda, E. Fujii, A. Tsujimura, Dynamic observation of brain-like learning in a ferroelectric synapse device, Jpn. J. Appl. Phys. 52 (4S) (2013) 04CE06.

[147] H. Mulaosmanovic, et al., Novel ferroelectric FET based synapse for neuromorphic systems, in: 2017 Symposium on VLSI Technology, Kyoto, 2017, pp. T176–T177.

[148] H. Mulaosmanovic, E. Chicca, M. Bertele, T. Mikolajick, S. Slesazeck, Mimicking biological neurons with a nanoscale ferroelectric transistor, Nanoscale 10 (2018) 21755–21763.

[149] H. Mulaosmanovic, T. Mikolajick, S. Slesazeck, Random number generation based on ferroelectric switching, IEEE Electron Device Lett. 39 (1) (2018) 135–138.

[150] G. Catalan, D. Jimenez, A. Gruverman, Negative capacitance detected, Nat. Mater. 14 (2015) 137.

[151] V.V. Zhirnov, R.K. Cavin, Negative capacitance to the rescue? Nat. Nanotechnol. 3 (2008) 77.

[152] C.M. Krowne, S. Kirchoefer, W. Chang, J.M. Pond, L.M.B. Alldredge, Nano Lett. 11 (2011) 988.

[153] A. Khan, D. Bhowmik, P. Yu, S.J. Kim, X. Pan, R. Ramesh, S. Salahuddin, Appl. Phys. Lett. 99 (2011) 113501.

[154] J.R. Daniel, N.K. Ponon, K.S.K. Kwa, B. Zou, P.K. Petrov, T. Wang, M. Neil, Alford, and anthony O'Neill, Nano Lett. (14) (2014) 3864–3868.

[155] W.W. Gao, A. Khan, X. Marti, C. Nelson, C. Serrao, J. Ravichandran, R. Ramesh, S. Salahuddin, Nano Lett. 14 (2014) 5814–5819.

[156] Y.J. Kim, M.H. Park, Y.H. Lee, H.J. Kim, W. Jeon, T. Moon, K.D. Kim, D.S. Jeong, H. Yamada, C.S. Hwang, Sci. Rep. 6 (2016) 19039.

[157] Y.J. Kim, M.H. Park, W. Jeon, H.J. Kim, T. Moon, Y.H. Lee, K.D. Kim, S.D. Hyun, C.S. Hwang, J. Appl. Phys. 118 (2015) 224105.

[158] Y.J. Kim, H. Yamada, T. Moon, Y.J. Kwon, C.H. An, H.J. Kim, K.D. Kim, Y.H. Lee, S.D. Hyun, M.H. Park, C.S. Hwang, Nano Lett. 16 (7) (2016) 4375–4381.

[159] A. Khan, K. Chatterjee, B. Wang, S. Drapcho, L. You, C. Serrao, S.R. Bakaul, R. Ramesh, S. Salahuddin, Nat. Mater. 14 (2015) 182.

[160] P. Zubko, J.C. Wojdeł, M. Hadjimichael, S. Fernandez-Pena, A. Sené, I. Luk'yanchuk, J.-M. Triscone, J. Íñiguez, Nature 534 (2016) 524–528. 7608.

[161] Y.J. Kim, H.W. Park, S.D. Hyun, H.J. Kim, K.D. Kim, Y.H. Lee, T. Moon, Y.B. Lee, M.H. Park, C.S. Hwang, Nano Lett. 17 (12) (2017) 7796–7802.

[162] M. Hoffmann, M. Peši'c, K. Chatterjee, A.I. Khan, S. Salahuddin, S. Slesazeck, U. Schroeder, T. Mikolajick, Adv. Funct. Mater. 26 (47) (2016) 8643–8649.

[163] D. Jimenez, E. Miranda, A. Godoy, IEEE Trans. Electronic Devices 57 (10) (2010) 2405–2409.

[164] H.P. Chen, V.C. Lee, A. Ohoka, J. Xiang, Y. Taur, IEEE Trans. Electronic Devices 58 (8) (2011).

[165] A. Rusu, G.A. Salvatore, D. Jiménez, A.M. Ionescu, IEDM Tech. Dig. (2010) 395–398.

[166] C.H. Cheng, A. Chin, IEEE Electron Device Lett. 35 (2) (2014) 274–276.
[167] M.H. Lee, Y.-T. Wei, K.-Y. Chu, J.-J. Huang, C.-W. Chen, C.-C. Cheng, M.-J. Chen, H.-Y. Lee, Y.-S. Chen, L.-H. Lee, M.-J. Tsai, IEEE Electron Device Lett. 36 (4) (2015) 294–296.
[168] S.A. Sherrill, P. Banerjee, G.W. Rubloff, S.B. Lee, High to ultra-high power electrical energy storage, Phys. Chem. Chem. Phys. 13 (2011) 20714–20723.
[169] L.C. Haspert, E. Gillette, S.B. Lee, G.W. Rubloff, Perspective: hybrid systems combining electrostatic and electrochemical nanostructures for ultrahigh power energy storage, Energ. Environ. Sci. 6 (2013) 2578.
[170] K. Yao, S. Chen, M. Rahimabady, M. S. Mirshekarloo, S. Yu, F.E.H. Tay, T. Sritharan, L. Lu (2011) 'Nonlinear dielectric thin films for high-power electric storage with energy density comparable with electrochemical supercapacitors' IEEE Trans. Ultrasonics Ferroelectr. Freq. Control 58(9): 1968.
[171] M.H. Park, H.J. Kim, Y.J. Kim, T. Moon, K.D. Kim, C.S. Hwang, Thin $Hf_xZr_{1-x}O_2$ films: a new lead-free system for electrostatic supercapacitors with large energy storage density and robust thermal stability, Adv. Energy Mater. 4 (2014) 1400610.
[172] F. Ali, X. Liu, D. Zhou, X. Yang, J. Xu, T. Schenk, J. Mueller, U. Schroeder, F. Cao, X. Dong, Silicon-doped hafnium oxide anti-ferroelectric thin films for energy storage, J. Appl. Phys. 122 (2017) 144105.
[173] P.D. Lomenzo, C.-C. Chung, C. Zhou, J.L. Jones, T. Nishida, Doped $Hf_{0.5}Zr_{0.5}O_2$ for high efficiency integrated supercapacitors, Appl. Phys. Lett. 110 (2017) 232904.
[174] G. Sebald, S. Pruvost, D. Guyomar, Energy harvesting based on Ericsson pyroelectric cycles in a relaxor ferroelectric ceramic, Smart Mater. Struct. 17 (2007) 015012.
[175] M. Hoffmann, U. Schroeder, C. Künneth, A. Kersch, S. Starschich, U. Böttger, T. Mikolajick, Ferroelectricphasetransitions in nanoscale HfO_2 films enable giant pyroelectric energy conversion and highly efficient supercapacitors, Nano Energy 18 (2015) 154–164.
[176] M.H. Park, H.J. Kim, Y.J. Kim, T. Moon, K.D. Kim, C.S. Hwang, Toward amultifunctional monolithic device based on pyroelectricity and the electrocaloric effect of thin antiferroelectric $Hf_xZr_{1-x}O_2$ films, Nano Energy 12 (2014) 131–140.
[177] C.R. Bowen, H.A. Kim, P.M. Weaver, S. Dunn, Piezoelectric and ferroelectric materials and structures for energy harvesting applications, Energ. Environ. Sci. 7 (2014) 25–44.
[178] Y. Yang, W. Guo, K.C. Pradel, G. Zhu, Y. Zhou, Y. Zhang, Y. Hu, L. Lin, Z.L. Wang, Pyroelectric nanogenerators for harvesting thermoelectric energy, Nano Lett. 12 (2012) 2833.
[179] B. Peng, H. Fan, Q. Zhang, A giant electrocaloric effect in nanoscale antiferroelectric and ferroelectric phases coexisting in a relaxor $Pb_{0.8}Ba_{0.2}ZrO_3$ thin film at room temperature, Adv. Funct. Mater. 23 (23) (2013) 2987–2992.

Advances in nanowire PCM

12

Massimo Longo
CNR—Institute for Microelectronics and Microsystems—Unit of Agrate
Brianza, Italy

12.1   Introduction

Interest in phase change random access memory (PCM) devices as future high per-forming, nonvolatile and non-charge-based memory devices is rapidly increasing. Those devices offer high-speed operation, endurance, and downscaling, even beyond lithographic limits [1–5], so that they are emerging as a leading contender for storage class memories that should fill the performance gap between volatile dynamic random access memory (DRAM) and nonvolatile flash memories [6]. The first PCM devices were commercialized in 2008 with a 128 Mbit device based on a 90-nm technology node [7] and miniaturization is in progress [8].

The information in a PCM device is stored in the different structural phases of the active material, typically chalcogenide compounds, such as $Ge_2Sb_2Te_5$ (GST). Suitably different values for the electrical resistivity correspond to these phases (the resistivity of the amorphous and crystalline phases even differing by 3–4 orders of magnitude) and can be used to assign binary codes. Reversible structural changes are induced in the compound by proper energy transfer, in the form of ns current pulses that induce Joule heating. A critical aspect is related to the crystalline to amorphous transition (reset operation), for which the material has to be melted and then rapidly quenched. Melting temperatures as high as around 600°C are necessary for this process in the case of GST; which requires relatively high reset currents (fractions of mA), thereby high power (fractions of mW) is required in PCM devices [9], in comparison with mature flash memory devices. The opposite process, namely the amorphous to crystalline transition, requires less energy, since the crystallization temperatures can be relatively low (~160°C for GST), but a reduced thermal stability of the amorphous phase leads to a poor data retention of the memory device.

Although PCM devices are promising in terms of fabrication costs, important issues for future development of PCM devices are the increase of the storage capacity per cell and the reduction of power consumption and costs.

Scaling down PCM cells from thin films to nanowires (NWs) may provide a good chance to increase the PCM performances beyond flash memory and their family. However, it requires specific solutions to each of the reliability problems introduced by the reduced length scale. As often occurs at the nanoscale, new physical effects, negligible at the larger length scale, have to be taken into account [10]. After basic descriptions of PCM mechanism, fabrication, and characterization techniques for phase change NWs, this chapter reviews the main properties of NW phase change materials

Advances in Non-volatile Memory and Storage Technology. https://doi.org/10.1016/B978-0-08-102584-0.00013-9

and how they affect device performance and reliability. It examines crystallization kinetics, phase transitions, thermal and electrical properties, as well as the properties of core-shell structures.

12.2   Strategies for improving the PCM performance

Storage capacity and power reduction of PCMs can be basically achieved by acting on the phase change material and/or on the programming volume size, meaning scaling down the memory cells.

12.2.1   Modifying phase change materials

Phase change alloys with a higher electrical resistivity and a lower thermal conductivity can reduce the switching currents, whereas different resistance levels can enable multilevel data storing capability (thus increased storage density); however, acting on the phase change material composition also means understanding the failure mechanisms, affecting its reliability [10].

First, alloys are required that can compromise between the high crystallization rate and the stability of the amorphous phase. Typical PCMs have a 10-year data retention at 105°C [11]. The ability to retain data and maintain the memory behavior at nanoscale geometry is determined by the controllability and stability of the amorphous-phase change material. For example, variations of the composition within largely employed Ge-Sb-Te system are known to have an impact to the speed/retention trade-off. The crystallization temperature can be increased by increasing the content of GeTe within the $(GeTe)_x(Sb_2Te_3)_y$ subsystem, thus improving the stability of the amorphous phase, although this entails some drawbacks for the crystallization speed. It has been reported that there are great improvements to the data retention when nitrogen doping around 10% is employed in GeTe, yielding data retention of 10 years at 241°C [12]. However, such a trade-off has not been clearly assessed yet. Currently, GST is used in PCMs, due to its fast crystallization dynamics and its high stability at operating temperatures. Memories based on using GST have shown 10 years retention for temperatures up to 110°C [13], which is sufficient for several applications, including mobile phone cards.

For automotive applications, the 10-year retention is required at temperatures higher than 125°C [14]. A higher retention temperature implies higher stability of the amorphous phase, thereby higher crystallization temperature. For this application, GST has high melting temperature ($T_M \sim 620$°C), but too low crystallization temperature ($T_X \sim 160$°C), therefore, insufficient for use of multi-bit storage capability. A way to increase the thermal stability of GST is to introduce a significant amount of Ge, thus achieving a crystallization temperature around 360°C, although segregation effects occur in the switched cells [15]. "In-based" chalcogenide materials, namely those belonging to the In-Ge-Te and In-Sb-Te systems, are also attracting attention, since they display higher thermal stability: In-Ge-Te (IGT) has a very high crystallization temperature ($T_C \sim 276$°C) and a 10-year retention at 170°C [14]. $In_3Sb_1Te_2$ (IST) has an even higher crystallization temperature ($T_C \sim 280$°C) [16]. Recently, In-based chalcogenide materials are increasingly being considered for PCM application and some reports of In-based PCMs can be found in the literature [14, 17–21].

12.2.2   Reducing volume size

A very effective method of power reduction is to reduce the size of the chalcogenide-based memory cells to the nanoscale (~100 nm). Lowering the active material volumes that are to be programmed allows for reduced power consumption and, at the same time, increases the cell density. Of course, this needs to be done by assuring the conservation of the basic material characteristics, even for very small dimensions. The scalability of PCM cells has been demonstrated down to 1.8 nm for GeTe nanoparticles [22] and the same group [23] has obtained writing currents as low as 1.4 µA, 1000 times lower than in ordinary PCM devices, by using carbon nanotube (CNT) electrodes, with good prospects for high-volume manufacturability and reliability.

Actually, shrinking the linear dimensions of PCM cells has an impact on the cell parameters, such as electrothermal resistance, current, voltage, and power dissipation, thus determining "scaling rules" [2]; so that the scaling of the active material/bottom electrode contact area [24], cell confinement in trenches/pores [25], and three-dimensional (3D) cross-bar architectures [26] have been explored for their potential in power consumption limitation. The best programming energy reduction and fast switching have been obtained in interfacial phase change memory (IPCM) devices by the use of ultrathin $GeTe/Sb_2Te_3$ superlattices [27]. In interfacial PCM, the switching energy is reduced, because the movement of the atoms is limited to a single dimension. The phase change would not occur by conventional melting and therefore, would require much less energy. The electrical energy required to set the IPCM cell has been reported to be as low as 11 pJ, whereas for conventional GST cells the same authors have reported 90 pJ [27].

Most of the GST thin films used in current PCM applications are obtained by physical deposition, such as sputtering, which is limited by its low deposition conformality, that is, it is unsuited to provide uniform filling of structured substrates, especially for vertical, high aspect ratio ones. In order to fabricate ultrascaled memory cells, more conformal techniques are needed to uniformly deposit GST inside structured substrates with trenches tens of nanometers wide. Among the growth methods that can provide highly conformal deposition, atomic layer deposition (ALD) and metal organic chemical vapor deposition (MOCVD) are being explored. Scaled PCM devices have already been realized by MOCVD and mixed ALD/chemical vapor deposition (CVD) processes [18, 28, 29].

An alternative method for reducing programming volumes and contact size areas is to employ phase change nanomaterials, synthesized either by top-down or bottom-up approaches. Some different approaches to achieve this are:

- the use of metal sidewall pattern technology [30];
- Sb-doped Te-rich nanotubes synthesized by the vapor transport method [31];
- sputtered GST in CNTs [32];
- GST nano-pillars fabricated by nano-imprint lithography [33];
- GST nano-rings created by focused laser beam followed by differential chemical etching [34];

- phase change nanodot arrays by the lift-off technique on a self-assembled di-block copolymer template [35];
- nano-PCM cell element obtained by using a focused ion beam (FIB) [36];
- colloidal GeTe nanocrystals [22];
- metal chalcogenide arrays created by using block copolymer-derived nanoreactors [37]; and
- Ge-Sb-Te and Sb_2Te_3 nanostructures deposited by selective area epitaxy and solvo-thermal synthesis [38].

12.3   The use of NWs

The use of 1D phase change structures, such as NWs, makes it possible to fully exploit the scaling properties of the PCM device. The phase change NW growth and its diameter, composition, and crystallinity can be controlled. When metal contacted, a NW can form a nano-PCM cell itself and, if dense arrays of NWs are obtainable, a further, novel highly performing device could be formed: a NW PCM cell, with a relatively low manufacturing cost and complementary metal-oxide semiconductor (CMOS) compatibility. The energy required to induce the phase transition is lowered, not only because of the good structural properties of these nanostructures (which can be defect free when the NWs are self-assembled) and the high geometrical scaling, but also due to other aspects related to size effects (Section 12.8).

The idea of NW-PCM device is very simple, based on the transformation of the active chalcogenide layer contained in a planar cell, into a one-dimensional (1D) cell. A small volume in the region close to the bottom electrode heater is crystallized and reamorphized during the switching cycles of a planar PCM cell. The 1D cell is formed by a NW, whose metallic contacts at both tips form the top and bottom electrodes, respectively. A nanocell is, therefore, obtained in which the NW is a self-heating resistor. A portion of it undergoes a phase transition from amorphous to crystalline, as the current flows through it. A proper dielectric material is needed to passivate the NW, not only to insulate it, but also to protect it from evaporating during the phase transitions. Fig. 12.1 displays the schemes of a planar PCM cell and a NW-based PCM cell.

The alloys used for phase change NW are essentially the same as those used for planar PCM cells. They belong to the IV, V, and VI groups of the periodic table and

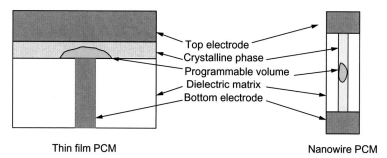

Thin film PCM   Nanowire PCM

Fig. 12.1 Schematics of the thin-film PCM device and the 1D nanowire PCM.

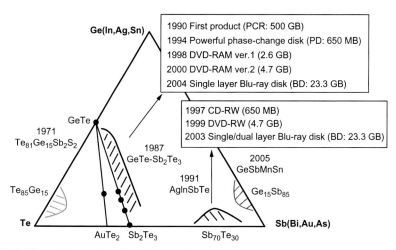

Fig. 12.2 Phase diagram showing different phase change alloys, and their year of discovery, for optical memories; those for electronic memories are mainly on the GeTe-Sb_2Te_3 tie line. Reprinted by permission from Macmillan Publishers Ltd.: Nature Materials M. Wuttig, N. Yamada, Phase-change materials for rewriteable data storage, Nat. Mater. 6 (2007) 824–32, copyright (2007).

aim at high-speed transitions, a large resistance window between the amorphous and crystalline phases, thermal stability of the amorphous phase (for data retention), and a high number of transitions before material degradation (cyclability). GeTe, Sb_2Te_3, and the alloys of the pseudobinary GeTe-Sb_2Te_3 tie line (in particular, the stable phases $Ge_2Sb_2Te_5$, $Ge_1Sb_2Te_4$, and $Ge_1Sb_4Te_7$) are the main alloys that have been investigated, along with In_2Se_3, Ge-Sb, Bi_2Te_3, InSb, and $In_3Sb_1Te_2$. The following section will illustrate examples of the most commonly used materials for phase change NWs and their corresponding synthesis techniques. In Fig. 12.2, the different alloys used for phase change optical and electronic memories are shown.

In the case of optical data storage, different material reflectivity of the amorphous and crystalline states is looked at, instead of electrical resistivity.

12.4 Fabrication of phase change NWs: Top-down approaches

As in the case of NWs based on III–V and nitride compounds, it is possible to obtain phase change chalcogenide NWs either via the "top-down," or the "bottom-up" approaches. The top-down approach involves patterning processes. It mainly employs lithographic techniques, based on ultraviolet (UV), X-ray and electron beams, or reactive ion etching (RIE), so that the NWs are obtained by properly removing material until the desired structures are exposed. Due to the high geometry control, this is the most compatible method with the current CMOS technology even in 3D architecture, thanks to the precise controllability of the fabrication process with high reproducibility.

The bottom-up approach is characterized by the use of the basic constituent blocks (often the elements themselves), which are built up to form complex NW structures, using chemical reactions and/or self-assembly mechanisms on nano-structured or chemically active surfaces. The NW size, morphology, and growth can thus be controlled in a single-step process. Moreover, structures such as core-shell NWs can be synthesized, which are not possible when employing top-down technologies.

Both the top-down and bottom-up approaches present positive and negative aspects. In terms of the top-down approach, the resolution of the chip-compatible lithographic tools is limited and the fabricated nanostructures may present some structural damages or defects; as a matter of fact, it has been reported that the presence of grain boundaries can enhance the electromigration effects (an atomic compositional variation on reset switching), thus leading to device failure [39]. Equally, the defect-free NWs obtained by the bottom-up approach cannot be easily located at desired positions or interfaced in a device and their technical implementation can be an issue. Future applications will need to consider using a combination of top-down and bottom-up techniques. For example, an ordered array of phase change NWs could be obtained by utilizing self-assembly through selective area deposition on favorable surface areas, by using nano-patterning.

The use of PCM line cells based on a single Sb-Te nanostrip, comparable to a NW because of the 1D design, was first reported in 2005. These have been obtained by top-down fabrication, in particular by electron beam lithography (EBL) and RIE (Fig. 12.3).

In this case, the reset current depends on a function of the strip cross section, but it was clearly advantageous that no heater integration was required. Analogously, the bridge cell concept has been reported, in which a $Ge_{15}Sb_{85}$ NW was deposited on top of TiN electrodes by EBL coupled to ion milling (Fig. 12.4), yielding a cross-sectional area of 60 nm^2. In this case, a minimum 90-µA reset current was obtained.

In subsequent studies, Huang et al. [40] have reported an e-beam-free process for the fabrication of GST phase change NWs based on an anisotropic ion-beam spacer etch process, yielding NWs with dimensions of 50 nm (width) × 100 nm (height). The process reproducibility was high, although a significant NW roughness, related to the photolithography process, was observed.

Ma et al. [41] have reported a so-called "self-aligned process" to obtain phase change NWs (Fig. 12.5).

The NWs are self-confined in properly processed electrode nano-gaps which, after EBL, lift-off and inductively coupled plasma (ICP), form lateral PCM devices [42]. Lateral phase change $Ge_{15}Sb_{85}$ NWs were fabricated by Gholipour et al. [43], starting from CVD-deposited Ge-Sb thin films and then defining the electrodes both by EBL or FIB. Whereas in the case of FIB fabrication some Ga contamination could have had a detrimental effect, in the case of EBL a sensible reset current consumption reduction down to a 1–2 mA was observed when reducing the NW cell width to 50 nm. In the same work, electrothermal modeling showed how the central spot of the NW reaches the crystallization temperature of Ge-Sb during the observed resistance change in the set process.

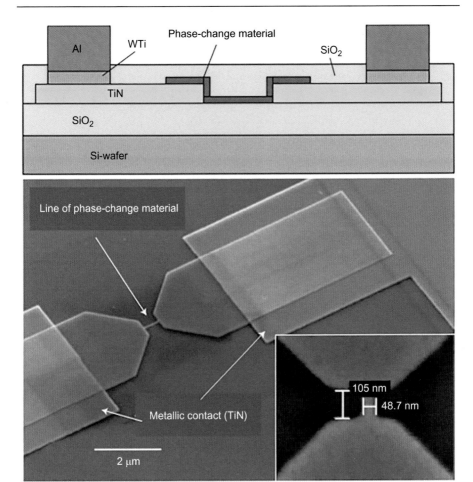

Fig. 12.3 *Top*: cross-section scheme of a line cell formed by TiN contacts and Al bond pads on a SiO$_2$/Si wafer. *Bottom*: scanning electron micrograph of the same cell (length=500 nm, width=50 nm), after contact and electron beam lithography. Inset: detail of similar cell with length reduced to 100 nm.

Reprinted by permission from Macmillan Publishers Ltd.: M.H.R. Lankhorst, B.W.S.M.M. Ketelaars, R.A.M. Wolters, Low-cost and nanoscale non-volatile memory concept for future silicon chips, Nat. Mater. 4 (2005) 347, copyright (2005).

12.5 Fabrication of phase change NWs: Bottom-up approaches

The bottom-up fabrication of phase change NWs is generally based on a deposition process, coupled to self-assembly mechanisms. Bottom-up NWs are typically single crystalline, defect-free structures, and thus ideal for studying size-dependent properties of the constituent materials. Physical deposition processes have been used, such as in the case of the Sb$_2$Te$_3$ NWs shown in Fig. 12.6.

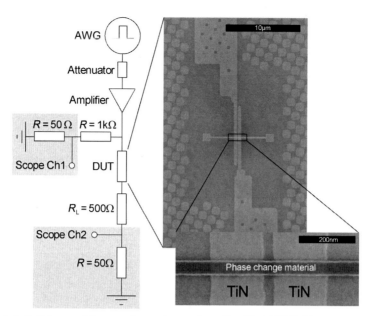

Fig. 12.4 *Left*: Scheme of the experimental setup for a Ge$_{15}$Sb$_{85}$ DUT bridge cell (device under test). *Right*: plan view SEM images of the bridge cell device, where the phase change material is suspended over TiN contacts. Inset: zoomed image of the bridge region.
Reprinted from D. Krebs, D.S. Raoux, C.T. Rettner, R.M. Shelby, G.W. Burr, M. Wuttig, Set characteristics of phase change bridge devices, Proc. Mater. Res. Soc. Symp. 1072 (2008) G06–G07, copyright (2008), with permission from Materials Research Society.

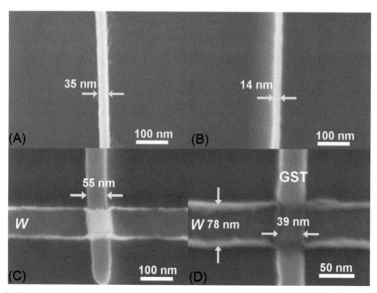

Fig. 12.5 Scanning electron microscopy images displaying: (A) etched GST layer with PMMA negative resist; (B) GST NW emerging after removal of the negative resist; (C) tungsten electrode deposition (W); and (D) final PCM device after resist removal, where GST is confined within the 39-nm metal electrode gap.
Reprinted with permission from H. Ma, X. Wang, J. Zhang, X. Wang, C. Hu, X. Yang, Y. Fu, X. Chen, Z. Song, S. Feng, A. Ji, F. Yang, A self-aligned process for phase-change material nanowire confined within metal electrode nanogap, Appl. Phys. Lett. 99 (2011) 173107, copyright (2011), AIP Publishing LLC.

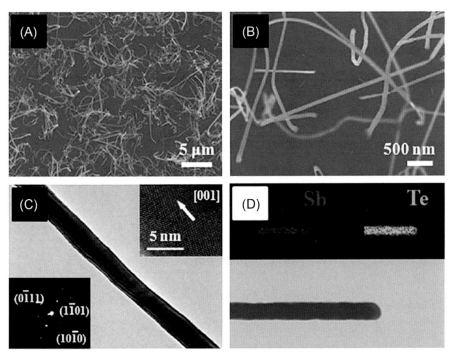

Fig. 12.6 (A) Low- and (B) high-resolution field emission scanning electron microscopy (FESEM) images of Sb_2Te_3 NWs obtained by a catalyst-free method from $Al_{22.6}Ge_{22.1}Sb_{17.7}Te_{37.7}$ layers. (C) Selected area electron diffraction (SAED) pattern and high-resolution transmission electron microscopy (HRTEM) images of a single NW showing its structure and orientation and (D) energy-dispersive spectroscopy (EDS) mapping images, showing the distribution of elements in the NW.
Reprinted from B.G. Kim, B.-S. Kim, S.-M. Jeong, S.-M. Choi, D. Whang, H.-L. Lee, Catalyst-free growth of Sb2Te3 nanowires, Mater. Lett. 65 (2011) 812–814, copyright (2010), with permission from Elsevier.

A 200-nm Al-Ge-Sb-Te thin layer (AGST, Al = 6.5–22.6 at.%) was deposited at room temperature (RT) by DC magnetron sputtering on a glass and on a Si(100) wafer by co-sputtering. Sb-Te NWs, 100 nm in width, were extruded spontaneously on the surface of thin AGST films for thermal annealing at around 300°C and above 12.4 at.% Al content. The hypothesis was that the "thermal stress-induced growth," namely the thermal mismatch between the substrate and the AGST layer inducing mass flow along grain boundaries on thermal annealing which forms the NWs, is the dominating mechanism. Notably, this approach does not require the use of metal catalysts; on the other hand, many NWs appeared as kinked or curved.

The most diffused technique for defect-free NW self-assembly makes use of CVD, coupled with vapor-solid (VS), and/or vapor-liquid-solid (VLS) mechanisms [44]. For the VLS mechanism, a metal catalyst is required to form a eutectic alloy at the substrate surface with the reactant vapors. The eutectic temperature is lower than each element's melting temperature, so that the alloy melts into a liquid droplet. Under

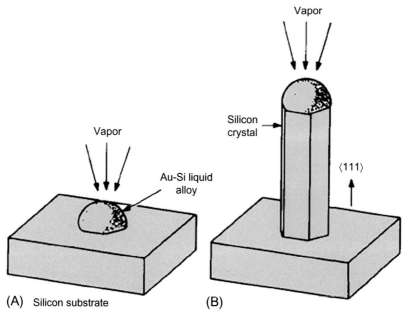

Fig. 12.7 Scheme illustrating the VLS mechanism for the growth of a silicon NW:
(A) formation of a liquid nanodroplet on the substrate and (B) growing crystalline NW with
the liquid droplet remaining at the tip.
Reprinted with permission from R.-S. Wagner, W.C. Ellis, Vapor-liquid-solid mechanism of
single crystal growth, Appl. Phys. Lett. 4 (1964) 89, copyright (1964), AIP Publishing LLC.

supersaturation conditions, the droplet forms a site for NW epitaxial growth; the NW
diameter is generally related to the catalyst size [45, 46]. This mechanism is shown
schematically in Fig. 12.7.

The phase diagram for Au (the most used catalyst) and GeTe-Sb$_2$Te$_3$ in Fig. 12.8A
from Meister et al. [47] shows that it is possible to reach the eutectic point at temperatures as low as 454°C in the VLS deposition of Sb-Te (ST) NWs.

This work showed how it is possible to grow Sb$_2$Te$_3$ and GeTe NWs using a VLS
mechanism with different morphologies: from straight to helical to curly, even coexisting (Fig. 12.8B–D).

This mechanism is actually employed for phase change NW self-assembly. This is
shown in Fig. 12.9A [48], which depicts the typical scheme of a horizontal furnace
that uses chemical vapor transport methods for the synthesis of NWs.

By tuning the temperature profile in the furnace and the position of sources (in
this case Te and Sb), it is possible to ensure that they thermally evaporate at higher
temperatures, followed by allowing the Sb-Te NW self-assembly by VLS at a lower
temperature (Fig. 12.9B). Different morphologies for phase change Sb$_2$Te$_3$ NWs could
be chosen by changing the furnace temperature profile (Fig. 12.9C), the location of
the reactants and the carrier gas flow, from straight to zigzag NWs, nanobelts, and
nanotubes (Fig. 12.9D).

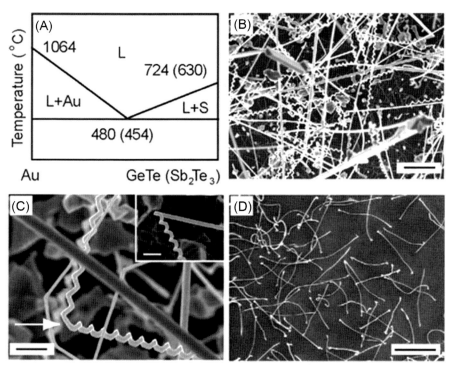

Fig. 12.8 (A) Quasi-binary phase diagram between Au and GeTe or Sb_2Te_3 (L = liquid; S = solid; drawing not to scale); (B) scanning electron microscopy (SEM) image of GeTe NWs with three types of morphologies (scale bar = 2 μm); and (C) SEM image of helical GeTe NWs with two different pitches (white arrow indicates the pitch change). Inset: SEM image of straight-helical NW (scale bar = 0.5 μm); and (D) SEM image of thin and curly GeTe NWs (scale bar = 2 μm).
Reprinted with permission from S. Meister, H. Peng, K. McIlwrath, K. Jarausch, X. Feng Zhang, et al., Synthesis and characterization of phase-change nanowires, Nano Lett. 6 (2006) 1514–1517, copyright (2006), American Chemical Society.

In general, the growth conditions deeply influence the NW length and diameter, especially the growth temperature time and reactant concentration, as in the case of GeTe NWs. Complete control of the resulting NW is not necessarily straightforward, as it depends on the deposition method, conditions and the complex reactions that may take place [49]. Once the reactant concentration is fixed, the growth temperature in the vapor transport methods can influence the composition of the NWs, as demonstrated by Philipose et al. [50] for InSb NWs and Longo et al. [51] for $Ge_1Sb_2Te_4$ NWs.

Many studies have reported the synthesis of phase change NWs by vapor transport methods in horizontal tube furnaces. These include single-crystal GST [52–55], Ge-Sb [56, 57], Sb_2Te_3 [47], GeTe [58], In_2Se_3 NWs [59, 60], Sb_2Te_3/GeTe [48], Ga-doped InO [61], and InSb NWs [50]. Although controlling the composition of the NW was not straightforward in vapor transport techniques, it was finally achieved in GST NWs [62], producing peculiar superlattice structures in $Ge_8Sb_2Te_{11}$, $Ge_3Sb_2Te_6$, $Ge_3Sb_8Te_6$,

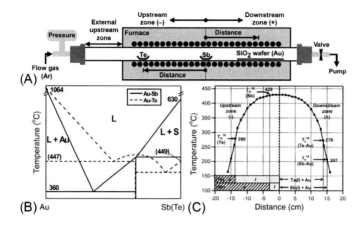

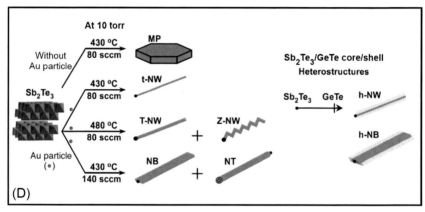

Fig. 12.9 (A) Scheme of the horizontal furnace for the growth of Sb$_2$Te$_3$ nanostructures; (B) binary phase diagram between Au and Sb or Te (L = liquid; S = solid); (C) temperature profile of the furnace at 430 °C (Tm^{10} = melting temperature at 10 Torr; Te^{10} = eutectic temperature with Au at 10 Torr); and (D) morphologies of Sb$_2$Te$_3$ nanostructures depending on the growth temperature and the Ar carrier gas flow rate (MP = microplate; t-NW = thin NW; T-NW = thick NW; Z-NW = zigzag NW; NB = nanobelt; NT = nanotube; h-NW = core/shell NW heterostructure; h-NB = core/shell NB heterostructure).
Reprinted with permission from J.S. Lee, S. Brittman, D. Yu, H. Park, Vapor–liquid–solid and vapor–solid growth of phase-change Sb$_2$Te$_3$ nanowires and Sb$_2$Te$_3$/GeTe nanowire heterostructures, J. Am. Chem. Soc. 130 (2008) 6252–6258, copyright (2008), American Chemical Society.

and Ge$_2$Sb$_7$Te$_4$. The VLS method was also adopted to synthesize doped NWs, analogously to the case of phase change layers. Zhang et al. [63] have investigated Bi-doped GeTe NWs, with a Bi content of around 3 at.%, in which the crystallization time is strongly reduced by a factor of 20 with respect to undoped GeTe NWs. In order to take advantage of the benefits provided by the introduction of Bi, Jung et al. [64] have reported the growth of a series of pseudobinary alloys formed by mGeTe·Bi$_2$Te$_3$ (m = 3–8) NWs by varying the Bi/(Ge + Bi) content in the GeTe and Bi$_2$Te$_3$ powders, to investigate the composition-dependent thermal stability and electrical properties of the NWs; different superlattice structures were obtained, with an evolution from the cubic to the hexagonal phase and increased thermal stability as m decreases.

The self-assembly approach is fundamental for the realization of core-shell NW structures. These were demonstrated by Lee et al. [48], who first grew Sb_2Te_3 core NWs by using VLS vapor transport and then uniformly coating, in situ, the core NWs with GeTe shell, through a VS growth. The results are shown in Fig. 12.10.

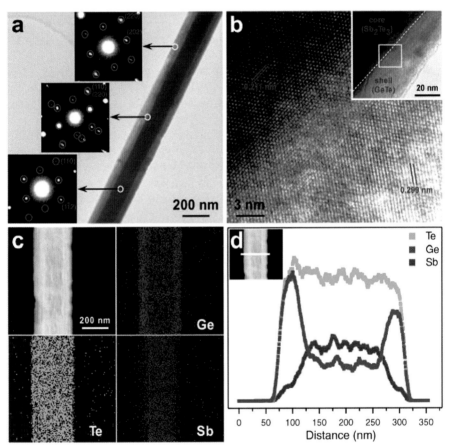

Fig. 12.10 (A) Transmission electron microscopy (TEM) image of a core-shell NW formed by a uniform GeTe shell and a Sb_2Te_3 core. Insets: selected-area electron diffraction (SAED) patterns with red *(blue)* circles indicating the Sb_2Te_3 (GeTe) diffraction spots. (B) High-resolution (HR)TEM image of a core/shell NW taken from the area in the inset, where the core-shell interface is shown. (C) Bright-field scanning TEM image and corresponding energy-dispersive X-ray (EDX) spectroscopy elemental mapping and (D) cross-sectional EDX line scans of a Sb_2Te_3/GeTe core-shell nanobelt.

Reprinted with permission from J.S. Lee, S. Brittman, D. Yu, H. Park, Vapor–liquid–solid and vapor–solid growth of phase-change Sb_2Te_3 nanowires and Sb_2Te_3/GeTe nanowire heterostructures, J. Am. Chem. Soc. 130 (2008) 6252–6258, copyright (2008), American Chemical Society.

Such structures might not seem to have special applications, but that is not the case. They make it possible to design novel multilevel PCM devices, in which further information can be stored beyond the reset (0) and set (1) binary codes. The possibility of obtaining multilevel PCM was already explored in thin films by Rozenberg et al. [65]. The drawbacks were narrow programming window and unsatisfactory reproducibility, due to poor control of the crystalline and amorphous states. A multilayer phase change material stack, formed by nitrogen-doped GST and GST layers sandwiching Ta_2O_5 barrier layer, was reported by Gyanathan and Yeo [66] to exhibit a resistance window of one order of magnitude.

The GST/GeTe core-shell NWs by Jung et al. [67] (Fig. 12.11) demonstrated that the way to remove such limitations is to couple different phase change materials which have different electrical and thermal properties.

The deposition was performed by VLS growth of the GST core NW, followed by the conformal deposition of the GeTe shell by VS growth. This kind of structure can yield multilevel programming, due to the introduction of intermediate resistive states (see Section 12.12).

The vapor transport growth is limited by factors including low flexibility, difficult compositional tuning, and industrial transfer. Other possibilities for improving this are offered by advanced chemical deposition techniques, such as the MOCVD technique. This provides good control of the material composition, high purity, industrial scalability on $12''$ substrates, and relatively high deposition rates. In MOCVD, the vapor phase transport is achieved using metalorganic compounds (precursors) and/or hydrides of the atomic species to be deposited. The deposition occurs through the pyrolysis (endothermic dissociation) of the precursors, which makes the atomic species available when under thermodynamic super-saturation conditions. The technique is highly flexible in terms of the choice of reactants and deposition parameters and allows large area deposition, thus potentially enabling large-scale NW assembly.

Notably, only one group has published on the self-assembly of GeTe, Ge-doped Sb-Te, and $Ge_1Sb_2Te_4$ NWs using the MOCVD coupled with the VLS technique [51, 68, 69]. The same group reported self-assembly of In-Sb-Te and In-doped Sb_4Te_1 NWs both by MOCVD [70]; the obtained defect-free NWs reached the minimum diameter of 15 nm. For GeTe, single-crystal NWs, with a length up to 4 μm, were self-assembled [69]; the effect of the catalyst size was studied and longer NWs were obtained by using 10 nm islands, whereas 50-nm islands induced NW self-assembly with a larger diameter and more dispersed length. The NW aspect ratio was in the range of 10–16. The same synthesis mechanism was found in $Ge_1Sb_2Te_4$ NWs by MOCVD and VLS [51] (Fig. 12.12).

However, in this case, the length distribution was centered at around 400 nm. Since they were deposited on SiO_2, NWs showed different orientations with respect to the substrate. Furthermore, background crystals were formed where larger Au islands did not induce NW formation. Notably, this is the only report of $Ge_1Sb_2Te_4$ NWs, which is independent of the deposition technique. Both GeTe and $Ge_1Sb_2Te_4$ NWs could be grown at 400°C, 50°C below the lowest eutectic point of the phase diagram [47], indicating the effect of their size, namely that a sub-eutectic growth was possible.

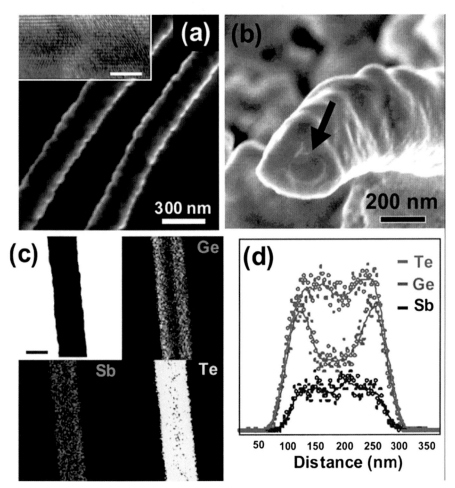

Fig. 12.11 (A) Dark-field transmission electron microscopy (TEM) image of GST(core)/ GeTe(shell) NWs. Inset: high-resolution (HR)TEM image of polycrystalline GeTe on the surface of the GST core NW (scale bar = 2 nm). (B) Scanning electron microscopy (SEM) image of a GST/GeTe core/shell NW cross section obtained by focused ion beam; the arrow indicates the core-shell interface. (C) Scanning (S)TEM image showing the elemental mapping in the core-shell NW (scale bar = 200 nm). (D) EDX line-scan profile across the NW in (C). Reprinted with permission from Y. Jung, S.-H. Lee, A.T. Jennings, R. Agarwal, Core-shell heterostructured phase change nanowire multistate memory, Nano Lett.Nano. Lett. 8 (2008) 2056–2062, copyright (2008), American Chemical Society.

Sb_2Se_3/Sb_2S_3 NWs were synthetized by pulsed VLS and CVD techniques by Yang et al. [71]. The molecular precursors were alternatively delivered to the Au nanocatalyst area in the form of pulses, to prevent pre-reactions and ensure better deposition control, in an analogous method to ALD. No catalyst was used for the MOCVD synthesis of phase change In-Sb-Te NWs on the planar $TiAlN/SiO_2/Si$ substrates displayed in Fig. 12.13 [72], so that even lower temperatures of 250°C could be used.

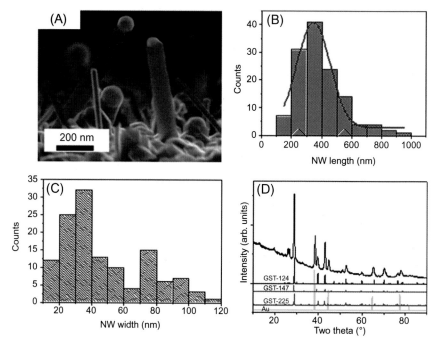

Fig. 12.12 (A) Scanning electron microscopy (SEM) cross-section image showing the GST NWs deposited by metalorganic chemical vapor deposition (MOCVD) coupled to vapor-liquid-solid (VLS) mechanism at 400°C. The Au droplets on top of the NWs are clearly visible; (B and C) statistical size distribution for the NWs in (A) (continuous line = best Gaussian data fit) and (D) X-ray diffraction (XRD) analysis of the sample in (A), along with the expected spectra from the different GST stable phases ($Ge_1Sb_2Te_4$, $Ge_1Sb_4Te_7$, $Ge_2Sb_2Te_5$) and from Au.

Reprinted with permission from M. Longo, R. Fallica, C. Wiemer, O. Salicio, M. Fanciulli, et al., Metal organic chemical vapor deposition of phase change $Ge_1Sb_2Te_4$ nanowires, Nano Lett. 12 (2012) 1509–1515, copyright (2012), American Chemical Society.

Here the driving mechanism was the protrusion under high supersaturation of InTe, caused by the relatively high concentration of the reactants in the growth chamber, to which Sb was incorporated, forming IST NWs.

Lee et al. [73] have reported on the liquid-phase synthesis of orthorhombic GeTe NWs. Their assembly was induced by the presence of dispersed Bismuth seeds, demonstrating the potential for this method in terms of low production costs, low deposition temperature (below 300°C), and high throughput. The NWs are shown in Fig. 12.14.

Due to the absence of the Bi seeds at the NW tips, Ostwald ripening was proposed as the growth mechanism, rather than solution-liquid-solid (SLS). The NW diameter ranged from 50 to 100 nm and their length was around 2 μm. Finally, Sb_2Se_3 NWs were deposited by microwave-activated solvo-thermal reaction and reported by Metha et al. [74].

The self-assembly of single crystalline, core-shell Ge/(In-Te) NWs was obtained for the first time in a single-step MOCVD process by Cecchini et al. [75]. The NWs

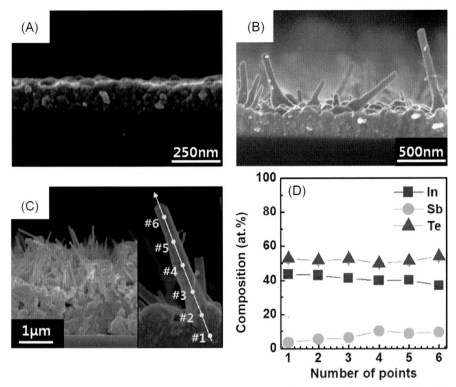

Fig. 12.13 In-Sb-Te NWs synthetized by metalorganic chemical vapor deposition (MOCVD): scanning electron microscopy (SEM) cross-sectional image of the samples grown at (A) 3.9×10^2, (B) 9.1×10^2 Pa, and (C) 13×10^2 Pa (left side). Right side image in (C) shows the SEM image of a NW selected for local compositional analysis in the numbered points. (D) Measured composition at each point along the NW length.

Reprinted with permission from J.K. Ahn, K.W. Park, H.-J. Jung, S.-G. Yoon, Phase-change InSbTe nanowires grown in situ at low temperature by metal-organic chemical vapor deposition, Nano Lett. 10 (2010) 472–477, copyright (2010), American Chemical Society.

Fig. 12.14 Scanning electron microscopy (SEM) image showing the formation of GeTe NWs and Te microparticles obtained by liquid-phase deposition.

Reprinted with permission from M.-H. Lee, T.G. Kim, B.-K. Ju, Y.-M. Sung, Bismuth seed-assisted liquid-P phase synthesis of germanium telluride nanowires, Cryst. Growth Des. 9 (2009) 938–941, copyright (2009), American Chemical Society.

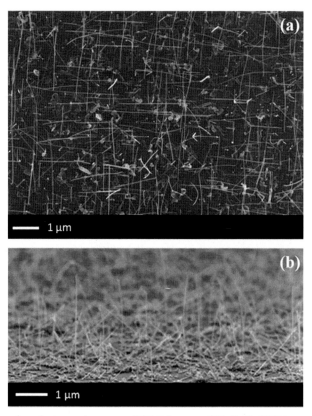

Fig. 12.15 Scanning electron microscopy (SEM) images of Ge/In-Te NWs grown on different substrates and catalyzed by 20 nm Au nanoparticles: (A) top view on Si (001) and (B) tilted view on Si (110).
Reproduced from R. Cecchini, S. Selmo, C. Wiemer, R. Rotunno, L. Lazzarini, M. De Luca, I. Zardo, M. Longo, Single-step Au-catalysed synthesis and microstructural characterization of core–shell Ge/In–Te nanowires by MOCVD, Mater. Res. Lett. 6 (2018) 29–35, which is an Open Access article distributed under the terms of the Creative Commons Attribution License (http://creativecommons.org/licenses/by/4.0/).

were as thin as 15 nm, were preferentially oriented with respect to the Si substrates, as shown in Fig. 12.15 and the core-shell formation mechanism was ascribed to a possible adding of phase separation between Ge and InTe, to the VLS process.

12.6 Fabrication of phase change NWs: Other techniques

12.6.1 Self-aligned nanotube-NW PCMs

Xiong et al. [76] introduced a nanofabrication method called the self-aligned nanotube-NW PCM, using the aid of CNT electrodes. This method does not require lithography to self-align the NWs and the CNTs. The nanogap in the CNTs is created

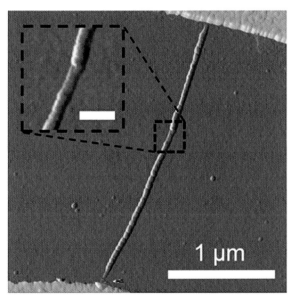

Fig. 12.16 Atomic force microscopy (AFM) image of a self-aligned NW with carbon nanotube electrodes of diameter ≈ 2.5 nm. The GST NW is ~40 nm wide, ~10 nm tall and capped by ~10 nm of SiO_2. The nanogap size is indirectly estimated to be ~30–60 nm. Inset: zoom of the nanogap region (scale bar = 150 nm).
Adapted with permission from F. Xiong, M.-H. Bae, Y. Dai, A.D. Liao, A. Behnam, E.A. Carrion, S. Hong, D. Ielmini, E. Pop, Self-aligned nanotube-nanowire phase change memory, Nano Lett. 13 (2013) 464–469, copyright (2013), American Chemical Society.

by inducing electrical breakdown. In all, 10 nm of GST are sputtered in the PMMA nanotrench (previously induced by Joule heating); when the remaining PMMA has been removed, a tiny GST NW results, perfectly aligned with the CNT ends, which can be used as electrodes (Fig. 12.16).

A process that did not require lithography was also developed by Fu et al. [77]. They prepared GST NWs using an SiN_x spacer as an etch mask, which formed a NW that was fully confined in a metal nano-gap. Such a device showed very low currents (16 μA) and set voltages (80 mV).

12.6.2   Template-based deposition

Further growth methods have been employed to self-assembly phase change NWs, in particular the template-based deposition (electrochemical deposition). This is a relatively simple method to fabricate low-dimensional structures. It uses an ionic solution containing the elements to be deposited on the surface of a conductive template by electrolysis. Anodized aluminum oxide (AAO) membranes containing nanopores are used to form the NWs. The NWs can be electrodeposited within the nanopores. By this method, arrays of Bi_2Te_3 NWs were deposited with diameters of around 25, 30, and 75 nm [78] (Fig. 12.17).

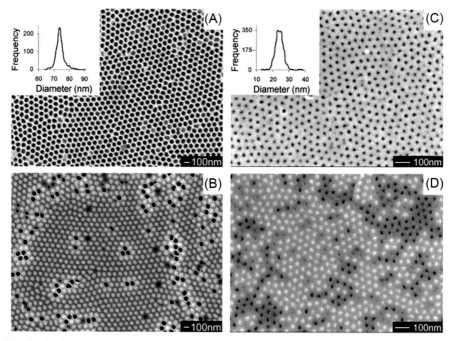

Fig. 12.17 Scanning electron microscopy (SEM) images of: (A) empty anodic alumina template with average pore size of 75 nm (see inset) and (B) corresponding electrodeposited Bi_2Te_3 NW array composite; (C) empty template with average pore size of 25 nm (see inset) and (D) corresponding electrodeposited Bi_2Te_3 NW array composite. Dark spots indicate empty pores and bright spots indicate the Bi_2Te_3 NWs.
Reprinted with permission from M.S. Sander, R. Gronsky, T. Sands, A.M. Stacy, Structure of bismuth telluride nanowire arrays fabricated by electrodeposition into porous anodic alumina templates, Chem. Mater. 15 (2003) 335–339, copyright (2003), American Chemical Society.

Bi_2Te_3 NWs were also deposited into 70 nm pores of AAO membranes at a 10 mV voltage, using an electrolyte of HNO_3, Bi_2O_3, and TeO_2 [79]. To analyze single NWs, the AAO membrane could be dissolved in NaOH. In order to obtain a PCM cell device formed by NW arrays in the membrane, Au/Ti and Pt electrodes were deposited at the base and top of the NWs for electrical analysis. Notably, Bi-Te NWs with different compositions from Bi_2Te_3 did not exhibit the phase change behavior.

Sb-Te NWs with variable compositions and aspect ratios were also electrodeposited into AAO membranes and amorphous $Sb_{37}Te_{63}$ NWs with an average diameter of 80 nm and length of 1.2 μm were grown by electrodeposition on homemade AAO templates, with a good vertical alignment [80].

Te-rich Sb-Te NWs have been reported by Jeong et al. [81], as deposited in nano-sized trench structures by atomic vapor deposition methods. The method was catalyst free and based on the selection of deposition conditions to induce the NW growth inside high aspect ratio trenches. Fig. 12.18 shows the scanning electron microscopy (SEM) images of the deposited Te-rich Sb-Te NWs inside the trenches, whereas a

Fig. 12.18 Scanning electron microscopy (SEM) images of Sb-Te NWs deposited by a catalyst-free vapor-solid (V-S) mechanism at (A) the border between the trench and the planar surface (tilted view), (B) the inside of the trench (tilted view), and (C) top view of the NWs. Reproduced with permission from J.H. Jeong, S.J. Park, S.B. An, D.J. Choi, Vapor-solid growth of Te-rich SbTe nanowires on a template of nano-size trenches, J. Cryst. Growth 410 (2015) 47–52, copyright (2014), Elsevier.

very low-density NW deposition occurred in the planar regions. Although the original scope was to obtain void-free trench devices, this work is one of the few attempts toward the realization of a regular array of phase change NWs.

12.7 Characterization of PC-NWs

The typical front-end characterization to check the NW formation is performed by SEM. This can also give an immediate indication of the possible VLS or VS mechanisms. The structure of the NWs is basically determined by their fabrication process. In the case of the top-down approach, NWs can be formed by either amorphous or crystalline material, depending on the starting layer. The attention is focused on the NW morphological properties, rather than the structure. The majority of phase change NWs are actually synthesized by bottom-up processes, often leading to crystalline NWs. In this case, the NW composition and structure is carefully analyzed, since the synthesis method can lead to significantly different structural and compositional properties, and therefore, to different phase change behavior. Moreover, the crystalline structures of the NWs may be different from those of the equivalent layers or bulk-like materials, because of their size.

A critical deposition parameter is the temperature which influences the crystalline phase of the NW when it is higher than the crystallization temperature of the involved material. It is useful to consider the case of GST, which is characterized by two transition temperatures:

1. amorphous to face-centered cubic (fcc) metastable phase transition around 150–170°C and
2. fcc to stable hexagonal close packed (hcp) transition, around 260–300°C, depending on the rate of temperature elevation [82].

Normally, the first transition is the one exploited in PCM devices. In the case of the largely diffused VLS mechanism, the NW synthesis temperature is higher than 300°C, meaning that the GST NWs are expected to exhibit hexagonal crystal phases. Analogous issues can be examined once it is considered that GeTe has a crystallization temperature of 145°C [83] and Sb_2Te_3 has 77°C [84].

The study of the structural and compositional analysis of phase change NWs is mainly based on high-resolution transmission electron microscopy (HR-TEM), energy-dispersive spectroscopy (EDS), and selected area electron diffraction (SAED) measurements that allow a local area analysis on single NWs. Other techniques can be of use, such as X-ray fluorescence (XRF) and X-ray diffraction (XRD). However, XRD and XRF can only collect information over relatively large areas, which can be a limitation when nonuniformity in the NW structure or when other synthesis products are present, such as undesired background crystals. This is the case for GeTe NWs grown by MOCVD on SiO_2 substrates [69]. Although the VLS mechanism is clear from the presence of the Au tip, different NW orientations and background crystals can be observed by SEM, when using Au nano-islands within average sizes of 10 nm (1 L) and 20 nm (2 L) as metal catalysts (Fig. 12.19).

Total reflection XRF (TXRF) gave an overall composition of $Ge_{0.46}Te_{0.54}$ for the grown material, including the presence of the Au XRF peak. However, the XRD analysis indicated a compatibility of the recorded NW spectra with both rhombohedral and cubic GeTe structures. More reliable information came from the local area analysis, performed by TEM (see Fig. 12.20, related to a 50-nm wide NW), which showed a dark Au seed containing around 10% of GeTe. Notably, the VLS mechanism occurred in a sub-eutectic condition at 400°C.

Moreover, size effects induced the NW metastable cubic phase at 400°C, when it is normally expected at 446°C [45]. The GeTe NWs appeared to be single crystal and defect free and to grow along the [110] direction; the EDX analysis confirmed the stoichiometric composition. Another case in which cubic GeTe NWs orientated along the [110] direction were obtained, is reported by X. Sun et al. [85]. In other cases, rhombohedral GeTe NWs were observed to grow along the [220] direction [86], along the [−111] direction [87], along the [202] direction [58] and along the [220], [003], and [202] directions [88]. It is worth noting that in the case of Bi-seed assisted GeTe NWs, synthetized by liquid-phase growth, orthorhombic NWs were obtained at very low temperatures (<300°C).

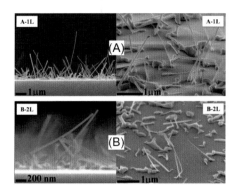

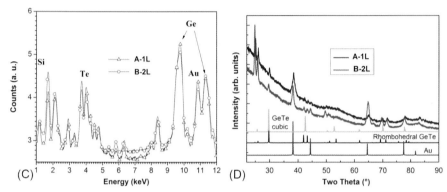

Fig. 12.19 Scanning electron microscopy (SEM) cross-sectional image (*left*) and tilted views (*right*) of GeTe NWs grown at 400°C with: (A) Te/Ge precursor ratio = 3.8, 10 nm Au-seeds (in A-1L); (B) Te/Ge precursor = 1.7, 20 nm Au-seeds (in B-2L); (C) comparison of the total X-ray fluorescence (TXRF) spectra obtained from the same samples; and (D) comparison of the X-ray diffraction (XRD) spectra obtained from the same samples.

Reprinted with permission from M. Longo, C. Wiemer, O. Salicio, M. Fanciulli, L. Lazzarini, et al., Au-catalyzed self assembly of GeTe nanowires by MOCVD, J. Cryst. Growth, 315 (2011) 152–156, copyright (2010), Elsevier.

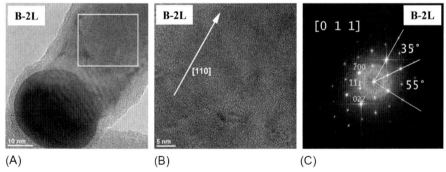

Fig. 12.20 High-resolution transmission electron microscopy (HRTEM) analysis of a single GeTe NW from sample B-2L of Fig. 12.19B: (A) Au-seed appears dark and the seed/NW body interface is featured by the Moiré fringes. The squared area is zoomed in (B), where the growth direction is displayed and in (C), which is a fast Fourier transform of the HRTEM image in (B), showing the identification of the diffraction spots and angles for the GeTe cubic phase.

Reprinted with permission from M. Longo, C. Wiemer, O. Salicio, M. Fanciulli, L. Lazzarini, Au-catalyzed self assembly of GeTe nanowires by MOCVD, J. Cryst. Growth 315 (2011) 152–156, copyright (2010), Elsevier.

A peculiar oxidation effect was observed in GeTe NWs grown by VLS. Sun et al. [85] found a 1–3-nm amorphous $GeO_2.TeO_2$ outer layer, after exposure to oxygen. This effect of oxidation was also observed by CNR-IMEM, Parma, Italy, in MOCVD-grown GeTe NWs. It was related to aging, as the oxide layer thickness increases with the time exposure to air (Fig. 12.21).

The Moiré fringes in Fig. 12.21 are the result of overlapping of the inner GeTe crystalline region of the NWs and of nanocrystals embedded in the amorphous layer. The nanocrystals, visible along the NW axis in the figure, were accurately identified to be mostly composed of Te and the amorphous layer mainly formed of GeO_2. On the basis of these results, it can be concluded that oxidation tends to occur on Ge, with the Te excess aggregating in Te-rich crystal clusters.

A further study by Rotunno et al. [89] allowed the precise identification of the aging mechanism in such GeTe NWs and the Te-rich clusters were identified as $GeTe_4$ nanocrystals; in Fig. 12.22A the TEM image of a NW after 12 weeks of aging is shown, where the thickness of the oxide amorphous layer is 20 nm; the Fourier transform (FT) in the inset displays the spatial frequencies caused either by the inner crystalline region or the Te-rich particles. In Fig. 12.22B, the distribution of the spatial frequencies is mapped in a false color image. This was obtained by performing the inverse FT (FT^{-1}) process for each frequency at a time and then by summing up all the obtained images. The intensities produced by the inner crystalline region and the outer Te-rich clusters were very well separated. The aging process is sketched out in Fig. 12.22C.

Most of the work on the structural analysis of single-crystalline $Ge_2Sb_2Te_5$ NWs has been done by the group of R. Agarwal at the Penn University of Pennsylvania, USA. They have worked on demonstrating the hcp structure and the [001] growth direction [54]. In another report, two dominant growth directions were identified, namely the [110] and [001] of the hcp structure [55]. The only reported work related to the (MOCVD) synthesis and structural characterization of the stable phase for single-crystal $Ge_1Sb_2Te_4$ NWs was by Longo et al. [51]. The homogeneity of the alloy was evidenced by the uniform EDX mapping and profiles. The rhombohedral phase and growth directions were found to correspond to the [1 −1 0 14] and

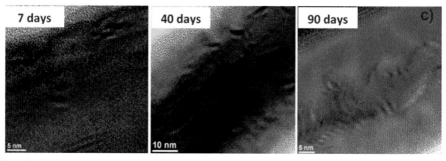

Fig. 12.21 Formation of an amorphous layer of increasing thickness (GeO_2) as a function of exposure time to air.
Courtesy of Enzo Rotunno and Laura Lazzarini, CNR-IMEM, Parma, Italy.

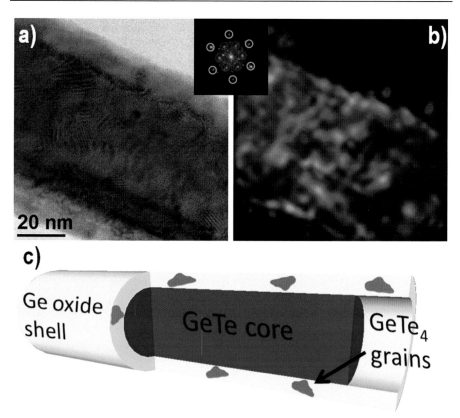

Fig. 12.22 (A) Transmission electron microscopy (TEM) image of a 12 weeks aged GeTe NW. (B) Map representing, in false colors, the region where the GeTe *(green)* and GeTe₄ *(red)* atomic structures are observed. The map is color coded according to the fast Fourier transform (FFT) reported as inset. (C) A graphical representation of the aging process.
Adapted from E. Rotunno, M. Longo, L. Lazzarini, Ageing of GeTe nanowires, Microscopie 26 (2016) 58–63, which is licensed under a Creative Commons Attribution-NonCommercial 4.0 International License.

[1 0 −1 0] directions of the hexagonal structure, corresponding to the [100] and [112] growth directions of the rock-salt cubic structure (Fig. 12.23).

The atomic arrangement of the Ge-Sb-Te compounds is a matter for debate. Many different configurations have been proposed. Rotunno et al. [90] have made progress with regard to the atomic arrangement of the rhombohedral structures in the stable phases of $Ge_2Sb_2Te_5$ and $Ge_2Sb_2Te_5$ NWs. They have identified the stacking sequence in their crystal structure through the comparison of experimental high-angle annular dark field scanning transmission electron microscopy (HAADF-STEM) results and STEM simulations. Different sets of theoretical models, proposing possible stacking sequences, were chosen for comparison. Fig. 12.24 shows the three models for the $Ge_2Sb_2Te_5$ structure proposed by Kooi and De Hosson [92], Matsunaga et al. [93], and Petrov et al. [91].

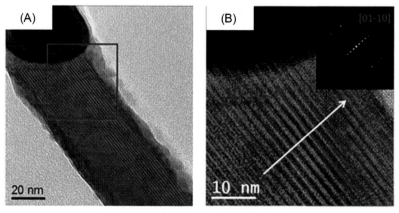

Fig. 12.23 (A) Transmission electron microscopy (TEM) image of a $Ge_1Sb_2Te_4$ NW deposited at 400°C, 50 mbar. The Au droplet on top is featured by a darker contrast; (B) [1 0 −1 0] zone axis high-resolution (HR)TEM image of the square in (A), showing that the lattice periodicity along the *c*-axis (arrow pointing to right) is orthogonal to the growth direction (arrow pointing to left). Inset: corresponding diffractogram.

Reprinted with permission from M. Longo, R. Fallica, C. Wiemer, O. Salicio, M. Fanciulli, et al., Metal organic chemical vapor deposition of phase change $Ge_1Sb_2Te_4$ nanowires, Nano Lett. 12 (2012) 1509–1515, copyright (2012), American Chemical Society.

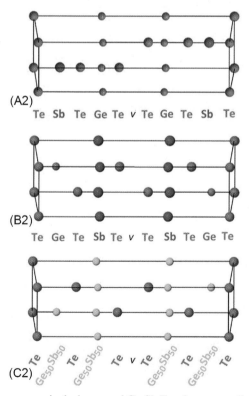

(A2)
Te Sb Te Ge Te v Te Ge Te Sb Te

(B2)
Te Ge Te Sb Te v Te Sb Te Ge Te

(C2)
Te Ge₅₀Sb₅₀ Te Ge₅₀Sb₅₀ Te v Te Ge₅₀Sb₅₀ Te Ge₅₀Sb₅₀ Te

Fig. 12.24 The layer sequences in the hexagonal $Ge_2Sb_2Te_5$ phase, according to the models proposed by Petrov et al. [91] (A), Kooi and De Hosson [92] (B), and Matsunaga et al. [93] (C). Reproduced from E. Rotunno, L. Lazzarini, M. Longo, V. Grillo, Crystal structure assessment of Ge–Sb–Te phase change nanowires, Nanoscale 5 (2013) 1557–1563 with permission from The Royal Society of Chemistry.

Matsunaga et al. [93] introduced the possibility of disorder in the Ge/Sb site occupation, since Ge and Sb are allowed to share the same lattice site in precise rates. "v" is referred to as the ordered vacancy layer. However, its possibility had been ruled out by theoretical works.

Fig. 12.25 displays the experimental HAADF-STEM image from a $Ge_2Sb_2Te_5$ NW and the simulations obtained from the models of Fig. 12.24. The darkest lines, separated by approximately 1.7 nm, are the vacancy layers. On the right, the corresponding line profile intensities are shown.

The highest cross-correlation coefficient, 0.866 (where 1 signifies coincidence), was obtained by the Matsunaga model, confirming that the structure in which Ge and Sb randomly share the same lattice sites does occur. Analogous conclusions were drawn in the case of the rhombohedral $Ge_1Sb_2Te_4$ NW structure. This was found to crystallize with mixed occupation in the Ge/Sb planes, according to a similar model, also from Matsunaga, despite the adverse theoretical predictions [94].

Jung et al. [62] grew alternative Ge-Sb-Te NWs by chemical vapor transport. They controlled the composition of GeTe and increased the Sb content, so that the value for x (calculated by: $x = [Sb]/([Ge] + [Sb])$) was varied between 0 and 0.8, in turn by varying the deposition temperature and the elemental sources ratio. Fig. 12.26 shows the main results for the nanostructures.

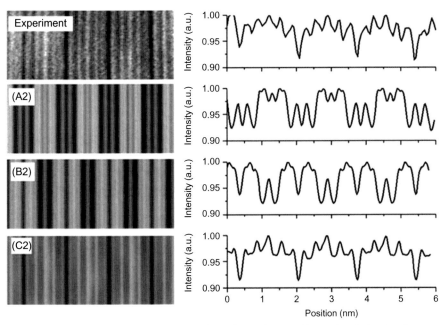

Fig. 12.25 The simulation results: left-hand panels are the experimental and simulated images of the cell structures in Fig. 12.24, labeled from (A) to (C). Right-hand graphs are the corresponding intensity line profiles.

Reproduced from E. Rotunno, L. Lazzarini, M. Longo, V. Grillo, Crystal structure assessment of Ge–Sb–Te phase change nanowires, Nanoscale 5 (2013) 1557–1563 with permission from the Royal Society of Chemistry.

GeSbTe nanowires

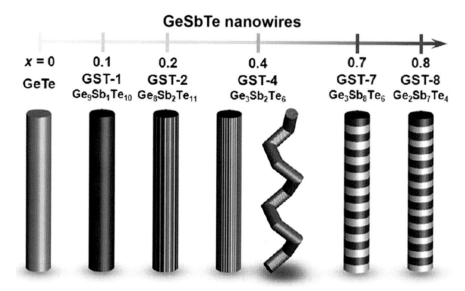

Fig. 12.26 Schematic diagram for the composition/phase tuned growth of the Ge-Sb-Te NWs: GeTe ($x=0$), $Ge_9Sb_1Te_{10}$ (GST-1; $x=0.1$), $Ge_8Sb_2Te_{11}$ (GST-2; $x=0.2$), $Ge_3Sb_2Te_6$ (GST-4; $x=0.4$), $Ge_3Sb_8Te_6$ (GST-7; $x=0.7$), and $Ge_2Sb_7Te_4$ (GST-8; $x=0.8$). For $x=(0.2-0.8)$, superlattice structures appeared.
Reprinted with permission from C.S. Jung, H.S. Kim, H.S. Im, Y.S. Seo, K. Park, S.H. Back, Y.J. Cho, C.H. Kim, J. Park, J.-P.Ahn, Polymorphism of GeSbTe superlattice nanowires, Nano Lett. 13 (2013) 543, copyright (2013), American Chemical Society.

The interesting outcome was that the increase of the Sb content, substituting Ge, lowered the transition temperature from the rhombohedral to the cubic phase and introduced a phase evolution yielding polymorphism. In particular, the latter resulted in rhombohedral/cubic phase coexistence, zigzag NWs, and superlattice structures, whose properties are summarized in Table 12.1.

Pure rhombohedral structures were again obtained when the Sb content was above 0.7, supporting the expectation that the Sb content increase lowers the transition temperature from the cubic to the rhombohedral phase.

Meister et al. [47] reported uniform, single-crystalline Sb_2Te_3 NWs, growing in a rhombohedral structure. Rhombohedral Sb_2Te_3 NWs were also obtained by Lee et al. [48], growing along the [110] direction. The case of Ge-doped SbTe NWs, reported by Longo et al. [68], should be noted. Here the Ge content was varied (x_{Ge}, ranging in the 1%–13% interval). For $x_{Ge}<5\%$, the NWs keep the Sb_2Te_3 structure, while for $5\%<x_{Ge}<13\%$, the NWs exhibit a mixed disordered structure and finally for $x_{Ge}>14\%$, the NWs crystallize as $Ge_1Sb_2Te_4$ (Fig. 12.27).

Further, the case of a Ge incorporation below 3 at.% and NW diameter below 40 nm, leading to a novel metastable Sb_2Te_3 polymorph, unstable in the bulk form, was reported by Rotunno et al. [95]. The stabilization of the primitive trigonal lattice (SG#164 instead of the expected SG#166 for bulk Sb_2Te_3) was identified by

Table 12.1 Properties of rhombohedral/cubic phase coexistence, zig-zag NWs, and superlattice structures

No.	Diameter[a] (nm)	x[b]	Composition	Crystal structure[c]	Growth direction	Period[d] (nm)	No. of layers[e]
GeTe	100	0	GeTe	R	$[0111]_R$		
GST-1	100	0.1	$Ge_9Sb_1Te_{10}$	R and C	$[110]_C$		
GST-2	150	0.2	$Ge_8Sb_2Te_{11}$	C	$[110]_C$	3.8	21(1)
GST-4	200	0.4	$Ge_3Sb_2Te_6$	C and R	$[110]_C$	2.2	11(1)
GST-7	400	0.7	$Ge_3Sb_8Te_6$	R	$[0001]_R$	3.4	17(0)
GST-8	400	0.8	$Ge_2Sb_7Te_4$	R	$[0001]_R$	2.7	13(0)

[a] Average diameter.
[b] $x = [Sb]/([Ge] + [Sb])$.
[c] R = rhombohedral, C = cubic (fcc).
[d] Average length of slabs in superlattice structures.
[e] Average number of atomic layers in each superlattice slab (number in parentheses is the number of vacancy layers per slab).
Adapted with permission from C.S. Jung, H.S. Kim, H.S. Im, Y.S. Seo, K. Park, S.H. Back, Y.J. Cho, C.H. Kim, J. Park, J.-P. Ahn, Polymorphism of GeSbTe superlattice nanowires, Nano Lett. 13 (2013) 543, copyright (2013), American Chemical Society.

advanced TEM techniques and turned out to be related to the particular sidewall faceting. Fig. 12.28 shows that the atomic positions inside the unit cell were assessed by STEM-HAADF analysis in two wire orientations. In (A) the dark lines corresponding to the van der Waals gaps are visible and the contrast allows the determination of the stacking sequence along the c-axis, identified as reported in the labels. The atomic positions in the two projections, compatible with the #164 space group, were confirmed in comparison with the related image simulations, reported in the inset of Fig. 12.28.

Han et al. [79] reported that Bi_2Te_3 NWs obtained by electrodeposition on an aluminum membrane at room temperature had a rhombohedral crystalline structure. Similar NWs were deposited by pulsed electrodeposition with a diameter of 40 nm, equal to the pore size. These exhibited the preferred [015] direction (Fig. 12.29).

Ge-Sb phase change NWs have been studied by Jung et al. [56] and are of interest for their good data retention and fast switching properties. In particular, the two eutectic compositions of Ge-rich Ge-Sb and Sb-rich Ge-Sb, with $Sb \geq 86\%$, can be achieved. For Sb-rich NWs, the expected structure was that of pure Sb, since they form a solid solution of Ge in Sb, with the Sb content varying between 0.86% and 0.90%. However, the structure determined by HR-TEM was that of hexagonal Sb, with growth direction along [012], where some stacking faults were also found, attributed to temperature accelerated growth kinetics. In the case of Ge-rich NWs, a tetragonal structure was obtained for single-crystalline Ge-Sb NWs growing along the [001] direction and exhibiting rectangular cross sections, as observed by SEM.

Karthik et al. [96] noted a threshold switching and differential negative resistance for $Sb_2Se_xS_{1-x}$ NWs, with diameter of approximately 30–150 nm, synthetized by a microwave-stimulated solvo-thermal synthesis technique. The structure of the single-crystal NWs was orthorhombic, with the growth direction aligned along the high electron mobility one, namely [010].

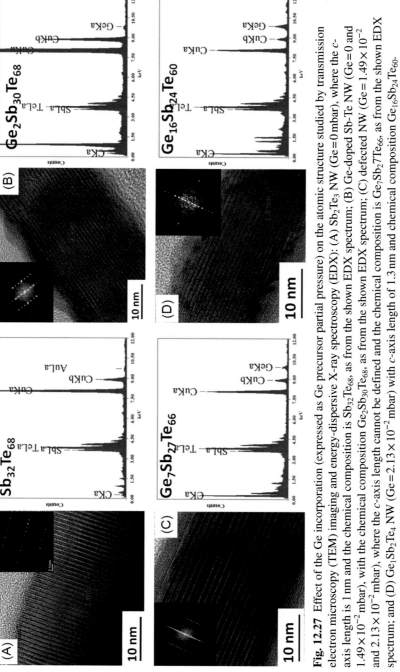

Fig. 12.27 Effect of the Ge incorporation (expressed as Ge precursor partial pressure) on the atomic structure studied by transmission electron microscopy (TEM) imaging and energy-dispersive X-ray spectroscopy (EDX): (A) Sb_2Te_3 NW (Ge=0 mbar), where the c-axis length is 1 nm and the chemical composition is $Sb_{32}Te_{68}$, as from the shown EDX spectrum; (B) Ge-doped Sb-Te NW (Ge=0 and 1.49×10^{-2} mbar), with the chemical composition $Ge_2Sb_{30}Te_{68}$, as from the shown EDX spectrum; (C) defected NW (Ge=1.49×10^{-2} and 2.13×10^{-2} mbar), where the c-axis length cannot be defined and the chemical composition is $Ge_7Sb_2Te_{66}$, as from the shown EDX spectrum; and (D) $Ge_1Sb_2Te_4$ NW (Ge=2.13×10^{-2} mbar) with c-axis length of 1.3 nm and chemical composition $Ge_{16}Sb_{24}Te_{60}$. Reprinted from M. Longo, T. Stoycheva, R. Fallica, C. Wiemer, L. Lazzarini, E. Rotunno, Au-catalyzed synthesis and characterisation of phase change Ge-doped Sb–Te nanowires by MOCVD, J. Cryst. Growth, 370 (2013) 323, copyright (2012), with permission from Elsevier.

Intensity (a.u.)

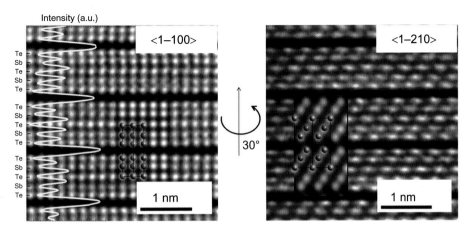

Fig. 12.28 High-resolution scanning transmission electron microscopy-high-angle annular dark field (STEM-HAADF) images of an Sb_2Te_3 NW oriented along two different zone axes, as labeled in the image. An intensity line profile is reported along the $\langle 1 -1\ 0\ 0 \rangle$ oriented image to identify the stacking sequence. Inset: the atomic models of the crystal structure and the image simulations.

Reprinted with permission from E. Rotunno, M. Longo, C. Wiemer, R. Fallica, D. Campi, M. Bernasconi, A.R. Lupini, S.J. Pennycook, L. Lazzarini, A novel Sb2Te3 polymorph stable at the nanoscale, Chem. Mater. 27 (2015) 4368–4373. Copyright (2015) American Chemical Society.

Finally, the case of In-based NWs should be cited. In_2Se_3 is a phase change material of interest, due to its large resistivity contrast between the amorphous and crystalline phase and relatively low programming current, even in the form of NWs, as it will be shown in Section 12.11. In order to study the phase transformation behavior in such NWs, a preliminary structural characterization was reported by Huang et al. [97]. Fig. 12.30 shows the TEM and EDS analysis of 50-nm-thick In_2Se_3 single-crystal NWs, indicating they grow along the [301] direction of the γ-phase.

Among the In-based phase change materials, it is worth mentioning the Ga-doped In_2O_3 NWs reported by Jin et al. [61], of interest for the dependence of their switching behavior on the Ga concentration and ultrafast switching speed. The average NW diameter was 40 nm, and the structure cubic. Last, an increasing interest is devoted to the synthesis of In-Sb-Te NWs, thanks to their relevant thermal stability. Ahn et al. [72] have obtained uniform, single-crystal $In_3Sb_1Te_2$ NWs with length 1–2 μm and diameter 50–70 nm. These NWs exhibited the expected fcc lattice structure. Sun et al. [98] have shown that the single-crystal $In_2 Se_3$ NWs, with diameter of 40–80 nm and length up to 100 μm, exhibited the β-phase of the hexagonal lattice structure, with a growth direction along [110]. Selmo et al. [70] analyzed the structure of $In_3Sb_{1.3}Te_{2.3}$ NWs with diameter down to 15 nm, exhibiting a rock salt phase; Fig. 12.31 shows the SEM and TEM characterization, along with the electron diffraction pattern.

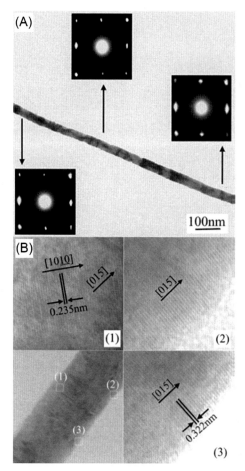

Fig. 12.29 (A) Transmission electron microscopy (TEM) image and selected-area electron diffraction (SAED) patterns of a single Bi_2Te_3 NW with a diameter of 40 nm. (B) The corresponding high-resolution (HR)TEM images: (1), (2), and (3) are the magnified HRTEM images in the areas marked by the three squares.
Source: L. Li, Y. Yang, X. Huang, G. Li, L. Zhang, Pulsed electrodeposition of single-crystalline Bi_2Te_3 nanowire arrays, Nanotechnology 17 (2006) 1706–1712, https://doi.org/10.1088/0957-4484/17/6/027, © 2006 IOP Publishing. Reproduced by permission of IOP Publishing. All rights reserved.

12.8 Size effects

As previously mentioned, one important mechanism in the phase change process is given by the crystalline to amorphous transition. In a NW, we can introduce the melting temperature (or melting point) T_m, causing the disappearance of the associated electron diffraction patterns. As in the case of III–V semiconductor NWs, the downscaling in chalcogenide NWs implies a melting point lowering. This has significant

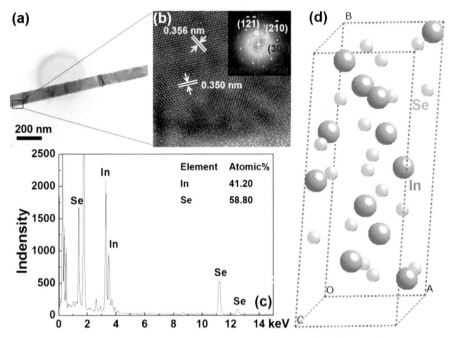

Fig. 12.30 (A) Transmission electron microscopy (TEM) image of a 50-nm In₂Se₃ NW. (B) High-resolution (HR)TEM image of the square in (A). Inset: corresponding fast Fourier transform (FFT) electron diffraction pattern with [−1 −2 3] zone axis. (C) Electron diffraction spectroscopy (EDS) spectrum and the corresponding atomic percentage of In and Se. (D) Schematic crystal structure of γ-phase In₂Se₃ NWs.
Reproduced with permission from Y.-T. Huang, C.-W. Huang, J.-Y. Chen, Y.-H. Ting, K.-C. Lu, Y.-L. Chueh, W.-W. Wu, Dynamic observation of phase transformation behaviors in indium(III) selenide nanowire based phase change memory, ACS Nano 8 (2014) 9457–9462, copyright (2014), American Chemical Society.

advantages for the amorphization (reset) currents, since the energy required to melt the material is reduced, provided that conventional melt-fast quenching is the actual mechanism occurring in NW amorphization. A reduction of the melting point in NWs is theoretically expected and can be expressed as a function of the NW diameter (D) by the formula [99]:

$$\frac{T_{\mathrm{m}}(D)}{T_{\mathrm{m}}(\infty)} = \exp\left(-\frac{2S_{vib}(\infty)}{3R}\frac{1}{(D/D_0-1)}\right), \tag{12.1}$$

where $S_{\mathrm{vib}}(\infty)$ is the bulk vibrational melting entropy; $T_{\mathrm{m}}(\infty)$ is the bulk melting point; R is the ideal gas constant; and D_0 is the critical radius at which almost all atoms of the particle are located on its surface.

Lee et al. [55] carried out in situ TEM experiments to detect the diffraction pattern on single 80-nm wide GeTe and GST NWs. They discovered a 38% reduction of T_{m} for GST NWs and a 44% reduction of T_{m} for GeTe NWs. Analogous results collected from some other reports are summarized in Table 12.2.

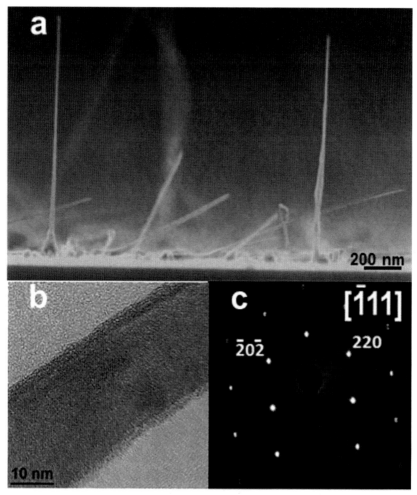

Fig. 12.31 Morphological and structural analysis performed on In-Sb-Te NWs grown on Si(001) substrates at metalorganic chemical vapor deposition (MOCVD) chamber pressure = 300 mbar and temperature = 325°C: (A) scanning electron microscopy (SEM) cross-view image of $In_3Sb_1Te_2$ NWs on Si(001), (B) high-resolution transmission electron microscopy (HRTEM) image of a $In_3Sb_1Te_2$ NW, and (C) its electron diffraction pattern. Reproduced with permission from S. Selmo, S. Cecchi, R. Cecchini, C. Wiemer, M. Fanciulli, E. Rotunno, L. Lazzarini, M. Longo, MOCVD growth and structural characterization of In-Sb-Te nanowires. Phys. Status Solidi A 213 (2016) 335–338, copyright (2015) WILEY-VCH Verlag GmbH & Co. KGaA, Weinheim.

The phase transition modes of NWs also depend on material crystallization kinetics, in particular on the crystallization mechanism, related to the amorphous to crystalline state transition. For example, in bulk-like materials belonging to the GeTe-Sb_2Te_3 tie line, especially regarding the three stable phases $Ge_1Sb_2Te_4$, $Ge_1Sb_4Te_7$, and $Ge_2Sb_2Te_5$, nucleation is the dominant mechanism over growth. It, therefore, results in

Table 12.2 Effect of melting temperature reduction in different NWs compared to their equivalent bulk materials

Material	T_m bulk (°C)	T_m nanowire (°C)	Reference
GeTe	725	410	Lee et al. [55]
		390	Sun et al. [85]
In_2Se_3	890	680	Sun et al. [98]
		722.5	Jin et al. [100]
$Ge_2Sb_2Te_5$	616	375	Lee et al. [55]
Ge	930	650	Wu and Yang [101]

a high crystallization temperature and low crystallization speed (thus high crystallization time, t_c) for alloys close to the GeTe composition, but in opposite properties in the case of alloys close to Sb_2Te_3. In materials such as Sb-rich Ge-Sb, Ge-doped $Sb_{70}Te_{30}$, or In-doped $Sb_{70}Te_{30}$, the growth of crystalline material from the amorphous regions dominates over nucleation. This gives high crystallization speeds, thermal stability, and lower threshold voltages, but lower cyclability [2, 102]. In determining the PCM device speed, the crystallization (set) process is generally expected to be the limiting step, since it is slower than amorphization (reset). Scaling down to 1D devices, phase change NWs have already turned out to yield reversible and fast phase switching (programming pulses <500 ns) in the case of GeTe (set = 100 ns, reset = 500 ns, [55]), Bi-doped GeTe (set = 50 µs, [63]), $Ge_2Sb_2Te_5$ (set = 100 ns, reset = 300 ns, [55]), $Ge_1Sb_2Te_4$ (reset = 300 ns, [51]), Bi_2Te_3 (reset = 40 ns, set 100 ns, [79]), Sb_2Te_3 (set = 500, reset = 100 ns, [48]), In_2Se_3 (set = 100 µs, reset = 100 ns, [100]), and In-Sb-Te (reset = 100 ns, [103]). An important effect of their size observed in phase change NWs is related to the reduction of the programming (reset) currents (see Section 12.11) with NW dimensions. Lee et al. [55] demonstrated this effect (Fig. 12.32) for PCM devices based on GeTe and $Ge_2Sb_2Te_5$ NWs.

Scaling down from 200 to 30-nm-thick NWs implied a current reduction from 3.70 to 0.43 mA for GeTe and from 1.31 to 0.15 mA for $Ge_2Sb_2Te_5$ NWs. The power consumption was similarly reduced from approximately 6 mW (thickness = 200 nm) to 0.63 mW (thickness = 30 nm) for $Ge_2Sb_2Te_5$ NWs. These effects could be interpreted by the increased heat localization and reduced melting temperature in diameter scaled NWs.

It is worth noting that the current and power reduction depends on material composition, which becomes a critical parameter in the switching behavior of nano-PCM cells. Yim et al. [104] studied both the melting temperature reduction and sublimation effects in GeTe NWs coated with SiO_2 layers. Their hypothesis was that the presence of the Au catalyst might affect the reduction of the melting temperature in a sort of reverse VLS mechanism. Hence, they evaluated the vaporization coefficient of the NWs as $\alpha = 10^{-3}$, which did not appear to depend on the NW diameter. Furthermore, they considered the sublimation of the remaining GeTe particles in the final stage of the NW sublimation without the Au seed and found that α is increased by up to 2×10^{-2} when the GeTe nanoparticle is reduced to sufficiently small dimensions. This suggests an interesting parameter to consider in the evaluation of nanoscaled phase change devices.

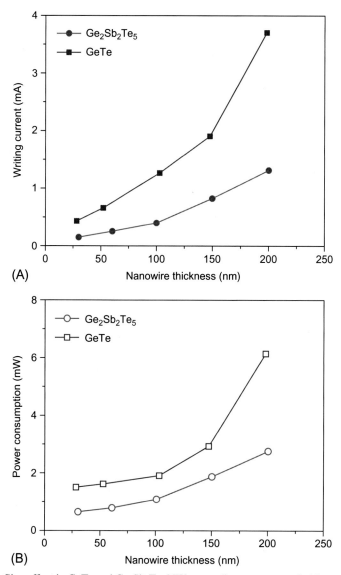

Fig. 12.32 Size effect in GeTe and $Ge_2Sb_2Te_5$ NWs regarding memory-switching characteristics. In (A) the dependence of writing current on NW thickness and in (B) the dependence of power consumption.

Reprinted from S.-H. Lee, Y. Jung, H.-S. Chung, A.T. Jennings, R. Agarwal, Comparative study of memory-switching phenomena in phase change GeTe and $Ge_2Sb_2Te_5$ nanowire devices, Phys. E. 40 (2008) 2474–2480, copyright (2007), with permission from Elsevier.

The recrystallization of the amorphized state is instantaneously induced, as in the PCM operation mode, by heating the material using current pulses to a temperature above its crystallization temperature but below its melting temperature, long enough to allow the structural rearrangement into the crystalline phase. Recrystallization kinetics influence the data retention capability of phase change NWs. Given that the crystalline state is the thermodynamically favored phase over the amorphous one, there exists a time after which the amorphous phase is spontaneously transformed into the crystalline one at a given (even low) temperature. This essentially determines the thermal stability and thus the data retention time for a phase change NW. The recrystallization time, or data retention time, is defined as the time required by an amorphized phase change device to lose its stored information without any applied current pulse. This time can be calculated by studying the temporal evolution of the NW resistance, starting from the amorphized state and maintaining a fixed temperature within the system.

Lee et al. [54] have examined the recrystallization times of GST NWs with different diameters (or thickness), by looking at amorphous → crystalline transformations occurring at 160°C (Fig. 12.33A).

The recrystallization time can be determined by the time at which the steep resistance drop occurs and it is evident that smaller NWs exhibited reduced crystallization times. Recrystallization is a thermally activated process, and thus related to activation energy, E_a. The Arrhenius plots of recrystallization times as a function of $1/kT$ (k=Boltzman constant, T=absolute temperature) show a straight line, whose slope provides the E_a value for different NW diameters (Fig. 12.33B). Size also clearly had an effect, such that reducing the NW size implied reduced E_a values. However, all NWs are indefinitely stable at room temperature, with a data retention time of the order of approximately 4.8×10^5 years for 30-nm NW-based devices.

A difficult consideration issue is the determination of the smallest NW size that can be favorably exploited to obtain both low programming currents and sufficient data retention. It must be taken into account that the operation mode of nonvolatile memories can be required at higher temperatures, such as in automotive applications. Lee et al. [54] reported that the data retention time changes from approximately 1800 years (E_a=2.34 eV) for a 200-nm-thick NW-based device operating at 80°C, to approximately 3 years (E_a=1.98 eV) for a 30-nm-thick GST NW-based device at 80°C.

The E_a data of Fig. 12.33B are plotted in Fig. 12.33C as a function of the NW thickness. By extrapolating the 20 nm thickness, a 1-year retention at 80°C (E_a=1.92 eV) is obtained, which corresponds to approximately 1.22×10^5 years at room temperature. Such results reflect the importance of size effects in NWs for high performing PCM devices. This is particularly true for homogeneous and defect-free self-assembled structures, where the very low presence of compositional nonuniformity and structural or surface damages have little influence on the recrystallization process. To summarize the size effects on recrystallization time, when the GST NW size shrinks, NWs crystallize faster (higher crystallization speed), and E_a is decreased; such effect was ascribed to an enhanced heterogeneous crystal nucleation at the NW surface, as related to the increasing surface to volume ratio; here it has to be noted that GST is featured by a nucleation-dominated crystallization behavior.

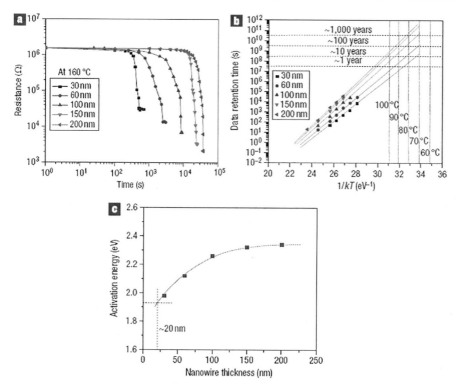

Fig. 12.33 (A) Size effect in Ge$_2$Sb$_2$Te$_5$ NWs on resistance change induced by recrystallization as a function of time at $T = 160°C$: recrystallization is faster in thinner NWs. (B) Arrhenius plots showing the NW Data retention time t (corresponding to recrystallization) versus $1/kT$ of amorphous NWs with different thicknesses, measured in the $T = (140–220°C)$ range. The slopes (activation energy E_a) exhibit a clear thickness dependence; the intercepts are basically independent of thickness. The Ge$_2$Sb$_2$Te$_5$ NW have unlimited data retention time at room temperature. (C) Size-dependent E_a values for NW recrystallization obtained from the slopes in (B).

Reprinted by permission from Macmillan Publishers Ltd.: S.-H. Lee, Y. Jung, R. Agarwal, Highly scalable non-volatile and ultra-low-power phase-change nanowire memory, Nat. Nanotechnol. 2 (2007) 626, copyright (2007).

Li [99] has derived analytical equations, which do not require arbitrary parameters, to describe the behavior of the crystallization activation energy $E_a(D)$ and the nucleation rate $N(D)$ in the amorphous phase for PCM nanomaterials, as a function of the characteristic dimension D, which corresponds to the diameter of NWs. The model proposed was based on the theory of heterogeneous nucleation, in which surfaces, interfaces, and defects catalyze the phase transition by lowering the energy barrier and increasing the transformation rate. This was found to fit well with experimental results in the case of Ge$_2$Sb$_2$Te$_5$ NWs. Fig. 12.34 shows a plot of the $E_a(D)$ theoretical and experimental values.

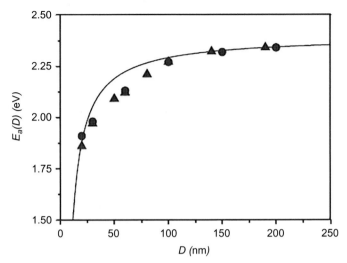

Fig. 12.34 Model prediction (*solid line*) for E_a as a function of the GST NW characteristic dimension D. The symbols ● [54] and ▲ [55] represent the experimental results. The size effect is evidenced.

Reprinted from M. Li, Size-dependent nucleation rate of $Ge_2Sb_2Te_5$ nanowires in the amorphous phase and crystallization activation energy, Mater. Lett. 76 (2012) 138, copyright (2012), with permission from Elsevier.

The decrease of E_a with decreasing NW size was ascribed to the high surface-to-volume ratio of NWs with decreasing diameter, where surface atoms play an important role.

The model is also useful for predicting the behavior of the experimental nucleation rate $N(D)$ values (Fig. 12.35).

The $N(D)$ increases as NW size decreases by seven orders of magnitude from a 190-nm-thick to a 20-nm-thick NW. The dashed line represents where the surface-to-volume effect was not considered. It reproduces the trend, but is more apart from the experimental data. The $N(D)$ size effect was attributed to both the increasing surface-to-volume effect and to the $E_a(D)$ decrease.

Khonic et al. [105] have showed that the E_a value is proportional to the crystallization temperature T_c, finding that $E_a = CT_c$, where C is a size-independent constant. The reduction in E_a with the NW size reduction also indirectly indicates that the latter contributes to thermal stability reduction. The opposite behavior was observed in phase change films and nanoparticles with size <10 nm. Caldwell et al. [22] reported an increase in the crystallization temperature of GeTe nanoparticles from 320°C (diameter = 3.4 nm) to 400°C (diameter = 1.8 nm), at least 150°C higher than the corresponding bulk value (170°C). The enhanced amorphous-phase stability of such nanoparticles was attributed to surface effects and to a lower surface energy of the amorphous phase than the crystalline one. Raoux et al. [3] reported an increase in T_c in GeTe thin films deposited on SiO_2 and capped in situ by 10 nm SiO_2 and a decrease in melting temperature, T_m, with respect to the bulk values (Fig. 12.36).

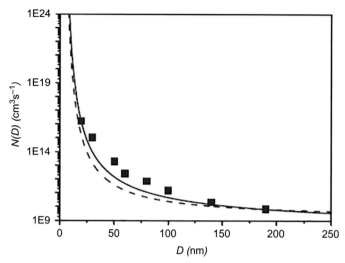

Fig. 12.35 Model prediction (*solid line*) for the nucleation rate N as a function of the GST NW characteristic dimension D. The dashed line represents the predictions ignoring the effect of the surface to volume ratio (see source for details). The symbol ■ denotes the experimental results. The size effect is evidenced.

Reprinted from M. Li, Size-dependent nucleation rate of $Ge_2Sb_2Te_5$ nanowires in the amorphous phase and crystallization activation energy, Mater. Lett. 76 (2012) 138, copyright (2012), with permission from Elsevier.

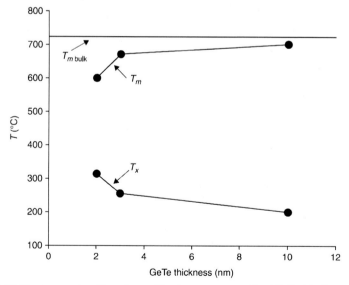

Fig. 12.36 Melting temperature T_m and crystallization temperature T_x of GeTe thin films, as a function of film thickness; the melting temperature of bulk GeTe (T_m bulk) is shown for comparison. Reprinted from S. Raoux, R.M. Shelby, J. Jordan-Sweet, B. Munoz, M. Salinga, et al., Phase change materials and their application to random access memory technology, Microelectron. Eng. 85 (2008) 2330–2333, copyright (2008), with permission from Elsevier.

Last, it should be noted that, in the case of encapsulated GST films with dimensions below 10 nm, the presence of the matrix or cladding layers induces compressive stress capable of increasing the crystallization temperature [106]. Therefore, the size effects of encapsulated chalcogenide nanostructures may exhibit different trends. However, the size effects of nanoscaled GeTe could indicate something more, if one recalls that this material is featured by a growth-dominated recrystallization behavior when the amorphous region is in the 20 nm range [107]. As a matter of fact, Gabardi et al. [108] performed atomistic simulations of 9-nm-thick GeTe NW, amorphizing and recrystallizing a small portion at its center. In the amorphous to crystalline transition that small portion is conceived as a supercooled liquid in contact with the neighboring crystalline phase, so that the dominating mechanism would be crystal growth proceeding from the crystal/liquid interface, rather than the heterogeneous nucleation at the NW surface. As a result, the melting temperature of the NW is decreased by 100°C and the maximum crystallization speed is reduced by a factor of 2 with respect to bulk GeTe. Fig. 12.37 shows the simulated crystal growth velocities in the case of bulk GeTe and GeTe NW, along with the results of fitting obtained on the basis of the classical nucleation theory.

Therefore, in this case, the crystallization process would be slowed down by size reduction, with beneficial effects in terms of the material thermal stability, thus of the device data retention capability.

PCM devices are affected by a drift of the resistance and threshold voltage in time. The threshold voltage is the typical minimum voltage at which the structural phase

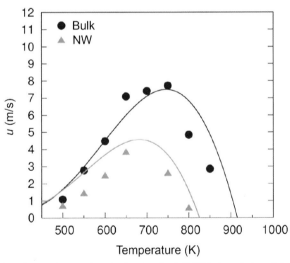

Fig. 12.37 Simulated crystal growth velocity for GeTe NW (*triangles*) and the (110) crystal/liquid interface in the bulk (*circles*). The solid lines are the results of fitting with equations from classical nucleation theory.

Reprinted with permission from S. Gabardi, E. Baldi, E. Bosoni, D. Campi, S. Caravati, G.C. Sosso, J. Behler, M. Bernasconi, Atomistic simulations of the crystallization and aging of GeTe nanowires, J. Phys. Chem. C 121 (2017) 23827–23838, copyright (2017), American Chemical Society.

change is induced, recalled more specifically in Section 12.11. Temporal drift is a detrimental effect on the PCM device reliability, especially when multilevel memory is involved. Some researchers ascribe this effect to compressive stress relaxation of the solidified amorphous phase [109] and some to an intrinsic structural relaxation, reducing the defects of the amorphous phase after phase transition [110]. Mitra et al. [111] investigated amorphous-phase change GST NWs and found that, when the NWs are exposed to free surfaces, their drift coefficients are extremely low in comparison with those of thin-film devices. The drift of amorphous-phase resistance, R, as a function of time, t, can be written as

$$R(t) = R(t_0)(t/t_0)^\alpha \tag{12.2}$$

where α is the resistance drift coefficient and is in the range of 0.01–0.1 in thin-film devices. A similar law involves the threshold voltage $V_{th}(t)$, with a logarithmic dependence on time:

$$\frac{V_{th}(t) - V_{th}(t_0)}{V_{th}(t_0)} = v \log\left(\frac{t}{t_0}\right) \tag{12.3}$$

where v is the threshold voltage drift coefficient and is in the range of 0.01–0.05.

The GST NWs in the work of Mitra et al. [111] exhibited a size-dependent resistance drift up to the thinnest 45 nm NW (Fig. 12.38) for which the drift coefficient was reduced up to one order of magnitude with respect to thin films.

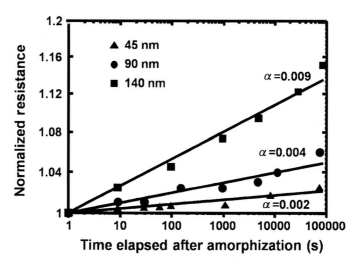

Fig. 12.38 Size effect in drift of normalized resistance for $Ge_2Sb_2Te_5$ NW devices. Drift is reduced in smaller diameter devices. Solid lines show results fitting to Eq. (12.2).
Reprinted with permission from M. Mitra, Y. Jung, D.S. Gianola, R. Agarwal, Extremely low drift of resistance and threshold voltage in amorphous phase change nanowire devices, Appl. Phys. Lett. 96 (2010) 22111, copyright (2010), AIP Publishing LLC.

This size effect is again attributed to the high surface to volume ratio and exposed surfaces in the NWs. In fact, the presence of 300-nm dielectric films of SiO_2 or Si_3N_4, embedding the NWs, induced stress effects (by a factor of 20 with respect to unembedded NWs), which makes the drift coefficient similar to those of thin-film PCMs. The conclusion supported the hypothesis that the built-in stress relaxation induced by the amorphization process is the major factor responsible for resistance drift.

Jin et al. [112] reported the decrease of resistance drift at reset state by lowering the diameter of In_2Se_3 NW-PCM devices encapsulated with a 400-nm-thick SiO_2 layer. The drift coefficient scaled down with the NW size, from $\alpha = 0.008$ for a 145-nm-thick NW, to $\alpha = 0.003$ for a 60-nm-thick NW.

In their atomistic simulation for aging of amorphous GeTe NWs, Gabardi et al. [108] examined the fraction of Ge-Ge homopolar bonds, whose presence they previously related to resistance drift in amorphous GeTe. Such a fraction turned out to be lower in NWs than in the bulk GeTe and tends to decrease with time, thus accounting for the lower resistance drift effect in the NWs.

Finally, NW-like line cells have been proposed by Yoon et al. [113]. These are based on the patterned nanoscale self-heating channels of $Ge_{18}Sb_2Te_5$. The voltage necessary for cell reset is decreased from 4.2 to 2.6 V, as the channel length was reduced from 500 to 100 nm, demonstrating the remarkable scaling effect on the power consumption.

In summary, the list below gives the size effects that can be observed in phase change NWs:

1. reduction of melting temperature;
2. reduction of reset currents and power consumption;
3. higher crystallization speed, reduction of activation energy for recrystallization (in nucleation-dominated NWs);
4. lower data retention (in nucleation-dominated NWs); and
5. reduction of resistance drift.

12.9 Phase transition mechanisms

Surprising discoveries have been made by in situ observation of structural transformations of phase change NWs during their phase transitions. Furthermore, this would give rare opportunity to directly observe the phase change mechanisms, since similar observation of current PCM devices are almost impossible. This is because the active phase change material is confined by many layers of different materials, making it impossible to observe the device during operation. Meister et al. [87] have studied the structural transformations during operation of GeTe NW-based devices, with NW diameters in the range of 60–200 nm. Single Pt electrodes were placed on electron transparent silicon nitride membranes to allow direct in situ TEM observation. To switch them back and forth, the NWs were encapsulated in a 20-nm SiO_2 sputtered layer. Fig. 12.39 illustrates the results of the real-time TEM observation of phase transition under a voltage scan, correlating the structural and electrical analysis.

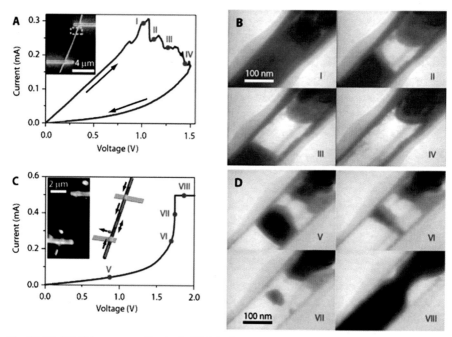

Fig. 12.39 (A) Voltage scan of a single NW device. The inset shows the location of the transmission electron microscopy (TEM) observation in panel (B). (B) TEM images taken in situ during the voltage scan in the panel (A), at times I, II, III, and IV. Note the correlation of resistance with void size. (C) Second voltage scan on the same wire showing switching to the initial low resistance state. The inset displays a scanning electron microscopy (SEM) image of the same NW after repeated pulsing and voltage scanning. On the NW, the dark area indicates the hollow tube and the bright area indicates the existence of GeTe materials. The inset also shows a schematic representation of the material flow inside the NW. (D) TEM images taken in situ. At time VIII, the void is suddenly refilled and the current jumps past compliance in the voltage scan.

Reprinted with permission from S. Meister, D.T. Schoen, M.A. Topinka, A.M. Minor, Y. Cui, Void formation induced electrical switching in phase-change nanowires, Nano Lett. 8 (2008) 4562–4567, copyright (2008), American Chemical Society.

Fig. 12.39A and B shows the transformation toward the highly resistive state, starting from the deposited crystalline NW (with a two-point resistance of 3.7 kΩ). As the resistance was increased by the applied current, the sudden formation of nanovoids close to the contact regions was observed. The current maintained through a preferential channel, due to the small shell of GeTe remaining on the SiO_2 nanotube walls. The final resistance value was 42 kΩ, with a void length of about 1 µm. The reverse transition mechanism also produced surprising results (Fig. 12.39, panels C and D). Starting from a state of high resistance due to large voids in the NW, the voltage scan induced movement of the material, until the molten phase began, so that the shell tube could be refilled, resulting in removing the voids and restoring the initial NW resistance.

Void formation was, therefore, assumed to be the mechanism accounting for the very large measured resistance window, which was of the order of 10^7 between the ON and OFF states. Notably, in the process of molten material filling, the voids were even observed coming from regions outside the contact areas. Structural transformations were observed in small regions of the NW, often located close to the contacts (due to the heat concentration for high contact resistance) or at necks in the NW, caused by switching. Rather than electromigration, the possible dominant mechanism causing the motion of liquid GeTe was identified as the expansion due to material melting and overpressure in the SiO_2 shell, due to Te evaporation.

Jung et al. [53] performed in situ HR-TEM observations of reversible phase transitions in proper PCM devices based on single GST NWs. They applied 50 ns voltage pulses with increasing amplitude and examined both the structural and electrical properties. A "notch" structure was achieved for the NW suspended along the observation trench, in order to obtain a "hot spot" in which the phase change was locally forced to occur, so that TEM analysis could be focused there. Fig. 12.40 shows the TEM images related to the different switching states of the NW close to the notch, where the phase transition is observable by the presence of white/dark contrast regions.

As well as full crystallization and full amorphization of the NWs, intermediate amorphous to crystalline transformations and vice versa were studied. Intermediate resistance states were reached by gradually increasing the voltage pulses. The case of HR-TEM analysis performed on partially recrystallized NWs should be noted, in which randomly oriented crystalline grains of approximately 5–7 nm were detected in the hot spot region, thus confirming the recrystallization mechanism is dominated by nucleation.

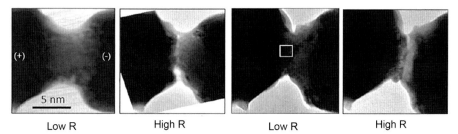

| Low R | High R | Low R | High R |

Fig. 12.40 The transmission electron microscopy (TEM) images show the single GST NW device during repeated switching cycles, yielding the high and low resistance (R) states. Low R is a set state and high R is a reset state; the amorphous region is evidenced by white contrast at the notch.

Adapted with permission from Y. Jung, S.-W. Nam, R. Agarwal, High-resolution transmission electron microscopy study of electrically-driven reversible phase change in $Ge_2Sb_2Te_5$ nanowires, Nano Lett. 11 (2011) 1364–1368, copyright (2011), American Chemical Society.

The same group [114] has also examined the phase change transitions from crystalline to amorphous states in a similar single-crystalline GST NW device, by in situ HR-TEM in real time. As in the previous case, GST NW-based PCM devices were suspended on proper 2 µm trenches. This allowed the TEM analysis under the application of a few 100 ns voltage pulses with increasing amplitude. After a time interval between pulses, which was sufficient to allow for device thermalization, the device resistance was measured. The first surprising result was in the NW resistance change behavior. While still in the crystalline state, a small decrease in the resistance ("resistance dip") took place just before amorphization, followed by an abrupt increase by two orders of magnitude. This effect was reproducibly observed in all NWs. Simultaneous HR-TEM observations with the appearance of the resistance dip allowed the identification of the presence of line defects, that is, dislocations. Such defects were characterized by a Burgers vector of ½ $\langle 1\ 1\ -2\ 0 \rangle$, meaning that the dislocation loops are in the $\langle 1\ 0\ -1\ 0 \rangle$ prismatic plane of the hcp GST NW, growing along the $[1\ 0\ -1\ 0]$ direction. The resistance dip was related to the line defects that propagate due to wind force breaking the NW symmetry, so that dislocations were observed to move in parallel to hole (majority) carriers, thereby under the influence of bias polarity. At the bottom of the resistance dip, the dislocations were stopped because of heavy accumulation, until amorphization took place. Amorphization was accompanied by a sharp resistance increase, with a bright line featuring the small amorphized region and a "cloud" of dislocations left behind it. Further experiments were carried out using the "notch" structure and real-time HR-TEM observations during the crystalline to amorphous transition, in order to better localize the amorphized region, yielding analogous results. Dark field (DF) TEM analysis clearly showed the presence of a cloud of dense dislocations generated by the heat shocks of the voltage pulses in the monocrystalline NW, in which vacancy condensation process is in progress and the presence of electrical wind force causes the dislocation motion. Thus, further vacancies can be "consumed," reducing the carrier scattering and decreasing the NW resistivity.

Generation, oriented motion and jamming of dislocations appear to be the basis of the observed process in the studied GST NWs. However, the authors could not exclude the possibility that a form of transient melting occurs in the highly dislocated region before the resistance increase for amorphization. Nevertheless, the work from Nam et al. [114] shed a new light on the role of defects in the transition mechanisms of phase change NWs and possibly of phase change materials in general. Defect-assisted amorphization consumes less power than the melt-quench process, since it has a lower activation energy. This implies lower phase separation effect on switching is expected, with beneficial effects on the device reliability. It is here useful to remind that the failure mechanisms in PCMs are mainly related to phase separation, electromigration, void formation, and thermal instability of the amorphous phase [10]. Moreover, in their work, Nam et al. [114] reported no appreciable change in the bottom-up, defect-free GST NW composition after phase transition, indicating that electromigration is rather related to the presence of grain boundaries, typically induced by the use of top-down techniques. The thermal behavior of the same GST NWs of the work by Nam et al. [114] was simulated by Ji and Feng [115]; it turned out that the NWs never reach the GST melting temperature, so confirming the route of amorphization following a lower energy, rapid state transformation, assisted by defect nucleation and clustering. This finding does not rule out the hypothesis of the "transient" situation, in which the dislocation jamming could just depress the NW melting temperature, so that a lower voltage becomes sufficient to induce a "premelting" condition that facilitates the melt-quench process. In any case, it is desirable that such favorable conditions can be obtained and controlled in PCM devices, to get advantages for power consumption limitations.

The role of disorder and electronic instability in NWs was studied by applying voltage pulses to single-crystalline GeTe NW-based devices [116], inducing the formation and pileup of dislocations and antiphase boundaries, until the material is transformed into a dirty metal first, and then to an insulator, even if it keeps its crystalline structure, thus proving a new electronic state of GeTe. So the progressive addition of disorder was demonstrated to be able to induce a metal-insulator transition (MIT) in GeTe NWs, before amorphization takes place, although a slight variation of pulse voltage conditions can lead to a disordered amorphous phase.

An analogous study was presented by Mafi et al. [117], related to the role of dislocations in In_2Se_3 NWs and confirming through TEM and scanning Kelving probe microscopy, coupled to DFT calculations, that dislocations (i) are generated via vacancy condensation, (ii) result mostly immobile, and (iii) cause an electrical resistivity increase by several orders of magnitude, still leaving the material in its single-crystalline phase.

It is here worth mentioning the analysis on the phase transition of GST NWs, performed by Lee et al. [118] on the basis of direct imaging and electrical characterization. Abundant evidence emerged that as-grown GST NWs are in the hcp phase and the direct phase transition hcp to amorphous implies a large volume change. The TEM analysis by Lee et al. [118] evidenced that, after iterative electrical switching, the NWs in the set state remain in the metastable fcc phase, so that the fcc-amorphous-fcc transition is involved in the phase change process. This is a kind of "metastabilization," featured by the Ge and Sb atoms migrating from intrinsic octahedral sites of Ge/Sb mixed layers into the tetrahedral sites of vacancy layers on electrical pulses, transforming the stable hcp structure into a metastable fcc structure. The structural hcp to fcc transformation was confirmed by an increased set resistance, which also indicates that, in addition to Joule heating, the set pulses of metastabilization induce lattice disordering (atomic migration) in the as-grown NWs. The electrical switching process offers the benefits of lower power consumption and volume change. Starting from the fcc phase, the reset pulses form an amorphous region around the longitudinal center of the NWs, made by an uneven distribution of partial crystallites. Fig. 12.41 sketches the core-centered phase transition.

Since the simulation results indicated that the central region does not attain the GST melting temperature, the reset operation is not based on fully melted liquid state. Rather, it was demonstrated to be related to a change in the atomic concentration (atomic migration even in this process) close to the core-centered region of the NW, followed by lattice disorder in the form of prismatic dislocation loops, eventually leading to the defect-assisted amorphization, as observed by Nam et al. [114].

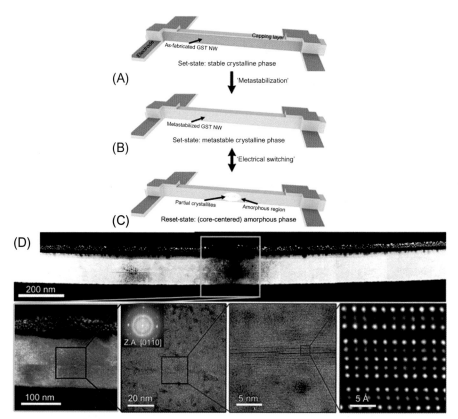

Fig. 12.41 Schematic representation of a core-centered phase transition in GST NWs, represented at different states: (A) as-fabricated, (B) after set-operation, and (C) after reset-operation. (D) (*top*) Scanning transmission electron microscopy (STEM)—high-angle annular dark-field (HAADF) scanning transmission electron microscopy images with sequential zooming of: the GST NW after reset operation (*bottom*), the NW center region (*green square*), a partially crystalline region (*blue square*), and the core part of the center region (*red* and *magenta square*).

Adapted with permission from J.-Y. Lee, J.-H. Kim, D.-J. Jeon, J. Han, J.-S. Yeo, Atomic migration induced crystal structure transformation and core-centered phase transition in single crystal Ge$_2$Sb$_2$Te$_5$ nanowires, Nano Lett. 16 (2016) 6078–6085, copyright (2016), American Chemical Society.

12.10   Thermal properties

The reversible phase transition between the amorphous and the crystalline phase in PCM-NWs is based on the Joule heating effect induced by proper electrical pulses. Furthermore, we have shown that only a portion of the NW is involved in the structural change. The dynamic behavior of PCM cells is strongly related to heat dissipation. Therefore, the thermal conduction properties in phase change NWs and the surrounding environment (in particular the thermal resistance or the inversely proportional thermal conductivity) are important for understanding the device operation and reliability. Therefore, the thermal analysis of nanostructured phase change systems, such as NWs and related size effects, needs to be explored in detail.

A molecular dynamics (MD) study conducted by Volz and Chen [119] has demonstrated that the thermal conductivity of semiconductor NWs can be sensibly lowered with respect to bulk values (up to two orders of magnitude in the case of Si NWs), mainly caused by phonon boundary scattering: due to the higher surface/volume ratio in NWs, the number of phonon collisions at the NW surface is much higher than in corresponding bulk material, thus increasing the heat flow resistance and decreasing the thermal conductivity. It is, therefore, expected that in chalcogenide NWs a similar effect takes place, but also benefits from heat localization and thus more efficient use of the transferred energy for phase switch. Heat localization in NWs caused by the Joule effect can be described in terms of the heat transport equation as [120]

$$k\frac{d^2T}{dX^2} = \rho J^2 \qquad (12.4)$$

where k is the thermal conductivity of the NW; ρ is the electrical resistivity of the NW; r is the radius of the NW; and J is the current density $= I/\pi r^2$. The writing current is given by

$$I = 2\pi\sqrt{\frac{2k\Delta T}{\rho}}\frac{r^2}{L} \qquad (12.5)$$

where $\Delta T =$ temperature difference between the transition temperature and room temperature. This equation shows the high impact of the NW size on the writing current. Reducing thermal conductivity lowers the writing currents, along with the decrease of the maximum temperature that can reach the "hot spot" of the NW.

A compact model was presented by Chen and Pop [121], in order to examine the temperature distribution in cylindrical PCM-GST NWs. This model solves the heat diffusion equation, taking into account the effect of thermal boundary resistance (TBR) at the interfaces formed by the contact area between the NW and the TiN electrodes, as well as the NW sidewalls and the SiO_2 embedding matrix [121, 122]. The model explains the programming current dependence on geometric properties. As regards to the dependence of programming current on the GST properties (stoichiometry) and interfaces with different insulators (SiO_2, SiN_x) and heaters (TiN), it came out that, if the thermal conductivity of the materials is lowered, the programming current is reduced, due to better thermal confinement (similar results were obtained for thin-film-based PCM devices, as in [123]); furthermore, better thermal confinement was obtained by higher TBR values, with the effect of reducing the programming current by up to a factor of 3.

When considering the experimental analysis, typical methods to measure the thermal conductivity in NWs are the 3ω method, Raman spectroscopy, and a combination of scanning thermal microscopy (SThM) and the 3ω method [124].

The group of Battaglia developed the SThM within the 3ω mode to measure the thermal conductivity of Sb_2Te_3 NWs with diameters larger than 100 nm [125]; the method does not require processing or NW suspension, and allows the determination of the thermal conductivity along the transverse direction of the NWs. The same method was then employed to measure the thermal resistance R_{Th} along the transverse direction (or thermal conductivity $k_{NW} = \Phi_{NW}/R_{Th}$, where Φ_{NW} is the NW thickness) in In$_3$SbTe$_2$ NWs with thickness of 13 and 23 nm [126]. Fig. 12.42 shows the atomic force microscopy (AFM) and SThM measurements amplitude and phase (SThM mode), performed on a 23-thick NW at the frequency of 523 Hz.

The measured temperature and phase, as a function of the frequency, averaged over the A and B areas in Fig. 12.42 were used to extract the thermal resistance value, taking into account the effect of the SiO$_2$/Si substrate. It turned out that the thermal conductivity was as low as $3.82 \, W \, m^{-1} \, K^{-1}$ for 13-nm-thick NWs, which is by a factor ~5 lower than for the bulk material. This offers confirmation of the phonon confinement along the NWs thickness Φ_{NW}, since the phonon mean free path in In$_3$SbTe$_2$ is comparable with the thickness of the examined NWs.

Park et al. [127] studied, by Raman spectroscopy, the thermal resistance of 10 μm long, crystalline GeTe and GST NWs, with diameters in the range of 80–160 nm for GeTe and 120–220 nm for GST. Their measured values of thermal conductivity are $1.44 \, W \, m^{-1} \, K^{-1}$ for GeTe and $1.13 \, W \, m^{-1} \, K^{-1}$ for GST NWs, respectively, which are clearly smaller than in corresponding thin films. Bipolar switching was observed in such NWs, due to the absence of a real phase change transition. Rather, a structural modulation might be present in the set/reset operations with morphological variations (voids and hillocks). This indicates a compositional change (electromigration), to which the bipolar switch could be ascribed. The conclusion was that the coupling of low thermal conductivity and long NWs makes it difficult to generate and transfer enough heat to the middle region of the NW, in order to induce a structural transition. This poses a warning for the design of PCM NWs, in which the benefits from reduced thermal conductivity have to be compared with the device length and compositional changes.

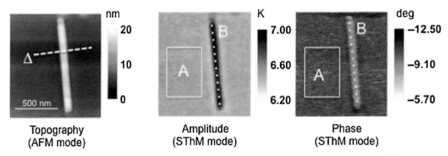

Topography	Amplitude	Phase
(AFM mode)	(SThM mode)	(SThM mode)

Fig. 12.42 Atomic force microscopy (AFM) topography, scanning thermal microscopy (SThM) amplitude and phase, measured in a 23-nm-thick In$_3$SbTe$_2$ NW at 523 Hz; the Δ line is used for the cross-section profile. Areas A and B are used to average the amplitude when the probe is in contact with the SiO$_2$/Si substrate and the NW.

Reproduced with permission from J.-L. Battaglia, A. Saci, I. De, R. Cecchini, S. Selmo, M. Fanciulli, S. Cecchi, M. Longo, Thermal resistance measurement of In$_3$SbTe$_2$ nanowires, Phys. Status Solidi A 214 (2017) 16000500, copyright (2016), WILEY-VCH Verlag GmbH & Co. KGaA, Weinheim.

Jin et al. [100] have studied the size dependence of the thermal resistance in In_2Se_3-based phase change NW PCM devices. The NWs had a diameter in the range of 60–200 nm and were encapsulated in a 400-nm-thick SiO_2 layer. The thermal resistance as a function of the NW diameter was investigated on the basis of the programming power, P_{prog}, measurement and the transition temperatures T_C (T_s for crystallization and T_r for amorphization). By integrating the Fourier equation along the heat flux direction, the following relation can be established:

$$T_C = T_{298K} + R_{Th} \times P_{Prog} \tag{12.6}$$

where R_{Th} is the NW thermal resistance and T_{298K} is the thermal chuck temperature. The Wiedemann-Franz law was then applied:

$$k/\sigma = LT \tag{12.7}$$

where k is the thermal conductivity, σ is the electrical conductivity, and L is the Lorentz number $= 2.44 \times 10^{-8}\,W\,\Omega\,K^{-2}$. By using this, it was possible to correlate the electrical resistance (R_e) and thermal resistance (R_{Th}) of the amorphous (A) and crystalline (C) phases:

$$\frac{R_e(A)}{R_e(C)} = \frac{R_{Th}(A)}{R_{Th}(C)} \tag{12.8}$$

The results produced by the above calculation are shown in Fig. 12.43, for the thermal resistance and power consumption of the NWs as a function of the NW diameter.

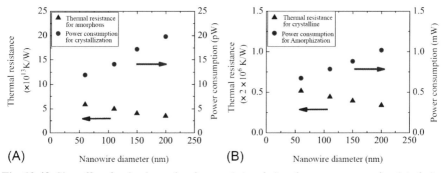

Fig. 12.43 Size effect for the thermal resistance (*triangles*) and power consumption (*circles*) of In_2Se_3 NWs as a function of the NW diameter. (A) The minimum programming power of 11.9 pW is enough for set (crystallization) in a 60-nm NW memory cell with 5.86×10^{13} K/W amorphous thermal resistance and (B) 0.673 mW are required for reset (amorphization) with 1.04×10^6 K/W crystalline thermal resistance.

Reprinted with permission from B. Jin, D. Kang, J. Kim, M. Meyyappan, J.-S. Lee, Thermal efficient and highly scalable In_2Se_3 nanowire phase change memory, J. Appl. Phys. 113 (2013) 164303, copyright (2013), AIP Publishing LLC.

Notably, the thermal resistance increases with decreasing NW diameter, while the power consumption decreases in In_2Sb_3 NWs. In this case, the thermal resistance ratio between the NW and bulk phase was around 10^5 for the amorphous and around 10 for the crystalline phase.

In PCM operation, undesirable "proximity disturbance" effects are related to "thermal cross talk," which is the probability that the heat generated by a programmed cell can be propagated to a nonprogrammed adjacent cell. This has the effect of destroying the correct information and storing incorrect data encoding [128]. The problem can also affect NW-based PCM devices. An analytical model for isotropic downscaling of phase change NW size by a factor of $1/k$ $(k > 1)$ showed that the thermal crosstalk effect is reduced in NWs; the operation speed is increased by a factor k^2 and the programming current is reduced by at least a factor k^3, demonstrating a further advantage of phase change NWs [129].

We can summarize the effects of employing phase change NWs, in comparison with their corresponding bulk materials, in the following points:

1. reduced thermal conductivity and
2. reduced proximity disturbance.

The combination of the transition temperature reduction (Section 12.8), reduced thermal conductivity and reduced volume to be programmed, would make the phase change NW a very efficient, low-power consuming PCM cell.

12.11   Electrical properties

The configuration used to measure the electrical properties of phase change NWs is given by the NWs contacted to proper electrodes as single nanocells, often horizontally placed on an insulating substrate (typically SiO_2). Metallization is usually brought about by using FIB techniques, coupled with nanolithography to achieve a whole circuit, up to the macro pads. The single NW can be exposed to air, although accompanied by the effect that oxidation and material evaporation rapidly degrade the nanocell and very few switching cycles are possible. In order to avoid this problem, the NW can be encapsulated in a proper dielectric material. Ideally, a regular array of vertical NWs buried in a dielectric matrix should be built, which allows the deposition of the contacts to the NW ends, thus yielding collective information on a large number of nanocells [79].

The electrical characterization of phase change NWs is used to study the electrically triggered transition of the material from the amorphous (reset, high resistance) to the crystalline (set, low resistance) phase and vice versa. The pulsed I/V (current vs voltage) technique is employed, similar to conventional PCM device measurement, which can determine the threshold voltage of the device, V_{th}. representing the minimum voltage required for the operation of the NW PCM device [2].

The amorphization of the NW devices (reset) is achieved by an intense, short voltage pulse, which is believed to rapidly melt and quench a portion of the NW or generate defects whose evolution/motion leads to very high resistance state. Crystallization (set) is achieved by a less intense, but longer voltage pulse, which triggers a more complex process. The increase of the electric field across the amorphous region causes the density of charge carriers to increase and temperature increases due to inelastic scattering. As a result, the resistance of the amorphous chalcogenide decreases, which means that there is sufficient current flow to heat the material above the crystallization temperature by the Joule effect [130].

The electrical analysis for NW-based PCM devices is intended to determine the same performance indicators as are used for PCM cells, as listed below (essentially depending on the phase change material properties):

- *Set and reset currents*: required for crystallization and amorphization; reset current is the highest current needed to allow the cell operation, and thus directly related to the cell power consumption.
- *Electrical resistance*: of the cell in the high conductivity (set) and low conductivity (reset) states.
- *Programming window*: defined by the resistance gap between the set and reset state. A programming window of at least two orders of magnitude is required for the proper operation of the device, to prevent ambiguities in the cell status data reading.
- *The number of switching cycles*: capable of keeping the width of the programming window constant, also defined as "endurance."
- *The threshold voltage*: voltage above which the processes leading to crystallization start.
- *The data retention time*: related to the retention of the amorphous phase (reset) state (as explained previously); the conventional requirement for nonvolatile storage is a retention >10 years at 85°C.

In order to program a NW cell, appropriate values for the voltage pulses have to be selected by considering the programming curve, which is the plot of the cell resistance (R) value as a function of the voltage pulse amplitude. This analysis provides the set (V_{set}) and reset (V_{reset}) voltage values (generally a few V), once the duration of the pulses has been selected (on the scale of hundreds of ns, necessary to transfer enough energy for crystallization). In Fig. 12.44 [55], the pulsed I-V (A) characteristics under a voltage sweeping and R-I programming curves (B) are reported for GeTe and GST NW devices (NW thickness = 100 nm). The as-deposited NWs were single crystalline and the corresponding I-V curve in Fig. 12.44(A) shows a linear dependence. In the amorphous to crystalline transition under increasing voltage pulses, the V_{th} was evident at 0.8 V for GeTe and 1.2 V for GST, after which the crystallization of the two materials was gradually completed.

Subsequently, the system returns to its initial situation and another cycle can be performed, indicating the reversibility of this process. The R-I programming curve in Fig. 12.44B shows how the current pulses of different amplitude could selectively induce either the crystallization or the amorphization of the phase change material. This in turn determines a crystalline state resistance at approximately 1.5 kΩ for GeTe and approximately 17 kΩ for GST and an amorphous state resistance of 3.8 MΩ for GeTe and 1.58 MΩ for GST. Minimum current amplitude of approximately 1.2 mA for 100 ns was necessary to start the amorphization of GeTe NWs, as opposed to that required for GST of 0.24 mA for 300 ns. GST NWs, therefore, created memory devices with a higher threshold voltage, but a lower current consumption needed to achieve

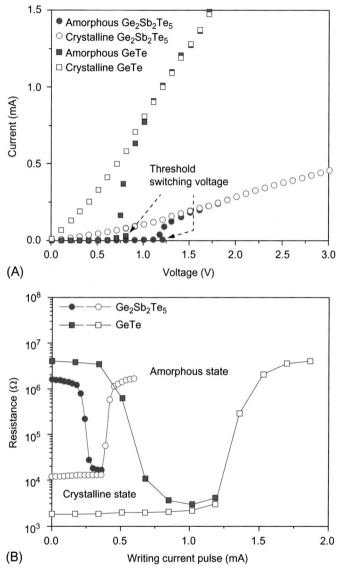

Fig. 12.44 Comparison of memory switching characteristics in GeTe and $Ge_2Sb_2Te_5$ NWs (100 nm thickness): (A) current-voltage (*I-V*) plots at amorphous (GeTe solid squares; $Ge_2Sb_2Te_5$ solid circles), and crystalline state (GeTe open squares; $Ge_2Sb_2Te_5$ open circles). The threshold voltage and the two amorphous and crystalline programming intervals are visible; and (B) programming *R-I* curves of the NWs memory device under 100 ns write and 500-ns erase pulses for GeTe and 100 ns write and 300-ns erase pulses for $Ge_2Sb_2Te_5$. The amorphous and crystalline regions appear well separated.

Reprinted from S.-H. Lee, Y. Jung, H.S. Chung, A.T. Jennings, R. Agarwal, Comparative study of memory-switching phenomena in phase change GeTe and $Ge_2Sb_2Te_5$ nanowire devices, Phys. E. 40 (2008) 2474–2480, copyright (2007), with permission from Elsevier.

crystallization, when compared with GeTe NW-based devices. However, GeTe NW devices provided a larger programming window.

Resistance-voltage (R-V) curves were used to evaluate the operating voltage for a PCM cell based on In_2Se_3 NWs, with and without an SiO_2 passivating layer of 400 nm by Baek et al. [59], by applying increasing voltage pulses, while maintaining pulse duration in all experiments (100 μs for set and 200 ns for reset); the results are shown in Fig. 12.45.

With the presence of a SiO_2 passivating layer, the values of V_{set} decreased from 6 to 3.8 V and V_{reset} from 8 to 4.8 V, when the programming window was approximately 2.5×10^5. This showed a 40% reduction in the programming voltage and a 27% reduction in the programming power. This effect was attributed to the reduced heat loss from the passivated NWs to the surroundings.

Longo et al. [51] have analyzed a $Ge_1Sb_2Te_4$ NW-based PCM device, for which the SEM image is shown in Fig. 12.46A.

The single NW (80 nm thick, 1 μm long) was contacted using FIB-Pt contacts touching the aluminum pads to pick up the signal. In Fig. 12.46B, the I-V characteristics are shown for the device. The NW was deposited as crystalline, with approximately 6 kΩ resistance. A threshold switching value of 1.35 V was recorded for amorphization, lower than in thin-film devices based on $Ge_1Sb_2Te_4$ (1.41 V). The amorphized NW state exhibited a resistance of approximately 450 kΩ and thus a programming window of about 10^2, which is smaller than in $Ge_2Sb_2Te_5$ NWs. The corresponding values from a 60 nm GST NW [55] are also shown in Fig. 12.46B, for comparison. The device could only be reversibly switched for nine times, which is much fewer than the 10^5 write/rewrite cycles reported in Lee et al. [54] for GST NWs. However, the $Ge_1Sb_2Te_4$ NWs were not embedded and so were exposed to air during cycling, meaning that oxidation and evaporation quickly degraded the test device during operation. Jung et al. [56] have reported on memory switching in Ge-Sb NWs with two different eutectic compositions: (i) Sb-rich Ge-Sb (Sb > 86 at.%) and (ii) Ge-rich GeSb. Electrical analysis was performed by examining their I-V, R-V, and cyclability curves. For 80-nm-thick Sb-rich Ge-Sb NWs, the switching behavior was observed with a V_{th} of approximately 4 V. However, after the first crystallization sweep, the initial resistance value of approximately 36 kΩ was not recovered, but ended up to higher values. Similar results were obtained for NWs up to 120 nm in thickness, indicating that electrical reversibility was not achieved. TEM results suggested that the partial irreversibility was due to phase separation of Ge under thermal annealing. Similar electrical measurements performed on Ge-rich Ge-Sb NWs did not show any switching behavior.

The switching behavior of $In_3Sb_1Te_2$ (IST) NWs with diameters as small as 20 nm has been studied by Selmo et al. [103] to demonstrate the low-power and low-programming voltage of phase change cells based on such NWs. The device was formed by a single NW (crystalline as deposited, thickness = 20 nm, gap between electrodes = 700 nm), contacted by Ni/Au microelectrodes fabricated by EBL on top of the NW. Fig. 12.47 shows the characteristic I-V curve for an IST NW device.

Starting from the set state (resistance ~50 kΩ), a voltage pulse of 3 V for 100 ns ($I_{res} = 60$ μA) was applied to induce the reset state (resistance ~ 12 MΩ). The set state was re-obtained by reaching a threshold voltage $V_{th} \sim 0.78$ V and completing the

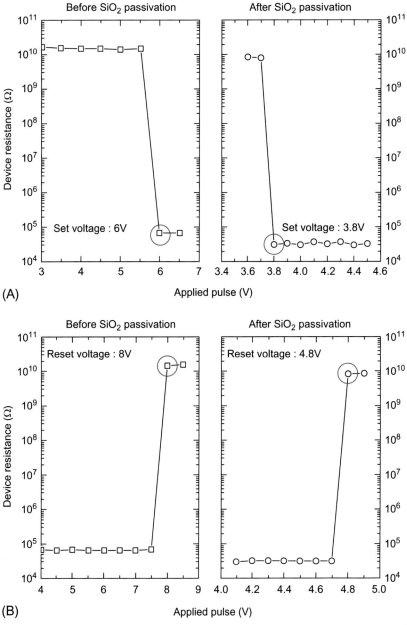

Fig. 12.45 Resistance as a function of applied pulses in In_2Se_3 NW devices before and after SiO_2 passivation: (A) set voltages with 100 ns pulses, amorphous phase to crystalline phase; and (B) reset voltages with 200 ns pulses.

Reprinted from C.-K. Baek, D. Kang, J.S. Kim, B. Jin, T. Rim, S. Park, M. Meyyappan, Y.-H. Jeong, J.-S. Lee, Improved performance of In_2Se_3 nanowire phase-change memory with SiO_2 passivation, Solid State Electron. 80 (2013) 10–13, copyright (2012), with permission from Elsevier.

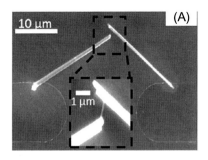

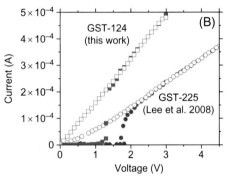

Fig. 12.46 (A) Scanning electron microscopy (SEM) image of a $Ge_1Sb_2Te_4$ (GST-124) NW, contacted by focused ion beam (FIB); the NW area is shown in the zoomed image; (B) pulsed *I-V* measurement of the device in (A) (solid squares for amorphous $Ge_1Sb_2Te_4$, open squares for crystalline $Ge_1Sb_2Te_4$) in comparison with analogous measurement from a $Ge_2Sb_2Te_5$ (GST-225) NW-based device (solid circles for amorphous $Ge_2Sb_2Te_5$, open circles for crystalline $Ge_2Sb_2Te_5$, as from [55]).
Reprinted with permission from M. Longo, R. Fallica, C. Wiemer, O. Salicio, M. Fanciulli, et al., Metal organic chemical vapor deposition of phase change $Ge_1Sb_2Te_4$ nanowires, Nano Lett. 12 (2012) 1509–1515, copyright (2012), American Chemical Society.

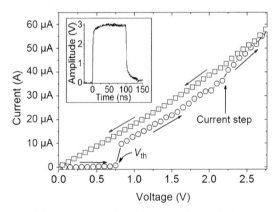

Fig. 12.47 *I-V* characteristics of a 20-nm $In_3Sb_1Te_2$ NW device in the amorphous (*circles*) and crystalline (*squares*) states. The threshold voltage (V_{th}) for the amorphous to crystalline transition and an additional step in the measured current at about 2.25 V are clearly observed. Inset: representative reset pulse directly measured at the device.
Reprinted from S. Selmo, R. Cecchini, S. Cecchi, C. Wiemer, M. Fanciulli, E. Rotunno, L. Lazzarini, M. Rigato, D. Pogany, A. Lugstein, M. Longo, Low power phase change memory switching of ultra-thin $In_3Sb_1Te_2$ nanowires, Appl. Phys. Lett. 109 (2016) 213103, with the permission of AIP Publishing.

crystallization (with the remarkably low resistance of ~50 kΩ) at a voltage of 2.75 V. By analyzing the I-V curves in the reset state within the subthreshold range, that is, for $V < V_{th}$, it was possible to extract the information that only a few tens of nm of the NW length are amorphized in the reset operation; therefore, just a small fraction of the NW is involved in the switching process. The presence of the current step in the I-V curve was attributed to a possible successive structural transformation of the crystalline state, that could, in principle, be useful for multilevel memory devices. A cycling test was performed on unpassivated 22-nm-thick NWs (gap = 480 nm) up to 8 cycles, with a resistance ratio of 10^2, as shown in Fig. 12.48A; the average reset current was 41 μA, the current density was 13 MA cm^{-2} and the corresponding reset power was 130 μW, much lower than in Ge-Sb-Te NWs.

Fig. 12.48B shows the programming R-V curve for the 20-nm-thick NW device, starting from both the initially high and low resistance state, with transition voltages at 2.0 V for set and 2.9 V for reset.

The effect of Bi-doping (~3 at.%) on the electrical properties of GeTe NWs has been studied by Zhang et al. [63]. The analysis of the I-V characteristics showed an increase in the set and reset resistances of around two orders of magnitude and the programming currents were decreased with respect to the undoped material; the crystallization time was also reduced by a factor of 20, as mentioned in Section 12.5.

Nukala et al. [131] has demonstrated how defect-templated amorphization can be even exploited for defect engineering to obtain ultralow-power switching in GeTe NWs. The crystalline NWs were He$^+$ ion irradiated, in order to induce extended defects; the induced defects caused carrier localization in the material, enhancing the carrier-lattice coupling, that is, local heating so that the energy provided by current pulses could work distorting the lattice with very small heat losses, until amorphization was induced. In Fig. 12.49, the current density (0.13–0.6 MA cm^{-2}) necessary for the amorphization of the defect-engineered NW-based devices decreases with decreasing device length, and is significantly smaller as compared with the melt-quench process (~50 MA cm^{-2}). Once reaching such a low-power consumption, the undesirable thermal cross talk and possible chemical segregation can be limited.

Noteworthy, it was also demonstrated that, after a few switching cycles, the recrystallized phase keeps the initial resistance value of the defect-engineered crystalline phase and, after 10^{20} switching cycles, the reset state resistance is stable.

The role of defect-assisted amorphization has been also invoked by Hwang et al. [132] to correlate the reset switching current in GeTe and GST NWs with their device contact resistance, for different NW diameters. The device/contact resistance and size-normalized reset currents I_{reset} were correlated to the NW diameter and an inversely proportional dependence between contact resistance and I_{reset} emerged. The conclusion was that the interfaces with high contact resistance contribute to the reset operation via a defect-assisted amorphization process, acting as dislocation generation centers (being lattice vacancy sources), thus opening new device optimization chances based on engineering of the interfaces between the NWs and the metal contact.

Finally, an extremely low reset current in very small GST NWs, 40 nm wide and approximately 10 nm long, was accomplished in a nanogap aligned to CNT electrodes, with a contact area of a few nm^2. A 10-nm passivating layer of SiO$_2$ was also deposited.

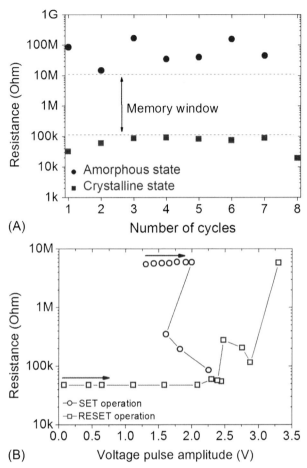

Fig. 12.48 (A) Set-reset cycling test of a 22-nm $In_3Sb_1Te_2$ NW device. The memory window defined by the resistance difference between the amorphous (*circles*) and crystalline (*squares*) states is indicated. (B) Resistance of a 20-nm $In_3Sb_1Te_2$ NW device after writing pulses of 100 ns with increasing voltage amplitudes, obtained for initially high (*circles*) and initially low (*squares*) resistance states.
Reproduced with permission from S. Selmo, R. Cecchini, S. Cecchi, C. Wiemer, M. Fanciulli, E. Rotunno, L. Lazzarini, M. Rigato, D. Pogany, A. Lugstein, M. Longo, Low power phase change memory switching of ultra-thin $In_3Sb_1Te_2$ nanowires, Appl. Phys. Lett. 109 (2016) 213103, copyright (2016), AIP Publishing.

The *I-V* and *R-I* curves in Fig. 12.50 show that the threshold switches between the reset state at 2.5 GΩ resistivity and the set state at 1.3 MΩ.

After the first amorphization, the threshold voltage value fell by 20%–30%, until it was stabilized at lower values of approximately 3.2 V, after 100 switches. The set current was approximately 0.4 μA and the reset current approximately 1.9 μA, which constitutes the smallest values ever reported (Fig. 12.50A). The programming window was approximately 2×10^3 and around 1500 switching cycles were demonstrated (Fig. 12.50B).

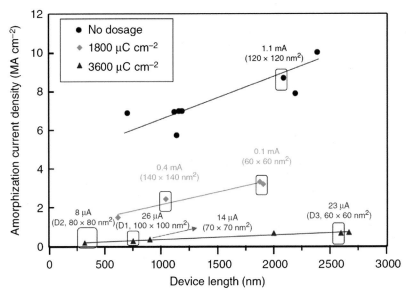

Fig. 12.49 Size effects on switching properties of GeTe NW devices with pre-induced defects by ion irradiation at different dosages. Amorphization current density is plotted as a function of the device length; the switching currents and device cross sections are also indicated. Reproduced from P. Nukala, C.-C. Lin, R. Composto, R. Agarwal, Ultralow-power switching via defect engineering in germanium telluride phase-change memory devices, Nat. Commun. 7 (2016) 10482, which is an open Access article under the Creative Commons Attribution 4.0 International License (http://creativecommons.org/licenses/by/4.0/).

Since early attempts, NW devices have shown a power reduction of one to two orders of magnitude, when compared to conventional thin-film PCMs [48, 55]. Table 12.3 summarizes the main electrical properties of phase change NW-based PCM devices, compared with equivalent devices based on the planar and confined cells. The potentialities of the NWs in terms of the programming window and programming currents can be seen.

12.12   Phase change properties of core-shell NWs

In the case of designing multilevel cell (MLC) storage, which would considerably increase data storage density, double layers of GST and Sb_2Te_3 have been considered [139]. Endurance was measured for up to 10^5 programming cycles. Due to the very close melting temperatures of the two alloys, severe interdiffusion occurs during the reset process. This limits the stability of the middle resistance level. Improved performances were reached by intercalating a TiN layer between Sb_2Te_3 and GST [140]; higher endurance was therefore possible. However, due to the low thermal conductivity of TiN, more heat was dissipated within the cell, and higher reset voltages were required, so that increased power consumption was necessary. Multilevel data storage

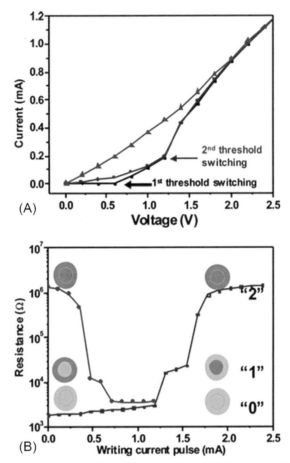

Fig. 12.50 (A) Electrical characteristics of the 1st, 10th, and 100th set switches, showing that the threshold voltage stabilizes at $V_{th} \approx 3.2\,V$. (B) Resistance switching after a series of current pulses with increasing amplitude. Set (reset) pulses have 300 ns (100 ns) width and rising (falling) edges of 50 ns (2 ns). The set (reset) current is $\sim 0.4\,\mu A$ ($\sim 1.9\,\mu A$). The ratio $R_{OFF}/R_{ON} = 2.5\,G\Omega/1.3\,M\Omega$, nearly $\sim 2000\times$.

Adapted with permission from F. Xiong, M.-H. Bae, Y. Dai, A.D. Liao, A. Behnam, E.A. Carrion, S. Hong, D. Ielmini, E. Pop, Self-aligned nanotube-nanowire phase change memory, Nano Lett. 13 (2013) 464–469, copyright (2013), American Chemical Society.

has been also investigated using the In_3SbTe_2 compound, which provided four different resistance levels, enabled by the distinct crystallization of the binary InSb and InTe compounds [20].

Multilevel data storage based on core-shell NWs has been investigated by Jung et al. [67], who used GST (core) and GeTe (shell) structures. Three different resistance levels were resolved in the R-I characteristics, with two slightly different middle resistance states, corresponding to the core chalcogenide material in the amorphous phase and the shell material in the crystalline phase or vice versa, although their endurance data are not available.

Table 12.3 Comparison of the main electrical switch properties in chalcogenide NWs and planar/confined PCM devices

Material and NW diameter (D)	Set resistance (Ω)	Reset resistance (Ω)	Switching cycles	Programming current	Threshold voltage (V)	Reference
GeTe NW, $D=60$ nm	10^4	10^8	10	NA	0.7	Fallica et al. [133]
GeTe NW, $D=100$ nm	2×10^3	$\sim4\times10^6$	NA	~0.5 mA for set, ~1.5 mA for reset[a]	0.8	Lee et al. [55]
GeTe NWs, ion irradiated	$\sim1\times10^4$	$\sim5\times10^6$	2×10^5	26 µA for reset	<1 V	Nukala et al. [131]
GST NW, $D=100$ nm	$\approx10^4$	$\sim2\times10^6$	NA	~0.24 mA for set, ~0.45 mA for reset[a]	1.2	Lee et al. [55]
GST NW, $D=60$ nm	$\approx10^4$	1.8×10^6	$>10^5$	0.25 mA for set	1.8	Lee et al. [54]
Ge$_1$Sb$_2$Te$_4$ NWs, $D=80$ nm	1.4×10^4	4.5×10^5	9	NA	1.35	Longo et al. [51]
Ge$_{0.2}$Sb$_3$Te$_{6.8}$ NWs, 50 nm	1.4×10^4	9.10×10^6	A few	NA	NA	Longo et al. [68]
GST NW, in a nanogap aligned to CNT, $D=40$ nm + SiO$_2$ passivation	$\sim1\times10^6$	$\sim7\times10^9$	$>10^3$	0.1 µA for set, 1.6 µA for reset	3.2	Xiong et al. [76]
In$_2$Se$_3$ NWs, $D=100$–350 nm No NW passivation	6.57×10^4	1.5×10^{10}	>10	NA	NA	Baek et al. [59]
In$_2$Se$_3$ NWs, $D=100$–350 nm + SiO$_2$ passivation	3.25×10^4	8.3×10^9	>20	NA	NA	Baek et al. [59]
Sb$_2$Te$_3$ NWs, $D=99$ nm	$\sim28.9\times10^3$	$\sim28\times10^6$	10	NA	0.75	Lee et al. [48]
Bi$_2$Te$_3$ NW, array	$\sim2.3\times10^3$	$\sim2\times10^7$	≈50	NA	1.2	Han et al. [79]
Bi$_2$Te$_3$ NW, single	0.6×10^3	7.5×10^6	NA	NA	≈0.35	Han et al. [79]
Ge-Sb NWs, $D=70$–100 nm	$\sim1\times10^5$	$\sim4\times10^6$	200	0.24 mA for set	4	Jung et al. [56]
In$_3$Sb$_1$Te$_2$ NWs, $D=22$ nm	$\sim1\times10^5$	$\sim1\times10^7$	8	41 µA for reset	2.25	Selmo et al. [103]
In$_3$Sb$_1$Te$_2$ NWs, $D=70$ nm	$\sim1\times10^8$	5×10^9	NA	NA	1.6	Ahn et al. [72]
11.5% Ga-doped In$_2$O$_3$ NWs, $D=40$ nm	$\sim6\times10^4$	$\sim1\times10^7$	60	NA	0.5 V	Jin et al. [61]

Sb$_2$Te line memory cell	~1×10^4	~5×10^6	10^7	280 μA	NA	Jedema et al. [134]
Doped Ge-Sb bridge cell	~3×10^4	~4×10^5	>10^4	60 μA for set 90 μA for reset	~1	Chen et al. [42]
GST NW by electron beam lithography—width 39 nm	~1×10^4	~1×10^7	NA	2 μA for set	1.81	Ma et al. [41]
GST planar cell	~5×10^3	~1×10^6	10^{12}	~100 μA for set, ~700 μA for reset	0.6	Lacaita and Wouters [2]
7.5-nm dash-type confined PCM cell	Programming window circa 1 order of magnitude		2×10^{10}	160 μA for reset	0.65	Im et al. [135]
Lateral PCM/CNT	~2×10^6	~5×10^7	200	1 μA for set, 5 μA for reset	3.5	Xiong et al. [136]
GeTe cross-point PCM with CNT electrodes	~2×10^6	~2×10^7	>100	1.4 μA for reset	5–13	Caldwell et al. [23]
45-nm Microtrench GST PCM cell (Numonyx)	~1×10^4	2×10^6	>10^8	200 μA for reset	NA	Servalli [137]
Cross spacer	1×10^4	2×10^5	>10^6	230 μA	NA	Chen et al. [138]
Interfacial PCM	~1×10^4	~1.5×10^6	>10^7	~0.2 mA for set, ~0.6 mA for reset	NA	Simpson et al. [27]

NA = not available.
[a] Estimated value.

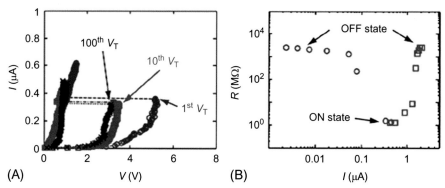

Fig. 12.51 (A) Current-voltage characteristics of a 200-nm Ge$_2$Sb$_2$Te$_5$/GeTe core/shell NW device to compare electrical behavior of the NW starting from fully crystalline *(blue)*, partially amorphized *(red)*, and fully amorphized *(black)* states. The two-step threshold switching (marked by *arrows*) is clearly resolved in the *I-V* sweep of the fully amorphized NW (amorphization current pulse; 2.0 mA, 100 ns). (B) Variation of resistance of the same core-shell NW device as a function of current pulses with varying amplitudes. Pulse durations are 100 ns for amorphization and 300 ns for crystallization. Three different resistive states (low, intermediate, and high) achieved with the application of current pulses are distinguishable. The schematics represent the cross section of the core-shell NW at each stage of transition, light regions represent crystalline phase and dark regions the amorphous one.
Reprinted with permission from Y. Jung, S.-H. Lee, A.T. Jennings, R. Agarwal, Core-shell heterostructured phase change nanowire multistate memory, Nano Lett. 8 (2008) 2056–2062, copyright (2008), American Chemical Society.

Programming could be performed by acting on the crystalline/amorphous phases of the involved materials in order to exploit low programming volumes, high cell density, and defect-free nanostructures. Multilevel data "0," "1," and "2" encoding was achieved by introducing different offsets of phase transitions, in a core-shell/shell-core sequence. This made it possible to obtain at least an intermediate, mixed resistive state, between the high and low resistive states of the core and shell materials (Fig. 12.51A and B).

12.13   Further applications of phase change NWs

Nanoscaled phase change materials in the form of NWs have become of interest for different additional applications. In the field of energy harvesting, the conversion of thermal energy into electrical energy is related to the thermoelectrical effect. The related efficiency is expressed by the figure of merit $ZT = \sigma S^2 T/k$, which depends on the Seebeck coefficient S, the thermal conductivity k, the electrical conductivity σ, and temperature T. It has been shown how thermal conductivity decreases with the NW size, still leaving the NW highly conductive, so an improvement in the thermoelectric effect for phase change NWs is, in principle, expected. Indeed, such expectations were fulfilled by different phase change NWs and related thermoelectric test modules in the case of Bi_2Te_3, Bi-Sb-Te, Sb_2Te_3, and Sb_2Se_3 NWs [122, 141–143]. Phase change NWs are also promising as topological insulator nanomaterials, where applications for spintronics and quantum computing are sought. Several reports are already present for Sb_2Te_3, Bi_2Te_3, and Bi_2Se_3 NWs [144–147].

A special mention goes to the use of a GeTe NW (diameter ~300 nm) embedded in a nonvolatile nanophotonic circuit, to study and demonstrate the reversible switching operations between the amorphous and crystalline states in both the electrical and optical modes; Fig. 12.52 shows the operation principle of the mixed-mode NW device.

Transitions were induced by evanescent coupling between light through a Si_3N_4 waveguide placed orthogonally to the NW and triggered with 50 ns optical pulses;

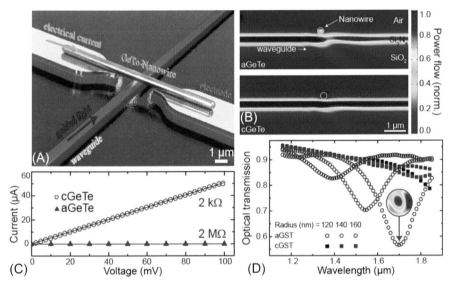

Fig. 12.52 Operation principle of the phase change NW mixed-mode measurement. (A) Sketch of the on-chip mixed-mode device. The GeTe NW is both electrically contacted and evanescently coupled to a photonic waveguide. This enables simultaneous measuring with optical and radio frequency (RF) signals. (B) Simulated interaction between the GeTe NW and the guided optical light for both the amorphous (upper panel) and crystalline (lower panel) phase state. The high refractive index contrast between the two states results in a significant change of transmitted optical power on switching which enables sensitive monitoring of the NW phase state by optical means. (C) Measured current-voltage (*I-V*) characteristics of a GeTe NW. Upon crystallization, the device resistance drops by over three orders of magnitude. (D) Simulated optical transmission spectrum of the GeTe photonic device for different dimensions of the GeTe NW. Resonant Mie scattering is observed which drastically enhances the evanescent interaction at on-resonance wavelengths in the amorphous-phase state. The enhancement of the electromagnetic energy density in the NW (inset) additionally increases the absorbed energy within the NW.
Reproduced with permission from Y. Lu, M. Stegmaier, P. Nukala, M.A. Giambra, S. Ferrari, A. Busacca, W.H.P. Pernice, R. Agarwal, Mixed-mode operation of hybrid phase-change nanophotonic circuits, Nano Lett. 17 (2017) 150–155, copyright (2017), American Chemical Society.

noteworthy, these transitions could be monitored optically (by a fiber-coupled pump-probe light) and electronically (by a radio frequency, RF, setup). On switching, the optical transmission was expected to change between 0.7 (amorphous GeTe) and 0.89 (crystalline GeTe), whereas the GeTe electrical resistance changed by three orders of magnitude (from a few MΩ to 6 kΩ); up to 8 mixed-mode cycling switches were reported [148]. These findings opened new possibilities for high-performance on-chip multifunctional optoelectronic circuits (MOC).

12.14   Conclusion

Phase change NWs are attractive 1D systems for PCM applications, since they offer programmable volume and cross-sectional area reduction; moreover, size effect allow confinement of heat, electrical current, and melting point depression, resulting in a reduction of the reset current, switching time, and power consumption. Further, phase change NWs are unique test devices for the direct analysis of a PCM cell and its failure mechanisms, not to mention alternative applications, from energy harvesting to photonics and spintronics.

Different methods are being explored to obtain phase change NWs, in order to realize highly scaled memory devices. Although some technological obstacles have to be overcome, the NW self-assembly based on CVD is one of the most studied, since it offers the possibility to realize peculiar NW geometries, compositional control, and defect-free nanostructures. In particular, the MOCVD technology should be highlighted, featured by high process control and large area deposition capability, thus allowing the potential for relatively easy industrial transfer.

Given the possibility that even core-shell structures can be implemented for multilevel memory devices, the performances shown in Table 12.4 can reasonably be expected for future phase change NW-based PCM devices.

At the same time, phase change NWs pose important challenges with regard to their technological implementation into PCM devices. Many approaches are being considered, both involving top-down and bottom-up or indeed mixed techniques and also employing CNTs. While lithography-based methods can result in compatibility with CMOS technology, bottom-up methods can lead to denser and defect-free NWs, with further benefit of providing an ideal means to study intrinsic nanostructure properties and size effects. It is likely that a mix of the above approaches will be necessary in future.

Table 12.4   Expected performances for future NW-PCM devices

Property	PCM state-of-the-art	NW-based PCM
Power consumption	~ 20 pJ/bit	5× reduction
Scalability	45 nm	10 nm
Density	~ 40 Gbits/Sq. in.	1 Tbits/sq.in.
Program/access time	50–100 ns	40–80 ns
Stability	10 years at 85°C	10 years at 85°C
Endurance	> 10^8 cycles	> 10^8 cycles

12.15   Sources of further information and advice

12.15.1   Books/publications

Bai, G., Liu, Z., Li, R., Xia, Y. and Yin, J. (2013), 'Phase change behaviour and critical size of $Ge_2Sb_2Te_5$ nanowires and nanotubes', *Physica B,* 411: 68–71.

Raoux, S. and Wuttig, M. eds (2009), *Phase Change Materials – Science and Applications,* Ed. Springer.

Yu, D., Brittman, S., Lee, J.S., Falk, A.L. and Park, H. (2008), 'Minimum voltage for threshold switching in nanoscale phase-change memory', *Nano Lett.,* 8(10): 3429–33.

Andrea Redaelli Ed. "Phase Change Memory", Device Physics, Reliability and Applications, Springer, ISBN 978-3-319-69052-0, 2018.

12.15.2   Research groups

Agarwal Group Nanoscale Phase-Change and Photonics: http://agarwal.seas.upenn.edu/.
CNR-IMM: http://www.imm.cnr.it/.
University of Milano Bicocca: http://www.unimib.it/go/102/Home/English.
Politecnico di Milano: http://www.polimi.it/en/english-version/.
RTWH Aachen: http://www.rwth-aachen.de/.
Paul-Drude-Intitut für Festkörperelektronik: http://www.pdi-berlin.de/.
University of Groningen: https://www.rug.nl/.
University of Bordeaux: https://www.u-bordeaux.com/.

12.15.3   Trade bodies

STMicroelectronics: www.st.com
Micron: www.micron.com
Samsung: www.samsung.com
IBM: www.ibm.com

12.15.4   Web sites

http://www.epcos.org/dcfault.htm
www.eetimes.com/
http://www.chipworks.com/en/

References

[1]   R. Bez, Chalcogenide PCM: a memory technology for next decade, in: IEDM Tech. Dig, 2009, pp. 89–92.
[2]   A.L. Lacaita, D.J. Wouters, Phase-change memories, Phys. Status Solidi A 205 (2008) 2281–2297.

[3] S. Raoux, R.M. Shelby, J. Jordan-Sweet, B. Munoz, M. Salinga, et al., Phase change materials and their application to random access memory technology, Microelectron. Eng. 85 (2008) 2330–2333.

[4] M. Terao, T. Morikawa, T. Ohta, Electrical phase-change memory: fundamentals and state of the art, J. Appl. Phys. 48 (2009) 080001.

[5] W. Welnic, M. Wuttig, Reversible switching in phase-change materials, Mater. Today 11 (2008) 20–27.

[6] S.W. Fong, M. Neuman, Wong H.-S- P., Phase-change memory—towards a storage – class memory, IEEE Trans. Electron Devices 11 (2017) 4374–4384.

[7] F. Bedeschi, R. Fackenthal, C. Resta, E.M. Donze, M. Jagasivamani, E. Buda, F. Pellizzer, D. Chow, A. Cabrini, G.M.A. Calvi, R. Faravelli, A. Fantini, G. Torelli, D. Mills, R. Gastaldi, G. Casagrande, A multi-level-cell bipolar-selected phase-change memory, in: Proc. of ISSCC2008 Tech. Dig., Institute of Electrical and Electronics Engineers (IEEE), 2008, pp. 428.

[8] Z. Song, S. Song, M. Zhu, L. Wu, K. Ren, W. Song, S. Feng, From octahedral structure motif to sub-nanosecond phase transitions in phase change materials for data storage, Sci. China Inf. Sci. 61 (2018) 081302.

[9] G.W. Burr, M.J. Breitwisch, M. Franceschini, D. Garetto, K. Gopalakrishnan, B. Jackson, et al., Phase change memory technology, J. Vac. Sci. Technol. B 28 (2010) 223.

[10] Q. Zheng, Y. Wang, J. Zhu, Nanoscale phase-change materials and devices, J. Phys. D. Appl. Phys. 50 (2017) 243002.

[11] A. Redaelli, D. Ielmini, A.L. Lacaita, A. Pirovano, F. Pellizzer, et al., Impact of crystallization statistics on data retention for phase change memories, in: IEDM Tech. Dig, 2005, pp. 742–745.

[12] C. Peng, L. Wu, F. Rao, Z. Song, X. Zhou, M. Zhe, et al., Nitrogen incorporated GeTe phase change thin film for high-temperature data retention and low-power application, Scr. Mater. 65 (2011) 327–330.

[13] A. Pirovano, A.L. Lacaita, A. Benvenuti, F. Pelizzer, S. Hudgens, Scaling analysis of phase-change memory technology, in: Electron Devices Meeting, IEDM'03 Technical Digest. IEEE International, 2003, pp. 29.6.1–29.6.4.

[14] T. Morikawa, K. Kurotsuchi, M.N. Kinoshita, N. Matsuzaki, Y. Matsui, et al., Doped In-Ge-Te phase change memory featuring stable operation and good data retention, in: Technical Digest—International Electron Devices Meeting, IEDM, 2007, pp. 307–310.

[15] V. Sousa, G. Navarro, N. Castellani, M. Coué, O. Cueto, C. Sabbione, P. Noé, L. Perniola, S. Blonkowski, P. Zuliani, R. Annunziata, Operation fundamentals in 12 Mb phase change memory based on innovative Ge-rich GST materials featuring high reliability performance, in: Symposium on VLSI Technology, Kyoto, 2015, 2015, pp. T98–T99.

[16] Y. Maeda, H. Andoh, I. Ikuta, H. Minemura, Reversible phase-change optical data storage in InSbTe alloy films, J. Appl. Phys. 64 (1988) 1715.

[17] K. Daly-Flynn, D. Strand, InSbTe phase-change materials for high performance multi-level recording, Jpn. J. Appl. Phys. 42 (2003) 795–799.

[18] R. Fallica, T. Stoycheva, C. Wiemer, M. Longo, Structural and electrical analysis of In–Sb–Te-based PCM cells, Phys. Status Solidi RRL 7 (2013) 1009–1013.

[19] E.T. Kim, J.Y. Lee, Y.T. Kim, Investigation of electrical characteristics of the $In_3Sb_1Te_2$ ternary alloy for application in phase-change memory, Phys. Status Solidi RRL 3 (2009) 103–105.

[20] Y.I. Kim, E.T. Kim, J.Y. Lee, Y.T. Kim, Microstructures corresponding to multilevel resistances of $In_3Sb_1Te_2$ phase-change memory, Appl. Phys. Lett. 98 (2011) 091915.

[21] P.D. Szkutnik, M. Aoukar, V. Todorova, L. Angélidès, B. Pelissier, D. Jourde, P. Michallon, C. Vallée, P. Noé, Impact of In doping on GeTe phase-change materials thin films obtained by means of an innovative plasma enhanced metalorganic chemical vapor deposition process, J. Appl. Phys. 121 (2017) 105301.

[22] M.A. Caldwell, S. Raoux, R.Y. Wang, H.-S.P. Wong, D.J. Milliron, Synthesis and size-dependent crystallization of colloidal germanium telluride nanoparticles, J. Mater. Chem. 20 (2010) 1285–1291.

[23] M.A. Caldwell, R.G.D. Jeyasingh, H.-S.P. Wong, D.J. Milliron, Nanoscale phase change memory materials, Nanoscale 4 (2012) 4382–4392.

[24] Y.H. Ha, J.H. Yi, H. Horii, J.H. Park, S.H. Joo, et al., An edge contact type cell for phase change RAM featuring very low power consumption, in: VLSI Symp. Tech. Dig, 2003, pp. 175–176.

[25] A. Pirovano, F. Pellizzer, I. Tortorelli, A. Riganó, R. Harrigan, et al., Phase-change memory technology with self-aligned trench cell architecture for 90 nm node and beyond, Solid State Electron. 52 (2008) 1467–1472.

[26] Y. Sasago, M. Kinoshita, T. Morikawa, K. Kurotsuchi, S. Hanzawa, et al., Cross-point phase change memory with $4F^2$ cell size driven by low-contact-resistivity poly-Si diode, in: Symposium on VLSI Technology Digest of Technical Papers, 2009, pp. 24–25.

[27] R.E. Simpson, P. Fons, A.V. Kolobov, T. Fukaya, M. Krbal, et al., Interfacial phase-change memory, Nat. Nanotechnol. 6 (2011) 501–505.

[28] B.J. Choi, S.H. Oh, S. Choi, T. Eom, Y.C. Shin, et al., Switching power reduction in phase change memory cell using CVD $Ge_2Sb_2Te_5$ and ultrathin TiO_2 films, J. Electrochem. Soc. 156 (2009) H59–H63.

[29] J.I. Lee, H. Park, S.L. Cho, Y.L. Park, B.J. Bae, et al., Highly scalable phase change memory with CVD GeSbTe for sub 50nm generation, in: Symposium on VLSI Technology Digest of Technical Papers, 12–14 June, 2007, pp. 102–103.

[30] H. Lv, Y. Lin, P. Zhou, T.B. Tang, B. Qiao, et al., A nano-scale-sized 3D element for phase change memories, Semicond. Sci. Technol. 21 (2006) 1013–1017.

[31] Y. Jung, R. Agarwal, C.-Y. Yang, R. Agarwal, Chalcogenide phase-change memory nanotubes for lower writing current operation, Nanotechnology 22 (2011) 254012.

[32] F. Xiong, A. Liao, E. Pop, Inducing chalcogenide phase change with ultra-narrow carbon nanotube heaters, Appl. Phys. Lett. 95 (2009) 243103.

[33] B.-J. Bae, S.-H. Hong, S.-Y. Hwang, J.-Y. Hwang, K.-Y. Yang, H. Lee, Electrical characterization of Ge–Sb–Te phase change nano-pillars using conductive atomic force microscopy, Semicond. Sci. Technol. 24 (2009) 075016.

[34] C.H. Chu, M.L. Tseng, C.D. Shiue, S.W. Chen, H.-P. Chiang, M. Mansuripur, D.P. Tsai, Fabrication of phase-change $Ge_2Sb_2Te_5$ nano-rings, Opt. Express 19 (2011) 12652–12657.

[35] Y. Zhang, S. Raoux, D. Krebs, L.E. Krupp, T. Topuria, et al., Phase change nanodots patterning using a self-assembled polymer lithography and crystallization analysis, J. Appl. Phys. 104 (2008) 074312.

[36] L. Liu, Z.-T. Song, S.L. Feng, B. Chen, Reversible phase change for C-RAM nano-cell-element fabricated by focused ion beam method, Chin. Phys. Lett. 22 (2005) 758.

[37] D.J. Milliron, M.A. Caldwell, H.-S.P. Wong, Synthesis of metal chalcogenide nanodot arrays using block copolymer-derived nanoreactors, Nano Lett. 7 (2007) 3504–3507.

[38] H. Hardtdegen, M. Mikulics, S. Rieß, M. Schuck, T. Saltzmann, U. Simon, M. Longo, Modern chemical synthesis methods towards low-dimensional phase change structures in the Ge–Sb–Te material system, Prog. Cryst. Growth Charact. Mater. 61 (2015) 27–45.

[39] Y.-S. Huang, C.-H. Hang, Y.-J. Huang, T.-E. Hsieh, Electromigration behaviors of Ge$_2$Sb$_2$Te$_5$ chalcogenide thin films under DC bias, J. Alloys Compd. 580 (2013) 449–456.

[40] R. Huang, K. Sun, K.S. Kiang, R. Chen, Y. Wang, B. Ggolipour, D.W. Hewak, C.H. De Groot, Contact resistance measurement of Ge$_2$Sb$_2$Te$_5$ phase change material to TiN electrode by spacer etched nanowire, Semicond. Sci. Technol. 29 (2014) 095003.

[41] H. Ma, X. Wang, J. Zhang, X. Wang, C. Hu, X. Yang, Y. Fu, X. Chen, Z. Song, S. Feng, A. Ji, F. Yang, A self-aligned process for phase-change material nanowire confined within metal electrode nanogap, Appl. Phys. Lett. 99 (2011) 173107.

[42] Y.C. Chen, C.T. Rettner, S. Raoux, et al., Ultra-thin phase-change bridge memory device using GeSb, in: International Electron Devices Meeting, IEDM 2006, 2006, pp. 1–4.

[43] B. Gholipour, C.-C. Huang, J.-Y. Ou, D.W. Hewak, Germanium antimony lateral nanowire phase change memory by chemical vapor deposition, Phys. Status Solidi B 250 (2013) 994–998.

[44] R.-S. Wagner, W.C. Ellis, Vapor-liquid-solid mechanism of single crystal growth, Appl. Phys. Lett. 4 (1964) 89.

[45] S. Noor Mohammad, Why droplet dimension can be larger than, equal to, or smaller than the nanowire dimension, J. Appl. Phys. 106 (2009) 104311.

[46] K.W. Schwartz, J. Tersoff, From droplets to nanowires: dynamics of vapor-liquid-solid growth, Phys. Rev. Lett. 102 (2009) 206101.

[47] S. Meister, H. Peng, K. McIlwrath, K. Jarausch, X. Feng Zhang, et al., Synthesis and characterization of phase-change nanowires, Nano Lett. 6 (2006) 1514–1517.

[48] J.S. Lee, S. Brittman, D. Yu, H. Park, Vapor–liquid–solid and vapor–solid growth of phase-change Sb$_2$Te$_3$ nanowires and Sb$_2$Te$_3$/GeTe nanowire heterostructures, J. Am. Chem. Soc. 130 (2008) 6252–6258.

[49] S.W. Jung, S.-M. Yoon, Y.-S. Park, S.-Y. Lee, B.-G. Yu, Control of the thickness and the length of germanium-telluride nanowires fabricated via the vapor-liquid-solid method, J. Korean Phys. Soc. 54 (2009) 653–659.

[50] U. Philipose, G. Sapkota, J. Salfi, H.E. Ruda, Influence of growth temperature on the stoichiometry of InSb nanowires grown by vapor phase transport, Semicond. Sci. Technol. 25 (2010) 075004.

[51] M. Longo, R. Fallica, C. Wiemer, O. Salicio, M. Fanciulli, et al., Metal organic chemical vapor deposition of phase change Ge$_1$Sb$_2$Te$_4$ nanowires, Nano Lett. 12 (2012) 1509–1515.

[52] Y. Jung, S.-H. Lee, D.-K. Ko, R. Agarwal, Synthesis and characterization of Ge$_2$Sb$_2$Te$_5$ nanowires with memory switching effect, J. Am. Chem. Soc. 128 (2006) 14026–14027.

[53] Y. Jung, S.-W. Nam, R. Agarwal, High-resolution transmission electron microscopy study of electrically-driven reversible phase change in Ge$_2$Sb$_2$Te$_5$ nanowires, Nano Lett. 11 (2011) 1364–1368.

[54] S.-H. Lee, Y. Jung, R. Agarwal, Highly scalable non-volatile and ultra-low-power phase-change nanowire memory, Nat. Nanotechnol. 2 (2007) 626.

[55] S.-H. Lee, Y. Jung, H.-S. Chung, A.T. Jennings, R. Agarwal, Comparative study of memory-switching phenomena in phase change GeTe and Ge$_2$Sb$_2$Te$_5$ nanowire devices, Phys. E. 40 (2008) 2474–2480.

[56] Y. Jung, C.-Y. Yang, S.-H. Lee, R. Agarwal, Phase-change Ge-Sb nanowires: synthesis, memory switching, and phase-instability, Nano Lett. 9 (2009) 2103–2108.

[57] X. Sun, B. Yu, G. Ng, M. Meyyappan, S. Ju, D.B. Janes, Germanium antimonide phase-change nanowires for memory applications, IEEE Trans. Electron Devices 55 (2008) 3131–3135.

[58] A. Jennings, Y. Jung, J. Engel, R. Agarwal, Diameter-controlled synthesis of phase-change germanium telluride nanowires via the vapor–liquid–solid mechanism, J. Phys. Chem. C 113 (2009) 6898–6901.

[59] C.-K. Baek, D. Kang, J.S. Kim, B. Jin, T. Rim, S. Park, M. Meyyappan, Y.-H. Jeong, J.-S. Lee, Improved performance of In$_2$Se$_3$ nanowire phase-change memory with SiO$_2$ passivation, Solid State Electron. 80 (2013) 10–13.

[60] S. Yu, X. Ju, G. Sun, T.D. Ng, M. Nguyen, et al., Indium selenide nanowire phase-change memory, Appl. Phys. Lett. 91 (2007) 133119.

[61] B. Jin, T. Lim, S. Ju, M.I. Latypov, H.S. Kim, M. Meyyappan, J.-S. Lee, Ga-doped indium oxide nanowire phase change random access memory cells, Nanotechnology 25 (2014) 055205.

[62] C.S. Jung, H.S. Kim, H.S. Im, Y.S. Seo, K. Park, S.H. Back, Y.J. Cho, C.H. Kim, J. Park, J.-P. Ahn, Polymorphism of GeSbTe superlattice nanowires, Nano Lett. 13 (2013) 543.

[63] J. Zhang, R. Huang, L. Shi, L. Wang, F. Wei, T. Kong, G. Cheng, Bi doping modulating structure and phase-change properties of GeTe nanowires, Appl. Phys. Lett. 102 (2013) 063104.

[64] C.S. Jung, H.S. Kim, H.S. Im, K. Park, J. Park, J.-P. Ahn, S.J. Yoo, J.-G. Kim, J.N. Kim, J.H. Shim, In situ temperature-dependent transmission electron microscopy studies of pseudobinary mGeTe·Bi$_2$Te$_3$ (m = 3–8) nanowires and first-principles calculations, Nano Lett. 15 (2015) 3923–3930.

[65] M.J. Rozenberg, I.H. Inoue, M.J. Sanchez, Nonvolatile memory with multilevel switching: a basic model, Phys. Rev. Lett. 92 (2004) 178302.

[66] A. Gyanathan, Y.-C. Yeo, Multi-level phase change memory devices with Ge$_2$Sb$_2$Te$_5$ layers separated by a thermal insulating Ta$_2$O$_5$ barrier layer, J. Appl. Phys. 110 (2011) 124517.

[67] Y. Jung, S.-H. Lee, A.T. Jennings, R. Agarwal, Core-shell heterostructured phase change nanowire multistate memory, Nano Lett. 8 (2008) 2056–2062.

[68] M. Longo, T. Stoycheva, R. Fallica, C. Wiemer, L. Lazzarini, E. Rotunno, Au-catalyzed synthesis and characterisation of phase change Ge-doped Sb–Te nanowires by MOCVD, J. Cryst. Growth 370 (2013) 323.

[69] M. Longo, C. Wiemer, O. Salicio, M. Fanciulli, L. Lazzarini, et al., Au-catalyzed self assembly of GeTe nanowires by MOCVD, J. Cryst. Growth 315 (2011) 152–156.

[70] S. Selmo, S. Cecchi, R. Cecchini, C. Wiemer, M. Fanciulli, E. Rotunno, L. Lazzarini, M. Longo, MOCVD growth and structural characterization of In-Sb-Te nanowires, Phys. Status Solidi A 213 (2016) 335–338.

[71] R.B. Yang, R.B. Yang, J. Bachmann, E. Pippel, A. Berger, J. Woltersdorf, U. Gösele, K. Nielsch, Pulsed vapor-liquid-solid growth of antimony selenide and antimony sulfide nanowires, Adv. Mater. 21 (2009) 3170–3174.

[72] J.K. Ahn, K.W. Park, H.-J. Jung, S.-G. Yoon, Phase-change InSbTe nanowires grown in situ at low temperature by metal-organic chemical vapor deposition, Nano Lett. 10 (2010) 472–477.

[73] M.-H. Lee, T.G. Kim, B.-K. Ju, Y.-M. Sung, Bismuth seed-assisted liquid-P phase synthesis of germanium telluride nanowires, Cryst. Growth Des. 9 (2009) 938–941.

[74] R.J. Metha, R.J. Mehta, C. Karthik, W. Jiang, B. Singh, Y. Shi, R.W. Siegel, T. Borca-Tasciuc, G. Ramanath, High electrical conductivity antimony selenide nanocrystals and assemblies, Nano Lett. 10 (2010) 4417–4422.

[75] R. Cecchini, S. Selmo, C. Wiemer, R. Rotunno, L. Lazzarini, M. De Luca, I. Zardo, M. Longo, Single-step Au-catalysed synthesis and microstructural characterization of core–shell Ge/In–Te nanowires by MOCVD, Mater. Res. Lett. 6 (2018) 29–35.

[76] F. Xiong, M.-H. Bae, Y. Dai, A.D. Liao, A. Behnam, E.A. Carrion, S. Hong, D. Ielmini, E. Pop, Self-aligned nanotube-nanowire phase change memory, Nano Lett. 13 (2013) 464–469.

[77] Y. Fu, X. Wang, J. Zhang, X. Wang, C. Chang, H. Ma, K. Cheng, X. Chen, Z. Song, S. Feng, A. Ji, F. Yang, A lithography-independent and fully confined fabrication process of phase-change materials in metal electrode nanogap with 16-μA threshold current and 80-mV SET voltage, Appl. Phys. A Mater. Sci. Process. 110 (2013) 173–177.

[78] M.S. Sander, R. Gronsky, T. Sands, A.M. Stacy, Structure of bismuth telluride nanowire arrays fabricated by electrodeposition into porous anodic alumina templates, Chem. Mater. 15 (2003) 335–339.

[79] N. Han, S.I. Kim, J.D. Yang, K. Lee, H. Sohn, H.-M. So, C.W. Ahn, K.-H. Yoo, Phase-change memory in Bi_2Te_3 nanowires, Adv. Mater. 23 (2011) 1871–1875.

[80] C. Ihalawela, R. Cook, X. Lin, H. Wang, G. Chen, Sb-Te phase-change nanowires by templated electrodeposition. MRS Proc. 1431 (2012) https://doi.org/10.1557/opl.2012.1105. Mrss12-1431-f05-03.

[81] J.H. Jeong, S.J. Park, S.B. An, D.J. Choi, Vapor-solid growth of Te-rich SbTe nanowires on a template of nano-size trenches, J. Cryst. Growth 410 (2015) 47–52.

[82] N. Kato, I. Konomi, Y. Seno, T. Motohiro, In situ X-ray diffraction study of crystallization process of GeSbTe thin films during heat treatment, Appl. Surf. Sci. 244 (2005) 281–284.

[83] N. Yamada, E. Ono, K. Nishiuki, N. Akahira, Rapid-phase transitions of $GeTe$-Sb_2Te_3 pseudobinary amorphous thin films for an optical disk memory, J. Appl. Phys. 69 (1991) 2849.

[84] V. Damodara Das, N. Soundararajan, M. Pattabi, Electrical conductivity and thermoelectric power of amorphous Sb_2Te_3 thin films and amorphous-crystalline transition, J. Mater. Sci. 22 (1987) 3522–3528.

[85] X. Sun, B. Yu, G. Ng, M. Meyyappan, One-dimensional phase-change nanostructure: germanium telluride nanowire, J. Phys. Chem. C 111 (2007) 2421–2425.

[86] S.-H. Lee, D.-K. Ko, Y. Jung, R. Agarwal, Size-dependent phase transition memory switching behavior and low writing currents in GeTe nanowires, Appl. Phys. Lett. 89 (2006) 223116.

[87] S. Meister, D.T. Schoen, M.A. Topinka, A.M. Minor, Y. Cui, Void formation induced electrical switching in phase-change nanowires, Nano Lett. 8 (2008) 4562–4567.

[88] D. Yu, J. Wu, Q. Gu, H. Park, Germanium telluride nanowires and nanohelices with memory-switching behavior, J. Am. Chem. Soc. 128 (2006) 8148–8149.

[89] E. Rotunno, M. Longo, L. Lazzarini, Ageing of GeTe nanowires, Microscopie 26 (2016) 58–63.

[90] E. Rotunno, L. Lazzarini, M. Longo, V. Grillo, Crystal structure assessment of Ge–Sb–Te phase change nanowires, Nanoscale 5 (2013) 1557–1563.

[91] I. Petrov, R.M. Imamov, Z.G. Pinsker, Electron-diffraction determination of the structures of $Ge_2Sb_2Te_5$ and $GeSb_4Te_7$, Sov. Phys. Crystallogr. 13 (1968) 339–342.

[92] B.J. Kooi, J.T. De Hosson, Electron diffraction and high-resolution transmission electron microscopy of the high temperature crystal structures of $Ge_xSb_2Te_{3+x}$ ($x = 1, 2, 3$) phase change material, J. Appl. Phys. 92 (2002) 3584.

[93] T. Matsunaga, M. Yamada, Y. Kubota, Structures of stable and metastable $Ge_2Sb_2Te_5$, an intermetallic compound in $GeTe$–Sb_2Te_3 pseudobinary systems, Acta Crystallogr. Sect. B: Struct. Sci. 60 (2004) 685–691.

[94] T. Matsunaga, Y. Yamada, Structural investigation of $GeSb_2Te_4$: a high-speed phase-change material, Phys. Rev. B: Condens. Matter Mater. Phys. 69 (2004) 104111.

[95] E. Rotunno, M. Longo, C. Wiemer, R. Fallica, D. Campi, M. Bernasconi, A.R. Lupini, S.J. Pennycook, L. Lazzarini, A novel Sb$_2$Te$_3$ polymorph stable at the nanoscale, Chem. Mater. 27 (2015) 4368–4373.

[96] C. Karthik, R.J. Mehta, W. Jiang, E. Castillo, T. Borca-Tasciuc, G. Ramanath, Threshold conductivity switching in sulfurized antimony selenide nanowires, Appl. Phys. Lett. 99 (2011) 103101.

[97] Y.-T. Huang, C.-W. Huang, J.-Y. Chen, Y.-H. Ting, K.-C. Lu, Y.-L. Chueh, W.-W. Wu, Dynamic observation of phase transformation behaviors in indium(III) selenide nanowire based phase change memory, ACS Nano 8 (2014) 9457–9462.

[98] X. Sun, B. Yu, G. Ng, T.D. Nguyen, M. Meyyappan, III–VI compound semiconductor indium selenide (In$_2$Se$_3$) nanowires: synthesis and characterization, Appl. Phys. Lett. 89 (2006) 233121.

[99] M. Li, Size-dependent nucleation rate of Ge$_2$Sb$_2$Te$_5$ nanowires in the amorphous phase and crystallization activation energy, Mater. Lett. 76 (2012) 138.

[100] B. Jin, D. Kang, J. Kim, M. Meyyappan, J.-S. Lee, Thermally efficient and highly scalable In$_2$Se$_3$ nanowire phase change memory, J. Appl. Phys. 113 (2013) 164303.

[101] Y. Wu, P. Yang, Melting and welding semiconductor nanowires in nanotubes, Adv. Mater. 13 (2001) 520–523.

[102] L. van Pieterson, M. van Schijndel, J.C.N. Rijpers, M. Kaiser, Te-free, Sb-based phase-change materials for high-speed rewritable optical recording, Appl. Phys. Lett. 83 (2003) 1373.

[103] S. Selmo, R. Cecchini, S. Cecchi, C. Wiemer, M. Fanciulli, E. Rotunno, L. Lazzarini, M. Rigato, D. Pogany, A. Lugstein, M. Longo, Low power phase change memory switching of ultra-thin In$_3$Sb$_1$Te$_2$ nanowires, Appl. Phys. Lett. 109 (2016) 213103.

[104] J.W.L. Yim, B. Xiang, J. Wu, Sublimation of GeTe nanowires and evidence of its size effect studied by in situ TEM, J. Am. Chem. Soc. 131 (2009) 14526–14530.

[105] V.A. Khonic, K. Kitagawa, H. Morii, On the determination of the crystallization activation energy of metallic glasses, J. Appl. Phys. 87 (2000) 8440.

[106] R.E. Simpson, M. Krbal, P. Fons, A.V. Kolobov, J. Tominaga, T. Uruga, H. Tanida, Toward the ultimate limit of phase change in Ge$_2$Sb$_2$Te$_5$, Nano Lett. 10 (2010) 414–419.

[107] G. Bruns, P. Merkelbach, C. Schlockermann, M. Salinga, M. Wuttig, T.D. Happ, J.B. Philipp, M. Kund, Nanosecond switching in GeTe phase change memory cells, Appl. Phys. Lett. 95 (2009) 043108.

[108] S. Gabardi, E. Baldi, E. Bosoni, D. Campi, S. Caravati, G.C. Sosso, J. Behler, M. Bernasconi, Atomistic simulations of the crystallization and aging of GeTe nanowires, J. Phys. Chem. C 121 (2017) 23827–23838.

[109] I.V. Karpov, M. Mitra, D. Kau, G. Spadini, Fundamental drift of parameters in chalcogenide phase change memory, J. Appl. Phys. 102 (2007) 124503.

[110] M. Boniardi, D. Ielmini, Physical origin of the resistance drift exponent in amorphous phase change materials, Appl. Phys. Lett. 98 (2011) 243506.

[111] M. Mitra, Y. Jung, D.S. Gianola, R. Agarwal, Extremely low drift of resistance and threshold voltage in amorphous phase change nanowire devices, Appl. Phys. Lett. 96 (2010) 22111.

[112] B. Jin, J. Kim, D. Kang, M. Meyyappan, J.-S. Lee, Size-dependent characteristics of highly-scalable In$_2$Se$_3$ nanowire phase-change random access memory, in: Proceedings of the 13th IEEE International Conference on Nanotechnology, Beijing, China, August 5–8, 2013, 2013, pp. 849–851.

[113] S.-M. Yoon, S.-W. Jung, S.-Y. Lee, Y.-S. Park, B.-G. Yu, Fabrication and electrical characterization of phase-change memory devices with nanoscale self-heating-channel structures, Microelectron. Eng. 85 (2008) 2334–2337.

[114]   S.-W. Nam, H.-S. Chung, Y.C. Lo, L. Qi, J. Li, Y. Lu, A.T.C. Johnson, Y. Jung, P. Nukala, R. Agarwal, Electrical wind force-driven and dislocation-templated amorphization in phase-change nanowires, Science 336 (2012) 1561–1566.
[115]   X.-Y. Ji, X.Q. Feng, Dislocation-templated amorphization of $Ge_2Sb_2Te_5$ nanowires under electric pulses: a theoretical model, J. Appl. Phys. 113 (2013) 243507.
[116]   P. Nukala, R. Agarwal, X. Qian, M.H. Jang, D. Dhara, K. Kumar, A.T.C. Johnson, J. Li, R. Agarwal, Direct observation of metal−insulator transition in single-crystalline germanium telluride nanowire memory devices prior to amorphization, Nano Lett. 14 (2014) 2201–2209.
[117]   E. Mafi, X. Tao, W. Zhu, Y. Gao, C. Wang, Y. Gu, Generation and the role of dislocations in single-crystalline phase-change In_2Se_3 nanowires under electrical pulses, Nanotechnology 27 (2016) 335704.
[118]   J.-Y. Lee, J.-H. Kim, D.-J. Jeon, J. Han, J.-S. Yeo, Atomic migration induced crystal structure transformation and core-centered phase transition in single crystal $Ge_2Sb_2Te_5$ nanowires, Nano Lett. 16 (2016) 6078–6085.
[119]   S.G. Volz, G. Chen, Molecular dynamics simulation of thermal conductivity of silicon nanowires, Appl. Phys. Lett. 75 (1999) 2056.
[120]   B. Piccione, R. Agarwal, Y. Jung, R. Agarwal, Size-dependent chemical transformation, structural phase change, and optical properties of nanowires, Philos. Mag. 93 (2013) 2089–2121.
[121]   I.-R. Chen, E. Pop, Compact thermal model for vertical nanowire phase-change memory cells, IEEE Trans. Electron Devices 56 (2009) 1523.
[122]   Y.M. Zuev, J.S. Lee, C. Galloy, H. Park, P. Kim, Diameter dependence of the transport properties of antimony telluride nanowires, Nano Lett. 10 (2010) 3037–3040.
[123]   J.P. Reifenberg, D.L. Kencke, K.E. Goodson, The impact of thermal boundary resistance in phase-change memory devices, IEEE Electron Device Lett. 29 (2008) 1112–1114.
[124]   M.M. Rojo, C.O. Caballero, A.F. Lopeandia, J. Rodriguez-Viejo, M. Martìn-Gonzalez, Review on measurement techniques of transport properties of nanowires, Nanoscale 5 (2013) 11526–11544.
[125]   A. Saci, J.-L. Battaglia, A. Kusiak, R. Fallica, M. Longo, Thermal conductivity measurement of a Sb_2Te_3 phase change nanowire, Appl. Phys. Lett. 104 (2014) 263103.
[126]   J.-L. Battaglia, A. Saci, I. De, R. Cecchini, S. Selmo, M. Fanciulli, S. Cecchi, M. Longo, Thermal resistance measurement of In_3SbTe_2 nanowires, Phys. Status Solidi A 214 (2017) 16000500.
[127]   S. Park, D. Park, K. Jeong, T. Kim, S. Park, M. Ahn, W.J. Yang, J.H. Han, H.S. Jeong, S.G. Jeon, J.Y. Song, M.-H. Cho, Effect of the thermal conductivity on resistive switching in GeTe and $Ge_2Sb_2Te_5$ nanowires, ACS Appl. Mater. Interfaces 7 (2015) 21819–21827.
[128]   A.L. Lacaita, Phase change memories: state-of-the-art, challenges and perspectives, Solid State Electron. 50 (2006) 24–31.
[129]   J. Liu, B. Yu, M.P. Anantram, Scaling analysis of nanowire phase-change memory, IEEE Electron Device Lett. 32 (2011) 1340.
[130]   A. Pirovano, A.L. Lacaita, A. Benvenuti, F. Pellizzer, R. Bez, Electronic switching in phase-change memories, IEEE Trans. Electron Devices 51 (2004) 452–459.
[131]   P. Nukala, C.-C. Lin, R. Composto, R. Agarwal, Ultralow-power switching via defect engineering in germanium telluride phase-change memory devices, Nat. Commun. 7 (2016) 10482.
[132]   I. Hwang, Y.-J. Cho, M.-J. Lee, M.-H. Jo, The role of contact resistance in GeTe and $Ge_2Sb_2Te_5$ nanowire phase change memory reset switching current, Appl. Phys. Lett. 106 (2015) 193106.

[133] R. Fallica, M. Longo, C. Wiemer, O. Salicio, M. Fanciulli, Electrical characterization of MOCVD-grown chalcogenide nanowires for phase change memory applications, in: MRS2011, San Francisco, CA, 24–28 April (oral presentation R5.8), 2011.

[134] F. Jedema, M. 't Zandt, R. Wolters, D. Castro, G. Hurkx, R. Delhougne, Scaling properties of doped Sb_2Te phase change line cells, in: Non-Volatile Semiconductor Memory Workshop, 2008 and 2008 International Conference on Memory Technology and Design. NVSMW/ICMTD 2008, 2008, pp. 43–45.

[135] D.H. Im, J.I. Lee, S.L. Cho, H.G. An, D.H. Kim, I.S. Kim, H. Park, D.H. Ahn, H. Horii, S.O. Park, U.I. Chung, J.T. Moon, A unified 7.5 nm dash-type confined cell for high performance PRAM device. in: 2008 IEEE International Electron Devices Meeting, San Francisco, CA, 2008, 2008, pp. 1–4, https://doi.org/10.1109/IEDM.2008.4796654.

[136] F. Xiong, A.D. Liao, D. Estrada, E. Pop, Low-power switching of phase-change materials with carbon nanotube electrodes, Science 332 (2011) 568.

[137] G. Servalli, A 45nm generation phase change memory technology, in: Electron Devices Meeting, IEDM 2009, IEEE International, 2009, pp. 5.7.1–5.7.4.

[138] W.S. Chen, C.M. Lee, D.S. Chao, Y.C. Chen, et al., A novel cross-spacer phase change memory with ultra-small lithography independent contact area, in: Electron Devices Meeting, IEDM 2007, IEEE International, 2007, pp. 319–322.

[139] F. Rao, Z. Song, L. Wu, B. Liu, S. Feng, Investigation on the stabilization of the median resistance state for phase change memory cell with doublelayer chalcogenide films, Appl. Phys. Lett. 91 (2007) 123511.

[140] F. Rao, Z. Song, L. Wu, Y. Gong, S. Feng, B. Chen, Phase change memory cell based on $Sb_2Te_3/TiN/Ge_2Sb_2Te_5$ sandwich-structure, Solid State Electron. 53 (2009) 276–278.

[141] S. Bäßler, T. Böhnert, J. Gooth, C. Schumacher, E. Pippel, K. Nielsch, Thermoelectric power factor of ternary single-crystalline Sb_2Te_3- and Bi_2Te_3-based nanowires, Nanotechnology 24 (2013) 495402.

[142] T.-Y. Ko, M. Shellaiah, K.W. Sun, Thermal and thermoelectric transport in highly resistive single Sb_2Se_3 nanowires and nanowire bundles, Sci. Rep. 6 (2016) 35086.

[143] I.K. Ng, K.Y. Koka, C.Z. Che Abd Rahman, T.F. Chooa, N.U. Saidina, Bismuth telluride based nanowires for thermoelectric power generation, Mater. Today Proc. 3 (2016) 533–537.

[144] Y.C. Arango, L. Huang, C. Chen, J. Avila, M.C. Asensio, D. Grützmacher, H. Lüth, J.C. Lu, T. Schäpers, Quantum transport and nano angle-resolved photoemission spectroscopy on the topological surface states of single Sb_2Te_3 nanowires, Sci. Rep. 6 (2016) 29493.

[145] B. Hamdou, J. Gooth, A. Dorn, E. Pippel, K. Nielsch, Aharonov-Bohm oscillations and weak antilocalization in topological insulator Sb_2Te_3 nanowires, Appl. Phys. Lett. 102 (2013) 223110.

[146] H.-S. Kim, H.S. Shin, J.S. Lee, C.W. Ahn, J.Y. Song, Y.-J. Doh, Quantum electrical transport properties of topological insulator Bi_2Te_3 nanowires, Curr. Appl. Phys. 16 (2016) 51–56.

[147] D. Kong, J.C. Randel, H. Peng, J.J. Cha, S. Meister, K. Lai, Y. Chen, Z.-X. Shen, H.C. Manoharan, Y. Cui, Topological insulator nanowires and nanoribbons, Nano Lett. 10 (2010) 329–333.

[148] Y. Lu, M. Stegmaier, P. Nukala, M.A. Giambra, S. Ferrari, A. Busacca, W.H.P. Pernice, R. Agarwal, Mixed-mode operation of hybrid phase-change nanophotonic circuits, Nano Lett. 17 (2017) 150–155.

Further Reading

[149] M. Wuttig, N. Yamada, Phase-change materials for rewriteable data storage, Nat. Mater. 6 (2007) 824–832.

[150] M.H.R. Lankhorst, B.W.S.M.M. Ketelaars, R.A.M. Wolters, Low-cost and nanoscale non-volatile memory concept for future silicon chips, Nat. Mater. 4 (2005) 347.

[151] D. Krebs, D.S. Raoux, C.T. Rettner, R.M. Shelby, G.W. Burr, M. Wuttig, Set characteristics of phase change bridge devices, Proc. Mater. Res. Soc. Symp. 1072 (2008) G06–G07.

[152] B.G. Kim, B.-S. Kim, S.-M. Jeong, S.-M. Choi, D. Whang, H.-L. Lee, Catalyst-free growth of Sb_2Te_3 nanowires, Mater. Lett. 65 (2011) 812–814.

[153] L. Li, Y. Yang, X. Huang, G. Li, L. Zhang, Pulsed electrodeposition of single-crystalline Bi_2Te_3 nanowire arrays, Nanotechnology 17 (2006) 1706–1712.

Flexible and transparent ReRAM devices for system on panel (SOP) application

Asal Kiazadeh, Jonas Deuermeier
CENIMAT/i3N, Department of Materials Science, Faculty of Sciences and Technology, Universidade NOVA de Lisboa and CEMOP/UNINOVA, Campus de Caparica, Caparica, Portugal

13.1   Introduction

The system on panel (SOP) or system on glass approach has been proposed to integrate all the components, such as sensors, memory, drivers, and controller of an electronic device on a single display panel leading to a transparent ultrathin device for cell phones, automobile, and other applications. The concept results in high performance and low power consumption and is very cost effective for the manufacturers due to the ease of the fabrication process.

To achieve the SOP technology integration, the electronic components should share similar materials. In the last decade, display technology has been developed rapidly toward applying various transparent oxide materials in oxide thin film transistors (TFTs): indium-gallium-zinc oxide (IGZO), zinc-tin oxide (ZTO), ZnO [1–5]. In parallel, new designs of flash memory using oxide materials was attempted by different research groups [6–10]. Transparent charge trapping memory devices suffer from issues, such as difficult erasing process and fast retention loss. The most efficient erasing process occurs only under light exposure [10]. Furthermore, typical flash memory cannot be further scaled down due to the intrinsic physical limits. They need comparatively large structures for a good performance [6]. On the other hand, IGZO engineered charge trapping memories (based on metal-oxide-nitride-oxide-semiconductor MONOS) [7] face the issues, such as low endurance of the nitride layer and certainly, to guarantee an ever-increasing memory density and ever-decreasing memory cost, a new technology is required. Resistive switching random access memories (ReRAM) are new candidates, owing to a simple structure of metal/resistive switching material/metal. They combine the best features of current memories, such as the fast speed of SRAM (few ns) with the high density of DRAM and the non-volatile properties of flash. Moreover, the ReRAM has some intrinsic characteristics like a gradual and continuous switching to different resistance states (analog mode) or abrupt switching behavior showing logical states which are encoded as different resistances (digital mode) [11].

Discovery of the resistive switching properties of metal oxides dates back to the 1960s [12–14]. One of the first reviews on resistive switching behavior including the concept of

Advances in Non-volatile Memory and Storage Technology. https://doi.org/10.1016/B978-0-08-102584-0.00014-0

memory characteristics on amorphous metal oxide was published in 1970 [15]. After the invention of the first physical memristor device based on TiO_2 by HP labs in 2008 [16], a large range of various materials appeared in the literature investigating the resistive switching properties. Clearly, it is necessary to focus on the resistive switching properties of specific materials for a certain application. Hence, this chapter is organized in line with the transparent and flexible metal oxide ReRAM devices.

13.1.1   Overview of the involved mechanisms

Flexible, transparent devices with ReRAM characteristics are reported for both unipolar and bipolar types. Basically, the difference between the resistive switching types relies on the parameters which trigger the reset process. A temperature effect or Joule heating is the main reason for the reset process in unipolar resistive switching and leads to the dissolution of the conductive filament (CF). On the contrary, the charge migration is believed to be the main reason for the reset in bipolar resistive switching. However, the temperature effect cannot be ignored in a bipolar mode [17]. Generally, the physics behind the resistive switching properties of the metal oxide transparent active materials (TAMs) are discussed in terms of the oxygen deficiency of the switching material, and the resistive switching can be based on a filamentary mechanism or an interface-type mechanism. When the switching is controlled by the CF, the low resistance state (LRS) is not dependent on the electrode area. Whereas, in the interface-type resistive switching both LRS and high resistance state (HRS) are influenced by a change of contact area. Interface-type resistive switching is designed by either selection of a Schottky and an ohmic electrode [17] or by modulation of asymmetric barrier heights induced by accumulation/depletion of charge carriers within defective states in a heterostructure device [18]. Interestingly, the CF-based and interface-type resistive switching may occur in a single-device structure depending on the electrical operation. Basically, a unipolar filamentary resistive switching can be altered to an interface-type by application of a reverse bias at sufficiently high current compliance. It generates an extra oxygen-deficient layer next to the bottom contact [19]. The interface-type resistive switching is homogeneously distributed over the device area. Thus, the current states are more stable and remain well defined when the device is scaled down. However, poor retention for devices based on crystalline active material is the major drawback of the interface type, especially when they are under bending condition. This drawback is interpreted due to back diffusion of oxygen vacancies through grain boundaries [20].

13.2   Technical challenges and recent developments— Conductivity, transparency, flexibility

The synthesis and design of TAMs sandwiched between transparent conducting electrodes are the key topics for the construction of high-performance transparent ReRAM devices. A sufficiently high sheet conductance is crucial for SOP technology, especially on large areas. The main challenge for a transparent SOP approach is to maintain the high sheet conductance when using transparent electrodes. The first transparent conducting oxide (TCO) was CdO which was reported around 100 years ago [21]. Nowadays, high-performance TCOs (such as indium-tin oxide (ITO) as the most common one) are obtained in low-cost manufacturing methods [22]. The active material and electrodes require at least ~85% optical transparency owing to a wide optical bandgap in the range of 3 eV or higher. For application in various low-cost, flexible, and high-end applications, they need to be processed at low temperatures. These constraints complicate the optimization of reliability and stability of the ReRAM characteristics compared to a nontransparent device.

The selection of the electrodes has a high impact on the resistive switching behavior. Inert electrodes with high work functions result in a high on/off ratio. However, the idealized scenario for the anode side of a single active layer is a metal with a high oxygen affinity, which provides a reservoir of oxygen vacancies by the formation of an ultrathin oxide layer between the electrode and the active material. This generally leads to higher endurance [23, 24]. To obtain flexible ReRAM devices with effective electrode conductivity, the electrode must be mechanically stable. In addition, the surface should be smooth with high physical contact with the TAM within the abundant pathways. One of the approaches is to add a thin metallic interlayer sandwiched between TCOs. This multilayer electrode configuration is more flexible than plain TCOs such as ITO at a small bending radius. The resistance of the multilayer contact remained unchanged under bending condition with the radius equal to 10 mm, and the switching performance is almost the same compared with the ReRAM device before the bending test [25]. Using percolated networks of silver nanowires is another proper choice for transparent and flexible applications.

Ag nanowires (AgNWs) embedded in polymer nanocomposites fabricated by conventional processing were reported as transparent electrodes for textile electronics but still with the rough surface of nanowire-based contacts [27] inducing leakage paths. Using the inverted layer processing, the coverage of the conductive pathways at the surface of the electrode can be enlarged leading to much better ReRAM performance, compared to the conventional method [26]. Two separate procedures of fabrication of the AgNWs/colorless polyimide (cPI) nanocomposite transparent and flexible electrode are shown in Fig. 13.1.

Another electrode used in flexible and transparent resistive switching devices is graphene which holds great promise due to outstanding chemical, mechanical, and electrical properties. Generally, excellent conductivity, high transparency and flexibility, cost-effectiveness, sustainability, and environmental friendliness satisfy all the requirements of next-generation electronics. Solution-based fabrication of graphene electrodes as compared to chemical vapor deposition (CVD) shows higher production yield and does not require extreme conditions, such as high temperature and vacuum [28]. Fig. 13.2 represents highly transparent, flexible SiO_x ReRAM devices with graphene contacts in crossbar arrays [29]. No degradation in memory characteristics is observed under bending conditions with a curvature radius of 6 mm.

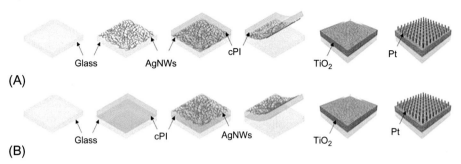

Fig. 13.1 Schematic representation of the fabrication procedure for AgNWs/cPI-based ReRAMs: (A) inverted layer processing and (B) normal (or conventional) processing. Reprinted from S.-W. Yeom, B. You, K. Cho, H.Y. Jung, J. Park, C. Shin, B.-K. Ju, J.-W. Kim, Silver nanowire/colorless-polyimide composite electrode: application in flexible and transparent resistive switching memory, Sci. Rep. 7 (2017) 3438, https://doi.org/10.1038/s41598-017-03746-1, under CC BY 4.0 license.

Besides the electrodes, constructing a ReRAM device with well-controlled defect states is crucial, rather than relying on the intrinsic defects. However, excessive defect concentrations lead to a nonreversible breakdown. In addition, as a rule of thumb, higher TAMs thicknesses require higher electroforming voltages or higher set voltages. To achieve a fully integrated transparent and flexible system consisting of memristors and transistors, both transistor semiconductor and resistive switching layers can be made of amorphous metal oxide semiconductors (AOS, see Section 13.3.1). Transparent TFTs based on an IGZO semiconductor show reliable performance and have already brought several favorable characteristics to the display technology. Zinc-tin-oxide (ZTO) is an interesting and nontoxic alternative since it avoids the usage of rare earth elements. Not only semiconductor materials but also high k dielectrics based on TaO_x, AlO_x, and HfO_x can be considered as TAMs for ReRAM applications (see Section 13.3.2) [30]. Here, a reversible dielectric breakdown, which leads to reliable resistance states, is a key parameter. Like the AOSs, high k materials are used for fabrication of the TFTs, which facilitates the integration into the already mature TFT technology for advanced transparent SOP applications.

Other examples for flexible TAMs are ultrathin 2D nanomaterials, such as graphene oxide and other derivatives, which exhibit appealing functionality, such as tunable electronic states by controlling the number of functional groups on the surface and excellent transparency and flexibility. The progress in the fabrication of ReRAM devices based on a high-quality ultrathin film on a large scale via simple techniques can be found in a tutorial review in Ref. [28]. In addition to oxide-based ReRAM devices, organic ReRAMs based on polymers [31] and cellulose derivatives [32] are candidates for fabrication of transparent and flexible devices. However, the environmental instability of organic electronic materials limits the possible applications of this class of materials [33].

For achieving a flexible device, the electrical properties of the ReRAM under bending condition and crack formation at a certain strain is a key parameter.

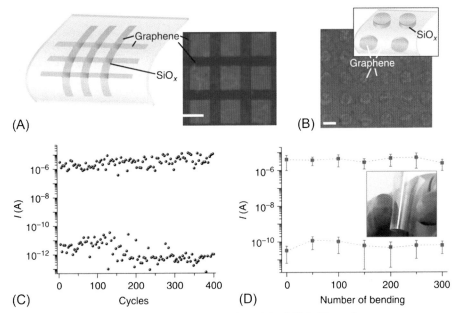

Fig. 13.2 (A) (Left panel) Schematic representation of the G/SiO$_x$/G crossbar structures on a plastic (fluoropolymer) substrate and (right panel) the optical image of the structures. Scale bar, 20 µm. (B) Optical image of the G/SiO$_x$/G pillar structures with the inset showing the schematic image. Scale bar, 100 µm. (C) Memory cycles from one of the crossbar devices using +5 and +14 V as set and reset voltages, respectively. The programming current is not shown here and the memory states (current) were recorded at +1 V. (D) Retention of both ON and OFF memory states (read at +1 V) from a crossbar device is shown on bending the plastic substrate around a ~1.2-cm diameter curvature; the central devices on the sheet being tested throughout the bending cycles. The inset shows the actual transparent memory devices using the pillar structures on the plastic substrate.

Reprinted from J. Yao, J. Lin, Y. Dai, G. Ruan, Z. Yan, L. Li, L. Zhong, D. Natelson, J.M. Tour, Highly transparent nonvolatile resistive memory devices from silicon oxide and graphene, Nat. Commun. 3 (2012) 1101, https://doi.org/10.1038/ncomms2110, with permission from Springer Nature.

Applying a ductile conductive material, such as silver in the ITO/Ag/ITO multilayered electrode structure results in a reliable electrical performance under continuous bending condition compared with only ITO [34]. Two flexibility factors—bending radius and bending times (cycle of endurance under bending condition)—are required to be analyzed. Polycrystalline materials (for both the electrodes and the TAMs) do not withstand excessive bending conditions without deterioration of the electrical properties. As shown in the plot of Fig. 13.3, the on/off ratio of the ReRAM based on ITO/NiO-ZnO/ITO under bending test condition is decreased >10 times after the first 100 cycles. Furthermore, the required voltage to set and reset the device increases linearly with the evolution of the bending time [35].

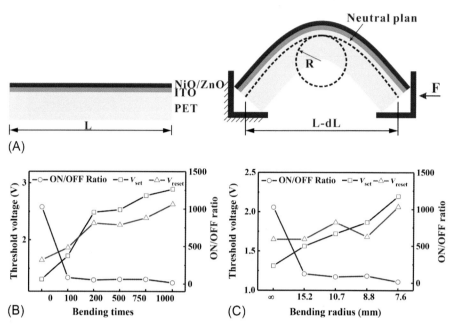

(A)

(B)   Bending times   (C)   Bending radius (mm)

Fig. 13.3 (A) Schematic illustration of the bending test for the NiO-ZnO nanocomposite films. (B) The dependence of V_{set}, V_{reset} and on/off ratio on the bending times, respectively. (C) Effect of bending radius on the on/off ratio, V_{set} and V_{reset}, respectively. Reprinted from H.L. Yuan, J.C. Li, Effect of bending on resistive switching of NiO/ZnO nanocomposite thin films, J. Alloys Compd. 709 (2017) 752–759, https://doi.org/10.1016/j. jallcom.2017.03.196, with permission from Elsevier.

Intrinsic defects and grain boundaries can provide an easy path for crack propagation. The induced stresses in ITO and the functional layer (NiO-ZnO) are increased by the relatively large Young's modulus compared to the organic substrate.

The NiO films are amorphously embedded with highly crystallized ZnO nanoparticles. The switching is forming free due to the distribution of random defects in the TAMs. No metallic trend is observed in temperature dependence analysis of the LRS under bending conditions. Hence, intrinsic defects, such as oxygen vacancies are responsible for the resistive switching, which can form CFs along the grain boundaries. The current at the HRS shows a superlinear increase with the voltage with different slopes (1.3–1.7) depending on the bending condition. Hence, the electrical behavior of the HRS is analyzed at different temperatures to understand the physics behind the bending effect. It is observed that relatively stable electrical properties are obtained under different radius and bending times. This suggests that the conduction mechanism of the device in HRS is not strongly influenced by the bending condition.

To investigate the on/off ratio degradation under bending condition, a simulation based on the Extended Finite Element Method is applied. It is revealed that cracks created will propagate in a critical bending condition of <9 mm (onset of crack: 1.77 GPa).

The crack may grow along the boundaries and it limits the charge transport. If it holds, the decrease of current at LRS is explainable. The oxygen can enter the film and, in addition, bending-induced heating effects would weaken the oxygen vacancies of CFs. Furthermore, the electric field of cracked ITO is changed, which limits the CF formation [35].

13.3 Literature survey on transparent and/or flexible electronic devices

13.3.1 Semiconductor oxide-based ReRAM devices (ZnO, IGZO, ZTO)

Semiconducting TAMs such as zinc oxide are widely reported using different fabrication techniques, such as radio frequency (RF) sputtering, CVD, atomic layer deposition (ALD), electrodeposition, and solution processing. Sputter deposition is preferred for good uniformity on large areas and a precise control of the oxygen amount. The latter allows to control the oxygen deficiency of the material, which causes a higher carrier concentration via an increase in donor density (zinc interstitials and oxygen vacancies) [36]. Note that the endurance of the resistive switching behavior clearly depends on controlling of defect states. Thus, different thermal treatments or metal stacks or adding an additional doping layer is crucial to develop better ReRAM performance [37]. High oxygen flow annihilates more oxygen vacancies in favor of less leakage current of the device, leading to better reset stability and higher on/off ratio. Thermal postdeposition treatment in air or another oxygen-rich atmosphere can have the same impact on the switching layer due to the decrease of native donor defects. When the active materials are polycrystalline, the excess of branches of CFs may result in stochastic resistive switching properties, thus the amorphous film is preferred. The elements In, Ga, and Sn are among the doping elements for ZnO film [37]. Due to the success of amorphous oxide, quaternary compounds, such as IGZO-based TFTs in display technology and integration of IGZO-TFTs with ReRAMs obtained from the same material, AOS are an excellent choice for SOP applications. The main motivation for the implementation of electronic circuits on amorphous oxides is related to the significantly reduced manufacturing costs. The first ReRAM for compatible usage dates back to 2010 [38]. In the following years, in order to improve the IGZO amorphous oxide ReRAM performance, more studies were done to understand the working mechanism [39–43] and to develop better device configurations. There appears to exist an agreement on an oxygen-vacancy mediated switching mechanism in IGZO.

Wang et al. show a unipolar resistive switching behavior of transparent flexible IGZO-based ReRAM. However, the optical transparency of the film by selecting Cu as the electrode is reduced to ~65%. Temperature analysis of the LRS reveals the temperature coefficient of $5.8 \times 10^{-4} \, K^{-1}$ which is one order of magnitude smaller than that of Cu ($\sim 10^{-3} \, K^{-1}$). One possible origin of the CF is oxygen deficient defects similar to transition metal oxide-based ReRAM. To prove the assumption, the temperature analysis of an oxygen-deficient film can be compared with the result of LRS. The resistance increases linearly with the temperature and the temperature coefficients of both films are similar and in the same order of magnitude ($\sim 10^{-4}$). This comparison shows that the formation of oxygen deficient defects under electric field results in LRS and not the formation of a metallic copper filament. The HRS and the initial resistance state of the device at pristine state exhibit the opposite trend with temperature, meaning the resistance decreases by increasing the temperature (semiconducting-like behavior). Due to Joule heating, a partial oxidation of oxygen deficient defects controls the transition between LRS to HRS [44].

Inspired by the low-cost amorphous oxide solution-processed thin-film transistors [45], in 2012, Moon-Seok Kim and his team developed IGZO solution-based ReRAM devices with a low on/off ratio [46]. In 2014, Wei Hu et al. [47] showed a better performance of unipolar ReRAM devices fabricated by a low- temperature solution-based process. In 2017, solution processed IGZO based bipolar ReRAM with low programming voltages of ± 1 V, a good retention time of up to 10^4 s and high endurance cycle was achieved at a low temperature of 200°C [24], suitable for SOP applications. Fig. 13.4 demonstrates the resistive switching properties of the devices. These devices are applicable to flexible substrates.

Ternary oxide ReRAMs (ZTO-based) are also fabricated by either physical method or low-cost solution processing techniques [48, 49]. ZTO does not contain expensive scarce elements (In, Ga), which is the main advantage of this group of AOSs.

13.3.2   High k dielectric ReRAM

High k dielectrics are another type of active materials, which can be applied to transparent and flexible ReRAM for SOP applications. Intrinsic defect states and the reversible dielectric breakdown motivate many research groups to study ReRAM properties of high k dielectrics. Furthermore, these materials are widely used as a gate dielectric in Si CMOS technology, which is an advantage to reduce the processing steps. To obtain the resistive switching device from an insulating active material, an electro-

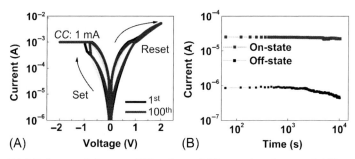

Fig. 13.4 (A) *I-V* characteristics over 100 cycles and (B) retention time at −0.1 V at room temperature for Ag/IGZOnps/TiO$_x$/Ti annealed at 200°C.
Reprinted from J. Rosa, A. Kiazadeh, L. Santos, J. Deuermeier, R. Martins, H.L. Gomes, E. Fortunato, Memristors using solution-based IGZO nanoparticles, ACS Omega 2 (2017) 8366–8372, https://doi.org/10.1021/acsomega.7b01167, with permission from the American Chemical Society.

forming process (soft breakdown) is required to activate resistive switching behaviors. The migration of oxygen ions induces an oxygen-deficient layer with respect to the pristine state and the created oxygen vacancies set the device to LRS. HfO_x is one of the most representative active material in literature with a broad range of studies: from the role of oxygen concentration, interface engineering, the impact of electrodes [50–64] to the presence of residual carbon on device reliability [65]. However, many reports target the data storage capability only using a rigid substrate. Understanding failure mechanisms due to mechanical change (bending/stretching) is of importance for further optimization. For transparent ReRAM devices consisting of $ITO/HfO_x/ITO$ on the flexible substrate polyethylene terephthalate (PET), metallic-type LRSs with the temperature coefficient of $4.0 \times 10^{-3} \, K^{-1}$ close to the value reported for hafnium nanowires is obtained [66]. The metallic nature of the conducting filament is due to the migration of non-lattice oxygen anions and the subsequent reduction of the Hf^{n+} cations. Furthermore, the resistance states are measured at different bending radii between 12 and 2 mm for 300 cycles. When the bending radius reached 4 mm, cracks appeared. The crack density increased with reducing bending radius. A decreased on/off ratio is also observed. Nevertheless, at a radius of 3 mm, still two resistance states are distinguishable. The set and reset voltages are also increased with decreasing bending radius and eventually at a radius of 2 mm, the ReRAM cannot be switched off [66]. The enhanced mechanical properties are related to the controlled formation of pseudo straight CFs in the amorphous film.

Generally, in comparison with ZnO-based memories including the NiO-ZnO (Section 13.2), HfO_x ReRAMs exhibit outstanding switching properties in a harsh environment, such as high oxygen partial pressure, corrosive agent exposure, and high moisture [67]. Fig. 13.5 shows the schematic and memory cycling properties of HfO_x

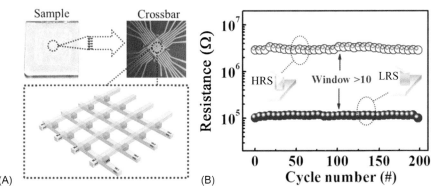

Fig. 13.5 (A) A schematic representation of crossbar configuration of HfO_2 transparent and flexible ReRAM. Image of the as-fabricated device is shown on the upper left. The device area of each transparent ReRAM cell inside the crossbar array is $1 \, \mu m^2$. (B) Endurance characteristics at −0.1 V of HfO_2 ReRAM cell in crossbar array.
Reprinted from P.K. Yang, C.H. Ho, D.H. Lien, J.R. Durán Retamal, C.F. Kang, K.M. Chen, T.H. Huang, Y.C. Yu, C.I. Wu, J.H. He, A fully transparent resistive memory for harsh environments, Sci. Rep. 5 (2015) 1–9, https://doi.org/10.1038/srep15087, under CC BY 4.0 license.

ReRAM. By applying Al as top contact to HfO_x, a thin layer of Al_2O_3 is formed, which increases the oxygen vacancies underneath the contact. A bilayer structure device shows lower set/reset voltages and more uniform resistive switching properties due to a confinement of CF formation and rupture [68].

ReRAMs based on Al_2O_3, as the most abundant material have been already reported via deposition methods such as sputtering, ALD, and thermal oxidation, and recently low-cost solution processing [69–79]. One of the promising flexible and transparent ReRAM device is the all-sputtered crossbar array of indium-zinc oxide (IZO)/Al_2O_3/IZO on a plastic substrate, which showed high stability under bending conditions [80]. Towards low-cost fabrication approaches, ReRAM devices are also developed from solution-based aluminum oxide in a combustion synthesis. Fig. 13.6. shows the schematic representation of the ReRAM based on AlO_x and typical electrical characteristics in unipolar and bipolar resistive switching mode. Transparency of the ReRAMs stacked between ITO and IZO is 80%. The fully solution-processed ReRAMs demonstrate encouraging results to be applied in cost-effective SOP technology [81].

13.3.3 Integration

ReRAM devices are suitable for high-density integration of three-dimensional (3D) crossbar architecture arrays [82]. However, the word line/ReRAM/bit line configuration exhibits an inherent readout margin issue due to the presence of sneak paths and the interference of neighboring cells (the high current of LRS deteriorates the readout margin of the HRS). Thus, an active element like a transistor or a nonlinear passive component, such as a diode should be added to each ReRAM cell as a selector device. The majority of the integration approaches are reported based on 1 transistor-1 ReRAM (1T-1R) similar to DRAM block consisting of one transistor and

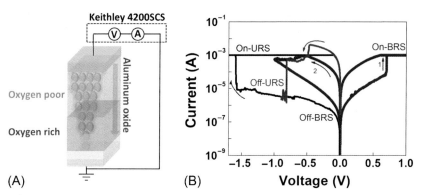

Fig. 13.6 (A) Device structure schematic Pt (bottom electrode)/Al_2O_3/Ti (top electrode). The oxygen-poor region in Al_2O_3 is a result of the presence of TiO_x at the interface; (B) typical *I-V* characteristic in unipolar resistive switching (URS) mode and bipolar resistive switching (BRS) mode.

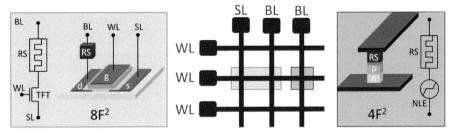

Fig. 13.7 Schematic representation of crossbar arrays (middle). 1T-1R structure (left). The size of the device is dominated by the size of the selecting transistor. The 1T-1R can be in NOR (lower latency) and NAND (higher density) configurations, 1D-1R (right). *WL*, word line; *BL*, Bit line; *SL*, source line; *RS*, resistive switching; *NLE*, nonlinear element (here NLE is a p-n diode as an example).

one capacitor, possibly because of the mature and reliable transistor technology [83–86]. Fig. 13.7 demonstrate the integration approaches.

Clearly, applying diodes compared with the transistor maximizes the memory density in the minimum active area and improves the stacking scalability towards the $4F^2$ (F: minimum feature size) integration density. The crossbar arrays of unipolar resistive switching devices based on ZnO in different stacking combined with the heterostructure p-n diode of NiO/ZnO and WO_3/ZnO tunnel barrier switching diodes showed stable resistive switching [87]. The current is efficiently suppressed on the other polarity where no resistive switching occurs. In the NiO/ZnO diode, the conduction is attributed to both holes and electrons whereas in WO_3/ZnO, only electrons participate in the charge transport [87]. A similar approach is shown in the Fig. 13.8 for devices based on only silicon layers (p-Si/SiO_2/n-Si crossbar array) [88]. Different doping types in the silicon contacts form a self-assembled diode within each junction. This built-in selector effectively suppress sneak paths in 3D crossbar arrays.

For a bipolar resistive switching, bidirectional selector devices [89] or threshold resistive switches [90] are required. However, obtaining the low off-state leakage and large rectifying ratio on one hand and realizing the high current density on the other hand, is challenging.

Another approach is fabricating a single selector-less ReRAM device with nonlinear resistive switching nature. This approach is feasible by interface engineering of the active material [91] and it has an advantage of less process complexity. The nonlinear resistive switching effectively prevents the sneak path current because the LRS is obtained at high voltages and the conductivity at low voltages ($V_{read}/2$) is efficiently reduced [92]. A typical example is shown in Fig. 13.9. Besides this, complementary resistive switching (CRS) is another selector-less ReRAM configuration. The conventional CRS consists of two symmetric resistive switching connected back to back. The CRS device requires an external resistor to adjust the switching voltages of two resistive switching elements. Currently, this issue is solved by the design of the CRS cells in a single-device Pt/HfO_2/HfO_x/Pt structure [93].

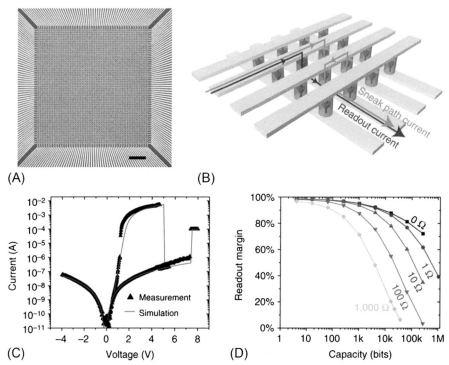

Fig. 13.8 Single-layer all-silicon crossbar array and array size evaluation. (A) A 64×64 p-Si/SiO$_2$/n-Si crossbar array. The devices in the arrays have a junction area of 5 mm $\times$ 5 mm. (B) Schematic representation of a crossbar array in which the sneak path problem is alleviated by intrinsic diodes at each cell. The *blue* line is one exemplary sneak path that includes one reverse biased cell, which significantly reduces the sneak path current. (C) The simulated single-device DC sweep *I-V* (solid line) curve from the SPICE simulation matches the measurement data (triangles), validating the SPICE model. (D) The normalized readout margin is larger than 39% for a 1 Mbits crossbar array if we consider wire resistance between each cell to be 1 vΩ, 27% for a 64 kbits array with 100 Ω, and 10% for 30 kbits array with 1000 Ω. Scale bar, 100 mm.
Reprinted from C. Li, L. Han, H. Jiang, M.H. Jang, P. Lin, Q. Wu, M. Barnell, J.J. Yang, H.L. Xin, Q. Xia, Three-dimensional crossbar arrays of self-rectifying Si/SiO$_2$/Si memristors, Nat. Commun. 8 (2017) 1–9, https://doi.org/10.1038/ncomms15666, under CC BY 4.0 license.

13.4   Conclusion

Modern electronics for SOP applications have rapidly developed towards flexible and transparent devices in the last decade. It is possible to process many TAMs for resistive switching applications at sufficiently low temperature to allow the use of flexible substrates. Among the active materials, the amorphous ones are preferred because of significant benefits from the stability of the electrical properties under bending conditions, compared to polycrystalline films.

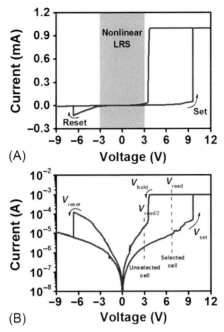

Fig. 13.9 Typical resistive switching behavior of Pt/TiO_2 nanorods/FTO device in linear scale and in semi-log scale.
Reprinted from C.-H. Huang, T.-S. Chou, J.-S. Huang, S.-M. Lin, Y.-L. Chueh, Self-selecting resistive switching scheme using TiO_2 nanorod arrays, Sci. Rep. 7 (2017) 2066, https://doi.org/10.1038/s41598-017-01354-7, under CC BY 4.0 license.

The main challenge in the implementation of flexible and transparent ReRAM for SOP technology is the electrode sheet conductance. New approaches of electrode configuration, such as ultrathin metallic layers between transparent conductive oxides (TCOs) and contacts based on metallic nanowires and graphene are promising candidates. Nevertheless, the integration of ReRAM in transparent and flexible SOP technology has not yet reached a sufficient state of development for a commercial application.

References

[1] K. Nomura, H. Ohta, A. Takagi, T. Kamiya, M. Hirano, H. Hosono, Room-temperature fabrication of transparent flexible thin-film transistors using amorphous oxide semiconductors, Nature 432 (2004) 488–492, https://doi.org/10.1038/nature03090.
[2] T. Kamiya, K. Nomura, H. Hosono, Present status of amorphous In–Ga–Zn–O thin-film transistors, Sci. Technol. Adv. Mater. 11 (2010) 44305, https://doi.org/10.1088/1468-6996/11/4/044305.
[3] P. Barquinha, A.M. Vila, G. Gonçalves, L. Pereira, R. Martins, J.R. Morante, E. Fortunato, Gallium–indium–zinc-oxide-based thin-film transistors: influence of the source/drain material, IEEE Trans. Electron Devices 55 (2008) 954–960, https://doi.org/10.1109/TED.2008.916717.

[4]   G. Bahubalindruni, V.G. Tavares, P. Barquinha, C. Duarte, R. Martins, E. Fortunato, P.G. de Oliveira, IEEE, Basic analog circuits with a-GIZO thin-film transistors: modeling and simulation. in: 2012 International Conference on Synthesis, Modeling, Analysis and Simulation Methods and Applications to Circuit Design (SMACD), 2012, pp. 261–264, https://doi.org/10.1109/SMACD.2012.6339389.

[5]   D. Salgueiro, A. Kiazadeh, R. Branquinho, L. Santos, P. Barquinha, R. Martins, E. Fortunato, Solution based zinc tin oxide TFTs: the dual role of the organic solvent. J. Phys. D Appl. Phys. 50 (2017) 65106, https://doi.org/10.1088/1361-6463/50/6/065106.

[6]   H. Yin, S. Kim, H. Lim, Y. Min, C.J. Kim, I. Song, J. Park, S.-W. Kim, A. Tikhonovsky, J. Hyun, Y. Park, Program/erase characteristics of amorphous gallium indium zinc oxide nonvolatile memory. IEEE Trans. Electron Devices 55 (2008) 2071–2077, https://doi.org/10.1109/TED.2008.926727.

[7]   N.C. Su, S.J. Wang, A. Chin, A nonvolatile InGaZnO charge-trapping-engineered flash memory with good retention characteristics. IEEE Electron Device Lett. 31 (2010) 201–203, https://doi.org/10.1109/LED.2009.2037986.

[8]   H. Ji, Y. Wei, X. Zhang, R. Jiang, Homogeneous-oxide stack in IGZO thin-film transistors for multi-level-cell NAND memory application. Appl. Phys. Lett. 111 (2017) 202102, https://doi.org/10.1063/1.4998207.

[9]   L.-F. He, H. Zhu, J. Xu, H. Liu, X.-R. Nie, L. Chen, Q.-Q. Sun, Y. Xia, D.W. Zhang, Light-erasable embedded charge-trapping memory based on MoS_2 for system-on-panel applications. Appl. Phys. Lett. (2017) 111, https://doi.org/10.1063/1.5000552.

[10] S.-B. Qian, Y. Shao, W.-J. Liu, D.W. Zhang, S.-J. Ding, Erasing-modes dependent performance of a-IGZO TFT memory with atomic-layer-deposited Ni nanocrystal charge storage layer. IEEE Trans. Electron Devices 64 (2017) 3023–3027, https://doi.org/10.1109/TED.2017.2702702.

[11] S. Chang Lee, Q. Hu, Y.-J. Baek, Y. Jin Choi, C. Jung Kang, H. Ho Lee, T.-S. Yoon, Analog and bipolar resistive switching in pn junction of n-type ZnO nanowires on p-type Si substrate. J. Appl. Phys. 114 (2013) 64502, https://doi.org/10.1063/1.4817838.

[12] T.W. Hickmott, Low-frequency negative resistance in thin anodic oxide films. J. Appl. Phys. 33 (1962) 2669–2682, https://doi.org/10.1063/1.1702530.

[13] T. Mukerjee, J. Allan, Negative resistance in $Al-Al_2O_3-Se-Au$ sandwich cells. J. Appl. Phys. 35 (1964) 2270–2271, https://doi.org/10.1063/1.1702839.

[14] J.F. Gibbons, W.E. Beadle, Switching properties of thin Nio films. Solid State Electron. 7 (1964) 785–790, https://doi.org/10.1016/0038-1101(64)90131-5.

[15] G. Dearnaley, A.M. Stoneham, D.V. Morgan, Electrical phenomena in amorphous oxide films. Rep. Prog. Phys. 33 (1970) 306, https://doi.org/10.1088/0034-4885/33/3/306.

[16] D.B. Strukov, G.S. Snider, D.R. Stewart, R.S. Williams, The missing memristor found. Nature 453 (2008) 80–83, https://doi.org/10.1038/nature06932.

[17] H. Akinaga, H. Shima, Resistive random access memory (ReRAM) based on metal oxides. Proc. IEEE 98 (2010) 2237–2251, https://doi.org/10.1109/JPROC.2010.2070830.

[18] Z. Xu, L. Yu, X. Xu, J. Miao, Y. Jiang, Effect of oxide/oxide interface on polarity dependent resistive switching behavior in ZnO/ZrO_2 heterostructures. Appl. Phys. Lett. 104 (2014) 192903, https://doi.org/10.1063/1.4878402.

[19] C.-H. Huang, J.-S. Huang, C.-C. Lai, H.-W. Huang, S.-J. Lin, Y.-L. Chueh, Manipulated transformation of filamentary and homogeneous resistive switching on ZnO thin film memristor with controllable multistate. ACS Appl. Mater. Interfaces 5 (2013) 6017–6023, https://doi.org/10.1021/am4007287.

[20] H.Y. Peng, G.P. Li, J.Y. Ye, Z.P. Wei, Z. Zhang, D.D. Wang, G.Z. Xing, T. Wu, Electrode dependence of resistive switching in Mn-doped ZnO: filamentary versus interfacial mechanisms. Appl. Phys. Lett. 96 (2010) 192113, https://doi.org/10.1063/1.3428365.

[21] K. Bädeker, Über die elektrische Leitfähigkeit und die thermoelektrische Kraft einiger Schwermetallverbindungen. Ann. Phys. 327 (1907) 749–766, https://doi.org/10.1002/andp.19073270409.

[22] H. Hosono, K. Ueda, Springer International Publishing, Springer Handbook of Electronic and Photonic Materials. 1–1 2017https://doi.org/10.1007/978-3-319-48933-9_58.

[23] W.-Y. Chang, H.-W. Huang, W.-T. Wang, C.-H. Hou, Y.-L. Chueh, J.-H. He, High uniformity of resistive switching characteristics in a Cr/ZnO/Pt device. J. Electrochem. Soc. 159 (2012) G29–G32, https://doi.org/10.1149/2.092203jes.

[24] J. Rosa, A. Kiazadeh, L. Santos, J. Deuermeier, R. Martins, H.L. Gomes, E. Fortunato, Memristors using solution-based IGZO nanoparticles. ACS Omega 2 (2017) 8366–8372, https://doi.org/10.1021/acsomega.7b01167.

[25] M. Kim, K.C. Choi, Transparent and flexible resistive random access memory based on Al$_2$O$_3$ film with multilayer electrodes. IEEE Trans. Electron Devices 64 (2017) 3508–3510, https://doi.org/10.1109/TED.2017.2716831.

[26] S.-W. Yeom, B. You, K. Cho, H.Y. Jung, J. Park, C. Shin, B.-K. Ju, J.-W. Kim, Silver nanowire/colorless-polyimide composite electrode: application in flexible and transparent resistive switching memory. Sci. Rep. 7 (2017) 3438, https://doi.org/10.1038/s41598-017-03746-1.

[27] K.-H. Ok, J. Kim, S.-R. Park, Y. Kim, C.-J. Lee, S.-J. Hong, M.-G. Kwak, N. Kim, C.J. Han, J.-W. Kim, Ultra-thin and smooth transparent electrode for flexible and leakage-free organic light-emitting diodes. Sci. Rep. 5 (2015) 9464, https://doi.org/10.1038/srep09464.

[28] C. Tan, Z. Liu, W. Huang, H. Zhang, Non-volatile resistive memory devices based on solution-processed ultrathin two-dimensional nanomaterials. Chem. Soc. Rev. 44 (2015) 2615–2628, https://doi.org/10.1039/C4CS00399C.

[29] J. Yao, J. Lin, Y. Dai, G. Ruan, Z. Yan, L. Li, L. Zhong, D. Natelson, J.M. Tour, Highly transparent nonvolatile resistive memory devices from silicon oxide and graphene. Nat. Commun. 3 (2012) https://doi.org/10.1038/ncomms2110.

[30] M. Lanza, Mario, A review on resistive switching in high-k dielectrics: a nanoscale point of view using conductive atomic force microscope. Materials 7 (2014) 2155–2182, https://doi.org/10.3390/ma7032155.

[31] H.C. Yu, M.Y. Kim, M. Hong, K. Nam, J.Y. Choi, K.H. Lee, K.K. Baeck, K.K. Kim, S. Cho, C.M. Chung, Fully transparent, non-volatile bipolar resistive memory based on flexible copolyimide films. Electron. Mater. Lett. 13 (2017) 1–8, https://doi.org/10.1007/s13391-017-6148-z.

[32] N. Raeis Hosseini, J.-S. Lee, Resistive switching memory based on bioinspired natural solid polymer electrolytes. ACS Nano 9 (2015) 419–426, https://doi.org/10.1021/nn5055909.

[33] M. Nikolka, I. Nasrallah, B. Rose, M.K. Ravva, K. Broch, A. Sadhanala, D. Harkin, J. Charmet, M. Hurhangee, A. Brown, S. Illig, P. Too, J. Jongman, I. McCulloch, J.-L. Bredas, H. Sirringhaus, High operational and environmental stability of high-mobility conjugated polymer field-effect transistors through the use of molecular additives. Nat. Mater. 16 (2017) 356–362, https://doi.org/10.1038/nmat4785.

[34] J. Won Seo, J.-W. Park, K.S. Lim, S.J. Kang, Y.H. Hong, J.H. Yang, L. Fang, G.Y. Sung, H.-K. Kim, Transparent flexible resistive random access memory fabricated at room temperature. Appl. Phys. Lett. 95 (2009) 133508, https://doi.org/10.1063/1.3242381.

[35] H.L. Yuan, J.C. Li, Effect of bending on resistive switching of NiO/ZnO nanocomposite thin films. J. Alloys Compd. 709 (2017) 752–759, https://doi.org/10.1016/j.jallcom.2017.03.196.

[36] P. Erhart, K. Albe, A. Klein, First-principles study of intrinsic point defects in ZnO: role of band structure, volume relaxation, and finite-size effects. Phys. Rev. B 73 (2006) 205203, https://doi.org/10.1103/PhysRevB.73.205203.

[37] M. Laurenti, S. Porro, C.F. Pirri, C. Ricciardi, A. Chiolerio, Zinc oxide thin films for memristive devices: a review. Crit. Rev. Solid State Mater. Sci. 42 (2017) 153–172, https://doi.org/10.1080/10408436.2016.1192988.

[38] M.-C. Chen, T.-C. Chang, S.-Y. Huang, S.-C. Chen, C.-W. Hu, C.-T. Tsai, S.M. Sze, Bipolar resistive switching characteristics of transparent indium gallium zinc oxide resistive random access memory. Electrochem. Solid State 13 (2010) H191, https://doi.org/10.1149/1.3360181.

[39] Z.Q. Wang, H.Y. Xu, X.H. Li, H. Yu, Y.C. Liu, X.J. Zhu, Synaptic learning and memory functions achieved using oxygen ion migration/diffusion in an amorphous InGaZnO memristor. Adv. Funct. Mater. 22 (2012) 2759–2765, https://doi.org/10.1002/adfm.201103148.

[40] C.-C. Lo, T.-E. Hsieh, Forming-free, bipolar resistivity switching characteristics of fully transparent resistive random access memory with IZO/α-IGZO/ITO structure. J. Phys. D Appl. Phys. 49 (2016) 385102, https://doi.org/10.1088/0022-3727/49/38/385102.

[41] Y.H. Kang, T.I. Lee, K.-J. Moon, J. Moon, K. Hong, J.-H. Cho, W. Lee, J.-M. Myoung, Observation of conductive filaments in a resistive switching nonvolatile memory device based on amorphous InGaZnO thin films. Mater. Chem. Phys. 138 (2013) 623–627, https://doi.org/10.1016/j.matchemphys.2012.12.029.

[42] Y.-S. Fan, P.-T. Liu, C.-H. Hsu, Investigation on amorphous InGaZnO based resistive switching memory with low-power, high-speed, high reliability. Thin Solid Films 549 (2013) 54–58, https://doi.org/10.1016/j.tsf.2013.09.033.

[43] Y. Pei, B. Mai, X. Zhang, R. Hu, Y. Li, Z. Chen, B. Fan, J. Liang, G. Wang, Forming free bipolar ReRAM of Ag/a-IGZO/Pt with improved resistive switching uniformity through controlling oxygen partial pressure. J. Electron. Mater. 44 (2015) 645–650, https://doi.org/10.1007/s11664-014-3547-x.

[44] Z.Q. Wang, H.Y. Xu, X.H. Li, X.T. Zhang, Y.X. Liu, Y.C. Liu, Flexible resistive switching memory device based on amorphous InGaZnO film with excellent mechanical endurance. IEEE Electron Device Lett. 32 (2011) 1442–1444, https://doi.org/10.1109/LED.2011.2162311.

[45] S.J. Kim, S. Yoon, H.J. Kim, Review of solution-processed oxide thin-film transistors. Jpn. J. Appl. Phys. 53 (2014) https://doi.org/10.7567/JJAP.53.02BA02. 02BA02.

[46] M.-S. Kim, Y. Hwan Hwang, S. Kim, Z. Guo, D.-I. Moon, J.-M. Choi, M.-L. Seol, B.-S. Bae, Y.-K. Choi, Effects of the oxygen vacancy concentration in InGaZnO-based resistance random access memory. Appl. Phys. Lett. 101 (2012) 243503, https://doi.org/10.1063/1.4770073.

[47] W. Hu, L. Zou, X. Chen, N. Qin, S. Li, D. Bao, Highly uniform resistive switching properties of amorphous InGaZnO thin films prepared by a low temperature photochemical solution deposition method. ACS Appl. Mater. Interfaces 6 (2014) 5012–5017, https://doi.org/10.1021/am500048y.

[48] J.S. Rajachidambaram, S. Murali, J.F. Conley, S.L. Golledge, G.S. Herman, Bipolar resistive switching in an amorphous zinc tin oxide memristive device. J. Vac. Sci. Technol. B: Microelectron. Nanometer Struct.-Process. Meas. Phenom. 31 (2013) https://doi.org/10.1116/1.4767124. 01A104.

[49] S. Murali, J.S. Rajachidambaram, S.Y. Han, C.H. Chang, G.S. Herman, J.F. Conley, Resistive switching in zinc-tin-oxide. Solid State Electron. 79 (2013) 248–252, https://doi.org/10.1016/j.sse.2012.06.016.

[50] S.Z. Rahaman, Y. De Lin, H.Y. Lee, Y.S. Chen, P.S. Chen, W.S. Chen, C.H. Hsu, K.H. Tsai, M.J. Tsai, P.H. Wang, The role of Ti buffer layer thickness on the resistive switching properties of hafnium oxide-based resistive switching memories. Langmuir 33 (2017) 4654–4665, https://doi.org/10.1021/acs.langmuir.7b00479.

[51] X.P. Wang, Y.Y. Chen, L. Pantisano, L. Goux, M. Jurczak, G. Groeseneken, D.J. Wouters, Effect of anodic interface layers on the unipolar switching of HfO_2-based resistive RAM. in: Proceedings of 2010 International Symposium on VLSI Technology, System and Application, VLSI-TSA 2010, 2010, pp. 140–141, https://doi.org/10.1109/VTSA.2010.5488914.

[52] N.S. Avasarala, M. Heyns, J. Van Houdt, M.H. Van Der Veen, D.J. Wouters, M. Jurczak, Switching behavior of HfO_2-based resistive RAM with vertical CNT bottom electrode. in: 2017 IEEE 9th International Memory Workshop, IMW 2017, 2017, https://doi.org/10.1109/IMW.2017.7939107. 0–3.

[53] L. Zhao, Z. Jiang, H.Y. Chen, J. Sohn, K. Okabe, B. Magyari-Köpe, H.S.P. Wong, Y. Nishi, Ultrathin (~2 nm) HfO_x as the fundamental resistive switching element: thickness scaling limit, stack engineering and 3D integration. in: Technical Digest—International Electron Devices Meeting, IEDM 2015–Feb, 2015, pp. 6.6.1–6.6.4, https://doi.org/10.1109/IEDM.2014.7046998.

[54] S.U. Sharath, S. Vogel, L. Molina-Luna, E. Hildebrandt, C. Wenger, J. Kurian, M. Duerrschnabel, T. Niermann, G. Niu, P. Calka, M. Lehmann, H.J. Kleebe, T. Schroeder, L. Alff, Control of switching modes and conductance quantization in oxygen engineered HfO_x based memristive devices. Adv. Funct. Mater. 27 (2017) 1–13, https://doi.org/10.1002/adfm.201700432.

[55] C. Walczyk, C. Wenger, D. Walczyk, M. Lukosius, I. Costina, M. Fraschke, J. Dabrowski, A. Fox, D. Wolansky, S. Thiess, E. Miranda, B. Tillack, T. Schroeder, On the role of Ti adlayers for resistive switching in HfO_2-based metal-insulator-metal structures: top versus bottom electrode integration. J. Vac. Sci. Technol. B: Nanotechnol. Microelectron.: Mater. Process. Meas. Phenom. 29 (2011) https://doi.org/10.1116/1.3536524. 01AD02.

[56] K.L. Lin, T.H. Hou, J. Shieh, J.H. Lin, C.T. Chou, Y.J. Lee, Electrode dependence of filament formation in HfO_2 resistive-switching memory. J. Appl. Phys. 109 (2011) 084104, https://doi.org/10.1063/1.3567915.

[57] S.U. Sharath, J. Kurian, P. Komissinskiy, E. Hildebrandt, T. Bertaud, C. Walczyk, P. Calka, T. Schroeder, L. Alff, Thickness independent reduced forming voltage in oxygen engineered HfO_2 based resistive switching memories. Appl. Phys. Lett. 105 (2014) 073505, https://doi.org/10.1063/1.4893605.

[58] B. Ku, Y. Abbas, A.S. Sokolov, C. Choi, Interface engineering of ALD HfO_2-based RRAM with Ar plasma treatment for reliable and uniform switching behaviors. J. Alloys Compd. 735 (2018) 1181–1188, https://doi.org/10.1016/j.jallcom.2017.11.267.

[59] F.Y. Yuan, N. Deng, C.C. Shih, Y.T. Tseng, T.C. Chang, K.C. Chang, M.H. Wang, W.C. Chen, H.X. Zheng, H. Wu, H. Qian, S.M. Sze, Conduction mechanism and improved endurance in HfO_2-based RRAM with nitridation treatment. Nanoscale Res. Lett. 12 (2017) 3–8, https://doi.org/10.1186/s11671-017-2330-3.

[60] T. Cabout, J. Buckley, C. Cagli, V. Jousseaume, J.F. Nodin, B. De Salvo, M. Bocquet, C. Muller, Role of Ti and Pt electrodes on resistance switching variability of HfO_2-based resistive random access memory. Thin Solid Films 533 (2013) 19–23, https://doi.org/10.1016/j.tsf.2012.11.050.

[61] A.S. Sokolov, Y.R. Jeon, S. Kim, B. Ku, D. Lim, H. Han, M.G. Chae, J. Lee, B.G. Ha, C. Choi, Influence of oxygen vacancies in ALD HfO$_{2-x}$thin films on non-volatile resistive switching phenomena with a Ti/HfO$_{2-x}$/Pt structure. Appl. Surf. Sci. 434 (2018) 822–830, https://doi.org/10.1016/j.apsusc.2017.11.016.

[62]. S. Ban and O. Kim, Improvement of switching uniformity in HfO$_x$-based resistive random access memory with a titanium film and effects of titanium on resistive switching behaviors, *Jpn. J. Appl. Phys.*

[63] C. Cagli, J. Buckley, V. Jousseaume, T. Cabout, A. Salaun, H. Grampeix, J.F. Nodin, H. Feldis, A. Persico, J. Cluzel, P. Lorenzi, L. Massari, R. Rao, F. Irrera, F. Aussenac, C. Carabasse, M. Coue, P. Calka, E. Martinez, L. Perniola, P. Blaise, Z. Fang, Y.H. Yu, G. Ghibaudo, D. Deleruyelle, M. Bocquet, C. Muller, A. Padovani, O. Pirrotta, L. Vandelli, L. Larcher, G. Reimbold, B. De Salvo, Experimental and theoretical study of electrode effects in HfO$_2$ based RRAM. vol. 3, 2011 International Electron Devices Meeting, 201128.7.1–28.7.4, https://doi.org/10.1109/IEDM.2011.6131634.

[64] S.U. Sharath, T. Bertaud, J. Kurian, E. Hildebrandt, C. Walczyk, P. Calka, P. Zaumseil, M. Sowinska, D. Walczyk, A. Gloskovskii, T. Schroeder, L. Alff, Towards forming-free resistive switching in oxygen engineered HfO$_{2-x}$. Appl. Phys. Lett. (2014) 104, https://doi.org/10.1063/1.4864653.

[65] G. Niu, X. Cartoixà, A. Grossi, C. Zambelli, P. Olivo, E. Perez, M.A. Schubert, P. Zaumseil, I. Costina, T. Schroeder, C. Wenger, Mechanism of the key impact of residual carbon content on the reliability of integrated resistive random access memory arrays. J. Phys. Chem. C 121 (2017) 7005–7014, https://doi.org/10.1021/acs.jpcc.6b12771.

[66] J. Shang, W. Xue, Z. Ji, G. Liu, X. Niu, X. Yi, L. Pan, Q. Zhan, X.-H. Xu, R.-W. Li, Highly flexible resistive switching memory based on amorphous-nanocrystalline hafnium oxide films. Nanoscale 9 (2017) 7037–7046, https://doi.org/10.1039/c6nr08687j.

[67] P.K. Yang, C.H. Ho, D.H. Lien, J.R. Durán Retamal, C.F. Kang, K.M. Chen, T.H. Huang, Y.C. Yu, C.I. Wu, J.H. He, A fully transparent resistive memory for harsh environments. Sci. Rep. 5 (2015) 1–9, https://doi.org/10.1038/srep15087.

[68] M. Akbari, M.-K. Kim, D. Kim, J.-S. Lee, Reproducible and reliable resistive switching behaviors of AlO$_X$/HfO$_X$ bilayer structures with Al electrode by atomic layer deposition. RSC Adv. 7 (2017) 16704–16708, https://doi.org/10.1039/C6RA26872B.

[69] I.-J. Baek, W.-J. Cho, Resistive switching organic-inorganic blended films for flexible memory applications. Solid-State Electron. 140 (2018) 129–133, https://doi.org/10.1016/j.sse.2017.10.030.

[70] S. Kim, Y.-K. Choi, Resistive switching of aluminum oxide for flexible memory. Appl. Phys. Lett. 92 (2008) 223508, https://doi.org/10.1063/1.2939555.

[71] Y.W.Y. Wu, B.L.B. Lee, H.P. Wong, Based RRAM using atomic layer deposition (ALD) with 1- RESET current. IEEE Electron Device Lett. 31 (2010) 1449–1451, https://doi.org/10.1109/LED.2010.2074177.

[72] S. Yu, R. Jeyasingh, Y. Wu, H.S. Philip Wong, Understanding the conduction and switching mechanism of metal oxide RRAM through low frequency noise and AC conductance measurement and analysis. in: Technical Digest—International Electron Devices Meeting, IEDM, 2011, pp. 275–278, https://doi.org/10.1109/IEDM.2011.6131537.

[73] W. Kim, S.I. Park, Z. Zhang, Y. Yang-Liauw, D. Sekar, H.-S.P. Wong, S.S. Wong, Forming-free nitrogen-doped AlO$_X$ RRAM with sub-µA programming current, in: 2011 Symposium on VLSI Technology—Digest of Technical Papers 31, 2011, pp. 22–23. http://doi.org/978-4-86348-166-4.

[74] A. Prakash, S. Maikap, W. Banerjee, D. Jana, C.S. Lai, Impact of electrically formed interfacial layer and improved memory characteristics of IrOx/high-κx/W structures

containing AlOx, GdOx, HfOx, and TaOx switching materials. Nanoscale Res. Lett. 8 (2013) 1–12, https://doi.org/10.1186/1556-276X-8-379.

[75] Z. Chen, F. Zhang, B. Chen, Y. Zheng, B. Gao, L. Liu, X. Liu, J. Kang, High-performance HfO_x/AlO_y-based resistive switching memory cross-point array fabricated by atomic layer deposition. Nanoscale Res. Lett. 10 (2015) 70, https://doi.org/10.1186/s11671-015-0738-1.

[76] K. Park, J.-S. Lee, Reliable resistive switching memory based on oxygen-vacancy-controlled bilayer structures. RSC Adv. 6 (2016) 21736–21741, https://doi.org/10.1039/C6RA00798H.

[77] J. Jang, Y. Song, K. Cho, Y. Kim, W. Lee, D. Yoo, S. Chung, T. Lee, Non-Volatile Aluminum Oxide Resistive Memory Devices on a Wrapping Paper Substrate. vol. 1, Flexible and Printed Electronics, 201634001, https://doi.org/10.1088/2058-8585/1/3/034001.

[78] S.-W. Yeom, H.J. Ha, J. Park, J.W. Shim, B.-K. Ju, Transparent bipolar resistive switching memory on a flexible substrate with indium-zinc-oxide electrodes. Nanotechnology 69 (2016) L1613, https://doi.org/10.1088/0957-4484/27/7/07LT01.

[79] S. Chen, Z. Lou, D. Chen, G. Shen, An artificial flexible visual memory system based on an UV-motivated memristor. Adv. Mater. 30 (2018) 1705400, https://doi.org/10.1002/adma.201705400.

[80] S.-W. Yeom, S.-C. Shin, T.-Y. Kim, H.J. Ha, Y.-H. Lee, J.W. Shim, B.-K. Ju, Transparent resistive switching memory using aluminum oxide on a flexible substrate. Nanotechnology 27 (2016) 07LT01, https://doi.org/10.1088/0957-4484/27/7/07LT01.

[81] E. Carlos, A. Kiazadeh, J. Deuermeier, R. Branquinho, R. Martins, E. Fortunato, Critical role of a double-layer configuration in solution-based unipolar resistive switching memories. Nanotechnology 29 (2018) 345206, https://doi.org/10.1088/1361-6528/aac9fb.

[82] C. Kügeler, M. Meier, R. Rosezin, S. Gilles, R. Waser, High density 3D memory architecture based on the resistive switching effect. Solid State Electron. 53 (2009) 1287–1292, https://doi.org/10.1016/j.sse.2009.09.034.

[83] R. Aluguri, D. Kumar, F.M. Simanjuntak, T.-Y. Tseng, One bipolar transistor selector—one resistive random access memory device for cross bar memory array. AIP Adv. 7 (2017) 95118, https://doi.org/10.1063/1.4994948.

[84] I.-J. Baek, W.-J. Cho, Fabrication of IGZO-based 1T-1R ReRAMs for flexible and transparent system-on-panel (SoP) application. J. Nanosci. Nanotechnol. 17 (2017) 3065–3070, https://doi.org/10.1166/jnn.2017.14042.

[85] M. Zhang, S. Long, G. Wang, R. Liu, X. Xu, Y. Li, D. Xu, Q. Liu, H. Lv, E. Miranda, J. Suñé, M. Liu, Statistical characteristics of reset switching in $Cu/HfO_2/Pt$ resistive switching memory. Nanoscale Res. Lett. 9 (2014) 694, https://doi.org/10.1186/1556-276X-9-694.

[86] A. Rodriguez-Fernandez, C. Cagli, L. Perniola, J. Suñé, E. Miranda, Effect of the voltage ramp rate on the set and reset voltages of ReRAM devices. Microelectron. Eng. 178 (2017) 61–65, https://doi.org/10.1016/j.mee.2017.04.039.

[87] J. Won Seo, S.J. Baik, S.J. Kang, Y.H. Hong, J.H. Yang, K.S. Lim, A ZnO cross-bar array resistive random access memory stacked with heterostructure diodes for eliminating the sneak current effect. Appl. Phys. Lett. 98 (2011) 233505, https://doi.org/10.1063/1.3599707.

[88] C. Li, L. Han, H. Jiang, M.H. Jang, P. Lin, Q. Wu, M. Barnell, J.J. Yang, H.L. Xin, Q. Xia, Three-dimensional crossbar arrays of self-rectifying $Si/SiO_2/Si$ memristors. Nat. Commun. 8 (2017) 1–9, https://doi.org/10.1038/ncomms15666.

[89] J.-J. Huang, Y.-M. Tseng, C.-W. Hsu, T.-H. Hou, Bipolar nonlinear $Ni/TiO_2/Ni$ selector for 1S1R crossbar array applications. IEEE Electron Device Lett. 32 (2011) 1427–1429, https://doi.org/10.1109/LED.2011.2161601.

[90] M.-J. Lee, D. Lee, H. Kim, H.-S. Choi, J.-B. Park, H.G. Kim, Y.-K. Cha, U.-I. Chung, I.-K. Yoo, K. Kim, IEEE, Highly-scalable threshold switching select device based on chaclogenide glasses for 3D nanoscaled memory arrays. in: 2012 International Electron Devices Meeting, 2012, pp. 2.6.1–2.6.3, https://doi.org/10.1109/IEDM.2012.6478966.

[91] X.L. Shao, L.W. Zhou, K.J. Yoon, H. Jiang, J.S. Zhao, K.L. Zhang, S. Yoo, C.S. Hwang, Electronic resistance switching in the Al/TiO$_x$/Al structure for forming-free and area-scalable memory. Nanoscale 7 (2015) 11063–11074, https://doi.org/10.1039/C4NR06417H.

[92] C.-H. Huang, T.-S. Chou, J.-S. Huang, S.-M. Lin, Y.-L. Chueh, Self-selecting resistive switching scheme using TiO$_2$ nanorod arrays. Sci. Rep. 7 (2017) 2066, https://doi.org/10.1038/s41598-017-01354-7.

[93] M. Yang, H. Wang, X.M. Gao, Y. Hao, X. Ma, H. Gao, Voltage-amplitude-controlled complementary and self-compliance bipolar resistive switching of slender filaments in Pt/HfO$_2$/HfO$_x$/Pt memory devices. J. Vac. Sci. Technol., B 351 (2017) 32203–173504, https://doi.org/10.1116/1.4983193.

RRAM/memristor for computing

14

Rivu Midya, Zhongrui Wang, Mingyi Rao, Navnidhi Kumar
Upadhyay, J. Joshua Yang
Department of Electrical and Computer Engineering, University of Massachusetts,
Amherst, MA, United States

14.1   Introduction to a memristor

Resistors, capacitors, and inductors have been the sole examples of fundamental passive circuit elements until recently. A series of equations that were formulated as a result of combined efforts by physicists and electrical engineers describe the relationship between these basic circuit variables. However, in 1971, Leon Chua predicted that there exists a fourth fundamental passive circuit element that can be used to explain the relationship between the magnetic flux φ and charge q [1]. Chua named the fourth element "memristor" where the memristance is defined as

$$\mathrm{d}\varphi = M\mathrm{d}q \tag{14.1}$$

In 2008, Stanley Williams and colleagues from the Hewlett-Packard Labs linked this theory with a real-life prototype of memristor [2], which is a two terminal device often called resistive random access memory (RRAM) when used as memories. They used a coupled variable-resistor model to explain the internal mechanism during switching of the memristor, with mobile dopants being under the influence of an electric field [3]. Assuming the memristor has two regions—a doped, low resistance region and an undoped, high resistance region, an external bias $v(t)$ will move the boundary of the two regions by causing the charged dopant to drift. This phenomenon can be mathematically described by

$$v(t) = \left(R_{\mathrm{ON}} \frac{w(t)}{D} + R_{\mathrm{OFF}} \left(1 - \frac{w(t)}{D} \right) \right) i(t) \tag{14.2}$$

where $w(t)$ is the width of the doped area and D is the width of the entire region. R_{ON} and R_{OFF} represent the resistance of a fully doped device and an undoped device, respectively. If $R_{\mathrm{OFF}} \ll R_{\mathrm{ON}}$, the following equation can be deduced:

$$M(q) = R_{\mathrm{OFF}} \left(1 - \frac{\mu_v R_{\mathrm{ON}}}{D^2} q(t) \right) \tag{14.3}$$

where $M(q)$ is the magnetic flux and μ_v is the average dopant mobility. Hence, there is a linear relationship between magnetic flux and charge. A more easily captured behavior of such a device is the nonlinear relationship between voltage and current, which is denoted by Eq. (14.1). Furthermore, the memristor is static if no current or voltage is applied. If $I(t) = 0$, we find $q(t)$ and $M(t)$ are constant.

Advances in Non-volatile Memory and Storage Technology. https://doi.org/10.1016/B978-0-08-102584-0.00015-2

Although, resistive switching phenomena has been widely studied ever since the early 1960s [4], but the overwhelming success of Si technology since then had played a critical role in dulling the interest in other alternative technologies. This changed in the beginning of the millennium when it became clear that keeping up with Moore's law was going to become more challenging than ever as the transistor dimensions shrunk to the unprecedented atomic level. Memristive devices are all the rage now because they are scalable, their remarkable electrical performance and new functionalities enabled by such devices.

14.1.1   Switching modes

Most memristive/resistive switching devices possess either unipolar or bipolar switching as illustrated in Fig. 14.1 [5]. SET is the process by which a memristive device transitions from its high resistance state (HRS) to its low resistance state (LRS). Incidentally, when a device transitions from the LRS back to the HRS, it is referred to as the RESET process. Most reported memristors require some kind of electroforming process before they can exhibit "normal" resistive switching. The electroforming or FORMING process requires a larger voltage than that required for subsequent switching for the initial HRS to LRS transition. The reason some devices need this FORMING (or electroforming) process is due to the fact that a larger voltage or more time (for a given voltage stimulus) is required to build a complete filament from the pristine state of a dielectric film, whereas RESET and SET will just rupture or restore a part of the filament. Dielectric memristive materials (e.g., oxides or nitrides) which operate by manipulating native dopants (e.g., oxygen or nitrogen vacancies), the electroforming process involves creating native dopants electrically, while RESET and SET processes are dominated by moving the

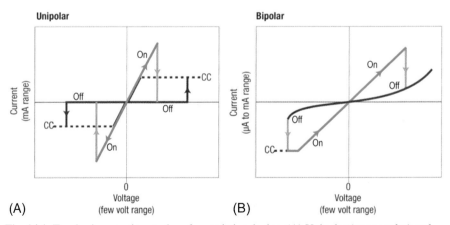

Fig. 14.1 Two basic operation modes of memristive device. (A) Unipolar (or nonpolar) and (B) bipolar. CC denotes compliance current controlled by external circuit.
Reproduced with permission from R. Waser, M. Aono, Nanoionics-based resistive switching memories, Nat. Mater. 6 (2007) 833–840, Copyright 2007, Springer.

native dopants electrically. Unipolar switching refers to switching where SET and RESET happen at the same polarity of applied voltages, while bipolar switching means SET and RESET happen at opposite polarities. If a device can be switched to either its HRS or its LRS at one polarity but can be switched back to its original resistance state at either polarity, the device is classified as a nonpolar memristor. Electroforming could be eschewed by building dopants or partial conducting paths into the switching film in the fabrication process [6].

14.1.2   Switching mechanism

Nonuniform (localized) switching and uniform switching are the two main identifiable types of redox-based resistive switching. The active region, in case of nonuniform switching devices is laterally localized rather than being uniformly distributed throughout the entire device area. The localized area is often referred to as the conductive filament (CF) or conduction channel. In case of uniform switching devices, the switching takes place across the entire device area and often at the interface between the electrode and the dielectric. Such kind of switching is often referred to as interfacial switching. The active region of nonuniform switching can also be vertically localized, that is, concentrated at the interface region. Nonuniform switching can be classified based on the mechanism of switching into three categories, namely, electrochemical metallization memory (ECM), valence change memory (VCM), and memristors which switch due to thermochemical activity (TCM). Nonredox resistive switching also occurs in phase-change memory (PCM).

14.1.2.1   Electrochemical metallization memory

Electrochemical metallization memristors [conductive bridge random access memory (CBRAM)] generally possess a dielectric film sandwiched between a relatively electrochemically inert electrode (e.g., Pt, Pd) and an electrochemically active metal electrode (e.g., Ag, Cu). ECM devices generally exhibit bipolar switching due to the formation and rupture of a CF composed of metal atoms. The applied electrical bias during the electroforming process ionizes the active metal atoms. Usually in ECM devices composed of electrolytic materials with high ion conductivity and low electron conductivity, the metal ions tend to move across the switching layer and accumulate on the inert electrode. They can nucleate and grow toward the active electrode side [7]. However, if the ion conductivity is low, the situation is different. In this case, usually, the filament grows toward the inert electrode after originating from the active electrode [8, 9]. The device is switched ON or to its LRS when the two electrodes are connected by a conductive channel. In the subsequent RESET process, an opposite polarity voltage again ionizes the metal atoms from the inert electrode side and drives them in the opposite direction, resulting in a partially ruptured filament. As a result of this, there is a residual portion of the ruptured filament next to an electrode. This makes the subsequent SET processes easier or in other words a smaller magnitude of voltage is required or a shorter time is needed to rebuild the filament.

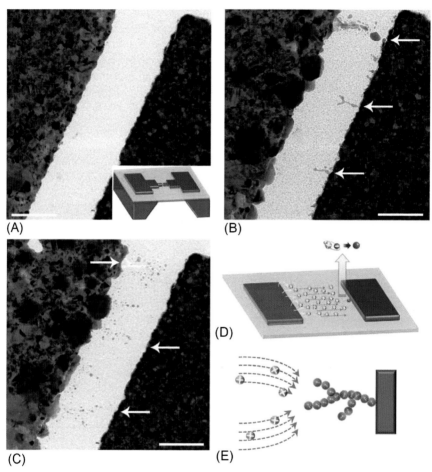

Fig. 14.2 (A) TEM image of as-fabricated resistive switching device on SiN$_x$ membrane. Inset: device structure schematic. Scale bar: 200 nm. (B) After FORMING process, some filaments have been observed as indicated by arrows. Scale bar: 200 nm. (C) After RESET process, the observation at the same site showed that the filament had dissolved. Scale bar: 200 nm. (D) Schematic representation showing the metal ionization, migration, and crystallization processes. (E) Schematic representation of cation nucleation at the electrode to form branch-shaped conductive filament.

Reproduced with permission from Y. Yang, et al. Observation of conducting filament growth in nanoscale resistive memories. Nat. Commun. 3 (2012) 732, Copyright 2012, Springer.

In 2012, Yang et al. observed that the conductive bridge of Ag filament is not necessarily a continuous wire. Instead, it can be composed of discrete nanoparticles (Fig. 14.2) [10]. The device is characterized by a lateral structure with Ag and Pt as the active electrode and inert electrode, respectively. The dynamics of the filament were observed by transmission electron microscopy (TEM). They also mathematically modeled the switching process by formulating a relationship of the wait time (time needed for ON-switching) t_W, electric field E, and temperature T. This shows that the filament growth can be limited by cation mobility in the dielectric film based on the following equation:

$$t_W = \frac{1}{v} e^{\frac{E_a'}{k_B T}} = \frac{1}{v} e^{\frac{E_a}{k_B T}} e^{\frac{-aE}{k_B T}} \tag{14.4}$$

where v is the attempt frequency and k_B is the Boltzmann constant. As a result of the applied electric field E, the real migration barrier E_a' is reduced from E_a to $E_a - aE$, where a is a constant. Similar TEM observation has been reported elsewhere [9, 11, 12].

Onofrio et al. carried out a detailed simulation of conductive bridge type resistive switching [13]. They used molecular dynamics simulation for this purpose. They concluded that a CF (in ECM devices) is responsible for the switching if and only if metallic clusters are formed as a result of metal ion aggregation. The resulting filament is stable and leads to nonvolatile switching.

14.1.2.2   Valence change memristor

The mobile species in VCM devices are oxygen (or nitrogen) vacancies, unlike ECM memristors, where active metal ions play a similar role as oxygen vacancies play in VCM memory [3]. Numerous oxides and a few nitrides have been used as thin dielectric films to demonstrate VCM resistive switching, including $SrTiO_3$ [14], TiO_2 [15], HfO_2 [16], Ta_2O_5 [17], and AlN [18]. Predominantly, the switching observed in such devices is of bipolar nature owing to the fact that such devices often possess an asymmetric structure. They generally, consist of a relatively inert electrode (e.g., Pt, Pd, TiN) serving as the anode during reset (OFF) switching operation, while the counter electrode is normally a more active and oxidizable metal.

The switching mechanism has been studied by Yang et al. [3] and can be described as follows. There is an excess of oxygen vacancies near the oxidizable electrode in the virgin state. Since oxygen vacancies serve as n-type dopants in TiO_2 [19], when a negative voltage is applied to the inert electrode, positively charged vacancies will be attracted to it. The device is switched ON when the conductive channel composed of oxygen vacancies penetrates through the film. When a positive voltage is applied to the inert electrode, oxygen vacancies are repelled away from that electrode and the original electronic barrier (e.g., a Schottky barrier between electrode and dielectric) is recovered. The device is then RESET or switched OFF.

A TEM study was conducted by Li et al. to observe the filament in a VCM memristor. The corresponding phase map is shown in Fig. 14.3. The phase map featuring $\Delta\varphi^{bias}(x,y)$ reflected the electric potential map because their difference is a constant [20]. Experimental results supported the conclusion that the FORMING and SET processes are a result of generated oxygen vacancies bridging both electrodes, while RESET is the result of recombination of oxygen vacancies in the filament and oxygen ions. A random circuit breaker network model was used to simulate the process and further supported their theory.

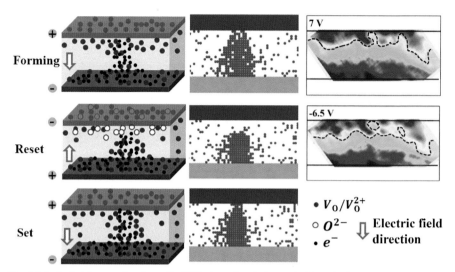

Fig. 14.3 Switching mechanism of a HfO₂-based resistive switching device. From top
to bottom are the forming, RESET, and SET process. The left column is the schematic
representation of switching dynamics. Mid column shows the simulation result of a conductive
filament using random circuit breaker network model. The right column is the potential map
of the resistive switching region, where dark red is for higher potential and dark blue is for
lower potential. The scalar bar means phase values, which is proportional to electric potential.
The dashed line is the boundary between positive and negative potential regions. A 7 V formed
the device, while −6.5 V RESET the device.

Reproduced with permission from C. Li, et al., Direct observations of nanofilament evolution
in switching processes in HfO₂-based resistive random access memory by in situ TEM studies,
Adv. Mater. (2017), Copyright 2017, Wiley.

14.1.2.3   Thermochemical effect memristor

Although thermochemical effects do coexist with valence change and electrochemical metallization mechanism, it alone can also cause resistive switching in dielectric
films. This kind of switching is a result of heat-induced formation and rupture of CFs.
Generally, it does not depend on external voltage polarity, so it demonstrates unipolar
or nonpolar behavior. Transition metal oxides (TMOs), including CuO [21], NiO [22],
and HfO₂ [23] can show such switching behavior.

Local heat is generated upon application of an electric field to the TMO. The
metal in the oxide is reduced to a lower valence or even metallic state as a result of
oxygen diffusing out of the high-temperature region due to thermophoresis. This
results in the device ON switching [7]. Due to Fick's diffusion, there is thermal dissolution of the metallic filament. This results in OFF switching. A reset simulation
performed by Russo et al. is illustrated in Fig. 14.4. They assume there is only one
filament and it is of a cylindrical shape. There is a local hot spot near the middle of
the filament from where the rupture of the filament begins. This hot spot eventually
melts the filament. Since a partially ruptured filament leads to a higher electric field
at the breaking point, the dissolution is accelerating itself, leading to an abrupt OFF
switching [24].

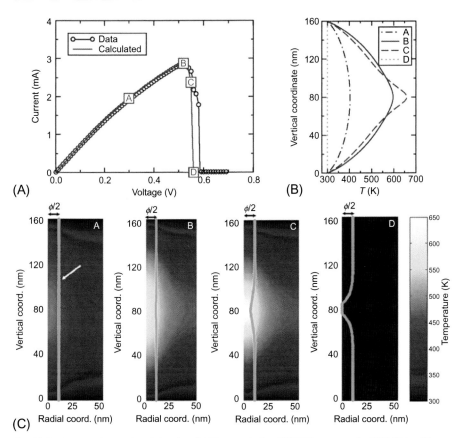

Fig. 14.4 (A) Experimental and simulated *I-V* curve during RESET process of a NiO film with n-Si as bottom electrode and top electrode. (B) and (C) Simulation of thermal dissolution of the filament corresponding to the four point A, B, C, and D in (A). High temperature is observed in the middle of the filament which accelerates the dissolution process. φ denotes the radius of filament. Reproduced with permission from U. Russo, et al., in: IEEE International Electron Devices Meeting. IEDM 2007, IEEE, 2007, pp. 775–778, Copyright 2007, IEEE.

Kwon et al. observed CFs in a unipolar switching TiO_2 device by performing in situ TEM [25]. The device has a structure of $Pt/TiO_2/Pt$. During the forming process, a blown-off region is observed, which is probably due to a sudden evolution of compressed oxygen gas released from the oxide during electroforming [26]. According to the in situ TEM observation, as voltage is applied to the device, some part in the TiO_2 film is changed to Ti_4O_7, a Magneli phase that is conductive at room temperature. The reversible conversion between such Magneli phase nanofilaments and the TiO_2 phase is responsible for the observed unipolar resistive switching in the TiO_2 thin film.

14.1.2.4 Interface switching memristor

Interface-type memristive switching is different from the aforementioned memristive switching mechanisms due to the fact that it does not form a localized filament to connect the top and bottom electrodes. The switching takes place uniformly across the entire device area at or near the interface between the dielectric layer and one of the electrodes. A FORMING process is usually not needed for such switching. It is worth noting that many filamentary types of switching can also occur in a region close to or at the interface between the inert electrode and the dielectric layer likely due to the relatively high electric field in this region.

One of the proposed explanations for interface switching involves the modulation of electronic transport by field-induced redistribution of oxygen vacancies. Resistance change in the stack is induced by the redistribution of oxygen vacancies which changes the electronic band structure and in turn the height or width of the energy barrier for electron transport. R. Meyer et al. studied interface-type switching based on a stack consisting of a tunnel oxide (TO) layer and a conductive metal oxide (CMO) [27]. The switching mechanism and a simulation on the oxygen vacancy drifting are shown in Fig. 14.5. Oxygen ions migrate from the CMO into TO, due to a high electric field caused by the application of an external voltage. The electron tunneling probability is reduced because the negatively charged oxygen ions lead to a high potential barrier. The device is thus programmed into a HRS. By changing the polarity of the voltage, oxygen vacancies are driven back from TO to CMO, which restores the tunneling barrier and the device switches back to its LRS.

It is worth noting that more than one switching mechanisms can be at play in a certain device. For instance, when an electric field drives metal ions or oxygen vacancies, a thermochemical reaction can also take place. This is the reason why different operation modes sometimes coexist in the same material system under different operation conditions [28]. Although unipolar switching usually shows a higher ON/OFF conductance ratio compared to bipolar switching, it typically consumes more energy and generates more heat, which is usually detrimental to the device's reliability in the long run.

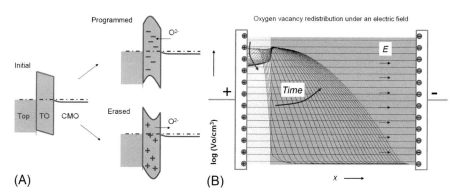

Fig. 14.5 (A) Sketched band structure of initial, programmed, and erased device. (B) Simulated time evolution of oxygen vacancy distribution.
Reproduced with permission from R. Meyer, et al., in: 9th Annual Non-Volatile Memory Technology Symposium, NVMTS 2008, IEEE, 2008, pp. 1–5, Copyright 2008, IEEE.

14.1.3   Figures of merit for resistive switching

There are a few indices which serve as a measure of the performance of memristors. Speed characterizes how fast the state of a memristor can change with time. Endurance measures how good a memristor could repeat a state transition cycle. Retention is a measure of the memristor's ability in maintaining its state, and uniformity depends on the performance variation from cycle to cycle or device to device.

14.1.3.1   Switching speed

One of the most important performance metrics of memory is the reading and writing speed. FLASH memory is slower compared to static random-access memory (SRAM) due to slow charging of the floating gate over a large electronic barrier via electron tunneling. SRAM, although much faster, is expensive and has a large-area footprint. Memristors have shown sub-100 ps switching [18], which is likely more than sufficient for most of the foreseeable applications in the near future.

The switching mechanism of memristors, and hence the factors that determine the switching speed of a memristor are debatable. Nevertheless, there are certain known factors. One of them is the mobility of mobile ions. As discussed in Section 14.1.2, most resistive switching is related to migration of either metal cations or oxygen (or nitrogen) anions. Ion migration can be described by Eq. (14.1). When an electric field is applied, charged ions are driven by electric force. If the electric force is large enough, the ions can overcome the energy barrier and hop to the next-neighboring site that is most favorable in terms of energy. Under a high electric field, the mobility of the ions can be exponentially enhanced. Thus, both the selected materials and the applied voltage affect the ion transport process. Temperature also has a complicated relationship with the switching process, not only with regards to ion migration but also in terms of formation of chemical bonds. Although it is generally true that by applying a higher voltage the device is able to switch exponentially faster [10], increasing voltage can have some undesired effects in terms of power consumption and device reliability.

There have been a few reports of encouraging speed performances recently. Choi et al. reported 85 ps switching in a TiN/AlN/Pt device (Fig. 14.6) [18]. In this work, nitride vacancies are used instead of oxygen vacancies as the migration species. The rationale behind this choice is the fact that nitride vacancies have a valence of +3, which might be subjected to a stronger electric force compared to +2 valued oxygen vacancies. In an earlier work, Pickett et al. observed that a very small volume of the dielectric acts as the switching site. Joule heating-induced phase change results in sub-nanosecond switching [29]. CBRAM is relatively slower compared to other types of switching mechanisms. It takes some time before a stable metallic filament is able to bridge the two electrodes [13]. In addition, multiple steps are involved in CBRAM switching, including oxidation, detaching ions from one electrode, ion transport, attaching ions to the other electrode and reduction, all of which contribute to a longer switching time. However, such switching can still take place within a few nanoseconds, which has been demonstrated experimentally [30].

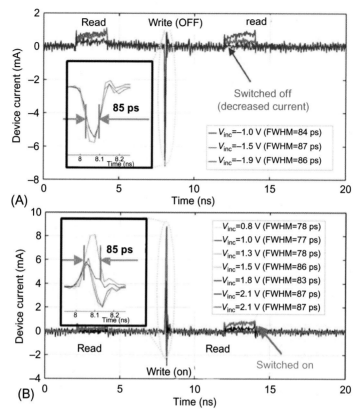

(A)

(B)

Fig. 14.6 (A) 85-ps voltage pulse OFF switching. (B) 85-ps voltage pulse ON switching. The insets to (A) and (B) are the zoom-in image of the switching pulses. A read pulse was applied before and after the switching pulse. Three attempts with increasing voltage magnitude were used. The devices were successfully switched only if the threshold voltage was reached, that is, −1.9 V for OFF switching and 2.1 V for ON switching. FWHM means full-width at half-maximum.

Reproduced with permission from B.J. Choi, et al., High-speed and low-energy nitride memristors, Adv. Funct. Mater. 26 (2016) 5290–5296, Copyright 2016, Wiley.

14.1.3.2 Endurance

The number of cycles a memory device can be programmed and erased before failing is called the "endurance" of a device. There are many factors that can influence the endurance of a device. The major factors are the constituents of the memristor and electrical operation schemes. After repeatedly switching a device, the device undergoes an irreversible change of the switching material at the active region of the device, forcing it to get stuck in its HRS or its LRS state.

In a study by Chen et al., three main mechanisms of endurance degradation in TMO-based VCM RRAM were discussed [31]. One of the proposed mechanisms states that oxygen species can react with an electrode and form a metal oxide layer, resulting in a highly resistive region, preventing the device from a successful SET operation. The second proposal states that excessive oxygen vacancies can make the filament too thick to be ruptured, leading to a RESET failure. The third proposal states due to the effect of a continuously applied electric field and joule heating, oxygen species tend to migrate away from the filament site. Therefore, available oxygen is inadequate to recombine with oxygen vacancies, also resulting in a RESET failure.

Changes in the distribution of metal are usually the reason for endurance degradation in CBRAM. Fig. 14.7 shows the material evolution after repeated SET and RESET [32]. The CBRAM device comprises of a layer of HfO_x sandwiched by Cu (active electrode) and Pt (inert electrode). The Cu electrode serves as a source of mobile Cu ions which can form a CF. According to the TEM observations and energy-dispersive X-ray spectroscopy (EDS) characterizations after switching the

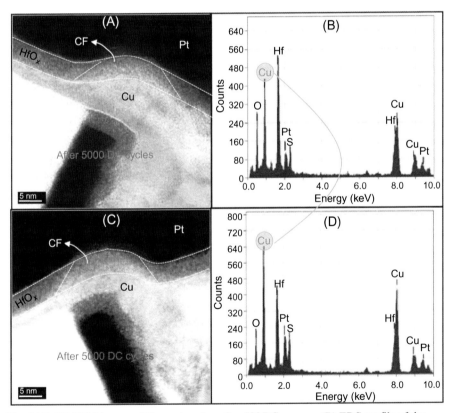

Fig. 14.7 (A) TEM image of filament region after 500 DC sweeps. (B) EDS profile of the filament region after 500 DC sweeps. (C) TEM image of filament region after 5000 DC sweeps. (D) EDS profile of the filament region after 5000 DC sweeps. CF is short for the conductive filament.

Reproduced with permission from H. Lv, et al., in: IEEE International Conference on Electron Devices and Solid-State Circuits (EDSSC), IEEE, 2015, pp. 229–232, Copyright 2015, IEEE.

device for a given number of cycles, Cu accumulates on the Pt electrode after being
excessively cycled. Subsequently, the gap between the electrode and the filament
edge becomes narrower and electrons can tunnel through, which causes the device to
be stuck in the ON state.

According to the 2015 International Technology Roadmap for Semiconductors
(ITRS), the valence change type memristor has the best endurance performance. An
important requirement of the high endurance is a thermodynamically stable filament.
For example, in TaO_x, the conductive Ta filament will not react with Ta_2O_5 even under
high temperature [33]. Similarly, the HfO_x-based memristor has also exhibited high
endurance [34], which is probably due to the fact that the Hf-O system only allows
two thermodynamically stable phases within the operating temperature range inside
the device.

Recently, a graphene/$MoS_{2-x}O_x$/graphene device has shown excellent switching
characteristics and endurance as high as 10^7 at high operating temperatures up to
340°C [35], suggesting the potential of using two-dimensional (2D) crystalline materials for memristors with desirable properties.

14.1.3.3   Retention

The amount of time a memory device can maintain its programed resistance state at
a certain temperature is known as its "retention." It is of particular importance for
nonvolatile memory applications. The standard retention requirement for nonvolatile
memory is 10 years at 85°C. Since the degradation time conforms with the Arrhenius
dependence on temperature, retention is usually tested at high stress temperatures so
that it can be performed within a reasonable timeframe, which also allows one to extract the activation energy for ion motion in the device.

In general, filament-based memristors have been reported to possess long retention
times, whereas electronic memristors possess relatively shorter retention times [36].
One of the possible reasons is as follows. The switching of an electronic device is
due to trapped-charge-induced modulation of the electron barrier height. At the same
temperature, it is easier for charge carriers to trap/detrap compared to the migration
of ions [37].

The retention failure in filament-based memristors is a result of unwanted migration
of oxygen or metal ions (Fig. 14.8) [38]. Oxygen vacancies or metal ions moving from
the filament site to elsewhere can make the device fail to maintain its LRS. It has been
observed that LRS failure usually happens before HRS failure when increasing testing
temperature, suggesting LRS retention is generally a more vulnerable issue [39].

One of the common ways to increase retention time in a filament-based memristor
is to limit ion migration during retention tests. After the introduction of a Hf cap layer,
an improvement of retention in HfO_2-based RRAM had been achieved. The additional
cap layer reduces available oxygen and slows down the recombination process [40].

It is, however, worth noting that reducing mobility of ions might lead to SET/
RESET failure or an increment of switching energy. Thus, a trade-off between retention and endurance needs to be achieved for specific applications. Furthermore,
different electrical conditions can benefit some of the performance parameters while

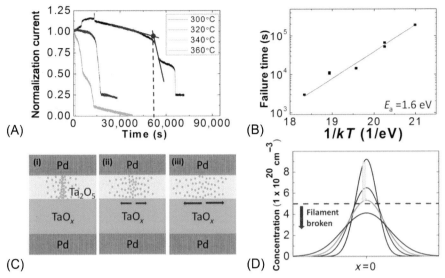

Fig. 14.8 (A) Retention time of Pd (top electrode)/Ta$_2$O$_5$/TaO$_x$/Pd (bottom electrode) memristor at 300°C½320°C, 340°C, and 360°C. (B) Fitted temperature dependence of the characteristic retention failure time (squares) following Arrhenius law. (C) Schematic representation of oxygen vacancy distribution at LRS (i), after oxygen vacancies diffused out (ii), and filament ruptured (iii). (D) A prediction of vacancy concentration profile as a function of time.

Reproduced with permission from S. Choi, J. Lee, S. Kim, W.D. Lu, Retention failure analysis of metal-oxide based resistive memory. Appl. Phys. Lett. 105 (2014) 113510, Copyright 2014, AIP.

harming others. It is important to optimize both parameters in a balanced way based on the application requirements [41].

14.1.3.4 Uniformity

Another critical figure of merit for a memristor is its uniformity. The uniformity of operating voltage, speed, resistance in HRS and LRS, and some other parameters will determine how easily a memristor can be accommodated into a large scale and multi-functional circuit.

Memristor devices show both device-to-device variation and cycle-to-cycle variation. There are several factors that can lead to cycle-to-cycle variation including randomness of ion migration, gradual changing of morphology of a filament or switching interface, current overshoot, and so on. Device-to-device variation can also arise due to limitations of fabrication conditions.

A twofold approach has been adopted for solving the uniformity problem. One is material oriented, while the other is electrical operation oriented. In the material-oriented method, one idea is to use doping methods to reduce formation energy and the migration barrier of oxygen vacancies so that the filament is easier to be either ruptured or restored under a reasonable amplitude of applied voltages [42]. Another effective approach is to insert a capping layer between the electrode and the resistive switching layer. Depending on the material stack, the capping layer may have different functions. Lv et al. inserted a GST ($Ge_2Sb_2Te_5$) layer in a Cu_xO-based memristor [43]. In the forming process, some conductive path is built in the GST layer and the filament tended to be confined to a localized region. Yu et al. used a buffer oxide layer which had a high oxygen migration barrier which impeded the filament from easily rupturing. The buffer layer is also useful to mitigate the over-RESET problem. (During reset, the filament is typically ruptured partially, leaving a narrow gap, but if an excessive voltage is applied the gap becomes wider.) This makes the HRS more uniform [44].

Electrical-oriented approaches can help enhance uniformity in a different manner. In general, the voltage applied to operate the device needs to be carefully balanced. Too low a voltage may lead to a thin, or even an unstable filament, while too high a voltage can cause the formation of unwanted filaments and excessive Joule heating. The electrical stimuli can affect the uniformity as well. In an earlier work, better uniformity was observed in current sweeping mode than in voltage sweeping mode. The reason could be that current sweeping could prevent formation of additional filaments once the device had been SET [45].

14.2   Applications

Memristive devices have many advantages over complementary metal-oxide-semiconductor (CMOS) devices such as increased throughput due to parallelism, low-power consumption, and reduced footprint and stackability. On top of that, memristive devices are intrinsically analog [46]. Due to the versatile nature of memristive devices, memristors can serve as the fundamental building blocks of many different kinds of networks, which in turn can reap the benefits of the rich properties possessed by these devices. Artificial neural networks or level-based computing rely heavily on accurate tuning of weights or conductance, whereas spiking neural networks rely on more complex memristor dynamics and threshold switching to function.

Being a novel circuit element, versatile memristive devices can be used for attractive alternative computing applications such as a true random number generator (TRNG) and for image compression tasks which are computationally expensive on similar CMOS platforms. Last but not the least, memristors, with proper passive memristor accessing devices, could lead to high-density three-dimensional (3D) packed array for ultra-high-density information storage. A certain kind of volatile memristive device can be used for this purpose, as they are threshold devices which possess high nonlinearity, making them good enough to suppress sneak path currents.

14.2.1   Neuromorphic systems

Neuromorphic computing represents several approaches for processing information that derives inspiration from neurobiological systems. This feature distinguishes neuromorphic systems from conventional computing systems. The brain is able to solve difficult engineering problems by using efficient low-power computation. The goal of neuromorphic engineering is to exploit the known properties of biological systems to design and implement efficient devices for engineering applications. Hardware neural emulators are useful to simulate large-scale neural models which could explain how intelligent behavior arises in the brain. The main advantages of neuromorphic emulators are that they are highly energy efficient, parallel and distributed, and require a small footprint.

In this section, we review efforts undertaken by the memristor community to build efficient neuromorphic systems. The discussion is divided into two parts, starting with a closer emulation of biological systems, namely, the bio-inspired approach and the versatile computation-oriented approach, namely, neural networks.

14.2.1.1 Bio-inspired approach

In this section, we review some advances made in emulating neurobiological systems closely, using memristive devices.

Artificial synapses

Chemical synapses act as junctions between neurons which can modulate the neural signal transmission via their synaptic weights, which can be updated through the so-called plasticity processes. Transistors have long been employed to implement synaptic functions [47–53]. Two-terminal circuit elements, such as memristors [1, 2, 5, 25, 54–56] have recently been utilized for this purpose in neuromorphic architectures for their simplicity and power efficiency [57–65].

Volatility of resistive switching in Ag-based ECM threshold switches has been reported for realizing neuronal functions like short-term synaptic plasticity (STP) [59, 66], short-term plasticity to long-term plasticity (LTP) transition [59, 63, 67], as well as spike-timing-dependent plasticity (STDP) [66, 67].

Ohno et al. reported short-term facilitation of a $Pt/Vac/Ag_2S/Ag$ system. The system showed high conductance upon the stimulation of each input pulse (80 mV/0.5 s) and decayed to the low conductance state during a relatively long (e.g., 20 s) interval between pulses, which mimicked the transmission weight relaxation in short-term facilitation of a biological synapse [59] (see Fig. 14.9A). With an increased stimulation frequency (2 s interval between 80 mV/0.5 s pulses), Ohno et al. identified a persistent conductance increment, or long-term facilitation, after seven spikes in the $Pt/Vac/Ag_2S/Ag$ system as shown in Fig. 14.9B.

Wang et al. reported similar Ag dynamics in a $Pt/SiO_xN_y:Ag/Pt$ threshold switch, sharing strong analogy with the Ca^{2+} dynamics of chemical synapses, which led to both short-term facilitation and depression depending on the time interval between spikes (denoted as t_{zero}) (see Fig. 14.9C). High-frequency stimulations increased the conductance of the threshold switch from its initial ON state conductance, while a slower conductance increase or even a decrease (short-term depression) from the same initial conductance was observed with a lower-frequency stimulation [66]. Another important feature of chemical synapses is that prolonged or excessive electrical spikes will lead to an inflection from initial facilitation to eventual depression at high-frequency

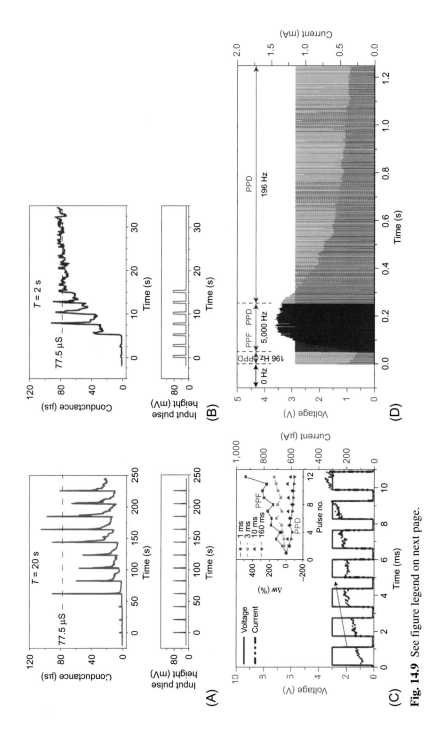

Fig. 14.9 See figure legend on next page.

stimulation [68–70], which has also been observed in Fig. 14.9D [66]. The device in its steady state had exhibited depression upon application of low-frequency (196 Hz) spikes, followed by facilitation at 5000 Hz simulation, then turned into depression with subsequent low-frequency (196 Hz) pulses, which eventually recovered the initial ON state conductance [66] (see Fig. 14.9D).

When combined with a nonvolatile Pt/TaO$_x$/Ta/Pt drift memristor, the Pt/SiO$_x$N$_y$:Ag/ Pt threshold switch was able to show LTP-based [71] learning protocols including the spike-rate-dependent potentiation [72] and STDP [71, 73, 74] (see Fig. 14.10A). Wang et al. reported the rate-dependent potentiation where the drift memristor weight (resistance) change was a function of the frequency of the applied pulses, as shown in Fig. 14.10B [66]. A higher frequency or shorter t_{zero} resulted insufficient time for the Pt/SiO$_x$N$_y$:Ag/Pt threshold switch to relax back to its HRS. Therefore, a larger voltage would drop across the drift memristor according to the voltage dividing between the drift and diffusive memristors. This induced a larger resistance change in the drift memristor, analogous to the LTP. On the other hand, a lower frequency of the stimulation spikes (a longer t_{zero}) led to a sufficient time for the threshold switch to relax back to its HRS and thus a smaller voltage drop across the drift memristor, inducing negligible resistance change of the drift memristor [66].

Wang et al. also demonstrated the STDP learning rule on the combined device in a similar fashion with nonoverlapping electrical pulses (see Fig. 14.10C for the waveform). The synaptic dynamics of the threshold switch offered an intrinsic timing mechanism for the combined device. The initial spike initiated the timer by turning on the threshold switch while not affecting the drift memristors [66]. The resistance of the diffusive memristor then gradually relaxed back, serving the role of a timed switch [66]. Whether the second spike was able to program the drift memristor or not depended on the time interval between two spikes (Δt) [66]. As shown in Fig. 14.10D, a narrower interval Δt corresponded to a smaller diffusive memristor resistance and a larger resistance change in the drift memristor and vice versa [66]. If the prespike

Fig. 14.9, Cont'd Synaptic applications with threshold switches (A) short-term facilitation and (B) long-term facilitation of a Pt/Vac/Ag$_2$S/Ag system. Change in the conductance of the inorganic synapse when the input pulses ($V = 80$ mV, $W = 0.5$ s) were applied with intervals of $T = 20$ s in (A) and 2 s in (B). (C) Short-term facilitation and depression of a Pt/SiO$_x$N$_y$:Ag/ Pt threshold switch. Device current response *(blue)* to multiple subsequent voltage pulses (3 V, 1 ms) at different inter-pulse time t_{zero}. The weight change slows down with increasing t_{zero} and eventually becomes negative. (D) Short-term depression following facilitation of a Pt/SiO$_x$N$_y$:Ag/Pt threshold switch. Device current *(blue)* responses to a train of voltage pulses (2.8 V, 100 μs) of the same amplitude but varying frequency. (D) Long-term spike-rate-dependent potentiation and bio-realistic spike-timing-dependent plasticity (SRDP and true STDP) behavior of a combined device consisting of a diffusive and a drift memristor Panels (A) and (B) Reproduced with permission T. Ohno, et al., Short-term plasticity and long-term potentiation mimicked in single inorganic synapses, Nat. Mater. 10 (2011) 591–595, Copyright 2011, Nature Publishing Group. Panels (C) and (D) Reproduced with permission from Z. Wang, et al., Memristors with diffusive dynamics as synaptic emulators for neuromorphic computing, Nat. Mater. 16 (2016) 101–108, Copyright 2017, Nature Publishing Group.

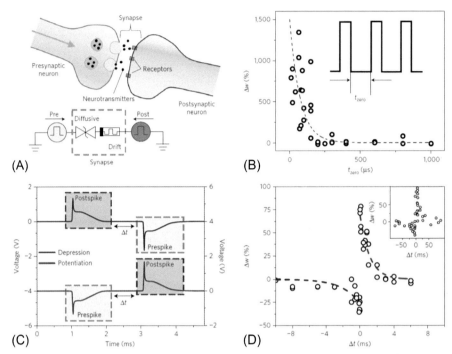

Fig. 14.10 (A) Illustration of a biological synaptic junction and a circuit diagram of the electronic counterpart consisting of a Pt/SiO$_x$N$_y$:Ag/Pt threshold switch connected in series with a Pt/TaO$_x$/Ta/Pt memristor. (B) Spike-rate-dependent potentiation of the combined device in (A). A longer t_{zero}, results in a lesser weight change of the memristor. (C) Schematic representation of the pulses applied to the combined device in (A) for STDP demonstration. (D) STDP of the combined device in (A). A shorter Δt corresponds to a larger weight change. The inset shows the spike-timing-dependent plasticity of a typical chemical synapse.
Panels (A)–(D) Reproduced with permission from Z. Wang, et al., Memristors with diffusive dynamics as synaptic emulators for neuromorphic computing, Nat. Mater. 16 (2016) 101–108, Copyright 2017, Nature Publishing Group.

appeared before the postspike, the drift memristor was SET by the post-spike (i.e., potentiation) [66]. If the prespike followed the postspike, depression would occur as the drift memristor was RESET by the prespike, which naturally met the STDP rule observed in chemical synapses without using engineered overlapping spikes [66].

Nociceptor
Imitating different aspects of the mammalian nervous system is an interesting and exciting direction of research. Memristors have been used mostly to emulate learning phenomenon exhibited by synapses and also to emulate certain behavior by neurons. However, there are still other instances where a memristor can be used to imitate the nervous system in other ingenious ways.

    Recently, in an article published by Yoon et al. [75], memristors were used to emulate a nociceptor. A nociceptor is a special kind of sensory neuron that specifically reacts to damaging or potentially damaging stimuli by sending a "warning" or a "threat" signal to the spinal cord or the brain which in turn initiates adequate motor response.

A diffusive memristor bears a one-to-one correspondence with a nociceptor (because these two elements both possess threshold dynamics) as illustrated in Fig. 14.11. Pulse measurements performed on a diffusive memristor revealed that the three key features of the nociceptor, namely, "threshold," "no adaptation," and "relaxation" were also characteristic of the diffusive memristor.

Apart from this, sensitization functions such as allodynia and hyperalgesia were demonstrated using the memristive nociceptor (Fig. 14.12).

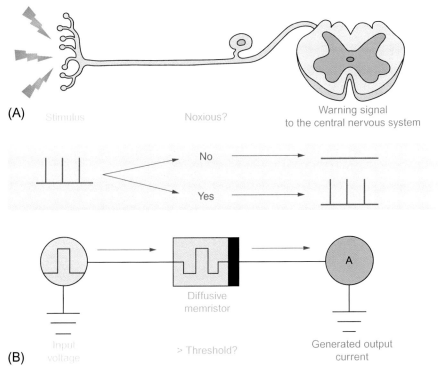

Fig. 14.11 One-to-one correspondence of the nociceptor system in the human body and an artificial nociceptor circuit consisting of a diffusive memristor (threshold switch). (A) When a noxious stimulus is received from a free nerve ending, the nociceptor compares the amplitude of the signal with a threshold value and decides whether an action potential should be generated and sent to the brain via the spinal cord (central nervous system). (B) An electrical pulse applied on the device emulates the external stimulus. When the pulse amplitude is higher than the threshold voltage of the diffusive memristor, the device is turned ON, and current pulses are generated.
Reproduced with permission from Yoon et al., An artificial nociceptor based on a diffusive memristor, Nat. Commun. 9 (2018) 417.

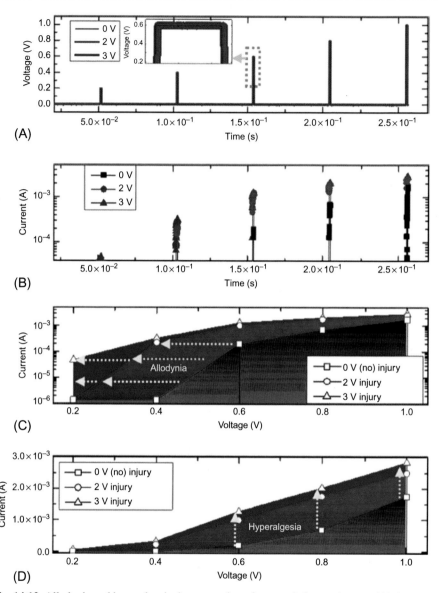

Fig. 14.12 Allodynia and hyperalgesia demonstration of a memristive nociceptor. (A) A train of input voltage pulses composed of a variable range of pulse amplitudes from 0.2 to 1.0 V, with a 1 ms pulse width applied on a nociceptor subjected to different set pulses (0 V as reference and 2, 3 V with a 1 ms width as injury cases) first and (B) the corresponding output currents, respectively. The maximum output currents at a different input voltage amplitudes (C) in log scale and (D) in linear scale, demonstrating the shift of the ON-switching voltage toward a lower threshold (allodynia) and the ON current toward higher currents (hyperalgesia). Reproduced with permission from Yoon et al., An artificial nociceptor based on a diffusive memristor, Nat. Commun. 9 (2018) 417.

14.2.1.2   Neural networks

The brain and the nervous system in animals is comprised of a complicated system of connected nerve cells or neurons which are interconnected by synapses. Neural networks are an attempt to replicate the nervous system to do computation in a similar way. Neural networks use a complex system of connected artificial neurons (or nodes) to perform various computational tasks. The neurons are interconnected with each other through artificial synapses which can be assigned a weight to modulate the strength of an input being received by a neuron [76]. The neurons are interconnected among each other to allow for the incoming stimulus to be processed and a corresponding output to be generated. Neural networks are generally organized into layers such as an input layer, intermediate hidden layers, and an output layer, where each layer is composed of a number of nodes to allow for data processing. Recently, there have been a lot of attempts to implement neural networks using memristive device arrays. In this section, we review some recent strides made in this area.

Pattern classification using passive memristor array

Researchers in UCSB have recently demonstrated a single perceptron in a passive memristor crossbar array for the first time [64]. They developed memristors with low variability by using a bilayer of Al_2O_3 and TiO_{2-x}. They fabricated an integrated 12×12 array of these devices as shown in Fig. 14.13A and B. The memristors in the array possessed sufficiently uniform current-voltage characteristics. The crossbar was used to implement a single-layer perceptron with 10 inputs and 3 outputs, which are fully connected with 30 synaptic weights as shown in Fig. 14.13C. This network was used to classify 3×3 pixel black and white images into three classes (Fig. 14.13D). The images were represented by nine network inputs (V_1, ..., V_9) which in turn corresponded to pixel values. Another input V_{10} was used as a source of bias for the nonlinear activation function. This network was tested on a set of 30 patterns. These pattern included the letters z, v, n and three sets of nine noisy versions of each letter which were generated by flipping one of the pixels of the original image (Fig. 14.13E).

Sparse coding

Hubel and Wiesel [77] had found in their iconic paper 50 years ago that some neurons in the primary visual cortex fire selectively only for certain specific orientations in their receptive fields. In other words, the neuronal representation in the primary visual cortex is highly input specific. They also observed the fact that for a given input, only a few neurons were active. In other words, active neurons for a given input are sparse. The sparse coding algorithm uses this concept to implement an input-specific neuronal representation by a small portion of the available neurons.

Memristor crossbar networks are naturally suitable for implementing neural network inspired algorithms due to the fact that they can be used to perform vector-matrix multiplication (VMM) operations with only a single read step (as explained in the previous section). Sheridan et al. [78] demonstrated sparse coding in a 32×32 WO_x-based VCM array (Fig. 14.14A and B). They used the array to implement the algorithm and performed natural image analysis using learned dictionaries. The prototypical method

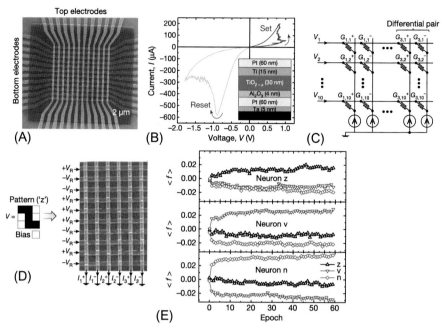

Fig. 14.13 (A) Integrated 12×12 crossbar with an Al_2O_3/TiO_{2-x} memristor at each crosspoint. (B) A typical current-voltage curve of a formed memristor. (C) An implementation of a single-layer perceptron using a 10×6 fragment of the memristive crossbar. (D) An example of the classification operation for a specific input pattern (stylized letter "z"), with the crossbar input signals equal to $+V_R$ or $-V_R$, depending on the pixel color. (The read and write biases were always $V_R = 0.1\,V$ and $V_W^{\pm} = \pm 1.3\,V$, respectively.) (E) The evolution of output signals, averaged over all patterns of a specific class.

Reproduced with permission from M. Prezioso, et al., Training and operation of an integrated neuromorphic network based on metal-oxide memristors, Nature 521 (2015) 61–64, Copyright 2015, Nature Publishing Group.

worked in two phases. In the first phase, the dictionary was mapped element wise to the weights (conductance) of the memristors in the crossbar array. The input vector was applied by encoding voltage pulses of fixed amplitude and variable width corresponding to the input data value. Due to the inherent dot product engine (DPE)-like nature of the memristor crossbar, the product of the input data with the dictionary mapped weight in the memristor corresponds to the charge generated in the memristor. In the same read step, all the elements in the same column sum their currents up due to Kirchoff's current law. The output neuron tied to the column is set active for the next phase, if the sum of the preceding single step dot product and the existing membrane potential of the neuron is higher than a preset threshold. In the next phase, the input image is reconstructed based on the currently active neurons and compared with the original image.

Fig. 14.14C shows that the dictionary comprises of a total of 19 features (where each feature is a 5×5 array). This dictionary is stored in a 32×32 array. 25×19 cells in the array were used because each 5×5 feature in the dictionary was vectorized.

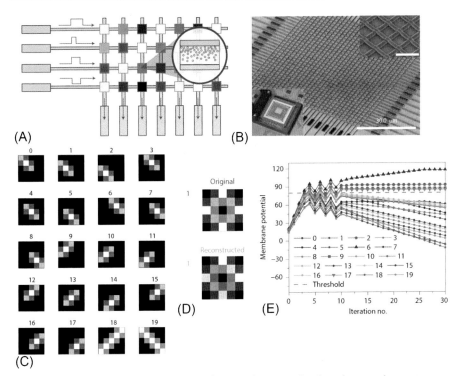

Fig. 14.14 (A) Schematic representation of a memristor crossbar-based computing system, showing the input neurons *(green)*, the memristor crossbar array, and the leaky integrating output neurons *(pink)*. A memristor is formed at each crosspoint, and can be programmed to different conductance states (represented in gray scale) by controlling the internal ion redistribution (inset). (B) Scanning electron micrograph (SEM) image of a fabricated memristor array used in this study. Upper right inset: magnified SEM image of the crossbar. Scale bar, 3 μm. Lower left inset: memristor chip integrated on the test board after wire bonding. (C) Dictionary elements programmed into the memristor crossbar array. Each dictionary element is stored in a single column and connected to an output neuron. The different gray scales represent four different levels. The neuron number is listed above each element. (D) The original image to be encoded and the reconstructed image after the memristor network settles. (E) Membrane potentials of the neurons as a function of iteration number during locally competitive algorithm (LCA) analysis. The *red* dashed horizontal line marks the threshold parameter λ. The same threshold, $\lambda = 80$, is used in all experiments in D–E. Reproduced with permission from P. M. Sheridan, et al., Sparse coding with memristor networks, Nat. Nanotechnol. 12 (2017) 784–789, Copyright 2017, Nature Publishing Group.

The activation function is illustrated in Fig. 14.14D. The activation function was acquired using the locally competitive algorithm. It has a threshold for the onset of nonzero activation. The change of membrane potential (u) for each neuron during iteration is plotted in Fig. 14.14E in which above threshold membrane potential converges after 30 iterations. The reconstructed data from the acquired activation were consistent with the original input data.

Analog computation and neural network classification with a memristor-based DPE

Hu et al. [79] used their 128×64 memristor crossbar array to do vector matrix multiplication. They used their one-transistor one-memristor (1T1R) network (refer to Section 14.2.2.2 for more details about 1T1R) to perform a single-layer neural network inference. Specifically, they performed a MNIST handwritten digit classification.

The authors train offline, a single-layer softmax neural network and convert it to the targeted conductance values and then program it into a 96×40 1T1R array (Fig. 14.15). A 19×20 pixel image is unwrapped into a 380 input vector which is converted into voltages and then applied to the matrix in four sets. The summed result of the four outputs yields the DPE classification which is illustrated in Fig. 14.9. Similarly, successful classification has been performed on the digits 4 and 5 and the experimental results (*red*) are in accordance with the ideal software results (*blue*). The total overall recognition accuracy for each digit for 10,000 images from the MNIST database is as high as 89.9%.

Fully memristive neural network

Artificial neurons and synapses could pave the way for more efficient implementation of neural network algorithms compared to traditional computation approaches. There have been several attempts at neuromorphic computing using traditional CMOS-based hardware [80, 81]. Although they have been successful in performing complicated cognitive applications, they have been severely inefficient with regards to power consumption. Looking toward alternative technologies such as memristors is the way to go for neuromorphic computing.

There have been a few reports of artificial neurons built using memristors, however, they possess limited bio-realistic dynamics and no direct interaction with the artificial synapses in an integrated network [82–85]. Recently, Wang et al. [86] have used Ag-based diffusive memristors to construct neural emulators that can show stochastic leaky integrate and fire dynamics with a tunable integration time. The authors went on to integrate these neurons with nonvolatile memristive synapses to build a fully memristive artificial neural network. The authors demonstrated unsupervised learning using the aforementioned integrated network.

A biological neuron has been functionally and physically emulated by a diffusive memristor neuron (Fig. 14.16A). Fig. 14.16B–E compares experimentally measured data with corresponding physics-based simulation results. The temporal behavior of the artificial neuron was characterized by applying a single super-threshold voltage pulse followed by a train of low-voltage pulses. The interaction of the RC time constant of the circuit with the internal Ag dynamics of the memristor led to the rise of a distinct delay time (τ_d) between the arrival of the voltage pulse and the rise of the output current. When a large circuit capacitance is chosen, the RC time constant is high as well and it dominates the delay time. With a smaller capacitance, the RC time becomes shorter and the internal Ag dynamics of the memristor dominates the delay time and thus the integrate-and-fire behavior, as shown in Fig. 14.16.

The integrated chip consists of 1T1R synaptic array and diffusive memristor neurons (Fig. 14.17A). The detailed structure of a single 1T1R cell is shown along with the associated connections (Fig. 14.17B). When all the transistors are turned on, the

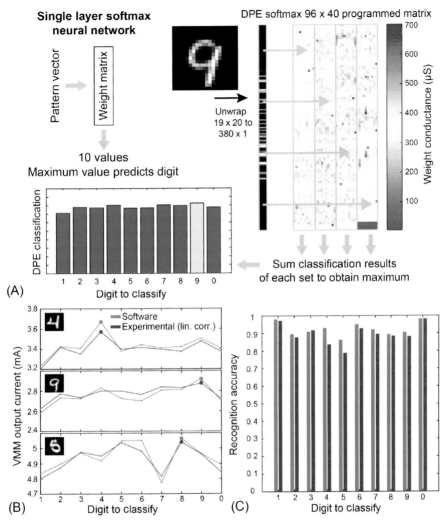

Fig. 14.15 Experimental demonstration of a single-layer neural network for MNIST handwritten digit classification. (A) Illustration of the computing procedure. A single-layer softmax neural network is trained offline, converted to target conductance values, and programmed into a 96×40 1T1M crossbar array. For a given input image, the 19×20 pixel image is unwrapped to a 380 input vector, converted to voltages, and applied to the memristor matrix in four sets. The summed result of the four outputs yields the DPE classification. This is illustrated here for the digit "9." (B) Some example classification results of the softmax single-layer neural network on the inset digits "4," "9," and "5" with ideal software results *(blue)* compared to the experimental results *(red)*. (C) Total recognition accuracy for each digit for 10,000 images from the MNIST database. Results are shown for the single-layer trained software result compared to the experimental system (same color legend as in B). The overall recognition accuracy is 89.9%. Reproduced with permission from M. Hu, et al., Memristor-based analog computation and neural network classification with a dot product engine, Adv. Mater. 30 (2018) 1705914, Copyright 2018, Wiley.

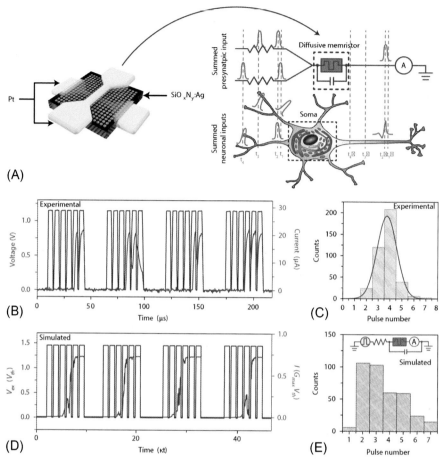

Fig. 14.16 Diffusive memristor artificial neuron. (A) Schematic illustration of a crosspoint diffusive memristor, which consists of a SiO_xN_y:Ag layer between two Pt electrodes. The artificial neuron receives software-summed presynaptic inputs via a pulsed voltage source and an equivalent synaptic resistor (20 μS conductance in this case). Both the artificial and biological neurons integrate input stimuli *(orange)* beginning at t_1 and fire when the threshold condition is reached (i.e., at t'_2). The integrated signal decays over time such that input stimuli spaced too far apart will fail to reach the threshold (i.e., the delay between t_3 and t_4). (B) Experimental response of the device to multiple subthreshold voltage pulses followed by a rest period of 200 μs (only 20 μs is shown for convenience). The device required multiple pulses to reach the threshold and "fire." (C) Histogram of the number of subthreshold voltage pulses required to successfully fire the artificial neuron *(red)* compared with a Gaussian distribution *(blue)*. (D) Simulated response of the device to multiple subthreshold voltage pulses as in B showing similar behavior to experiment, with the resting time between pulse trains chosen to allow the Ag in the device to diffuse back to the OFF state. (Only 10% of the rest period is shown for convenience.) The time is measured in temperature relaxation time, where κ is the heat transfer coefficient (see Methods). (E) Simulated switching statistics with respect to pulse numbers (within each train), consistent with the experimental results in C. The inset illustrates the circuit diagram used in the simulation.

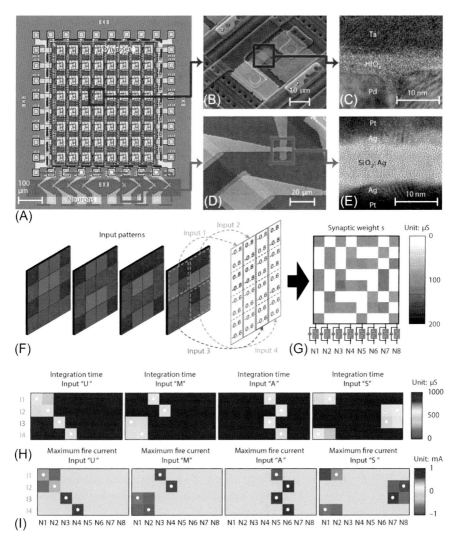

Fig. 14.17 Fully integrated memristive neural network for pattern classification. (A) Optical micrograph of the integrated memristive neural network, consisting of an 8×8 1T1R memristive synapse crossbar interfacing with eight diffusive memristor artificial neurons. (Each neuron used in this demonstration has an external capacitor not shown here.) (B) Scanning electron micrograph of a single 1T1R cell. Memristive synapses of the same row share bottom electrode lines while those of the same column share top electrode and transistor gate lines. (C) Cross-sectional transmission electron microscopy image of the integrated Pd/HfO$_x$/Ta drift memristor prepared by focused-ion-beam cutting. (D) Scanning electron micrograph of a single diffusive memristor junction. (E) High-resolution transmission electron micrograph of the cross section of the Pt/Ag/SiO$_x$:Ag/Ag/Pt diffusive memristor showing amorphous background SiO$_x$ with nanocrystalline thin Ag layers. (F) The input pattern consists of four letters, "UMAS," with artificially added noise. Each input pattern consists of 4×4 pixels, which are divided into four inputs (Input 1, Input 2, Input 3, and Input 4).

(See figure legend on next page)

1T1R array works as a fully connected memristor crossbar. The TEM image shows that an amorphous HfO_2 layer is sandwiched between Pd and Ta electrodes (Fig. 14.17C). A single diffusive memristor (Fig. 14.17D) and its cross-sectional TEM (Fig. 14.17E) showing the amorphous nature of the background SiO_x dielectric lattices and the nanocrystalline Ag layer. The synapses were initially programmed to have different weights. Letters "U," "M," "A," and "S" with artificially added noise were used as inputs. The *red* and *blue* squares in Fig. 14.17F represent the input differential voltages fed to the rows of the synaptic array. Every input pattern is divided into four sub-images of size 2×2, with a stride of two. A convolution filter is employed for each sub-image by using eight synapses in a column. Thus, there is a total of eight filters in the 8×8 array. Fig. 14.17G shows the measured weights after programming. The negative values of the convolution matrices are mapped to the conductance of memristor cells by grouping memristors from adjacent rows to form a differential pair. The results of the convolution of the eight filters to each sub-image are concurrently revealed by the firing of their corresponding diffusive memristor artificial neurons, which serve the role of ReLUs. This network can produce a unique response for each input pattern, as illustrated in Fig. 14.17H and I, in the form of integration time and the maximum fire current.

Integrated nanosystem

Schulaker et al. [87] presented a prototypical transformative nanosystem that consists of more than a million RRAM cells and more than two million carbon nanotube (CNT) field effect transistors. This prototypical system promises the use of emerging nanotechnology-based devices in digital logic circuits and for dense data storage. This system is different from conventional integrated circuit architectures in that it can be built in a 3D fashion thereby allowing for fine grained and dense vertical connectivity between layers (Fig. 14.18A) of computing, data storage and input/output. Thus, this prototypical system is able to capture massive amounts of data every second, store it on the chip, perform in situ processing of the stored data, and produce the desired output after extensive processing of the data. The prototypical system is built on silicon logic circuitry indicating that it is compatible with existing foundry infrastructure for silicon-based technologies. The authors have demonstrated the sensing and classification of ambient gases in this work (Fig. 14.18C–F).

Fig. 14.17, Cont'd Each input covers a sub-array of 2×2 size (four pixels) of the original pattern, using differential pairs as listed. Triangular voltage waveforms are fed to the eight rows of synapses of the network. (G) Measured conductance weights of the memristors after programming the eight convolutional filters (one filter per column) onto the 8×8 array using a differential pair scheme. Each of the eight columns interfaces with a diffusive memristor neuron at the end of the column. (H, I) Measured integration time and maximum amplitude of fire current of the artificial neurons as responses to the "UMAS" input patterns. Each individual input pattern is associated with a unique firing pattern of the eight artificial neurons. The ideal output patterns are marked by the white dots for neurons with positive fire current flowing out of the network.
Reproduced with permission from Z. Wang, et al., Fully memristive neural networks for pattern classification with unsupervised learning, Nat. Elect. 1 (2018) 137, Copyright 2018, Nature Publishing Group.

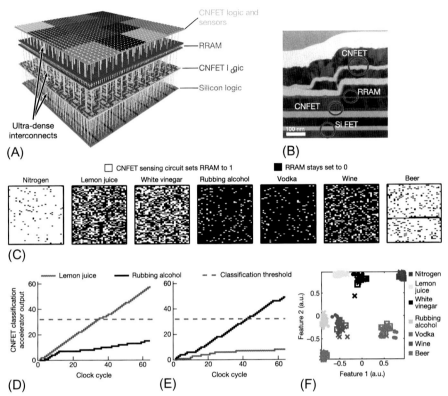

Fig. 14.18 (A) Illustration of the nanosystem. It consists of four monolithically integrated vertical layers, connected through dense vertical interconnects: fourth (top) layer, CNFET sensors and logic (including more than one million CNFET inverters, which operate as gas sensors); third layer, RRAM (1 Mbit); second layer, CNFET logic (the CNFET row decoders and CNFET classification accelerator); first (bottom) layer, silicon FET logic. (B) Cross-sectional transmission electron microscopy (TEM) image of the four-layer chip, highlighting each layer. The brighter sections of the TEM image are cross sections of wires and the darker sections are different oxides (such as gate dielectrics or interlayer dielectrics). Scale bar, 100 nm. (C–F) Results from the nanosystem: (C) sensor data (generated from the fourth layer) written into the RRAM (on the third layer) from a sample of 2048 monolithic 3D cells, measured under seven different ambient gases. A white pixel corresponds to the RRAM in that monolithic 3D cell setting to 1 during the sensing phase; a black pixel corresponds it staying set to 0. (D, E) Measured output from a CNFET-based classification accelerator upon exposure to vapors of lemon juice (D) and rubbing alcohol (E). The nanosystem compares the measured output (from 128 monolithic 3D cells) to previously learned vectors of weights (learned and stored off-chip). Each gas is correctly classified: in both cases, the only output that exceeds the positive classification threshold corresponds to the gas that was present during sensing. (F) Principal-component analysis (performed off-chip) of the sensor data shown in C demonstrates the ability to correctly classify nitrogen and six vapors, illustrating that our nanosystem functions correctly. a.u., arbitrary units. Reproduced with permission from M.M. Shulaker, et al., Three-dimensional integration of nanotechnologies for computing and data storage on a single chip, Nature 547 (2017) 74, Copyright 2017, Nature Publishing Group.

14.2.2   Unconventional computing

In this section, we discuss about the demonstration of some alternative computing applications implemented using memristive systems. Those functions were inefficiently
implemented with CMOS circuits. Being a novel circuit building block, memristors
provide an alternative to perform these tasks with projected better performance. We start
by discussing the TRNG implementation using memristors that harnesses the inherent
stochasticity of memristors, then we discuss a powerful image processing application
of memristors and finally discuss a Hamming distance comparator built using memristors. These applications speak volumes about the versatility of memristive devices.

14.2.2.1   True random number generator

TRNGs could potentially play a crucial role in modern day cryptography and security.
TRNGs rely on physical randomness of entropy sources for their operation.

Over the past few years, there have been a plethora of different random number generation schemes such as those based on the ring oscillators [88, 89], thermal noise [90,
91], and switching variability of memory devices [92–98]. There have been a handful
of reports on RRAM-based TRNG schemes which use the inherent variation of the programming amplitude and duration as the source of the desired randomness [92, 95–98].

Jiang et al. reported a TRNG based on the randomness of the programming duration of a Pt/SiO$_x$:Ag/Ag/Pt threshold switch diffusive memristor with a setup as shown
in Fig. 14.19A and B [99]. In each cycle, a fixed input voltage pulse (V_a), as illustrated in Fig. 14.19C(a), was applied to the threshold switch which was subsequently
turned on with a random delay time leading to the rise of the output voltage (V_b)
(Fig. 14.19C(b)) [99]. The rise of the voltage V_b flipped the outcome of the comparator
(V_c) and thus enabled the AND gate to transmit the clock signal (Fig. 14.19C(c–e)),
which activated the counter to record the effective ON state duration of the device in
binary format (Fig. 14.19C(f)). Using the last bit of the delay per cycle, as shown in
Fig. 14.19D, the TRNG was capable to generate continuous outputs of the random bit
stream under a train of input pulses (see Fig. 14.19E).

14.2.2.2   Analog signal and image processing

In general, analog weights are used by brain-inspired technologies for completing tasks
efficiently. CMOS approaches rely on on-chip weight storage using SRAM or off-chip
weight storage using DRAM. The total amount of memory available for utilization by the
network is limited due to chip space being allocated for on-chip memory. Moreover, this
approach is highly area inefficient. The DRAM approach which involves using off-chip
weight storage results in high-power consumption and also results in higher overall latency as data needs to be transferred from off-chip weight storage system to the processor.
Memristive devices provide an alternative as analog weights can be programmed into
specific conductance values in these devices. The 1T1R structure provides for the precise
control of conductance of each memristor in the crossbar array. Each individual memristor can be tuned using the transistor's gate linear I-V relation [100]. Moreover, sneak path
current issues can be eliminated in larger crossbars due to the inherent property of the
structure and also a compliance current can be applied which can prevent damage to the

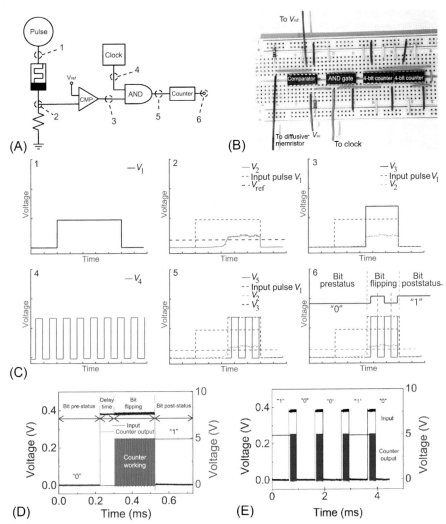

Fig. 14.19 TRNG based on the threshold switches. (A) Circuit diagram and (B) real circuit of the Pt/SiO$_x$:Ag/Ag/Pt threshold switch. (C) Schematic pulse waveforms at each stage of the circuit (as labeled in A), illustrating the working principle of the TRNG. The stochastic delay time of the threshold switch leads to variation of the pulse width shown in C, which enables the counter to effectively count the ON state duration (shown in F). Only 3 out of 8 bits are shown here as an example. (D) Monitored one counter output in response to input voltage pulse applied on our diffusive memristor. (E) Monitored one random binary output flipping from "1" → "0" → "0" → "1" → "0" over continuous switching cycles.

Panels (A)–(E) Reproduced with permission from H. Jiang, et al., A Novel True Random Number Generator Based on Stochastic Diffusive Memristors, 2017, Copyright 2017.

memristive devices. Recently, Yao et al. fabricated a neuromorphic network composed of a 1024-cell (128×8 array) for gray scale face classification from the Yale face database using analog, nonvolatile memristive devices as electronic synapses [65].

Fairly recently, Li et al. [101] reported an implementation of such hybrid memristor transistor crossbar arrays, for performing vector matrix multiplication in a faster and energy-efficient manner. They have been able to use their 128×64 array (Fig. 14.20) to multiply an analog voltage amplitude vector arriving at the ends of the rows with the analog conductance values of each memristor in a certain way, such that they get the dot product sum in the form of a current at the end of the columns. Such a large-scale VMM operation hasn't been reported before. They went on to use this array for performing certain image processing tasks as discussed in this section.

The authors used their array to do discrete cosine transform (DCT) by converting the mathematical definition of DCT into a matrix form and then using two approaches

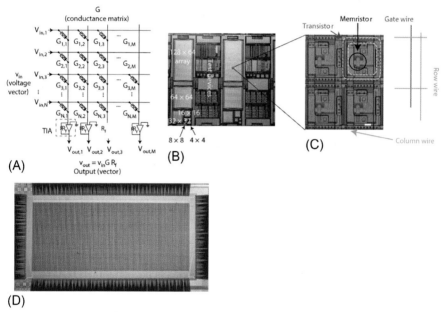

Fig. 14.20 Data stored in a 128×64 1T1R memristor crossbar, demonstrating conductance state linearity, write precision and accuracy, and read stability and reproducibility. (A) Schematic representation of the VMM operation. Multiplication is performed via Ohm's law, as the product of the voltage applied to a row and the conductance of a crosspoint cell yields a current injected into the column, while the currents on each column are summed according to the Kirchhoff's current law. The total current from each column is converted to a voltage by a TIA, which also provides a virtual ground for the column wires. (B) A 2 cm×2 cm detail from a photograph showing two dies of 1T1R memristor crossbars, each of which contains array sizes from 4×4 to 128×64 cells, along with various test devices. (C) Microscope image of four cells in a 1T1R array (scale bar, 10 μm). Crosses are memristors, and the transistors are ring shaped. Inset: schematic representation showing how the memristors and transistors are connected into an array. (D) Photograph of a probe card in contact with an operational 128×64 1T1R array (scale bar, 500 μm).

Reproduced with permission from C. Li, et al., Analogue signal and image processing with large memristor crossbars, Nat. Elect. 1 (2018) 52, Copyright 2018, Nature Publishing Group.

of (a) linear transformation and (b) differential mapping to accommodate negative values of M_{dct} into the memristor conductance map. Using this model of doing DCT, the authors performed image compression. They mapped the pixel intensities of the input image into voltage values to be applied to the rows and columns (Fig. 14.21).

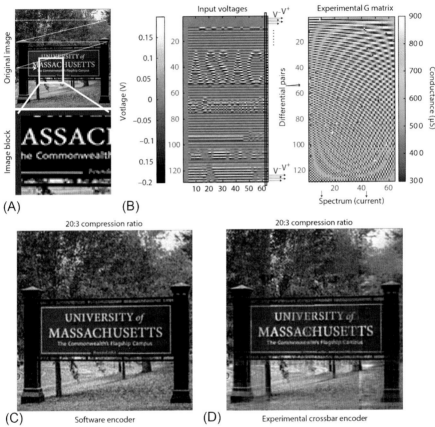

Fig. 14.21 Experimental 2D DCT demonstration using differential conductance pairs for image compression and processing. (A) The original image for compression was input into the crossbar for the 2D DCT, block by block. The *white arrow* shows the block processing sequence. The lower image shows a representative image block to be processed. (B) The image block was converted to voltages that were applied to the row wires of the crossbar (left), with neighboring wires having a voltage pair with the same amplitude, representing image pixel intensity, but opposite polarity. The paired voltages (represented by the horizontal arrows) were applied to a differential pair of memristor conductance, with the resulting net current representing the product of the absolute voltage value and the conductance difference of the differential pair. Right: differential DCT written into the 128×64 array, with the small number of stuck "on" or "off" memristors evident as disruptions in the pattern. (C, D) Images decoded from the 2D DCT by software (C) and experimentally (D). Before decoding, only the frequencies representing the top 15% of the spectral intensity were preserved (a 20:3 compression ratio).
Reproduced with permission from C. Li, et al., Analogue signal and image processing with large memristor crossbars, Nat. Elect. 1 (2018) 52, Copyright 2018, Nature Publishing Group.

Then using a similar approach, they performed DCT on these values and acquired the 2D cosine transform of the image. They retained the frequencies containing the top 15% of the spectral amplitudes, thereby attaining a compression ratio of 20:3. Then, they used the 2D inverse DCT function in MATLAB software to reconstruct the original image. On comparing the results obtained using the 2D DCT function of MATLAB using the results obtained using this approach, one can appreciate the memristor crossbars' ability to get a reasonable image.

The equation for matrix operation is as follows:

$$y = x \cdot M_{dct}$$

where x is the input signals vector, M_{dct} is the DCT matrix, and y is the output spectrum vector.

The authors then experimentally demonstrated convolution using differential memristor conductance pairs (Fig. 14.22). They added artificial Gaussian noise to a standard Lena image and then represented each color channel of the image with a floating-point number between 0 and 1. The pixel intensities were represented by two voltages with equal amplitude and opposite polarity, which were then input into a pair of memristors in adjacent rows. The difference of conductance of the memristor pair represents one matrix element in the convolution and the measured conductance after programming 10 convolutional filters into the 25×20 crossbar array has been demonstrated. In all, 10 different images were obtained in parallel by using the convolution operation.

14.2.14.3 Hamming distance comparator

Yoon et al. [102] used an array of unipolar memristors to demonstrate a Hamming distance comparator for two strings of analog states represented by voltages.

Specifically, two sets of voltages are applied to the array of memristors (Fig. 14.23) by the memristor-based comparator. The memristors are initially in their HRS and will switch to its LRS when the voltages applied on them differ by the switching threshold of the memristors. On applying a reading voltage to the memristor and then summing all the resultant currents, an accurate representation of the Hamming distance is obtained (Fig. 14.24).

14.2.3 Selecting device for larger array

The key to enabling both lateral scalability and vertical scalability at the same time is the passive assembly of ultra-high-density memristors. This would enable the use of memristors in a variety of applications in memory and neuromorphic computing. One of the advantages of 2D crossbar arrays is the ease of fabrication and high-density layout ($4F^2$). However, such crossbar arrays suffer from issues related to sneak path currents from neighboring cells when a target cell is read or programmed, leading to unwanted reduction of the read margin and an increased power consumption [103–107].

Different schemes have been proposed to get rid of sneak path current. Schemes such as employing metal oxide semiconductor field effect transistors (MOSFETs), or equipping memristors with I-V nonlinearity have been suggested to eliminate the sneak

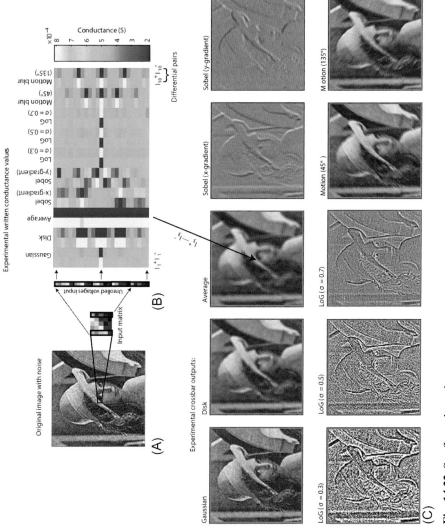

Fig. 14.22 See figure legend on next page.

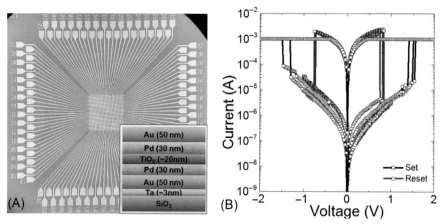

Fig. 14.23 Purpose-designed unipolar memristor and crossbar for Hamming distance computation. (A) Optical microscopic image of a 32×32 memristor crossbar. The inset is a schematic illustration of the device stack structure. (B) Semilog plot of the *I-V* curves for switching.
Reproduced with permission from N. Ge, et al., An efficient analog Hamming distance comparator realized with a unipolar memristor array: a showcase of physical computing, Sci. Rep. 7 (2017) 40135.

path current [7]. Among these approaches, a passive two-terminal nonlinear selector is a forerunner because of its simplicity, low-power consumption, scalability, and 3D stacking capability [103–107]. Schottky diodes [108], tunneling junctions [109, 110], ovonic threshold switches (OTS) [111–114], metal-insulator transitions (MIT) devices [115–119], mixed-ionic-electronic conduction (MIEC) devices [36, 120, 121], and FAST selectors [36, 120] have been proposed as the selecting devices to be integrated with memristors at each crosspoint. Recently, threshold switches based on Ag or Cu have been introduced and have demonstrated the potential of having high nonlinearity and steep transition slope [122–126].

Fig. 14.22, Cont'd Experimental convolution demonstration with differential memristor conductance pairs. (A) The input image was a standard Lena image with artificially added Gaussian white noise. Each color channel of the image is represented by a floating-point number between 0 and 1, and the added noise has a standard deviation of 0.004 and zero mean. (B) Measured conductance after programming 10 convolutional filters into the 25×20 crossbar. The pixel intensities, represented by two voltages with equal amplitude and opposite polarity, were input into the crossbar onto a pair of memristors in adjacent columns. The difference in the conductance of the memristor pair represents one matrix element of a convolution. (C) In all, 10 different filtered images obtaine in parallel by the convolution operation: Gaussian, disk, and average reduce noise by smoothing the image, Laplacian of Gaussian (LoG) with various parameters and Sobel (*x* and *y* gradient) were used to detect edges, and Motion to generate motion blur.
Reproduced with permission from C. Li, et al., Analogue signal and image processing with large memristor crossbars, Nat. Elect. 1 (2018) 52, Copyright 2018, Nature Publishing Group.

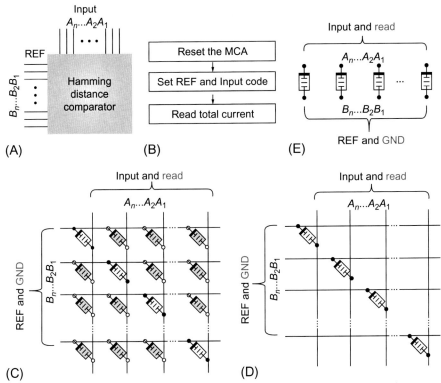

Fig. 14.24 Hamming distance comparator. (A) Schematic illustration of the core circuit architecture. (B) Steps for the comparator operation. (C) Active unipolar memristors on a full-populated crossbar with only diagonal devices utilized. The input and read conditions are indicated. The rest of the memristor devices highlighted in gray color are unused devices and they will remain in the HRS all the time, which will have minimum impact to the circuit operation. (D) Unipolar memristors only exist on the diagonal line of a crossbar with the input and read conditions indicated for practical application. (D) Unipolar memristors on the in-line architecture with the input and read conditions indicated for practical application. Reproduced with permission from N. Ge, et al., An efficient analog Hamming distance comparator realized with a unipolar memristor array: a showcase of physical computing, Sci. Rep. 7 (2017) 40135.

Luo et al. suggested the use of a Cu/Cu:HfO$_2$/HfO$_2$/Pt threshold switch as a selecting device in a memristor array [122]. Performance predictions have been made based on the observed switching characteristics [122]. In a hypothetical worst-case scenario, a 100 ON/OFF ratio of 1-selector-1-memristor combination device could guarantee a sufficient read margin in an array with a size of up to 10^5 Gb [122] (Fig. 14.25A). With both V/2 and V/3 writing scheme, the selector is predicted to effectively reduce the leakage current and thus lead to a lower power consumption [122] (see Fig. 14.25B).

Midya et al. fabricated a vertically integrated 1-selector-1-memristor cell which included a Pd/Ag/HfO$_x$/Ag/Pd threshold switch on top of a Pd/Ta$_2$O$_5$/TaO$_x$/Pd memristor

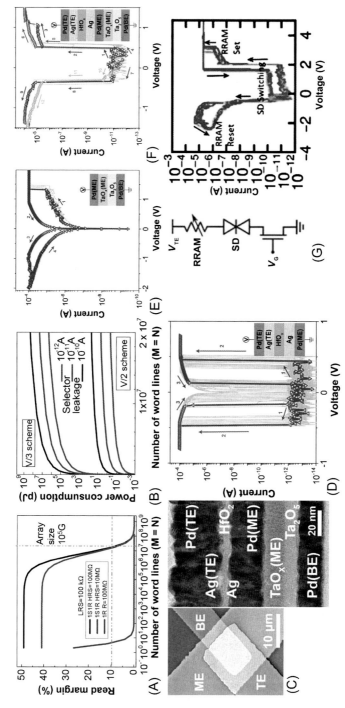

Fig. 14.25 See figure legend on next page.

with a Pd layer as a common middle electrode, as shown in Fig. 14.25C [127]. Electrical DC voltage sweeps had been applied to the individual Pd/Ag/HfO$_x$/Ag/Pd threshold selector as well as Pd/Ta$_2$O$_5$/TaO$_x$/Pd memristor, showing repeatable bidirectional threshold switching and bipolar memristive switching, respectively, as illustrated in Fig. 14.25D and E [127]. The *I-V* characteristics of the integrated 1-selector-1-memristor are shown in Fig. 14.25F. The Pd/Ag/HfO$_x$/Ag/Pd threshold switch was first turned on at ~0.5 V followed by the SET transition of the Pd/Ta$_2$O$_5$/TaO$_x$/Pd memristor at ~1.2 V. Programming of the memristor was confirmed by the subsequent reading sweep (see Fig. 14.25F) Under negative biasing, the threshold switch was turned on at ~−0.4 V followed by the RESET of memristor in the range from ~−0.8 to ~−1.4 V. The subsequent sweep verified the state of the memristor (see Fig. 14.25F).

Bricalli et al. reported characterization of a C/SiO$_x$/Ag threshold switch connected in series with a discrete C/SiO$_x$/Ti memristor, which is schematically shown in the left panel of Fig. 14.25G [128]. The right panel of Fig. 14.25G shows the switching of the selector (V_{th} ~2 V and ~−0.5 V) and the memristor (~3 V and ~−2 V), which proves its worth in storage array applications [128].

14.3 Conclusion

We introduce the concept of the memristor and then discuss electrical operation schemes of these devices. Then, we summarize the different reported physical mechanisms of memristive devices, including thermochemical effects responsible for the unipolar switching, valence change switching due to anion motion, and electrochemical cells relying on migration of cations. We then discuss some figures of merit and their correlations to the memristive switching mechanisms, such as speed, endurance,

---

Fig. 14.25, Cont'd Nonlinear selecting devices by threshold switches. (A) Read margin analysis in the worst-case condition of memristor array with Cu/Cu:HfO$_2$/HfO$_2$/Pt threshold switch as selecting devices. (B) Power consumption assessment of memristor array with Cu/Cu:HfO$_2$/HfO$_2$/Pt threshold switch as selecting devices with various writing schemes and array size. (C) SEM top view of the vertically integrated Pd/Ag/HfO$_x$/Ag/Pd selector and Pd/Ta$_2$O$_5$/TaO$_x$/Pd memristor (left) and TEM of the cross section. (D) and (E) Repeatable bipolar threshold switching (D) and memristive switching (E) of the individual Pd/Ag/HfO$_x$/Ag/Pd threshold switch and Pd/Ta$_2$O$_5$/TaO$_x$/Pd memristor. (F) DC *I-V* characteristics of the vertically integrated threshold switch and memristor. The threshold switch turned on at ~0.5 V/~−0.4 V (*blue/red* curve), followed by the SET/RESET of the memristor at ~1.2 V/~−0.8 V. The programming of memristor was verified by the subsequent sweep (*purple/magenta* curve). (G) Schematic illustration of the wired C/SiO$_x$/Ag threshold switch and a C/SiO$_x$/Ti memristor and measured *I-V* curve (B), indicating separate switching of threshold switch and memristor. Panels (A) and (B) Reproduced with permission Q. Luo, et al., in: IEEE International Electron Devices Meeting (IEDM), 2015, pp. 10.14.11–10.14.14, Copyright 2015, IEEE. Panels (C)–(F) Reproduced with permission R. Midya, et al., Anatomy of Ag/hafnia based selectors with 1010 nonlinearity, Adv. Mater. 29 (2017), Copyright 2017, Wiley-VCH. Panel (G) Reproduced with permission A. Bricalli, et al., in: IEEE International Electron Devices Meeting (IEDM), 2016, pp. 4.3.1–4.3.4, Copyright 2016, IEEE.

retention, and uniformity. Finally, we review some recent demonstrations of novel applications using these devices, consisting of brain-inspired systems, unconventional computing platforms, and large-scale integration for data storage.

References

[1] L. Chua, Memristor-the missing circuit element, IEEE Trans. Circuit Theory 18 (1971) 507–519.
[2] D.B. Strukov, G.S. Snider, D.R. Stewart, R.S. Williams, The missing memristor found, Nature 453 (2008) 80–83.
[3] J.J. Yang, et al., Memristive switching mechanism for metal/oxide/metal nanodevices, Nat. Nanotechnol. 3 (2008) 429–433.
[4] S.R. Ovshinsky, Reversible electrical switching phenomena in disordered structures, Phys. Rev. Lett. 21 (1968) 1450–1453.
[5] R. Waser, M. Aono, Nanoionics-based resistive switching memories, Nat. Mater. 6 (2007) 833–840.
[6] J.H. Yoon, et al., Truly electroforming-free and low-energy memristors with preconditioned conductive tunneling paths, Adv. Funct. Mater. 27 (2017) 1702010.
[7] R. Waser, R. Dittmann, G. Staikov, K. Szot, Redox-based resistive switching memories–nanoionic mechanisms, prospects, and challenges, Adv. Mater. 21 (2009) 2632–2663.
[8] Y. Yang, et al., Electrochemical dynamics of nanoscale metallic inclusions in dielectrics, Nat. Commun. 5 (2014).
[9] Q. Liu, et al., Real-time observation on dynamic growth/dissolution of conductive filaments in oxide-electrolyte-based ReRAM, Adv. Mater. 24 (2012) 1844–1849.
[10] Y. Yang, et al., Observation of conducting filament growth in nanoscale resistive memories, Nat. Commun. 3 (2012) 732.
[11] Z. Xu, Y. Bando, W. Wang, X. Bai, D. Golberg, Real-time in situ HRTEM-resolved resistance switching of Ag_2S nanoscale ionic conductor, ACS Nano 4 (2010) 2515–2522.
[12] X. Tian, et al., Bipolar electrochemical mechanism for mass transfer in nanoionic resistive memories, Adv. Mater. 26 (2014) 3649–3654.
[13] N. Onofrio, D. Guzman, A. Strachan, Atomic origin of ultrafast resistance switching in nanoscale electrometallization cells, Nat. Mater. 14 (2015) 440–446.
[14] K. Szot, W. Speier, G. Bihlmayer, R. Waser, Switching the electrical resistance of individual dislocations in single-crystalline $SrTiO_3$, Nat. Mater. 5 (2006) 312–320.
[15] K.M. Kim, et al., Electrically configurable electroforming and bipolar resistive switching in $Pt/TiO_2/Pt$ structures, Nanotechnology 21 (2010) 305203.
[16] H. Lee, et al., IEEE International Electron Devices Meeting. IEDM 2008, IEEE, 2008, pp 1–4.
[17] M.-J. Lee, et al., A fast, high-endurance and scalable non-volatile memory device made from asymmetric Ta_2O_{5-x}/TaO_{2-x} bilayer structures, Nat. Mater. 10 (2011) 625–630.
[18] B.J. Choi, et al., High-speed and low-energy nitride memristors, Adv. Funct. Mater. 26 (2016) 5290–5296.
[19] P. Knauth, H. Tuller, Electrical and defect thermodynamic properties of nanocrystalline titanium dioxide, J. Appl. Phys. 85 (1999) 897–902.
[20] Y. Yao, et al., In situ electron holography study of charge distribution in high-κ charge-trapping memory, Nat. Commun. 4 (2013).
[21] R. Yasuhara, et al., Inhomogeneous chemical states in resistance-switching devices with a planar-type Pt/CuO/Pt structure, Appl. Phys. Lett. 95 (2009) 012110.

[22] S. Seo, et al., Reproducible resistance switching in polycrystalline NiO films, Appl. Phys. Lett. 85 (2004) 5655–5657.
[23] M. Chan, T. Zhang, V. Ho, P. Lee, Resistive switching effects of HfO_2 high-k dielectric, Microelectron. Eng. 85 (2008) 2420–2424.
[24] U. Russo, et al., IEEE International Electron Devices Meeting. IEDM 2007, IEEE, 2007, pp. 775–778.
[25] D.-H. Kwon, et al., Atomic structure of conducting nanofilaments in TiO_2 resistive switching memory, Nat. Nanotechnol. 5 (2010) 148–153.
[26] D.S. Jeong, H. Schroeder, U. Breuer, R. Waser, Characteristic electroforming behavior in Pt/TiO_2/Pt resistive switching cells depending on atmosphere, J. Appl. Phys. 104 (2008) 123716.
[27] R. Meyer, et al., 9th Annual Non-Volatile Memory Technology Symposium. NVMTS 2008, IEEE, 2008, pp. 1–5.
[28] D.S. Jeong, H. Schroeder, R. Waser, Coexistence of bipolar and unipolar resistive switching behaviors in a Pt/TiO_2/Pt stack, Electrochem. Solid-State Lett. 10 (2007) G51–G53.
[29] M.D. Pickett, R.S. Williams, Sub-100 fJ and sub-nanosecond thermally driven threshold switching in niobium oxide crosspoint nanodevices, Nanotechnology 23 (2012) 215202.
[30] Y.C. Yang, F. Pan, Q. Liu, M. Liu, F. Zeng, Fully room-temperature-fabricated nonvolatile resistive memory for ultrafast and high-density memory application, Nano Lett. 9 (2009) 1636–1643.
[31] B. Chen, et al., IEEE International Electron Devices Meeting (IEDM), IEEE, 2011, pp. 12.13.11–12.13.14.
[32] H. Lv, et al., IEEE International Conference on Electron Devices and Solid-State Circuits (EDSSC), IEEE, 2015, pp. 229–232.
[33] J.J. Yang, et al., High switching endurance in TaO_x memristive devices, Appl. Phys. Lett. 97 (2010) 232102.
[34] H. Lee, et al., IEEE International Electron Devices Meeting (IEDM), IEEE, 2010, pp. 19.17.11–19.17.14.
[35] M. Wang, et al., Robust memristors based on layered two-dimensional materials, Nat. Elect. 1 (2018) 130–136.
[36] S.H. Jo, T. Kumar, S. Narayanan, H. Nazarian, Cross-point resistive ram based on field-assisted superlinear threshold selector, IEEE Trans. Electron Dev. 62 (2015) 3477–3481.
[37] J.H. Yoon, et al., Pt/Ta_2O_5/HfO_{2-x}/Ti resistive switching memory competing with multilevel NAND flash, Adv. Mater. 27 (2015) 3811–3816.
[38] S. Choi, J. Lee, S. Kim, W.D. Lu, Retention failure analysis of metal-oxide based resistive memory, Appl. Phys. Lett. 105 (2014) 113510.
[39] S. Yu, Y. Yin Chen, X. Guan, H.-S. Philip Wong, J.A. Kittl, A Monte Carlo study of the low resistance state retention of HfO_x based resistive switching memory, Appl. Phys. Lett. 100 (2012) 043507.
[40] Y.Y. Chen, et al., IEEE International Electron Devices Meeting (IEDM), IEEE, 2013, pp. 10.11.11–10.11.14.
[41] C. Nail, et al., IEEE International Electron Devices Meeting (IEDM), IEEE, 2016, pp. 4.5. 1–4.5. 4.
[42] H. Zhang, et al., Ionic doping effect in ZrO_2 resistive switching memory, Appl. Phys. Lett. 96 (2010) 123502.
[43] H. Lv, H. Wan, T. Tang, Improvement of resistive switching uniformity by introducing a thin GST interface layer, IEEE Electron Device Lett. 31 (2010) 978–980.
[44] S. Yu, X. Guan, H.-S.P. Wong, On the switching parameter variation of metal oxide RRAM—Part II: model corroboration and device design strategy, IEEE Trans.Electron Devices 59 (2012) 1183–1188.

[45] W. Lian, et al., Improved resistive switching uniformity in Cu/HfO$_2$/Pt devices by using current sweeping mode, IEEE Electron Device Lett. 32 (2011) 1053–1055.

[46] T. Chang, Y. Yang, W. Lu, Building neuromorphic circuits with memristive devices, IEEE Circuits Syst. Mag. 13 (2013) 56–73.

[47] L.O. Chua, L. Yang, Cellular neural networks: applications, IEEE Trans. Circuits Syst. 35 (1988) 1273–1290.

[48] U. Ramacher, SYNAPSE—a neurocomputer that synthesizes neural algorithms on a parallel systolic engine, J. Parallel Distrib. Comput. 14 (1992) 306–318.

[49] C. Diorio, P. Hasler, B.A. Minch, C.A. Mead, A single-transistor silicon synapse, IEEE Trans. Electron Dev. 43 (1996) 1972–1980.

[50] G. Indiveri, E. Chicca, R. Douglas, A VLSI array of low-power spiking neurons and bistable synapses with spike-timing dependent plasticity, IEEE Trans. Neural Netw. 17 (2006) 211–221.

[51] C. Bartolozzi, G. Indiveri, Synaptic dynamics in analog VLSI, Neural Comput. 19 (2007) 2581–2603.

[52] J.H.B. Wijekoon, P. Dudek, Compact silicon neuron circuit with spiking and bursting behaviour, Neural Netw. 21 (2008) 524–534.

[53] W. Xu, S.-Y. Min, H. Hwang, T.-W. Lee, Organic core-sheath nanowire artificial synapses with femtojoule energy consumption, Sci. Adv. 2 (2016) e1501326.

[54] A. Wedig, et al., Nanoscale cation motion in TaO$_x$, HfO$_x$ and TiO$_x$ memristive systems, Nat. Nanotechnol. 11 (2016) 67–74.

[55] W. Wu, et al., Improving analog switching in HfO$_x$-based resistive memory with a thermal enhanced layer, IEEE Electron Device Lett. 38 (2017) 1019–1022.

[56] H. Wu, et al., Resistive random access memory for future information processing system, Proc. IEEE (2017) 1–20.

[57] S.H. Jo, et al., Nanoscale memristor device as synapse in neuromorphic systems, Nano Lett. 10 (2010) 1297–1301.

[58] S. Yu, Y. Wu, R. Jeyasingh, D. Kuzum, H.S.P. Wong, An electronic synapse device based on metal oxide resistive switching memory for neuromorphic computation, IEEE Trans. Elect. Dev. 58 (2011) 2729–2737.

[59] T. Ohno, et al., Short-term plasticity and long-term potentiation mimicked in single inorganic synapses, Nat. Mater. 10 (2011) 591–595.

[60] Z.Q. Wang, et al., Synaptic learning and memory functions achieved using oxygen ion migration/diffusion in an amorphous InGaZnO memristor, Adv. Funct. Mater. 22 (2012) 2759–2765.

[61] J.J. Yang, D.B. Strukov, D.R. Stewart, Memristive devices for computing, Nat. Nanotechnol. 8 (2013) 13–24.

[62] H. Lim, I. Kim, J.S. Kim, C.S. Hwang, D.S. Jeong, Short-term memory of TiO$_2$-based electrochemical capacitors: empirical analysis with adoption of a sliding threshold, Nanotechnology 24 (2013) 384005.

[63] S. La Barbera, D. Vuillaume, F. Alibart, Filamentary switching: synaptic plasticity through device volatility, ACS Nano 9 (2015) 941–949.

[64] M. Prezioso, et al., Training and operation of an integrated neuromorphic network based on metal-oxide memristors, Nature 521 (2015) 61–64.

[65] P. Yao, et al., Face classification using electronic synapses, Nat. Commun. 8 (2017) 15199.

[66] Z. Wang, et al., Memristors with diffusive dynamics as synaptic emulators for neuromorphic computing, Nat. Mater. 16 (2016) 101–108.

[67] S. La Barbera, A.F. Vincent, D. Vuillaume, D. Querlioz, F. Alibart, Interplay of multiple synaptic plasticity features in filamentary memristive devices for neuromorphic computing, Sci. Rep. 6 (2016) 39216.

[68] J.I. Hubbard, Repetitive stimulation at the mammalian neuromuscular junction, and the mobilization of transmitter, J. Physiol. 169 (1963) 641–662.

[69] L.E. Dobrunz, C.F. Stevens, Heterogeneity of release probability, facilitation, and depletion at central synapses, Neuron 18 (1997) 995–1008.

[70] R.S. Zucker, W.G. Regehr, Short-term synaptic plasticity, Annu. Rev. Physiol. 64 (2002) 355–405.

[71] G.-Q. Bi, M.-M. Poo, Synaptic modifications in cultured hippocampal neurons: dependence on spike timing, synaptic strength, and postsynaptic cell type, J. Neurosci. 18 (1998) 10464–10472.

[72] T. Dunwiddie, G. Lynch, Long-term potentiation and depression of synaptic responses in the rat hippocampus: localization and frequency dependency, J. Physiol. 276 (1978) 353–367.

[73] L.I. Zhang, H.W. Tao, C.E. Holt, W.A. Harris, M.-m. Poo, A critical window for cooperation and competition among developing retinotectal synapses, Nature 395 (1998) 37–44.

[74] S. Song, K.D. Miller, L.F. Abbott, Competitive Hebbian learning through spike-timing-dependent synaptic plasticity, Nat. Neurosci. 3 (2000) 919–926.

[75] J.H. Yoon, et al., An artificial nociceptor based on a diffusive memristor, Nat. Commun. 9 (2018) 417.

[76] H.C. Anderson, Neural network machines, IEEE Potentials 8 (1989) 13–16.

[77] H.D. Hubel, T.N. Wiesel, Receptive fields, binocular interaction and functional architecture in the cat's visual cortex, J. Physiol. 160 (1962) 106–154.

[78] P.M. Sheridan, et al., Sparse coding with memristor networks, Nat. Nanotechnol. 12 (2017) 784–789.

[79] M. Hu, et al., Memristor-based analog computation and neural network classification with a dot product engine, Adv. Mater. 30 (2018) 1705914.

[80] D. Silver, et al., Mastering the game of Go with deep neural networks and tree search, Nature 529 (2016) 484–489.

[81] P.A. Merolla, et al., A million spiking-neuron integrated circuit with a scalable communication network and interface, Science 345 (2014) 668–673.

[82] M.D. Pickett, G. Medeiros-Ribeiro, R.S. Williams, A scalable neuristor built with Mott memristors, Nat. Mater. 12 (2013) 114–117.

[83] P. Stoliar, et al., A leaky-integrate-and-fire neuron analog realized with a mott insulator, Adv. Funct. Mater. (2017) 1604740.

[84] H. Lim, et al., Relaxation oscillator-realized artificial electronic neurons, their responses, and noise, Nanoscale 8 (2016) 9629–9640.

[85] T. Tuma, A. Pantazi, M. Le Gallo, A. Sebastian, E. Eleftheriou, Stochastic phase-change neurons, Nat. Nanotechnol. 11 (2016) 693–699.

[86] Z. Wang, et al., Fully memristive neural networks for pattern classification with unsupervised learning, Nat. Elect. 1 (2018) 137.

[87] M.M. Shulaker, et al., Three-dimensional integration of nanotechnologies for computing and data storage on a single chip, Nature 547 (2017) 74.

[88] M. Bucci, et al., A high-speed IC random-number source for SmartCard microcontrollers, IEEE Trans. Circuits Syst. I Fundam. Theory Appl. 50 (2003) 1373–1380.

[89] K. Yang, et al., IEEE International Solid-State Circuits Conference Digest of Technical Papers, ISSCC, 2014, pp. 280–281.

[90] C.Y. Huang, W.C. Shen, Y.H. Tseng, Y.C. King, C.J. Lin, A contact-resistive random-access-memory-based true random number generator, IEEE Electron Device Lett. 33 (2012) 1108–1110.

[91] Z. Wei, et al., Electron Devices Meeting (IEDM), IEEE International, 2016, pp. 4.8.1–4.8.4.

[92] S. Gaba, P. Sheridan, J. Zhou, S. Choi, W. Lu, Stochastic memristive devices for computing and neuromorphic applications, Nanoscale 5 (2013) 5872–5878.

[93] A. Fukushima, et al., Spin dice: a scalable truly random number generator based on spintronics, Appl. Phys. Express 7 (2014) 083001.

[94] C. Won Ho, et al., IEEE International Electron Devices Meeting, 2014, pp. 12.15.11–12.15.14.

[95] S. Balatti, S. Ambrogio, Z. Wang, D. Ielmini, True random number generation by variability of resistive switching in oxide-based devices, IEEE J. Emerg. Sel. Top. Circuits Syst. 5 (2015) 214–221.

[96] Y. Wang, W. Wen, H. Li, M. Hu, A Novel True Random Number Generator Design Leveraging Emerging Memristor Technology, Association for Computing Machinery (ACM), 2015, pp. 271–276.

[97] S. Balatti, et al., Physical unbiased generation of random numbers with coupled resistive switching devices, IEEE Trans. Elect. Dev. 63 (2016) 2029–2035.

[98] M.L. Gallo, T. Tuma, F. Zipoli, A. Sebastian, E. Eleftheriou, 46th European Solid-State Device Research Conference, ESSDERC, 2016, pp. 373–376.

[99] H. Jiang, et al., A Novel True Random Number Generator Based on Stochastic Diffusive Memristors, Nature Communications, 2017.

[100] M. Hu, et al., 53nd ACM/EDAC/IEEE Design Automation Conference (DAC), IEEE, 2016, pp. 1–6.

[101] C. Li, et al., Analogue signal and image processing with large memristor crossbars, Nat. Elect. 1 (2018) 52.

[102] N. Ge, et al., An efficient analog Hamming distance comparator realized with a unipolar memristor array: a showcase of physical computing, Sci. Rep. 7 (2017) 40135.

[103] G.W. Burr, et al., Access devices for 3D crosspoint memory, J. Vac. Sci. Technol. B 32 (2014) 040802.

[104] S. Kim, J. Zhou, W.D. Lu, Crossbar RRAM arrays: selector device requirements during write operation, IEEE Trans. Elect. Dev. 61 (2014) 2820–2826.

[105] J.Y. Seok, et al., A review of three-dimensional resistive switching cross-bar array memories from the integration and materials property points of view, Adv. Funct. Mater. 24 (2014) 5316–5339.

[106] J. Zhou, K.H. Kim, W. Lu, Crossbar RRAM arrays: selector device requirements during read operation, IEEE Trans. Elect. Dev. 61 (2014) 1369–1376.

[107] G.W. Burr, R.S. Shenoy, H. Hwang, in: D. Ielmini, R. Waser (Eds.), Resistive Switching, Wiley-VCH Verlag GmbH & Co. KGaA, 2016, pp. 623–660.

[108] G.H. Kim, et al., 32×32 Crossbar array resistive memory composed of a stacked schottky diode and unipolar resistive memory, Adv. Funct. Mater. 23 (2013) 1440–1449.

[109] A. Kawahara, et al., An 8 Mb multi-layered cross-point ReRAM macro with 443 MB/s write throughput, IEEE J. Solid State Circuits 48 (2013) 178–185.

[110] B.J. Choi, et al., Trilayer tunnel selectors for memristor memory cells, Adv. Mater. 28 (2016) 356–362.

[111] M. Anbarasu, M. Wimmer, G. Bruns, M. Salinga, M. Wuttig, Nanosecond threshold switching of GeTe$_6$ cells and their potential as selector devices, Appl. Phys. Lett. 100 (2012) 143505.

[112] M.J. Lee, et al., Electron Devices Meeting (IEDM), IEEE International, 2012, pp. 2.6.1–2.6.3.

[113] S. Kim, et al., Symposium on VLSI Technology (VLSIT), 2013 T240–T241.

[114] M.J. Lee, et al., A plasma-treated chalcogenide switch device for stackable scalable 3D nanoscale memory, Nat. Commun. 4 (2013) 2629.
[115] M. Son, et al., Excellent selector characteristics of nanoscale VO_2 for high-density bipolar ReRAM applications, IEEE Electron Device Lett. 32 (2011) 1579–1581.
[116] S. Kim, et al., Symposium on VLSI Technology (VLSIT), 2012, pp. 155–156.
[117] E. Cha, et al., IEEE International Electron Devices Meeting (IEDM), 2013, pp. 10.15.11–10.15.14.
[118] D. Lee, et al., IEEE International Electron Devices Meeting (IEDM), 2013, pp. 10.17.11–10.17.14.
[119] G.A. Gibson, et al., An accurate locally active memristor model for S-type negative differential resistance in NbOx, Appl. Phys. Lett. 108 (2016) 023505.
[120] S.H. Jo, T. Kumar, S. Narayanan, W.D. Lu, H. Nazarian, IEEE International Electron Devices Meeting (IEDM), 2014, pp. 6.7.1–6.7.4.
[121] P. Narayanan, et al., Exploring the design space for crossbar arrays built with mixed-ionic-electronic-conduction (MIEC) access devices, IEEE J. Electron Devices Soc. 3 (2015) 423–434.
[122] Q. Luo, et al., IEEE International Electron Devices Meeting (IEDM), 2015, pp. 10.14.11–10.14.14.
[123] J. Song, J. Woo, A. Prakash, D. Lee, H. Hwang, Threshold selector with high selectivity and steep slope for cross-point memory array, IEEE Electron Device Lett. 36 (2015) 681–683.
[124] J. Song, et al., Bidirectional threshold switching in engineered multilayer ($Cu_2O/Ag:Cu_2O/Cu_2O$) stack for cross-point selector application, Appl. Phys. Lett. 107 (2015) 113504.
[125] W. Chen, H.J. Barnaby, M.N. Kozicki, Volatile and non-volatile switching in Cu-SiO_2 programmable metallization cells, IEEE Electron Device Lett. 37 (2016) 580–583.
[126] J. Song, et al., Steep slope field-effect transistors with Ag/TiO_2-based threshold switching device, IEEE Electron Device Lett. 37 (2016) 932–934.
[127] R. Midya, et al., Anatomy of Ag/hafnia based selectors with 10^{10} nonlinearity, Adv. Mater. 29 (2017) 1604457.
[128] A. Bricalli, et al., IEEE International Electron Devices Meeting (IEDM), 2016, pp. 4.3.1–4.3.4.

Further reading

[129] C. Li, et al., Direct observations of nanofilament evolution in switching processes in HfO_2-based resistive random access memory by in situ TEM studies, Adv. Mater. (2017).

Emerging memory technologies for neuromorphic hardware

E. Vianello, L. Perniola, B. De Salvo
CEA-LETI Minatec Campus, Grenoble, France

15.1 Artificial intelligence systems: Status and challenges

With billions of easy-access and low-cost connected devices, in the past few decades, the world has entered the era of hyperconnectivity, enabling people and machines to interact in a symbiotic way—anytime, anywhere—with both the physical and cyber worlds. Artificial intelligence (AI) has been at the center of this revolution [1]. In recent years, we have seen a boost in the performance and applications of machine learning (ML), driven by several factors: (i) the enormous storehouses of data—images, video, audio, and text files strewn across the Internet—which have been essential to the dramatic improvement of learning/training approaches and algorithms; (ii) the increased computational power of modern computers (the advent of parallel computing for neural network processing having compensated the slowing down of Moore's law below the 10 nm node). Among the many fields of ML, deep learning (DL) is the most popular.

Today, for tasks such as image or speech recognition, ML applications are equaling or even surpassing expert human performance. Other tasks considered as extremely difficult in the past, such as natural language comprehension or complex games, have been successfully tackled. The particular case of the AlphaGo program from Google is remarkable in that it demonstrates how to increase performance by refining the algorithm architecture and combining several techniques of ML (DL techniques with reinforcement learning). For the time being, data are transmitted in hierarchical infrastructures, and applications have to deal with different levels of analysis, such as Cloud computing, the edge (networked mobile devices), and the end devices (wireless sensor nodes). Currently, most of the data processing for DL training, and even for the inference phase, happens in the Cloud (i.e., data are sent to a data center and then processed there, before pushing operational decisions back to the edge platform). AI algorithms are not useful in settings where connectivity is sparse. Moreover, training a DL network in the Cloud (with conventional processors or GPU) on extremely large datasets involves intensive computing tasks and can take several weeks [2]. The power limitations of servers used for DL are expected to slow down the pace of performance improvements. This poses a great challenge to computing platform designers.

Bringing intelligence to the *edge* or to *end devices* means doing useful processing of the data as close to the collection point as possible and allowing

Advances in Non-volatile Memory and Storage Technology. https://doi.org/10.1016/B978-0-08-102584-0.00016-4

systems to make some operational decisions locally, possibly semi-autonomously. Distributing intelligence over the network is important for a number of reasons. Safety will require local decision making, in real time, without having to rely on a connection that could be interrupted for different reasons. Running real-time DL locally is essential for many applications, from landing drones to navigating driverless cars. The delay caused by the round-trip to the Cloud could lead to disastrous or even fatal results. Privacy will require that key data do not leave the user's device, while transmission of high-level information, generated by local neural network algorithms, will be authorized. Raw videos generated by millions of cameras will have to be locally analyzed to limit bandwidth issues and communication costs. For all these reasons, new concepts and technologies that can bring AI closer to the edge and end devices are highly demanded. The main design goal, in distributed applications covering several levels of hierarchy similar to the brain, is to find a global optimum between performance and energy consumption. This imperatively requires a holistic research approach, where the technology stack—from device to applications—is redesigned. As shown in Fig. 15.1, to address embedded applications, major industrial players and start-ups have developed specialized edge platforms that can execute ML algorithms (inference) on embedded hardware (CPU and GPU), such as Movidius Myriad X, MobilEye EyeQ5, and Jetson TX2. Impressive power improvements (down to a few Watts) have been achieved by exploiting Moore's law (pushing the Finfet technology down to the 7 nm node) and with hardware-software co-optimization. Since many mobile applications are "always-on" (e.g., voice commands), low power is critical for mobile IoT [3]. In this context, several research groups have focused on hardware designs of convolutional neural network (CNN) accelerators. Precision-scalable processors (implemented in 40-nm LP CMOS) for deep neural networks have shown power consumption in the range of 70 mW [4]. The need for off-chip storage devices, like DRAMs, significantly increases power consumption. Recently, mobile-oriented applications (like keyword spotting and face detection) were demonstrated with a low-power programmable DL accelerator [5]—with on-chip weight storage—which consumed <300 µW.

It is worth mentioning that the challenges of bringing intelligence into low-power IoT connected end devices (with applications ranging from habitat monitoring to medical surveillance) are much more demanding than those associated with traditional networked mobile devices at the edge [6]. Most connected end devices are wireless sensor nodes containing microcontrollers, wireless transceivers, sensors, and actuators. The power requirement for these systems is extremely critical (<100 µW for normal workloads) as these devices often operate using energy harvesting sources or a single battery for several years. To improve implementation efficiency of ML, different approaches have been explored. Nevertheless, given the energy costs related to the memory system, and the constraints on both parallelism and technology scaling, it might seem like there is not much room for additional energy improvements [7]. Finding new, affordable, and energy efficient ways to implement inference and learning through new specialized low-power and distributed compute engines is thus key for future intelligent systems.

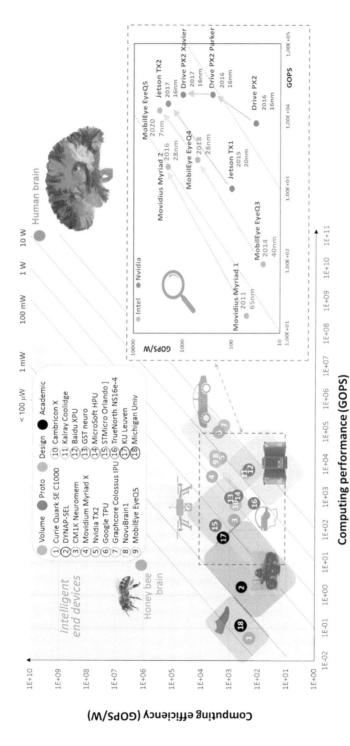

Fig. 15.1 Comparison of computing efficiency (GOPS/W)—during the inference phase—versus computing performance (GOPS) of several intelligent chips from literature and the web, showing the gap between the intelligent end device requirements and existing solutions. Note that we took the very coarse approximation of a 1:1 correspondence between OPS, FLOPS, IPS, SOPS (SOPS = firing rate × average active synapses) [1].

15.2   Advanced technologies for brain-inspired computing

Inspired by the human brain, whose computing performance and efficiency still remain unmatched (see Fig. 15.1), a radically different approach is being investigated. It consists of implementing *bio-inspired architectures in optimized neuromorphic hardware* to provide direct one-to-one mapping with the learning algorithm running on it. This approach, which originated with the pioneering work of Carver Mead [8], has yet to be fully demonstrated and industrialized. Implementation limitations are linked to several elements [9], such as the difficulties to emulate the behavior of the neural network elementary components (such as neurons and synapses) with standard CMOS technologies and to achieve a 3D brain-like high-density connectivity with 2D-layer technologies. In the following paragraphs, essential brain-inspired operating principles (such as spike coding and STDP) will be introduced, followed by a detailed discussion on how emerging technologies could lead to new neuromorphic hardware and change the rules of the game.

15.2.1   Spike coding and spike-timing-dependent-plasticity

The first brain-inspired operating principle to consider is the way neuron states are encoded in a system. In the past, neuron values were encoded using analog or digital values. A recent trend in neuromorphic computing is to encode neuron values as pulses or spikes [8, 10, 11]. This parsimonious signal coding was inspired by the way neurons of the central nervous system interact and could lead to higher energy efficiencies. It differs from the traditional signal rate coding (used in today's main industrial neural network applications), which employs the average frequency of spikes in a given time window. The values manipulated in those networks (i.e., inputs and outputs of neurons) are numbers representing the "cumulative" effect of spikes over time. However, if input/output signals are represented as pulses (spikes), the multiplication operation between input signals and synaptic weights is reduced to a gating operation at the synapse level. This typically produces a weighted current at the arrival of the presynaptic spike that is integrated by the postsynaptic neuron. The higher the frequency of the input spikes, the larger the value integrated by the neuron. Furthermore, if many synapses receive input spikes in parallel, the weighted sum operation is implemented directly at the input node of the postsynaptic neuron following Kirchhoff's current law. Power consumption can therefore be reduced by implementing this spike or event-based signal representation (called Address-Event Representation, AER) using asynchronous schemes. Given these features and because this representation is also optimal for transmitting signals across long distances or chip boundaries, most of the recent state-of-the-art neuromorphic computing approaches are using AER. Moreover, spiking neurons offer the additional advantage of being easily interfaced with low-power spiking sensors (be it image-, audio-, tactile-, or chemical sensors [12–16]).

The second brain-inspired principle essential to neuromorphic systems is the learning paradigm (i.e., the way the synaptic connections among neurons are created, modified, and preserved). The computation schemes to define the synaptic weights can be divided into two types: (1) supervised learning, where the inference process is based on training examples (this is the case of most neural-inspired ML algorithms, which show impressive performance—at the cost of huge power dissipation—for solving very specific tasks); (2) unsupervised learning, which does not use any feedback from an external teacher, but attempts to classify inputs based on the underlying statistics of the data. Spike-timing-dependent-plasticity (STDP) is a bio-inspired algorithm that enables unsupervised learning. The assumption underlying STDP is that synapses tend to reinforce causal links. That is, when the presynaptic neuron spikes just before the postsynaptic neuron spikes, the synapse between the two becomes stronger. Therefore, if the presynaptic neuron spikes again, the synapse will allow the postsynaptic neuron to spike faster or with a higher occurrence probability.

Optimized neuromorphic hardware for future ultralow-power cognitive systems can be successfully implemented by coupling the aforementioned novel brain-inspired paradigms with novel emerging technologies (such as advanced CMOS, 3D technologies, emerging resistive memories, and Silicon photonics) [1]. In particular, in the following sections, we will present the extraordinary potential of resistive memory technologies to provide intelligent features in neuromorphic hardware.

15.3   Emerging nonvolatile resistive memories (ReRAMs) for neuromorphic computing

Several large-scale neuromorphic systems have been proposed in the last years, taking advantage of the enormous potentialities of current Silicon technologies. Examples include the Heidelberg's HICANN [17], IBM's TrueNorth [18, 19], and ETH's ROLLS [20] chips. These approaches use standard CMOS technologies to implement both neurons and synapses. The synaptic weights are stored in analog or digital devices, such as capacitors or SRAM. Nevertheless, SRAM-based synapses are affected by the problems of area consumption and data volatility. When the network is turned off, the synaptic weights stored in the SRAM are lost, stressing the need for storage in nonvolatile memories (NVMs) during or after the learning process, but NVMs come with additional power and area consumption.

15.3.1   Main features of ReRAM

Recently, new memory technologies, called ReRAM (e.g., phase change memory—PCM, spin-transfer torque magnetic memory—STT-MRAM, metal oxide resistive switching memory—OxRAM, conductive-bridge memory—CBRAM, and vertical resistive memories—VRAM) have appeared [21, 22]. These memories offer several key features, such as low voltages (ranging from 1 to 3 V), fast programming, and reading time (few 10's of ns, even <1 ns), long data retention, single bit alterability, execution in place, good cycling performance (higher than Flash), density, and ease of integration in the Back-End-Of-Line of advanced CMOS. ReRAM are currently developed for applications such as microcontrollers [21], servers, and high-performance computers. Bringing the memory close to the processing unit will revolutionize the traditional memory hierarchy [23] and facilitate the implementation of in-memory computing architectures.

Looking into details at the aforementioned NVM technologies, we see that universal signatures exist, despite from the very different nature of the physics behind them. Gathering together the data we have obtained for several years in LETI on

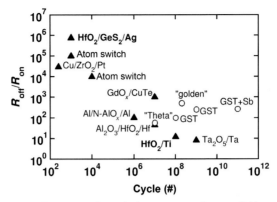

Fig. 15.2 Universal plot of programming window versus endurance [21].

the different emerging memory technologies, as well as results summarized by Stanford University in Ref. [24], we conclude that the programming window (ratio between high and low resistive states), endurance and data retention performances are linked together in a sort of global energy metric before device breakdown. Indeed, in Fig. 15.2 we can appreciate the link between the programming window and the endurance capabilities of the different technologies: high programming window is typically associated to poor endurance capabilities, (examples are certain class of chalco-based CBRAM [25] or nonpolar OxRAM [26]) and indeed for these technologies the overall filament is believed to participate to the creation and the dissolution events. On the contrary, small programming window associated with outstanding endurance can be ascribed to displacing very few atoms/ions, even spins, during the write or erase event allowing long-lasting device electrical integrity. Data also show that a universal trade-off exists between endurance and data retention itself. The better the performances in endurance, the lower the data retention capabilities at high temperature [21]. The "transformational" ability of PCM [27–29], adjusting the alchemy of phase-change materials, is known since roughly a decade. Looking at the major papers published so far, a global trade-off seems to exist between the programming time for the SET operation and the device data retention: the better the data retention is, the longer is the time needed to operate a fine SET state. This trade-off is represented in Fig. 15.3. Indeed it seems to be reasonable that a better stability of the RESET state is associated with a more stable amorphous phase. Such stability is experimented when devices are subjected to medium temperature stress (i.e., the environmental temperature during data retention tests) as well as high-temperature stress (i.e., temperature internally induced, thanks to the joule-heating effect during the SET process).

Holding these statements true, it is essential to target each technology or optimized memory stack in view of a specific application. Due to their low power consumption, multi-value resistance levels, and nonvolatility, ReRAM memories are also promising for implementing energy-efficient bio-inspired synapses in complex neural network systems [30–32] (see Fig. 15.4).

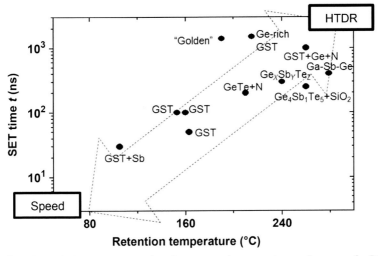

Fig. 15.3 Universal plot on programming time versus data retention performance for PCM devices [21].

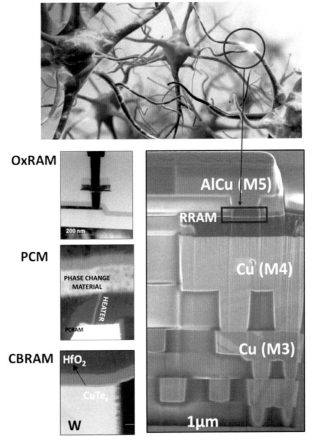

Fig. 15.4 Illustration of biological synapse and the concept of using ReRAM as synapses [1].

15.4   Example of ReRAM convolutional neural networks with supervised learning

The CNNs are among the most promising architectures for embedded applications due to their relatively small memory footprint. The organization of convolutional layers in CNNs is originally inspired by the structure of the visual system in mammals [33–35]. Software implementations of CNNs were applied with great success in applications such as traffic sign recognition [36], the analysis of biological images [37], and the detection of faces, complex text, pedestrians on the streets, and human bodies in natural images [38–40]. A major recent practical success of software implementations of CNNs is the face recognition software proposed by Facebook [41].

The power consumption to perform convolution operations is computationally expensive in the CNN implementation on CPUs and GPUs. This hinders their integration in portable devices. In recent years, a dedicated system on chip (SoC) solutions and FPGA platforms have been used to implement these networks for increasing performances while decreasing their power consumption. A hardware implementation of CNNs based on the ReRAM devices can further improve the power efficiency.

15.4.1   OxRAM-based CNN for image classification

In [42], a CNN spike-based architecture for pattern recognition, using HfO_2-OxRAM devices as synapses for convolution kernels was presented. Thanks to the use of ReRAM synapses to implement the kernel, the convolution operations are performed directly in memory, reducing the latency per image recognition with respect to software implementations on GPU. The proposed CNN architecture is composed of a feature extraction module (made of two cascaded convolutional layers) and a classification module (made of two fully connected layers) (see Fig. 15.5). While in the fully connected classification module the neurons of a given layer are connected to every neuron of the previous layer by a large number of synapses, in convolutional layers a small set of synapses (constituting several kernels) are shared among different neurons to connect layer N and $N-1$ through a convolutional operation. A convolutional layer is composed of several (output) feature maps, each of them being connected to some or all the features maps of the previous layer through a convolutional kernel. The kernel corresponds to a feature that has to be localized in the input image. In a layer, each feature map contains the results of the convolution of some or all the input maps (which are the output feature maps of the previous layer), each of them with a different convolution kernel. It contains information about the locations where the kernel features are present in the input map. The extraction module, therefore, transforms the input image into a simpler set of feature maps. The classification module connects the obtained set of feature maps to the output neuron layer. Each output neuron is associated to a category (the 10 digit categories for the MNIST database and the 43 German Traffic Signs for the GTSRB database): an output neuron spikes when the image presented to the input of the network belongs to its category. In the first convolutional layer, each output feature map is connected to the input image through a convolutional

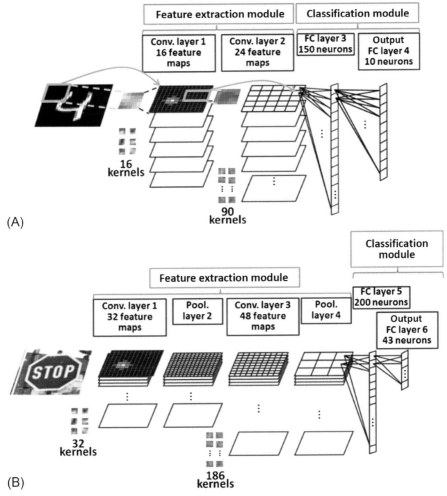

Fig. 15.5 CNN architecture for (A) handwritten digits recognition (MNIST database) and (B) traffic signs recognition (GTSRB database). For the second convolutional layer, a non-complete connection scheme (each output feature map is connected only to some of the input features maps) is adopted. The max-pooling layers in the GTSRB network reduce the size of the feature maps by a factor 2, thus reducing the complexity of the network. The stride is 2 and 1 for the MNIST and GTSRB networks respectively [42].

kernel; 16 and 32 feature maps are used in the first convolutional layer for the MNIST and GTSRB applications respectively. For the second convolutional layer, we adopted a noncomplete connection scheme; each output feature map is connected only to some of the input features maps through a convolutional kernel. For MNIST 16 input feature maps are connected to 24 output feature maps though 90 kernels, while for the complex GTSRB 32 input feature maps are connected to 48 output feature maps though 186 kernels. To produce smaller output volume data it is possible to increase the stride with which the kernel is sided on the input data or to add a max-pooling layer. For the MNIST and GTSRB the stride is 2 and 1, respectively. In the case of GTSRB network, the down-sampling operation is performed by adding a max-pooling layer. The estimated size of the OxRAM array needed to implement the CNNs is 600 kbit for MNIST and 1 Mb for GTSRB, respectively.

In order to study the impact of the OxRAM electrical performances and reliability on the network, we fully characterized a 16-kbit OxRAM demonstrator integrated into a 28-nm CMOS digital test-chip (Fig. 15.6) [43]. OxRAM devices feature a metal-insulator-metal (MIM) structure composed of a 5-nm-thick HfO_2 layer sandwiched between a Ti top electrode and a TiN bottom electrode. A bitcell is composed of 1 Transistor-1 Resistor (1T1R) structure. Fig. 15.6 reports the cumulative distributions of Low Resistance State (LRS) and High Resistance State (HRS) extracted from the 16-kbit OxRAM array statistics. No correction code or smart programming algorithms have been used. All the network simulations presented into the following take into account the real LRS and HRS distributions presented in Fig. 15.6.

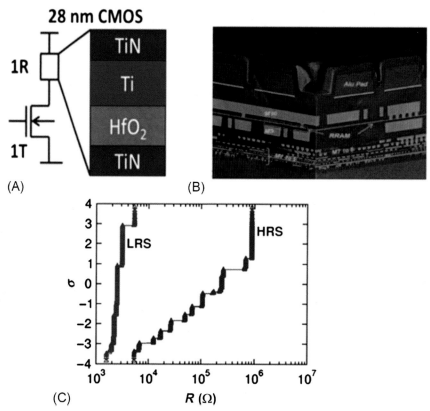

Fig. 15.6 (A) 1T1R bitcell schematic, (B) SEM cross section of CMOS 28 nm stack including HfO_2 based OxRAM cells, (C) cumulative distributions of LRS and HRS for 16-kbit demonstrator. Distributions are cut at 1 M due to the lower limit in current sensing [43].

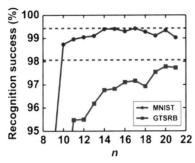

Fig. 15.7 Recognition success as a function of the number n of parallel OxRAM devices used to implement an equivalent synapse [42].

To implement a synapse in hardware, n OxRAM devices connected in parallel are used. In order to define the optimal value of the synaptic weight resolution, that is, the number (n) of OxRAM devices needed per synapse, simulations have been performed on both MNIST and GTSRB databases using the CNN architectures presented in Fig. 15.5. Simulation results in terms of recognition success as a function of the value n are shown in Fig. 15.7. The recognition success improves as n increases, for n higher than 12 the maximum network performance >99% is reached for MNIST. The *blue* curve in Fig. 15.7 reports the recognition success for the GTSRB database. More complex application tasks are more demanding in terms of number of cells per synapse.

15.5 Example of ReRAM fully connected neural networks with unsupervised learning

The use of ReRAM synapses also opens a path toward online real-time unsupervised learning (through continuous weight updating performed on local synaptic weights) and biological brain lifelong learning abilities (i.e., once learned, it is almost impossible to train the same algorithm or network on a different task without completely relearning all parameters). Plasticity will play an important role in achieving these goals.

Two main approaches (see Fig. 15.8) to emulate synaptic conductance modulation have been successfully demonstrated. In the *analog approach*, multiple low-resistance states for emulating long-term potentiation (cumulative increase of conductance, LTP) and multiple HRSs for long-term depression (cumulative and gradual decrease of conductance, LTD) are used. In the *binary approach*, only two distinct resistive states (LRS and HRS) are used per device, with probabilistic STDP bio-inspired learning rules. This approach is also motivated by biological studies which suggest that STDP learning might be a partially stochastic process in nature. In the case of the binary approach, in order to improve performance, a single synapse could be composed of n multiple binary cells in parallel. Several ideas have been proposed to implement STDP with memory devices.

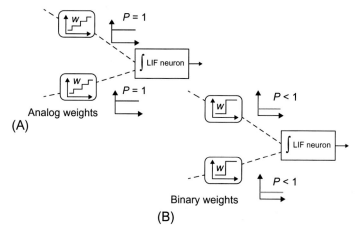

Fig. 15.8 Schematic illustrating (A) multilevel deterministic- and (B) binary probabilistic-synapses connected to a LIF neuron (*W*, weight; *p*, probability).

15.5.1   PCM based fully connected neural networks for image classification

A simplified version of STDP is presented in [44], where the analog time dependence of biological STDP is neglected and only two conditions (i.e., increasing or decreasing synaptic weight) are considered. This model requires technologies with multilevel capability. Phase-change memories show a strong asymmetry between the SET and RESET process: whereas the SET process is extremely gradual and resembles learning in neural networks, the RESET process is abrupt. In [44, 45] a 2-PCM synapse (see Fig. 15.9) that recreates artificial symmetry between SET and RESET by employing two devices per synapse has been proposed. This strategy has been shown to achieve unsupervised learning in a fully connected neural network for automobile tracking (see Fig. 15.10). An average detection rate of 92% and system power consumption for learning of 112 μW have been demonstrated by means of system level simulations.

15.5.2   CBRAM-based fully connected neural networks for auditory signal classification

In Ref. [46], an original methodology to use conductive bridge RAM (CBRAM) devices as easy to program and low-power binary synapses with stochastic learning rules is proposed. This learning scheme has been demonstrated on a fully connected neural network able to process asynchronous *analog data* streams for recognition and extraction of repetitive patterns in a fully unsupervised way. The demonstrated applications exhibit very good performance (auditory pattern sensitivity >2) and ultralow synaptic power dissipation (0.55 μW) in the learning mode.

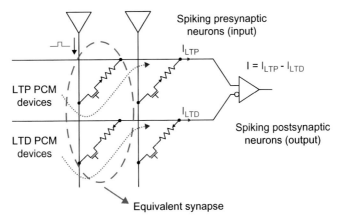

Fig. 15.9 Schematic representation showing the concept of 2-PCM synapse. The contribution of the current from the LTD device is subtracted at the postsynaptic neuron [44, 45].

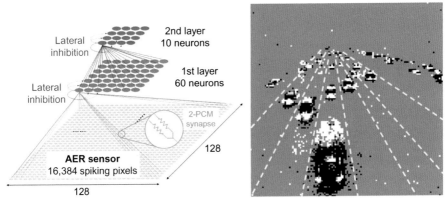

Fig. 15.10 Left: Illustration showing a two-layer feed forward fully connected spiking neural network. AER data is fed from a DVS artificial retina in the first layer of the neural network. Each neuron in the first layer is connected to every pixel by two synapses. Each neuron in the second layer is connected to neurons in the first layer by two synapses. Right: Screenshot from AER motion data of cars passing on a six lane freeway, recorded using the artificial silicon DVS retina sensor. The dotted lines marking the lanes were added. The difference in color indicates the difference in luminous intensity at a particular pixel [44, 45].

15.5.3 OxRAM based fully connected neural networks for real-time spike sorting of real biological data

Low-power neuromorphic computing systems can also be coupled with brain-computer-interfaces (BCI) to enable the design of autonomous implantable devices for rehabilitation purposes, capable of making decisions based on real-time online processing of in vivo recorded biological signals. In Ref. [47], a ReRAM-based two-layer fully connected neural network able to identify, learn, recognize, and distinguish between different spike shapes of measured biological signals without any supervision, was proposed. In order to illustrate the validity of the proposed spike sorting methodology, we measured the extracellular activity from in vitro Crayfish nerves recorded simultaneously with intracellular data of one motor or sensory neuron of the T5 ganglion (see Fig. 15.11). Fig. 15.12 shows the topological view of the network architecture. The biological signal is encoded by 32 frequency band-pass filters. The 32 filtered signals are then full-wave rectified and presented to the input layer of 32 neurons where the analog continuous signals are converted into spikes which are then propagated along the synapses to the five output neurons. To solve one of the main challenges of biological signal treatment in BCI, the high background noise level, a synaptic compound using HfO_2 based OxRAM cells; able to implement two different flavors of spike-based synaptic plasticity, the long-term and the short-term learning rules, has been presented [48]. Thanks to long-term plasticity, the system is capable of learning based on an unsupervised paradigm, while the short-term plasticity allows for improved accuracy despite the significant background noise in the input data.

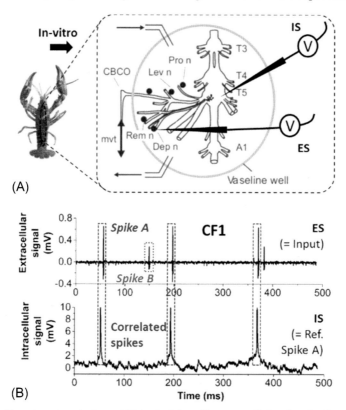

Fig. 15.11 Experiment to obtain real biological data. The extracellular signal reflects Spike A and B and in the following is called CF1. The intracellular activity reflects only Spike A and it can be used as the ground truth to assess the spike sorting capability for the detection of Spike A [47].

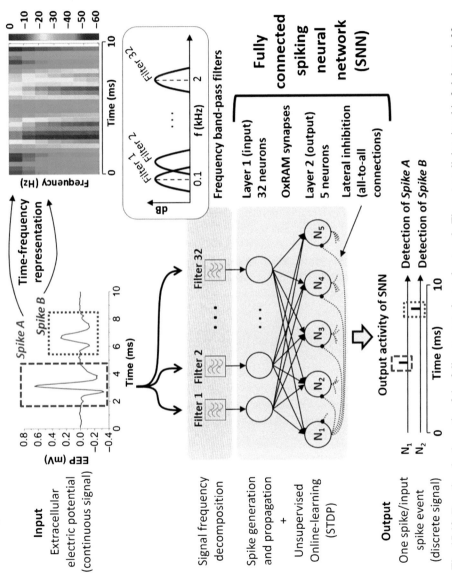

Fig. 15.12 Functional schematic of the fully connected neural network. The extracellular signal is fed through 32 frequency band-pass filters which are connected to the fully connected neural network. Synapses are based on HfO_2-based OxRAM devices. Output neurons become selective to different input spikes shapes [47].

15.6 Conclusions and perspectives

In this chapter, we showed how new disruptive technologies, and, in particular, the novel emerging memory technologies, coupled to new computing paradigms and algorithms, will allow for low-power *brain-inspired hardware*, which distributes intelligence over the whole IoT network, all-the-way down to ultralow-power end-devices. To map the embedded systems requirements, we presented a holistic research approach to develop low-power architectures inspired by the human brain, where process development and integration, circuit design, system architecture, and learning algorithms were simultaneously optimized.

Future artificial cognitive systems will be autonomous physical systems which will need to interact in real time with the environment and individuals everywhere. Physical constraints will shape the dynamics of these interactions. In such systems, as with biological organisms, *the link between the low-level sensory-motor processes, control systems, and cognition will play a key role*. The biological plausibility of artificial systems should not be a burden for engineers' creativity. Nonetheless, we believe that more interactions between AI engineers, neuroscientists, and biologists will be strongly beneficial from a fundamental point of view.

Bio-inspired approaches will force us to think differently. We would like to mention that biology teaches us that noise can improve the performance of biological sensory systems. Inspired by this assessment, several studies have been devoted to leveraging intrinsic device noise for neuromorphic computing. For example, the stochastic switching behavior of ReRAM under weak programming conditions was used to implement synapses with probabilistic STDP learning rules [49] and neuron circuits with stochastic firing [50].

Simpler biological systems, rather than the human brain, will be highly inspirational and instructive. For example, the use of *insects as templates for artificial intelligent systems* [51] highlights the need to think in a systemic way, as *organisms do not decouple sensors and signal treatment*. Future autonomous systems will have to perform intelligent tasks well beyond the possibilities of current ML systems (designed with a traditional input-output scheme and optimized to address classification tasks). The way they learn autonomously will be essential to define their predictive and interactive capabilities.

References

[1] B. DeSalvo, Brain inspired technologies: towards chips that think? in: IEEE ISSCC, 2018.
[2] S. Venkataramani, Efficient Embedded Learning for the IoT Devices, IEEE, 2016.
[3] N.D. Lane, et al., Can deep learning revolutionize mobile sensing? in: ACM International Workshop on Mobile Computing Systems and Applications, 2015.
[4] B. Moons, et al., A 0.3–2.6 TOPS/W Precision-Scalable Processor for Real-Time Large-Scale ConvNets, VLSI, 2016.
[5] S. Bang, et al., 14.7 A 288 µW programmable deep-learning processor with 270 kb on-chip weight storage using non-uniform memory hierarchy for mobile intelligence, in: ISSCC, 2017.

[6] A. Chandrakasan, Ultra-low-power networked systems, in: Nano Tera Workshop, 2015.

[7] M. Horowitz, Computing's energy problem (and what we can do about it), in: ISSCC, 2014.

[8] C. Mead, Analog VLSI and Neural Systems, Addison-Wesley VLSI Systems Series, 1989.

[9] F. Clermidy, et al., Advanced Technologies for Brain-Inspired Computing, IEEE, 2014.

[10] G. Indiveri, et al., Neuromorphic Architectures for Spiking Deep Neural Networks, IEDM, 2015.

[11] R. Brette, Philosophy of the spike: rate-based vs. spike-based theories of the brain, Front. Syst. Neurosci. 9 (2015) 151.

[12] P. Lichtsteiner, et al., A 128×128 120 dB 30 mW asynchronous vision sensor that responds to relative intensity change, in: IEEE ISSCC, 2006.

[13] C. Posch, et al., A QVGA 143 dB dynamic range asynchronous address-event PWM dynamic image sensor with lossless pixel-level video compression, in: IEEE ISSCC, 2010.

[14] P. Georgiou, et al., Low-power spiking chemical pixel sensor, Electron. Lett. 42 (23) (2006) 1331.

[15] M. Yang et al., "A 0.5 V 55 μW 64×2-channel binaural silicon cochlea for event-driven stereo-audio sensing", IEEE ISSCC 2016.

[16] S. Caviglia, et al., Spike-based readout of POSFET tactile sensors, IEEE Trans.Circuits Syst. 64 (6) (2017) 1421.

[17] J. Schemmel, et al., A wafer-scale neuromorphic hardware system for large-scale neural modeling, IEEE Int. Symp. Circuits Syst. (2010).

[18] P.A. Merolla, et al., A million spiking-neuron integrated circuit with a scalable communication network and interface, Science 345 (6197) (2014) 668.

[19] J. Sawada, et al., TrueNorth ecosystem for brain-inspired computing: scalable systems, software, and applications, in: ACM/IEEE International Conference for High Performance Computing, Networking, Storage and Analysis, 2016.

[20] N. Qiao, et al., A reconfigurable on-line learning spiking neuromorphic processor comprising 256 neurons and 128 K synapses, Front. Neurosci. 9 (2015) 141.

[21] L. Perniola, et al., Universal signatures from non-universal memories: clues for the future, in: IEEE 8th International Memory Workshop (IMW), 2016.

[22] E. Vianello, et al., Resistive memories for ultra-low-power embedded computing design, in: IEDM Technical Dig, 2014, pp. 144–147.

[23] P. Cappelletti, Non volatile memory evolution and revolution, in: IEDM, 2016.

[24] Stanford Memory Trends, https://nano.stanford.edu/stanford-memory-trends.

[25] G. Palma, et al., Interface engineering of Ag-GeS$_2$-based conductive bridge RAM for reconfigurable logic applications, IEEE Trans. Electron Devices 61 (3) (2014) 793–800.

[26] C. Cagli, et al., Experimental and theoretical study of electrode effects in HfO$_2$ based RRAM, in: IEDM Technical Dig, 2011.

[27] P. Zuliani, et al., Engineering of chalcogenide materials for embedded applications of Phase Change Memory, Solid-State Electr. 111 (2015) 27–31.

[28] G. Navarro, et al., Trade-off between SET and data retention performance thanks to innovative materials for phase-change memory, in: IEDM Technical Dig, 2013, pp. 570–573.

[29] V. Sousa, et al., Operation fundamentals in 12 Mb phase change memory based on innovative Ge-rich GST materials featuring high reliability performance, in: VLSI Tech. Symp, 2015, pp. 98–99.

[30] B. DeSalvo, From memory in our brain to emerging resistive memories in neuromorphic systems, in: IEEE IMW, 2015.

[31] D. Lee, et al., Oxide based nanoscale analog synapse device for neural signal recognition system, in: IEEE IEDM, 2015.

[32] G.W. Burr, et al., Large-scale neural networks implemented with non-volatile memory as the synaptic weight element: comparative performance analysis (accuracy, speed, and power), in: IEEE IEDM, 2015.

[33] D.H. Hubel, et al., Receptive fields, binocular interaction and functional architecture in the cats visual cortex, J. Physiol. 160 (1962) 106–154.

[34] D.J. Felleman, et al., Distributed hierarchical processing in the primate cerebral cortex, Cereb. Cortex 1 (1991) 1–47.

[35] K. Fukushima, Artificial vision by multi-layered neural networks: neocognitron and its advances, Neural Netw. 37 (2013) 103–119.

[36] D. Ciresan, et al., Multi–column deep neural network for traffic sign classification, Neural Netw. 32 (2012) 333–338.

[37] F. Ning, et al., Toward automatic phenotyping of developing embryos from videos, IEEE Trans. Image Process. 14 (2005) 1360–1371.

[38] P. Sermanet, et al., Pedestrian detection with unsupervised multi–stage feature learning, in: IEEE Conference on Computer Vision and Pattern Recognition (CVPR), 2013. 36263633.

[39] R. Vaillant, et al., A convolutional neural network hand tracker, IEEE Proc.-Vis. Image Signal Process. 141 (1994) 245–250.

[40] C. Garcia, et al., Convolutional face finder: a neural architecture for fast and robust face detection, IEEE Trans. Pattern Anal. Mach. Intell. 26 (2004) 14081423.

[41] Y. Taigman, et al., Deepface: closing the gap to human-level performance in face verification, in: IEEE Conference on Computer Vision and Pattern Recognition (CVPR), 2014, pp. 1701–1708.

[42] D. Garbin, et al., HfO$_2$-based OxRAM devices as synapses for convolutional neural networks, IEEE Trans. Electron Devices (2015).

[43] A. Benoist, et al., 28nm advanced CMOS resistive RAM solution as embedded non-volatile memory, in: Proceedings of the IEEE Reliability Physics Symposium, 2014, pp. 2E.6.1–2E.6.5.

[44] O. Bichler, et al., Synapses made by two phase-change memory devices for efficient spiking neural networks, IEEE Trans. Electron Devices (2012).

[45] M. Suri, et al., Phase Change Memory as Synapse for Ultra-Dense Neuromorphic Systems: Application to Complex Visual Pattern Extraction, IEDM, 2011.

[46] M. Suri, et al., Bio-inspired stochastic computing using binary CBRAM synapses, IEEE Trans. Electron Devices 60 (7) (2013) 2402.

[47] T. Werner, et al., Spiking neural networks based on OxRAM synapses for real-time unsupervised spike sorting, Front. Neurosci. 10 (2016) 474.

[48] E. Vianello, et al., Resistive Memories for Spike-Based Neuromorphic Circuits, IMW, 2017.

[49] N. Locatelli, et al., Spintronic devices as key elements for energy-efficient neuroinspired architectures, in: Design, Automation & Test in Europe Conference & Exhibition, 2015, p. 994.

[50] G. Palma, et al., Stochastic neuron design using conductive bridge RAM, in: IEEE/ACM International Symposium on Nanoscale Architectures, 2013.

[51] J. Casas, et al., Biomimetic flow sensors, in: Encyclopedia of Nanotechnology, Springer Verlag, 2012, p. 264.

Neuromorphic computing with resistive switching memory devices

Daniele Ielmini[a], Stefano Ambrogio[b]
[a]Dipartimento di Elettronica, Informazione e Bioingegneria, Politecnico of Milan and IU.NET, Milan, Italy, [b]IBM Research-Almaden, San Jose, CA, United States

16.1   Introduction

For the last 50 years, digital computing machines have improved their performance by scaling down the field-effect transistor (FET) according to the Moore's law, which predicts the doubling of the transistor count on the chip every 18 months [1]. Currently, the scaling trend in the microelectronics industry is facing the hard challenge of extreme dynamic power consumption, which is preventing the frequency increase due to the excessive chip heating [2]. To overcome the current limitations of scaling, novel devices have been proposed with the purpose of improving the subthreshold slope, thus allowing for the reduction of the dynamic power at constant static power consumption. New proposals include the tunnel FET [3], the negative-capacitance FET [4], and many other concepts [5], which are currently at the stage of demonstration in academic and industrial research labs.

On the other hand, the search for increasingly high operating frequency is questioned by the energy efficiency of the von Neumann architecture. Fig. 16.1 shows the power density of commercial central processing unit (CPU) chips as a function of the frequency bandwidth, demonstrating the relentless increase according to the Moore's law since 1971 [6]. In comparison, the human brain is located at the end of low power (about 20 W total consumed power) and low frequency (about 10 Hz). These data indicate that functionality may be improved not only by a mere increase of the number of operations per second, but also, and most importantly, by the architecture of the computing and memory elements.

The high performance and energy efficiency in the human brain can be explained by the colocation of the computing and memory elements, and the parallelism of the neural network architecture. Fig. 16.2 shows a simple feed-forward neural network with an input layer of neurons, two hidden layers, and an output layer with one neuron for classification. The network is fully connected (FC), in that each neuron in layer i is connected to each neuron in layer $i-1$ and layer $i+1$ through synapses. Each neuron in layer i collects the signals delivered by neurons in layer $i-1$, weighted by the corresponding synapse. The specific function of the neural network, for example, the type of pattern that can be recognized and

Advances in Non-volatile Memory and Storage Technology. https://doi.org/10.1016/B978-0-08-102584-0.00017-6

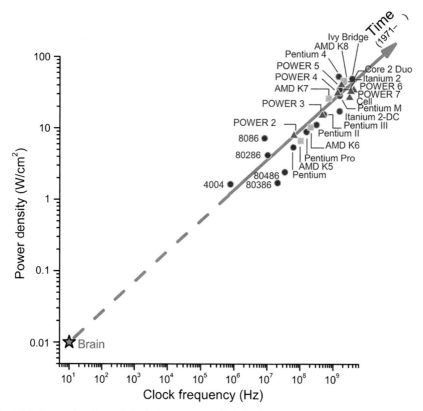

Fig. 16.1 Power density and clock frequency of digital microprocessors between 1971 and 2014. Moore's law improves both power density and clock frequency, however, battery and thermal limitations of power dissipation forced a saturation during the last few technology generations. Despite the low-frequency and low-power density, the brain can outperform a digital computer in many cognitive tasks.
Reproduced with permission from P.A. Merolla, J.V. Arthur, R. Alvarez-Icaza, A.S. Cassidy, J. Sawada, F. Akopyan, B.L. Jackson, N. Imam, C. Guo, Y. Nakamura, B. Brezzo, I. Vo, S.K. Esser, R. Appuswamy, B. Taba, A. Amir, M.D. Flickner, W.P. Risk, R. Manohar, D.S. Modha, A million spiking-neuron integrated circuit with a scalable communication network and interface, Science 345 (6197) (2014) 668–673. https://doi.org/10.1126/science.1254642. Copyright AAAS (2014).

distinguished by the network, is dictated by the weight of the synapses and the architecture, for example, the number of layers and the number of neurons in each layer. Synapses thus play a critical role for neuromorphic engineering that is the study and design of neural networks for developing cognitive circuits emulating the functioning of the brain. Note that memory and computing are sparsely distributed within a neural network, thus enabling enhanced performance with respect to von Neumann architectures, where the memory chip and the CPU are physically separated [7].

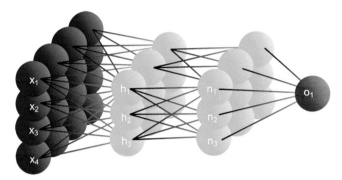

Fig. 16.2 Sketch of a feed-forward FC neural network with an input layer of neurons (x_1, x_2, etc.), two hidden layers (h_1, h_2, ..., n_1, n_2, ...), and an output layer with one neuron (o_1) for classification.

In this scenario, emerging memory technologies such as resistive switching memory (RRAM) [8–10], phase-change memory (PCM) [11–13], and magnetic memory (magnetoresistive random-access memory (MRAM)) [14–16] can play a pivotal role, thanks to possibility of multilevel storage, nanoscale dimension, back-end-of-line (BEOL) integration, and good reliability. Emerging memory devices can both store data and provide computing functionality, such as Boolean logic operations in RRAM devices [17–20] and PCM devices [21–23]. Accumulative crystallization in PCM allows for algebraic summation [24] and prime factorization [25]. Finally, RRAM and PCM allow for integration in cross-point arrays, which naturally provide matrix-vector multiplication (MVM) via physical computing through the Ohm's law and the Kirchhoff's law [26, 27].

As a result of the high computing capabilities in emerging memories, several schemes were proposed for artificial synapses and neurons in neuromorphic circuits. Synapses in cross-point arrays were demonstrated using PCM [28, 29] and RRAM [30, 31]. Brain-inspired synapses capable of changing their weights according to spike-timing-dependent plasticity (STDP) were demonstrated by PCM devices [32–34] and RRAM devices [35–40]. Electronic neuron circuits with integrate-and-fire operation were shown either relying on the accumulation of input spiking signals by crystallization in PCM devices [41] or using the threshold switching operation in volatile-type RRAM [42, 43]. Hardware implementations of neurons and synapses with PCM or RRAM technologies were demonstrated for supervised learning [29–31] or unsupervised learning [44]. Despite several challenges to upscale the proposed concepts to the higher level of complete systems capable of cognitive computing, the reported results are extremely promising for the development of neuromorphic circuits based on the emerging memories.

This chapter addresses RRAM devices, and most generally emerging memories such as PCMs, for neuromorphic circuits. RRAM-based neural networks will be reviewed with reference to supervised learning and unsupervised learning, discussing RRAM synaptic structures and the techniques for weight update in the network. Brain-inspired associative memory and error correction in recurrent Hopfield networks will be finally discussed.

16.2   Neural networks for supervised learning

Neural networks find extensive applications in pattern recognition, such as recognition of objects or faces within a picture. The most popular approach to this purpose is the deep learning concept, where FC-artificial neural networks (ANNs) with multiple-layer perceptron (MLP) structure are trained to provide an abstract representation of the submitted data [45]. Deep learning has recently led to several breakthroughs in various fields, including speech recognition [46], image recognition and object detection [47], machine translation [48], and drug discovery and genomics [49]. The use of two-terminal memory elements, such as PCM or RRAM, for the synaptic connections in the ANN of Fig. 16.2 has been early recognized as an attractive strategy for at least two reasons: first, the memory element can represent a multiple-bit value thanks to the analog nature of PCM and RRAM, thus offering the capability to replace many single-bit RAM cells for weight storage. For instance, up to eight levels, that is, three bits, have been demonstrated with RRAM devices [50, 51] and PCM devices [52], thus supporting the feasibility of a very high synaptic density in the ANN. The second added value of using two-terminal resistive memories is the cross-point architecture which enables physical computation of MVM by Ohm's and Kirchhoff's laws [26, 27]. In fact, the total current I_j of the cross-point column of index j is given by

$$I_j = \sum_i G_{ij} V_i \tag{16.1}$$

where V_i is the voltage applied to the cross-point row of index i, and G_{ij} is the synaptic conductance (or weight) connecting the row i and the column j. As a result, MVM is completed in situ within one single step, without any need for multiplication-accumulation (MAC) requiring multiple steps in a digital CPU with time-consuming exchange of input/output data with the memory chip. Supervised training supported by RRAM/PCM synaptic arrays thus seems a promising trend for saving power and speeding up the learning process for deep learning applications.

16.2.1   Network training by the backpropagation algorithm

Fig. 16.3 shows a typical ANN with MLP structure adopting PCM synapses [29]. For supervised learning, first a pattern, for example, a handwritten digit from the MNIST dataset, is submitted to the input layer, thus resulting in a feedforward propagation of the pattern across the several layers of neurons. In the feedforward propagation, neurons compute their output values as a proper function of the sum of the signals from the preceding layer, each signal multiplied by the corresponding synaptic weight, namely:

$$x_j^B = f\left(\sum_i w_{ij} x_i^A \right) \tag{16.2}$$

where x_i^A and x_j^B are the output signals of neuron i in layer A and neuron j in layer B, respectively, w_{ij} is the weight of the synapse connecting these two neurons, and

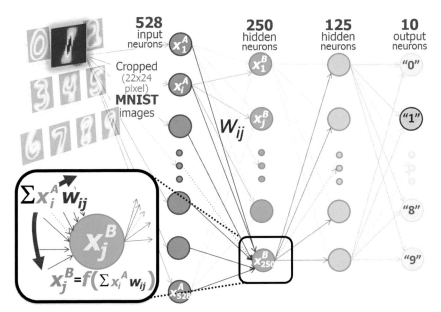

Fig. 16.3 Sketch of a feedforward FC-ANN with MLP structure adopting PCM synapses. For supervised learning of handwritten digits from the MNIST dataset, patterns are submitted to input neurons, and then propagated forward by the nonlinear transfer function of the neuron according to Eq. (16.2). Only the output neuron "1" should fire in response to the presentation of a digit "1" as input pattern.
Reproduced with permission from G.W. Burr, R.M. Shelby, S. Sidler, C. di Nolfo, J. Jang, I. Boybat, R.S. Shenoy, P. Narayanan, K. Virwani, E.U. Giacometti, B.N. Kurdi, H. Hwang, Experimental demonstration and tolerancing of a large-scale neural network (165 000 synapses) using phase-change memory as the synaptic weight element, IEEE Trans. Electron Devices 62 (11) (2015) 3498–3507. https://doi.org/10.1109/TED.2015.2439635. Copyright IEEE (2015).

f is a suitable nonlinear function representing the threshold-type behavior of the McCulloch-Pitts (MCP) neuron [53]. Fig. 16.4 shows some typical transfer functions adopted for ANNs, including the hyperbolic tangent function, the logistic function, and the rectifying linear unit (ReLU) function, which has been demonstrated to speed up the supervised learning in the backpropagation scheme and simplify the tuning of the parameters [54]. At the end of the forward propagation process, the synaptic weights are updated according to the backpropagation scheme shown in Fig. 16.5. Here, the output signals x_j^D are compared with the correct answer g_j provided by each pattern label, and the comparison leads to an error term δ_j^D given by

$$\delta_j^D = f'\left(x_j^D\right)\left(g_j - x_j^D\right) \tag{16.3}$$

where f' is the derivative of the neuron transfer function with respect to the neuron value x_j^D. The error term δ_j^D in Eq. (16.3) is then used to update the weights of

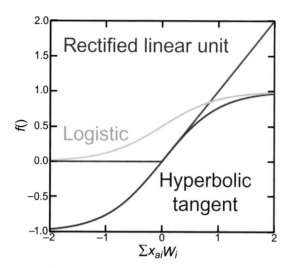

Fig. 16.4 Examples of nonlinear transfer functions adopted for neurons in ANNs, including the hyperbolic tangent function, the logistic function, and the rectifying linear unit (ReLU) function.

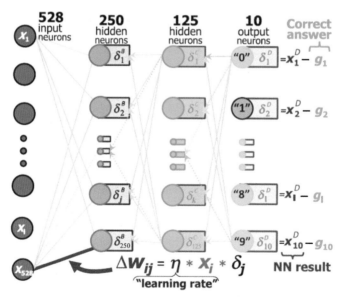

Fig. 16.5 Sketch of an ANN depicting the supervised learning process by the backpropagation scheme. The error from Eq. (16.3) is back propagated from output to input for weight update according to Eq. (16.4).

synapses connected to the output neuron layer according to the incremental update formula [55]:

$$\Delta w_{ij} = \eta x_i^C \delta_j^D \tag{16.4}$$

where η is a learning efficiency parameter dictating the speed of update of the back-propagation process. Note that the weight in Eq. (16.4) is changed proportionally to the error δ_j^D, which marks the distance of the actual weight from the ideal value for correctly recognizing the submitted pattern, and the input signal x_i^C, which is a figure of the importance of the synapse in the recognition process for the specific pattern. The learning efficiency η plays a key role for convergence and accuracy, therefore, its value must be carefully tuned to maximize the network ability to classify unknown input images. Errors in the preceding hidden layer C are calculated according to [55]

$$\delta_k^C = f'\left(x_k^C\right)\sum_j w_{ij}\delta_i^D \tag{16.5}$$

where f' is the derivative of the neuron transfer function with respect to the internal variable x_k. Also, Eq. (16.5) allows to iteratively compute errors across the ANN layers in the backpropagation direction.

16.2.2 Weight update of resistive switching devices

The weight update according to Eq. (16.4) poses a significant challenge in the design of PCM and RRAM synapses, in that both potentiation and depression must be achieved via incremental steps. However, in general, only one of the two operations can be gradually achieved: for instance, the crystallization process in PCM is incremental in that the application of sequential pulses causes more amorphous phase to change the crystalline state [21, 24]. On the other hand, amorphization process is only a function of the applied voltage and current in each pulse, without any significant dependence on the actual state of the PCM device. As a result, only incremental potentiation can be achieved in PCM devices, that is, Eq. (16.4) is only applicable for increasing the synaptic weight w_{ij}, or PCM conductance G_{ij}. Conversely, RRAM devices generally show incremental depression by repeating reset pulses [56], while incremental potentiation requires specific device materials, such as interfacial switching $Pr_{1-x}Ca_xMnO_3$ [57] or bilayer-stacked TaO_x-TiO_2 [58]. Another issue with resistive arrays is that a conductance G_{ij} can only map a positive weight w_{ij}, whereas both positive and negative weights are generally needed to represent input patterns in ANNs [55, 57].

These problems can be overcome by a differential approach, where each synapse is represented by two memory elements, for example, two PCM devices [28, 29] or two RRAM devices [57], and synaptic currents are obtained as the difference between the two paths in the differential scheme, namely:

$$I_j = \sum_i \left(G_{ij}^+ - G_{ij}^-\right)V_i \tag{16.6}$$

where G_{jj}^{+} and G_{jj}^{-} are the conductance values of the positive and negative resistive elements in the synaptic cell at position (i,j) in the array [57]. As a result, potentiation of weight w_{ij} can be achieved by either potentiation of conductance G_{ij}^{+}, or depression of conductance G_{ij}^{-}, depending on potentiation or depression being incremental in the adopted resistive switching synapse.

Another key issue for resistive switching synapse is the linearity and symmetry of the plasticity characteristics. Clearly, update according to Eq. (16.4) requires that the amount of potentiation/depression depends only on the number of pulses applied to the synapse element, not on the actual weight of the synapse which is generally not known. As a result, an ideal synapse should be linear, in that a certain amount of potentiation/depression can be achieved by a fixed amount of set/reset pulses irrespective of the resistive state of the device. On the other hand, real devices fail in satisfying linearity as shown by the weight update characteristics in Fig. 16.6 [59]: all the three reported cases show nonlinear change of conductance along both the potentiation and depression branches of the characteristics. For instance, potentiation is relatively steep if potentiating pulses are applied to a device with low conductance, whereas slow potentiation is obtained for relatively high conductance. Similarly, steep depression occurs for high-conductance states, with saturation taking place at low conductance. As a figure of merit, one can define a linearity factor α, which controls the power dependence of the normalized weight g on the number of pulses given by

$$g = \frac{G - G_0}{G_1 - G_0},$$
(16.7)

where g varies between 0 and 1 as the synaptic weight G changes from the lowest value G_0, corresponding to the high-resistance state (HRS), and the highest value G_1, corresponding to the low-resistance state (LRS) [57]. The normalized weight increases with the normalized number of pulses x, also changing between 0 and 1, according to the power law:

$$g = x^{\frac{1}{\alpha}},$$
(16.8)

with the nonlinearity factor α generally varying from 3 to 6 for potentiation and being equal to 1 only for ideally linear potentiation/depression synapses.

The lack of linear update causes inefficient learning in Fig. 16.7, showing the learning efficiency for training, that is, the probability for correct recognition of a pattern belonging to the data set and already submitted to the network during training, and for test, that is, the probability of correct recognition of a pattern which has not been submitted to the network during training [57]. The learning efficiency has been studied for various types of update characteristics, such as nonlinear/nonsymmetric characteristics with $\alpha > 1$ for potentiation and $\alpha < 1$ for depression (B), nonlinear/symmetric characteristics with $\alpha < 1$ relatively close to 1 (C), linear characteristics (D), nonlinear/symmetric characteristics with large $\alpha > 1$ (E), and nonlinear/symmetric characteristics with small $\alpha < 1$ (F). In general, the learning efficiency suffers from both nonlinearity and asymmetry, with the best performance for linear characteristics (D), and the worst behavior for nonlinear, nonsymmetric update (B). These results support the

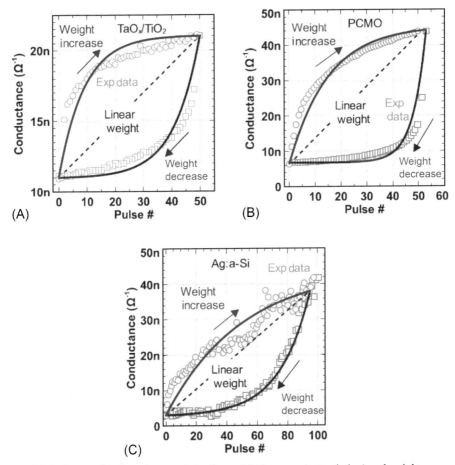

Fig. 16.6 Measured update characteristics for weight increase (potentiation) and weight decrease (depression) for TaO_x/TiO_2 (A), PCMO (B), and Ag:a-Si (C) resistive devices. Data show the measured conductance in response to the application of a sequence of pulses with constant positive/negative amplitude. Analog potentiation/depression is needed for backpropagation in the supervised learning of ANNs.
Reproduced with permission from S. Yu, P.-Y. Chen, Y. Cao, L. Xia, Y. Wang, H. Wu, Scaling-up resistive synaptic arrays for neuro-inspired architecture: challenges and prospect, in: IEDM Tech. Dig. 2015, 2015, pp. 451–454. https://doi.org/10.1109/IEDM.2015.7409718. Copyright IEEE (2015).

importance of a broad research scope aiming at linearity and symmetry in update characteristics, either by material engineering [57–59], or by circuit design [60]. Another important device property is the available dynamic range, namely the ratio between the highest and the lowest achievable conductance. A higher dynamic range allows to accommodate a larger number of intermediate steps, thus providing improved ANN performances. Toward this goal, an interesting approach is the periodic carry concept [61], where, instead of using one device for G^+ and one for G^-, more devices are

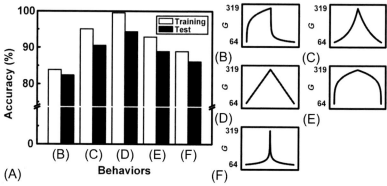

Fig. 16.7 (A) Learning efficiency for training, namely the probability for correct recognition of a pattern belonging to the data set and already submitted to the network during training, and for test, namely the probability of correct recognition of a pattern which was not submitted to the network during training. The learning efficiency was studied for various types of update characteristics, such as nonlinear/nonsymmetric characteristics with $\alpha > 1$ for potentiation and $\alpha < 1$ for depression (B), nonlinear/symmetric characteristics with $\alpha < 1$ relatively close to 1 (C), linear characteristics (D), nonlinear/symmetric characteristics with large $\alpha > 1$ (E), and nonlinear/symmetric characteristics with small $\alpha < 1$ (F).
Reproduced with permission from J.-W. Jang, S. Park, G.W. Burr, H. Hwang, Y.-H. Jeong, Optimization of conductance change in Pr1-xCaxMnO3-based synaptic devices for neuromorphic systems, IEEE Electron Device Lett. 36 (5) (2015) 457–459. https://doi. org/10.1109/LED.2015.2418342. Copyright IEEE (2015).

employed, each of them with a varying significance obtained by weighting the single device contributions with incremental coefficients (e.g., 1, 1/5, 1/25, and 1/125 [61]). With the periodic carry concept, the available number of levels can thus increase linearly with the number of devices employed.

16.2.3 Acceleration of ANNs with resistive switching devices

Crossbar architectures display large potential advantages for the calculation of MVM operations, thus being ideal to implement FC-ANNs. Several studies have been proposed to either perform training [29, 30] or directly program pretrained weights [62, 63]. The goal of the crossbar implementation is to achieve substantial speedup over conventional approaches to training and forward inference of deep learning networks, generally adopting the graphical processing unit (GPU) for fast MVM [29]. While GPUs show high performance in training convolutional neural networks (CNNs) [64], due to their ability to rapidly manipulate and move kernel weights during convolution operations, the performance on FC-ANNs is reduced because of the very large amount of weights to be trained. In this context, dense crossbar arrays of resistive devices represent an optimal solution, ideally reducing the duration of the weight update within an entire crossbar, hence ANN layer, to a single clock step [29].

To combine the fast training in cross-point architectures with the flexibility of GPUs, recently proposed analog/digital hybrid systems implement the crossbar array as an analog computational unit able to largely accelerate the calculation of the MVM [65–68], thus allowing to relax the computational load on the digital section [68]. For instance, a dot-product engine was proposed where the resistive crossbar array efficiently calculates MVM via vector scalar product $\mathbf{a} \cdot \mathbf{b} = \Sigma a_i b_i$, achieving in simulation software-like accuracy on MNIST digit recognition by implementing pretrained weights and efficiently performing forward inference [65, 66]. In a recently proposed hybrid scheme, ANN training is performed by storing the weight updates into a high-precision memory, and then transferring the cumulated weight change ΔW on a crossbar memory only when ΔW shows a size comparable with the device conductance change [68]. Based on these reports, hybrid systems incorporating analog/digital complementary metal-oxide-semiconductor (CMOS) circuits and cross-point arrays of emerging memories, such as PCM and RRAM, appear as the most promising approach to ANNs and other neural networks for deep learning applications.

16.3 Spiking neural networks for unsupervised learning

In a typical ANN, the input and output information are carried by the amplitude of a voltage or a current, which are linked to the synaptic weights by Eq. (16.1). While input/output signals might be either continuous or pulsed, there is no information carried by the time variables of the pulse, such as pulse width, or the frequency and the time of the occurrence of a stimulus. This is opposite to how the brain represents information, which is mapped by the specific neuron being active, and by the specific time of the neuron action, or spike. Such a spatiotemporal coding is what makes the brain highly energy efficient and highly functional in representing and elaborating complex information [69–71]. Another conjectured mode of information coding in the brain is rate coding [69–71], where the average rate of neuron spiking is used to describe the relevant input/output information.

Neuromorphic systems aiming at replicating the correct data processing in the brain usually adopt the spiking neural network (SNN) architecture with suitable spatiotemporal or rate coding of the information. Fig. 16.8 illustrates a typical SNN with single-layer perceptron structure, consisting of a first layer of neurons, each referred to as a presynaptic neuron (PRE), and a second layer with a single postsynaptic neuron (POST), which is connected to each PRE by a synapse [44]. The PRE spikes are transmitted through the synaptic connections to the POST for processing and learning.

16.3.1 Neurons and synapses in SNNs

In a SNN, neurons typically have a leaky integrate&fire (LIF) behavior, where input spikes are accumulated into an internal variable, usually referred to as membrane potential or internal potential V_{int}. As V_{int} reaches a threshold value, the neuron fires, that is, emits a spike toward the next layer of neurons, while resetting V_{int} back to zero. Fig. 16.9A shows the typical circuit of an electronic LIF neuron [44], including a first stage for leaky integration of an input current, with output V_{int}. The latter serves as input for the second "fire" stage, consisting of a voltage-controlled astable circuit for

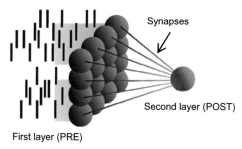

Fig. 16.8 Sketch of a feedforward FC SNNs with a single-layer perceptron architecture.
Reproduced from G. Pedretti, V. Milo, S. Ambrogio, R. Carboni, S. Bianchi, A. Calderoni,
N. Ramaswamy, A.S. Spinelli, D. Ielmini, Memristive neural network for on-line learning
and tracking with brain-inspired spike timing dependent plasticity, Sci. Rep. 7 (2017), 5288.
https://doi.org/10.1038/s41598-017-05480-0.

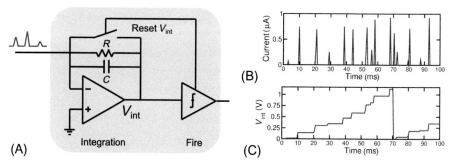

Fig. 16.9 (A) Sketch of an integrate&fire neuron, (B) input spikes, and (C) corresponding
internal potential V_{int} according to the simulations. At fire, the accumulation is reset and a
spike is generated toward the next layer of neurons.
Reproduced from G. Pedretti, V. Milo, S. Ambrogio, R. Carboni, S. Bianchi, A. Calderoni,
N. Ramaswamy, A.S. Spinelli, D. Ielmini, Memristive neural network for on-line learning
and tracking with brain-inspired spike timing dependent plasticity, Sci. Rep. 7 (2017), 5288.
https://doi.org/10.1038/s41598-017-05480-0.

generating a spike as V_{int} overcomes a given threshold. At the fire event, the fire stage
also sends a control signal back to the integration to reset the internal voltage V_{int} to
zero by short circuiting the feedback capacitor C. Fig. 16.9B shows a typical spiking
current signal, while Fig. 16.9C shows the calculated V_{int} from a circuit simulation
with the fire threshold $V_{th} = 1$ V [44].

Fig. 16.10A illustrates a biological system of two neurons connected by a synapse.
To replicate such elementary system in silico, the synapse can be implemented by a
hybrid CMOS/RRAM structure as represented in Fig. 16.10B. Here, the synapse combines a RRAM element with a FET in a 1-transistor/1-resistor (1T1R) structure, where
the HfO$_2$-based RRAM shows a bipolar switching behavior with a set transition from
HRS to LRS at positive voltage, and a reset transition from HRS to LRS for negative
voltage (Fig. 16.10C). The current during set transition is forced to remain below a
compliance current I_C, which can be controlled by the gate voltage of the series FET.

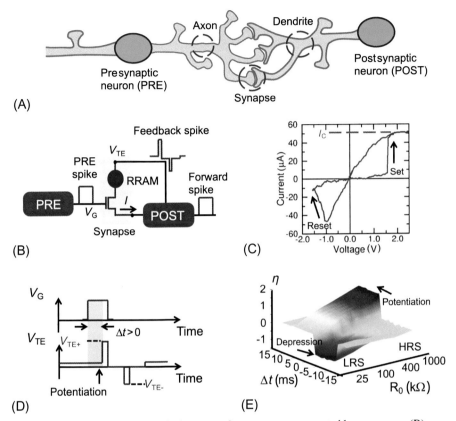

Fig. 16.10 (A) Sketch of a biological system of two neurons connected by a synapse, (B) synapse circuit implementation into a hybrid CMOS/RRAM 1T1R structure, (C) I-V curve of the 1T1R HfO$_2$-based RRAM with bipolar switching characteristics, (D) PRE and POST spikes overlapping for $\Delta t > 0$, and (E) measured STDP characteristics of η as a function of spike time delay and initial synaptic resistance R_0.

Reproduced from G. Pedretti, V. Milo, S. Ambrogio, R. Carboni, S. Bianchi, A. Calderoni, N. Ramaswamy, A.S. Spinelli, D. Ielmini, Memristive neural network for on-line learning and tracking with brain-inspired spike timing dependent plasticity, Sci. Rep. 7 (2017), 5288. https://doi.org/10.1038/s41598-017-05480-0.

When the PRE emits a spiking voltage to the gate of the FET, a small current proportional to the 1T1R conductance flows thanks to the constant voltage V_{TE} applied to the top electrode. The bottom electrode is connected to the input node of the POST, which thus integrates all spiking currents from the connected PREs. When the internal (membrane) potential V_{int} of the POST reaches a threshold, the POST fires, which consists of the generation of a feedforward spike to the next layer of neurons. Also, the POST generates a *feedback* spike composed by a positive voltage pulse and a negative voltage pulse, which can temporally overlap with the PRE spike. In case the PRE spike is temporally preceding the POST spike by a time delay $\Delta t > 0$ (Fig. 16.10D), the positive branch of the feedback spike partially overlaps with the PRE spike, thus causing a set transition of the RRAM, or synaptic potentiation. Conversely, if the PRE spike is temporally following the POST spike by $\Delta t < 0$, the negative branch of the feedback spike partially overlaps with the PRE spike, thus inducing a reset transition of the RRAM, or synaptic depression. The delay-dependent RRAM conductance changes reproduce STDP, where potentiation and depression occur for positive and negative Δt, respectively [72–76]. The PRE spike is 10 ms long, which is designed to cause no impact on the 1T1R conductance for relatively long delays, that is, $|\Delta t| > 10$ ms, according to the STDP learning rule. Fig. 16.10E shows the experimental STDP curve, reporting the ratio $\eta = \log_{10}(R_0/R)$ as a function of the delay time Δt for various initial resistances R_0 of the RRAM device. Data indicate depression and potentiation for $\Delta t < 0$ and $\Delta t > 0$, respectively [44]. Note that the 1T1R synapse is bistable, as a weight update always leads to a full set transition to LRS, or a full reset transition to HRS, irrespective of the initial resistance R_0. This implementation of STDP contrasts with the ANN trained with the backpropagation algorithm in Section 16.2, as it only requires two synaptic levels, thus being even more resilient to RRAM variability [77] and noise [78, 79]. On the other hand, the repeated request for full transitions may cause a stronger device degradation, with respect to the incremental conductance changes in backpropagation-based ANN.

16.3.2   Unsupervised learning by STDP

A key difference between brain-inspired SNNs and ANNs for deep learning is the type of learning taking place at synaptic level. Since there is no direct supervision in the brain, SNNs usually implement unsupervised learning by STDP, where the synaptic network autonomously learns a pattern that is stochastically submitted to the network. Unsupervised learning by STDP has been demonstrated in hardware SNNs with 1T1R synapses based on the feedforward structure shown in Fig. 16.8. Fig. 16.11 shows the experimental STDP behavior for a 1T1R synapse with the structure of Fig. 16.10, which was implemented in a 4×4 feedforward perceptron [44]. Based on the temporal overlap between the PRE and POST spikes in the 1T1R synapse (Fig. 16.11A), the synaptic weight undergoes potentiation for $\Delta t > 0$ ($\Delta t = +3$ ms in Fig. 16.11B) and depression for $\Delta t < 0$ ($\Delta t = -7$ ms in Fig. 16.11B). Fig. 16.11C summarizes the STDP response of the synapse, showing the correlation between the resistance $R(t_i)$ measured before the application of the spikes, and the resistance $R(t_{i+1})$ measured after the application of the spikes, for various pulse combinations. Cases with $0 < \Delta t < 10$ ms reveal clear transitions to the LRS, while the synapse switches to the HRS for -10 ms $< \Delta t < 0$. Finally, cases for $|\Delta t| > 10$ ms show no resistance variation, as the pulses are never applied to the device at the same time [44].

Fig. 16.12A shows the circuit implementation of perceptron network (Fig. 16.8), evidencing an architecture where many PREs are connected to the POST by 1T1R synapses. To achieve unsupervised specialization on an input pattern, the pattern (e.g., the "X" in Fig. 16.12B) is repeatedly submitted by the PREs through the synaptic channels, and on purpose randomly alternated with noise (Fig. 16.12C), as indicated by the raster plot in Fig. 16.12D, which shows the pixels of the 64 channels as a

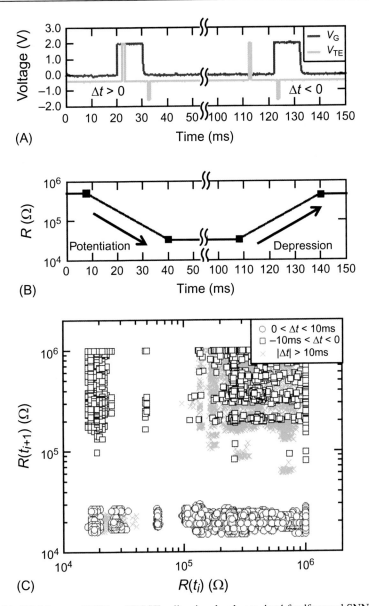

Fig. 16.11 (A) Measured PRE and POST spikes in a hardware 4×4 feedforward SNN, (B) corresponding change of resistance indicating potentiation ($\Delta t = +3$ ms) and depression ($\Delta t = -7$ ms), and (C) correlation plot of final resistance $R(t_{i+1})$ and initial resistance $R(t_i)$ for various delays in the STDP synapse, leading to potentiation (positive Δt), depression (negative Δt), and no change (large Δt).

Reproduced from G. Pedretti, V. Milo, S. Ambrogio, R. Carboni, S. Bianchi, A. Calderoni, N. Ramaswamy, A.S. Spinelli, D. Ielmini, Memristive neural network for on-line learning and tracking with brain-inspired spike timing dependent plasticity, Sci. Rep. 7 (2017), 5288. https://doi.org/10.1038/s41598-017-05480-0.

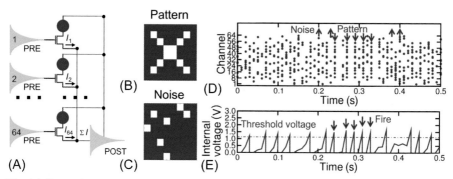

Fig. 16.12 (A) Sketch of a feedforward SNN with 1T1R synapses, (B) pattern and (C) noise adopted for unsupervised learning, (D) input spikes, and (E) corresponding V_{int} indicating fire in response to the presentation of the pattern after learning.
Reproduced with permission from S. Ambrogio, S. Balatti, V. Milo, R. Carboni, Z. Wang, A. Calderoni, N. Ramaswamy, D. Ielmini, Neuromorphic learning and recognition with one-transistor-one-resistor synapses and bistable metal oxide RRAM, IEEE Trans. Electron Devices 63 (4) (2016) 1508–1515. https://doi.org/10.1109/TED.2016.2526647. Copyright IEEE (2016).

function of time. Thanks to the unsupervised learning, once the network has specialized on the input pattern, the internal voltage V_{int} in Fig. 16.12E shows a clear correlation between fire events and input pattern [40].

Fig. 16.13A shows the simulated behavior of the synaptic weights during the unsupervised training described in Fig. 16.12. Simulations were carried out according to a stochastic model of RRAM devices, where set/reset events led to a statistically distributed resistance values replicating the experimentally observed distributions [40]. Starting from a random distribution of synaptic conductances, the synapses within the pattern channels converge to full LRS in around 50 epochs, while synapses in the other channels, also known as background synapses, show a gradual depression, as evidenced by synaptic conductance maps in Fig. 16.13B–D. The contrast between the abrupt potentiation and the gradual depression origins from the different roles of input pattern and noise. In fact, the pattern submission generates an immediate potentiation of pattern synapses, because pattern PRE spikes are correlated in time, thus are followed by a POST fire which causes potentiation according to the STDP rule. On the other hand, PRE noise spikes are not correlated, therefore, are much more likely to follow a POST fire, rather than anticipate a POST fire. Synapse depression thus occurs according to the STDP rule. However, since noise spiking density is relatively small compared to the pattern density, the depression rate is lower than the potentiation rate, the latter approaching the one- or few-shot learning speed [80]. Higher noise density leads to a faster background depression, hence increased overall training speed. However, excessive noise prevents the initial pattern potentiation, due to the competition between pattern and noise in inducing POST fire. The maximum amount of injected noise, hence maximum learning rate, is thus dictated by a trade-off between learning speed and training robustness and stability [81].

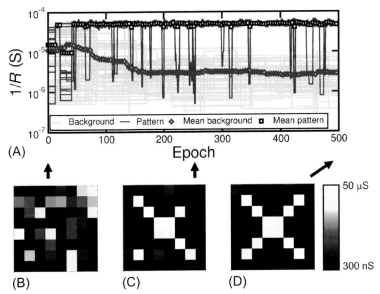

Fig. 16.13 (A) Measured conductance of the 1T1R synapses during unsupervised learning, and color plots of the synaptic weights for (B) initial state, (C) after 250 epochs, and (D) after 500 epochs. While potentiation of pattern synapses is almost immediate, the depression of background synapses is more gradual due to the uncorrelated, low-density noise spikes. Reproduced with permission from S. Ambrogio, S. Balatti, V. Milo, R. Carboni, Z. Wang, A. Calderoni, N. Ramaswamy, D. Ielmini, Neuromorphic learning and recognition with one-transistor-one-resistor synapses and bistable metal oxide RRAM, IEEE Trans. Electron Devices 63 (4) (2016) 1508–1515. https://doi.org/10.1109/TED.2016.2526647. Copyright IEEE (2016).

16.3.3 Hardware demonstration of unsupervised learning

A FC SNN with perceptron architecture was experimentally demonstrated by implementing PRE/POST neurons and synapses on a printed circuit board (PCB) [44]. An Arduino Due microcontroller (μC) was adopted to describe the spiking neurons, while the 1T1R HfO$_2$ RRAM synapses were wired together to build a 4×4 perceptron network. Fig. 16.14A shows the schematic circuit layout, while Fig. 16.14B shows the practical hardware implementation on the PCB [44].

Fig. 16.15 shows the measured synaptic weights during the unsupervised learning by stochastic spikes. The visual pattern consisted of a diagonal from lower-left to upper-right of the 4×4 image and was alternated with random noise functional for depression. Starting from a random distribution of synaptic conductance, synapses automatically adapt to the submitted pattern, reaching a complete pattern potentiation and background depression in about 1000 epochs, as shown by the color plot of the synaptic weights in Fig. 16.15A–D. Fig. 16.15E shows the spiking activity alternating pattern and noise, while Fig. 16.15F shows the detailed measured conductance evolution with time. The noise density was relatively low (about 5%), due to the small size of the visual pattern. Despite the programming variability and resistance fluctuations which characterize the RRAM, Fig. 16.15F indicates a marked resistance window

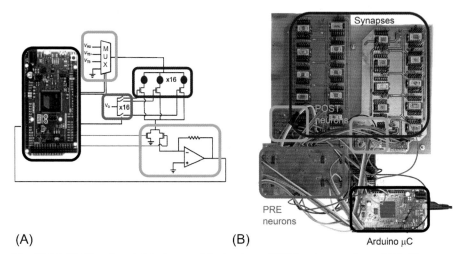

(A)   (B)   Arduino μC

Fig. 16.14 (A) Sketch and (B) picture of the hardware SNN implemented on a PCB. The 1T1R synapses and the PRE/POST neurons controlled by the microcontroller are indicated. Reproduced from G. Pedretti, V. Milo, S. Ambrogio, R. Carboni, S. Bianchi, A. Calderoni, N. Ramaswamy, A.S. Spinelli, D. Ielmini, Memristive neural network for on-line learning and tracking with brain-inspired spike timing dependent plasticity, Sci. Rep. 7 (2017), 5288. https://doi.org/10.1038/s41598-017-05480-0.

between the pattern synapses at LRS and background synapses at HRS, thus supporting the robustness of the learning algorithm.

In addition to pattern specialization and recognition, one of the advantages of unsupervised algorithm is the ability to adapt to input variations, at least in case input variations are slower than the learning time. This adaptive behavior contrasts with ANNs trained with backpropagation algorithm, where the "catastrophic forgetting" [82] prevents the network to adapt to modifications of the input information, such as the addition of a new class, thus forcing to retrain the entire network rather than just adapt a portion of the weights.

To demonstrate the adaptability of the STDP algorithm, Fig. 16.16 shows the successive learning of four different patterns, which were submitted during presentation phases at increasing times. During each phase, a different pattern was applied (Fig. 16.16A), leading to the adaptation of the synaptic weights to the submitted pattern (Fig. 16.16B). Fig. 16.16C shows the four visual patterns, consisting of four pixels gradually shifting from the top bar to the left bar in three steps. Fig. 16.16D shows the color plot of the synaptic weights, revealing a fast adaptability to the variation of the input pattern. Interestingly, the synapses not involved in the change of the input pattern do not significantly change their conductance after input shift, thus confirming the robustness of STDP to input variations [81].

STDP networks can also enable the classification and recognition of more than one pattern by implementing inhibitory synapses. Fig. 16.17A shows a perceptron SNN where the input layer is FC to two POST neurons, defined as POST1 and POST2. The two POSTs are mutually connected by inhibitory synapses, namely nonadaptive

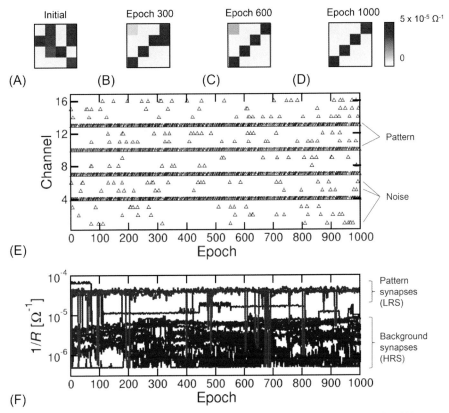

Fig. 16.15 Color plot of synaptic conductance values for (A) initial states, (B) after 300 epochs, (C) after 600 epochs, and (D) after 1000 epochs, (E) corresponding input spikes, and (F) real-time behavior of the synaptic conductance. Pattern synapses are potentiated to LRS, while background synapses are depressed to HRS.
Reproduced from G. Pedretti, V. Milo, S. Ambrogio, R. Carboni, S. Bianchi, A. Calderoni, N. Ramaswamy, A.S. Spinelli, D. Ielmini, Memristive neural network for on-line learning and tracking with brain-inspired spike timing dependent plasticity, Sci. Rep. 7 (2017), 5288. https://doi.org/10.1038/s41598-017-05480-0.

synapses which provide a negative current spike from a firing POST to the other. For instance, when POST1 fires in response to submitted pattern 1, POST2 receives a negative spike, thus inducing a decrease of the internal potential and preventing POST2 to fire in response to the same pattern. This allows each neuron to specialize on separate patterns, thus maximizing the information storage and recognition capability of the network. Two 3×3 RRAM synaptic arrays were physically implemented to fully connect the nine input axon channels to POST1 and POST2. To demonstrate learning and synaptic adaptation in the network, the two moving patterns in Fig. 16.17B, consisting of top/bottom horizontal stripes shifting to the clockwise or counterclockwise direction by one pixel every submission phase, were submitted. The patterns were randomly presented one at a time at the PRE channels, and were alternated with noise

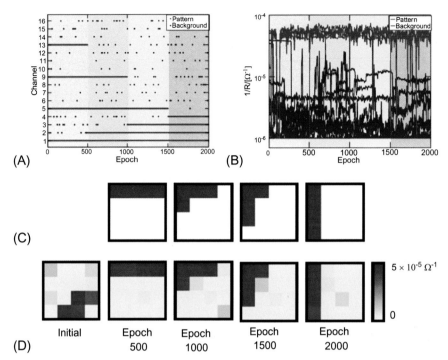

Fig. 16.16 (A) Input spikes, (B) synaptic weight evolution with time, (C) submitted patterns for each phase, and (D) corresponding weights for the initial state and after each training phase. The submitted pattern was initially a top bar, and then shifted to the left by one pixel at a time at every epoch.

Reproduced from G. Pedretti, V. Milo, S. Ambrogio, R. Carboni, S. Bianchi, A. Calderoni, N. Ramaswamy, A.S. Spinelli, D. Ielmini, Stochastic learning in neuromorphic hardware via spike timing dependent plasticity with RRAM synapses, IEEE J. Emerging Sel. Top. Circuits Syst. 8 (1) (2018) 77–85. https://doi.org/10.1109/JETCAS.2017.2773124.

according to the previously discussed stochastic approach. Every 1000 epochs the patterns were shifted as indicated by arrows. Fig. 16.17C and D shows the color plots of the measured synaptic weights for POST1 and POST2, respectively, at the end of each training phase. Note that each POST randomly learns one of the two patterns initially, and then remains locked to the same pattern during the subsequent phase, due to the tendency to minimize the synaptic weight change from one phase to the other. Fig. 16.17E and F shows the measured weight conductance for increasing epochs during each phase for POST1 and POST2, respectively. Pattern synapses are almost immediately potentiated to enable specialization of POST1 and POST2 to learn different patterns and track them during subsequent phases, thus keeping memory of the previous stored pattern.

Bistability of 1T1R synapses allows for robust learning thus providing strong advantages in a real implementation, however, the amount of information that can be encoded is only one bit, while several cases need more capacity. For instance, gray-scale images require a larger amount of information during synapse

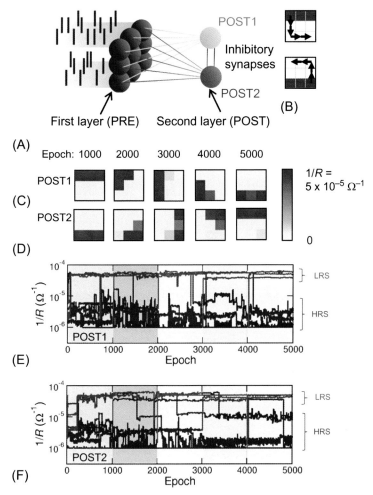

Fig. 16.17 (A) SNN for multi-pattern learning by POST1 and POST2, (B) submitted pattern 1 (top bar with counterclockwise shift) and pattern 2 (bottom bar with counterclockwise shift), color plot of the synaptic conductance for synapses pointing to (C) POST1 and (D) POST2, measured synaptic weights for (E) POST1 and (F) POST2 as a function of epoch. Reproduced from G. Pedretti, V. Milo, S. Ambrogio, R. Carboni, S. Bianchi, A. Calderoni, N. Ramaswamy, A.S. Spinelli, D. Ielmini, Memristive neural network for on-line learning and tracking with brain-inspired spike timing dependent plasticity, Sci. Rep. 7 (2017), 5288. https://doi.org/10.1038/s41598-017-05480-0.

training, namely, not only the position of the pattern, but also the signal amplitude at that position. Fig. 16.18 shows an experimental demonstration of grayscale image recognition, where the pattern was characterized by white and gray pixels. Fig. 16.18A–D shows the conductance color plots of experimental RRAMs during pattern learning. Information was stored into the 1T1R synapses changing the compliance current I_C proportional to pixel brightness, where gray correspond

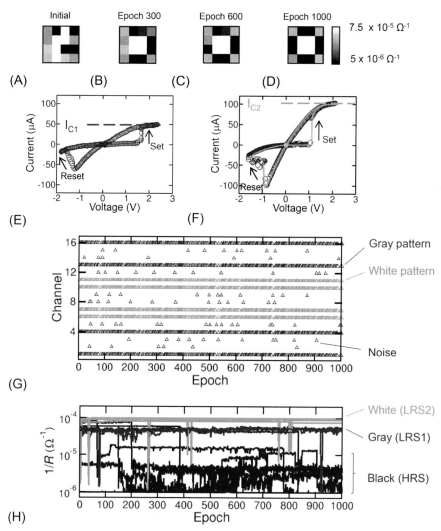

Fig. 16.18 Synaptic weights in a gray-scale color plot for (A) initial states, (B) after 300 epochs, (C) after 600 epochs, (D) after 1000 epochs, (E) measured I-V curve of the HfO$_2$ RRAM for a low compliance current $I_{C1} = 50\,\mu A$, corresponding to gray color, and (F) $I_{C2} = 100\,\mu A$ corresponding to white color, (G) input spikes, and (H) measured synaptic weights as a function of time.

Reproduced from G. Pedretti, V. Milo, S. Ambrogio, R. Carboni, S. Bianchi, A. Calderoni, N. Ramaswamy, A.S. Spinelli, D. Ielmini, Memristive neural network for on-line learning and tracking with brain-inspired spike timing dependent plasticity, Sci. Rep. 7 (2017), 5288. https://doi.org/10.1038/s41598-017-05480-0.

to $I_C = 50\,\mu A$ (Fig. 16.18E), while white corresponds to the highest $I_C = 100\,\mu A$ (Fig. 16.18F). The black background was achieved by noise-induced depression. Fig. 16.18G shows the raster plot of input spikes, while Fig. 16.18H shows the measured synaptic conductances for increasing epochs, supporting gray-scale learning with gray and white patterns reaching separate LRS conductance values LRS1 and LRS2, respectively [44].

16.4   Conclusions and outlook

This chapter reviews the state of the art regarding neuromorphic computing using RRAM devices. Two computing approaches have been covered, namely the ANN for supervised training and the SNN for brain-inspired unsupervised learning via STDP. ANNs find application in computer vision and can improve the training speed and energy efficiency with respect to CMOS-based systems, such as the GPU, thanks to physical computation of MVM within the memory array. The main challenge is the control of conductance in the RRAM, due to variability and nonlinearity of weight update characteristics. On the other hand, RRAM seems a promising solution for synaptic elements in brain-inspired SNNs, due to robust learning and relatively easy implementation of bio-realistic learning schemes, such as STDP and spike-rate-dependent plasticity (SRDP) rules. Unsupervised learning via STDP in SNNs was demonstrated both in simulation and hardware experiments, thus supporting the feasibility of RRAM-based SNNs for brain-inspired neuro-computing.

Although RRAM seems promising for synaptic elements in neural networks, there are still several challenges to reach commercial viability with respect to other existing technologies. In ANN applications, RRAM synapses are still affected by relatively slow programming compared to static random access memory (SRAM) and dynamic random-access memory (DRAM), strong nonlinearity of weight update [57–60], and variability issues [31, 77]. In particular, gradual set and reset processes are usually observed in a limited class of RRAM devices, which still requires optimization to improve the symmetry and linearity and thus achieve robust learning in supervised training. Materials engineering, for example, adopting interface-type switching in bilayer structures [58], or synaptic circuit engineering [60] are needed to improve and optimize RRAM characteristics for analog ANNs.

Regarding brain-inspired SNNs, the application scenario for this technology is not clear yet. The implementation of cognitive primitives similar to the human brain is extremely promising for realizing image and speech recognition in hardware, with relatively high-energy efficiency and large information storage density. However, these applications still require synaptic devices capable of closely resembling the biological learning rules in the brain, such as short-term activation in synaptic plasticity, which might enable bio-realistic learning by the physics of the device [83]. The architecture to achieve this goal might require both feedforward and recurrent SNNs, such as Hopfield networks with attractor-based dynamics [84]. A broadly interdisciplinary approach involving device engineering, circuit design, and neuroscience is thus needed to create a research roadmap toward the development of truly brain-inspired neuromorphic computing.

Acknowledgments

This chapter has received funding from the European Research Council (ERC) under the European Union's Horizon 2020 research and innovation program (grant agreement no. 648635).

References

[1] G.E. Moore, Cramming more components onto integrated circuits, Electronics (1965) 114–117.

[2] K.J. Kuhn, Considerations for ultimate CMOS scaling, IEEE Trans. Electron Devices 59 (7) (2012) 1813–1828, https://doi.org/10.1109/TED.2012.2193129.

[3] A.M. Ionescu, H. Riel, Tunnel field-effect transistors as energy-efficient electronic switches. Nature 479 (7373) (2011) 329–337, https://doi.org/10.1038/nature10679.

[4] S. Salahuddin, S. Datta, Use of negative capacitance to provide voltage amplification for low power nanoscale devices, Nano Lett. 8 (2) (2008) 405–410, https://doi.org/10.1021/nl071804g.

[5] D.E. Nikonov, I.A. Young, Overview of beyond-CMOS devices and a uniform methodology for their benchmarking, Proc. IEEE 101 (12) (2013) 2498–2533, https://doi.org/10.1109/JPROC.2013.2252317.

[6] P.A. Merolla, J.V. Arthur, R. Alvarez-Icaza, A.S. Cassidy, J. Sawada, F. Akopyan, B.L. Jackson, N. Imam, C. Guo, Y. Nakamura, B. Brezzo, I. Vo, S.K. Esser, R. Appuswamy, B. Taba, A. Amir, M.D. Flickner, W.P. Risk, R. Manohar, D.S. Modha, A million spiking-neuron integrated circuit with a scalable communication network and interface, Science 345 (6197) (2014) 668–673, https://doi.org/10.1126/science.1254642.

[7] G. Indiveri, S.-C. Liu, Memory and information processing in neuromorphic systems. Proc. IEEE 103 (8) (2015) 1379–1397, https://doi.org/10.1109/JPROC.2015.2444094.

[8] R. Waser, M. Aono, Nanoionics-based resistive switching memories, Nat. Mater. 6 (2007) 833–840, https://doi.org/10.1038/nmat2023.

[9] H.-S.P. Wong, H.-Y. Lee, S. Yu, Y.-S. Chen, Y. Wu, P.-S. Chen, B. Lee, F.T. Chen, M.-J. Tsai, Metal-oxide RRAM. 100 (6) (2012) 1951–1970, https://doi.org/10.1109/JPROC.2012.2190369.

[10] D. Ielmini, Resistive switching memories based on metal oxides: mechanisms, reliability and scaling, Semicond. Sci. Technol. 31 (2016) 063002, https://doi.org/10.1088/0268-1242/31/6/063002.

[11] M. Wuttig, N. Yamada, Phase-change materials for rewriteable data storage, Nat. Mater. 6 (2007) 824–832, https://doi.org/10.1038/nmat2009.

[12] S. Raoux, W. Welnic, D. Ielmini, Phase change materials and their application to nonvolatile memories, Chem. Rev. 110 (1) (2010) 240–267, https://doi.org/10.1021/cr900040x.

[13] G.W. Burr, M.J. Breitwisch, M. Franceschini, D. Garetto, K. Gopalakrishnan, B. Jackson, B. Kurdi, C. Lam, L.A. Lastras, A. Padilla, B. Rajendran, S. Raoux, R.S. Shenoy, Phase change memory technology, J. Vac. Sci. Technol. B 28 (2) (2010) 223–262, https://doi.org/10.1116/1.3301579.

[14] C. Chappert, A. Fert, F.N. Van Dau, The emergence of spin electronics in data storage, Nat. Mater. 6 (2007) 813–823, https://doi.org/10.1038/nmat2024.

[15] B. Dieny, R.C. Sousa, J. Hérault, C. Papusoi, G. Prenat, U. Ebels, D. Houssameddine, B. Rodmacq, S. Auffret, L.D. Buda-Prejbeanu, Spin-transfer effect and its use in spintronic components, Int. J. Nanotechnol. 7 (2010) 591, https://doi.org/10.1504/IJNT.2010.031735.

[16] A.D. Kent, D.C. Worledge, A new spin on magnetic memories, Nat. Nanotechnol. 10 (2015) 187–191, https://doi.org/10.1038/nnano.2015.24.

[17] J. Borghetti, G.S. Snider, P.J. Kuekes, J.J. Yang, D.R. Stewart, R.S. Williams, 'Memristive' switches enable 'stateful' logic operations via material implication, Nature 464 (2010) 873–876, https://doi.org/10.1038/nature08940.

[18] S. Balatti, S. Ambrogio, D. Ielmini, Normally-off logic based on resistive switches—Part I: Logic gates, IEEE Trans. Electron Devices 62 (6) (2015) 1831–1838, https://doi.org/10.1109/TED.2015.2422999.

[19] B. Chen, F. Cai, J. Zhou, W. Ma, P. Sheridan, W.D. Lu, Efficient in-memory computing architecture based on crossbar arrays, in: IEDM Tech. Dig, 2015, pp. 17.5.1–17.5.4, https://doi.org/10.1109/IEDM.2015.7409720.

[20] P. Huang, J. Kang, Y. Zhao, S. Chen, R. Han, Z. Zhou, Z. Chen, W. Ma, M. Li, L. Liu, X. Liu, Reconfigurable nonvolatile logic operations in resistance switching crossbar array for large-scale circuits, Adv. Mater. 28 (44) (2016) 9758–9764, https://doi.org/10.1002/adma.201602418.

[21] M. Cassinerio, N. Ciocchini, D. Ielmini, Logic computation in phase change materials by threshold and memory switching, Adv. Mater. 25 (41) (2013) 5975–5980, https://doi.org/10.1002/adma.201301940.

[22] Y. Li, Y.P. Zhong, Y.F. Deng, Y.X. Zhou, L. Xu, X.S. Miao, Nonvolatile "AND," "OR," and "NOT" Boolean logic gates based on phase-change memory, J. Appl. Phys. 114 (2013) 234503, https://doi.org/10.1063/1.4852995.

[23] D. Loke, J.M. Skelton, W.-J. Wang, T.-H. Lee, R. Zhao, T.-C. Chong, S.R. Elliott, Ultrafast phase-change logic device driven by melting processes, Proc. Natl. Acad. Sci. U. S. A. 111 (2014) 13272–13277, https://doi.org/10.1073/pnas.1407633111.

[24] C.D. Wright, Y. Liu, K.I. Kohary, M.M. Aziz, R.J. Hicken, Arithmetic and biologically-inspired computing using phase-change materials, Adv. Mater. 23 (30) (2011) 3408–3413, https://doi.org/10.1002/adma.201101060.

[25] P. Hosseini, A. Sebastian, N. Papandreou, C.D. Wright, H. Bhaskaran, Accumulation-based computing using phase-change memories with FET access devices, IEEE Electron Device Lett. 36 (9) (2015) 975–977, https://doi.org/10.1109/LED.2015.2457243.

[26] S.N. Truong, K.-S. Min, New memristor-based crossbar array architecture with 50% area reduction and 48% power saving for matrix-vector multiplication of analog neuromorphic computing, J. Semicond. Technol. Sci. 14 (3) (2014) 356–363, https://doi.org/10.5573/JSTS.2014.14.3.356.

[27] P. Gu, B. Li, T. Tang, S. Yu, Y. Cao, Y. Wang, H. Yang, Technological exploration of RRAM crossbar array for matrix-vector multiplication, in: 2015 20th Asia and South Pacific Design Automation Conference (ASP-DAC), 2015, pp. 1–6, https://doi.org/10.1109/ASPDAC.2015.7058989.

[28] O. Bichler, M. Suri, D. Querlioz, D. Vuillaume, B. DeSalvo, C. Gamrat, Visual pattern extraction using energy-efficient "2-PCM synapse" neuromorphic architecture, IEEE Trans. Electron Devices 59 (8) (2012) 2206–2214, https://doi.org/10.1109/TED.2012.2197951.

[29] G.W. Burr, R.M. Shelby, S. Sidler, C. di Nolfo, J. Jang, I. Boybat, R.S. Shenoy, P. Narayanan, K. Virwani, E.U. Giacometti, B.N. Kurdi, H. Hwang, Experimental

demonstration and tolerancing of a large-scale neural network (165 000 synapses) using phase-change memory as the synaptic weight element, IEEE Trans. Electron Devices 62 (11) (2015) 3498–3507, https://doi.org/10.1109/TED.2015.2439635.

[30] M. Prezioso, F. Merrikh-Bayat, B.D. Hoskins, G.C. Adam, K.K. Likharev, D.B. Strukov, Training and operation of an integrated neuromorphic network based on metal-oxide memristors, Nature 521 (7550) (2015) 61–64, https://doi.org/10.1038/nature14441.

[31] P. Yao, H. Wu, B. Gao, S.B. Eryilmaz, X. Huang, W. Zhang, Q. Zhang, N. Deng, L. Shi, H.-S.P. Wong, H. Qian, Face classification using electronic synapses, Nat. Commun. 8 (2017) 15199, https://doi.org/10.1038/ncomms15199.

[32] D. Kuzum, R.G.D. Jeyasingh, B. Lee, H.-S.P. Wong, Nanoelectronic programmable synapses based on phase change materials for brain-inspired computing, Nano Lett. 12 (5) (2012) 2179–2186, https://doi.org/10.1021/nl201040y.

[33] S. Kim, M. Ishii, S. Lewis, T. Perri, M. BrightSky, W. Kim, R. Jordan, G.W. Burr, N. Sosa, A. Ray, J.-P. Han, C. Miller, K. Hosokawa, C. Lam, NVM neuromorphic core with 64k-cell (256-by-256) phase change memory synaptic array with on-chip neuron circuits for continuous in-situ learning, in: IEDM Tech. Dig. 2015, 2015, pp. 443–446, https://doi.org/10.1109/IEDM.2015.7409716.

[34] S. Ambrogio, N. Ciocchini, M. Laudato, V. Milo, A. Pirovano, P. Fantini, D. Ielmini, Unsupervised learning by spike timing dependent plasticity in phase change memory (PCM) synapses, Front. Neurosci. 10 (2016) 56, https://doi.org/10.3389/fnins.2016.00056.

[35] S.H. Jo, T. Chang, I. Ebong, B.B. Bhadviya, P. Mazumder, W. Lu, Nanoscale memristor device as synapse in neuromorphic systems, Nano Lett. 10 (4) (2010) 1297–1301, https://doi.org/10.1021/nl904092h.

[36] S. Yu, Y. Wu, R. Jeyasingh, D. Kuzum, H.-S.P. Wong, An electronic synapse device based on metal oxide resistive switching memory for neuromorphic computation, IEEE Trans. Electron Devices 58 (8) (2011) 2729–2737, https://doi.org/10.1109/TED.2011.2147791.

[37] K. Seo, I. Kim, S. Jung, M. Jo, S. Park, J. Park, J. Shin, K.P. Biju, J. Kong, K. Lee, B. Lee, H. Hwang, Analog memory and spike-timing-dependent plasticity characteristics of a nanoscale titanium oxide bilayer resistive switching device. Nanotechnology 22 (25) (2011) 254023, https://doi.org/10.1088/0957-4484/22/25/254023.

[38] S. Ambrogio, S. Balatti, F. Nardi, S. Facchinetti, D. Ielmini, Spike-timing dependent plasticity in a transistor-selected resistive switching memory, Nanotechnology 24 (2013) 384012, https://doi.org/10.1088/0957-4484/24/38/384012.

[39] Z.-Q. Wang, S. Ambrogio, S. Balatti, D. Ielmini, A 2-transistor/1-resistor artificial synapse capable of communication and stochastic learning for neuromorphic systems, Front. Neurosci. 8 (2015) 438, https://doi.org/10.3389/fnins.2014.00438.

[40] S. Ambrogio, S. Balatti, V. Milo, R. Carboni, Z. Wang, A. Calderoni, N. Ramaswamy, D. Ielmini, Neuromorphic learning and recognition with one-transistor-one-resistor synapses and bistable metal oxide RRAM, IEEE Trans. Electron Devices 63 (4) (2016) 1508–1515, https://doi.org/10.1109/TED.2016.2526647.

[41] T. Tuma, A. Pantazi, M. Le Gallo, A. Sebastian, E. Eleftheriou, Stochastic phase-change neurons. Nat. Nanotechnol. 11 (2016) 693–699, https://doi.org/10.1038/nnano.2016.70.

[42] M.D. Pickett, G. Medeiros-Ribeiro, R.S. Williams, A scalable neuristor built with Mott memristors, Nat. Mater. 12 (2013) 114–117, https://doi.org/10.1038/nmat3510.

[43] X. Zhang, W. Wang, Q. Liu, X. Zhao, J. Wei, R. Cao, Z. Yao, X. Zhu, F. Zhang, H. Lv, S. Long, M. Liu, An artificial neuron based on a threshold switching memristor, IEEE Electron Device Lett. 39 (2) (2018) 308–311, https://doi.org/10.1109/LED.2017.2782752.

[44] G. Pedretti, V. Milo, S. Ambrogio, R. Carboni, S. Bianchi, A. Calderoni, N. Ramaswamy, A.S. Spinelli, D. Ielmini, Memristive neural network for on-line learning and tracking

with brain-inspired spike timing dependent plasticity, Sci. Rep. 7 (2017) 5288, https://doi.org/10.1038/s41598-017-05480-0.

[45] Y. LeCun, L. Bottou, Y. Bengio, P. Haffner, Gradient-based learning applied to document recognition, Proc. IEEE 86 (11) (1998) 2278–2324, https://doi.org/10.1109/5.726791.

[46] A. Graves, A. Mohamed, G. Hinton, Speech recognition with deep recurrent neural networks, in: 2013 IEEE International Conference on Acoustics, Speech and Signal Processing, 2013, pp. 6645–6649, https://doi.org/10.1109/ICASSP.2013.6638947.

[47] A. Krizhevsky, I. Sutskever, G.E. Hinton, ImageNet classification with deep convolutional neural networks, in: Advances in Neural Information Processing Systems, NIPS'12 Proceedings of the 25th International Conference on Neural Information Processing Systems, vol. 1, Lake Tahoe, Nevada, December 03–06, Curran Associates, Inc., USA, 2012, pp. 1097–1105.

[48] https://www.nytimes.com/2016/12/14/magazine/the-great-ai-awakening.html.

[49] Y. LeCun, Y. Bengio, G. Hinton, Deep learning, Nature 521 (2015) 436–444, https://doi.org/10.1038/nature14539.

[50] S. Balatti, S. Larentis, D. Gilmer, D. Ielmini, Multiple memory states in resistive switching devices through controlled size and orientation of the conductive filament, Adv. Mater. 25 (10) (2013) 1474–1478, https://doi.org/10.1002/adma.201204097.

[51] A. Prakash, J. Park, J. Song, J. Woo, E.-J. Cha, H. Hwang, Demonstration of low power 3-bit multilevel cell characteristics in a TaO$_x$-based RRAM by stack engineering, IEEE Electron Device Lett. 36 (1) (2015) 32–34, https://doi.org/10.1109/LED.2014.2375200.

[52] A. Athmanathan, M. Stanisavljevic, N. Papandreou, H. Pozidis, E. Eleftheriou, Multilevel-cell phase-change memory: a viable technology, IEEE J. Emerging Sel. Top. Circuits Syst. 6 (1) (2016) 87–100, https://doi.org/10.1109/JETCAS.2016.2528598.

[53] W.S. McCulloch, W. Pitts, A logical calculus of the ideas immanent in nervous activity, Bull. Math. Biophys. 5 (4) (1943) 115–133.

[54] X. Glorot, A. Bordes, Y. Bengio, Deep sparse rectifier neural networks, in: Proc. 14th International Conference on Artificial Intelligence and Statistics, vol. 15, 2011, pp. 315–323.

[55] S. Haykin, Neural Networks and Learning Machines, third ed., Prentice Hall, 2009.

[56] S. Yu, B. Gao, Z. Fang, H. Yu, J. Kang, H.-S.P. Wong, A low energy oxide-based electronic synaptic device for neuromorphic visual systems with tolerance to device variation, Adv. Mater. 25 (12) (2013) 1774–1779, https://doi.org/10.1002/adma.201203680.

[57] J.-W. Jang, S. Park, G.W. Burr, H. Hwang, Y.-H. Jeong, Optimization of conductance change in Pr$_{1-x}$Ca$_x$MnO$_3$-based synaptic devices for neuromorphic systems, IEEE Electron Device Lett. 36 (5) (2015) 457–459, https://doi.org/10.1109/LED.2015.2418342.

[58] I.-T. Wang, Y.-C. Lin, Y.-F. Wang, C.-W. Hsu, T.-H. Hou, 3D synaptic architecture with ultralow sub-10 fJ energy per spike for neuromorphic computation, in: IEDM Tech. Dig, 2014, pp. 665–668, https://doi.org/10.1109/IEDM.2014.7047127.

[59] S. Yu, P.-Y. Chen, Y. Cao, L. Xia, Y. Wang, H. Wu, Scaling-up resistive synaptic arrays for neuro-inspired architecture: challenges and prospect, in: IEDM Tech. Dig. 2015, 2015, pp. 451–454, https://doi.org/10.1109/IEDM.2015.7409718.

[60] K. Moon, M. Kwak, J. Park, D. Lee, H. Hwang, Improved conductance linearity and conductance ratio of 1T2R synapse device for neuromorphic systems, IEEE Electron Device Lett. 38 (8) (2017) 1023–1026, https://doi.org/10.1109/LED.2017.2721638.

[61] S. Agarwal, R.B.J. Gedrim, A.H. Hsia, D.R. Hughart, E.J. Fuller, A.A. Talin, C.D. James, S.J. Plimpton, M.J. Marinella, Achieving ideal accuracies in analog neuromorphic computing using periodic carry, in: Symp. VLSI Tech. Dig, 2017, pp. 174–175, https://doi.org/10.23919/VLSIT.2017.7998164.

[62] F. Merrikh-Bayat, M. Prezioso, B. Chakrabarti, H. Nili, I. Kataeva, D. Strukov, Implementation of multilayer perceptron network with highly uniform passive memristive crossbar circuits, Nat. Commun. 9 (1) (2018) 2331, https://doi.org/10.1038/s41467-018-04482-4.

[63] S. Yu, Z. Li, P.-Y. Chen, H. Wu, B. Gao, D. Wang, W. Wu, H. Qian, Binary neural network with 16 Mb RRAM macro chip for classification and online training, in: IEDM Tech. Dig. 2016, 2016, pp. 416–419, https://doi.org/10.1109/IEDM.2016.7838429.

[64] A. Coates, B. Huval, T. Wang, D. Wu, A.Y. Ng, B.C. Catanzaro, Deep learning with COTS HPC systems, in: Proceedings of the 30th International Conference on Machine Learning, vol. 28 (3), 2013, pp. 1337–1345.

[65] M. Hu, J.P. Strachan, Z. Li, R.S. Williams, Dot-product engine as computing memory to accelerate machine learning algorithms, in: 17th International Symposium on Quality Electronic Design (ISQED), 2016, pp. 374–379, https://doi.org/10.1109/ISQED.2016.7479230.

[66] M. Hu, J.P. Strachan, Z. Li, E.M. Grafals, N. Davila, C. Graves, S. Lam, N. Ge, J.J. Yang, R.S. Williams, Dot-product engine for neuromorphic computing: programming 1T1M crossbar to accelerate matrix-vector multiplication, in: 53rd Annual Design Automation Conference (DAC), 2016, pp. 1–6, https://doi.org/10.1145/2897937.2898010.

[67] M. Le Gallo, A. Sebastian, R. Mathis, M. Manica, H. Giefers, T. Tuma, C. Bekas, A. Curioni, E. Eleftheriou, Mixed-precision in-memory computing, arXiv:1701.04279v4, 2017.

[68] S.R. Nandakumar, M. Le Gallo, I. Boybat, B. Rajendran, A. Sebastian, E. Eleftheriou, Mixed-precision training of deep neural networks using computational memory, arXiv:1712.01192v1, 2017.

[69] D. Saha, K. Leong, C. Li, S. Peterson, G. Siegel, B. Raman, A spatiotemporal coding mechanism for background-invariant odor recognition, Nat. Neurosci. 16 (2013) 1830–1839, https://doi.org/10.1038/nn.3570.

[70] J. Humble, S. Denham, T. Wennekers, Spatio-temporal pattern recognizers using spiking neurons and spike-timing-dependent plasticity, Front. Comput. Neurosci. 6 (2012) 84, https://doi.org/10.3389/fncom.2012.00084.

[71] T. Deneux, A. Kempf, A. Daret, E. Ponsot, B. Bathellier, Temporal asymmetries in auditory coding and perception reflect multi-layered nonlinearities, Nat. Commun. 7 (2016) 12682, https://doi.org/10.1038/ncomms12682.

[72] G.-Q. Bi, M.-M. Poo, Synaptic modifications in cultured hippocampal neurons: dependence on spike timing, synaptic strength, and post synaptic cell type, J. Neurosci. 18 (24) (1998) 10464–10472.

[73] G.M. Wittenberg, S.S.-H. Wang, Malleability of spike-timing-dependent plasticity at the CA3-CA1 synapse, J. Neurosci. 26 (24) (2006) 6610–6617, https://doi.org/10.1523/JNEUROSCI.5388-05.2006.

[74] L.F. Abbott, S.B. Nelson, Synaptic plasticity: taming the beast. Nat. Neurosci. 3 (Suppl) (2000) 1178–1183, https://doi.org/10.1038/81453.

[75] S. Song, K.D. Miller, L.F. Abbott, Competitive Hebbian learning through spike-timing dependent synaptic plasticity, Nat. Neurosci. 3 (9) (2000) 919–926, https://doi.org/10.1038/78829.

[76] N. Caporale, Y. Dan, Spike-timing dependent plasticity: a Hebbian learning rule, Annu. Rev. Neurosci. 31 (2008) 25–46, https://doi.org/10.1146/annurev.neuro.31.060407.125639.

[77] S. Ambrogio, S. Balatti, A. Cubeta, A. Calderoni, N. Ramaswamy, D. Iclmini, Statistical fluctuations in HfO$_x$ resistive-switching memory: Part I—Set/reset variability, IEEE Trans. Electron Devices 61 (8) (2014) 2912–2919, https://doi.org/10.1109/TED.2014.2330200.

[78] S. Ambrogio, S. Balatti, A. Cubeta, A. Calderoni, N. Ramaswamy, D. Ielmini, Statistical fluctuations in HfO_x resistive-switching memory: Part II—Random telegraph noise, IEEE Trans. Electron Devices 61 (8) (2014) 2920–2927, https://doi.org/10.1109/TED.2014.2330202.

[79] S. Ambrogio, S. Balatti, V. McCaffrey, D. Wang, D. Ielmini, Noise-induced resistance broadening in resistive switching memory—Part II: Array statistics, IEEE Trans. Electron Devices 62 (11) (2015) 3812–3819, https://doi.org/10.1109/TED.2015.2477135.

[80] G. Pedretti, S. Bianchi, V. Milo, A. Calderoni, N. Ramaswamy, D. Ielmini, Modeling-based design of brain-inspired spiking neural networks with RRAM learning synapses, in: IEDM Tech. Dig, 2017, pp. 653–656, https://doi.org/10.1109/IEDM.2017.8268467.

[81] G. Pedretti, V. Milo, S. Ambrogio, R. Carboni, S. Bianchi, A. Calderoni, N. Ramaswamy, A.S. Spinelli, D. Ielmini, Stochastic learning in neuromorphic hardware via spike timing dependent plasticity with RRAM synapses, IEEE J. Emerging Sel. Top. Circuits Syst. 8 (1) (2018) 77–85, https://doi.org/10.1109/JETCAS.2017.2773124.

[82] R.M. French, Catastrophic forgetting in connectionist networks, Trends Cogn. Sci. 3 (4) (1999) 128–135, https://doi.org/10.1016/S1364-6613(99)01294-2.

[83] Z. Wang, S. Joshi, S.E. Savel'ev, H. Jiang, R. Midya, P. Lin, M. Hu, N. Ge, J.P. Strachan, Z. Li, Q. Wu, M. Barnell, G.-L. Li, H.L. Xin, R.S. Williams, Q. Xia, J.J. Yang, Memristors with diffusive dynamics as synaptic emulators for neuromorphic computing, Nat. Mater. 16 (2017) 101–108, https://doi.org/10.1038/nmat4756.

[84] V. Milo, D. Ielmini, E. Chicca, Attractor networks and associative memories with STDP learning in RRAM synapses, in: IEDM Tech. Dig, 2017, pp. 263–266, https://doi.org/10.1109/IEDM.2017.8268369.

Index

Note: Page numbers followed by *f* indicate figures and *t* indicate tables.